BIOFUELS

Production and Future Perspectives

Edited by

RAM SARUP SINGH

Punjabi University, Patiala, India

ASHOK PANDEY

CSIR - National Institute for
Interdisciplinary Science & Technology, Trivandrum, India

EDGARD GNANSOUNOU

EPFL, Lausanne, Switzerland

CRC Press
Taylor & Francis Group
Boca Raton London New York

CRC Press is an imprint of the
Taylor & Francis Group, an **informa** business

CRC Press
Taylor & Francis Group
6000 Broken Sound Parkway NW, Suite 300
Boca Raton, FL 33487-2742

First issued in paperback 2019

ISBN-13: 978-1-4987-2359-6 (hbk)
ISBN-13: 978-0-367-87311-0 (pbk)

Library of Congress Cataloging-in-Publication Data

Names: Singh, Ram Sarup, editor. | Pandey, Ashok, editor. | Gnansounou, Edgard, editor.
Title: Biofuels : production and future perspectives / editors, Ram Sarup Singh, Ashok Pandey, Edgard Gnansounou.
Description: Boca Raton, FL : CRC Press/Taylor & Francis Group, 2016. | Includes bibliographical references and index.
Identifiers: LCCN 2016012567 | ISBN 9781498723596 (alk. paper)
Subjects: | MESH: Biofuels | Biomass | Biotechnology
Classification: LCC TP339 | NLM QV 241 | DDC 662/.88--dc23
LC record available at http://lccn.loc.gov/2016012567

Visit the Taylor & Francis Web site at
http://www.taylorandfrancis.com

and the CRC Press Web site at
http://www.crcpress.com

Contents

Contents

SECTION III   Biofuels from Algae

SECTION IV   Future Perspectives of Biofuels

Preface

Amidst the backdrop of declining energy resources and dwindling fossil fuels, there has been a rise in global concern regarding the availability of food for the next generation with respect to finite natural resources. Ever since the Industrial Revolution took off in the eighteenth century, an ever-increasing demand for energy by modern society has been met by fossil fuels, signifying total dependence of humankind on fossil fuels. Fossil fuels have taken millions of years to become available, and, thus, their supply is finite and might become scarce, or even run out, in the near future. Fossil fuel burning has led to the emission of enhanced carbon dioxide into the atmosphere, a major cause of global warming, which apparently is leading to long-term drastic changes in climate and sea level. To provide both power and fuel for transport, alternative energy supplies are required. On all these frontiers, biofuels appear to be part of the solution as an alternative, "greener" energy substitute for fossil fuels, which can be replenished within a short time with additional benefits to the environment. These renewable and sustainable biofuel resources are enough to feed the energy-hungry civilization.

In recent times, liquid biofuels have attracted much interest, and there is an unprecedented increase in biofuels production and utilization. The biofuels market has been growing since the early 2000s owing to the need to enhance energy security and promote agriculture/rural development, affecting global agricultural commodity markets. Nations around the world require staggering amounts of energy for various sectors, and meeting this ever-increasing demand in a way that minimizes energy disruptions is a key challenge of the twenty-first century. Considering the energy source, feedstock production system sustainability is a matter of grave importance as biofuels feedstock come from agriculture. The importance of biofuels in transportation is not the question today; rather, their implications on the economy, environment, and health of society are of major concern. To replace the bulk of transport fossil fuels, instead of first-generation, second- and third-generation biofuels should be utilized so as to not compromise food crops. Furthermore, various policy decisions would impact and determine the broader social and economic impacts of biofuels. Domestically produced biofuels have been favored by various national policies, which, at times at the expense of import, nurture biofuels. Of late, various government agencies have been promoting and investing in research on biofuels utilization to reduce oil dependence and greenhouse gas emissions. Furthermore, it will be a "challenge" for the coming generation of scientists to develop more sustainable ways to fight energy crises.

The aim of this book is to provide in-depth information on the most recent developments in the area of biofuels. It leans toward the latest research-based information sandwiched by fundamentals, principles, and practices. The book is divided into four sections consisting of 21 chapters. Section I presents an overview of biofuels, comprising chapters on historical perspectives; public opinion on and global demand for biofuels; economic aspects, market, and policies of biofuels; and sustainability criteria for liquid biofuels. Section II focuses on the unification of biofuel production methods, including "second-generation biofuels" from various feedstock. Within the last decade, there has been a spectacular reawakening of interest in algal fuel. Section III is devoted to biofuels from algae, focusing on issues pertaining to their production, design of photobioreactors, and sustainability. Section IV critically explores the future perspectives of biofuels, including enzymes involved and their immobilization in biofuels production, proteomics of biofuel crops and cyanobacteria, and biofuel cells. All forgoing scientific, ecological, economic, and technological aspects of biofuels have been dealt with comprehensively by well-known experts in their respective fields. The text in each chapter is supported by numerous clear, informative tables and figures. Each chapter contains relevant references of published articles, which offer a potentially large amount of primary information and further links to a nexus of data and ideas.

This book is intended for postgraduate students and researchers from industry and academia who are working in the area of biofuels. Its purpose is to usher readers with enhanced knowledge and serve as an up-to-date reference source. The authors have provided a novel framework to illuminate interactions between food, feed, and fuel synergies in relation to sustainable development.

The editors sincerely thank all the contributors for their outstanding efforts to provide state-of-the-art information on the subject matter of their respective chapters. Their efforts have certainly enhanced our knowledge of biofuels. We also acknowledge the help from the reviewers, who, in spite of their busy schedules, helped us by evaluating the manuscripts and gave their critical inputs to refine and improve the chapters. We thank the publishers/authors of various articles whose works have been cited/included in the book. We warmly thank Dr. Michael Slaughter, Jennifer Ahringer, and the team at CRC Press/Taylor & Francis Group for their cooperation and effort in producing this book. We place on record our deep sense of appreciation for consistent support from Rupinder Pal Singh, Hemantpreet Kaur, Amandeep Kaur Walia, Shivani Rani, and Navpreet Kaur from Punjabi University, Patiala, for their help in preparing this book.

We hope that the book will help readers to find the needed information on the latest research and advances, especially innovations, in biofuels.

Ram Sarup Singh

Ashok Pandey

Edgard Gnansounou

Editors

Professor Ram Sarup Singh, MSc, MEd, MPhil, PhD, FBRS, is professor and former head, Department of Biotechnology, Punjabi University, Patiala, Punjab, India. He earned his master's degree in botany and PhD in biotechnology from Punjabi University, Patiala. He joined as assistant professor in the Department of Biotechnology at the same university in 1994 and since then has been working there in different faculty positions. He is a recipient of many national and international awards and fellowships: MASHAV-UNESCO fellowship (2006), Rehovat Campus, Hebrew University of Jerusalem, Israel; INSA Visiting Scientist (2008), National Institute of Pharmaceutical & Educational Research, Mohali, India; Fellow (2012), Biotech Research Society, India; Popularization of Science Award (2013), Baba Farid University of Health Sciences, Faridkot and Punjab State Council for Science & Technology, Chandigarh, India; and visiting professor (2014), Swiss Federal Institute of Technology, Lausanne (EPFL), Switzerland. Dr. Singh has more than 22 years of teaching and research experience in industrial biotechnology. His current focus areas of research are biofuels, microbial polysaccharides, industrial enzymes, microbial lectins, fructooligosaccharides, high fructose/maltose syrups, etc. He has to his credit more than 200 publications/communications, which include 2 patents, 3 books, 20 book chapters, 35 popular articles, 120 original research and review papers with h-index of 24. He is on the Advisory Board of the National Institute of Ayurvedic Pharmaceutical Research, Patiala, India. Dr. Singh was guest editor of special issues (2013) of three national/international journals: *Biologia, Journal of Scientific & Industrial Research*, and *Indian Journal of Experimental Biology*. He is on the editorial board of *International Journal of Food & Fermentation Technology* and *Journal of Environmental Sciences & Sustainability*. He is honorary consultant to various regional food industries and has successfully completed a few consultancy projects.

Professor Ashok Pandey, MSc, DPhil, FBRS, FNASc, FIOBB, FISEES, FAMI, is eminent scientist at the Center of Innovative and Applied Bioprocessing, Mohali (a national institute under the Department of Biotechnology, Ministry of Science and Technology, Government of India) and former chief scientist and head of the Biotechnology Division at CSIR—National Institute for Interdisciplinary Science and Technology, Trivandrum, Kerala, India. He is adjunct professor at MACFAST, Thiruvalla, Kerala, and Kalaslingam University, Krishnan Koil, Tamil Nadu. His major research interests are in the areas of microbial, enzyme, and bioprocess technology, which span over various programs, including biomass to fuels and chemicals, probiotics and nutraceuticals, industrial enzymes, solid-state fermentation, etc. He has more than 1100 publications/communications, which include 16 patents, 50+ books, 125 book chapters, 423 original and review papers, with an h-index of 74 and more than 23,700 citations (Google Scholar). He has transferred four technologies to industries and served as a consultant on about a dozen projects for Indian/international industries. He is editor-in-chief of the book series Current Developments in Biotechnology and Bioengineering, comprising nine books published by Elsevier.

Professor Pandey is a recipient of many national and international awards and fellowships, which include elected member of the European Academy of Sciences and Arts, Germany; fellow of the International Society for Energy, Environment and Sustainability; fellow of the National Academy of Science (India); fellow of the Biotech Research Society, India; fellow of the International Organization of Biotechnology and Bioengineering; fellow of the Association of Microbiologists of India; honorary doctorate degree from Univesite Blaise Pascal, France; Thomson Scientific India Citation Laureate Award, USA; Lupin visiting fellowship, visiting professor at the University Blaise Pascal, France; Federal University of Parana, Brazil and EPFL, Switzerland; Best Scientific Work Achievement award, Government of Cuba; UNESCO professor; Raman Research Fellowship Award, CSIR; GBF, Germany and CNRS, France Fellowship; and Young Scientist Award. He was chairman of the International Society of Food, Agriculture and Environment, Finland (Food & Health), during 2003–2004. He is founder-president of the Biotech Research Society, India (www.brsi.in); International Coordinator of the International Forum on Industrial Bioprocesses, France (www.ifibiop.org), Chairman of the International Society for Energy, Environment and Sustainability (www.isees.org), and vice-president of the All India Biotech Association (www.aibaonline.com). Professor Pandey is editor-in-chief of *Bioresource Technology* and an honorary executive advisor of two international journals, namely, *Journal of Water Sustainability* and *Journal of Energy and Environmental Sustainability*. He is an editorial board member of several international and Indian journals and also member of several national and international committees.

Professor Edgard Gnansounou, MS, PhD, is professor of modeling and planning of energy systems at the Swiss Federal Institute of Technology, Lausanne (EPFL), Switzerland, where he is director of the Bioenergy and Energy Planning Research Group. His current research work comprises techno-economic and environmental assessment of biorefinery schemes based on the conversion of agricultural residues. He is leading research projects in this field in several countries, including Brazil, Colombia, and South Africa. Dr. Gnansounou is credited with numerous papers in high-impact scientific journals. He is a member of the editorial board of *Bioresource Technology*. He graduated with an MS in civil engineering and a PhD in energy systems from the Swiss Federal Institute of Technology, Lausanne. He was a visiting researcher at the Thayer College, Dartmouth School of Engineering, with Professor

Charles Wyman (USA); at Polytech of Clermont-Ferrand, University Blaise Pascal (France); and at the Center of Biofuels, CSIR-National Institute for Interdisciplinary Science and Technology, Trivandrum, India, with Professor Ashok Pandey. He was also a visiting professor at the African University of Science and Technology (Abuja, Nigeria). He is a citizen of Benin (Africa) and Switzerland.

Contributors

David Benson
Department of Politics
University of Exeter
Exeter, United Kingdom

Ranjeeta Bhari
Department of Biotechnology
Punjabi University
Patiala, Punjab, India

Thallada Bhaskar
Bio-Fuels Division
Indian Institute of Petroleum
Council of Scientific and Industrial
    Research
Dehradun, Uttarakhand, India

Parameswaran Binod
National Institute for Interdisciplinary
    Science and Technology
Council of Scientific and Industrial
    Research
Trivandrum, Kerala, India

Bijoy Biswas
Bio-Fuels Division
Indian Institute of Petroleum
Council of Scientific and Industrial
    Research
Dehradun, Uttarakhand, India

P. Chiranjeevi
Bioengineering and Environmental
    Sciences
Indian Institute of Chemical Technology
Council of Scientific and Industrial
    Research
Hyderabad, Telangana, India

and

Academy for Scientific and Industrial
    Research
New Delhi, India

Somayeh Farzad
Department of Process Engineering
Stellenbosch University
Stellenbosch, South Africa

Johann F. Görgens
Department of Process Engineering
Stellenbosch University
Stellenbosch, South Africa

Lalitha Devi Gottumukkala
Department of Process Engineering
University of Stellenbosch
Stellenbosch, South Africa

Vaibhav V. Goud
Department of Chemical Engineering
Indian Institute of Technology
Guwahati, Assam, India

Tobias Ide
Research Group Climate Change and
    Security
University of Hamburg
Hamburg, Germany

and

Georg Eckert Institute for International
    Textbook Research
Braunschweig, Germany

Benan İnan
Department of Bioengineering
Yıldız Technical University
Istanbul, Turkey

Manpreet Kaur
Department of Biotechnology
Punjabi University
Patiala, Punjab, India

Simran Preet Kaur
Department of Biotechnology
Maharishi Markandeshwar University
Mullana, Haryana, India

Jasvirinder Singh Khattar
Department of Botany
Punjabi University
Patiala, Punjab, India

Anıl Tevfik Koçer
Department of Bioengineering
Yıldız Technical University
Istanbul, Turkey

Bhavya B. Krishna
Bio-Fuels Division
Indian Institute of Petroleum
Council of Scientific and Industrial
    Research
Dehradun, Uttarakhand, India

Jitendra Kumar
Bio-Fuels Division
Indian Institute of Petroleum
Council of Scientific and Industrial
    Research
Dehradun, Uttarakhand, India

Gill Malin
Centre for Ocean and Atmospheric
    Sciences
School of Environmental Sciences
University of East Anglia
Norwich, United Kingdom

Mohsen Ali Mandegari
Department of Process Engineering
Stellenbosch University
Stellenbosch, South Africa

Manpreet Kaur Mann
Department of Biotechnology
Punjabi University
Patiala, Punjab, India

Anil K. Mathew
Centre for Biofuels
National Institute for Interdisciplinary
    Science and Technology
Council of Scientific and Industrial
    Research
Trivandrum, Kerala, India

Umesh Mishra
Separation and Conversion
    Technologies
VITO—Flemish Institute for
    Technological Research
Boeretang, Belgium

and

Department of Civil Engineering
National Institute of Technology,
    Agartala
Agartala, Tripura, India

S. Venkata Mohan
Bioengineering and Environmental
  Sciences
Indian Institute of Chemical Technology
Council of Scientific and Industrial
  Research
Hyderabad, Telangana, India

Gunda Mohanakrishna
Separation and Conversion
  Technologies
VITO—Flemish Institute for
  Technological Research
Boeretang, Belgium

Gaurav Mundada
Bio-Fuels Division
Indian Institute of Petroleum
Council of Scientific and Industrial
  Research
Dehradun, Uttarakhand, India

M. Muthukumaran
Department of Botany
Ramakrishna Mission Vivekananda
  College (Autonomous)
Chennai, Tamil Nadu, India

Somkiat Ngamprasertsith
Department of Chemical Technology
Faculty of Science
and
Center of Excellence on Petrochemical
  and Materials Technology
Chulalongkorn University
Bangkok, Thailand

Didem Özçimen
Department of Bioengineering
Yıldız Technical University
Istanbul, Turkey

Ashok Pandey
National Institute for Interdisciplinary
  Science and Technology
Council of Scientific and Industrial
  Research
Trivandrum, Kerala, India

Deepak Pant
Separation and Conversion
  Technologies
VITO—Flemish Institute for
  Technological Research
Boeretang, Belgium

Brenda Parker
Department of Biochemical Engineering
University College
London, United Kingdom

Shahnaz Parveen
Department of Botany
Punjabi University
Patiala, Punjab, India

C. Nagendranatha Reddy
Bioengineering and Environmental
  Sciences
Indian Institute of Chemical Technology
Council of Scientific and Industrial
  Research
Hyderabad, Telangana, India
and
Academy for Scientific and Industrial
  Research
New Delhi, India

Ali Shemsedin Reshad
Department of Chemical Engineering
Indian Institute of Technology
Guwahati, Assam, India

Zubaidai Reyimu
Department of Bioengineering
Yıldız Technical University
Istanbul, Turkey

M.V. Rohit
Bioengineering and Environmental
    Sciences
Indian Institute of Chemical Technology
Council of Scientific and Industrial
    Research
Hyderabad, Telangana, India
and
Academy for Scientific and Industrial
    Research
New Delhi, India

Meena Sankar
Centre for Biofuels
National Institute for Interdisciplinary
    Science and Technology
Council of Scientific and Industrial
    Research
Trivandrum, Kerala, India

Ruengwit Sawangkeaw
The Institute of Biotechnology and
    Genetic Engineering
Chulalongkorn University
Bangkok, Thailand

Nirmala Sehrawat
Department of Biotechnology
Maharishi Markandeshwar University
Mullana, Haryana, India

Kirsten Selbmann
Institute for Education, Culture and
    Sustainable Development
Bochum University of Applied Sciences
Bochum, Germany

Raveendran Sindhu
National Institute for Interdisciplinary
    Science and Technology
Council of Scientific and Industrial
    Research
Trivandrum, Kerala, India

Davinder Pal Singh
Department of Botany
Punjabi University
Patiala, Punjab, India

Ram Sarup Singh
Department of Biotechnology
Punjabi University
Patiala, Punjab, India

Rawel Singh
Bio-Fuels Division
Indian Institute of Petroleum
Council of Scientific and Industrial
    Research
Dehradun, Uttarakhand, India

Yadvinder Singh
Department of Botany and
    Environmental Science
Sri Guru Granth Sahib World University
Fatehgarh Sahib, Punjab, India

V. Sivasubramaian
Department of Botany
Ramakrishna Mission Vivekananda
    College (Autonomous)
Chennai, Tamil Nadu, India

Balwinder Singh Sooch
Department of Biotechnology
Punjabi University
Patiala, Punjab, India

Vartika Srivastava
Bio-Fuels Division
Indian Institute of Petroleum
Council of Scientific and Industrial
    Research
Dehradun, Uttarakhand, India

Rajeev Kumar Sukumaran
Centre for Biofuels
National Institute for Interdisciplinary
    Science and Technology
Council of Scientific and Industrial
    Research
Trivandrum, Kerala, India

Shivani Thakur
Department of Biotechnology
Punjabi University
Patiala, Punjab, India

Pankaj Tiwari
Department of Chemical Engineering
Indian Institute of Technology
Guwahati, Assam, India

Amandeep Kaur Walia
Department of Biotechnology
Punjabi University
Patiala, Punjab, India

Mukesh Yadav
Department of Biotechnology
Maharishi Markandeshwar University
Mullana, Haryana, India

Pallavi Yadav
Bio-Fuels Division
Indian Institute of Petroleum
Council of Scientific and Industrial
    Research
Dehradun, Uttarakhand, India

Section I

Overview of Biofuels

1

Biofuels
Historical Perspectives and Public Opinions

Ram Sarup Singh and Amandeep Kaur Walia

Contents

Abstract

Since the dawn of the mankind, biofuels have been used and have marked their presence from the time fire was discovered. Today, with rising crude oil prices and various environmental challenges and political instability, there is an urgent need to explore new alternatives to fossil fuels. Over the past 15 years, biofuels have been extensively researched, produced, and used in solid, liquid, and gaseous forms. During its long history, biofuel production had many peaks and valleys, and currently has the potential to replace the supply of fossil fuels for an energy-hungry civilization. Further, the acceptance of a technology depends on public opinion. Over the years, despite large-scale investments in biofuel expansion and research, little is known regarding public opinions of biofuels. How the public perceives this alternative fuel technology and on what basis their opinions are formed is necessary to determine its future stability. The current chapter focuses on historical perspectives of biofuels and public opinion regarding biofuels considering various risk–benefit perceptions. Public opinions in different regions and regarding second-generation biofuels has also been discussed.

1.1 Introduction

A biofuel is a fuel derived from plant biomass including materials from organisms that died relatively recently and from the metabolic by-products of living organisms (Bungay, 1982; Demirbas, 2009). Through thermal, chemical, and biochemical conversion, this biomass can be converted into biofuels. Throughout man's long history, biomass fuels have been used. Conventional biofuels or "first-generation biofuels" are made from starch, sugar, or vegetable oil, whereas advanced or "second-generation biofuels" are made from lignocellulosic biomass or woody crops and agricultural residues or waste. This makes it harder to extract the required advanced biofuel and thus requires various physical and chemical treatments for its conversion to liquid fuels required for transportation (IEA Bioenergy Task, 2015; Ramirez et al., 2015).

The U.S. government has made development of renewable energies a top priority, owing to instability of oil prices in large part and to lessen dependence on foreign oil. "Renewable energy is a pivotal aspect of policy for the USA" is a noteworthy statement by President Barack Obama that signifies the relative importance of biofuels. Further biofuel production has been made mandatory by the United States, making hundreds of millions of dollars by investing in fledgling industry (Chu and Vilsack, 2009). The biofuel industry is used as a supply of energy as well as basic chemicals, thus serving dual purpose in the economy (Zaborsky, 1982). On an industrial scale, only biodiesel and bioethanol, including ethyl tertiary butyl ether (ETBE), make up more than 90% of biofuel market (Antoni et al., 2007). The use of lignocellulosics as alternative substrates has yet to be exploited (Demain et al., 2005). However, development is in progress, and various companies and publicly funded institutions are involved in the development of feasible processes. Against a backdrop of increasing political instability in oil-producing countries and rising prices for crude oil, the use of biobased alcohols as basic chemicals or solvents is back under consideration (Antoni et al., 2007). Major focus areas of biofuel research (Figure 1.1) include improvement in biomass conversion to biofuels, optimizing biomass resource production, and biofuel testing for engine compatibility and emissions modeling (Bente, 1984).

To determine the future of scientific energy innovation, it is crucial for policymakers and industry leaders alike to understand how the public evaluates risks and benefits for emerging innovations like biofuels (Cacciatore et al., 2012). Further strategies can be designed for an effective communication between the general public and those in scientific community, based on better understanding. The current chapter focuses on history of biofuels to date and further giving insight into public opinions regarding biofuels with its varied pros and cons in different domains including environmental, economic, ethical, and political.

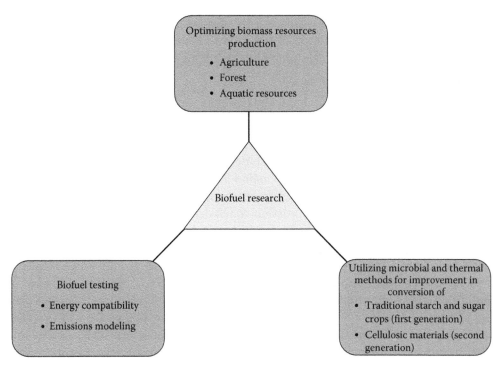

Figure 1.1 Focus areas of biofuel research.

1.2 History of Biofuels

Various renewable sources such as wood, waterwheels, and windmills have been utilized by mankind, for most of its existence. By the beginning of the twentieth century, up to 30% of arable land was planted with crops, to feed horses and oxen used in transportation. Table 1.1 describes use of biofuels during four historical epochs and their major concerns.

To promote rural development and self-sufficiency, Germany created the world's first large-scale biofuel industry decades before World War I (*London Times*, 1902).

TABLE 1.1   Biofuel Use and Chief Concerns during Four Historical Eras

Time Period	Chief Concerns	Biofuel Utilization
Mid-nineteenth century	—	As primary lamp and cooking fuel
Around twentieth century	Fuel quality and petroleum depletion	Internal combustion engine
Mid- to late twentieth century	International oil politics chiefly Arab oil embargoes (1970s) spurring national energy security investment	Internal combustion engine
Early twenty-first century	Climate change, biodiversity, and sustainability broadly framing energy research and policy debates	Internal combustion engine

The German program, which began in 1899, involved farm distillery construction, tariffs on imported oil, promotion of ethanol-fueled appliances (Tweedy, 1917), and research involving ethanol fuel automobiles. An early approach to household and small-scale energy systems was through Germany's "Materialbrennereien" program. In the early 1900s, an exhibit was developed in Germany, that was devoted to alcohol-powered automobiles, household appliances, and alcohol-powered agricultural engines (Automobile Club de France, 1902), which traveled between 1901 and 1904 to France, Italy, and Spain and then between 1907 and 1908 to the United States (Lucke and Woodward, 1907). As part of British defense research effort, an alcohol motor fuel committee was created in 1914 (*London Times*, 1914), which was charged with considering supply source, manufacture method, and production costs for alcohol fuel (Fox, 1924). In 1921, the commission concluded making alcohol a likely fuel in terms of cost as compared to petroleum in tropical and remote areas of the world (*London Times*, 1921). Prior to World War I, the French Agricultural Ministry promoted its ethanol fuel program (concerned with increasing oil import from Russia and the United States and about rising surplus of other crops), which led to rise in French ethanol fuel production from 2.7 to 8.3 million gallons from 1900 to 1905. In February 1923, after World War I, Article Six was passed based on the French committee's recommendation, which required gasoline importers to buy alcohol for 10% blends from state alcohol service (Fox, 1924). In 1935, biofuel utilization rate was at its peak (406 million liters), which accounted for 7% of all fuel use but, owing to poor harvest, declined to 194 million liters by 1937 (Egloff, 1939). Thus, a worldwide impact occurred due to German, French, and British biofuel law and research. Historical perspectives of various biofuels (solid, liquid, and gaseous) have been tabulated in Table 1.2 and discussed further.

1.2.1 Liquid Biofuels

1.2.1.1 Bioethanol

Commercially, bioethanol is produced mostly from sugarcane, sugar beet, and corn with other sources being cassava or cellulosic materials (grasses, trees, or waste product from crops). It can be blended with conventional fuels to at least 10% (10% ethanol to 90% gasoline). Around 4000 BC, humans used to make alcohol as a beverage from cereals, berries and grapes, etc., using fermentation technology. Since antiquity, olive oil and various other oils derived from plants and animals have been used for lamp oil (Appolonia and Sussman, 1983). During the early 1700s, lamps were fueled by vegetable oil and fats, and during the 1800s, whale oil was preferred until a modern method for refining kerosene was developed by Abraham Gesner in 1846 (Russell, 2003). By the late 1830s, owing to diminishing supply of expensive whale oil, it was replaced by ethanol blended with turpentine. Before the discovery of petroleum by Edwin Drake in 1859, ethanol was developed as an alternative fuel (Kovarik et al., 1998). To fund the Civil War, a $2.08 tax per gallon of ethanol was imposed in 1862 by the U.S. Congress, which continued well past the war's conclusion. This tax made ethanol more expensive than gasoline, thus favoring gasoline's use for the internal

TABLE 1.2   History of Biofuels

Time Period	Events
Biofuel for illumination	
Early 1700s	Lamps fueled by vegetable oil and fats.
1800	Whale oil, lard oil, and camphene preferred as lamp oil.
1830	Whale oil replaced by ethanol blended with turpentine as an illuminant.
1862	U.S. Congress increased tax on alcohol, leading to a boost for petroleum industry.
Germany	
1899	Research into ethanol-fueled trucks and automobiles started.
1902	Government promoted ethanol-fueled household appliances.
1906	Germany's "Materialbrennereien" program started.
Around 1900	An exhibit devoted to alcohol-powered automobiles, farm machinery, lamps, stoves, heaters, and other household appliances debuts.
1901–1904	Exhibit travels to France, Italy, and Spain.
1907–1908	Exhibit travels to USA.
USA	
1917	Alexander Graham Bell quoted ethanol as "it makes a beautiful, clean and efficient fuel" that can be manufactured from any vegetable matter capable of fermentation.
Britain	
1914–1921	Scarcity of oil resources led to interest in biofuels; alcohol motor fuel committee was created, which after research concluded that cost of alcohol in comparison with petroleum made it a likely fuel in tropical and remote areas of the world.
France	
1900–1905	French ethanol fuel program led to rise in French ethanol fuel production.
1923 (After World War I)	Article Six requiring gasoline importers to buy alcohol for 10% blends from state alcohol service was passed.
1935	Biofuel use peaked at 406 million gallons.
Biofuel for Internal combustion engine (ICE)	
1826	Samuel Morey developed the first authentic ICE that ran on ethanol and turpentine.
1860	Nicholas August Otto used ethanol fuel blend as fuel in engine.
1893	Rudolph Diesel envisaged potential of pure vegetable oil to power machines and invented compression-ignited diesel engines.
1906–1925	Henry Ford mentioned ethyl alcohol as "fuel of the future" and used it to power tractors and his model T cars.
1917	Alexander Graham Bell highlighted abundance of potential feedstock for ethanol production.
1917–1919 (World War I)	Increase in ethanol demand owing to raw material rationing.
1941–1945 (World War II)	Methanol used as fuel in Germany during World War II.

(Continued)

TABLE 1.2 (*Continued*) History of Biofuels

Time Period	Events
1930s	Farm chemurgy movement started in the United States, with its primary agenda being use of ethanol as octane booster in gasoline.
1937	Belgian scientist patented procedure for vegetable oil transformation for its use as fuel.
1943	Ethanol used to produce 77% of synthetic rubber in the United States.
1960s–1970s	In USSR, biobutanol (0.1 million tons per year) produced by fermentation.
1970s–2000s	Energy crisis.
1973	Initial energy crisis took place when Arab countries dropped oil production by 5%, leading to rise in oil prices.
1977	The first industrial-scale process for biodiesel production was developed.
1978	Next energy crisis occurred when Iranian dissent grew into strike in nation's oil refineries, shutting down 5% of the world's oil exports.
1980s–2000s	Lead gas phaseout
1989	Rapeseed used for biodiesel production at the world's first biodiesel industrial plant.
1996	Waste grease processing into biodiesel on commercial scale.
2008–2011	U.S. production of ethanol and biodiesel increased more than 40%.
2000–2013	Commercial global bioethanol production increased from 4.5 to 23.4 billion gallons. Biodiesel production increased from 213 to 628 million gallons.
2013	The first commercial-scale cellulosic ethanol plant entered into full operation worldwide.

combustion engine (ICE) (Dimitri and Effland, 2007). Even after the tax's repeal in 1906, it became difficult for ethanol infrastructure to compete with gasoline.

An authentic and first ICE (U.S. patent 4378 issued April 1, 1826) fueled by ethanol and turpentine was designed by American inventor Samuel Morey. In 1860, another ICE was developed by German engineer Nikolaus August Otto, which ran on ethanol fuel blend (Songstad et al., 2009). The next era (late nineteenth to early twentieth century), where biofuels became prominent, was linked with the invention of the automobile. The world's first practical automobile to be powered by an IC engine was designed and built by German engineer Karl Benz in 1885 (Loeb, 2004). To power gasoline engines, ethanol can be used as a gasoline substitute; thus, in 1913, ethanol was tested as an engine fuel. Henry Ford, another American industrialist, constructed ethanol-powered tractors. In 1906, Ford said carburetors on his Model T cars will use gasoline or alcohol. In 1925, Ford mentioned ethyl alcohol as the "fuel of the future." Alexander Graham Bell highlighted the abundance of potential feedstocks intended for the ethanol production by stating in a 1917 *National Geographic* interview (Anonymous, 1917) that "alcohol makes a beautiful, clean and efficient fuel that can be manufactured from any vegetable matter capable of fermentation such as crop residues, grasses, farm waste and city garbage." In the past, various ethanol and ethanol–gasoline blends have been used as automotive fuels (Hunt, 1981; Kovarik et al., 1998). During the mid-1930s, Alcolene and Agrol (alcohol–gasoline blends) brands were sold in the market.

During the 1920s and 1930s, Henry Ford promoted a new movement called chemurgy (Finlay, 2004). Chemurgy focused on crop utilization in biobased materials, and as evident from the need to produce synthetic rubber during the onset of World War II, it had ethanol as its prominent agenda. In 1943, ethanol was used to produce nearly 77% of the synthetic rubber in the United States (Finlay, 2004). Owing to rationing of raw materials and natural resources during world war years (1917–1919 and 1941–1945), an increase in ethanol demand was witnessed in the United States. Domestic use employed ethanol as a substitute for gasoline. Tremendous socioeconomic and environmental impacts have been generated owing to heavy dependence on crude oil. To counter the rapid escalation of oil prices, which has continued through the early twenty-first century, Arab oil embargoes of the 1970s were remembered by those who experienced it as the first vocal call for a domestic source of renewable energy. In the 1973 oil crisis, there was a revival of interest in ethanol as fuel. In 1979, at South Dakota University, the first pilot bioethanol plant with a distillation column was established (Songstad et al., 2009). Due to the OPEC oil export embargo (in 1973) and Iranian revolution (in 1979), substantial shortages and soaring prices of crude oil severely affected the economies of major industrial countries including the United States, Western Europe, Japan, Canada, and Australia (Lifset, 2014).

The U.S. Environmental Protection Agency's leaded gasoline phaseout in the 1980s caused an increase in ethanol usage as an octane booster and volume extender, yet methyl *tert*-butyl ether (MTBE) dominated most oxygenated gasoline markets during the 1990s (Soloman et al., 2007). However, increasing restrictions on MTBE as a fuel oxygenate led to rapid growth in U.S. ethanol production since 2002 (Soloman et al., 2007).

Commercially, starch-/sugar-based crops are used for production of bioethanol. In European countries, the predominant feedstock for bioethanol is wheat and sugar beet, whereas in Brazil, sugarcane is the primary feedstock. The global commercial bioethanol production has increased from 4.0 billion gallons in 1990 to 4.5 billion gallons in 2000 and 23.4 billion gallons in 2013 (RFA, 2014). In ethanol utilization, Brazil and the United States are world leaders. Bioethanol competes with human food and animal feed for source material and is thus regarded as "first-generation" biofuel. Further, to reduce their adverse effects, nonfood lignocellulosic plant materials (crop residues, food processing wastes, forest slashes, yard trimmings, and municipal organic refuses) have been explored for manufacture of "second-generation" bioethanol. Various lignocellulosic materials widely available can be used as feedstock for bioethanol, and for extraction of simple sugars into lignocellulosic biomass, acid and enzymatic hydrolysis can be used (Badger, 2002). Through pilot plant operations, the nation produced 218,000 gal of cellulosic ethanol in 2013 (IER, 2013). On October 9, 2013, the first commercial-scale cellulosic ethanol plant (the Crescentino Bio-refinery, Crescentino, Vercelli, Italy) entered into full operation worldwide, with expected expansion of commercial production of cellulosic ethanol in the future, thus boosting bioethanol production globally as a gasoline alternative.

1.2.1.2 Biodiesel

The process involving the combination of organically derived oils with alcohol (ethanol or methanol) in the presence of a catalyst to form ethyl or methyl ester leads to the formation of biodiesel, which can be used as neat fuel or blended with conventional diesel (Maniatis, 2003). A wide range of feedstocks such as mustard seed oil, palm oil, sunflower, rapeseed, soybean and jatropha, peanut, and cotton seed could be used for biodiesel production. In 1988, the term "biodiesel" was initially coined (Wang, 1988); however, use of vegetable oil as fuel in place of diesel dates back to 1900. In the late eighteenth century, before the invention of electric and gas lights, vegetable oil and animal fats were used for lighting oil lamps (Thomson, 2003). In 1893, German engineer Rudolph Diesel envisioned the potential of pure vegetable oil for powering machines in agriculture and thus invented the compression-ignited diesel engine. At the 1900 World's Fair in Paris, a peanut oil–fueled diesel engine was demonstrated by French Otto Company as described by Knothe (2001). In China, to produce a version of gasoline and kerosene, tung oil and other vegetable oils are used (Cheng, 1945; Chang and Wan, 1947). India conducted research on vegetable oil conversion to diesel as necessitated by fuel shortages during World War II (Chowhury et al., 1942). However, low prices of petrol diesel quenched the developing attention on vegetable oil–based fuels. Further, high viscosity of vegetable oils as compared to diesel engine fuel was regarded as a key problem. Transesterification of vegetable oil, animal fats, waste grease, or algal lipids in the presence of an alcohol or alkaline catalyst leads to formation of a yellowish liquid biodiesel. The earliest description of biodiesel generation was in 1937, when Belgian scientist George Chavanne patented the "Procedure for the vegetable oil transformation for their uses as fuels" (Knothe, 2001). Since the 1950s, owing to geographic and economic factors rather than a fuel shortage, there has been keen interest in converting vegetable oil into biodiesel. The first industrial-scale process for biodiesel production was developed in 1977 by Brazilian scientist Expedito Parente. In 1989, rapeseed was used for production of biodiesel at the world's first industrial-scale biodiesel plant, operated in Asperhofen, Austria. Processing of waste grease into biodiesel was done commercially (Pacific Biodiesel, Maui, Hawaii) by the United States in 1996. In the global fuel market, biodiesel rose in popularity as evident by high production (213 million–6.289 billion gallons from 2000 to 2013) due to increased awareness of energy security, government tax subsidies, and tremendously high foreign oil prices after 2001 (EIA, 2014).

1.2.1.3 Other Biofuels

Since 1916, biobutanol has been in almost continuous production, mostly used as a solvent as well as a basic chemical and often misdescribed as a "new" fuel. By 1927, butanol was increasingly used for the synthetic rubber industry and in lacquer solvent butyl acetate production. During World War II, butanol was used as an aviation fuel by Japan and possibly other combatants, when their fossil fuel supply was exhausted. In the USSR, more than 100,000 tons per year of biobutanol was produced by fermentation during the 1960s and 1970s. Lignocellulosic hydrolysates from agricultural waste materials were used as substrate (Zverlov et al., 2006). However, its production stopped in the early 1990s when the USSR dissolved.

During World War II, methanol was widely used as fuel in Germany (Demirbas, 2007). By the thermochemical conversion of wood, the traditional process produced renewable methanol. Methanol separation as well as conversion from sugar beet pulp was studied; however, the designated industrial process has not been yet realized (Anonymous, 2004; Wang, 2006).

Pyrolysis is a technique that involves heating (300°C–900°C) plant biomass in the absence of air leading to the formation of three products: biochar, bio-oil, and syngas. Crude pyrolysis bio-oil contains significant amounts of colloidal char particles, and further, moderately upgraded bio-oil has been applied as a heavy fuel oil substitute to power various static appliances. Various drop-in biofuels (biomass-derived liquid hydrocarbons) include butanol, sugar hydrocarbons, liquefied biomass, and syngas complexes (AFDC, 2012). These can meet the specifications regarding existing petrol distillate fuel and be ready to "drop in" to the existing fuel supply with intensive research on its production and utilization as directed in future renewable fuel development.

1.2.2 Solid Biofuels

Ever since man discovered fire, biofuels in solid form have been in use. For domestic purposes, firewood was the predominant fuel. Before the nineteenth century, firewood was the primary fuel of the whole world for cooking and heating. Prior to the early twentieth century, the woodburning fireplace, heater, and cookstove were standard household equipment in the United States. To date, with an annual consumption of 1730 million m³, nearly 2.6 billion (40%) of world's population (mainly in the rural areas of developing countries of Asia and sub-Saharan Africa) rely on firewood to satiate its energy requirements (FAO, 2013; IEA, 2013). The world consumption of firewood has thus increased slightly (3%) over the past 13 years, and it is projected that global consumption of firewood will remain relatively constant in the future.

Woodchips (small pieces of wood from cutting tree trunks and branches) have been increasingly used for bioheat (heating) and biopower (electricity generation) since the beginning of the twenty-first century (DOE, 2000). More processed biofuel products are wood pellets made by grinding wood chunks into sawdust through a hammer mill. In 2012, the global production of wood pellets was 19.1 million tons and estimated to increase to 45.2 million tons in 2020 (van Tilburg, 2013). Charcoal, another solid biofuel, has been used by humans in metallurgy to smelt ores for copper and iron since the Bronze Age back in 3000 BC. During AD 700, in China's Tang dynasty, charcoal was the designated governmental fuel for cooking and heating. During World War II in Europe, wood gas (generated by partially burning charcoal) was used to power automobiles as gasoline was scarce (de Decker, 2010).

1.2.3 Gaseous Biofuels

Biogas, a renewable gaseous fuel generated by anaerobic digestion of organic wastes, is an alternative fuel to natural gas. At commercial scale, natural gas was initially exploited in 1821 at Fredonia, New York, for lighting (DOE, 2013). As early as the tenth century BC in ancient Assyria, biogas was utilized to heat bathwater by humans. Anaerobic digestion of animal manure for flammable gas was already practiced way back 2000 years ago, by people in India and China (Lusk, 1998). The English chemist Sir Humphry Davy identified methane as the flammable gas from cattle manure ponds in 1808. In 1859 at Bombay, the first recorded anaerobic digestion plant was constructed. In 1895, biogas was collected from a sewage treatment facility in Exeter, England to light streetlamps. In the 1950s, to provide biogas for cooking and lighting, China built around 3.5 million low-technology anaerobic digesters in the rural areas, which have further increased to 45 million in 2012 (Xia, 2013). Germany generated 18.2 billion kWh of electricity by operating 6800 biogas plants and thus became the leading country to meet the nation's natural gas demand (Stephan, 2013). In 2012, to convert agricultural, industrial, and municipal wastes to biogas, Europe had installed 8960 biowaste digesters (predominantly farm based). In the late 1970s with financial incentives from the federal government for biogas production, the United States started to install manure-based digester systems on livestock farms. There has been a rise in the number of operating farm biogas plants since 2000, generating 9.1 billion kWh of electricity in 2012 in the United States (Stephan, 2013). With generation of 6 billion N m^3 of biomethane per annum, China is currently the largest biogas producer (Rajgor, 2013). Thus, global potential for biogas is quite remarkable, and with increasing public awareness of biogas and anaerobic digestion, it could play a crucial role in the overall bioenergy sector. Gasification or pyrolysis of plant materials produces syngas, another gaseous biofuel, and with 144 gasification plants operating globally in 2010, its further research and development would lead to improvement in cost-effectiveness of feedstock preparation and help in achieving economic viability for biogas production.

1.3 Public Opinions Regarding Biofuels

Global concerns were raised for the first time during 2008 when the cost of crude oil surpassed $100 a barrel, and with continuous near-record high prices and depleting fossil fuel resources, interest in biofuel production has risen as an alternative source of energy (Raswant et al., 2008). National energy security of developing countries could be enhanced by biofuel production, which can reduce dependence and expenditure of oil import and international oil price volatility exposure. In the late 1970s, Brazil initiated its biofuel program to counteract rising oil prices, which has made Brazil today the world's largest exporter of bioethanol.

It is of vital importance as to how public opinion develops regarding emerging science and technology. Public acceptance influences the success of a technology. Lack of public support for genetically modified foods in Europe can lead to rejection based on its fluctuating risk–benefit perceptions (Gaskell et al., 1999, 2004; Sjoberg, 2004; Wohlers, 2010). In order to pass judgment about science and technology, risk and benefit considerations (Siegrist, 2000) and party identification (Kim, 2011) are common opinion-forming cues. To influence public opinion about biofuels, these two perceptions may work in tandem (Fung et al., 2014).

1.3.1 Risk–Benefit Perceptions

Even though there has been considerable increase in production and utilization of biofuels, research regarding public opinion is rare, and there has been general public support for biofuels as shown by various polling results (Bolsen and Cook, 2008; Pew Research Center, 2008; Rabe and Borick, 2008; Wegener and Kelly, 2008). Respondents had a fair amount of knowledge regarding biofuels, but were less informed about its policies when a focus group study was conducted in a bioenergy-producing state (Delshad et al., 2010). Supporters recognize biofuels as environmentally friendly and economically affordable, whereas opponents deemed them as unsafe and did not support cap–trade policies (Delshad et al., 2010). Thus, important determinants regarding public attitudes are their perceptions about the advantages and disadvantages of biofuels.

Prior experience with using biofuels also influences attitudes as suggested by some studies. For example, truck fleet operators who used biodiesel in their vehicles are supportive of biodiesel, whereas non-biodiesel users are not agreeing to it in an interview conducted according to a National Biodiesel Board commissioned study (ASG Renaissance, 2004).

A favorable view by some studies (Bolsen and Cook, 2008; Rabe and Borick, 2008) and concerns/doubts by other researchers (Belden et al., 2011) have led to a mixed response about biofuels. An extensive knowledge of science is required to make an informed assessment of risk and benefit of a technology. People tend to believe sources they trust due to their lack of knowledge (Siegrist et al., 2000). Thus, biofuels have become a controversial science issue and are associated with promise and peril influencing four major domains:

1.3.1.1 Environmental Consequences

A positive impact of biofuels in the environment domain is its potential to reduce greenhouse emissions and slow down global warming (Forge, 2007). Biofuels are renewable and clean burning (bioethanol, biodiesel) and thus can be stored and distributed using existing infrastructure and, in turn, are easier to commercialize than other alternatives (SciDev.Net). In an attempt to lessen costs and cut down on greenhouse gas emissions, several airlines have begun testing biofuel blends in their jet fuel (Krauss, 2008). With the goals of increasing fuel security and reducing fuel cost and harmful environmental impacts, the U.S. military is testing biofuel blends (Biello, 2009). Carbon neutrality is the most debated point of biofuel, meaning that carbon is absorbed by growing plants and they release carbon they absorbed when harvested, thus emitting fewer greenhouse gases as compared to fossil fuels. The debate over the *net* carbon saving will give varied results depending on feedstock type, cultivation methods, conversion technology, and energy efficiency (Hazell, 2007). Potential of some biofuel crops such as Jatropha and Pongamia growing on marginal lands to improve soil quality and reduce erosion while their oil cake provides for soil-improving organic nutrients is considerable (Kartha, 2006). Further, biofuels as fuelwood and dung substitute can increase energy efficiency and decrease health risks.

However, expansion of biofuel crops by clearing land for planting biofuel crops can harm the environment by displacing other crops. It might threaten biodiversity and wildlife (Groom et al., 2008) in terms of conversion of natural forest and grassland into new cropland for growing feedstock meant for biofuel production (Searchinger et al., 2008). Biofuel opponents like Jean Ziegler, United Nations Special Rapporteur, on the right to food called it a "crime against humanity to divert arable land to the production of crops which are then burned for fuels" (Mathews, 2008). The current utilization of 1% of the world's arable land for biofuel development might increase up to 20% by 2050 by FAO estimation. In Indonesia and Malaysia, for the development of oil palm plantations, around 14–15 million hectares of peatlands have been cleared. Shifting of biodiverse ecosystem and farming systems to industrial monocultures might pose a threat to ecosystem integrity. Water pollution (Hoekman, 2009) and air pollution (Pimentel et al., 2009) caused by production and combustion of corn-based ethanol might have harmful impacts on human health and the environment. In some cases, dramatic decline in soil fertility and structure through biomass production is another environmental concern.

1.3.1.2 Economic Consequences

Through employment creation, biofuels could contribute to alleviating poverty as it is a labor-intensive job. Around 0.4 million, new jobs (in different public and private sectors) were created directly or indirectly in terms of production, construction, and research in ethanol industry that led to the addition of $42 billion to the U.S. economy in 2011 (U.S. Department of Energy). Biofuel production was stimulated in sugarcane-producing regions in Brazil, and through agro-industrial activities, the income generated further helped "capitalize" agriculture (Zarrilli, 2006). In many developing countries where land patterns, conditions, and uses are varied from those in the developed world, the competition for land uses among food and fuel might not be an overriding issue (Peskett et al., 2007). Economic development of the foreign exchange–strapped economies of many developing countries could be improved either by substituting for oil imports or by generating revenue through biofuel export.

However, concerns regarding rising food prices (owing to the need of corn for ethanol production) in the United States are contributing to a quick increase in agricultural commodity demands (Tangermann, 2008). Utilization of food crops as feedstocks for biofuel production has led to a rise in their prices, which might have an adverse impact on many food-importing countries. For example, there has been a hike in maize price mainly because of the U.S. ethanol program (WDR, 2008). Due to biodiesel production, prices of oil crops such as palm, soybean, and rapeseed have increased. Thus, owing to the increase in food prices, net purchasers of food and urban consumers are expected to suffer. Further, aviation fuel cost might rise if switched to biofuels (Newcomb, 2012), and thus, biofuel cost may become unaffordable without government subsidies (FAO, 2008).

1.3.1.3 Ethical/Social Implications

Utilization of cereal crop residues (low-cost and abundant agricultural waste) as potential feedstocks would neither require additional land nor increase greenhouse gas emissions during bioenergy production process (Hay, 2010). Developing biofuels will help the United Nations to maintain its global leadership in science and technology (*Scientific American*, 2012).

However, food supply might shrink because of increasing use of croplands for biofuels, which can drive up food prices to non-affordable levels, leading to world hunger (Ho, 2011). Further concerns regarding quality of life in local communities and a decline in residential property value surrounding biofuel plants have been raised by citizen groups (Hoyer and Saewitz, 2007).

1.3.1.4 Political Ramifications

The ongoing war in Iraq, anti-Americanism in the Middle East, and instability in the Persian Gulf risking oil tankers have led to new urgency regarding issue of energy independence. This U.S. energy independence can be met by developing biofuels that are domestically produced making a more reliable supply (Hoekman, 2009). To enable biofuels to compete with conventional gasoline and diesel, governments have provided substantial support including consumption and production incentives and mandatory blending standards (Raswant et al., 2008). The downfall in this arena includes the allocation of a large proportion of government funding (for domestic energy) toward biofuel development as compared to other renewable energies.

1.3.2 Party Identification and Public Opinion

An individual attachment to a political party based on a sense of closeness varying in degree of intensity is defined as party identification (Green et al., 2002). As opinion formation remains grounded in social identities, an individual's opinion regarding science and technology might be influenced by partisanship (Smith and Hogg, 2008). According to Hogg and Turner (1987), when individuals categorize themselves as members of a group, they behave consistently with the group prototype as they internalize the prototypical attributes of the group such as attitude endorsement. Party identification is thus an enduring, stable, and important predisposition (Goren, 2005).

In American political context, the Democratic Party's platform is based on modern liberalism (McGowan, 2007), whereas the Republican Party's platform is by and large grounded on conservative principles (Regnery, 2012). Public attitude regarding a mainstream renewable energy technology (biofuels and related policies) with interactive effect of party identification and temporarily accessible risk–benefit perception has been explained by Fung et al. (2014). Their results suggested that party identification likely influences individual attitude toward biofuels, after controlling risk–benefit priming. Ideologically liberal individuals (Democrats) tend to hold positive

attitudes toward biofuels, being more supportive than Republicans. After controlling for risk priming, biofuel utilization was more likely supported by Democrats.
However after controlling for biofuel benefit perceptions and considering biofuel
risks, there was significant effect on their attitudes. Their results also suggested that
when individuals thought of environmental risks, they were less likely to support
biofuel production and use. As compared to political risks, people tend to put more
emphasis on environmental and economic risks (Fung et al., 2014). Thus, the strength
of party identification considering different risk/benefits of biofuels can vary attitude
toward biofuels and related policies.

1.3.3 Public Perceptions in Varied Regions

Analysis of the public acceptance of biofuels in Greece shows that there is significant
lack of information regarding biofuels among the young and people of low education
with general concern regarding energy dependence on fossil fuels (Savvanidou et al.,
2010). As compared to other renewable resources, very few preferred biofuels, and the
chief concern of the majority of respondents was to first save energy than the use of
alternative energy. However, despite all hesitations, around half of the people considered biofuel usage as an effective solution for energy problems and against climatic
changes (Savvanidou et al., 2010).

In the United Kingdom, biofuel studies have shown that about half of UK farmers plan to grow biofuel sugar beet and 34% biofuel oilseed rape, whereas one-third
remain undecided (Mattison and Nottis, 2007). Only 55% of UK citizens know what
biofuels are, with 9 out of 10 not well informed about biofuels (FOE, 2008). Around
44% of UK citizens believe that improved public transport could reduce emissions
from road transport, whereas only 14% believe in biofuels as an effective solution to
these emissions (FOE, 2008).

In the United States, although citizens are not well informed about ethanol production, a majority of them consider "using biofuels is a good idea" (Wegener and Kelly,
2008). Around 60% of U.S. citizens are somewhat knowledgeable regarding biodiesel,
and if biodiesel prices are comparable to conventional diesel, majority (98%) of U.S.
farmers would use it (Lahmann, 2005). A survey conducted in 2008 reported that
around 67% of U.S. respondents were keen to learn more regarding the alternative
fuel, while a majority of those surveyed agreed that biofuels can help in cutting down
greenhouse gas emissions and help reduce U.S. dependence on foreign oil (Survey,
2009). However, reservations were expressed regarding corn-based ethanol with 43%
fearing that ethanol production will create pressure on local water supplies and 44%
voicing concern that it will put pressure on the food supply.

In a survey conducted in Canada (West, 2007), high to very high knowledge about
ethanol and biodiesel was shown by only 56% of the respondents, whereas the rest
had low to very low knowledge. Majority of Canadians (73%–85%) were in favor of
information actions, whereas 42%–56% were in favor of mandatory ethanol-blended
gasoline (West, 2007).

1.3.4 Second-Generation Biofuels and Public Opinion

In terms of sustainable growth in production and consumption, scientists are developing second-generation biofuels. Owing to higher yield per hectare of cellulose as compared to starch/sugar cops, second-generation biofuels will likely depend on cellulosic matter for feedstock. Second-generation biofuels can provide benefits including short-term gasoline price dampening, in terms of trade improvement and wealth transfers, and thus could make biofuels more beneficial to society (Rajagopal et al., 2007). A significantly higher EROI and carbon sequestration could be achieved with use of cellulosic biomass for energy production as compared to sugar-/starch-based biofuels (Sheehan et al., 2003; Farrell et al., 2006; Tilman et al., 2006). In terms of transportation sector, second-generation biofuels are considered one of the key mitigation technologies according to Intergovernmental Panel on Climate Change's fourth assessment report on climate change (IPCC, 2007). A change in agricultural landscape and the sources, levels, and variability of farm income will occur with large-scale production of new types of crops meant for energy production. Further, how well the process of technology adoption by farmers and processors is managed, will affect the socio-economic impact of biofuel production; thus, confluence of agricultural policy with environmental and energy policies is expected (Rajagopal et al., 2007).

Second-generation biofuel proponents claim they address many of the corn-based ethanol criticisms such as, compared to first-generation biofuels, cellulosic fuels that have a higher net energy balance, and as they are made from inedible feedstocks, they do not contribute to food price inflation (Bush, 2006; EAA, 2008). The development of an advanced biofuel industry that does not utilize food crops as its principal feedstock has been the focus of U.S. biofuel policy in the past 10 years (Mondou and Skogstad, 2012). To support the development of biofuel industry, the federal government of Canada has also pursued policies (Mondou and Skogstad, 2012), and some Canadian provinces imposed gasoline content mandates (Evans, 2013). Even though none of these policies in Canada has focused on specific targets for cellulosic ethanol (Mondou and Skogstad, 2012), Canadian industry has been much more supportive toward second-generation biofuel resources from agricultural waste products and forest biomass (Ackom et al., 2010). However, owing to the lack of specific policy support, the robustness of consumer demand for more sustainable biofuels will only ensure the future success of the industry in Canada (Dragojlovic and Einsiedel, 2015). Currently, in the United States, consumer perceptions of advanced biofuels are much more positive as compared to corn-based ethanol (Delshad et al., 2010; Delshad and Raymond, 2013) and strongly support biofuel policies in Canada (Dragojlovic and Einsiedel, 2014). However in both the United States and Canada, public attitude is relatively weak and subject to change as it is based on relatively low levels of information about technology (Wegener et al., 2014). Over time, the initially positive attitude of U.S. consumers toward corn-based ethanol has declined owing to potential economic, social, and environmental costs of the technology, which were pointed out by biofuel policy opponents (Talamini et al., 2012; Delshad and Raymond, 2013). Recently it has been suggested that although at the moment, cellulosic biofuels are being viewed as extremely positive in Canada, potential exists that if exposed to arguments pointing to negative environmental effects upon adoption of cellulosic biofuels, Canadians attitudes might become much more negative in the future (Dragojlovic and Einsiedel, 2015). Thus, public support is potentially vulnerable to various arguments focusing on unintended consequences regarding producing advanced biofuels.

1.4 Conclusions

Against a backdrop of the whole world facing crises of fossil fuel depletion and rapid rise in petroleum prices, it is imperative to explore newer alternatives that can help reduce oil import dependence and reduce emissions of greenhouse gases. Biofuel history dates back to its utilization in three forms: from solid biofuels (including firewood, wood charcoal, woodchips, and wood pellets) to liquid (bioethanol, biodiesel, biobutanol, etc.) and gaseous biofuels (syngas/methane gas). Among these, bioethanol and biodiesel have been researched and utilized extensively. In recent years, biofuel production and consumption have increased considerably; however, little is known regarding public attitude toward them. Second-generation biofuels are preferred over first-generation biofuels owing to various limitations associated. Thus, there is a need for strategic planning, careful assessment, and transparency in case of investment before rapidly advancing toward the route for biofuels as a solution for sustaining our energy demands. Sustainable standards must be developed for biofuel production, and it should be ensured that global diversity required to sustain life on earth is not affected by biofuel production.

References

Ackom, E., Mabee, W., and J. Saddler. 2010. Industrial sustainability of competing wood energy options in Canada. *Appl Biochem Biotechnol.* 162:2259.

AFDC. 2012. *Drop-in Biofuels.* Washington, DC: Alternative Fuel Data Center, Department of Energy. http://www.afdc.energy.gov/fuels/emerging_dropin_biofuels.html. Accessed on January 2, 2016.

Anonymous. 1917. *National Geographic.* 31:131.

Anonymous. 2004. Biomethanol from sugar beet pulp. *Energ Sust Dev Mag.* 3:15.

Antoni, D., Zverlov, V.V., and W.H. Schwarz. 2007. Biofuels from microbes. *Appl Microbiol Biotechnol.* 77:23–35.

Appolonia, A. and V. Sussman. 1983. The samaritan oil lamps from Apolonia—Arsuf, *Tel Aviv.* 10(4):71–96.

ASG Renaissance. 2004. Biodiesel end-user survey: Implications for industry growth. Final report out. The National Biodiesel Board, Jefferson City, MI. http://www.agmrc.org/media/cms/20040202_fle029_0994F38B6A5CC.pdf. Accessed on January 2, 2016.

Automobile Club de France. 1902. *Congres des applications de L'alcool denature, Paris.* Paris, France: Automobile Club de France.

Badger, P.C. 2002. Ethanol from cellulose: A general review. In: *Trends in New Crops and New Uses*, eds. Janick, J. and Whipkey, A., pp. 17–21. Alexandria, VA: ASHS Press.

Belden, R. and Stewart. 2011. Public opinion on federal farm and biofuels policy: Highlights from the 2010 survey on agriculture and the environment. Technical report. http://www.farmsfoodandfuel.org/system/files/BRS%202010%20Poll%20Highlights_2.pdf. Accessed on January 2, 2016.

Bente, P. 1984. *International Bioenergy Handbook*. Washington, DC: The Bioenergy Council, New York.

Biello, D. 2009. Navy green: Military investigates biofuels to power its ships and planes. *Scientific American*. http://www.scientificamerican.com/article/navy-investigates-biofuels-to-power-ships-airplanes. Accessed on January 2, 2016.

Bolsen, T. and F. Cook. 2008. The polls-trends: Public opinion on energy policy: 1974–2006. *Public Opin Q*. 72:364–388.

Bungay, H.R. 1982. Biomass refining. *Science*. 218:643–646.

Cacciatore, M.A., Binder, A.R., Scheufele, D.A., and B.R. Shaw. 2012. Public attitudes toward biofuels. Effects of knowledge, political partisanship, and media use. *Polit Life Sci*. 31:36–51.

Chang, C.-C. and S.-W. Wan. 1947. China's motor fuels from rung oil. *Ind Eng Chem*. 39:1543–1548.

Cheng, F.-W. 1945. China produces fuels from vegetable oils. *Chem Metall Eng*. 52:99.

Chowhury, D.H., Mukerji, S.N., Aggarwal, J.S., and L.C. Verman. 1942. Indian vegetable fuel oils for diesel engines. *Gas Oil Power*. 37:80–85.

Chu and Vilsack. 2009. Secretaries Chu and Vilsack announce more than $600 million investment in advanced biorefinery projects. Washington, DC: United States Department of Energy. http://energy.gov/articles/secretaries-chu-and-vilsack-announce-more-600-million-investment-advanced-biorefinery. Accessed on January 5, 2016.

De Decker, K. 2010. Wood gas vehicles: Firewood in the fuel tank. *Low-Tech Magazine*. http://www.lowtechmagazine.com/2010/01/wood-gas-cars.html. Accessed on January 8, 2016.

Delshad, A. and L. Raymond. 2013. Media framing and public attitudes toward biofuels. *Rev Policy Res*. 30:190.

Delshad, A.B., Raymond, L., Sawicki, V., and D.T. Wegener. 2010. Public attitudes toward political and technological options for biofuels. *Energy Pol*. 38:3414–3425.

Demain, A.L., Newcomb, M., and J.H.D. Wu. 2005. Cellulase, *Clostridia*, and ethanol. *Microbiol Mol Biol Rev*. 69:124–154.

Demirbas, A. 2007. Progress and recent trends in biofuels. *Progr Energ Combust Sci*. 33:1–18.

Demirbas, A. 2009. Political, economic and environmental impacts of biofuels: A review. *Appl Energy*. 86:S108–S117.

Dimitri, C. and A. Effland. 2007. Fueling the automobile: An economic exploration of early adoption of gasoline over ethanol. *J Agric Food Indus Org*. 5:1–17.

DOE. 2000. *Biomass Co-Firing: A Renewable Alternative for Utilities*. Washington, DC: US Department of Energy.

DOE. 2013. *The History of Natural Gas*. Washington, DC: US Department of Energy. http://fossil.energy.gov/education/energylessons/gas/gas_history.html. Accessed on January 5, 2016.

Dragojlovic, N. and E. Einsiedel. 2014. The polarization of public opinion on biofuels in North America: Key drivers and future trends. *Biofuels* 5:233.

Egloff, G. 1939. *Motor Fuel Economy of Europe*. Washington, DC: American Petroleum Institute.

EIA. 2014. *Monthly biodiesel production report*. Washington, DC: US Energy Information Administration.

Evans, B. 2013. Biofuels Canada: USDA report. Global Agricultural Information Network (GAIN) report. Ottawa, Canada: USDA Foreign Agricultural Service.

FAO. 2008. Biofuels: Prospects, risks and opportunities. Rome, Italy: Food and Agriculture Organization of the United Nations. http://www.fao.org/docrep/011/i0100e/i0100e00.htm. Accessed on January 5, 2016.

FAO. 2013. FAOStat—Forestry database. Geneva, Switzerland: Food and Agriculture Organization of the United Nations.

Farrell, A.E., Plevin, R.J., Turner, B.T., Jones, A.D., O'Hare, M., and D.M. Kammen. 2006. Ethanol can contribute to energy and environmental goals. *Science*. 311:506–508.

Finlay, M.R. 2004. Old efforts at new uses: A brief history of chemurgy and the American search for biobased materials. *J Ind Ecol*. 7:33–46.

FOE. 2008. Public in the dark about biofuels in their petrol. Friends of the Earth. www.foe.co.uk/resource/press_releases/public_in_the_dark_about_b_15042008. Accessed on January 5, 2016.

Forge, F. 2007. *Biofuels—An Energy, Environmental or Agricultural Policy?* Ottawa, Ontario, Canada: Library of Parliament, Science and Technology Division, Government of Canada. http://www.parl.gc.ca/information/library/PRBpubs/prb0637-e.htm#why. Accessed on January 7, 2016.

Fox, H.S. 1924. Alcohol motor fuels. In: *Supplementary Report to World Trade in Gasoline*, Bureau of Domestic and Foreign Commerce, US Department of Commerce Monograph, Trade Promotion Series No. 20. Washington, DC: Government Printing Office.

Fung, T.K.F., Choi, D.H., Scheufele, D.A., and B.R. Shaw. 2014. Public opinion about biofuels: The interplay between party identification and risk/benefit perception. *Energ Pol*. 73:344–355.

Gaskell, G., Allum, N., Wagner, W. et al. 2004. GM foods and the misperception of risk perception. *Risk Anal*. 24:185–194.

Gaskell, G., Bauer, M.W., Durant, J., and N. Allum. 1999. Worlds apart? The reception of genetically modified foods in Europe and the US. *Science*. 285:384–387.

Goren, P. 2005. Party identification and core political values. *Am J Polit Sci*. 49:882–897.

Green, D., Palmquist, B., and E. Schickler. 2002. *Partisan Hearts and Minds: Political Parties and the Social Identities of Voters*. New Haven, CT: Yale University Press.

Groom, M., Gray, E., and P. Townsend. 2008. Biofuels and biodiversity: Principles for creating better policies for biofuels production. *Conserv Biol*. 22:602–609.

Hay, J. 2010. Value of crops residues as biofuels. http://cropwatch.unl.edu/web/bioenergy/crop-residues. Accessed on January 7, 2016.

Hazell, P. 2007. Bioenergy: Opportunities and challenges. In: Presentation made for the *Sweet Sorghum Consultation at IFAD*, Rome, Italy.

Ho, M. 2011. Biofuels and world hunger. *Third World Resurgence*. 247:26–28.

Hoekman, S. 2009. Biofuels in the U.S.—Challenges and opportunities. *Renew Energ*. 34:14–22.

Hogg, M. and J. Turner. 1987. Social identity and conformity: A theory of referent informational influence. In: *Current Issues in European Social Psychology*, eds. Doise, W. and Moscovici, S., pp. 139–182. Cambridge, U.K.: Cambridge University Press.

Hoyer, M. and M. Saewitz. 2007. Study denounces project's impact. November 20, 2007, News paper *Virginia-Pilot*, USA.

Hunt, D.V. 1981. *The Gasohol Handbook*. New York: Industrial Press.

IEA. 2013. *World Energy Outlook 2013*. Paris, France: International Energy Agency.

IEA Bioenergy Task. 2015. The potential and challenges of dropin fuels (members only). IEA Bioenergy Task 39 Commercializing liquid biofuels. task39.sites.olt.ubc.ca. Accessed on January 7, 2016.

IER. 2013. EPA ignores reality with 2014 ethanol mandates. Washington, DC: Institute for Energy Research. http://instituteforenergyresearch.org/analysis/epa-ignores-reality-with-2014-ethanol-mandate. Accessed on January 7, 2016.

IPCC. 2007. Summary for policymakers climate change 2007: Mitigation. Contribution of working group III to the fourth assessment, report of the Intergovernmental Panel on Climate Change. Cambridge, U.K.: Cambridge University Press.

Kartha, S. 2006. Environmental effects of bioenergy. In: *Bioenergy and Agriculture: Promises and Challenges*, eds. Hazell, P. and Pachauri, R., Focus 14. Washington, DC: IFPRI.

Kim, K. 2011. Public understanding of the politics of global warming in the news media: The hostile media approach. *Public Underst Sci.* 20:690–705.

Knothe, G. 2001. Historical perspective on vegetable oil-based diesel fuels. *AOCS Inform.* 12:1103–1107.

Kovarik, B., Ford, H., and C. Kettering. 1998. The "fuel of the future". *Automot Hist Rev.* 32:7–27.

Krauss, C. 2008. Taking flight on jatropha fuel. *New York Times.* http://green.blogs. nytimes.com/2008/12/09/taking-flight-on-jatropha-fuel/. Accessed on January 7, 2016.

Lahmann, E.L. 2005. Biodiesel in Oregon: An agricultural perspective. Honors baccalaureate of science in environmental economics, policy, and management. University of Oregon, Eugene.

Lifset, R. 2014. *American Energy Policy in the 1970s.* Norman, OK: University of Oklahoma Press.

Loeb, A.P. 2004. Steam versus electric versus internal combustion: Choosing vehicle technology at the start of the automotive age. *Transport Res Rec.* 1885:1–7.

London Times. 1902. The Kaiser's new scheme. Reprint, *New York Times*, April 24, 1902, p. 9.

London Times. 1914. Alcohol motor fuel. *London Times*, January 22, 1914, p. 10.

London Times. 1921. Fuel from waste. *London Times*, December 24, 1921, p. 14.

Lucke, C.E. and S.M. Woodward. 1907. The use of alcohol and gasoline in farm engines. USDA Farmers Bulletin No. 277. Washington, DC: GPO.

Lusk P. 1998. Methane recovery from animal manures: The current opportunities casebook. NREL/SR-580-25145. Golden, CO: National Renewable Energy Laboratory.

Maniatis, K. 2003. Biofuels in the European Union. Paper presented to *F.O. Lichts World Biofuels*, Seville, Spain.

Mathews, J.A. 2008. Opinion: Is growing biofuel crops a crime against humanity? *Biofuels Bioproduct Bioref.* 2:97–99.

Mattison, E.H. and K. Nottis. 2007. Intentions of UK farmers toward biofuels crop production: Implications for policy targets and land use change. *Environ Sci Technol.* 41:5589–5594.

McGowan, J. 2007. *American Liberalism: An Interpretation for Our Time.* Chapel Hill, NC: The University of North Carolina Press.

Mondou, M. and G. Skogstad. 2012. The regulation of biofuels in the United States, European Union and Canada. University of Toronto, Canada. http://www.ag-innovation.usask. ca/Mondou%20&%20Skogstad-CAIRN%20report-30%20March.pdf. Accessed on January 10, 2016.

Newcomb, T. 2012. *Jet Green Times.* http://business.time.com/2012/09/13/biofuel-industry-scales-up-for-airlines. Accessed on January 7, 2016.

Peskett, L., Slater, R., Stevens, C., and A. Dufey. 2007. Biofuels, agriculture and poverty reduction. Overseas Development Institute (ODI). Programme of advisory support services for rural livelihoods. Department for International Development, London, England, pp. 1–24.

Pew Research Center. 2008. Ethanol research loses ground, continued division on ANWR: Public sends mixed signals on energy policy. Washington, DC. http://www.people-press.org. Accessed on January 8, 2016.

Pimentel, D., Marklein, A., Toth, M. et al. 2009. Food versus biofuels: Environmental and economic costs. *Hum Ecol.* 37:1–12.

Rabe, B. and C. Borick. 2008. Global warming and climate policy options: Key findings report. Policy report, 11. Ann Arbor, MI: Center for Local, State, Urban Policy, University of Michigan.

Rajagopal, D., Sexton, S.E., Roland-Holst, D., and Zilberman, D. 2007. Challenge of biofuel: Filling the tank without emptying the stomach? *Environ Res Lett.* 2:044004.

Rajgor, G. 2013. Renewable power generation—2012 figures. Part six: Biomass. *Renew Energ Focus.* 14:24–25.

Ramirez, J., Brown, R., and R. Thomas. 2015. A review of hydrothermal liquefaction biocrude properties and prospects for upgrading to transportation fuels. *Energies.* 8:6765–6794.

Raswant, V., Hart, N., and M. Romano. 2008. Biofuel expansion: Challenges, risks and opportunities for rural poor people. How the poor can benefit from this emerging opportunity. Paper presented at the Round Table organized during the thirty-first session of IFAD's Governing Council, Rome, Italy.

Regnery, A. 2012. The pillars of modern American conservatism. *Intercoll Rev.* 47:3–12.

RFA. 2014. *Ethanol Industry Statistics.* Washington, DC: Renewable Fuel Association. http://www.ethanolrfa.org/pages/statistics. Accessed on January 8, 2016.

Russell, L.S. 2003. *A Heritage of Light: Lamps and Lighting in the Early Canadian Home.* University of Toronto Press, Toronto, Canada.

Savvanidou, E., Zervas, E., and K.P. Tsagarakis. 2010. Public acceptance of biofuels. *Energ Pol.* 38:3482–3488.

Scientific American. 2012. Obama and Romney Tackle 14 Top Science Questions. *Scientific American.* http://www.scientificamerican.com/article.cfm?id=obama-romney-science-debate. Accessed on January 8, 2016.

Searchinger, T., Heimlich, R., Houghton, R. et al. 2008. Use of U.S. croplands for biofuels increases greenhouse gases through emissions from land use change. *Science.* 319(5867):1238–1240. http://www.bionica.info/biblioteca/Searchinger2008CroplandsIncreaseGreenhouse.pdf. Accessed on January 10, 2016.

Sheehan, J., Aden, A., Paustian, K. et al. 2003. Energy and environmental aspects of using corn stover for fuel ethanol. *J Ind Ecol.* 7:117–146.

Siegrist, M. 2000. The influence of trust and perceptions of risks and benefits on the acceptance of gene technology. *Risk Anal.* 20:195–204.

Siegrist, M., Cvetkovich, G., and C. Roth. 2000. Salient value similarity, social trust, and risk/benefit perception. *Risk Anal.* 20:353–362.

Sjoberg, L. 2004. Principles of risk perception applied to gene technology. *EMBO Rep.* 5:S47–S51.

Smith, J. and M. Hogg. 2008. Social identity and attitudes. In: *Attitudes and Attitude Change,* eds. Crano, W. and Prislin, R., pp. 337–360. New York: Psychology Press.

Songstad, D.D., Lakshmanan, P., Chen, J., Gibbons, W., Hughes, S., and R. Nelson. 2009. Historical perspective of biofuels: Learning from the past to rediscover the future. *In Vitro Cell Dev Biol Plant.* 45:189–192.

Stephan, D. 2013. *Germany Remains the World's Leading Biogas Energy Producer.* Würzburg, Germany: Vogel Business Media. http://www.processworldwide.com/management/markets_industries/articles/391244. Accessed on January 10, 2016.

Survey. 2009. Survey shows high interest in biofuels. Madison, WI: University of Wisconsin-Madison. http://www.news.wisc.edu/releases/15175. Accessed on January 10, 2016.

Talamini, E., Caldarelli, C.E., Wubben, E.F.M., and H. Dewes. 2012. The composition and impact of stakeholders' agendas on US ethanol production. *Energy Pol.* 50:647.

Tangermann, S. 2008. What's causing global food price inflation? http://www.voxeu.org/index.php?q=node/1437. Accessed on January 10, 2016.

Thomson, J. 2003. *The Scot Who Lit the World—The Story of William Murdoch Inventor of Gas Lighting.* Glasgow, U.K.: Janet Thomson.

Tilman, D., Hill, J., and C. Lehman. 2006. Carbon-negative biofuels from low-input high-diversity grassland biomass. *Science.* 314:1598.

Tweedy, R.N. 1917. *Industrial Alcohol.* Dublin, Ireland: Plunkett House.

US Department of Energy. 2013. Biofuels create green jobs. http://www1.eere.energy.gov/bioenergy/pdfs/green_jobs_factsheet2.pdf. Accessed on January 10, 2016.

US Patent No. 4378, Issued April 1, 1826.

van Tilburg, M. 2013. More pellets, please. Peachtree Corners, GA: *Site Selection Magazine.* http://www.siteselection.com/issues/2013/jul/world-reports.cfm. Accessed on January 10, 2016.

Wang, Q. 2006. Biomethanol conversion from sugar beet pulp with pectin methyl esterase. Master thesis, University of Maryland, USA.

Wang, R. 1988. Development of biodiesel fuel. *Taiyangneng Xuehao.* 9:434–436.

WDR. 2008. World Bank, World Development Report (WDR) 2008, Washington, DC.

Wegener, D. and J.R. Kelly. 2008. Social psychological dimensions of bioenergy development and public acceptance. *Bioenergy Res.* 1:107–117.

Wegener, D.T., Kelly, J.R., Wallace, L.E., and V. Sawicki. 2014. Public opinions of biofuels: Attitude strength and willingness to use biofuels. *Biofuels.* 5:249.

West, G. 2007. Canadian attitudes toward government intervention policies in the biofuel market. In: *Agricultural Institute of Canada Conference,* Edmonton, Alberta, Canada.

Wohlers, A.E. 2010. Regulating genetically modified food: Policy trajectories, political culture, and risk perceptions in the U.S., Canada, and EU. *Polit Life Sci.* 29:17–39.

Xia, Z. 2013. *Domestic Biogas in a Changing China: Can Biogas Still Meet the Energy Needs of China's Rural Households?* London, U.K.: International Institute for Environment and Development.

Zaborsky, O.R. 1982. Chemicals from renewable resources: An endorsement for biotechnology. *Enzyme Microbiol Technol.* 4:364–365.

Zarrilli, S. 2006. Trade and sustainable development implications of the emerging biofuels market in international center for trade and sustainable development linking trade, climate change and energy: Selected issue briefs. www.ictsd.org. Accessed on January 10, 2016.

Zverlov, V.V., Berezina, O., Velikodvorskaya, G.A., and W.H. Schwarz. 2006. Bacterial acetone and butanol production by industrial fermentation in the Soviet Union: Use of hydrolyzed agricultural waste for biorefinery. *Appl Microbiol Biotechnol.* 71:587–597.

2

Climate Change, Biofuels, and Conflict

Tobias Ide and Kirsten Selbmann

Contents

Abstract

The potential links between climate change and violent conflict gained much attention from scientists and policy makers alike, while few studies have discussed the possible impact of climate change mitigation measures on conflict dynamics. This chapter addresses this gap by investigating whether biofuel production, frequently discussed as a contribution to reduce greenhouse gas emissions and mitigate climate change, has the potential to influence the dynamics of violent conflicts. In order to do so, it first highlights the potential negative impacts of biofuel production in the local cultivation areas. The conflict relevance of these impacts is then assessed from an environmental conflict and a political ecology perspective. The environmental conflict perspective identifies food insecurity, socioeconomic decline, and land evictions as drivers of violent conflict under conditions such as preexisting tensions, recent political changes, unequal resource distribution patterns, low economic growth, or high poverty rates. The relevance of large-scale resource utilization schemes for material and symbolic conflicts within local communities as well as between local communities and more powerful national/international players is emphasized by the political ecology perspective. An in-depth case study of Brazil, and particularly of the locations of Mato Grosso do Sul and Bico do Papagaio, illustrates the theoretical claims and empirical results of both perspectives. This chapter concludes with four suggestions for future research on biofuels and conflict: avoid selecting cases on the dependent variable, trace whether conflicts are related to biofuels or rather to similar developments, put more emphasis on intragroup conflicts, and study the indirect effects of biofuel production.

2.1 Introduction

Possible links between climate change and violent conflict have gained broader attention in the political and scholarly community since 2007 (McDonald 2013). The rapidly expanding research field of climate change and conflict could build on previous studies on socioenvironmental conflicts (e.g., Homer-Dixon and Blitt 1998; Peluso and Watts 2001) and civil war onset (e.g., Blattman and Miguel 2010; Dixon 2009). But this literature has so far largely ignored the potential links between conflict and climate change adaptation and mitigation measures (Nordås and Gleditsch 2007), such as hydroenergy projects, higher energy prices, or geoengineering (Maas and Scheffran 2012; Zhouri and Oliveira 2009). Similarly, there is a very large amount of studies dealing with the potential negative impacts of biofuels, but few of them use concepts and insights developed by peace and conflict studies or socioenvironmental conflict research (Dietz et al. 2015). This chapter addresses this gap by investigating whether biofuel production, frequently discussed as a contribution to reduce greenhouse gas emissions and mitigate climate change (WBGU 2009), has the potential to influence the dynamics of violent conflicts. In doing so, it will draw on and connect the literature on (1) the biofuel production, (2) the political ecology of conflict, and (3) the causes and triggers of violent conflicts.

Existing research has claimed that biofuels can contribute to conflicts, including violent conflicts, in two ways: either they aggravate tensions in the production areas or they contribute to a rise of world food prices, thus stimulating grievances and eventually violence (Webersik and Bergius 2013). Due to the limited scope of this chapter, we focus solely on the local (or regional) tensions caused by biofuels in or around the production areas. In addition, we only deal with the links between violent conflict and the cultivation of crops for biofuel production. A conflict is defined in this study as a manifest clash of two or more social groups' interests that are perceived as contradictive or incompatible by the respective groups (Ide et al. 2014). We use the term violence in order to describe all forms of physical, direct violence against human beings or property (Galtung 1969).

This chapter proceeds as follows. In the next section, we outline the possible local- and regional-level negative impacts of biofuel cultivation. We then go on to assess the relevance of these impacts for violent conflict by building on two research perspectives, namely, environmental conflict and political ecology. Afterwards, the case of Brazil as one of the largest producers and exporters of biofuels is discussed in greater detail before a conclusion is drawn.

2.2 Biofuel Cultivation and Its Consequences

The issue of biofuels is controversially discussed in the public sphere and the scientific community. Nearly every aspect of biofuels is a matter of intensive debate, including their carbon footprint, influence on tropical rain forests, impact on workers' rights and indigenous communities, and relation to food prices and to the industrial agriculture (e.g., Fast 2009; McMichael 2009; Venghaus and Selbmann 2014). With regard to the consequences in the local production areas, the following negative impacts of biofuel cultivation are frequently mentioned in the literature (e.g., Anseeuw et al. 2012; Cotula et al. 2008; Garcez and Vianna 2009; Obidzinski et al. 2012):

1. Crops for the production of biofuels are usually cultivated in large-scale monoculture plantations, supported by huge amounts of fertilizers, pesticides, and fungicides as well as large-scale irrigation systems. These characteristics can, alone or in combination, contribute to the scarcity of land or good-quality water.
2. There are several reports about the expulsion of smallholders and indigenous people from their land in order to grow sugarcane, oil palms, or soy. This is especially the case if these people have no formal land titles, even though they live on the land for a long time and their titles are sanctioned by customary law. In addition, the land that was considered as unproductive and thus available for land reform is now used for biofuel cultivation.
3. Deforestation can occur as a direct consequence of biofuel production if forests are cleared in order to cultivate soy or sugarcane. However, many experts are more concerned about indirect land use changes, including deforestation, caused by biofuels. These occur when farmers or pastoralists are displaced from or forced to sell their land and then move to forest areas in order to establish new agricultural or pastoral plots.
4. Biofuels can also contribute to local food insecurity if smallholders lose the plot of land securing their livelihood or if food cultivation is abandoned in favor of soy, oil palm, or sugarcane plantations for biofuel production.
5. Biofuels are accused of producing serious health hazards in the cultivation regions due to the contamination of soil and water by agrochemicals. A related problem is air pollution caused by the burning of sugarcane fields in order to make the harvesting easier or by the burning of forests in order to gain new land for cultivation.
6. There is evidence that biofuels cause socioeconomic decline in the cultivation regions, for example, because their production offers less job opportunities than traditional farming or because working in sugarcane and soy plantations is dangerous and poorly paid.
7. Biofuels are said to cause two types of migration. First, rural-to-rural migration of poor landless people occurs when workers move to the producing regions in order to work at the plantations. These people are often hardly socially integrated in the receiving regions and are frequently denied basic worker rights. Second, there is also an out-migration from the soy- and sugarcane-producing regions to urban areas because people have lost their land, job, and/or income. The consequence is rural-to-urban migration of individuals, often to urban slums.

That does not mean, however, that biofuel cultivation has these impacts everywhere and every time. We rather claim that the mentioned effects of biofuel production can be found in a significant number of cases in the past and very likely also in the future, thus making it worthy to investigate their effect on violent conflicts.

2.3 Biofuel Cultivation, Conflict, and Violence: Theoretical and Empirical Insights

There is a lack of research dealing explicitly with the links between biofuel cultivation and violent conflict, but existing research on environment-conflict links can be used to gain some insights into the conflict potential of biofuels in the local production areas. Especially the environmental conflict and political ecology perspectives are important in this context.

2.3.1 Environmental Conflict Perspective

The environmental conflict perspective gained ground in peace and conflict research during the 1990s (Homer-Dixon and Blitt 1998) and inspired a large number of studies conducted in the last decade (Meierding 2013). The core claim made by this perspective is that environmental changes can trigger renewable resource scarcity and migration if the sensitivity of the affected communities is high and no adaptation measures are conducted. Resource scarcity and migration can then, in turn, raise the risk of violent conflict onset. Resource scarcity, for instance, might stimulate direct conflict over access to important resources or fuel general grievances due to a reduction of livelihood security (Deligiannis 2012; Homer-Dixon 1999). But a scarcity of renewable resources can also lead to lower opportunity costs for joining a violent group vis-à-vis remaining in the formal economy (Barnett and Adger 2007) or to better opportunities for elite-driven mobilization of resource-scarce and thus economically deprived social or ethnic groups (Kahl 2006). Migration is suspected to cause ethnic or social tensions or conflicts over scarce resources in the receiving areas (Reuveny 2008). However, the onset of such violent environmental conflicts is always dependent on several scope conditions, such as preexisting tensions, recent political changes, unequal resource distribution patterns, low economic growth, or high poverty rates (Carius et al. 2006; Ide 2015).

There are hardly any studies testing the claims of the environmental conflict perspective in the context of biofuel production (Backhouse 2015; Gerber 2011). This is surprising given that the cultivation of sugarcane, oil palms, and soy is claimed to cause the scarcity of land, forest resources, food, and good-quality freshwater as well as migration. However, one can use studies on the conflict relevance of each of these factors in order to gain insights on the conflict relevance of biofuel cultivation (see Table 2.1 for an overview).

Large-N studies concerning the conflict relevance of a reduced availability of freshwater show no conclusive results. For example, Raleigh and Urdal (2007) and Gizelis and Wooden (2010) claim a weak but significant connection between lower freshwater availability and civil conflicts. The results of Theisen (2008) show no significant correlation, which is in line with the study of Melander and Sundberg (2011). Hendrix and Glaser (2007) and Salehyan and Hendrix (2014) even find a low freshwater availability to be weakly (but significantly) correlated with a lower violent conflict risk. The literature on links between rainfall changes (which could

TABLE 2.1 Conflict Relevance of the Assumed Impacts of Biofuel Cultivation

Assumed Impact of Biofuel Cultivation	Robust Link to Violent Conflict
Water scarcity	No
Food insecurity	Yes
Socioeconomic decline	Yes
Land evictions	Yes
Deforestation	No
Reduced health status	No
Internal migration	No

be used as a proxy for freshwater shortages) and violent conflict is similarly inconclusive (Theisen et al. 2013).

Food insecurity is widely acknowledged to be a driver of violent conflicts in the literature. This is confirmed by a large number of quantitative studies (e.g., Raleigh et al. 2015; Rowhani et al. 2011; Smith 2014; Wischnath and Buhaug 2014) and also by several case studies (e.g., Engels 2014; Schilling et al. 2012) and theoretical considerations (e.g., Barnett and Adger 2007). The same is true of socioeconomic decline, which is identified as a cause or trigger of violent conflict in nearly all studies available. This effect is independent of the concrete operationalization of socioeconomic decline, being it low economic growth, low per capita income, or high infant mortality (Dixon 2009; Esty et al. 1999; Hegre and Sambanis 2006).

To the authors' knowledge, there is no systematic evaluation of the conflict relevance of the eviction of small-scale farmers, herders, and indigenous people from their land. However, there is a similarity between expulsion of rural people from their land and soil degradation because the most relevant consequence of both processes is the inability of rural people to sustain their livelihoods due to a lack of sufficient agricultural land. Despite the exact magnitude and causal chains are still disputed, there is an increasing consensus in the literature that land degradation is a driver of violent conflict (e.g., Esty et al. 1999; Raleigh and Urdal 2007; Theisen 2008). This is supported by a case study evidence that finds that unequal land distribution and the expulsion of people living on their land for a long time can lead to violent conflict (Howard and Homer-Dixon 1998; Kahl 2006). Still, one should have in mind these are at best indications and by no means evidence for a general correlation between land evictions and violent conflicts.

Regarding the conflictivity of deforestation, there are only four more general studies available. Two of the three studies supporting a deforestation-conflict nexus are quite old (Esty et al. 1999; Hauge and Ellingsen 1998), and the results of Hauge and Ellingsen (1998) could not be reproduced by Theisen (2008). Only the results of Gerber (2011) are more recent and robust. Therefore, there is no empirical basis to make any claims about the relevance of deforestation for violent conflict. Similarly, Pinstrup-Andersen and Shimokawa (2008) provide some evidence that a poor health status is indeed positively correlated with the risk of civil war onset, but there are not enough studies on this issue yet. Finally, there are some studies conforming a correlation between international migration and violent conflict (Reuveny 2007; Salehyan and Gleditsch 2006), but the relevance of internal migration in this context is not well understood, probably because of a lack of adequate data (Reuveny 2008).

Therefore, we can conclude from the environmental conflict literature that food insecurity, socioeconomic decline, and land evictions are those possible consequences of large-scale biofuel cultivation which are most likely to drive violent conflicts.

2.3.2 Political Ecology Perspective

The political ecology tradition also focuses on the relations between the environment and conflicts but chooses a theoretical and empirical focus that is quite different from the one taken by the environmental conflict perspective. In general, political ecologists emphasize that there exists a natural environment with defined biophysical characteristics but that this environment is always structured by human activities. There exist societal rules defining who can control, assess, and use natural resources, and these rules are often shaped by unequal power relationships (Peluso and Watts 2001; Verhoeven 2011). Especially the enclosure or appropriation of natural resources by actors with previously low access to or control over these resources often provides the starting point for struggles between social groups. The scarcity and importance of natural resources is therefore culturally, historically, and socially produced, and the resulting socioenvironmental struggles are driven by local and site-specific and regional, national, and transnational interest constellations, cultures, and power relations (Horowitz 2009; Vandergeest 2003).

Political ecology has done extensive research on the consequences of large-scale natural resource utilization schemes. Examples include the Bougainville copper mine (Regan 1998), nickel mining in New Caledonia (Horowitz 2009), sugarcane plantations in the Philippines (Montefrio 2013), lumbering in Cambodia (Le Billon 2000), and industrial agriculture in India (Jewitt 2008). Two insights from this research can be used to assess the relevance of biofuel cultivation for violent conflict onset in Brazil. First, nearly all case studies agree that large-scale (capitalist) environmental projects cause tensions between local inhabitants, which are affected by evictions and the negative ecological side effects such as water pollution, and the state and/or private companies, which are executing these resource utilization schemes. These conflicts are especially likely to escalate into violence if they occur in regions inhabited by traditionally marginalized groups or if the local people feel their culture or livelihood is severely threatened (Allen 2013). In these cases, people are often ready to defend and unwilling to leave their traditional land if they are not (violently) forced to do so.

If separatist or rebel groups already exist in the respective region, there is even a risk for the onset of a civil war (Regan 1998), although this scenario is unlikely to become a reality in comparatively strong states such as Brazil or Malaysia. But other violent expressions of conflicts associated with large-scale resource utilization schemes are likely to occur around biofuel plantations, also because some facilitating conditions such as a high symbolic relevance of land, preexisting identity-based cleavages (e.g., indigenous vs. nonindigenous), and concrete negative impacts on the local population can be detected in many biofuel-producing countries (Dietz et al. 2015; Wright and Wolford 2003). Such forms of violence usually encompass destruction of production equipment and facilities, imprisonments, and violent evictions of local inhabitants (especially activists and resistant political leaders) as well as violent clashes between the affected local populations and the security forces of the state and/or company responsible for the utilization scheme (Watts 2004). Indeed, there are several reports about the occurrence of these forms of violence related to the cultivation of sugarcane, oil palms, and soy in countries as diverse as Brazil, Indonesia, or Mozambique (Cotula et al. 2008).

A second key finding of political ecology research is that large-scale natural resource utilization schemes create tensions within local communities, sometimes even escalating into violence. This is the case because large-scale resource utilization schemes tend to divide the local communities into several people benefiting from the project (people who can sell their land for good prices or find jobs in the extraction site) and usually a larger group of people who suffers from the negative impacts of the utilization scheme without having access to jobs, adequate compensation, improved healthcare, or other benefits (Gerber 2011). The latter group of people often campaigns against the project, while the former group of beneficiaries tends to support the project (Horowitz 2009; Montefrio 2013; Watts 2004). To quote one example mentioned by Reportér Brasil (2009) in the context of soy farming:

> In the Ligeiro IL, in the state of Rio Grande do Sul, mechanised monoculture has created income concentration among the 1.9 thousand Kaigang who live in the area and a divide in the villages. In 2005, 300 Indians who did not agree with the use of the territory for commercial production of soybean were forcefully withdrawn from the area by leaders of the negotiation with farmers.

2.4 Case Study: Brazil

In this section, we are going to apply and illustrate the theoretical and empirical insights discussed in the previous section by focusing on the case of Brazil. We have chosen this case for two reasons. First, Brazil is the second largest producer and first largest exporter of ethanol to be used in car engines worldwide (Kaup and Selbmann 2013; WBGU 2009). While this ethanol is almost completely extracted out of sugarcane, in 2002, the Brazilian government launched a large-scale biodiesel program, utilizing mainly soy (Pousa et al. 2007). Second, there have been various claims about the conflict-stimulating effects of biofuels in Brazil for years (Backhouse 2015; Fernandes et al. 2010; Venghaus and Selbmann 2014).

Brazil does not currently experience a civil war and is unlikely to do so in the near future given its comparatively high political stability, economic growth, and standard of living. But with regard to low-level violent conflict, Brazil is a most-likely case for a biofuel-conflict link according to the environmental conflict and political ecology perspectives. Large parts of the country are still suffering from social tensions, high poverty rates, a highly unequal distribution of land/wealth, and the discrimination of indigenous people. Sugar and soy are mainly produced at large, agro-industrial plantations. Few people (and even fewer locals) benefit from these plantations, while they often cause a loss of land, food insecurity, and socioeconomic decline in the surrounding regions. This causes local grievances and resistance (especially if the symbolic value of land is considered high), which in turn might lead to the use of force by those who benefit from biofuel cultivation (e.g., landowners and the companies involved in sugarcane/soy cultivation and processing) (Avritzer 2010; OMCT 2008). The resulting escalation process has the potential to erupt into violence, although it is also possible that one of the involved groups backs down or the conflict is mediated before turning violent.

Since Brazil is a large country characterized by cultural, economic, political, and ecological diversity, it is impossible to specify these claims further given the lack of comparable, high-quality data for conflicts (or their absence) around sugarcane or soy plantations. In the following, we therefore provide evidence from two cases that are documented rather well and that are located in different states of Brazil.

2.4.1 Guarani-Kaiowá in Mato Grosso do Sul

This case study was produced by drawing on several reports (Anaya 2009; Suárez et al. 2008; Survival International 2010) and more recent information on the case provided by the *BBC News* (2011, 2012) and Survival International (2009, n.d.).

The Guarani-Kaiowá are indigenous people consisting out of several large family groups, each of them owning their own land (*tekohá*). The Guarani's understanding of owning the land refers not only the physical space but also the resources (e.g., forests, waters, animals) located on it and to the spiritual dimension of land as a home for the group and the heritage of honored ancestors. But since the 1970s, land in the Guarani territory has been allocated to large-scale cattle ranches and more recently to sugarcane plantations. Today, there are twenty sugarcane factories surrounded by large sugarcane fields in Mato Grosso do Sul. Of these factories, 13 are located on land claimed by the Guarani. Several other sugarcane factories are planned to open on land considered as ancestral by the Guarani (ten Kate 2011).

The consequences of the establishment of this industrial sugarcane agriculture include the pollution of rivers from which the Guarani extract drinking water, rapid deforestation, and increasing land pressure. More and more Guarani families are forced to live either in overcrowded reserves or on roadsides. The amount of land available to the Guarani is constantly shrinking while they are suffering the pollution of water, soil, and crops by agrochemicals. As a consequence, most Guarani are unable to continue their traditional lifestyle of hunting, fishing, and cultivating crops, have lost their self-sufficiency, and became dependent on plantation work or government support. The consequences of this dire situation are increasing poverty, alcoholism, bad health conditions and malnutrition, low life expectancy (approximately 45 years), high suicide rates and internal conflict.

Next to these structural forms of violence (Galtung 1969), three physical manifestations of violent conflicts related to biofuels can be distinguished:

1. *Violence related to evictions*: Violence is often used to evict Guarani who are unwilling to leave plots of land (to be) used for sugarcane plantations. In September 2009, for example, a community of 130 Guarani called Laranjeira Ñanderuwas was forced to leave their village, which was later set on fire by unidentified people (presumably private security guards paid by a landowner). In 2005, a Guarani activist was shot by a private security guard during a forced eviction of indigenous people from the Nanderu Marangatu territory, which was about to be used for sugarcane fields.

2. *Violence related to reoccupations (retomadas)*: Due to land pressure, some members of the Guarani try to reoccupy parts of their former land. This is opposed by the new/formal landowners, often leading to violent struggles. The Guarani community of Kurusu Mba, for instance, tried to reoccupy the land they were evicted from in 1975 three times since 2005, but each attempt was successfully countered by the owners and their gunmen, firing on the occupiers and assassinating several leaders and activists. To give another example, in November 2011, forty masked gunmen attacked a Guarani camp near the town of Amambai several weeks after the reoccupation of their ancestral land and killed at least one leader of the occupants, while injuring other group members. It would be mistaken, however, to consider the Guarani as merely passive victims. In April 2006, for instance, "two civic policemen were murdered by indigenous people, because they entered the community of Passo Piraju without identifying themselves" (Suárez et al. 2008).

3. *Conflicts related to the demarcation process*: In November 2007, the Brazilian government decided to demarcate territories for the Guarani-Kaiowá. The state government and local landowners were able to delay the process until now by using legal means, but also by hiring gunmen and security guards to prevent government employees from accessing their property and determining which parts of it can be classified as ancestral Guarani land. However, no casualties or deaths related to this struggle have been reported yet.

Many of the large landowners in Mato Grosso do Sul are either influential local or regional politicians or have good relations to large national and transnational companies. This is why in several cases, the landowners and their security personnel receive no or only minor punishments. Another relevant factor is that Guarani often have only customary, but no formal land titles. In such cases, evictions are usually supported or even carried out by the police.

2.4.2 Bico do Papagaio

Information on the Bico do Papagaio case were obtained from Bickel (2004), the Comissão Pastoral da Terra (2007), the Procuradoria da Repúliva Federal do Tocantins (2012), and the Reportér Brasil (2008). Bico do Papagaio is part of the Tocantins state, which is well known for its high soil fertility. Consequentially, soy production increased rapidly since 2003 and was further accelerated by the introduction of biodiesel production in the area in 2007. In 2012, the areas dedicated to biodiesel production in Tocantins cover 130,935 m^3 and are very likely to grow in size in the future (ABIOVE 2013). There are various negative impacts of the industrial soy agriculture in the region affecting the local inhabitants, mainly small farmers and indigenous people. The pesticides and herbicides used for cultivating soy spill often over to other fields, causing a contamination of crops and soils and a yield reduction of up to 50%. Many locals furthermore complain about a reduction in fish stocks in the rivers and an increase in skin diseases due to the agrochemicals used on the soy plantations. Since land scarcity as well as river and soil pollution undermine traditional subsistence farming, many people of Bico do Papagaio are forced either to migrate or to work on the soy plantations. There are various reports about inhumane and at times even slave-like working conditions on these plantations.

In contrast to the Guarani case, reoccupation and demarcation processes play only a minor role in Bico do Papagaio. But there are conflicts related to land rights and forced evictions that are similar to the ones in Mato Grosso do Sul. Even before the start of the soy boom, 105,000 hectares of land were allocated to the Companhia de Promoção Agrícola, which intended to grow soy in Bico do Papagaio. This allocation was possible because the land was considered unproductive or not in use despite the fact that it sustained the livelihoods of many local families for generations. Since they had no officially registered land titles, these people were evicted from the land without proper compensation. If people resisted leaving their land, violence was used against them by unidentified actors, including beatings and the destruction of property. Fortunately, no deaths are reported. Similarly, residents of the Barra do Ouro municipality and the indigenous Krahô people living in the North of Tocantins report that they have been physically threatened and that part of their cattle has been killed by local soy farmers during land disputes.

Due to the close connections (and often personal union) between local or regional political and economics elites and the soy farmers, the political and administrative support for the smallholders and indigenous people in these struggles is rather low. This power inequality is aggravated by the smallholders' lack of official land titles and the missing recognition of customary land use practices.

2.5 Conclusions

The goal of this chapter was to explore the potential links between biofuel cultivation and violent conflict. It has been shown that the large-scale cultivation of sugarcane, oil palms, and soy has several negative side effects. Some of these are, according to the environmental conflict and the political ecology perspectives, likely to contribute to the onset of violent conflicts. In the case of Brazil, and more specifically in Mato Grosso do Sul and Bico do Papagaio, both perspectives were largely confirmed. In line with the environmental conflict perspective, socioeconomic deprivation, food insecurity, and land scarcity all contributed to land disputes that escalated into violence. As predicted by the political ecology perspective, tensions were exacerbated by the large-scale cultivation of soy and sugarcane, the high cultural relevance of land for indigenous people, the involvement of nonlocal actors (e.g., large companies), and the power inequalities between biofuel farmers and smallholders. This is in line with other recent studies on biofuels and violent conflicts in Brazil (Hermele 2009; OMCT 2008) and other regions of the world (Marin-Burgos 2015; Menguita-Feranil 2013).

However, there are several issues that need to be addressed in order to gain a better understanding of the dynamics of violence triggered by biofuel cultivation. First, most existing studies are conducted in areas where violence occurred or intensified simultaneously with the industrial cultivation of sugarcane, oil palms, or soy. But this is a "selecting on the dependent variable" strategy already criticized by Gleditsch (1998). What about the locations were biofuel expansion coincidences with the absence of violence? What factors inhibit the escalation of conflicts there? Second, a research needs to work out the links between biofuels and violent conflict more precisely. Not every conflict related to sugarcane, oil palms, or soy is automatically related to biofuels, since these crops are used for other purposes as well. And some conflicts about sugarcane, oil palms, and soy plantations used for biofuel production have other roots, such as land disputes or forests clearing that precede biofuel cultivation by several years. Third, following the political ecology perspective, research should focus stronger on intragroup conflicts related to biofuel cultivation (e.g., between biofuel supporters and opponents in local communities). And finally, we do not know much yet about the indirect and spatially distant effects of increasing biofuel cultivation. Especially rural-to-urban migration seems to be relevant in this context since the links between growing urban conglomerates marked by poor living conditions and increasing levels of violence are discussed in the literature for quite some time (e.g., Rodgers 2009). If these issues are addressed, we are confident that a better understanding of biofuel-related conflicts is achieved and can be used to prevent their violent escalation and to address the human security challenges that the industrial biofuel production poses.

References

ABIOVE. 2013. Biodiesel: Entrega e produção—Fevereiro 2013. Dados de produção e entrega de biodiesel no Brasil. http://www.abiove.org.br/site/index.php?page=estatistica&area=NC0yLTE (accessed March 26, 2013).

Allen, M.G. 2013. Melanesia's violent environments: Towards a political ecology of conflict in the western Pacific. *Geoforum* 44:152–161.

Anaya, J. 2009. Report of the special rapporteur on the situation of human rights and fundamental freedom of indigenous people. Report on the situation of indigenous people in Brazil. New York: United Nations.

Anseeuw, W., Wily, L.A., Cotula, L., and M. Taylor 2012. *Land Rights and the Rush for Land.* Rome, Italy: ILC.

Avritzer, L. 2010. Living under a democracy: Participation and its impact on the living conditions of the poor. *Lat Am Res Rev* 45:166–185.

Backhouse, M. 2015. Green grabbing—The case of palm oil expansion in so-called degraded areas in the eastern Brazilian Amazon. In *The Political Ecology of Agrofuels*, eds. K. Dietz, B. Engels, O. Pye, and A. Brunnengräber, pp. 167–185. New York: Routledge.

Barnett, J. and W.N. Adger 2007. Climate change, human security and violent conflict. *Polit Geogr* 26:639–655.

BBC News. 2011. Brazil indigenous Guarani leader Nisio Gomes killed. http://www.bbc.co.uk/news/world-latin-america-15799712 (accessed April 07, 2012).

BBC News. 2012. Brazil biofuel: Shell axes 'illegal' sugar cane plan. http://www.bbc.co.uk/news/world-latin-america-18433008 (accessed April 07, 2012).

Bickel, U. 2004. *Brasil: Expansão da soja, conflitos sócio-ecológicos e segurança alimentar.* Bonn, Germany: University of Bonn.

Blattman, C. and E. Miguel. 2010. Civil war. *J Eco Lit* 48:3–57.

Carius, A., Tänzler, D., and J. Winterstein 2006. *Weltkarte von Umweltkonflikten.* Berlin, Germany: Adelphi.

Comissão Pastoral da Terra. 2007. Conflitos no campo Brasil 2006. http://www.cptnacional.org.br/index.php/component/jdownloads/finish/43-conflitos-no-campo-brasil-publicacao/244-conflitos-no-campo-brasil-2006?Itemid=23 (accessed April 22, 2013).

Cotula, L., Dyer, N., and S. Vermeulen 2008. *Fuelling Exclusion? The Biofuels Boom and Poor People's Access to Land.* London, U.K.: IIED.

Deligiannis, T. 2012. The evolution of environment-conflict research: Toward a livelihood framework. *Global Environ Polit* 12:78–100.

Dietz, K., Engels, B., Pye, O., and A. Brunnengräber. 2015. *The Political Ecology of Agrofuels.* New York: Routledge.

Dixon, J. 2009. What causes civil war? Integrating quantitative research findings. *Int Stud Rev* 11:707–735.

Engels, B. 2014. Contentious politics of scale: The global food price crisis and local protest in Burkina Faso. *Soc Mov Stud* 13:180–194.

Esty, D.C., Goldstone, J.A., Gurr, T.R. et al. 1999. State failure task force report: Phase II findings. *Environ Change Secur Proj Rep* 5:49–72.

Fast, S. 2009. The biofuels debate: Searching for the role of environmental justice in environmental discourse. *Environ J* 37:83–100.

Fernandes, F.M., Welch, C.A., and E.C. Gonçalves 2010. Agrofuel policies in Brazil: Paradigmatic and territorial disputes. *J Peasant Stud* 37:793–819.

Galtung, J. 1969. Violence, peace, and peace research. *J Peace Res* 6:167–191.

Garcez, C.A.G.G. and J.N.D.S. Vianna. 2009. Brazilian biodiesel policy: Social and environmental considerations of sustainability. *Energy* 34:645–654.

Gerber, J.-F. 2011. Conflicts over industrial tree plantations in the south: Who, how and why? *Global Environ Change* 21:165–176.

Gizelis, T.-I. and A.E. Wooden. 2010. Water resources, institutions, & intrastate conflict. *Polit Geogr* 29:444–453.

Gleditsch, N.P. 1998. Armed conflict and the environment: A critique of the literature. *J Peace Res* 35:381–400.

Hauge, W. and T. Ellingsen. 1998. Beyond environmental scarcity: Causal pathways to conflict. *J Peace Res* 35:299–317.

Hegre, H. and N. Sambanis. 2006. Sensitivity analysis of empirical results on civil war onset. *J Conflict Resolut* 50:508–535.

Hendrix, C.S. and S.M. Glaser. 2007. Trends and triggers: Climate, climate change and civil conflict in Sub-Saharan Africa. *Polit Geogr* 26:695–715.

Hermele, K. 2009. *Agrofuels, Brazil and Conflicting Land Uses.* Lund, Sweden: University of Lund.

Homer-Dixon, T. 1999. *Environmental Scarcity and Violence.* Princeton, NJ: Princeton University Press.

Homer-Dixon, T. and J. Blitt. 1998. *Ecoviolence: Links among Environment, Population, and Security.* Lanham, MD: Rowman & Littlefield.

Horowitz, L.S. 2009. Environmental violence and crises of legitimacy in New Caledonia. *Polit Geogr* 28:248–258.

Howard, P. and T. Homer-Dixon. 1998. The case of Chiapas, Mexico. In *Ecoviolence: Links among Environment, Population, and Security*, eds. T. Homer-Dixon and J. Blitt, pp. 19–65. Lanham, MD: Rowman & Littlefield.

Ide, T. 2015. Why do conflicts over scarce renewable resources turn violent? A qualitative comparative analysis. *Global Environ Change* 33:61–70.

Ide, T., Schilling, J., Link, J.S.A., Scheffran, J., Ngyruya, G., and T. Weinzierl. 2014. On exposure, vulnerability and violence: Spatial distribution of risk factors for climate change and violent conflict across Kenya and Uganda. *Polit Geogr* 43:68–81.

Jewitt, S. 2008. Political ecology of Jharkhand conflicts. *Asia Pacific Viewpoint* 49:68–82.

Kahl, C.H. 2006. *States, Scarcity, and Civil Strife in the Developing World.* Princeton, NJ: Princeton University Press.

Kaup, F. and K. Selbmann. 2013. The seesaw of Germany's biofuel policy—Tracing the evolvement to its current state. *Energ Pol* 62:513–521.

Le Billon, P. 2000. The political ecology of transition in Cambodia 1989–1999: War, peace and forest exploitation. *Dev Change* 31:785–805.

Maas, A. and J.S. Scheffran. 2012. Climate conflicts 2.0? Geoengineering as a challenge for international peace and security. *Secur Peace* 30:193–200.

Marin-Burgos, V. 2015. Social-environmental conflicts and agrofuel crops: The case of oil palm expansion in Columbia. In *The Political Ecology of Agrofuels*, eds. K. Dietz, B. Engels, O. Pye, and A. Brunnengräber, pp. 148–166. New York: Routledge.

McDonald, M. 2013. Discourses of climate security. *Polit Geogr* 33:42–51.

McMichael, P. 2009. The agrofuels project at large. *Crit Sociol* 35:825–839.

Meierding, E. 2013. Climate change and conflict: Avoiding small talk about the weather. *Int Stud Rev* 15:185–203.

Melander, E. and R. Sundberg. 2011. Climate change, environmental stress, and violent conflict—Tests introducing the UCDP Georeferenced Event Dataset. *51st ISA Annual Convention*, Montreal, Quebec, Canada, March 16–19, 2011.

Menguita-Feranil, M.L. 2013. Contradictions in palm oil promotion in the Philippines. In *The Palm Oil Controversy in Southeast Asia: A Transnational Perspective*, eds. O. Pye and J. Bhattacharya, pp. 97–119. Pasir Panjang, Singapore: ISAS.

Montefrio, M.J.H. 2013. The green economy and land conflict. *Peace Rev* 25:502–509.

Nordås, R. and N.P. Gleditsch 2007. Climate change and conflict. *Polit Geogr* 26:627–638.

Obidzinski, K., Andriani, R., Komarudin, H., and A. Andrianto. 2012. Environmental and social impacts of oil palm plantations and their implications for biofuel production in Indonesia. *Ecol Soc* 17:25–44.

OMCT. 2008. List of issues arising from the second periodic report of Brazil to the committee on economic, social and cultural rights. Geneva, Switzerland: OMCT.

Peluso, N.L. and M. Watts. 2001. Violent environments. In *Violent Environments*, eds. N.L. Peluso and M. Watts, pp. 3–38. Ithaca, NY: Cornell University Press.

Pinstrup-Andersen, P. and S. Shimokawa. 2008. Do poverty and poor health and nutrition increase the risk of armed conflict onset? *Food Pol* 33:513–520.

Pousa, G.P., Santos, A.L.F.S., and P.A.Z.S. Suarez. 2007. History and policy of biodiesel in Brazil. *Energ Pol* 35:5393–5398.

Procuradoria da Repúliva Federal do Tocantins. 2012. Conflitos agrários são tema de audiência pública no norte do estado. http://www.prto.mpf.gov.br/news/

conflitos-agrarios-sao-tema-de-audiencia-publica-no-norte-do-estado   (accessed April 19, 2013).

Raleigh, C., Choi, H.J., and D. Kniveton 2015. The devil is in the details: An investigation of the relationship between conflict, food price and climate across Africa. *Global Environ Change* 32:187–199.

Raleigh, C. and H. Urdal 2007. Climate change, environmental degradation and armed conflict. *Polit Geogr* 26:674–694.

Regan, A.J. 1998. Causes and course of the Bougainville conflict. *J Pac Hist* 33:269–285.

Reportér Brasil. 2008. *O brasil dos agrocombustíveis. Os impactos das lavouras sobre a terra, o meio e a sociedade.* São Paulo, Brazil: Reportér Brasil.

Reportér Brasil. 2009. *Brasil on Biofuels: Soybean, Castor Bean 2009.* São Paulo, Brazil: Reportér Brasil.

Reuveny, R. 2007. Climate change-induced migration and violent conflict. *Polit Geogr* 26:656–673.

Reuveny, R. 2008. Ecomigration and violent conflict: Case studies and public policy implications. *Hum Ecol* 36:1–13.

Rodgers, D. 2009. Slum wars of the 21st century: Gangs, mano dura and new urban geography of conflict in Central America. *Dev Change* 40:949–976.

Rowhani, P., Degomme, O., Guha-Sapir, D., and E. Lambin. 2011. Malnutrition and conflict in East Africa: The impacts of resource variability on human security. *Climatic Change* 105:207–222.

Salehyan, I. and K.S. Gleditsch. 2006. Refugees and the spread of civil war. *Int Org* 60:335–366.

Salehyan, I. and C. Hendrix. 2014. Climate shocks and political violence. *Global Environ Change* 28:239–250.

Schilling, J., Opiyo, F., and J. Scheffran 2012. Raiding pastoral livelihoods: Motives and effects of violent conflict in north-eastern Kenya. *Pastoralism* 2:1–16.

Smith, T. 2014. Feeding unrest: Disentangling the causal relationship between food price shocks and sociopolitical conflict in urban Africa. *J Peace Res* 51:679–695.

Suárez, S.M., Bickel, U., Garbers, F., and L. Goldfarb. 2008. Agrofuels in Brazil: Fact-finding mission report on the impacts of the agrofuels expansion on the enjoyment of social rights of rural workers, indigenous peoples and peasants in Brazil. Heidelberg, Germany: FIAN.

Survival International. 2009. Village torched as Guarani Indians evicted from their land. http://www.survivalinternational.org/news/4949 (accessed April 04, 2012).

Survival International. 2010. Violations of the rights of the Guarani of mato grosso do sul state, Brazil: A survival international report to the un committee on the elimination of racial discrimination. London, U.K.: Survival International.

Survival International. n.d. The Guarani: Fighting back. http://www.survivalinternational.org/tribes/guarani/fightingback#main (accessed April 04, 2012).

ten Kate, A. 2011. *Royal Dutch Shell and Its Sustainability Troubles.* Amsterdam, the Netherlands: Milieudefensie.

Theisen, O.M. 2008. Blood and soil? Resource scarcity and internal armed conflict revisited. *J Peace Res* 45:801–818.

Theisen, O.M., Gleditsch, N.P., and H. Buhaug. 2013. Is climate change a driver of armed conflict? *Climatic Change* 117:613–625.

Vandergeest, P. 2003. Racialization and citizenship in Thai forest politics. *Soc Nat Res* 16:19–37.

Venghaus, S. and K. Selbmann. 2014. Biofuel as social fuel: Introducing socio-environmental services as a means to reduce global inequity? *Ecol Econ* 97:84–92.

Verhoeven, H. 2011. Climate change, conflict and development in Sudan: Global neo-malthusian narratives and local power struggles. *Dev Change* 42:679–707.

Watts, M. 2004. Resource curse? Governmentality, oil and power in the Niger Delta, Nigeria. *Geopolitics* 9:50–80.

WBGU. 2009. *Future Bioenergy and Sustainable Land Use*. London, U.K.: Earthscan.

Webersik, C. and M. Bergius. 2013. The biofuel transition. In *Backdraft: The Conflict Potential of Climate Change Adaptation and Mitigation*, eds. G.D. Dabelko, L. Herzen, S. Null, M. Parker, and R. Sticklor, pp. 34–36. Washington, DC: Woodrow Wilson Center.

Wischnath, G. and H. Buhaug. 2014. Rice or riots: On food production and conflict severity across India. *Polit Geogr* 43:6–15.

Wright, A. and W. Wolford. 2003. *To Inherit the Earth: The Landless Movement and the Struggle for a New Brazil*. Oakland, CA: Food First Books.

Zhouri, A. and R. Oliveira. 2009. Development projects and violence in rural Brazil: The case of hydroelectric dams. In *Connected Accountabilities: Environmental Justice & Global Citizenship*, ed. S. Vemuri, pp. 197–217. Oxford, U.K.: Inter-Disciplinary Press.

3

Global Demands of Biofuels
Technologies, Economic Aspects, Market, and Policies

Ram Sarup Singh and Shivani Thakur

Contents

Abstract

Currently, the world is confronted with the twin crisis of fuel depletion and global environmental degradation, which resulted in rapid climate change. Indiscriminate extraction and lavish consumption of fossil fuels have led to the reduction of underground carbon resources. Moreover, the price of crude oil keeps on fluctuating and rising on a daily basis. Meanwhile, the scenario war for oil has already been started due to subsequent environmental concerns and political events in the Middle East. Over the globe, scientists have explored several fossil fuel alternatives (biofuels) of bio-origin, which have the potential to quench the ever-increasing thrust of today's population. Within few years from now, biofuels will transform from a niche energy source. In the foreseeable future, biofuel policies still require motivation by a plethora of political concerns, which can reduce dependence on fossil fuels and improve the environmental concerns. The shift from traditional fossil fuels to more exotic biofuels would constitute a very revolutionary change. In this chapter, an attempt has been made to collate the information on current biofuel scenario, technologies, economic aspects, market, and policies.

3.1 Introduction

Long back in 1925, Henry Ford stated, "We can get fuel from fruit, from that shrub by the roadside, or from apples, weeds, sawdusts almost everything! There is fuel in every bit of vegetable matter that can be fermented. There is enough alcohol in 1 year's yield of a hectare of potatoes to drive machinery necessary to cultivate the field for a hundred years, and it remains from someone's to find out how this fuel can be produced commercially better fuel at a cheaper price than we know now."

Today, energy security is becoming a serious issue as fossil fuels are non-renewable energy and will deplete eventually in the near future (Masjuki et al. 2013). The increased use of fossil fuels has caused greenhouse gas (GHG) emissions, which have subsequently caused undesirable damage to the environment. The volatility of oil prices along with major concerns about the climate change, oil supply security, and depleting reserves has sparked renewed interest in the production of fuels from renewable resources (Dellmonaco et al. 2010). Then, a big challenge is to find new fields and to implement energy from the projected natural gases and non-conventional sources. As for now, future energy demand is expected to grow at an annual growth rate of 5%–7.9% for the next 20 years. The total biofuel demand is expected to meet approximately 27% of the total transport fuel demand in 2050 (Muller-Langer et al. 2014).

Biofuels are promoted as the best means to meet the prospected increase in energy demand in the coming years. The concept of biofuels was conceived in the 1970s, when the world faced a large-scale oil crisis. Biofuels can assist international development and poverty alleviation, because most people in the developing countries participate in agriculture to increase agricultural income, which strongly improves the overall welfare and also increases food security (Dale and Ong 2014). Biofuels are non-polluting, locally available, sustainable, and reliable fuels that are obtained from renewable sources (Demirbas 2008). Renewable resources are more evenly distributed than fossil and nuclear resources. There are other niche of renewable resources such as biogas, which have been derived by anaerobic treatment of manure and other biomass materials. However, the volumes of biogas used for transportation are relatively very small today (Naik et al. 2010).

The expanding biofuel sector involves both opportunities and threats for development (Bindraban et al. 2009). The increased pressure on arable land currently used for food production can lead to severe food shortages, in particular for the developing world, where already more than 800 million people are suffering from hunger and malnutrition (Dragone et al. 2010). So, increasing the production rate of biofuels within available land can be achieved through intensive research in biotechnology, plant agronomy, and use of precision agricultural techniques (Masjuki et al. 2013). An entire branch of biotechnology, referred to as "white biotechnology," embraces the bioproduction of fuels and chemicals from renewable sources. The major factors that account for the explosive growth in biofuel sector and widespread enthusiasm for the technology are (1) opportunity to reduce dependence on fossil fuels through renewable energy, (2) search for energy security and energy dependence as emerging economies like the United States, (3) as a potential to reduce the net emission of CO_2 into the atmosphere and to reduce global warming, (4) to raise commodity prices, to improve income, and, most importantly, to increase rural employment opportunities (FAO 2006). Recent advances in synthetic biology can provide new tools for metabolic engineers to direct their strategies and construct optimal biocatalysts for the sustainable production of biofuels (Dellomonaco et al. 2010). India is the world's sixth largest consumer of energy. The demand for energy is estimated to grow by eight times by the year 2030 at the present growth rate of 4.8% per annum. Every year, India is losing a substantial amount of foreign exchangers through import of crude fossil fuel, which caters to about 70% of the country's requirement (Kaushik and Biswas 2010). To date, little research has specifically addressed biofuels in Asian context.

Transportation and agricultural sectors are the major consumers of fossil fuels and are the biggest contributors to environmental pollution that can be reduced by replacing fossil fuels by renewable fuels of bio-origin (Agarwal 2007). Nevertheless, the use of fossil fuels in transportation sector is growing faster and the trend appears to be moving up dramatically (Masjuki et al. 2013). The efficacy of alternative biofuel policies in achieving energy, environmental, and agricultural policy goal can be assessed using economic cost–benefit analysis (de Gorter and Just 2010). Biofuel policies could be motivated by a plethora of political concerns related to reducing dependence on oil, improving the environment, and increasing the agricultural income (Rajagopal et al. 2007). Policy makers should realize the future crisis to make a short-, medium-, and long-term policy considering all the views, aspects, and alternatives (Masjuki et al. 2013). The main approaches should be (1) tax reduction/exemption for biofuels and (2) biofuel obligations. Unfortunately, the broader picture is not so attractive. A number of concerns are raised by these developments without subsidies, because most of the biofuels have to compete price-wise with petroleum products in most regions of the world (Doornbosch and Steenblik 2007). In order to replace existing fossil fuels and to compete with them, technology development is a prerequisite. In order to enhance the production rate and oil yield, there is no alternative regardless of the advanced technologies.

3.2 End-Use Biofuel Technologies

Since the Arab oil embargo of the 1970s, Brazil has made an incomparable effort in the reduction of its energy dependency by intensifying and extending ethanol production from sugarcane. Today, Brazil is the second largest worldwide ethanol producer.

Renewable fuel standard has mandated biofuel production in the United States, since the establishment of the Energy Policy Act of 2005. Accordingly, 36 billion gallons of biofuels are supposed to be supplied to the market by 2022. Advanced biofuels need to constitute 58.3% of the total mandate (Ziolkowska 2014). The fundamental problem for the advanced biofuel industry is that, despite many attempts, none of them succeeded in identifying a commercially viable way to produce advanced biofuels at a cost-competitive level with petroleum fuels.

Several studies have been undertaken to address this problem and to provide a viable solution. One of the possible solutions is to use two fungal strains (*Thielavia terrestris* and *Myceliophthora thermophila*) because their enzymes are active at high temperatures between 40°C and 75°C (Berka et al. 2011). This will accelerate the biofuel production process and reduce the cost of second-generation biofuels. This will also improve the efficiency of biofuel production in large-scale biorefineries. Moreover, fungi can be exposed to genetic manipulation in order to increase the enzyme efficiency even more than its wild types (Stephanopoulos 2007; Vinuselvi et al. 2011). A similar solution has been given by the scientists from the U.S. Department of Energy (DOE), the BioEnergy Science Center, and the University of California, where they utilized the bacterium *Clostridium celluloyticum* capable of breaking down the cellulose that enabled the production of isobutanol in a single inexpensive step (Casey 2012). Isobutanol can be burned in car engines with a heat value higher than that of ethanol and similar to gasoline. Thus, the economics of using *C. celluloyticum* to break down cellulose is very promising in the near future. Furthermore, DOE researchers have found engineered strains of *Escherichia coli* having the ability to break down cellulose and hemicellulose of plant cell walls. In this way, necessary expensive processing steps in conventional systems can be eliminated that could subsequently reduce the final biofuel price and allow a faster commercialization process for the second-generation biofuels (Casey 2011).

3.2.1 Spark-Ignition Engines

For the combustion of petroleum, usually spark-ignition engines are used. These are internal combustion engines, where the fuel–air mixture is ignited with the spark. Spark-ignition engines can be either two strokes or four strokes. Generally, spark-ignition engines can also run with bioethanol if 10%–25% ethanol is mixed with gasoline, and no modifications in engines are required. But higher ethanol component blended in petroleum lowers its sustainability for standard car engines due to its certain characteristics. In general, E10 blends do not require engine tuning or vehicle modifications. In many countries, vehicles have been made adaptive by using ethanol-compatible materials in the fuel system and by turning engines for a midrange point, usually at the 22% ethanol level (E22). For using fuels with higher ethanol blends (E20–E100), conventional engines have to be refitted. This is due to characterization of ethanol to dissolve certain rubber and plastics. Larger carburetor jets have to be installed; moreover, for the temperature below 13°C, cold-starting systems are needed for maximizing the combustion and minimizing non-vaporized ethanol. Depending upon the particular customization requirement, refitting costs may run from few Euros to more than €500. Recently, a number of vehicles are manufactured with engines that can run on any petroleum/bioethanol ratio ≤85%. The sensors of flexible-fuel vehicles (FFVs), that is, E85 FFV, can automatically detect the type of the fuel and adapt engine running. They adjust the air/fuel ratio and the ignition tuning to compensate the different octane levels of fuels in the engines' cylinders. Dedicated ethanol vehicles are more efficient in using pure ethanol due to better combustion characteristics than FFVs. They retain dual-fuel capacity. Moreover, ethanol, in particular, ensures complete combustion, reducing carbon monoxide emissions (U.S. EPA 2010). Volkswagen, Fait, General Motors, and Ford have all produced dedicated ethanol versions for more than 25 years with full warranty coverage.

3.2.2 Compression-Ignition Engines

Compression-ignition engines are designed for being fueled with diesel. Ethanol is difficult to ignite in a compression-ignition engine. The only option for ethanol ignition is to blend it with an additive to enhance fuel ignition. Other approaches of using ethanol in diesel engines are either to use diesel and ethanol simultaneously by "fumigation" or to convert the diesel engine into a spark-ignition engine.

3.2.3 Fuel Cells

Direct ethanol fuel cell (DEFC) systems hold several advantages, as when bioethanol is fed directly into DEFC and no catalytic reforming is required as such. Moreover, storage of ethanol is much easier than that of hydrogen, and storage of liquid ethanol does not need to be done at high pressure. This will enable vehicles to use a combination of conventional and lower-cost fueling systems (WWI 2006).

3.2.4 Lipid Biofuels

Biodiesel can cause damages to conventional engines. Because of their solvent properties, fuel supply systems and fuel filters may clog due to breakdown deposits. The appropriate blending ratio of biodiesel with fossil diesel depends on appropriate measures. Typical blends are B5, B20, and B30 with 5%, 20%, and 30% biodiesel concentration, respectively. For older models, rubber and plastic components must be replaced with more resistant materials. The use of lower blends requires no or only minor technological modifications, whereas the use of higher blends, such as B100, needs more efforts in modification of engine and fuel component system.

3.2.5 Compression-Ignition Engines for Pure Plant Oil

Pure plant oil (PPO) cannot be used in diesel engines due to its high viscosity, so the engines should be modified accordingly. When PPO is used in unmodified engines, results can be poor atomization of the fuel in combustion chamber, incomplete combustion, coking of the injectors, and accumulation of soot deposition in the piston crown, rings, and lubricating oil (WWI 2006). Several refitting concepts have been developed, which include either preheating the fuel injection system or using the pre-equipment with a two-tank system. By using the latter technology, the engine is first started with diesel and is switched off to PPO, when the operating temperature is achieved. Shortly before being turned off, it should be ensured that it does not contain the traces of PPO. PPO should not be used in neither pure form nor mixed with diesel in updated engines, as its combustion properties differ so widely from those of diesel and can cause damage to the injection systems and deposits in the engine may also occur (Paul and Kemnitz 2006).

3.3 Economic Aspects of Biofuels

Interest in biofuels began with oil shocks in the 1970s, but the more rapid developments of biofuel industry in recent years has been primarily driven by mandates, subsidies, climate change concerns, emissions targets, and energy security. Broadly speaking, currently biofuels are more expensive than fossil fuels. From 2004 to 2006, the production of ethanol fuel grew by 26% and biodiesel grew by 172%. In 2004, 3.4 billion gallons of ethanol fuel was produced from 10% of the corn crop. Ethanol demand is expected to more than double in the next 10 years. To meet this demand, new technologies must move out from the laboratories to commercial reality (Bothast 2005). The world ethanol production is about 60% from sugar crop feedstocks. At present, biodiesel accounts for less than 0.2% of the diesel consumed for transport (UN 2006). According to one assessment, biodiesel is about US$0.27 per liter more expensive than regular diesel (Duncan 2003; OECD-FAO 2007). The major economic factor to consider for input costs of biodiesel production is the feedstock, which is about 75%–80% of the total operating cost. On an energy basis, ethanol is currently more expensive than gasoline in all regions of the world except Brazil. Ethanol produced from corns and grains in the United States and Europe, respectively, is more expensive than from sugarcane in Brazil.

Generally, biofuels are more costly than fossil fuels, and consumers will only use them if the cost is compensated by the government or if they are forced to use the biofuels. Currently, biofuels require subsidies, tariffs, fuel mandates, and other government supports for economic viability. Thus, government and consumers are both paying a significant premium to gain the expected benefits of biofuels. A myriad of policies are employed for U.S. biofuels including consumption subsidies, mandated minimum levels of consumption, and production subsidies including feedstock, import barrier, and sustainability standards (Gardner and Tyner 2007).

In Asia, biofuel production requires additional use of land, water, and fertilizers. Additional fertilizers are required to significantly increase the biofuel crop production. In India, where Jatropha plantation has been promoted for biodiesel production, it will require an additional 14.9 mt of organic manure and 2.6 mt of fertilizer per year to meet the production target. Technological advancements are required for the oil extraction, transesterification, and fermentation processes for the production of biodiesel and bioethanol to meet the requirements of biofuels. Moreover, the development of kinetic models that include accurate regulatory network parameters to facilitate the identification of enzymatic bottlenecks in metabolic pathways can be harnessed to achieve production of biofuels (Dellomonaco et al. 2010).

The demand for energy is increasing everyday due to rapid outgrowth of population and urbanization (Demibas 2008). Several countries have already enacted laws that mandate the production of biofuels to meet future demands (Kojima et al. 2007). Total cost of biofuels is composed of the cost of biomass production, biomass transportation, biomass conversion, and labor. Reduction of our demand for petroleum products could also reduce its price and will also generate economic benefits for consumers (Huang et al. 2013). Economic advantages of a biofuel industry would include an increased number of rural manufacturing jobs, an increased income taxes, reduced GHG emissions, and reduction of country's reliance on crude oil imports.

3.4 Biofuel Policies: World Scenario

Commercially available biofuels are almost entirely produced from food crops like sugarcane, sugar beet, corn, and oil seeds; therefore, policies encouraging biofuel production have repercussions on the markets of goods related with biofuel production (Sorda et al. 2010). A recent report issued by the Joint Research Center of the European Commission in Seville (2010) provides an extensive overview of current policy actions promoted by countries across the world to foster both biofuel production and consumption. The main reasons behind the countries' decision for *green energy* production are the will (or the need) to reduce dependence on fossil fuels (energy security), to reduce GHG emissions (climate change mitigation), and to increase demand for certain agricultural products that suffer from production surplus (support to farmer's income). Each of these three reasons has been criticized. Energy security, for example, could be achieved not only by encouraging biofuel use, but also through other forms of domestically produced renewable energy such as solar and wind power. The contribution of biofuels in reducing GHG emissions has been contested as well. Currently, biofuels are mainly produced from agricultural commodities such as ethanol from corn and biodiesel from rapeseed or palm oil. Farmers can be induced, by higher commodity prices, to put more land under cultivation or to make their production processes more intensive. This may result in an increase of CO_2 emissions from the agriculture sector that can eventually offset the GHG emission reduction obtained from an increase in biofuel consumption. Finally, it is true that an increased demand for food and non-food agricultural products can raise farmer's income; however, this might come at the expenses of food consumer worldwide and the environment. Biofuel policies are making an upward pressure on agricultural prices and can also undermine the environment since they encourage the expansion of agricultural areas at the expenses of rainforests and wilderness (direct and indirect land-use changes). It is important to fully understand the policies issued by the major biofuel producer and consumer countries, because their decisions can have a substantial impact on world markets of both bioenergy and agricultural products.

3.4.1 Brazil

Since the 1970s, Brazil has been at the lead to produce biofuels, in particular ethanol from sugarcane. Because of a combination of climate, soil, and 45 years of sustainable technological research and development, Brazil is currently the lowest-cost producer of sugarcane to date and, consequently, of ethanol for automotive transport. In 2006, there were 320 combined sugar mills and bioethanol distilleries in the country, with a total installed processing capacity in excess of 430 million tons of sugarcane. Including 51 new plants and expansion of those existing, together they could produce up to 30 million tons of sugar and 18 billion liters of ethanol per year (GBEP 2007). The largest plant in Brazil has a production of just below 330 million liters of ethanol per year. There are about 250 separate producers, but most of them are grouped in two associations that make up 70% of the market.

Unlike other countries with substantial biofuel production, Brazil does not offer production subsidies for bioethanol. However, the government has made it mandatory since 1977 for light vehicles to have the E20 blend, with vehicles running also on using up to E25 blends. Brazilian sugarcane ethanol is the only ethanol that is competitive with petroleum, and the E20 mandate causes minimum distortion, because it requires ethanol up to the cost-equivalent level. As a result, Brazil has been hailed as an example of successful biofuel subsidization, and its current mandate is purportedly for environmental rather than economic reasons. However, Brazil's current ethanol infrastructure is extremely costly to set-up for the government and taxpayers, and it requires decades of taxpayer subsidies, before it became economically viable (Xavier 2007). Though Brazil has a comparative advantage in ethanol production, it still suffers substantial drawbacks through subsidies that serve as an interesting lesson, especially for countries with less cost-effective biofuels such as the United States and European Union (EU) nations.

Although bioethanol is currently not subsidized, the government has given tax breaks to company producers of biodiesel to support domestic production and the research and development of biodiesel. The Brazilian government created the Brazilian Biodiesel Program in 2003 in order to encourage domestic production of biodiesel from SVO and to limit the import of biodiesel. Companies compete for the distribution and sale of produced biodiesel and are evaluated for social sustainability plans. The ministry of agrarian development claimed that 30,000 families are employed in the raw material production of biodiesel, although Brazil has recently come under criticism that projects do not contribute significantly to rural development and job creation from biodiesel is far lower. By the mid-1980s, more than three-quarters of all cars in Brazil were running on hydrous ethanol. A surge in sugar prices at the end of the 1980s, coupled with lower oil prices, led to a slump in ethanol production as growers diverted their production to the export market and to a loss of public confidence in the security of ethanol supply. By the end of the 1990s, sales of ethanol-fueled cars had almost dried up. Interest in ethanol rebounded in the early 2000s with higher oil prices, and the first flex-fuel cars were introduced.

Rising demand for oxygenates has also driven up ethanol prices, boosting the profitability of ethanol production, and has stimulated investment in new sugarcane plantations and biorefineries. Less than 3 years after they were introduced, Flex-fuel vehicles (FFVs) now make up more than 70% of the vehicles sold in Brazil. Vehicle prices are no higher than for conventional gasoline cars. All refueling stations in Brazil sell near-pure hydrous ethanol (E95) and anhydrous gasohol, and about a quarter also sell a 20% anhydrous ethanol blend (E20). In total, almost two-thirds of the ethanol currently consumed in Brazil is anhydrous. The price of ethanol has risen faster than that of gasoline in the past years, mainly due to high international sugar prices. This has prompted the government to lower the minimum ethanol content in gasoline blends from 25% to 20% in order to prevent an ethanol shortage. Gasoline that does not contain ethanol can no longer be marketed in Brazil.

3.4.2 Argentina

Argentina started to subsidize biofuels in 2007 mainly to diversify the energy supply, to reduce the environmental impact, and to promote the rural development. The program focuses on conventional biofuels, since Argentina already has a large biodiesel industry based on soybean oil and a rapidly increasing ethanol industry based on sugarcane and grains. Starting from January 2010, gasoline and diesel sold in Argentina must contain 5% of biofuel. Prices for ethanol and biodiesel sold into the domestic market are established by the law. Biofuel producers do not enjoy tax incentives if they sell their products abroad. Conversely, producers who sell in the domestic market can ask for the reimbursement of the value-added tax (VAT). In addition, the government assures that all biofuel produced will be purchased for the 15-year period. However, the incentives are not guaranteed (they are renewed annually) and prices are set by the government. Argentinean biodiesel is produced from soybean, and major plants are located in the Rosario area. Argentina is still a small player in the world biofuel market, but its production is growing rapidly. Ethanol production is much more limited and is linked to the sugar industry. Biofuel produced in Argentina, despite the domestic incentives, is almost entirely exported because it enjoys more favorable export tariffs (15% effective levy) than the raw materials (soybean and soybean oil) form which it is produced. However, the new, stricter EU standards for biofuels (minimum of 35% reduction in GHG emissions) might represent a constraint for Argentinean exports in the coming years. The fact that Spain banned imports of biodiesel from extra-EU countries might negatively affect the Argentinean biodiesel industry in the coming years.

3.4.3 United States

Subsidies in the United States range from US$5.5 billion to US$7.3 billion annually and support the exponentially growing production of corn ethanol. Driven by subsidies, U.S. ethanol production has grown from 16.2 billion liters in 2005 to an estimated 24.5 billion liters in 2007. Given the current subsidies and support, production of biofuels is expected to grow over the next several years. The figure will reach 36 billion US gallons of renewable fuel used by 2022. Ethanol production in 2006 represented about 3.5% of motor vehicle gasoline supplies in the country. Most ethanol is used in low-percentage gasoline blends, but sales of high-percentage blends are rising. About 6 million FFVs are now running on E85 (a blend of 85% ethanol and 15% gasoline). The United States has a long history of tax reductions for biofuels, and it exempted gasohol (E10) in 1978 from the US$0.04/gal fuel-excise tax. This was replaced by an income tax credit in 2004. However, many U.S. states still retain fuel-excise tax reductions on pure biofuels and blends, with a value of about US$0.20/gal. These tax reductions are complemented by biofuel mandates that further support biofuel consumption. Energy Policy Act 2005 established a mandate that requires renewable resources account for at least 4.2% of transport fuel distributed to U.S. motorists. U.S. subsidies, like those of other OECD countries, boost supply at every step of the production process. Investors in biofuels also benefit from tax credits and grants from local, state, and federal governments, a trend called "subsidy stacking." Municipal governments can offer free land and utility; the state offers tax credits for investment and economic development grants; and the federal agency provides support through environmental, agricultural, and regional development program. Energy Policy Act 2005 expanded grants for capital inputs, authorizing an average of US$250 million over 2 years in grants for cellulosic ethanol plants, as well as loans for ethanol production from cellulosic biomass or municipal solid waste. Municipalities and states have offered further support through similar grants and investment incentives. See the Global Subsidies Initiative (2007) report for more details.

The production capacity of the U.S. ethanol industry is rising sharply as new plants have been built or are under construction. By the end of 2007, over 126 ethanol plants were in operation and another 100 were under construction. Most of them are dry mills, which produce ethanol as the primary output; wet mills are designed to produce a range of products alongside ethanol, including maize oil, syrup, and animal feed. Production capacity in the industry is expected to exceed a staggering 36 billion liters (10 billion gallons) by 2008, but even this addition will not be sufficient to meet all new demands. The U.S. ethanol demand is outstripping the supply, with about 2.3 billion liters imported in 2006, mostly from Brazil. As a result, there are calls for import tariffs to be removed to prevent domestic ethanol prices from rising further, which would push up gasoline prices at the pump, and for fuel standards to be eased. The price of ethanol has risen sharply in recent years in absolute terms and relative to gasoline.

Ironically, despite significant support, the U.S. biofuel policy appears to have little net impact on the nation's oil use. This is because the amount of fuel displaced by ethanol is more than offset by increased gasoline consumption due to less energy stringent vehicle efficiency standards permitted by a loophole in legislation promoting flex-fuel vehicles (Childs et al. 2007). While ethanol's share in the overall gasoline market is relatively small, its importance to the corn market is comparatively large. About 14% of corn use went to ethanol production in the 2005–2006 crop year. Carryover stocks of corn represented about 17.5% of use at the end of 2006, but expanded use of corn to produce ethanol in the 2006–2007 crop year will leave the ending stocks-to-use ratio at 7.5% (USDA 2006). With continued strong ethanol expansion, the USDA's 2007 long-term projections indicate that more than 30% of the corn crop will be used to produce ethanol by 2009–2010, remaining near that share in subsequent years. The United States also produces a small volume of biodiesel, mainly from soybeans; output totaled 220 kilotonne of oil equivalent (ktoe) in 2005, which is less than half of 1% of that of ethanol, although production capacity is growing rapidly. Support for biodiesel is much more recent. Minnesota was the first state to introduce a requirement that diesel must contain at least 2% biodiesel in 2005. A federal excise tax credit of 1 cent per gallon of crop-based biodiesel for each percentage point share in the fuel blend was introduced in January 2005. Soybean producers also receive hefty subsidies from the federal government.

3.4.4 Canada

Almost 70% of Canadian ethanol is produced from corn, while the remaining 30% from wheat. Biodiesel is from animal fats like tallow grease, yellow grease (used oil from deep fryers), and canola (rapeseed oil). Similar to the United States, Canada also mandates the blending of regular gasoline with ethanol. The Environmental Protection Act Bill C-33 established that gasoline must contain 5% of renewable fuel in 2010—similarly, addition of 2% renewable content in diesel and heating oil by 2012. It has been estimated that, in order to meet these targets, at least 1.9 billion liters of ethanol should be produced at current gasoline sale trends. To achieve the diesel and heating oil target, an additional 520 million liters of ethanol must be either produced or imported (Sorda et al. 2010). Federal mandates are not the only policy in place to support ethanol production. Ethanol manufacturers have been enjoying a CAN$0.10/L incentive rate since April 2008, thanks to the *ecoENERGY for Biofuels Program*. Starting from January 2011, however, the incentive rate started to decline by CAN$0.01 per year until it will reach CAN$0.04 in 2015 and 2016. A similar incentive exists also for biodiesel producers. In this case, the incentive rate is CAN$0.20, also declining in the next few years, until it reaches CAN$0.06 in 2016. Several found schemes are also in place to expand biofuel production through an increase of production capacity (new infrastructure). Similar to what happens in the United States, also in Canada, local government (provinces) integrates federal measures with their own policies. Finally, trade protection is much lower than that of the United States and the EU. Renewable fuels from NAFTA countries can be imported duty-free, while there is an import tariff of CAN$0.05/L on Brazilian ethanol.

3.4.5 European Union

Over recent years, the EU has significantly increased its consumption of biofuels. According to the first estimates for 2006, biofuel consumption in the EU grew from just below 3 million tonnes of oil equivalent (Mtoe) in 2005 to approximately 6 Mtoe in 2006, growth of 86.5%, and reaching a 1.9% share of fuels used in transport (EurObserv'ER 2007). Biodiesel predominates, representing 71.6% of the energy content of biofuels dedicated to transport, significantly ahead of bioethanol (16.3%) and the other biofuels (12.1%, i.e., 629,809 tons oil equivalent [toe] of vegetable oil and 13,940 toe of biogas) (EurObserv'ER 2007). Consumption of biodiesel increased by 71.4% between 2005 and 2006, compared with 57.5% growth for bioethanol. Data for 2005 show that the total area used for energy crop production was around 2.8 million hectares, representing about 3% of the total EU-25 arable land (EC 2006). Biodiesel and ethanol are mainly used blended with diesel and gasoline, respectively, in low proportions, but high-proportion blends (e.g., ethanol used for FFVs) and pure forms are also available in some countries, such as Sweden. Most of the ethanol is processed into ethyl tertiary butyl ether (ETBE) that is used as an additive to gasoline. Biodiesel is produced primarily from rapeseed. In 2004, an estimated 4.1 million tons of rapeseed was used, equal to slightly more than 20% of EU-25 oilseed production. Germany is the main producer, followed by France, Italy, and the Czech Republic. Since, EU is by far the world's largest producer of biodiesel, there is no significant external trade. Import duties on biodiesel and vegetable oils are between 0% and 5% (EC 2006). EU production of bioethanol is estimated to have used around 1.2 million tons of cereals and 1 million tons of sugar beet from 2004's raw materials. This represents 0.4% of the total EU-25 cereals and 0.8% of sugar beet production. Apart from France, where three-quarters of bioethanol is obtained from sugar beet, the majority of EU plants process grains mainly maize, wheat, and barley. The leading EU producers are Spain, Germany, and Sweden. In Europe, biofuels have been championed as an energy source that can provide new incomes for farmers both domestically and abroad, increase security of energy supply, and reduce GHG emissions from transport. EU currently does not have a community-wide excise tax on transport fuels, and member states can grant tax preferences according to their individual needs. However, there are coordinated efforts to increase the use of biofuels to meet a proposed mandate to fill 10% of transportation energy needs with biofuels by 2020. At the European Council summit on March 8–9, 2007, the EU's member states formally endorsed the 10% biofuel target but made it clear that such a goal must be subject to sustainable biofuel production and that the so-called second-generation biofuels become commercially viable (EC 2007). This conditionality is linked to increasing concerns about the sustainability of the first-generation biofuels currently available (e.g., biodiesel, bioethanol), which are made from agricultural crops. In early 2008, the European Commission proposed a mandatory sustainability certification scheme for both imported and domestically produced biofuels, requiring at least a 35% reduction in GHG emissions compared with fossil fuels. While Europe lags behind the United States and Brazil in ethanol production, it has provided support to its growing biodiesel industry. Energy crops in EU member states are heavily subsidized, and farmers are compensated for setting aside land. Set-aside land makes up about 10% of the total EU farmland, and it is used 95% of the time to grow energy crops. Energy crops further qualify for set-aside payments and energy crop aid, and they are excluded from production quotas. Nine member states have further set mandatory blending requirements, and the majority couple the mandate with fuel-excise tax exemptions. While information on capital investment support is difficult, given individual member programs, available data show that state aid to industry may account for up to 60% of initial investment, with governments regularly providing grants that account for 15%–40% of capital infrastructure investment.

3.4.6 India

India's biofuel production efforts are centered on second-generation biodiesel made from Jatropha. While mandatory blends are currently E5, there are discussions to raise the standard to E10 and eventually E20 blends as biodiesel from Jatropha becomes more cost effective. Individual states in India have adopted various policies to support the growing of Jatropha and research into biofuel production. The state of Andhra Pradesh formed a public–private partnership with the firm Reliance Industries, giving the firm 200 ac of land for Jatropha cultivation for biodiesel use. Similarly, the states of Karnataka, Chhattisgarh, and Rajasthan are promoting the planting of Jatropha saplings. In particular, Chhattisgarh will become self-reliant on energy by 2015, using biodiesel and selling Jatropha seeds for profit. In addition to encouraging Jatropha planting, the state of Tamil Nadu has abolished the purchase tax on Jatropha in order to promote its distribution and use.

3.4.7 China

China's policies focus especially on ethanol since the country is a net importer of vegetable oil. In 2002, the government launched the *Ethanol Promotion Program* in order to reduce corn excessive stocks. In 2004, the National Development and Reform Commission (NDRC) initiated the *State Scheme of Extensive Pilot Projects on Bioethanol Gasoline for Automobiles* with which the government controlled both production and distribution of ethanol. In 2006, some pilot projects implemented in 5 provinces and 27 cities reached the 10% blending target. Later on in the same year, the NDRC proposed to set a 6.6 billion liter target for 2010, but the proposal was rejected by the State Council because of high food prices. In 2007, the NDRC launched the *Medium- and Long-Term Development Plan for Renewable Energy*, which establishes that renewable energy's share of total primary energy consumption must rise to 10% by 2010 and to 15% by 2020. Of course, biofuels will play a key role to reach these targets. Ethanol production is projected to reach 2 million tons by 2010 and 10 million tons by 2020. Biodiesel consumption targets were also established to be 200,000 tons by 2010 and 2 million tons by 2020. The government is fully committed in meeting these objectives without affecting food prices. To do so, it does not allow factories to employ corn in ethanol production but instead encourages the employment of crops such as cassava, sorghum, and sweet potatoes. These restrictions, however, reduce the potential of Chinese production, which is nevertheless forecasted to reach 1.7 million tons in 2009.

The government also controls ethanol price, which is maintained at a level that would make production not economically feasible without external financial assistance.

In 2007, producers were granted a US$200/ton subsidy, which was replaced in 2008 by payments based on the evaluation of individual plant's performance. Additionally, ethanol producers do not have to pay the 5% consumption tax and the 17% VAT. Intermediate inputs like grains and fertilizers were also granted financial assistance. Similar to what happens in both the United States and the EU, also in China, research on second-generation biofuels is supported. The subsidy is about US$438/ha for Jatropha plantations and US$394 for Cassava. No direct subsidies are given for biodiesel. Biofuel production in China is directed by the state through the state-owned industry. Production and demand are stringently planned and controlled. The Chinese government has recognized the importance of using sustainable energy, and the NDRC is directing increased production of biofuels, with a target to produce 2 million tons of biodiesel by 2020. China has a large variety of feedstock options for biodiesel production as well, with promise in Jatropha, Rapeseed, and Soybean. The State Forestry Administration recently allocated 7000 ha in Hebei province for biodiesel production. Hebei is one of seven regions that will be used as biofuel demonstration forests. In 2007, the NDRC signed a *memorandum of understanding* with the U.S. Departments of Energy and Agriculture to facilitate further development of biofuels and facilitate transfer of scientific and technical knowledge on feedstocks and biofuel production. Although widespread mandates have not yet been established in China, there are mandatory E10 blends in five provinces, that is, Heilongjiang, Jilin, Liaoning, Anhui, and Henan.

3.4.8 Australia

Australia started subsidizing the biofuel sector in 2001 and set a non-binding target of 350 million liters to be reached by 2010. In 2006, two Australian states set two even more stricter targets: New South Wales a 10% binding share of ethanol in gasoline by 2011 and Queensland a 5% one. Australia's biofuel production in 2007 was 83 million liter for ethanol and 77 million for biodiesel. The statistical data from U.S. EIA, 2015, were related to Asia–Oceania biofuel production pattern during 2000–2012. Despite these small figures, biofuels are highly subsidized if compared to other industries. The most important policy is the tax rebate, which has been established to offset the fuel-excise duty of A$0.38143 per liter for both ethanol and biodiesel until 2011. In July 2011, the tax rebate for ethanol was abolished even though the excise was lowered to A$0.125. The *Energy Grant–Cleaner Fuel*, however, kept guaranteeing an alternative subsidy for ethanol: A$0.1/L decreasing by A$0.025/year until 2015. Biodiesel underwent a similar treatment when the excise duty dropped to A$0.191/L in 2011 and the *Energy Grant–Cleaner Fuel* program introduced a A$0.153/L subsidy, also decreasing until 2015, when it will be eliminated.

3.4.9 Thailand

Thailand's biofuel policy incentivizes both biodiesel through mandatory blending and ethanol consumption through tax exemptions, which allow ethanol blends to be cheaper than regular gasoline. Ethanol is produced from sugarcane and molasses, but tapioca-based production is expanding. Gasoline blended with ethanol is called "gasohol," which has contributed to a significant decrease of standard gasoline and currently accounts for nearly 50% of total gasoline consumption. Gasohol is exempted from the excise tax and this allows ethanol blends to be 10%–15% cheaper than regular gasoline. The government is active in further promoting biofuel production, especially through the diffusion of E20 and E85 blends as well as of flex-fuel vehicles. E20 and E85 blends will be substantially cheaper than regular gasoline (−20% and −50%, respectively), thanks to excise duty exemptions and additional state subsidy from the *State Oil Fund*. Biodiesel with 2% methyl ester content has completely replaced regular diesel in the whole country in 2008. A B5 blend should start to be enforced in 2011. Additionally, the government set up the *Committee on Biofuel Development and Promotion* in order to increase domestic palm oil production, the main biodiesel feedstock, of which currently Thailand is a net importer (Sorda, Banse and Kemfert 2010).

3.4.10 Malaysia

Malaysia is one of the greatest palm oil producers in the world, and this gives the country a big competitive advantage in biodiesel production. The government first intervened in 2005 with the *National Biofuel Policy* that introduced a 5% biodiesel blending (B5) mandate, which has been implemented recently. At the moment, biofuel production in Malaysia is not economically viable yet, mainly because of high palm oil prices, which makes palm oil producers (that are also biofuel manufactures) more convenient to directly sell palm oil instead of further processing it. The Malaysian biodiesel sector is suffering from competition from neighboring Indonesia. As a result, Malaysian exports have substantially decreased in the recent years (USDA 2012). Malaysia produces two types of biodiesel, that is, envodiesel and palm methyl esters (PME) biodiesel. The latter is the result of the blending of regular diesel with raw palm oil and is used only domestically, despite car manufacturers discourage its use. The former is exported and represents the largest share of total biodiesel production (75%). In 2007, Malaysia exported slightly less than 100,000 tons of PME biodiesel. Producers can be eligible for financial support through two aid schemes, that is, the *pioneer status* and the *incentive tax allowance*. The first provides a 70% tax reduction on the statutory income obtained from biodiesel production for 5 years. The second is for the companies with high investment costs in equipment and machinery, where allowances spent for fixed assets can be detracted from taxable income for a 5-year period. Trade restriction measures are not present. Exports of processed palm oil or biodiesel are duty-free, while crude palm oil exports are taxed.

3.4.11 Indonesia

In 2008, the Indonesian government established mandatory levels of biofuel consumption. Biofuel must reach 2.5% of total energy consumption by 2010 and 20% by

2025. The ethanol component of gasoline was of 3% by 2010 and increase to 15% by 2025. Later on these targets were reformulated and raised to 10% of biofuel share by 2010. Indonesian Ministry of Energy and Mineral Resources and Parliament come to an agreement to provide biofuel subsidies at 3000 rupiah per liter for biodiesel and 3500 rupiah per liter of ethanol in 2013. Moreover, Indonesia coal and mineral mining companies are required to consume 2% of biofuels in their total fuel consumption starting from July 2012. Biofuel blending policies, however, have been hampered by high feedstock prices. In 2006, Pertamina, the state-owned oil and fuel distribution company and only biofuel supplier, started selling a gasoline blended with 5% biodiesel (B5); however, in 2009, it was forced to reduce the blend to 1% due to higher palm oil prices. Pertamina, starting from February 2012, started blending conventional diesel with 7.5% biodiesel. New production facilities are under construction for both biodiesel and ethanol. In 2010, biodiesel production capacity was more than 4 million tons/year. Ethanol capacity was much lower, but the new facilities under construction will make it possible to reach higher outputs. In 2011, almost 90% of Indonesian biodiesel production was exported. Most of biodiesel goes to the EU. In 2011, 39% of total EU biodiesel imports came from Indonesia. The government provides fuel subsidies for almost 15 billion dollars, which are used to allow the selling of ethanol and biofuel blends at the same price of standard gasoline. Of course, this implies heavy losses for Pertamina, which account for about 40 million dollars/year.

3.5 Conclusions

Because of the strong and growing demand of crude oil and stricter emission standards, the demand for alternative fuels is increasing. Biofuels are the right solution as an alternative to petroleum fuels to bridge the gap. Biofuels are steadily gaining recognition as an important part of agricultural and energy sectors. Within a couple of years, biofuels have transformed from a niche energy source. All the countries hope that biofuels will provide a win–win strategy that can simultaneously promote energy security and economic developmental protection. Production of biofuels is still expensive; moreover, the fuel quality is not yet constant and conversion technologies of certain biofuels are still immature.

There is an urgent need to review existing biofuel policies in an international context in order to protect the poor and to promote rural and agricultural development while ensuring environmental sustainability. Undoubtedly, the multidisciplinary research efforts that combine new molecular approaches for strain development and process integration in the framework of process engineering will allow expansion and commercial implementation of innovative technologies to exploit the vast resources. There is lack of knowledge on biofuels in general public. Public acceptance for biofuels will be the last challenge to be addressed once all the systems are in place. Since public is the major user of fossil fuel in transportation sector, lack of public support for new transportation fuels can eventually lead to catastrophic failure.

References

Agarwal, A.K. 2007. Biofuels (alcohols and biodiesel) applications as fuels for internal combustion engines. *Prog Energ Combust Sci.* 33:233–271.

Berka, R.M. Grigoriev, I.V., Otillar, R. et al. 2011. Comparative genomic analysis of the thermophilic biomass-degrading fungi *Myceliophthora thermophila* and *Thielavia terrestris. Nat Biotechnol.* 29:922–927.

Bindraban, P. Bulte, E. Conijn, S. Eickhout, B. Hoogwijk, M., and Londo, M. 2009. Can biofuels be sustainable by 2020? An assessment for an obligatory blending target of 10% in the Netherlands. Climate Change, Scientific Assessment and Policy Analysis. Netherlands Environmental Assessment Agency, Bilthoven, Utrecht.

Bothast, R.J. 2005. New technologies in biofuel production. Presentation at USDA, Washington, DC. Agricultural Outlook Forum (AOF), February 24, 2005. Accessed on January 2, 2016.

Casey, T. 2011. Cargill, Shell and Honda team up to make gasoline from pine and corn waste. *Cleantechnica.* http://cleantechnica.com/2011/06/06/cargill-shell-and-honda-team-up-to-make-gasoline-from-pine-and-corn-waste/. Accessed on January 2, 2016.

Casey, T. 2012. U.S. Department of energy announces new biofuel to replace gasoline. *Cleantechnica.* http://cleantechnica.com/2011/03/08/u-s-department-of-energy-announces-new-biofuel-to-replace-gasoline/. Accessed on January 2, 2016.

Childs, S. Britt and Bradley, R. 2007. Plants at the pump: Biofuels, climate change and sustainability. World Resources Institute, Washington, DC.

Dale, B.E. and Ong, R.G. 2014. Design, implementation and evaluation of sustainable bioenergy production system. *Biofuels Bioprod Biorefin.* 8:487–503.

de Gorter, H. and Just, D.R. 2010. The social costs and benefits of biofuels: The intersection of environmental, energy and agricultural policy. *Appl Econ Perspect Pol.* 32:4–32.

Dellomonaco, C. Fava, F., and Gonzalez, R. 2010. The path to next generation biofuels: Successes and challenges in the era of synthetic biology. *Microb Cell Fact.* 9:3. http://www.microbialcellfactories.com/content/9/1/3The path to next generation biofuels. Accessed on January 2, 2016.

Demirbas, A. 2008. Biofuels sources, biofuels policy, biofuels economy and global biofuels projections. *Energ Convers Manage.* 49:2016–2116.

Doornbosch, R. and Steenblik, R. 2007. Biofuels: Is the cure worse than the disease? Technical report, OECD, Paris, France.

Dragone, G. Fernandes, B. Vicente, A.A., and Teixeira, J.A. 2010. Third generation biofuels from microalgae. In *Current Research Technology and Education Topics in Applied Microbiology and Microbial Biotechnology*, ed. A. Mendez-Vilas, pp. 1355–1366. Formatex, Spain.

Duncan, J. 2003. Costs of biodiesel production. *Energy Efficiency and Conservation Authority*, Wellington, New Zealand, pp. 1–26.

EC (Commission of the European Communities). 2006. Biofuels in the European Union: A vision for 2030 and beyond. *EUR22066*, Brussels, Belgium, p. 33.

EC (Commission of the European Communities). 2007. EU energy summit: A new start for Europe. *AGRI-G-2/WWD*, Brussels, Belgium, p. 10.

FAO. 2006. *Agricultural Market Impact of Future Growth in the Production of Biofuels*. Paris, France: Organisation for Economic Cooperation and Development (OECD).

Gardner, B. and Tyner, W. 2007. Exploration in biofuels, economics, policy and history: Explorations in biofuels economics, policy, and history: Introduction to the special issue. *J Agric Food Ind Organ.* 5(2): 1–7.

GBEP. 2007. http://www.globalbioenergy.org/events1/events-2007/task-force-on-ghg-2007/en/. Accessed on January 2, 2016.

Huang, H., Khanna, M., Onal, H., and Chen, X. 2013. Stacking low carbon policies on the renewable fuels standard: Economic and greenhouse gas implications. *Energ Pol.* 56:5–15.

Kaushik, N. and Biswas, S. 2010. New generation biofuels—Technology & economic perspectives. Technology Information, Forecasting & Assessment Council (TIFAC), Department of Science & Technology (DST), Government of India, New Delhi, India.

Kojima, M. Donald, M., and William, W. 2007. *Considering Trade Policies for Liquid Biofuels.* Washington, DC: The International Bank for Reconstruction and Development/The World Bank.

Masjuki, H.H., Kalam, M.A., Mofijur, M., and Shahabuddin, M. 2013. Biofuel: Policy standardization and recommendation for sustainable future energy supply. *Energ Procedia.* 42:577–586.

Muller Langer, F., Majer, S., and O'Keffee, S. 2014. Benchmarking biofuels—A comparison of technical, economic and environmental indicators. *Energ Sustain Soc.* 4:20.

Naik, S.N., Goud, V.V., Rout, P.K., and Dalai, A.K. 2010. Production of first and second generation biofuels: A comprehensive review. *Renew Sustain Energ Rev.* 14:578–597.

OECD-FAO. 2007. *OECD-FAO Agricultural Outlook 2007–2016.* Paris, France: OECD and Rome, Italy: FAO.

Paul, N. and Kemnitz, D. 2006. Biofuels-plants, raw materials, products. Fachagentur Nachwachsende Rohstoffe eV (FNR). WPR Communication, Berlin, Germany, p. 43.

Rajagopal, D., Sexton, S., Hochman, G., Roland-Holst, D., and Zilberman, D. 2007. Challenge of biofuel: Filling the tank without emptying the stomach. *Environ Res Lett.* 2:1–9.

Sorda, G. Banse, M., and Kemfert, C. 2010. An overview of biofuel policies across the world. *Energ Pol.* 38:6977–6988.

Stephanopoulos, G. 2007. Challenges in engineering microbes for biofuels production. *Science.* 315:801–804.

UN (United Nations). 2006. The emerging biofuels market: Regulatory, trade and development implications. United Nations Conference on Trade and Development, New York.

US Environmental Protection Agency. 2010. Renewable fuel standard program (RFS2) regulatory impact analysis, EU biofuels annual 2010 GAIN report number BR 10006, Washington, DC, USDA.

USDA. 2007. Census of agriculture United States. Summary and state data. http://www.agcensus.usda.gov/Publications/2007/. Accessed on January 2, 2016.

USDA. 2012. EU biofuels annual 2012 GAIN report number NL2020. Washington, DC: USDA.

Vinuselvi, P., Park, J.M., Lee, J.M., Oh, K., Ghim, J.M., and Lee, S.K. 2011. Engineering microorganisms for biofuel production. *Biofuels.* 2:153–166.

WWI (Worldwatch Institute). 2006. Biofuels for transportation, global potential and implications for sustainable agriculture and energy in the 21st century. Submitted report prepared for BMELV in cooperation with GTZ and FNR. Washington, DC, June 7, 2006, p. 398.

Xavier, M.R. 2007. *The Brazilian Sugarcane Ethanol Experience.* Washington, DC: Competitive Enterprise Institute.

Ziolkowska, J.R. 2014. Prospective technologies, feedstocks and market innovations for ethanol and biodiesel production in the US. *Biotechnol Rep.* 4:94–98.

4

Trends and Sustainability Criteria for Liquid Biofuels

Abdul-Sattar Nizami, Gunda Mohanakrishna,
Umesh Mishra, and Deepak Pant

Contents

Abstract

Anthropogenic greenhouse gas (GHG) emissions are changing our Earth's climate very rapidly and causing global warming phenomenon. There is scientific, social, and political consensus that 20% of global GHG emissions are due to the transport sector that is also blamed for increasing oil demand worldwide. The growth in the transport sector is estimated to increase by 1.3% per year until 2030. The increase in GHG emissions and high demand for fuel in the transport sector can be reduced significantly by replacing fossil fuels with liquid biofuels, which are derived from plant materials and appear to be carbon-neutral, renewable, and capable for cultivation under harsh environments. The plant materials used in producing liquid biofuels are also a potential source of value-added products such as feed, materials and chemicals, in addition to biofuels. This chapter reviews the current trends in liquid biofuel systems on global platform and criteria for sustainability pertaining to liquid biofuels. The three types of sustainability criteria for liquid biofuels, including economic sustainability, environmental sustainability, and social sustainability are discussed in detail.

4.1 Introduction

4.1.1 Feedstocks and Characteristics of Biofuels

Biofuels are commonly classified as primary and secondary biofuels. Unprocessed forms of biomass such as fuelwood, wood chips, and pellets are categorized as primary biofuels (Lee and Lavoie 2013), whereas secondary biofuels are produced by processing of biomass into ethanol, biodiesel, and dimethyl ether (Cherubini and Ulgiati 2010). Moreover, based on raw materials, biofuels are divided into first-,

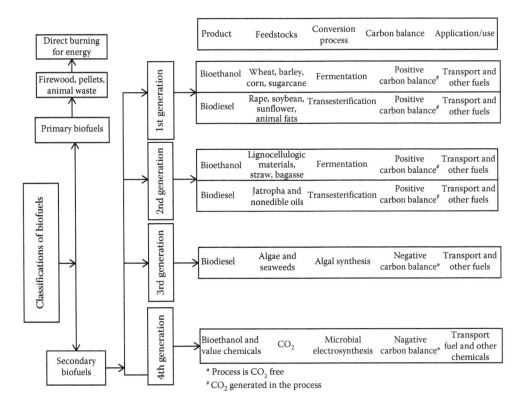

Figure 4.1 The classification of liquid biofuels and the production routes for bioethanol and biodiesel. (Adapted from Naik, S. et al., *Renew. Energy*, 35(8), 1624, 2010.)

second-, and third-generation biofuels (Ragauskas et al. 2006). The classification of liquid biofuels and the potential production routes for bioethanol and biodiesel are shown in Figure 4.1. Raw materials used in biofuel production may come from forest, agriculture, and municipal or industrial organic wastes (Larson 2008; Nizami et al. 2015a,b,c; Ouda et al. 2016).

Biofuels can be in the form of solid such as fuelwood, charcoal, and wood pellets; liquid such as biodiesel, ethanol, and pyrolysis oil; or gas such as methane and hydrogen (Nigam and Singh 2011; Demirbas et al. 2016; Miandad et al. 2016a,b). However, liquid biofuels are advantageous over solid and gaseous biofuels as they replace the most abundant conventional fuels such as diesel and petrol. Moreover, liquid biofuels are proved to be economically competitive with fossil fuels. However, the production of liquid biofuels in sufficient quantities is still required to make a meaningful impact on energy supply and demand (Ahmad et al. 2011). Figure 4.2 depicts the net energy gain over the energy sources used to produce ethanol and biodiesel from soybean and corn feedstocks individually (Hill et al. 2006).

4.1.1.1 First-Generation Liquid Biofuels

First-generation liquid biofuels are mainly produced from sugars, and carbohydrate-rich grains and seeds (Nigam and Singh 2011). This category of biofuels requires

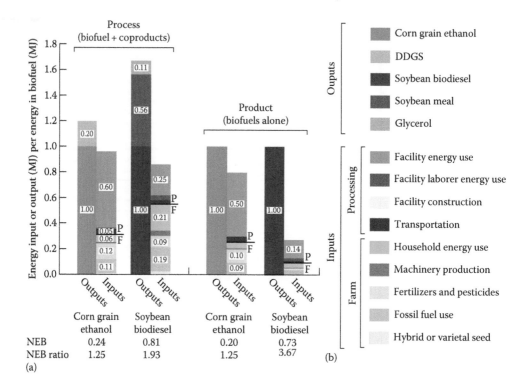

Figure 4.2 An example study showing that the production of both corn grain ethanol and soybean biodiesel results in positive net energy balance (NEB). Energy inputs and outputs are expressed per unit energy of a biofuel. All nine input categories are consistently ordered in each set of inputs, as in the legend, but some are so small as to be nearly imperceptible. Individual inputs and outputs of ≥0.05 are labeled. The NEB (energy output − energy input) and NEB ratio (energy output/energy input) of each biofuel are presented for both the entire production process (a) and the biofuel only (i.e., after excluding coproduct energy credits and energy allocated to coproduct production) (b) (From Hill, J. et al., *Proc. Natl. Acad. Sci. USA*, 103, 11206, 2006.)

a relatively simple process for the production of final fuel product (Nigam et al. 1997). Ethanol is the most popular first-generation biofuel that is produced by the fermentation of sugars (Larson 2008). Since biocatalysts are used in the conversion of organic matter to ethanol, it is denoted as bioethanol (Demirbas 2011). Yeast and enzymes are widely used as biocatalysts in bioethanol production, which convert simple carbohydrates such as glucose and starch into bioethanol (Suresh and Rao 1999). Distillation and dehydration are the steps involved to bring bioethanol to the desired concentration, which is suitable for direct use or to be mixed with conventional liquid fuels (Love et al. 1998). Hydrolysis is applied as a pretreatment step to convert carbohydrate into glucose, if the feedstock is made from complex carbohydrates (Zhao et al. 2009).

Biodiesel is another first-generation liquid biofuel that is preferred for sustainable energy management. Vegetable oils and oleaginous plants are the major sources in biodiesel production through the transesterification process, which uses alkaline, acidic, or enzymatic catalyzers and ethanol or methanol to convert lipids into fatty acids (Zhao et al. 2009). Glycerin or glycerol is the major by-product of biodiesel production (Venkata Mohan et al. 2011). Among available feedstocks for bioethanol production, sugarcane is the most widely used feedstock. However, various studies have highlighted the use of first-generation biofuels as an unviable option due to competition of feedstocks with food supply (Patil et al. 2008; Singh et al. 2010; Rathore et al. 2016).

Perennial grasses, sweet sorghum, corn or maize, and cassava are also major feedstocks for bioethanol production. Brazil and other tropical countries such as Australia, Peru, South Africa, and India are the most suitable regions for sugarcane production (Clay 2004; Xavier 2007). The quality of sugarcane determines the bioethanol yield (De Oliveira et al. 2005). Besides sugarcane juice, by-products of the sugarcane industry such as molasses are also used for bioethanol production. Sweet sorghum is widely grown in China, India, the United States, Australia, Brazil, Zimbabwe, and European Union (EU). India, the United States, and China are already making bioethanol from sweet sorghum. The sugar content of sweet sorghum is extracted and fermented to produce bioethanol. The range of bioethanol production from sweet sorghum is from 0.21 to 0.6 tons/ha (Elbehri et al. 2013).

Maize is a highly productive temperate crop that is cultivated in several parts of the world. Maximum agricultural output of maize is from 7 to 11 tons/ha, which is higher than any other cereal crop, and bioethanol yield can reach up to 3500 L/ha of corn. For both corn feedstock production and ethanol conversion, water consumption is relatively low with 4 L of water per liter of ethanol produced (Aden 2007). According to McKendry (2002), 8.4% of global maize production is used in the conversion of bioethanol that makes maize the dominant feedstock for liquid biofuel production. In recent years, a sharp rise in the U.S. ethanol production has resulted in high demand for maize as the feedstock of choice. The United States is producing almost all the biofuel from maize/corn. However, corn ethanol has been criticized for its lower greenhouse gas (GHG) emission savings (Farrell et al. 2006).

Cassava is a well-considered feedstock for bioethanol production. It is a perennial tuber crop that mostly grows in Africa but also grows in Asia and Latin America. It is a drought-resistant crop with high capacity to overcome diseases. Tropical and subtropical regions, where rainfall is above 600 mm in 2–3 months' cycle, are optimum for the cultivation of cassava (Lokko et al. 2007). Nigeria, Brazil, Thailand, and Indonesia are the four largest producers of cassava, accounting for about 50% of global cassava production. The production of cassava tuber per unit land is found to be higher (up to 90 tons of roots per ha). The major use of cassava is in starch production, which can be used as a raw material in the food industry and as a source of sweeteners and citric acid. Cassava chips or pellets have gained business worldwide as commercial products. Bioethanol yield from cassava varies from 137 to 190 L/ton cassava that is equivalent to 3.7 to 6.3 kL/ha (Lokko et al. 2007; Xavier 2007).

The biodiesel feedstocks have been found to have lower yields per a specific crop than the bioethanol feedstocks. However, the yield is found to be higher for palm oil grown in tropical regions. The EU is the largest producer of biodiesel in the world, which is followed by the United States (Bozbas 2008). Rapeseed is an annual herb, which grows in temperate regions ideally under 500 mm of annual rainfall. Rapeseed is also referred as canola (Christou and Alexopoulou 2012). Rapeseed oil conventionally has several food applications. Nowadays, it is identified as a promising feedstock for the production of biodiesel and value-added products such as soaps and lubricants. The widespread cultivation of rapeseed crop in the EU is due to the dominant position of EU in biodiesel production. Among available feedstocks for biodiesel production, rapeseed alone accounts for approximately 59% of biodiesel production (Pahl and McKibben 2008).

Oil palm is an important oil crop for biodiesel production and an important source of human and animal feed. Indonesia, Malaysia, and Nigeria are the largest oil palm–producing countries in the world (Beckman 2006). Palm oil accounts for about 10% of total biodiesel production, which is increasing rapidly in Indonesia and Malaysia (Pahl and McKibben 2008). Soybean is also an important feedstock for biodiesel production. Soy is a legume crop that has 17.5% of oil content that is lower in comparison to oil palm (Elbehri et al. 2013). Soy accounts for 75%–90% of total biodiesel production (Carriquiry 2007). Moreover, many other countries such as Brazil, Argentina, Paraguay, and Bolivia are using soy-based fuels to blend with liquid fossil fuels (Beckman 2006; Xavier 2007).

4.1.1.2 Second-Generation Liquid Biofuels

Second-generation liquid biofuels are produced from either inedible residues of food crops or inedible plant biomass such as grasses or trees explicitly grown for energy production (Korres et al. 2011; Nizami and Ismail 2013). Two fundamental processes such as biological and thermochemical processes are used to produce liquid biofuels from forests, agricultural, and lignocellulosic biomass (Sims et al. 2010). The main advantage of second-generation biofuels is the elimination of direct food versus fuel competition, which is a major disadvantage associated with first-generation biofuels (Sims et al. 2010). Feedstocks of second-generation biofuels can be produced specifically for energy purposes that enable higher production per unit land. They also increase the land use efficiency in comparison to first-generation biofuels. Moreover, it is believed that second-generation biofuels have potential for lower costs, and significant energy along with several environmental benefits (Larson 2008). However, the raw material is less complex than the feedstocks of first-generation biofuels that require more sophisticated and complex processing and production equipment. Furthermore, it is manifest from the literature that the production of second-generation biofuels needs more investment per unit of production at large-scale facilities (Stevens et al. 2004). To achieve second-generation biofuels as the potential energy source, extensive research and development (R&D) on feedstock production and conversion technologies is required to be explored (Nigam and Singh 2011).

The future production of ethanol is foreseen to include both first- and second-generation feedstocks (Stevens et al. 2004). Second-generation biofuels such as ethanol and butanol are produced using biochemical conversion processes, whereas all other biofuels are generated using thermochemical processes (Barron et al. 1996). Most of second-generation fuels produced through thermochemical processes are currently being produced from fossil fuels sources that include methanol, refined Fischer–Tropsch liquid fuels, and dimethyl ether (Lange 2007). Pyrolysis oils are unrefined fuels that are also produced using thermochemical processes; however, they require additional refining before being used in engines (Larson 2008). The utilization of aboveground biomass for the production of second-generation liquid biofuels provides better land-use efficiency than that of first-generation liquid biofuels (Lange 2007).

Jatropha is one of the major feedstocks for second-generation biofuels, especially for biodiesel. It is a drought-resistant and inedible perennial crop (Fairless 2007). Similar to cassava, it can grow in arid and harsh environments with limited water, and agrochemicals supply. The oil content of jatropha seed for biodiesel production is extracted mechanically by means of oil presses or chemical methods (Antizar-Ladislao and Turrion-Gomez 2008). The by-products of the process such as husks and cake can be used as fertilizers or briquettes for power and heat generation, while fatty acids can be used in the production of soaps. Other by-products of the transesterification process can be used as a potassium fertilizer and glycerin. Glycerin alone accounts for 10% of total process output. Toxic substances are removed during the process in order to make protein-rich animal feed free from toxins (Reinhardt et al. 2007). The biodiesel yield from jatropha crop ranges from 340 to795 L/ha on unfertile land and from 795 to 2840 L/ha on normal soils (Weyerhaeuser et al. 2007).

4.1.1.3 Third-Generation Liquid Biofuels

Third-generation biofuels are the latest generation of biofuels that are explicitly derived from microbes and microalgae (Chisti 2007). They are now widely accepted as promising alternative energy resources that are positively lacking the major drawbacks associated with first- and second-generation biofuels (Dragone et al. 2010). Microalgae are capable to produce 15–300 times more oil for biodiesel production in comparison to traditional crops per area. Moreover, conventional crop plants are generally harvested only once or twice a year, whereas microalgae have a very short harvesting cycle of 10–30 days. This will allow round the year harvesting of microalgae for multiple cycles per crop that will significantly increase the total algae biofuel yields (Schenk et al. 2008). Integrated productions of bioethanol and biodiesel were depicted in Figure 4.3.

Microalgae have chlorophyll *a* as their primary photosynthetic pigment. They lack a sterile covering around the reproductive cells (Brennan and Owende 2010). Prokaryotic and eukaryotic cells are the two types of cells present in algae. Prokaryotic cells are without membrane-bound organelles such as nuclei, mitochondria, plastids, and Golgi bodies and they are present in the group cyanobacteria. The remaining algal species are eukaryotic with membrane-bound organelles. Microalgae can be autotrophic or heterotrophic or both, that is, mixotrophic (Lee et al. 2007). Autotrophic microalgae use inorganic compounds as a source of carbon. The photoautotrophic microalgae use sunlight as a source of energy, whereas heterotrophic microalgae use oxidizing inorganic compounds as a source of energy. Some photosynthetic microalgae follow mixotrophic metabolism that is a combination of heterotrophy and autotrophy (Liang et al. 2009).

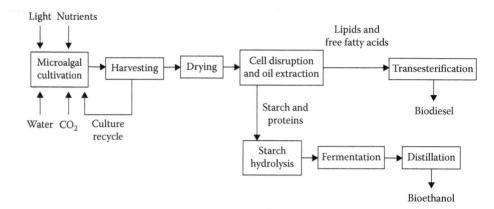

Figure 4.3 Integrated process for the production of biodiesel and bioethanol from micro-algae. (Adapted from Naik, S. et al., *Renew. Energy*, 35(8), 1624, 2010.)

Photosynthesis is a key component for survival of autotrophic algae, where they convert solar radiation and CO_2 into oxygen and adenosine triphosphate that are usable energy currency at the cellular level (Brennan and Owende 2010). This energy is converted to biomass and lipid material in microalgae. Lipid accumulation means an increased concentration of lipids within the cells of microalgae (Brennan and Owende 2010). An integrated system for biofuel production from microalgae includes the cultivation step with separation of the biomass from the growth medium along with succeeding lipid extraction for biodiesel production using a thermochemical process of transesterification. Following oil extraction, enzymatic hydrolysis leads to formation of fermentable sugars that can be converted into bioethanol through fermentation and distillation using the conventional ethanol distillation technology (Dragone et al. 2010).

4.1.2 Biorefining Process for Conversion Process

The dramatic rise in petroleum prices in the last decade has promoted liquid biofuels to become cost-competitive fuels with petroleum-based fuels. Biodiesel and bioethanol are the two main types of commercial liquid fuels among first-generation biofuels. Bioethanol is a replacement of petrol or gasoline in flexi-fuel vehicles that have the capability of blending bioethanol with petrol (Naik et al. 2010).

4.1.2.1 Transesterification for Biodiesel Production

Fatty acid methyl esters (FAMEs) of vegetable oils are commonly referred as biodiesel that is gaining importance as an eco-friendly fuel substitute. Biodiesel is produced from renewable agro-sources such as vegetable oils and animal fats, and residual oil or fat by chemical reaction with alcohols by using a catalyst (Kulkarni et al. 2006; Gardy et al. 2014). Glycerol is a by-product from the process that has several commercial applications (Meher et al. 2006). The esterification occurs between free fatty acids (R-COOH) and methanol (CH$_3$OH), whereas transesterification takes place between triglyceride (R-COOR′) and methanol that is adsorbed on the acidic site of the catalyst. The reaction of the carbonyl oxygen of free fatty acid including monoglyceride with the acidic site of the catalyst forms carbocation. As a result, tetrahedral intermediate is produced by the nucleophilic reaction of carbocation with alcohol (Kulkarni et al. 2006).

4.1.2.2 Fermentation for Ethanol Conversion Processes

An array of carbohydrate-based raw materials have been examined for the production of ethanol using the fermentation process. These raw materials can be categorized into three major types: (1) sugar crops such as sugarcane, beetroot, wheat, palm juice, and fruits; (2) starch crops such as grains such as sweet sorghum, corn, wheat, barley, and rice, and root plants such as potato and cassava; and (3) cellulosic crops such as wood and wood waste and agricultural residues (Ragauskas et al. 2006). The ethanol produced from food grain crops is known as grain alcohol, whereas ethanol produced from lignocellulosic biomass is called biomass-ethanol or bioethanol. The both of these alcohols are produced through fermentation or biochemical processes (Minteer 2006; Jin et al. 2015).

Structurally, starch is a long-chain polymer of glucose. First, the macromolecular structure of starch breaks down into simpler and smaller molecules such as glucose. To achieve this starch, biomass is ground and mixed with water to produce 15%–20% starch mash. The mash is then hydrolyzed using a biochemical via enzymatic or microbial processes or thermochemical process (Minteer 2006). The produced glucose then undergoes the fermentation process, where hexoses such as glucose are converted into ethanol. The anaerobic bacteria or specific enzymes act on this conversion. If the raw feedstock is agro-based biomass, it contains complex structures such as cellulose and lignin. To break down cellulosic materials into fermentable molecules such as glucose, energy-intensive chemical and enzymatic processes are used. Therefore, the process cost becomes higher due to energy-intensive processes (Zhang 2008).

4.1.3 Types of Biorefineries for Liquid Biofuel Production

4.1.3.1 Green Biorefinery

A green biorefinery is a multiproduct system that utilizes abundantly available green biomass as the raw material for the production of industrial products (Xiu and Shahbazi 2015). If a biorefinery processes its refinery products and fractions according to the physicochemical properties of the plant material, it is called a green biorefinery (Figure 4.4). In the past, much attention had been paid toward the presence of cellulose, hemicellulose, and lignin in cellulosic materials, though there are considerable amounts of other biomolecules such as protein available in these biological materials (Kamm and Kamm 2004). For example, Spirulina has high-protein content (>60%) that is more valuable than an equal weight of carbohydrate and lipid. This leads to the development of a protein-recovery-based

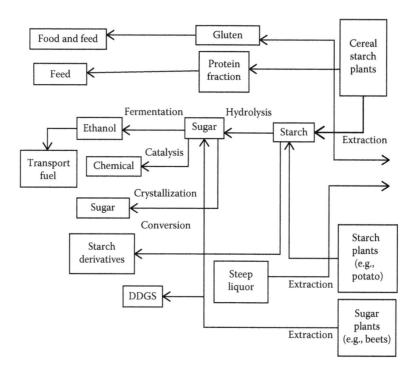

Figure 4.4 Schematic of a green biorefinery that also represents the biorefinery concept for first-generation biofuels. (Adapted from Naik, S. et al., *Renew. Energy*, 35(8), 1624, 2010.)

biorefinery. At present, the feedstocks used in the green biorefinery are mainly green grasses such as alfalfa, clover, and immature cereals from extensive land cultivation (Fernando et al. 2006).

The value-added intermediates from the green biorefinery are proteins, soluble sugars, celluloses, hemicelluloses, and lignin parts (Xiu and Shahbazi 2015). During the first step, the refinery processes the green biomass from their natural form into a fiber-rich press cake and nutrient-rich green juice using a wet fractionation process (Kamm and Kamm 2007). The press cake contains celluloses, starches, crude drugs, dyes, pigments, and other organics, whereas the green juice contains proteins, amino acids, organic acids, enzymes, hormones, dyes, organic substances, and minerals (Kamm et al. 2010). The press cake is suitable for the production of green feed pellets and also as a potential raw material for the production of chemicals, syngas, and synthetic fuels (Rahman et al. 2015).

4.1.3.2 Lignocellulosic-Based Biorefinery

Lignocellulosic plant materials contain two types of polysaccharides that are cellulose and hemicellulose (Zhang 2008). These two polysaccharides are bound together by a third component called lignin. A lignocellulosic-based biorefinery consists of three main chemical fractions: (1) hemicellulose and sugar molecules, mainly pentoses; (2) cellulose (polymer of glucose); and (3) lignin (polymers of phenols) (Tyson et al. 2005). The potential products of the lignocellulosic-based biorefinery are shown

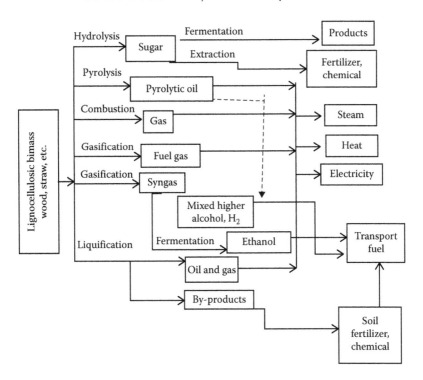

Figure 4.5 Schematic of a lignocellulose-based biorefinery that also represents the biorefinery concept for second-generation biofuels. (Redrawn based on Naik, S. et al., *Renew. Energy*, 35(8), 1624, 2010.)

in Figure 4.5. Biological conversion of lignocellulose has the following three main steps: (1) lignocellulosic pretreatment with the conversion of recalcitrant lignocellulosic structure to reactive cellulosic intermediates; (2) enzymatic hydrolysis with hydrolysis of cellulose by cellulases, where reactive intermediates are converted into fermentable sugars such as glucose; and (3) fermentation, where final products such as lactic acid, succinic acid, and ethanol or other bio-based chemicals are produced (Demain et al. 2005; Hong et al. 2013).

The pretreatment of lignocellulose is one of the costliest steps and has a major influence on lignocellulose particle size reduction and enzymatic hydrolysis, and finally in the process of fermentation (Wooley et al. 1999). Decomposing the locked polysaccharides of the recalcitrant structure of lignocellulose is one of the key research priorities for the emerging biofuels and bio-based industry (Zhang 2008). If lignin scaffolds are isolated in an economical way, considerable amounts of monoaromatic hydrocarbons can be extracted. This will add significant value to the primary processes (Zhang et al. 2006). The primary technologies in the lignocellulosic-based biorefinery generate primary chemicals that could make a variety of fuels, valuable chemicals, and useful materials along with power (Wyman et al. 2005). In this direction, five stages have been demonstrated, including sugar, thermochemical or syngas, biogas, carbon-rich chains, and plant products stages (Naik et al. 2010; Tahir et al. 2015).

4.1.3.3 Algae-Based Biorefinery

Biological fixation of CO_2 using photosynthetic microorganisms particularly micro-algae is a potential source of bioenergy production and mitigation of global warming (Chisti 2007). Algae perform three types of metabolisms such as autotrophic, heterotrophic, and mixotrophic. Besides lipids, algae also accumulate carbohydrate biomass. The lipids contain several components such as neutral lipids, polar lipids, sterols, wax esters, and hydrocarbons (Venkata Mohan et al. 2015a; Rehan et al. 2016a). Many algae species also produce phenyl derivatives such as carotenoids, terpenes, tocopherols, quinones, and phenylated pyrroles (Venkata Mohan et al. 2011).

Algae can grow under diverse and varying environmental conditions that allow them to be used for wastewater treatment (Figure 4.6). Since algae have high CO_2 tapping and fixation ability, it can be utilized to reduce CO_2 emissions from power plants and other industries (Brown and Zeiler 1993). The algae biomass can be used for extracting oil and producing biodiesel. Algae synthesize different fatty acids such as medium-chain, long-chain, and very-long-chain species of fatty acids. Algae oil from the stagnant water bodies that were receiving domestic sewage showed 33 different types of saturated and unsaturated fatty acids along with food and fuel features (Venkata Mohan et al. 2011). Oil content of some algae species is found to be 80% of

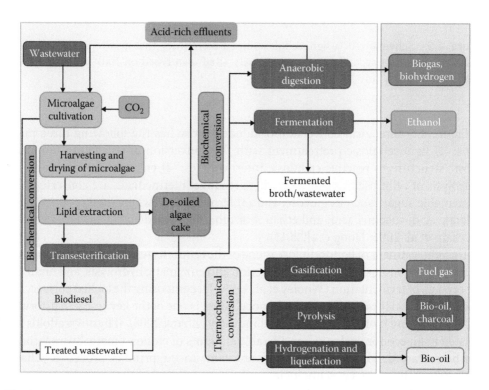

Figure 4.6 Schematic presentation of the microalgae-based biorefinery concept for third-generation biofuels. (Reused from Venkata Mohan, S. et al., Algal biorefinery, in *Biomass and Biofuels: Advanced Biorefineries for Sustainable Production and Distribution*, 2015a, pp. 199–216; Venkata Mohan, S. et al., *Bioresour. Technol.*, 102(2), 1109, 2015b. With permission.)

the total-body dry weight (Bridgewater et al. 1999). Microalgae produce much higher yields of biodiesel production in comparison to widely used agricultural oil crops including soybean, rapeseed, and oil palm (Harun et al. 2010).

The algae biorefinery concept is based on the petroleum refineries that yield multiple fuels and products of petro-chemicals from crude petroleum oil (Brennan and Owende 2010). The main products of an algae biorefinery are lipids, proteins, carbohydrates, and other value-added components such as pigments, antioxidants, fatty acids, and vitamins (Schmid-Straiger 2009; Singh and Gu 2010). According to the report of the U.S. Department of Energy on "National Algal Biofuels Technology Roadmap," the future of algae biofuel is bright in many ways (US-DOE 2009; USDA 2012). The positive factors of microalgae include fast growth potential, noninterference with food crops, large production of neutral lipids, survival and good growth in saline and brackish water, tolerance under desert, arid, and semiarid conditions, potential CO_2 sequestration agent, and production of value-added products (US-DOE 2009; Naik et al. 2010). In this perspective, the commercialization of the algae biorefinery has witnessed several developments and technological innovations in synthetic biology, metabolic engineering and genomics, the advances in bioreactor designs, and large-scale raceway ponds (Harun et al. 2010; Singh and Gu 2010).

4.1.4 Integrated Biorefineries

The biorefineries discussed earlier are based on single conversion technology to produce different chemicals and individual fuel. Integrated biorefineries use different bio-based feedstocks for the production of multiple biofuels, energy, and chemicals (Nizami et al. 2016; Sadef et al. 2015). Typically, all or some of the energy generated by the refinery can be used in its operation, and rest of the energy or product can be considered for revenue generation. The challenges of climate change, energy supply security, and increasing costs of conventional fuels are the main drivers in the development of integrated biorefineries (Azapagic 2014).

Integrated biorefineries basically integrate three different platforms such as thermochemical, sugar, and non-platform or existing technologies. The integrated biorefinery generates electricity from thermochemical processes, whereas fermentation technologies produce value-added products and bioethanol. Since a biorefinery is a capital-intensive project, integration of several conversion technologies including thermochemical and biochemical will reduce the overall cost. It will have more flexibility in products production and will generate its own electricity (Naik et al. 2010).

Bio-oil production from pyrolysis is an emerging process in the biorefinery development that could be directed via a conventional petrochemical refinery to generate multiple value-added chemicals (Miandad et al. 2016c; Rehan et al. 2016b). This is very advantageous for the existing petroleum refineries, as all the required infrastructures for separation and purification of generated products are in place (Bridgwater and Cottam 1992). Bio-oil chemical characteristics vary with the type of feedstock, however a woody biomass typically produces a mixture of water (30%), phenolics (30%), aldehydes and ketones (20%), alcohols (15%), and miscellaneous compounds

(Bridgwater and Cottam 1992). Hydrodeoxygenation (HDO) is for the hydrogenation of raw bio-oils. The bio-oil can be converted into a liquid hydrocarbon with similar characteristics of crude petroleum oil, after several HDO treatment steps.

4.2 Recent Trends in Liquid Biofuels

4.2.1 Blending of Biofuels with Liquid Fossil Fuels

The 1970s global fuel crises triggered awareness of oil restrictions and shortages. It helped to focus on alcohols as alternative liquid fuel sources. For the first time, a blend of 10% ethanol and unleaded gasoline (E10) was brought into the commercial market of the United States and in the Midwestern States. Later on in the 1980s, the use of ethanol blended with diesel was found to be technically acceptable for existing diesel engines. In the beginning of the twenty-first century, the commitment by the U.S. government to increase bioenergy threefold has added momentum to the search for viable biofuels. Similarly, the EU also adopted a directive for the use of biofuels. Blends such as ethanol blended with gasoline and biodiesel blended with diesel are being introduced to many countries in the world to reduce the hydrocarbon fuel's demand. However, certain technological limitations do not allow blending of ethanol and biodiesel with liquid fossil fuels at higher percentages. Therefore, it is important to quantify ethanol in blended gasoline and biodiesel in blended diesel.

4.2.1.1 Blending of Bioethanol with Petrol or Gasoline

There are different common ethanol fuel mixtures in place nowadays. It is possible to use these mixtures in IC engines, if the engines are designed or modified for this purpose. For example, ethanol can be blended with gasoline to be used in gasoline engines. In the case of high ethanol contents, it is possible to use after minor modifications in the engine. However, ethanol damages plastic fuel tanks and fuel lines of vehicles and airplanes. Ethanol fuel mixtures are denoted by "E" and followed by numbers that explain the volume percentage of ethanol in the mixture. For instance, E10 means 10% anhydrous ethanol and 90% gasoline. Low-ethanol blends such as E5, E10, and E25 are called as gasohols. E10 is the most common gasohol in use worldwide.

4.2.1.2 Blending of Biodiesel

Biodiesel blended with fossil-based diesel is the most commonly distributed fuel for use in the retail marketplace. The letter "B" is used to describe the amount of biodiesel in any fuel mix. The volume percentage of biodiesel in the mixture is suffixed to the "B." For example, B20 means 20% biodiesel and 90% hydrocarbon-based diesel. Various blends of biodiesels are available in the market ranging from B2 (2% biodiesel) to B100 (100% biodiesel). B20 or lower biodiesel grades can be used in diesel engines with no or minor modifications. The B6–B20 blends are enclosed under the ASTM D7467 standard. According to national renewable energy laboratory (NREL) report, B100 (100% biodiesel) can also be used, but the engine may require few modifications (Tyson et al. 2005).

4.2.2 Biofuels in Railway Transportation

Biodiesel is one of the most promising biofuels in the railway transportation sector. The worldwide consumption of liquid fuel by railways is around 34,000 million liters (UIC 2007). Only the EU consumes 15% diesel on railways among industrialized countries, whereas it is dominated in the United States with up to 70%. In case of the EU and Canada, liquid fuel consumption is decreasing (UIC 2007). On the contrary, it is increasing in the United States at a higher rate. Technologically, blending of biodiesel is feasible in diesel traction systems, but there is relatively little experience with biodiesel use in railways. Most engine manufacturers have shown their interest to include B5. However, less attention was paid in the case of higher blends such as B10–B100 (UIC 2007). Tax incentives for the use of biofuels in railways are less, as they often pay lower tax on fuel in comparison to road transport. Therefore, it will be a challenge to increase biodiesel share in the railway transport sector.

The EU has set 10 goals for the competitive and efficient transport system with an ambition to reduce 60% GHG emissions (Korres et al. 2010; Singh et al. 2011). For instance, 30% of road freight over 300 km will be shifted to railway or waterborne transport by 2030 and 50% by 2050. By 2050, the EU high-speed rail network will be completed, and most of the medium-distance passenger transport will be covered by rail only. In between 2010 and 2011, a 1-year trial was started by Amtrak with B20 in a diesel locomotive operation in the Southern States of the United States. Similar trials on the BNSF railway were carried out by Montana State University in 2011 (EBTP 2013). In August 2012, Indian Railways also announced plans for establishing four biodiesel plants. Virgin Trains, a British train-operating company, claimed to have run the United Kingdom's first biodiesel train with 80% diesel and 20% biodiesel. Since 2007, the Royal Train is successfully operating with 100% biodiesel fuel supplied by Green Fuels Ltd. During the summer of 2008, a similar state-owned railroad in eastern Washington ran a test on 25% biodiesel and 75% diesel mixture. In 2007, Disneyland began operating the park trains on 98% biodiesel (EBTP 2013). The International Union of Railways in association with the United Kingdom's Association of Train Operating Companies published a short report on railways and biofuel that concluded the test results from European Railways and Indian Railways. The report revealed that biodiesel is technically feasible for use in railway traction engines in lower biodiesel blends. However, potential disadvantages with higher blend are the increased fuel consumption and lower engine power.

4.2.3 Marine and Inland Waterways

In 2007, about 2300 Mtoe of energy was used in global transport, in which 10% was consumed by marine transport alone (EBTP 2013; McGill et al. 2013). In the EU, inland waterways consumed approximately 1.6% of total energy consumption (Eurostat 2014). In waterways, there are different modes of transport such as inland navigation, short-sea, and maritime transport. The EU alone has 40.6 thousand km of inland waterways and intra-EU maritime transport (EBTP 2013). Since electrification is not possible in water transport, gaseous and liquid biofuels are promising substitutes for fossil fuels. Many alternatives such as methanol, Bio-LNG, and biodiesel are promising substitutes for marine distillates and residual fuels (EBTP 2013; McGill et al. 2013; Eurostat 2014).

Biofuels are now widely considered to lower carbon intensity in the ships propulsion and to mitigate carbon emissions to local air. The waterways sector is still in early stage to orient toward biofuels. In diesel engines of ships, biodiesel (FAME), vegetable oil and hydrogenated vegetable oil, dimethyl ether, gas-to-liquid and biomass-to-liquid fuels potentially can be used. In March 2015, Stena Line launched a ferry (Stena Germanica) powered with methanol for the first time on the Kiel–Gothenburg route. The fuel system and engines of Stena Germanica have been adapted in the Remontowa shipyard, Gdansk, Poland, in collaboration with Stena Line and Wärtsilä. The project cost was 22 million Euros with 50% support from the EU's project of Motorways of the Seas. Dual fuel technology was used, with methanol as the main fuel along with an option to use marine gas oil (MGO) for backup (EBTP 2013). In the transport sector, water transport mode is the fastest growing and it claims to have relatively low-energy consumption for long-distance transportation (EBTP 2013).

4.2.4 Use of Biofuels in Industries

In domestic and commercial boilers, biodiesel can be used as a heating source. A standardized mixture of heating oil and biofuel was used for such purposes. This mixture is also called a bioheat that is a registered trademark of National Biodiesel Board (NBB) and National Oilheat Research Alliance (NORA) in the United States and Columbia Fuels in Canada (Anonymous 2014). ASTM 396 recognizes a mixture of up to 5% biodiesel as equivalent to virgin petroleum-based heating oil. Blends with a higher percentage of biofuel (up to 20%) are used by many consumers. Research and development (R&D) is underway to determine how such mixtures affect the performance of boilers (Anonymous 2014). The old furnaces that contain rubber parts can be affected by the solvent properties of biodiesel. Moreover, the used varnishes left behind by petroleum-based diesel can be released and they may clog pipes. Therefore, fuel filtering and regular filter replacement are needed. On the other hand, using biodiesel as a blend along with decreasing the petroleum ratio over time can permit the varnishes to come off more progressively and be less expected to clog. A law has been passed in Massachusetts, according to that diesel used for home heating in the state is to be increased to 2% biofuel by 2010 and 5% biofuel by 2013 (Anonymous 2014).

4.3 Criteria for Sustainability of Liquid Biofuels

4.3.1 Sustainable Development

Sustainability is defined as *development that meets the needs of the present without compromising the ability of future generations to meet their own needs* (Elbehri et al. 2013).

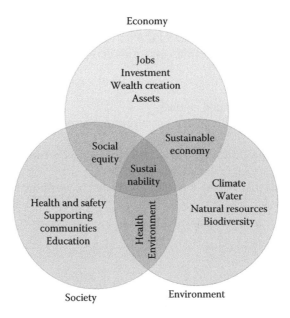

Figure 4.7   Venn diagram representing the three main components of criteria for sustainability of liquid biofuels. (Adapted from Remmen, A. et al., Life cycle management: A business guide to sustainability, UNEP and Danish Standards, 2007.)

Sustainability is an essential condition for long-term feasibility of the existing resources. Sustainable development strives to satisfy human needs with respect to economic affordability, environment friendliness, and social acceptability (Azapagic 2014). Sustainable development issues such as environmental impacts such as global warming, acidification, and loss of biodiversity; economic aspects such as costs and profits; and social concerns including employment, health, and human rights must be considered in evaluating any product, process, or system's sustainability (Figure 4.7). The following are important factors to make liquid biofuels the vectors for sustainable development (Elbehri et al. 2013; Buchholz et al. 2009):

- Encouragement for liquid biofuels production as a renewable energy and to reduce the dependency on fossil fuels.
- Liquid biofuels are intended to mitigate GHG emissions and improve climate change situation along with a more pleasant condition on earth.
- Major part of the land, labor and resources that are used in the production of liquid biofuels should not interfere with the existing resources of food and feed production.

4.3.2 Classification of Sustainability

Sustainability of liquid biofuels is generally categorized into the following three factors: economic sustainability, environmental sustainability, and social sustainability. Among these factors, environmental sustainability and economic sustainability are required to be preserved along with social sustainability properties. However, investments in liquid biofuels for long-term periods must pass the economic sustainability criteria that can be assessed based on the following points (Elbehri et al. 2013):

- Economical analysis of liquid biofuels
- Stable competitive conditions that encourage producers to choose liquid biofuels production
- Impacts with the production and feedstock competition between basic and biofuels application (i.e., as food and as a feed)
- Extent of biofuels that can make them a more reliable substitute for fossil fuels

In the case of environmental sustainability, the following criteria are required to be considered:

- Low GHG emissions
- No or low soil stress due to feedstock cultivation
- Ability to maintain high productive capacity
- Availability of water resources
- Pollution related to air and water
- No influence on the biodiversity

The third factor social sustainability should involve the following criteria in the production of raw materials along with liquid biofuels:

- Encouragement for rural development
- Gender main streaming
- Involvement of community
- Inclusiveness of small farmers with the industry
- Agreement with labor and land rights

4.4 Details of Economic Sustainability Criteria

An economic criterion is very comprehensive, as it is influenced by several factors including the aforementioned. The goal of a criterion is to ensure the short-term and long-term economic feasibility of the productive system (Elbehri et al. 2013).

4.4.1 Profitability and Efficiency

The important criterion for the long-term feasibility of a production system is utilizing resources to make a marketable product that also confirms the economic profitability of the product. Manufacturers will only be willing to start liquid biofuels production, if the liquid biofuel is economically affordable and profitable.

The following are key factors that can influence the profitability of liquid biofuels (Elbehri et al. 2013):

- Alternative usage of the feedstock
- The prices for liquid biofuels and others value-added products such as food, feed, materials, and timber
- Cost balance between feedstock cultivation cost and biofuel production cost
- Biofuel production costs should maintain below the price of oil equivalent.

Thus, oil prices set a price ceiling for biofuels. If their costs exceed the set oil prices value, liquid biofuels will be eliminated from the market (Schmidhuber 2007).

In recent years, biofuel subsidies have been regulated under the Global Subsidies Initiative (GSI) and by International Energy Agency (IEA). According to IEA, global renewable subsidies for biofuels was reached up to $22 billion in 2010 (Steenblik 2007). At the end, learning rates and cumulative production will determine the accomplishment of long-term costs (Elbehri et al. 2013). The economic cost-effectiveness also depends on the efficient use of coproducts and the feasibility and availability of markets for end-products.

4.4.2 Economic Equity

The perception of economic equity refers to justice in allocation of resources between simultaneous competing interests. Economic equity has established relatively less consideration in comparison to inter-generational equity between present and future generations (Jabareen 2008; Elbehri et al. 2013). Economic equity covers the following points, based on Elbehri et al. (2013):

- Quality of life, social and economic justice, democracy, public contribution and empowerment
- The occurrence and scale of unsustainable practices from power disparity
- The environmental limits of maintaining ecosystems
- The growing global demand for liquid biofuels and the associated environmental and socioeconomic changes. These changes might have a different impact on both men and women, living in the same household
- Male and female headed households; by means of their access and control to land and other productive assets
- Public participation in policy and decision making process
- Employment prospective
- Food security

Both the nature and the effect of these influences will be based on the socioeconomic and policy context and the specific technology. High land-use requirements for liquid biofuels might add extra pressure on marginal lands for the biofuels production (Elbehri et al. 2013). For example, the government of India, through the National

Mission on Biofuels program, converted around four million hectares of marginal lands to oil seed crops such as jatropha for producing biodiesel.

4.4.3 Competition with Food

Liquid biofuels production may compete with food and other resources such as land, labor, and water. Moreover, food security is a critical developmental goal that may cause conflicts with energy security at several levels. The extent of food security could delay large-scale liquid biofuels developments to balance between projected growth, size of population, availability or scarcity of land, and suitability of food or energy crops (Elbehri et al. 2013). The following are some more factors regarding competition with food:

- The projections of increased productivity and the implications for land availability.
- The relative profitability of feedstock for liquid biofuels in competition with alternative uses of resources such as land, water, and labor for food and other industrial purposes.
- Incentives or subsidies for feedstocks for bioenergy production in competition with food or other crop uses.
- Biofuels can play a negative role in destabilizing food supply, especially if agronomic prices are more linked to energy prices.
- The trade-offs between the positive benefits of climate change, secure energy, wealth generation, and coproducts.
- Competition with food is an important factor when investing in liquid biofuels.

If the policies which introduce sustainability criteria and standards are effectively implemented, they could significantly contribute to justifying the potential fuels versus food fight. Moreover, the assessment of sustainability issues related to biofuels and feedstocks for food is relatively higher in developing countries, as most of the people are involved with agriculture (Elbehri et al. 2013; Sorda et al. 2010).

4.4.4 Cost-Benefit Analyses

Cost-benefit analysis (CBA) is a widely accepted tool applied to evaluate financial and economic profitability of a project or process. In CBA, a net present value (NPV) is estimated by considering the expected inflows, outflows, and factors such as time and risk preferences of affected investors (Elbehri et al. 2013). If the NPV is positive, the project should be executed. CBA also estimates direct values of the project; however, it needs all the costs and benefits in monetary terms or facts including the following:

- In case of imperceptible impacts or products that are presently not traded on the market, methods based on revealed or stated preferences can be applied.
- CBA tries to quantify cost and benefits that do not essentially have any market cost.
- These costs are often known external costs or benefits. In such case, environmental benefits, security of supply benefits, and employment benefits are relevant.

- Environmental benefits of the different biofuels and their alternatives have been estimated using quantification of biofuels' GHG emission values through life cycle assessment (LCA) tools.
- CBA needs to make forecasts of the future.
- The decisions with respect to liquid biofuels will have a significant economic and financial impact for decades to come.

4.4.5 Full-Cost Pricing

According to Elbehri et al. (2013), full-cost pricing means "the price of a transaction that reflects information not only about its individual costs and benefits, but also the external cost it imposes on society through environmental damage". This is specifically important in the area of climate change, where such tools can effectively reduce using carbon emissions (Elbehri et al. 2013). However, the application of full-cost pricing goes beyond carbon and may include areas such as waste generation, local pollution, agriculture, and fisheries. Full-cost pricing also indicates phasing out damaging subsidies or incentives. Some developed countries have developed carbon pricing policies. For instance, Scandinavian countries, the Netherlands, and the United Kingdom have formulated carbon taxes but, carbon taxes in these countries are exempted for large emitters (Elbehri et al. 2013).

4.4.6 Change in Foreign Exchange Balance

Importing or exporting fossil fuels has a large impact on foreign exchange reserves. The impact doesn't change with a fossil fuel or biofuel. Moreover, locally produced biomass for biofuels can change the foreign exchange reserves once fossil fuels are substituted by domestic production and use of biofuels. The changes in the imported or exported and/or produced fuels have a direct effect on the foreign exchange reserves of countries both exporting and importing or producing fuels. Sometimes, it is possible for some countries that producing biofuels in a country itself may be more expensive than importing fossil fuels. However, economic development could have a positive impact on associated revenues, employment, and economic benefits to build a national biofuel industry. On the other hand, blending biofuels with fossil fuels would positively impact foreign exchange reserves. As foreign exchange reserves provide the means to purchase imports and protect the value of their currency, they are immensely important to the economic development of all countries (Fritsche et al. 2005; Sagisaka 2008).

Foreign exchange earnings and savings can be calculated according to Fritsche et al (2005) as follows:

$$\text{Foreign exchange earnings} = \text{Price per unit of convertible material} \times \text{Total volume of exports}$$

$$\text{Foreign exchange savings} = \text{Amount (in weight) of biomass} \times \text{Density of biomass}$$
$$\times \text{ Foreign exchange savings per fossil fuel displacement}$$

The aim of this indicator is to measure the contribution of the biofuels production and use the National Balance of Payments of the country that is an accounting record of all monetary transactions between a nation and the world. These monetary transactions include payments for the export and import of goods, services, and financial capital, as well as financial transfers of the country. Biofuel production and use help governments to estimate its contribution to the economic development of the country (Fritsche et al. 2005; Sagisaka 2008).

4.4.7 Productivity

Productivity is an indicator of resource availability and efficiency in biofuel production, processing, and distribution. This indicator primarily denotes to the efficiency of land use for biofuel production. The productivity indicator is formed by four values: (1) productivity of biofuel feedstocks, (2) the efficiency of feedstock processing, (3) the overall efficiency of production of the end-products, and (4) associated production costs per unit of biofuel (Fritsche et al. 2005).

The productivity indicator can be used at the farm scale, landscape, or country level by considering all other coproducts of the process. The benefits of more efficient use of resources will increase the availability of resources, reduce detrimental environmental impacts, and promote economic sustainability (Fritsche et al. 2005). Efficiency related to input parameters such as water, fertilizers, and labor that are used in biofuel production will not be directly addressed by this indicator but will assess the final productivity measurement and costs of production (Ball et al. 2001; Alston et al. 2010).

4.4.8 Net Energy Balance

Net energy balance is mainly related to the resource availability and efficiency in biofuel production, conversion, distribution, and end-use applications. To produce biofuels, input energy is required at different steps of the value chain that may be available through fossil or renewable fuels. The net energy ratio (NER) that is defined as the ratio between energy output and total energy input is a useful indicator for the relative energy efficiency of a specified pathway of biofuel production (Fritsche et al. 2005). If more energy is consumed during the biofuel lifecycle, less net energy will be available from the process. Therefore, the efficient use of energy is vital to improve energy security and optimize the use of available natural resources. If the energy required for biofuel production comes from hydrocarbons, a high NER will indicate efficient use of nonrenewable resources. This indicator also suggests using NER to measure the energy efficiency of a biofuel system. Farrell et al. (2006) suggested calculating and using the energy balance (net energy value = energy output − energy input) in the Energy and Resource Group Biofuel Analysis using Meta-Model (Fritsche et al. 2005; Bureau et al. 2010).

4.4.9 Gross Value Added

According to the World Bank, gross value added (GVA) is defined as qualitative change and restructuring in a country's economy in connection with technological and social progress (Fritsche et al. 2005). Gross domestic product (GDP) per capita measures the level of total economic output of a nation in relation to its population and the standard of living (Fritsche et al. 2005). Economic growth originates in two ways; economy can grow extensively by utilizing resources extensively or intensively by utilizing the same quantity of resources more productively or efficiently (Bolt et al. 2002). If the economic growth is attained with more labor, no growth can be seen in per capita income. However, when economic growth is attained by more productive utilization of all resources and labor, higher per capita income is expected (Domac et al. 2005). Furthermore, improved living standard of people is also expected. Intensive economic growth needs economic development. Mutual interaction of both intensive and extensive growth changes the economy of a nation. GVA provides the economic value of the total goods and services which have been generated. It also considers the input cost and raw materials that are directly associated to the production. This indicator also notifies the theme of economic sustainability and biofuels competitiveness (Fritsche et al. 2005; Arndt et al. 2008). GVA is calculated by the following formula:

$$\text{Gross value added} = \text{Total output value} - \text{Intermediate inputs}$$

4.4.10 Change in Consumption of Fossil Fuels

This indicator of change in consumption of fossil fuels and traditional use of biomass is primarily related to economic development, energy security, diversification of sources and supply, and rural and social development (Fritsche et al. 2005; Sagar and Kartha 2007). Utilization of locally produced biomass for biofuel can relocate the consumption of fossil fuels (Gehlhar et al. 2011). This will have a positive impact on the economic development and energy security of any country or region (Fritsche et al. 2005). Furthermore, public savings by avoiding fossil fuel imports can be diverted toward promotion of infrastructure development, education, sanitation, and other essential services (Sagisaka 2008). Substituting traditional use of biomass with a modern biofuel will benefit for socioeconomic development, especially in rural areas (Fritsche et al. 2005; Sagar and Kartha 2007).

4.4.11 Energy Diversity

The energy diversity indicator denotes to energy security and diversification of sources and supply. According to the United Nations development program (UNDP), energy security is defined as "the availability of energy at all times in various forms, in sufficient quantities, and at affordable prices without unacceptable or irreversible impact on the environment" (UNDP 2004; Chester 2010). Several inter-connected aspects are linked with energy security, including (1) availability of energy sources, (2) accessibility of energy supplies, (3) adequacy of capacity to produce, distribute, and use the energy, (4) affordability of energy prices, (5) and environmental sustainability of energy (Fritsche et al. 2005).

4.5 Criteria for Environmental Sustainability

Environmental sustainability is the second criterion for the assessment of liquid biofuels. It includes energy balances, GHG emission savings, soil stress, and its ability to preserve productive capacity, water resources, air and water pollution, and biodiversity (Elbehri et al. 2013). The environmental sustainability criteria are discussed in more detail in the following sections.

4.5.1 Energy Balance

Energy balance is explained as the relation or ratio between the renewable energy output of the produced biofuel and fossil energy input required in its production (Elbehri et al. 2013). This is a key factor in assessing the desirability of liquid biofuels that includes the following major issues:

- Energy balance measures the extent of biomass that can be qualified to substitute fossil fuels.
- Energy balance of 1.0 indicates no net energy gain or loss (Armstrong et al. 2002), whereas an energy balance of 2.0 indicates that a liter of biofuel contains twice the amount of input energy. Ethanol from sugarcane has the highest energy balance from 2 to 8.
- Biodiesel from sugar beet, temperate oilseeds, maize, and wheat has less energy balance. It is estimated to be in the range of 1–4 for rapeseed and soybean feedstocks.
- Variations in fossil energy balances across several feedstocks and fuels are based on feedstock productivity, agricultural practices, geographical location, sources of energy used for the conversion process, and conversion technologies.

4.5.2 Greenhouse Gas and Other Air Pollutants

Reduction of GHG emissions that cause global warming is one of the key drivers for liquid biofuels generation (Elbehri et al. 2013). The facts related to GHGs in relation to biofuels cover the following important issues:

- Intergovernmental panel on climate change (IPCC) suggested reducing 50%–85% of GHG emissions by 2050 to stabilize their concentration in the atmosphere (IPCC 2007).
- Fossil fuels use in transport sector, heating, and cooling systems are the largest contributors (70%–75% of CO_2 emissions) in global warming phenomenon.
- GHG assessments characteristically include CO_2, CH_4, N_2O, and HC. The gases released in the whole product life cycle of liquid biofuels depend on the cultivating and harvesting practices such as fertilizer use, pesticide application, harvesting, and conversion, distribution, and consumption processes.
- Production and consumption of liquid biofuels are required to prove that the net GHG emissions are lower than those of fossil fuels.
- Plants absorb CO_2 while they are growing. This offset the CO_2 produced on fuel combustion at other points in the process of liquid biofuels production.
- Liquid biofuels are one of the solutions to decarbonize transport.

4.5.3 Emissions of Non-GHG Air Pollutants

Non-GHG air pollutants are mainly related to the important area of air quality and human health and safety (Fritsche et al. 2005; Ali et al. 2016; Eqani et al. 2016). The application of machinery in the cultivation of feedstock for producing liquid biofuels may generate air pollutants or automobile smoke (Ahmad et al. 2016). Field burning and forest burning also generate fine dust particles in the local atmosphere. Both PM 2.5 and PM 10 were dominant in such smokes. Similarly, the refining process of biofuels produced from feedstocks may also contribute to the whole life cycle balance of non-GHG pollutants. In addition, transportation and distribution of feedstocks and processed biofuels to the storage points by transporting vehicles may release air pollutants (Gorham 2002). It is reported that use of fuel in transport contributes to the larger segment of the national air pollution inventories of many countries (EEA et al. 2007). Tailpipe emission from the transport sector is the principal component that changes air quality in many global cities. The use of liquid biofuels may decrease pollution relative to fossil fuels with a reduction in particulate matter (PM10 and PM2.5) (US-EPA 2002). A comprehensive and comparative data analysis of conventional fuels and biofuels stretches the practical viability of renewable biofuels in the future.

4.5.4 Land Use Change

After GHG, land use plays an important role in environmental sustainability. Remote-sensing imaging is the most common method used to estimate land use change (LUC), especially for monitoring deforestation. Different techniques are used to identify the factors involved in the LUC (dos Santos Silva et al. 2008). Furthermore, comparative analysis using primary and secondary data on areas from the past will help to estimate and predict future land use patterns (Nassar et al. 2008). The demand of new products such as biofuel feedstocks has increased on converted land that is described as direct LUC and included in the carbon accounting methods in most LCA tools. Indirect land use change (ILUC) denotes the second, third, and higher degrees of land substitutions, which is a complex analysis (Elbehri et al. 2013). There is an argument about measuring GHG emissions resulting from ILUC, which may occur when the increased demand for biofuel crops moves other crops to new land areas. The ILUC of biofuels refers to the unintentional results of releasing more CO_2 emissions, due to LUC induced by the development of croplands for increased biodiesel or bioethanol production (Searchinger et al. 2008; Elbehri et al. 2013).

4.5.5 Preservation of Biodiversity

Abundance of species in a habitat is called as biodiversity that is essential for an ecosystem's performance. Increased production of biomass for biofuels can results in both positive and negative impacts on the biodiversity (Sala et al. 2009). When unfertile lands are used, the diversity of species might be increased. While large cultivation of monoculture crops can be harmful to local biodiversity including expansion of invasive species, habitat loss, and pollution with the use of fertilizers and herbicides (Elbehri et al. 2013). In the twenty-first century, the reduction in biodiversity has considered as one of the most extreme environmental threats. On a global scale, biodiversity is critical for ecosystem functions that in return ensure hydrological cycles and diverse gene pools, favoring agriculture (UNEP 2008).

4.5.6 Water Quality

The objective of the water quality indicator is to assess the influence of biofuel feedstock production, transportation, and its conversion or processing on the water quality. Use of fertilizers for the plantation of biofuel feedstocks may increase the concentration of nitrogen and phosphorus in the soil that further enters into water bodies (Fritsche et al. 2005). Similarly, use of insecticides may contaminate waterways for which the water quality may degrade. Land ecosystems and water streams have the capability to absorb and release nitrogen through nitrification and denitrification processes (Vitousek et al. 2002). When nitrogen concentration from fertilizers, septic tanks, and atmospheric deposition exceeds the carrying capacity of terrestrial ecosystems to retain and recycle it, the additional concentration may enter into water bodies and create a "cascading" effect in the ecosystems (Galloway and Cowling 2002). Similarly, phosphorus is discharged from the phosphate fertilizers that may enter into the soil and may go beyond the permissible limit and ultimately reach waterways. An excess phosphate may help in the high growth rate of algal blooms and cyanobacteria that produce taste and odor problems (Evans et al. 2000). The residues of applied insecticides to the plantation of biofuel feedstocks are hazardous to the biota of water bodies. The increased concentration of these pesticides may change the biota of water bodies that may affect aquatic food chains (Rabalais and Turner 2001). Moreover, the cultivation of perennial energy crops can significantly contribute to mitigate the leakage of plant nutrients from the landscape into waterways (Rahmanian et al. 2014). Furthermore, some energy crops are able to remove heavy metals from the soils (Smith et al. 2003; Fritsche et al. 2005; Aulenbach 2006).

4.5.7 Water Use and Efficiency

A large quantity of water is required for producing and processing the biofuel feed-stocks. In certain areas, there may be a competing demand for surface water or groundwater. Other multiple impacts such as degradation in water quality, groundwater descending, modification of surface geochemistry, seasonal reduction of instream flow, fluctuations in water supply for domestic use, and variation of the existing agricultural yields were identified in water use and its efficiency (Berndes 2008). To ensure long-term biofuel production and processing, continuous accessibility to water resources with a large reserve of water is required round the year (Berndes 2008).

4.5.8 Life Cycle Assessment

LCA is a widely used technique to determine whether a liquid biofuel production process results in a net reduction of GHG emission savings with a positive energy balance (Nizami and Ismail 2013). According to ISO 14040, an LCA is a "compilation and evaluation of the inputs, outputs, and the potential environmental impacts of a product system throughout its life cycle ." In conducting an LCA, all input and output data in all phases of the product's life cycle is required, including (1) biomass production, (2) feedstock storage, (3) feedstock transportation, (4) liquid biofuel production, (5) its transportation and (6) final use (Rathore et al. 2016). Moreover, all outputs are accounted such as gases either leaked or captured and by-products. There are many LCA approaches that have been reported, however most of them focus on a few key input categories and two primary environmental criteria such as GHG emission savings and net energy balance (Singh et al. 2010, 2011; Shahzad et al. 2015).

4.6 Criteria for Social Sustainability

The social criteria of biofuel sustainability can address any interlinked problems, which results in several methodological difficulties such as the challenge of distinguishing between direct and indirect social concerns. In this case, three aspects of social sustainability are commonly discussed: (1) land ownership rights, (2) local stewardship of common property resources, and (3) labor rights (Elbehri et al. 2013). All these social sustainability criteria more or less address a common goal, which according to Elbehri et al. (2013) is "the need to integrate small-scale farmers within biofuel development and ensure inclusive benefit sharing, safeguarding of basic rights, and local means of livelihood consequent to the introduction of biofuels."

4.6.1 Labor/Employment

The growth in rural employment by biofuels production has played a significant role in many developing countries. Similarly, liquid biofuels production can increase the rural development and local employment by bringing capital to the agriculture sector. Moreover, the introduction of new technologies will result in better control and access to fertilizers, infrastructure, and high-yielding crops (Elbehri et al. 2013). The round table on sustainable biofuels recommended the non-violation of labor or human rights, assurance of decent work for workers, liberty to organize with no slave or child labor, and discrimination, along with respecting minimum wage, and international health and safety standards of stakeholders with respect to human and labor rights (Coelho 2008; Elbehri et al. 2013).

4.6.2 Price and Supply of a National Food Basket

There are numerous factors that affect the price and supply of national food basket along with biofuels use and domestic production. These factors include demand for foodstuffs, feed, and fiber, imports and exports of food, weather conditions, and energy prices (Fritsche et al. 2005). This indicator targets to estimate the impacts of biofuel use and their domestic production on the price and supply of food products. Food basket is defined at regional or national levels, and it includes staple crops that make the dominant part of the diet and supply a major share of the energy and nutrient requirements of the individuals in a given region or country (Elbehri et al. 2013). The indicator also aims to evaluate the influence of variation in the prices of the food basket components at several levels such as national, regional, and household welfare (Fritsche et al. 2005). In addition, the production of biofuel feedstock may change demand for land, water, and fertilizers. The price change can influence the final price of foodstuffs such as staple crops. However, changes in the prices of main staple crops will be at international and national levels (Diaz-Chavez et al. 2010; Minot 2010).

4.6.3 Change in Income

The wage and nonwage incomes are reported to change due to the biofuel production. Employment and wages are critical factors for rural and social development of many countries, especially in the developing world that is found to be influenced significantly by the biofuel sector (Fritsche et al. 2005). Moreover, wage levels shows an indication of labor conditions and employment in this sector in comparison to other related sectors. In addition, this indicator is focused on measuring the variation in income from the sale and own consumption of biofuels by self-employed individuals and households. Self-employment is another important income source which is linked with biofuel production (Fritsche et al. 2005). Furthermore, it can significantly affect rural and social development through improved purchasing power, livelihood options, and welfare (Madlener and Myles 2000). Net job creation and income generation due to the biofuel industry can lead to improved living standard in terms of household consumption levels, social cohesion, and stability (Walter et al. 2008).

4.6.4 Biofuel Used to Improve Access to Modern Energy Services

This indicator is mainly related to the access and improving the energy, especially to modern energy services provided by the modern biofuel for both households and businesses (Fritsche et al. 2005). The UN secretary general's Advisory Group on Energy and Climate Change defined universal energy access as "access to clean, reliable, and affordable energy services for cooking and heating, lighting, communication, and productive uses" (AGECC 2010). Practically, this needs inexpensive access to modern energy services such as electricity generation for lighting, communication and other household uses, mechanical power for irrigation and agricultural processing, and latest technologies for cooking and heating purposes (Fritsche et al. 2005; Bazilian and Nussbaumer 2010).

Modern biofuels play a vital role in providing or improving access to latest energy services. The total amount or percentage increase in access to modern energy services through modern biofuel and total quantity of biofuel use is an important indicator and sign of modern biofuel contribution to energy access (Fritsche et al. 2005). Access to energy is one of the strongest indices for socioeconomic development that includes GDP per capita, and the human development index (Bazilian and Nussbaumer 2010). According to UNDESA (2007), "modern energy services are an essential component of providing adequate food, shelter, water, sanitation, medical care, education, and access to communication. Lack of access to modern energy services contributes to poverty and deprivation and limits economic and human development. Adequate, affordable, and reliable energy services are necessary to guarantee sustainable economic and human development and also the achievement of the Millennium Development Goals." Intensifying access to energy and improving modern energy services using biofuels will help to forward several environmental and socioeconomic benefits (Vera and Langlois 2007).

Acknowledgments

Gunda Mohanakrishna gratefully acknowledges the Marie-Curie Intra-European Fellowship (IEF)–supported project BIO-ELECTRO-ETHYLENE (Grant No: 626959) from the European Commission. Prof. Umesh Mishra acknowledges the Department of Biotechnology (DBT) for Overseas Associate Fellowship for 6 months' research stay at VITO, Belgium.

References

Aden, A. 2007. Water usage for current and future ethanol production. *Southwest Hydrology* 6: 22–23.

AGECC. 2010. Energy for a sustainable future: Summary report and recommendations. The Secretary-General's Advisory Group on Energy and Climate Change, New York.

Ahmad, A. L., Yasin, N. M., Derek, C. J. C., Lim, J. K. 2011. Microalgae as a sustainable energy source for biodiesel production: A review. *Renewable and Sustainable Energy Reviews* 15(1): 584–593.

Ahmad, I., Rehan, M., Balkhyour, M. A., Abbas, M., Basahi, J. M., Almeelbi, T., Ismail, I. M. I. 2016. Review of environmental pollution and health risks at motor vehicle repair workshops: challenges and perspectives for Saudi Arabia. *International Journal of Agricultural and Environmental Research* 2(1): 1–22. ISSN: 2414-8245.

Ali, N., Ismail, I. M. I., Eqani, S. A. M. A. S., Malarvannan, G., Kadi, M. W., Rehan, M., Covaci, A. 2016. Brominated and Organophosphate Flame Retardants in Indoor Dust of Jeddah, Kingdom of Saudi Arabia: Implications for Human Exposure. *Science of the Total Environment* 569–570: 269–277. http://dx.doi.org/10.1016/j.scitotenv.2016.06.093.

Alston, J., Anderson, M., James, J., Pardey, P. 2010. *Persistence Pays: U.S. Agricultural Productivity Growth and the Benefits from R&D Spending.* Springer, New York.

Anonymous. 2014. Biodiesel and its capabilities. Where can we use biodiesel? https://sites.google.com/site/kosovoorg/3-resume. Accessed April 15, 2016.

Antizar-Ladislao, B., Turrion-Gomez, J. L. 2008. Second-generation biofuels and local bioenergy systems. *Biofuels, Bioproducts and Biorefining* 2(5): 455–469.

Armstrong, A., Baro, J., Dartoy, J., Groves, A., Nikkonen, J., Rickeard, D., Thompson, N. Larivé, J. 2002. Energy and greenhouse. Gas balance of biofuels for Europe—An update. Concave Ad Hoc. Group on Alternative Fuels. Report no. 2/02. Brussels, Belgium.

Arndt, C., Benfica, R., Tarp, F., Thurlow, J., Uaiene, R. 2008. Biofuels, poverty, and growth. IFPRI discussion paper 00803. International Food Policy Research Institute (IFPRI), Washington, DC.

Aulenbach, B. T. 2006. Annual dissolved nitrite plus nitrate and total phosphorus loads for Susquehanna, St. Lawrence, Mississippi-Atchafalaya, and Columbia River Basins, 1968–2004. USGS Open File Report 06-1087. http://pubs.usgs.gov/of/2006/1087/. Accessed May 28, 2016.

Azapagic, A. 2014. Sustainability considerations for integrated biorefineries. *Trends in Biotechnology* 32(1): 1–4.

Ball, V. E., Bureau, J. C., Butault, J. P., Nehring, R. 2001. Levels of farm sector productivity: An international comparison. *Journal of Productivity Analysis* 15(1): 5–29.

Barron, N., Brady, D., Love, G., Marchant, R., Nigam, P., McHale, L., McHale, A. P. 1996. Alginate-immobilized thermotolerant yeast for conversion of cellulose to ethanol. *Progress in Biotechnology* 11: 379–383.

Bazilian, M., Nussbaumer, P. 2010. UNIDO contribution to *the Fourth UN Conference on LDCs—Energy Services,* New York.

Beckman, C. 2006. Biodiesel. *Bi-Weekly Bulletin* 19(15): 1–4. http://dsp-psd.pwgsc.gc.ca/Collection/A27-18-19-15E.pdf. Retrieved April 28, 2009.

Berndes, G. 2008. Water demand for global bioenergy production: Trends, risks and opportunities. *Journal of Cleaner Production* 15(18): 1778–1786.

Bolt, K., Matete, M., Clemens, M. 2002. Manual for calculating adjusted net savings. Environment Department, World Bank, Washington, DC, pp. 1–23.

Bozbas, K. 2008. Biodiesel as an alternative motor fuel: Production and policies in the European Union. *Renewable and Sustainable Energy Reviews* 12(2): 542–552.

Brennan, L., Owende, P. 2010. Biofuels from microalgae—A review of technologies for production, processing, and extractions of biofuels and co-products. *Renewable and Sustainable Energy Reviews* 14(2): 557–577.

Bridgewater, A., Czernik, C., Diebold, J., Mekr, D., Radlein, P. 1999. *Fast Pyrolysis of Biomass: A Handbook*. CPL Scientific Publishing Services, Ltd., Newbury, U.K.

Bridgwater, A. V., Cottam, M. L. 1992. Opportunities for biomass pyrolysis liquids production and upgrading. *Energy & Fuels* 6(2): 113–120.

Brown, L. M., Zeiler, K. G. 1993. Aquatic biomass and carbon dioxide trapping. *Energy Conversion and Management* 34(9): 1005–1013.

Buchholz, T., Luzadis, V. A., Volk, T. A. 2009. Sustainability criteria for bioenergy systems: Results from an expert survey. *Journal of Cleaner Production* 17: S86–S98.

Bureau, J. C., Disdier, A. C., Gauroy, C., Tréguer, D. 2010. A quantitative assessment of the determinants of the net energy value of biofuels. *Energy Policy* 38(5): 2282–2290.

Carriquiry, M. 2007. U.S. biodiesel production: Recent developments and prospects. *Iowa Ag Review* 13(2): 8–11.

Cherubini, F., Ulgiati, S. 2010. Crop residues as raw materials for biorefinery systems—A LCA case study. *Applied Energy* 87(1): 47–57.

Chester, L. 2010. Conceptualising energy security and making explicit its polysemic nature. *Energy Policy* 38(2): 887–895.

Chisti, Y. 2007. Biodiesel from microalgae. *Biotechnology Advances* 25(3): 294–306.

Christou, M., Alexopoulou, E. 2012. The terrestrial biomass: Formation and properties (crops and residual biomass). In *Biorefinery: From Biomass to Chemicals and Fuels*, pp. 49–80. Walter De Gruyter Incorporated, Berlin, Germany.

Clay, J. 2004. *World Agriculture and the Environment—A Commodity-by-Commodity Guide to Impacts and Practices*. World Wildlife Fund. Island Press, Washington, DC.

Coelho, S. 2005. Biofuels—Advantages and trade barriers. UNCTAD/DITC/TED/2005/1, February 4, 2005. UNCTAD, São Paulo, Brazil.

De Oliveira, M. E. D., Vaughan, B. E., Rykiel, E. J. 2005. Ethanol as fuel: Energy, carbon dioxide balances, and ecological footprint. *BioScience* 55(7): 593–602.

Demain, A. L., Newcomb, M., Wu, J. H. D. 2005. Cellulase, clostridia, and ethanol. *Microbiology and Molecular Biology Reviews* 69: 124–154.

Demirbas, A. 2011. Competitive liquid biofuels from biomass. *Applied Energy* 88(1): 17–28.

Demirbas, A., Rehan, M., Al-Sasi, B. O., Nizami, A. S. 2016. Evaluation of natural gas hydrates as a future methane source. *Petroleum Science and Technology*. doi:10.1080/10916466. 2016.1185442.

Diaz-Chavez, R., Mutimba, S., Watson, H., Rodriguez-Sanchez, S., Nguer, M. 2010. Mapping food and bioenergy in Africa. A report prepared on behalf of FARA. Forum for Agricultural Research in Africa, Ghana, Africa. ERA-ARD, SROs, FARA, 3.

Domac, J., Richards, K., Risovic, S. 2005. Socio-economic drivers in implementing bioenergy projects. *Biomass and Bioenergy* 28(2): 97–106.

dos Santos Silva, M. P., Camara, G., Escada, M. I. S., de Souza, R. C. M. 2008. Remote-sensing image mining: Detecting agents of land-use change in tropical forest areas. *International Journal of Remote Sensing* 29: 4803–4822.

Dragone, G., Fernandes, B. D., Vicente, A. A., Teixeira, J. A. 2010. Third generation biofuels from microalgae. *Current Research, Technology and Education Topics in Applied Microbiology and Biotechnology* 2: 1355–1366.

EBTP (European Biofuels Technology Platform). 2013. Use of biofuels in shipping. http://www.biofuelstp.eu/shipping-biofuels.html. Accessed May 11, 2016.

EEA, EMEP, and CORINAIR. 2007. Atmospheric emission inventory guidebook. Technical report no. 16/2007. European Environment Agency, Copenhagen, Denmark.

Elbehri, A., Segerstedt, A., Liu, P. 2013. Biofuels and the sustainability challenge: A global assessment of sustainability issues, trends and policies for biofuels and related feedstocks. Food and Agriculture Organization of the United Nations (FAO), Rome, Italy.

Eqani, S. A. M. A. S., Khalid, R., Bostan, N., Saqib, Z., Mohmand, J., Rehan, M., Ali, N., Katsoyiannis, I. A., Shen, H. 2016. Human lead (Pb) exposure via dust from different land use settings of Pakistan: A case study from two urban mountainous cities. *Chemosphere* 155: 259–265. http://dx.doi.org/10.1016/j.chemosphere.2016.04.036.

Eurostat. 2014. Inland waterways freight transport—Quarterly and annual data. http://ec.europa.eu/eurostat/statistics-explained/index.php/Inland_waterways_freight_transport_-_quarterly_and_annual_data. Accessed May 28, 2016.

Evans, C. D., Jenkins, A., Wright R. F. 2000. Surface water acidification in the South Pennines I. Current status and spatial variability. *Environmental Pollution* 109: 11–20.

Fairless, D. 2007. Biofuel: The little shrub that could—Maybe. *Nature* 449: 652–655.

Farrell, A. E., Plevin, R. J., Turner, B. T., Jones, A. D., O'Hare, M., Kammen, D. M. 2006. Ethanol can contribute to energy and environmental goals. *Science* 311: 506–508.

Fernando, S., Adhikari, S., Chandrapal, C., Murali, N. 2006. Biorefineries: Current status, challenges, and future direction. *Energy & Fuels* 20: 1727–1737.

Fritsche, U. R., Hunecke, K., Wiegmann, K. 2005. Criteria for assessing environmental, economic and social aspects of biofuels in developing countries. Oko-Institute e.V., Copenhagen, Denmark.

Galloway, J., Cowling, E. 2002. Reactive nitrogen and the world: 200 years of change. *Ambio* 31: 64–71.

Gardy, J., Hassanpour, A., Lai, X. Rehan, M., 2014. The influence of blending process on the quality of rapeseed oil-used cooking oil biodiesels. *International Scientific Journal (Journal of Environmental Science)* 3: 233–240.

Gehlhar, M., Winston, A., Somwaru, A. 2011. Effects of increased biofuels on the U.S. economy in 2022. ERR-102. U.S. Department of Agriculture (USDA), Economic Research Service, Diane Publishing, Collingdale, PA.

Gorham, R. 2002. Air pollution from ground transportation. An assessment of strategies and tactics and proposed actions for the international community. The Global Initiative on Transport Emission, Division for Sustainable Development, UN Department of Economic and Social Affairs, United Nations, New York.

Harun, R., Singh, M., Forde, G. M., Danquah, M. K. 2010. Bioprocess engineering of microalgae to produce a variety of consumer products. *Renewable and Sustainable Energy Reviews* 14: 1037–1047.

Hill, J., Nelson, E., Tilman, D., Polasky, S., Tiffany, D. 2006. Environmental, economic, and energetic costs and benefits of biodiesel and ethanol biofuels. *Proceedings National Academy Science USA* 103: 11206–11210.

Hong, Y., Nizami, A. S., Pourbafrani, M., Saville, B. A., MacLean, H. L. 2013. Impact of cellulase production on environmental and financial metrics for lignocellulosic ethanol. *Biofuels, Bioproducts, Biorefinery* 7(3): 303–313.

IPCC. 2007. Climate change 2007: The physical science basis. In Solomon, S., Qin, D., Manning, M. et al. (Eds.), *Contribution of Working Group I to the Fourth Assessment Report of the Intergovernmental Panel on Climate Change*, pp. 1–996. Cambridge University Press, Cambridge, U.K.

Jabareen, Y. 2008. A new conceptual framework for sustainable development. *Environment, Development and Sustainability* 10: 197–192.

Jin, M., Slininger, P. J., Dien, B. S., Waghmode, S., Moser, B. R., Orjuela, A., Balan, V. 2015. Microbial lipid-based lignocellulosic biorefinery: Feasibility and challenges. *Trends in Biotechnology* 33: 43–54.

Kamm, B., Hille, C., Schönicke, P., Dautzenberg, G. 2010. Green biorefinery demonstration plant in Havelland (Germany). *Biofuels, Bioproducts and Biorefining* 4: 253–262.

Kamm, B., Kamm, M. 2004. Principles of biorefineries. *Applied Microbiology and Biotechnology* 64: 137–145.

Kamm, B., Kamm, M. 2007. Biorefineries—Multi product processes. In Ulber, R., Sell, D. (Eds.), *White Biotechnology*, pp. 175–204. Springer, Berlin, Germany.

Korres, N. E., Singh, A., Nizami, A. S., Murphy, J. D. 2010. Is grass biomethane a sustainable transport biofuel? *Biofuels, Bioproducts and Biorefining* 4: 310–325. doi:10.1002/bbb.

Korres, N. E., Thamsiriroj, T., Smyth, B. M., Nizami, A. S., Singh, A., Murphy, J. D. 2011. Grass biomethane for agriculture and energy. In Lichtfouse, E. (Ed.), *Sustainable Agriculture Reviews*, Vol. 7, pp. 5–49. Springer-Verlag, London, U.K. doi:10.1007/978-94-007-1521-9_2.

Kulkarni, M. G., Gopinath, R., Meher, L. C., Dalai, A. K. 2006. Solid acid catalyzed biodiesel production by simultaneous esterification and transesterification. *Green Chemistry* 8: 1056–1062.

Lange, J. P. 2007. Lignocellulose conversion: An introduction to chemistry, process and economics. *Biofuels, Bioproducts and Biorefining* 1: 39–48.

Larson, E. D. 2008. Biofuel production technologies: Status, prospects and implications for trade and development. Report no. UNCTAD/DITC/TED/2007/10. *United Nations Conference on Trade and Development*, New York.

Lee, R. A., Lavoie, J. M. 2013. From first-to third-generation biofuels: Challenges of producing a commodity from a biomass of increasing complexity. *Animal Frontiers* 3: 6–11.

Lee, S., Speight, J. G., Loyalka, S. K. 2007. *Handbook of Alternative Fuel Technologies*. CRC Taylor & Francis Group, Boca Raton, FL.

Liang, Y., Sarkany, N., Cui, Y. 2009. Biomass and lipid productivities of *Chlorella vulgaris* under autotrophic, heterotrophic and mixotrophic growth conditions. *Biotechnology Letters* 31: 1043–1049.

Lokko, Y., Okogbenin, E., Mba, C., Dixon, A., Raji, A., Fregene, M. 2007. Cassava. In Kole, C. (Ed.), *Pulses, Sugar and Tuber Crops*. Springer, Heidelberg, Germany.

Love, G., Gough, S., Brady, D., Barron, N., Nigam, P., Singh, D., McHale, A. P. 1998. Continuous ethanol fermentation at 45 C using *Kluyveromyces marxianus* IMB3 immobilized in calcium alginate and kissiris. *Bioprocess Engineering* 18: 187–189.

Madlener, R., Myles, H. 2000. Modelling socio-economic aspects of bioenergy systems: A survey prepared for IEA Bioenergy Task 29. In *IEA Bioenergy Task*, Vol. 29, IEA, France.

McGill, R., Remley, W. B., Winther, K. (2013). Alternative fuels for marine applications. A report from the IEA advanced motor fuels implementing agreement, International Energy Agency (IEA), France. Annex, Vol. 41, pp. 3–10.

McKendry, P. 2002. Energy production from biomass (part 1): Overview of biomass. *Bioresource Technology* 83: 37–46.

Meher, L. C., Sagar, D. V., Naik, S. N. 2006. Technical aspects of biodiesel production by transesterification—A review. *Renewable and Sustainable Energy Reviews* 10: 248–268.

Miandad, R., Rehan, M., Ouda, O. K. M., Khan, M. Z., Ismail, I. M. I., Shahzad, K., Nizami, A. S. 2016a. Waste-to-hydrogen energy in Saudi Arabia: Challenges and perspectives. In *Biohydrogen Production: Sustainability of Current Technology and Future Perspective*. doi:10.1007/978-81-322-3577-4_11. Springer India.

Miandad, R., Barakat, M., Rehan, M., Ismail, I. M. I., Nizami, A. S. 2016b. The energy and value-added products from pyrolysis of waste plastics. In *Recycling of Solid Waste for Biofuels and Bio-chemicals*. Environmental Footprints and Eco-design of Products and Processes Book Series. doi:10.1007/978-981-10-0150-5_12. Springer Science+Business Media, Singapore.

Miandad, R., Barakat, M. A., Aburiazaiza, A. S., Rehan, M., Nizami, A. S. 2016c. Catalytic Pyrolysis of Plastic Waste: A Review. *Process Safety and Environmental Protection* 102: 822–838. http://dx.doi.org/10.1016/j.psep.2016.06.022.

Minot, N. 2010. Transmission of world food price changes to markets in Sub-Saharan Africa. International Food Policy Research Institute, Washington, DC.

Minteer, S. 2006. *Alcoholic Fuels*. CRC Taylor & Francis Group, Boca Raton, FL.

Naik, S., Goud, V. V., Rout, P. K., Jacobson, K., Dalai, A. K. 2010. Characterization of Canadian biomass for alternative renewable biofuel. *Renewable Energy* 35(8): 1624–1631.

Nassar, A., Rudorff, B., Antoniazzi, L., Alves de Aguiar, D., Bacchi, M., Adami, M. 2008. Prospects of the sugar cane expansion in Brazil: Impacts on direct and indirect land use changes. In Zuurbier, P., van de Vooren, J. (Eds.), *Sugar Cane Ethanol. Contributions to Climate Change Mitigation and the Environment*. Wageningen Academic Publishers, Wageningen, the Netherlands.

Nigam, P., Banat, I. M., Singh, D., McHale, A. P., Marchant, R. 1997. Continuous ethanol production by thermotolerant *Kluyveromyces marxianus* IMB3 immobilized on mineral Kissiris at 45 C. *World Journal of Microbiology and Biotechnology* 13: 283–288.

Nigam, P. S., Singh, A. 2011. Production of liquid biofuels from renewable resources. *Progress in Energy and Combustion Science* 37: 52–68.

Nizami, A. S., Ismail, I. M. I. 2013. Life cycle assessment of biomethane from lignocellulosic biomass. In *Life Cycle Assessment of Renewable Energy Sources*. Green Energy and Technology Book Series, Vol. XVI, 293pp., 100 illus. Springer-Verlag, London, U.K. doi:10.1007/978-1-4471-5364-1_4.

Nizami, A. S., Ouda, O. K. M., Rehan, M., El-Maghraby, A. M. O., Gardy, J., Hassanpour, A., Kumar, S., Ismail, I. M. I. 2015a. The potential of Saudi Arabian natural zeolites in energy recovery technologies. *Energy* 1–10, doi:10.1016/j.energy.2015.07.030.

Nizami, A. S., Rehan, M., Ouda, O. K. M., Shahzad, K., Sadef, Y., Iqbal, T., Ismail, I. M. I 2015b. An argument for developing waste-to-energy technologies in Saudi Arabia. *Chemical Engineering Transactions* 45: 337–342. doi:10.3303/CET1545057.

Nizami, A. S., Rehan, M., Ismail, I. M. I., Almeelbi, T., Ouda, O. K. M. 2015c. Waste biorefinery in Makkah: a solution to convert waste produced during Hajj and Umrah Seasons into wealth. *15th Scientific Symposium for Hajj, Umrah and Madinah visit*. Madinah, Saudi Arabia. doi: 10.13140/RG.2.1.4303.6560.

Nizami, A. S., Shahzad, K., Rehan, M., Ouda, O. K. M., Khan, M. Z., Ismail, I. M. I., Almeelbi, T., Basahi, J. M., Demirbas, A. 2016. Developing waste biorefinery in Makkah: A way forward to convert urban waste into renewable energy. *Applied Energy*. http://dx.doi.org/10.1016/j.apenergy.2016.04.116.

Ouda, O. K. M., Raza, S. A., Nizami, A. S., Rehan, M., Al-Waked, R., Korres, N. E. 2016. Waste to energy potential: A case study of Saudi Arabia. *Renewable and Sustainable Energy Reviews* 61: 328–340. doi:10.1016/j.rser.2016.04.005.

Pahl, G., McKibben, P. 2008. *Biodiesel: Growing a New Energy Economy*, 2nd edn. Chelsea Green Publishing, White River Junction, VT.

Patil, V., Tran, K. Q., Giselrød, H. R. 2008. Towards sustainable production of biofuels from microalgae. *International Journal of Molecular Sciences* 9: 1188–1195.

Rabalais, N. N., Turner, R. E. 2001. Coastal hypoxia: Consequences for living resources and ecosystems. Coastal and estuarine studies 58. American Geophysical Union, Washington, DC.

Ragauskas, A. J., Williams, C. K., Davison, B. H., Britovsek, G., Cairney, J., Eckert, C. A., Tschaplinski, T. 2006. The path forward for biofuels and biomaterials. *Science* 311: 484–489.

Rahman, Q. M., Wang, L., Zhang, B., Xiu, S., Shahbazi, A. 2015. Green biorefinery of fresh cattail for microalgal culture and ethanol production. *Bioresource Technology* 185: 436–440.

Rahmanian, N., Ali, S. H. B., Homayoonfard, M., Ali, N. J., Rehan, M., Sadef, Y., Nizami, A. S. 2015. Analysis of physiochemical parameters to evaluate the drinking water quality in the State of Perak, Malaysia. *Journal of Chemistry*, 2015, Article ID 716125, 1–10, http://dx.doi.org/10.1155/2015/716125.

Rathore, D., Nizami, A. S., Pant, D., Singh, A. 2016. Key issues in estimating energy and greenhouse gas savings of biofuels: Challenges and perspectives. *Biofuel Research Journal* 10: 380–393.

Rehan, M., Nizami, A. S., Taylan, O., Al-Sasi, B. O., Demirbas, A. 2016a. Determination of wax content in crude oil. *Petroleum Science and Technology* 34(9): 799–804. http://dx.doi.org/10.1080/10916466.2016.1169287.

Rehan, M., Nizami, A. S., Shahzad, K., Ouda, O. K. M., Ismail, I. M. I., Almeelbi, T., Iqbal, T., Demirbas, A. 2016b. Pyrolytic liquid fuel: A source of renewable energy in Makkah. *Energy Sources, Part A: Recovery, Utilization, and Environmental Effects*. doi:10.1080/15567036.2016.1153753.

Reinhardt, G., Gärtner, S., Rettenmaier, N., Münch, J., von Falkenstein, E. 2007. Screening life cycle assessment of Jatropha biodiesel. Final report. IFEU – Institute for Energy and Environmental Research Heidelberg GmbH, Heidelberg, Germany.

Remmen, A., Jensen, A. A., Frydendal, J. 2007. Life cycle management: A business guide to sustainability. UNEP DTIE, France.

Sagar, A. D., Kartha, S. 2007. Bioenergy and sustainable development? *Annual Reviews of Environment and Resources* 32: 131–167.

Sagisaka, M. 2008. Guidelines to assess sustainability of biomass utilisation in East Asia. ERIA Research Project Report 2009. http://www.eria.org/RPR-2008-8-2.pdf. Accessed June 1, 2016.

Sadef, Y., Nizami, A. S., Batool, S. A., Chaudhary, M. N., Ouda, O. K. M., Asam, Z. Z., Habib, K., Rehan, M., Demibras, A. 2015. Waste-to-energy and recycling value for developing integrated solid waste management plan in Lahore. *Energy Sources, Part B: Economics, Planning, and Policy*. doi:10.1080/1556249.2015.105295.

Sala, O., Sax, D. Leslie, H. 2009. Biodiversity consequences of increased biofuel production. In Howarth, R., Bringezu, S. (Eds.), *Biofuels: Environmental Consequences and Interactions with Changing Land Use*. Proceedings of Scientific Committee on Problems of the Environment (SCOPE) International Biofuels Project Rapid Assessment, Gummersbach, Germany, September 22–25, 2008.

Schenk, P. M., Thomas-Hall, S. R., Stephens, E., Marx, U. C., Mussgnug, J. H., Posten, C., Hankamer, B. 2008. Second generation biofuels: High-efficiency microalgae for biodiesel production. *Bioenergy Research* 1: 20–43.

Schmidhuber, J. 2007. Biofuels: An emerging threat to Europe's food security? Impact of an increased biomass use on agricultural markets, prices and food security: A longer-term perspective. Policy paper 17/05/2007. Notre Europe, Paris, France.

Schmid-Straiger U. Algae biorefinery—Concept. *National German Workshop on Biorefineries*, Worms, Germany, September 15, 2009.

Searchinger, T., Heimlich, R., Houghton, R. A., Dong, F., Elobeid, A., Fabiosa, J., Yu, T. H. 2008. Use of US croplands for biofuels increases greenhouse gases through emissions from land-use change. *Science* 319(5867): 1238–1240.

Shahzad, K., Rehan, M., Ismail, I. M. I., Sagir, M., Tahir, M. S., Bertok, B., Nizami, A. S. 2015. Comparative life cycle analysis of different lighting devices. *Chemical Engineering Transactions* 45: 631–636. http://dx.doi.org/10.3303/CET1545106.

Sims, R. E., Mabee, W., Saddler, J. N., Taylor, M. 2010. An overview of second generation biofuel technologies. *Bioresource Technology* 101(6): 1570–1580.

Singh, A., Pant, D., Korres, N. E., Nizami, A. S., Prasad, S., Murphy, J. D. 2010. Key issues in life cycle assessment of ethanol production from lignocellulosic biomass: Challenges and perspectives. *Bioresource Technology* 10: 5003–5012.

Singh, A., Nizami, A. S., Korres, N. E., Murphy, J. D. 2011. The effect of reactor design on the sustainability of grass biomethane. *Renewable and Sustainable Energy Reviews* 15: 1567–1574.

Singh, J., Gu, S. 2010. Commercialization potential of microalgae for biofuels production. *Renewable and Sustainable Energy Reviews* 14: 2596–2610.

Smith, S. V., Swaney, D. P., Talaue-McManus, L., Bartley, J. D., Sandhei, P. T., McLaughlin, C. J., Dupra, V. C. et al. 2003. Humans, hydrology, and the distribution of inorganic nutrient loading to the ocean. *BioScience* 53: 235–245.

Sorda, G., Banse, M., Kemfert, C. 2010. An overview of biofuel policies across the world. *Energy Policy* 38: 6977–6988.

Steenblik, R. 2007. Biofuels at what cost? Government support for ethanol and biodiesel in selected OECD countries: A synthesis report addressing subsidies in biofuels in Australia, Canada, the European Union, Switzerland and the US, Global Subsidies Initiative. *IISD*, Rome, Italy, September 2007.

Stevens, D. J., Worgetten, M., Saddler, J. 2004. Biofuels for transportation: An examination of policy and technical issues. IEA Bioenergy Task 39, Liquid Biofuels Final Report 2001e2003, Paris, France.

Suresh, K., Rao, L. V. 1999. Utilization of damaged sorghum and rice grains for ethanol production by simultaneous saccharification and fermentation. *Bioresource Technology* 68(3): 301–304.

Tahir, M. S., Shahzad, K., Shahid, Z., Sagir, M., Rehan, M., Nizami, A. S. 2015. Producing methane enriched biogas using solvent absorption method. *Chemical Engineering Transactions* 45: 1309–1314. http://dx.doi.org/10.3303/CET1545219.

Tyson, K. S., Bozell, J., Wallace, R., Petersen, E., Moens, L. 2005. Biomass oil analysis: Research needs and recommendations. NREL Technical Report. http://www.eere.energy.gov/biomass/pdfs/34796.pdf. Accessed May 28, 2016.

UIC. 2007. Railways and biofuel, The International Union of Railways (UIC) Report in association with Association of Train Operating Companies (ATOC), Paris, France, July 2007.

UNDESA. 2007. *Indicators of Sustainable Development, Guidelines and Methodologies*, 3rd edn. United Nations, New York.

UNDP. 2004. World energy assessment overview: 2004 update. http://www.undp.org/content/undp/en/home/librarypage/environment-energy/sustainable_energy/world_energy_assessmentoverview2004update.html. Accessed May 27, 2016.

UNEP. 2008. The potential impacts of biofuels on biodiversity, Matters arising from SBSTTA recommendation XII/7. UNEP/CBD/ COP/9/26. *Conference of the Parties to the Convention on Biological Diversity Ninth Meeting*, Bonn, Germany, May 19–30, 2008.

USDA. 2012. USDA—Biofuel feedstock and coproduct market data. http://www.ers.usda.gov/topics/farm-economy/bioenergy/biofuel-feedstock-coproduct-market-data.aspx. Retrieved January 2013.

US-DOE. 2009. National algal biofuels technology roadmap. U.S. Department of Energy, Washington, DC. https://e-center.doe.gov/iips/faopor.nsf/UNID/79E3ABCACC9AC14 A852575CA00799D99/$file/AlgalBiofuels_Roadmap_7.pdf. Accessed July 30, 2015.

US-EPA. 2002. A comprehensive analysis of biodiesel impacts on exhaust emissions. Draft technical report. US EPA 420-P-02-001. https://www3.epa.gov/otaq/models/analysis/biodsl/p02001.pdf. Accessed June 1, 2016.

Venkata Mohan, S., Devi, M. P., Mohanakrishna, G., Amarnath, N., Babu, M. L., Sarma, P. N. 2011. Potential of mixed microalgae to harness biodiesel from ecological water-bodies with simultaneous treatment. *Bioresource Technology* 102(2): 1109–1117.

Venkata Mohan, S., Rohit, M. V., Chandra, R., Goud, K. R. 2015a. Algal biorefinery. In Shibu, J., Thallada, B. (Eds.), *Biomass and Biofuels: Advanced Biorefineries for Sustainable Production and Distribution*, pp. 199–216.

Venkata Mohan, S., Rohit, M. V., Chiranjeevi, P., Chandra, R., Navaneeth, B. 2015b. Heterotrophic microalgae cultivation to synergize biodiesel production with waste remediation: Progress and perspectives. *Bioresource Technology* 184: 169–178.

Vera, I., Langlois, L. 2007. Energy indicators for sustainable development. *Energy* 32: 875–882.

Vitousek, P. M., Cassman, K., Cleveland, C., Crews, T., Field, C. B., Grimm, N. B., Sprent, J. I. 2002. Towards an ecological understanding of biological nitrogen fixation. *Biogeochemistry* 57: 1–45.

Walter, A., Dolzan, P., Quilodrán, O., Garcia, J., Da Silva, C., Piacente, F., Segerstedt, A. 2008. A sustainability analysis of the Brazilian ethanol. Report submitted to the United Kingdom Embassy, Brazil.

Weyerhaeuser, H., Tennigkeit, T., Yufang, S., Kahrl, F. 2007. Biofuels in China: An analysis of the opportunities and challenges of *Jatropha curcas* in Southwest China. ICRAF working paper number 53. http://www.jatropha.pro/PDF%20bestanden/Biofuels%20in%20China.pdf. Accessed June 1, 2016.

Wooley, R., Ruth, M., Glassner, D., Sheehan, J. 1999. Process design and costing of bioethanol technology: A tool for determining the status and direction of research and development. *Biotechnology Progress* 15: 794–803.

Wyman, C. E., Dale, B. E., Elander, R. T., Holtzapple, M., Ladisch, M. R., Lee, Y. Y. 2005. Coordinated development of leading biomass pretreatment technologies. *Bioresource Technology* 96: 1959–1966.

Xavier, M. R. 2007. The Brazilian sugarcane ethanol experience. Competitive Enterprise Institute, Washington DC.

Xiu, S., Shahbazi, A. 2015. Development of green biorefinery for biomass utilization: A review. *Trends in Renewable Energy* 1: 4–15.

Zhang, Y. H. P. 2008. Reviving the carbohydrate economy via multi-product lignocellulose biorefineries. *Journal of Industrial Microbiology and Biotechnology* 35: 367–375.

Zhang, Y. H. P., Himmel, M. E., Mielenz, J. R. 2006. Outlook for cellulase improvement: Screening and selection strategies. *Biotechnology Advances* 24: 452–481.

Zhao, R., Bean, S. R., Wang, D., Park, S. H., Schober, T. J., Wilson, J. D. 2009. Small-scale mashing procedure for predicting ethanol yield of sorghum grain. *Journal of Cereal Science* 49: 230–238.

Section II

Production of Biofuels

5

Lipid-Based Biomasses as Feedstock for Biofuels Production

Somkiat Ngamprasertsith and Ruengwit Sawangkeaw

Contents

Abstract

Biofuels are defined in the beginning of this chapter, followed by the information about lipid-based biomasses and first-generation feedstocks for biofuel production. Fatty acids commonly found in lipid-based biomasses, molecular weight of general plant oils, and acidity of lipid-based biomasses are briefly described. Since the introduction of first-generation biomasses, mostly plant oils have been utilized as transportation fuels and their costs keep fluctuating due to food or fuel dilemma. Hence, many lipid-based biomasses were introduced as second-generation feedstock worldwide. The second-generation feedstock should not be aligned with the food industry. Wastes and industrial by-products are inexpensive feedstock, but their productivity and quality are not consistent. Sources of waste cooking oil (WCO), animal fat, oilseed, and other industrial and municipal wastes are mentioned. Moreover, wastes' and by-products' costs are low; however, pretreatment, collection, and handling might increase their costs. Among these wastes and by-products, WCO from restaurants acquired earliest attention. Thus, advantages and disadvantages of biofuel production using WCO as feedstock are summarized. In contrast, lipid-based biomasses derived from nonedible oils, insects, and oleaginous microorganisms were demonstrated to resolve problems on unstable productivity and feedstock quality. However, their costs when produced in a large scale remain questionable because they are in initial developmental stages in a small scale. Moreover, agricultural science research is necessary for increasing nonedible oil plants' productivity. Studies on insects having the abilities to convert several wastes to lipid-based biomasses have been recently reported. Although oleaginous microorganisms are promising due to high productivity and simplicity of genetic engineering, microalgal cultivation and harvesting systems have been debated. Carbon sources are used to cultivate oleaginous yeasts and fungi; however, resultant lipid-based biomasses are somewhat involved in food or fuel dilemma. In the conclusion section, the topics of practical importance for each lipid-based biomass are summarized.

5.1 Introduction

Alternative fuels derived from renewable resources, the first-generation feedstocks, are a promising new energy source for transportation in many countries. For example, bioethanol produced primarily from corn, sugarcane, and cassava is used as an automotive fuel in spark-ignition engines in the United States, Brazil, and Thailand. Furthermore, biodiesel synthesized from lipid-based biomasses is also a promising biofuel for compression-ignition engine. Nowadays, edible oils such as those from sunflower, soybean, and oil palm are the first-generation feedstocks used in the industrial production of biodiesel.

The term "biofuels" in this chapter includes biodiesel, fatty acid alkyl esters (FAAE), and other biofuels that are thermochemically produced via pyrolysis, catalytic decarboxylation, catalytic hydrodeoxygenation, and supercritical techniques. Specifically, the term "lipid-based biofuels" will be used within this chapter. Please bear in mind that the term "biofuels" in this chapter excludes other biofuels synthesized from sugar-based or lignocellulosic biomasses, such as methane (biogas), methanol, ethanol, and butanol. Additionally, the term "lipid-based biofuels" mostly refers to biodiesel that is chemically synthesized by transesterification and/or esterification reactions.

The 24 properties of biodiesel are strictly specified by international standards, such as EN14214 and ASTM D6751-14. Thermochemically produced lipid-based biofuels normally fail to meet the 96.5% ester content designated in the international standard for biodiesel (EN14214), so they cannot be classified as biodiesels. However, their properties, which are very similar to those of petrodiesel or gasoline, make them of interest as alternative transportation fuels. This chapter aims to emphasize the discovery of lipid-based feedstocks for biofuel production.

Because of the food or fuel dilemma, the use of the second-generation feedstocks, lignocellulosic biomasses, to produce bioethanol was demonstrated over 50 years ago (FitzPatrick et al. 2010). The second-generation feedstocks for lipid-based biomasses were discovered later than those for lignocellulosic biomasses. In this chapter, brief details concerning lipid-based biomasses are introduced first followed by the first-generation feedstocks for lipid-based biofuel production. Further, the second-generation feedstocks are categorized according to the consistency of their productivity. For example, the productivity of waste cooking oil (WCO) is inconsistent, whereas increasing cultivation could increase the consistency of nonedible oil plant production. Finally, the challenges involved in the production of lipid-based feedstocks for biofuel production are summarized.

5.2 Lipid-Based Biomasses

Lipids are a group of compounds found in living organisms that are completely miscible with nonpolar organic solvents. Thus, lipids naturally include many compounds, such as fatty acids, glycerides, fat-soluble vitamins, cholesterols, phospholipids, and glycolipids. The chemical structures of lipids depend on their source, the type of organism, and their biological function in the cell. However, the lipid-based biomasses, considered feedstocks for biofuel production, consist primarily of glycerides because their primary structure consists of C_8–C_{24} straight-chain fatty acids.

Glycerides, a collective name for the esterification products of glycerol and fatty acids, include triglycerides, diglycerides, and monoglycerides. The backbone of a glyceride molecule is glycerol, which consists of three hydroxyl groups (–OH) attached to three individual carbon atoms (see Figure 5.1). The carboxylic acid (–COOH) groups of fatty acids can be esterified, one by one, with the hydroxyl groups of glycerol. For example, an esterification between palmitic acid and glycerol is illustrated in Figure 5.1.

When the three hydroxyl groups of glycerol are completely esterified with three fatty acids, this molecule is called a "triglyceride." Molecules having two of the three

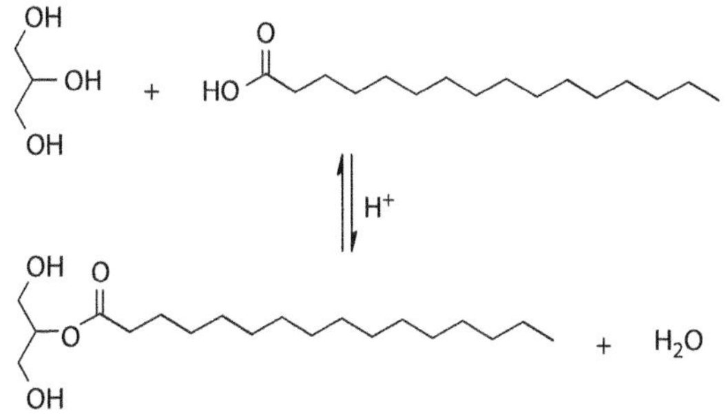

Figure 5.1 An acid-catalyzed esterification reaction between glycerol and palmitic acid (C16:0).

hydroxyl groups of glycerol esterified are called "diglycerides" and so on. Thus, the variety of glyceride structures originates from the number of fatty acids that can be esterified with the three hydroxyl groups of glycerol. It should be noted that the distribution of fatty acids in glycerides is not completely random; however, the detailed molecular structure of the glycerides is still unclear. The distribution of triacylglycerides can be estimated by reverse-phase HPLC, as described elsewhere (Christie et al. 2007).

Fatty acids are monocarboxylic acids with a straight chain of 4–22 carbon atoms, mostly in even numbers. The hydrocarbon chains of fatty acids can be categorized by the number of double bonds (C=C) between the carbon atoms in the chain. Consequently, they are divided into saturated chains, which contain no double bonds, and unsaturated chains, which contain at least one double bond. Furthermore, fatty acids containing multiple double bonds between carbon atoms are called "polyunsaturated fatty acids" or "PUFAs." The nomenclature of fatty acids is symbolized as CX:Y, where X and Y are the number of carbon atoms and C=C double bonds, respectively. For example, the symbol (shorthand name) for palmitic acid—a saturated fatty acid containing 16 carbon atoms—is C16:0. For additional details, consult the definitive work on fatty acids available elsewhere (Harwood and Scrimgeour 2007). The common fatty acids generally found in the lipid-based biomasses used for biofuel production are summarized in Table 5.1.

The distribution of fatty acids in glycerides, commonly called the fatty acid profile, fatty acid composition, or fatty acid component, is individually specified by the nature of each lipid-based biomass. For example, animal fat is typically composed of saturated fatty acids with high carbon numbers (up to 16 atoms), and thus it is a solid at ambient temperature. In addition, the linear structure of saturated fatty acids allows the individual molecules to get closer to each other than does the bent structure of unsaturated fatty acids, as illustrated in Figure 5.2. Consequently, the glycerides that contain a large number of PUFAs are liquids at ambient, even low temperature, because of the steric hindrance among the fatty acid chains. For instance, PUFAs are usually found in fish oils, particularly arctic marine fish oils. On the other hand, the oil plants such as oil palm (*Elaeis guineensis*) and coconut (*Cocos nucifera*) have a

TABLE 5.1   Fatty Acids Commonly Found in Lipid-Based Biomasses

Common Name	IUPAC Name	Number of Carbon	Number of Double Bond	Shorthand Name	Molecular Weight
Caprylic acid	Octanoic acid	8	0	C8:0	144.21
Capric acid	Decanoic acid	10	0	C10:0	172.26
Lauric acid	Dodecanoic acid	12	0	C12:0	200.32
Myristic acid	Tetradecanoic acid	14	0	C14:0	228.37
Palmitic acid	Hexadecanoic acid	16	0	C16:0	256.42
Stearic acid	Octadecanoic acid	18	0	C18:0	284.47
Oleic acid	Octadecenoic acid[a]	18	1	C18:1	282.46
Linoleic acid	Octadecadienoic acid[a]	18	2	C18:2	280.45
Linolenic acid	Octadecatrienoic acid[a]	18	3	C18:3	278.44

[a] The position(s) of the double bond(s) in this fatty acid molecule is(are) not specified.

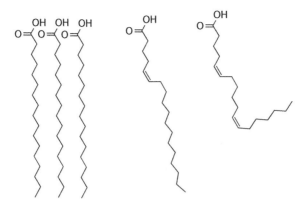

Figure 5.2 Chemical structures of saturated (three molecules of C18:0) and unsaturated fatty acids (C18:1 and C18:2).

high amount of saturated fatty acids, whereas some of them, such as olive (*Olea europaea*) and sunflower (*Helianthus annuus*), have a high amount of unsaturated fatty acids. It could be concluded that the exact molecular structure of each lipid-based biomass is unclear. For further analytical methods and discussions of triglycerides' structure, authoritative literature is available (Christie et al. 2007).

The fatty acid profiles of fats and oils can be conveniently analyzed using the two American Oil Chemists' Society (AOCS) standard methods (Christie et al. 2007). First, AOCS Ce 2-97 can be applied to prepare methyl esters of fatty acids using boron trifluoride (BF_3) in methanol (MeOH) as the reactant. The bottom layer, which contains glycerol, is removed by phase separation. Because of the high boiling point of the glycerides, conversion into a fatty acid methyl ester can prevent the thermal decomposition of glycerides before they are volatized. Second, the amount of each fatty acid in the top layer can be quantified by gas chromatography (GC) using the standard method of AOCS Ce 1-97. A simplified schematic diagram of the process used to determine a fatty acid profile is shown in Figure 5.3. Because all of the fatty acids in the glycerides are entirely converted to fatty acid esters, the origin of each fatty acid on the parent glycerides is not conclusively known.

It should be noted that the fatty acid profile is distinct from the triglyceride profile. The distribution of triglyceride species in lipid-based biomasses can be identified and quantified using chromatographic, spectrophotometric, and spectroscopic methods. A recent review estimated that there are 20–79 triglyceride species in the 10 edible oils (Andrikopoulos 2002). Additionally, analysis of the triglyceride profile is more complicated than analysis of the fatty acid profile. However, after the triglycerides are converted to biofuels, the properties of the resultant biofuels strongly depend on the fatty acid profile. Therefore, the triglyceride profiles have not commonly been reported during research on the topic of lipid-based biofuel production.

Even though the exact molecular structures of lipid-based biomasses are unknown, the molecular weight of each oil (MW_{oil}) can be experimentally estimated using the saponification value (SV). The molecular weights of lipid-based biomasses

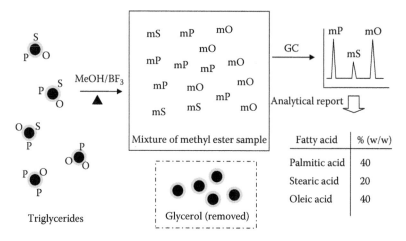

Figure 5.3 A simplified diagram of the standard method for fatty acid profile analysis. ● is glycerol; S, O, and P are stearic, oleic, and palmitic acids, respectively; and m refers to the methyl ester of the corresponding fatty acid; for example, mS is methyl stearate.

are required to calculate the molar ratio between oil and alcohol in conventional biodiesel production and to estimate the loss beyond the edible oil refining process. The SV is the amount of KOH required to completely neutralize all the fatty acids in a 1 g sample of a lipid-based biomass. Those fatty acids are derived from the hydrolysis of the lipid sample in aqueous KOH solution. Hence, light lipids, which contain mainly short-chain fatty acids, have a greater number of molecules per unit mass than heavy lipids (see Figure 5.4). In other words, a high SV reflects a low MW for

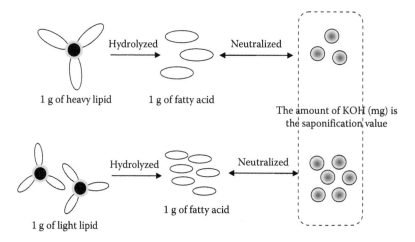

Figure 5.4 Simplified diagram demonstrating the relationship between molecular mass and saponification value for heavy and light lipids. ● is glycerol; ⬭ is a fatty acid and the ellipse size is relative to molecular mass; and ◉ is KOH.

TABLE 5.2   Literature Value for Relative Molecular Weight of General Plant Oils as Derived from Their Saponification Value

Oil	MW_{oil}	SV (mg KOH/g Oil)
Oilseeds		
Canola	890	189
Corn	887	190
Cottonseed	860	196
Peanut	887	190
Soybean	878	192
Sunflower	878	192
Tree fruits and kernels		
Coconut	674	248
Olive	887	190
Palm	851	198
Palm kernel	704	239

Source: Christie, W.W. et al., Analysis, in *The Lipid Handbook with CD-ROM*, 3rd edn., CRC Press, Boca Raton, FL, 2007, pp. 415–469.

lipid-based biomasses. In practice, the MW_{oil} can be estimated using the following equation:

$$MW_{oil} = \frac{3 \times 56.11}{SV} \times 1000 \tag{5.1}$$

The MW_{oil} and SV for selected plant oils are shown in Table 5.2.

On the other hand, a pseudotriglyceride molecular method has been devised to calculate the molecular weights of triglycerides using their fatty acid profiles (Espinosa et al. 2002; Bunyakiat et al. 2006). The lipid-based biomasses were represented with the following chemical formula: $[(CH_2COO)_2CHCOO](CH=CH)_m(CH_2)_n(CH_3)_3$. The parameters m and n are the degree of unsaturation and the molecular weight, as shown in Equations 5.2 and 5.3, respectively:

$$m = \sum_{i=1}^{N} m_i x_i \tag{5.2}$$

$$n = \sum_{i=1}^{N} n_i x_i \tag{5.3}$$

To calculate the values of m and n, the mole fraction of each fatty acid (x_i) is necessary. However, the fatty acid profile, commonly reported in terms of mass fraction, has to be converted before using the result to calculate the molecular weight. The example, a calculation of the molecular weight of palm oil, is illustrated in Table 5.3.

TABLE 5.3   Calculation of the Degree of Unsaturation (m) and Molecular Weight (n) for Palm Oil

Composing Fatty Acids in Pseudotriglyceride	% (w/w)	MW_{FA}	x_i	n_i	m_i	$x_i n_i$	$x_i m_i$
Myristic (C14:0)	2.25	228.37	0.027	36	0	0.49	0.00
Palmitic (C16:0)	38.50	256.42	0.406	42	0	20.91	0.00
Stearic (C18:0)	4.50	284.47	0.043	48	0	1.16	0.00
Oleic (C18:1)	45.00	282.46	0.430	42	3	17.58	1.26
Linoleic (C18:2)	1.75	280.45	0.017	36	6	1.55	0.26
Linolenic (C18:3)	8.00	278.44	0.078	30	9	0.09	0.03
					Σ	41.77	1.54

Source: Lee, K. and Ofori-Boateng, C., Production of palm biofuels toward sustainable development, in *Sustainability of Biofuel Production from Oil Palm Biomass*, Springer, Singapore, 2013, pp. 107–146.

After rounding the values of m and n to the nearest integer and inserting them into the pseudotriglyceride formula $[(CH_2COO)_2CHCOO](CH{=}CH)_m(CH_2)_n(CH_3)_3$, a molecular weight of 858 can be calculated for crude palm oil, which is close to its experimental value (Table 5.2). Additionally, the values of m_i and n_i for the pseudotriglyceride and condensed formulas of tripalmitin and triolein are illustrated in Figure 5.5.

The acidity of lipid-based biomasses is an important property that indicates the quality of the feedstock, particularly for conventional biodiesel production processes. It arises through the hydrolysis of glycerides caused by improper processing or storage conditions. Acidity can be measured using either the % free fatty acid content (%FFA) or the acid value (AV). For practical purposes, the %FFA is half of the AV when all FFAs are assumed to be oleic acid. The level of acidity for conventional biodiesel production processes is limited to below 0.5% (w/w) FFA. To use the high %FFA lipid-based biomasses as a biodiesel production feedstock, a suitable pretreatment process,

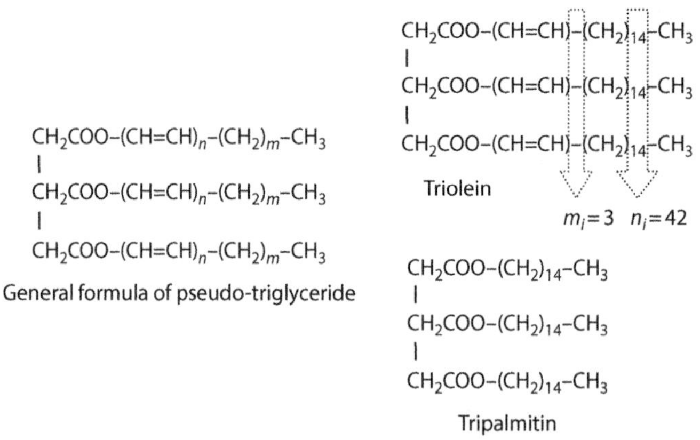

Figure 5.5   The chemical formulae of selected triglycerides based on the pseudotriglyceride assumption.

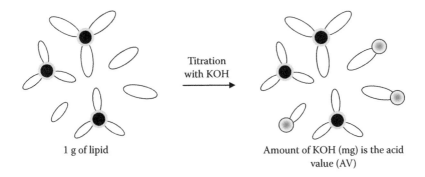

Figure 5.6 The reaction between free fatty acid and KOH. ⬤ is glycerol; ⬭ is fatty acid; and ⬤ is KOH.

particularly the esterification of FFAs, is necessary. When alternative feedstocks are discovered, it is essential that their %FFA be reported in the research article. An informative graphic for the definition of the AV is shown in Figure 5.6.

Regardless of fatty acid profile and physical properties, the productivity of lipid-based biomasses is the most important characteristic of a good feedstock candidate for biofuel production. For oil plants, the productivity of lipid-based biomasses is summarized into kilogram per hectare per year. For some micro-organisms, the productivity is calculated in terms of the number of kilograms of biomass or lipid per cubic meter. Nevertheless, the productivity of lipid-based biomasses derived from wastes, such as WCO and animal fat, is sporadic. To indicate high productivity, the lipid content per dry mass (%wt) of novel feedstocks is commonly reported.

5.3 First-Generation Lipid-Based Biomasses

Plant oils are first-generation lipid-based biomasses initially employed to produce biodiesel because of their availability. They are classified into two types, saturated and unsaturated oils, by their iodine value (IV). The standard IV of 86.13 is based on triolein, which contains three double bonds in each molecule. For saturated oils, such as coconut and palm kernel oils, the IV is lowered to 9–18. In contrast, the IV of unsaturated oils, such as soybean and corn oils, is in the range of 120–132. For highly unsaturated oils, such linseed and marine fish oils, the IV is approximately 185 (Christie et al. 2007). Indeed, the IV depends directly on the fatty acid profile, as shown in Table 5.4.

The fatty acid profile of a feedstock is reflected in the biodiesel properties in a few ways (Knothe et al. 2003; Dias et al. 2008). For example, biodiesel produced from unsaturated feedstocks has a slightly lower viscosity and density than that derived from saturated oils because of the steric hindrance of the fatty acids (see Figure 5.2). Conversely, increased unsaturation reduces the cetane number of the resultant bio-diesel, whereas increased chain length enhances the cetane number. Furthermore,

TABLE 5.4 Fatty Acid Profiles of General Edible Oils

Oil	C12:0	C14:0	C16:0	C18:0	C18:1	C18:2	C18:3	Other
Oilseeds								
Canola (rapeseed)	—	—	3.9	1.9	64.1	18.7	9.2	0.6 (C20:0)
Corn	—	—	9.2–11.0	1.1–2.6	19.5–25.8	39.4–65.6	0.5–2.6	—
Cottonseed	—	0.9	24.7	2.3	17.6	53.3	0.3	0.7 (C16:1)
Peanut	—	0.1	11.6	3.1	46.5	31.4	—	1.5 (C20:0) 1.4 (C20:1) 3.0 (C22:0) 1.0 (C24:0)
Soybean	—	0.1	11.0	4.0	23.4	53.2	7.8	0.1 (C14:0) 0.3 (C20:0)
Sunflower	0.5	0.2	6.8	4.7	18.6	68.2	0.5	0.4 (C20:0)
Tree fruits and kernels								
Coconut	46–50	17–19	8–10	2–3	5–7	1–2.5	—	0.2–0.8 (C6:0) 6–9 (C8:0) 9–10 (C10:0)
Olive	—	0.0–0.1	7.5–20.0	0.5–5.0	55.0–83.0	3.5–21	0.0–1.8	0.3–3.5 (C16:1) 0.0–0.8 (C20:0)
Palm	0.2	0.8–1.1	39.0–44.1	4.5–5.0	39.2–43.2	10.1–11.5	0.3–0.4	0.1–0.2 (C16:1) 0.2–0.4 (C20:0)
Palm kernel	44–51	15–17	7–10	2–3	12–19	1–2	—	0–1 (C6:0) 3–5 (C8:0) 3–5 (C10:0)

Source: Gunstone, F., *Vegetable Oils in Food Technology: Composition, Properties and Uses*, John Wiley & Sons, Chichester, U.K., 2011.

the oxidative stability of unsaturated fatty acids decreases significantly with increasing numbers of double bonds or increasing IV. The unsaturated fatty acids, particularly PUFAs, are sensitive to UV light, oxygen, and free radicals, which can lead to polymerization and epoxide/peroxide formation. On the other hand, the cold flow properties, such as pour point and cold filter plugging point, are improved when unsaturated oils are used as feedstocks (Dwivedi and Sharma 2014). The issues of oxidative stability and cold flow properties can be resolved by adding a trace amount of antioxidant and winterizer into the biodiesel.

The feedstock price is reported to represent approximately 70%–80% of the biodiesel production cost (Chhetri et al. 2008). Consequently, the economic feasibility of the lipid-based biofuel industry strongly depends on oil productivity, which is somewhat reflected in their price, as illustrated in Table 5.5.

Plant oils have been closely aligned with the food industry for many decades; however, their price has been somewhat influenced since they have been applied as sources of alternative energy. The sustainability of land use for food or fuel production has frequently been questioned and criticized. To avoid the economic and ethical competition between the food and fuel industries, second-generation feedstocks have been introduced.

As seen from Table 5.5, it is clear that palm oil is the cheapest edible oil because of its highest productivity. Furthermore, the 300–500 kg of palm kernel oil produced simultaneously with crude palm oil is a bonus. Despite having the cheapest price and the highest productivity, researchers from the top palm oil producer suggest that the

TABLE 5.5   Lipid Contents and Productivities of General Edible Oil Plants

Oil	Lipid Content (%wt)	Oil Productivity (kg/ha/Year)	Price[a] (US$/kg)
Oilseeds			
Canola (rapeseed)	40–45	590.7–663.8	0.75–0.84
Corn	3–6	241.9–438.8	0.70–1.30
Cottonseed	18–20	208.1–236.3	0.90–1.50
Peanut	45–50	1260.1–1400.8	1.36–1.39
Soybean	18–20	450–506.3	0.70–0.72
Sunflower	35–45	517.6–663.8	0.96–1.07
Tree fruits and kernels			
Coconut	65–68	731.3–978.8	1.18–1.22
Olive	15–35	101.3–292.5	3.35–3.53
Palm	45–50	3004.1–5006.8	0.63–0.67
Palm kernel	45–50	300.4–500.7	0.90–1.08

Sources: Tickell, J. and Tickell, K., *From the Fryer to the Fuel Tank: The Complete Guide to Using Vegetable Oil as an Alternative Fuel*, Tickell Energy, Hollywood, CA, 2003; O'Brien, R.D., *Fats and Oils: Formulating and Processing for Applications*, 3rd edn., CRC Press, Boca Raton, FL, 2008.

[a] Refers to the price range between September 2014 and February 2015 obtained from www.indexmundi.com, except the price of corn and cottonseed oils, which were obtained from www.alibaba.com.

conversion of wastes, which are both lipid-based and sugar-based biomasses, will be necessary to establish the sustainability of the biofuel industry. For instance, the palm oil industry generates approximately 190 million tons of solid and liquid residues worldwide; yet only 10% of them are commercially utilized as fertilizers. The conversion of those wastes, both lipid-based and sugar-based wastes, provides for the development of sustainable biofuel production from palm biomasses. For further details, an authoritative book is available elsewhere (Lee and Ofori-Boateng 2013). The concerns of alternative feedstocks in the palm oil industry could be applicable to other food industries.

Although the prices of some second-generation feedstocks, particularly those derived from wastes and by-products, are cheaper than those of the first-generation feedstocks, their productivity and quality are much lower than those of the first-generation feedstocks. The productivity of nonedible oil plants could be improved by increased research and development in agricultural science, but enhancing the productivity of lipid-based biomasses from wastes and industrial by-products will be difficult. Moreover, the lipid-based wastes and by-products generally have a high acidity that imparts negative effects on conventional biodiesel production. On the other hand, the second-generation feedstocks, such as insects and oleaginous microorganisms, or single-cell oils, are currently more expensive than plant oils. The insect oils are promoted because of their fast growth rate and waste-to-lipid conversion ability, whereas the oleaginous microorganisms are candidates for second-generation feedstocks because their high productivity has been reported in several reports (Meng et al. 2009; Subramaniam et al. 2010; Ahmad et al. 2011). However, both insect and microorganisms (single-cell oils) also have high acidity, just like lipid-based wastes and industrial by-products.

5.4 Second-Generation Lipid-Based Biomasses

5.4.1 Wastes and By-Products

5.4.1.1 Waste Cooking Oil from Households and the Food Industry

In the past, WCO from households was managed improperly, such as pouring it down the kitchen drain, or wiping it up with a paper towel, putting it in a sealable container, and then throwing it away. Recycling WCO by, for example, using it to lubricate snow shovels, to fuel oil lamps, and to make soap was proposed. Thus, the major problem with household WCO is that its availability is sporadic. From the food industry and restaurants, a high volume of WCO was sold as an additive for animal feed in poultry and swine farms. Because of the return of some harmful substances to the human food chain, the European Union has restricted the blending of WCO into animal feed since 2002 (Kulkarni and Dalai 2006). The collection of WCO from households, restaurants, and the food industry and converting this WCO into biodiesel is the most proper way to recycle WCO and reduce waste. The physical properties of used cooking oil depend on several factors, such as the type of frying pan, the type of fried food, and the duration of frying. Increases in viscosity, density, and AV have been commonly reported. However, the fatty acid profile of WCO is insignificantly different from that of fresh cooking oil. Furthermore, the physical contaminants that come from food processing, such as burned food bits, paper, and aluminum foil, are also found in WCO. Thus, WCO fluctuates in physical quality and requires testing methods for monitoring. The suggested testing methods and selected properties of WCO samples are illustrated in Table 5.6.

The properties of WCO that affect catalytic biodiesel production processes are the acidity and moisture content. For example, the allowable levels of FFA and moisture in feedstock are 0.5%–1.0% (w/w) and 0.05%–0.06% (v/v), respectively, for alkaline-catalyzed process (Canakci 2007; Mazubert et al. 2013). On the other hand, thermo-chemical processes, such as pyrolysis, catalytic cracking, and supercritical alcohols, are capable of dealing with WCO because of their high tolerance for FFA and moisture. The advantages and disadvantages of biofuel production processes are summarized in Table 5.7.

As a result of the high FFA and moisture content of WCO, suitable pretreatment processes are essential before using WCO as a feedstock in commercial biodiesel production (homogeneous base-catalyzed process). The pretreatment processes mainly include separation of suspended solids and reduction of FFA. The suspended solids can simply be separated by filtration or decanting, whereas FFA can be reduced using many approaches. For example, esterification with alcohols or spent glycerol using acidic catalysts such as H_2SO_4, H_3PO_4, and ion-exchange resins has been reported (Özbay et al. 2008). Moreover, vacuum distillation of WCO, similar to that in the edible oil refining process, is the effective method to simultaneously reduce FFA and moisture contents (Banerjee and Chakraborty 2009). In contrast, the use of vacuum distillation probably increases the capital and operating costs of biodiesel production from WCO to a level close to the costs of producing biofuels from refined edible oils. It should be noted that all of the processes presented in Table 5.7, as well as the pretreatment processes, are applicable to other lipid-based biomasses that have high FFA and moisture content.

Besides processing WCO to produce biofuels, policy research on the ways to recycle WCO is also important. A study of recycling methods in China, the United States, and Japan (Zhang et al. 2014) revealed that the third-party take-back (TPT) method was suitable for China and Japan, whereas the biodiesel enterprise take-back (BET) method was practical in the United States. The stakeholders of biodiesel production from WCO are the government, biodiesel enterprises, third-party providers, illegal manufacturers, and restaurants. The policies concerning WCO deposal and biodiesel production were set by the government for restaurants and biodiesel enterprises, respectively. For example, the government collects a deposal fee and implements penalties and subsidies for biodiesel enterprises and customers. The WCO could be delivered to the biodiesel enterprises by the third-party providers or by restaurants, largely depending on the size of the restaurants. Illegal manufacturers cannot be negligible because they are stimulated by poor control and interest guidance mechanisms of WCO. They also disrupt the supply of WCO to legal enterprises. The recycle rate of WCO could be enhanced by improving technical support for WCO collection, developing strict penalty mechanisms for restaurants, setting up compensatory financing for restaurants, and reforming the subsidy system.

TABLE 5.6 Selected Properties and Testing Methods for Waste Cooking Oil Samples

Physical and Chemical Properties	Sample					Testing Method
	United States WCO	Canadian WCO	European WCO	Vietnamese WCO		
% Solid portion	N/A	19	20	N/A		Filtration/centrifugal method
% Water content	<2.0	7.3	1.1–1.4	N/A		Karl Fischer method/ASTM E203
% Free fatty acid	8.8–25.5	1.4–5.6	2.4–3.1	0.4–1.2		AOCS Ca 5a-40
% Total polar compound	N/A	22	N/A	N/A		AOCS Cd 20-91
% Oxidized triglycerides	N/A	4.72	N/A	N/A		IUPAC, 2.508 (1987)
% Polymerized triglycerides	N/A	1.43	N/A	N/A		AOCS Cd 22-91 or IUPAC, 2.508 (1987)
Acid value (mg KOH/g oil)	17.5–51.0	2.5–11.2	5.3–6.3	0.74–2.0		AOCS Te 1a-64
Saponification value (mg KOH/g oil)	N/A	177.87	195.1–204.31	264.1–272.0		AOCS Cd 3b-76
Density at40 °C (kg/m³)	N/A	900	937–939	920		ASTM D1298
Peroxide value (m$_{eq}$/kg)	0.8–4.6	N/A	5.6–6.3	N/A		AOCS Cd 8b-90
Viscosity at 40°C (cSt)	N/A	44.7	190.2–201.3	27.42–30.05		ASTM D445
Iodine value (g I$_2$/100 g oil)	52–82	N/A	104.3–115.3	8.57–13.2		AOCS Cd 1b-87

Sources: Sawangkeaw, R. and Ngamprasertsith, S., *Renew. Sust. Energ. Rev.,* 25, 97, 2013; Canakci, M., *Bioresour. Technol.,* 98, 183, 2007.
Abbreviation: N/A, not available.

TABLE 5.7 Advantages and Disadvantages of Biofuel Production Processes Using Waste Cooking Oil as a Feedstock

Process	Advantage	Disadvantage
Homogeneous basic—catalyzed	• Very fast reaction rate • Mild reaction conditions • Inexpensive	• Sensitive to FFA and moisture contents • Reduced fuel yield and increased volume of wastewater due to soap formation
Heterogeneous basic—catalyzed	• Mild reaction conditions • Easy separation and regeneration of catalyst	• Poisoned at ambient air • Sensitive to FFA and moisture contents • Leaching of catalyst • Energy intensive
Homogeneous acidic—catalyzed	• Insensitive to FFA and moisture content • Simultaneous esterification and transesterification	• Very slow reaction rate • Corrosive • Difficult catalyst separation
Heterogeneous acid—catalyzed	• Insensitive to FFA and moisture content • Simultaneous esterification and transesterification • Easy separation and regeneration of catalyst	• Complication of catalyst synthesis • High reaction temperature, high alcohol to oil molar ratio, long reaction time • Leaching of catalyst • Energy intensive
Enzymatic catalyzed	• Low reaction temperature • Require only one purification step	• Very slow reaction rate • High costs of catalyst • Catalyst sensitive to alcohol
Thermal cracking or pyrolysis	• Insensitive to FFA and moisture content • Catalyst-free • Operate at atmospheric pressure • Applicable to produce kerosene, gasoline, and diesel	• High temperature (500°C–850°C) • Lower liquid yield because of char and gaseous product formation • Require product upgrading • Biofuel highly contains cyclic and aromatic hydrocarbons *(Continued)*

TABLE 5.7 (*Continued*) Advantages and Disadvantages of Biofuel Production Processes Using Waste Cooking Oil as a Feedstock

Process	Advantage	Disadvantage
Catalytic cracking	• Insensitive to FFA and moisture content • Operate at atmospheric pressure • Using commercial catalysts such as HZSM-5, MCM-41 • Applicable to produce kerosene, gasoline, and diesel	• Moderate temperature (450°C–500°C) • Require product upgrading • Biofuels contain cyclic and aromatic hydrocarbons • Deactivation of catalysts because of carbon deposition
Hydrotreating	• Biofuel has higher energy content than FAMEs, excellent combustion quality, and good low-temperature properties • Easy to adapt to the petroleum refinery • Using commercial catalysts such as NiMo, CoMo, and NiW/γ-Al$_2$O$_3$	• Moderate temperature • Require high volume of hydrogen • Hydrogen currently derived from nonrenewable source
Supercritical alcohols	• Very fast reaction rate • Insensitive to FFA and moisture content • Catalyst-free	• Moderate temperature (300°C–350°C) and high pressure (15–35 MPa) • High alcohol: oil molar ratio (40:1–42:1)
Combined subcritical water and supercritical alcohols	• Fast reaction rate • Insensitive to FFA and moisture content • Catalyst-free	• Moderate temperature (280°C–300°C) and high pressure (15–35 MPa) • Moderate water: oil and alcohol/oil molar ratios (24:1)

Sources: Lam, M.K. et al., *Biotechnol. Adv.*, 28, 500, 2010; Lee, J.S. and Saka, S., *Bioresour. Technol.*, 101, 7191, 2010.

5.4.1.2 Animal Fats from Slaughter

Several animal fats, including alligator fat, beef tallow, chicken fat, duck tallow, fish oil, lamb meat, and pork lard, have been reported as alternative feedstocks for lipid-based biofuel production. The only major problem with using animal fat as a biodiesel feedstock in conventional catalytic processes is the high FFA content of animal fat. The extraction of animal fat from meat and skin is commonly performed by melting at 60°C–120°C, followed by the removal of the solid residue by filtration (Chakraborty et al. 2014). It has been reported that 70% of the moisture content of animal fat is lost during the melting process at 110°C (Mata et al. 2011). The FFA removal processes mentioned for WCO could be used to solve the FFA problem of animal fats. The fatty acid profiles and FFA contents of selected animal fats are shown in Table 5.8.

Since the fatty acid profile of animal fats is only slightly different than that of the edible oils, which are the present feedstock for biodiesel production, the use of animal fats will not significantly influence the properties of the resulting biodiesel (Canakci and Sanli 2008; Hoekman et al. 2012). Because of its high saturated fatty acid profile, the resultant biodiesel is less prone to oxidation but has increased cloud point and cold filter plugging temperatures. Additional methods of biofuel production from animal fats, such as the supercritical transesterification of chicken fat and pork lard (Marulanda et al. 2010; Shin et al. 2012), the pyrolysis of waste fish oil (Wisniewski et al. 2010), and the catalytic cracking of beef tallow (Ito et al. 2012), have also been demonstrated. The biofuels obtained from these additional methods have properties similar to those of the biofuels produced from edible oils.

According to the data from the U.S. Department of Agriculture (USDA), world meat production and consumption were nearly 250 million tons in 2013. Those meats consisted of 43.2% swine, 33.4% chicken, and 23.4% beef. However, the worldwide productions of chicken, swine, and beef have grown by 4.1%, 2.5%, and 1.1%, respectively, annually over the past decades. By the way of contrast, the amount of animal fat depends on many factors, for example, meal industries, population, per capita income, religions, and food culture. For example, the amount of animal fat in India was relatively low because around 35%–40% of the population, at least 450 million people, are vegetarians. In contrast, the top meat producers of the world, including the United States, Brazil, and China, have abundant animal fat from slaughter.

5.4.1.3 Oilseeds from the Fruit Processing Industry

Production of fruit juice, canned fruit, dried fruit, and ketchup also ordinarily generates fruit seeds, skin, and pulp as by-products. In former times, these industrial by-products were generally disposed of in landfills, although some by-products, including citrus waste (seed and pulp) and olive and tomato pomaces (skin and seed), were considered for livestock. However, the fruit seeds, which usually contain oil, could be converted to lipid-based biofuel and other minor components. The oil content and fatty acid profiles of some fruit seeds are summarized in Table 5.9.

The oil content of fruit seeds has generally been determined according to the method ISO 659:1998, whereas the fatty acid composition was analyzed following the ISO standard ISO 5509:2000 or AOCS Ce 1-97 and Ce 2-97 as mentioned in the introduction. Like WCO and animal fats, the fatty acid profiles of all the seed oils in

TABLE 5.8 Physical Properties and Fatty Acid Profiles of Some Animal Fats

Sample	Fatty Acid Content (% w/w)						%FFA
	C16:0	C18:0	C18:1	C18:2	C18:3	Other	
Alligator fat	23.30	5.30	55.54			1.27 (C14:0) 12.1 (C16:1) 1.56 (C20:1) 0.28 (C20:3) 0.64 (C20:4)	8–11
Beef tallow	22.30–28.00	19.20–41.28	28.82–48.9	0.58–3.00	0.50–2.00	2.69–4.00 (C14:0)	3.6–15
Chicken fat	21.00–25.20	5.50–5.90	38.20–48.50	17.30–23.80	0.70–1.90	5.80–7.80 (C16:1)	5–25
Duck tallow	17.00	4.00	59.40	19.60			14.56
Waste fish oil	27.30–28.10	11.70–12.10	42.80–44.60	9.40–10.50	0.50–0.70	4.00–4.50 (C16:1)	4.90–10.70
Lamb meat	10.10–28.10	6.00–27.20	35.00–31.28	1.59–36.00	0.88	4.00 (C14:0) 2.00 (C20:0) 2.00 (C22:0)	N/A
Pork lard	24.7	12.1	44.9	11.9	1.5	1.8 (C14:0) 2.5 (C16:1)	0.5–1.5

Source: Sawangkeaw, R. and Ngamprasertsith, S., *Renew. Sust. Energ. Rev.*, 25, 97, 2013.
Abbreviation: N/A, not available.

TABLE 5.9 Fatty Acid Profiles and Oil Contents of Some Fruit Seeds

Seed Sample	Fatty Acid Content (% w/w)						%Oil Content
	C16:0	C18:0	C18:1	C18:2	C18:3	Other	
Apple (*Malus communis*)	6.8–7	1.9–2.0	36.2–40.4	48.1–51.7	0.3–0.6	0.1 (C16:1) 0.5 (C18:4)	21.9–25.6
Apricot (*Prunus armeniaca*)	4.9–6.4	1.0–1.1	59.6–69.7	23.1–31.4	0.1–0.2	1.1–1.2 (C20:1) 0.6–0.9 (C16:1) 0.1 (C20:1)	45.2–53.4
Blackcurrant (*Ribes nigrum*)	6.4–7.6	1.3–2.0	8.0–13.2	39.6–46.8	27.6–34.9	2.5–4.8 (C18:4)	16–27.6
Black cherry (*Prunus serotina*)	4	4	35	27	28	—	21.3
Grape	~5	~3	14–22	60–75	<1.0		7–20
Guava (*Psidium guajava* L.)	7.9	4.1	14.3	72.8	0.3	0.3 (C20:0)	10.5
Mandarin orange (*Citrus reticulata*) 0.25–1.09 %FFA	12.10–36.25	1.20–5.95	18.34–26.10	26.93–37.80	3.40–4.66	1.85–2.95 (C12:0)	24–41
Yellow melon (*Cucumis melo* var. *inodorus* Naudin)	8.7	5.3	26.4	59	—	0.2 (C20:0)	30.6
Kiwifruit (*Actinidia arguta*)	5	2	11	14	66	—	33
Passion fruit (*Passiflora edulis* Sims)	11.7	2.6	18.7	66.3	0.2	0.2 (C16:1) 0.1 (C20:0)	24.3

(Continued)

TABLE 5.9 (*Continued*)   Fatty Acid Profiles and Oil Contents of Some Fruit Seeds

Seed Sample	Fatty Acid Content (% w/w)						%Oil Content
	C16:0	C18:0	C18:1	C18:2	C18:3	Other	
Pear (*Pyrus communis*)	9	2.1	32.3	53.6	0.4	0.2 (C16:1) 1.2 (C20:1)	31.7
Pomegranate (*Punica granatum* L.), 0.6–1.4 %FFFA	4.3–4.6	2.3–2.6	7.1–8.0	6.6–8.2	76.1–78.2	0.2–0.4 (C20:0) 0.6–0.7 (C20:1)	12.7
Caravela pumpkin (*Cucurbita moschata*)	13.5	9.9	36.3	38.8	—	0.5 (C20:0)	33.5
Quince (*Cydonia vulgaris*)	6.8	1.5	34.4	55.4	0.2	0.2 (C16:1) 0.6 (C20:1)	16.9
Soursop (*Annona muricata* L.)	19.4	4.8	43.3	29.7	0.8	1.1 (C16:1) 0.5 (C20:0)	2.01
Tomato (*Solanum lycopersicum*)	12.3–13.4	4.6–5.5	20.0–22.2	56.1–59.1	2.0–2.8	—	20.2–35

Sources: Catchpole, O.J. et al., *J. Supercrit. Fluids*, 47, 591, 2009; Savikin, K.P. et al., *Chem. Biodivers.*, 10, 157, 2013; Sawangkeaw, R. and Ngamprasertsith, S., *Renew. Sust. Energ. Rev.*, 25, 97, 2013; Aguerrebere, I.A. et al., *Food Chem.*, 124, 983, 2011; Khoddami, A. and Roberts, T.H., *Lipid Technol.*, 27, 40, 2015; da Silva, A.C. and Jorge, N., *Food Res. Int.*, 66, 493, 2014.

Table 5.9 do not differ from those of edible oils; thus, they do not significantly impact the resultant biofuel. For example, the fuel properties of biodiesel from mandarin orange (Rashid et al. 2013) and tomato (Giannelos et al. 2005) seed oils were evaluated in lab-scale studies.

Because fruit seeds were considered waste in previous times, the productivity of fruit seed oil is still unclear. For instance, around 750,000 tons of citrus seed is produced globally per year, estimated using the worldwide citrus production and amount of seed contained in dried pulp. It has been reported that approximately 5% of tomato pomace is generated as waste during tomato paste manufacturing (Giannelos et al. 2005). The wet pomace contains 33% seed, 27% skin, and 40% pulp, whereas the dried pomace contains 44% seed and 56% combined pulp and skin (Kaur et al. 2005). Therefore, it can be deduced that 2%–3% tomato seed is a by-product of tomato paste manufacturing.

Although fruit seeds have a small and uncertain productivity, the value of the minor compounds that they contain is much higher than that of the minor compounds present in lipid-based biomasses used as biofuel feedstocks. For example, the antioxidants isolated from tomato seeds include sterols, tocopherols, lipopigments, and demethylsterol. They could be used by the pharmaceutical industry. Some bioactive phytochemicals are also found in the palm oil waste (Ofori-Boateng and Lee 2013). Therefore, the extraction of value-added components before converting the lipid-based biomasses to biofuel is a good strategy to improve the economic competitiveness and sustainability of biorefining.

5.4.1.4 Soap Stock and Fatty Acid Distillate from the Cooking Oils Industry

The refining process of edible oil production aims to reduce the acidity of crude vegetable oils to a certain level, generally 0.05 %FFA. By using the two types of equipment, soap splitters and vacuum distillation towers, the FFA content of crude vegetable oil is separated into two forms: soap stock and fatty acid distillate (FAD), respectively. Approximately 5%–6% soap stock or FAD is obtained per unit volume of refined edible oil (Shao et al. 2009). Formerly, soap stock was considered a waste, emulsified water containing FFA and soap. The state of soap stock, liquid or solid, depends on both the %FFA and the triglyceride content. Soap stock can be acidified to increase FFA content; the resulting material is called "acid oil." When obtained as FAD, the waste was utilized as an alternative fuel in boilers. It should be noticed that FAD was also considered a high potential source of phytochemicals (Piloto-Rodríguez et al. 2014). The properties of soap stock and FAD are summarized in Table 5.10.

Even though the fatty acid profiles of soap stock and FAD are somewhat dissimilar to that of their parent oils, except for the soap stock from marine fish oil, they do not significantly impact biofuel processing or the fuel properties, as mentioned earlier. The very high amount of PUFAs in marine fish oil may lead to the poor oxidative stability of biodiesel. Optional approaches, such as adding antioxidants, catalytic hydrogenation, and thermochemical methods to break carbon–carbon double bonds, can improve the oxidative stability of lipid-based biofuels as well.

Hence, thermochemical processes have been demonstrated to convert soap stock to biofuels because of their high FFA and water tolerance. It was reported that the biofuel

TABLE 5.10 Fatty Acid Profiles and % Free Fatty Acids of Some Soap Stock and Fatty Acid Distillate Samples

Sample	Fatty Acid Content (% w/w)						%FFA
	C16:0	C18:0	C18:1	C18:2	C18:3	Other	
Soap stock of marine fish	19.61	5.24	20.94	2.69	0.90	5.16 (C16:1) 1.82 (C17:0) 4.75 (C20:0) 2.54 (C20:4) 1.55 (C22:0) 3.86 (C22:4) 2.44 (C22:5) 15.91 (C22:6)	N/A
Olive soap stock	11.90	2.90	81.30	0.3			N/A
Palm soap stock	47.00	4.50	42.90	0.2		1.20 (C14:0)	N/A
Palm fatty acid distillate	42.9–51.5	4.1–4.9	32.8–39.8	8.6–11.3	0.2–0.6		72.7–92.9
Peanut soap stock	13.0	3.1	51.3	22.7		1.6 (C20:0) 1.7 (C20:1) 4.3 (C22:0) 2.3 (C24:0)	N/A
Rapeseed soap stock	4.80	2.90	87.60	0.6			N/A
Soybean soap stock	12.79	4.69	24.77	50.01	7.23	0.50 (C22:0)	51.6–59.3

Sources: Lee, K. and Ofori-Boateng, C., Production of palm biofuels toward sustainable development, in *Sustainability of Biofuel Production from Oil Palm Biomass*, Springer, Singapore, 2013, pp. 107–146; Lappi, H. and Alen, R., *J. Anal. Appl. Pyrol.*, 91, 154, 2011; Hilten, R. et al., *Bioresour. Technol.*, 102, 8288, 2011; Shao, P. et al., *Biosyst. Eng.*, 102, 285, 2009; Lin, C.Y. and Lin, Y.W., *Energies*, 5, 2370, 2012; Lin, C.Y. and Li, R.J., *Fuel Process. Technol.*, 90, 130, 2009; Piloto-Rodriguez, R. et al., *Braz. J. Chem. Eng.*, 31, 287, 2014.

Abbreviation: N/A, not available.

obtained from pyrolysis of soap stock comprises linear alkanes and alkenes, similar to those found in gasoline and diesel fuels (Lappi and Alen 2011). Furthermore, the catalytic pyrolysis of peanut soap stock was investigated in a fixed-bed reactor with HZSM-5 catalyst at temperatures of 450°C–550°C (Hilten et al. 2011). The results revealed that a yield in the range of 22%–35% (v/v of feed) could be obtained at the optimal temperature of 500°C. On the other hand, transesterification with super-critical methanol was applied to produce biodiesel from soybean soap stock (Lin and Lin 2012). It was observed that the amount of unsaturated fatty acid was significantly reduced after the soap stock was treated with supercritical methanol at 350°C for 30 min, at a methanol to oil ratio of 42:1.

5.4.1.5 Other Industrial and Municipal Wastes

Biological wastewater treatment in both industry and the household represents a high potential lipid-based biomass producer. It has been reported that the municipal wastewater contains up to 40% lipid within its total organic section (Muller et al. 2014). This lipid-based biomass can be physically separated by floatation in a grease trap and biologically accumulated by microorganisms during activated sludge treatment. In a grease trap, the grease could be characterized as yellow grease containing less than 15% (w/w) FFA and brown grease containing more than 15% (w/w) FFA (Canakci and Van Gerpen 2001). The sediment in wastewater treatment tank, known as activated sludge, also comprises oleaginous (oil-rich) microorganisms. In addition to wastewater, other wastes have been reported as potential feedstocks for lipid-based biofuel production, including leather waste, fish oil waste, spent coffee grounds from instant coffee manufacturers (Al-Hamamre et al. 2012), and tall oil from paper pulping (Keskin et al. 2010; White et al. 2011). The properties of some industrial and municipal wastes are illustrated in Table 5.11.

According to Table 5.11, palmitic (C16:0), oleic (C18:1), and linoleic (C18:2) acids are commonly found in industrial wastes, whereas stearic (C18:0) and linolenic (C18:3) acids are present in trace quantities. The fatty acid profiles of industrial and municipal wastes do not influence the properties of the resulting biofuel, as mentioned earlier. In addition, the high level of FFA could pose problems for conventional biodiesel production processes.

Formerly, these industrial wastes were considered worthless and represented environmental problems; thus, their cost is significantly lower than that of the first-generation feedstocks. However, extraction of the lipid-based biomass from these industrial wastes is the greatest challenge from a research point of view. In general, the lipid-based biofuel production process can be sequentially divided into extraction, reaction, and purification steps. The completeness of the extraction step influences the subsequent reaction and purification steps. Because the moisture content in some wastes is quite high, a drying process is required during the conventional solvent extraction. In some cases, the amount of energy consumed by this drying process is larger than the energy gained from the resulting lipid-based biofuels. For example, some approaches, such as *in situ* transesterification using enzymes and reactive extraction using supercritical methanol/CO_2, have been demonstrated to shorten the overall process (Burton et al. 2010; Calixto et al. 2011). Biological conversion using

TABLE 5.11 Fatty Acid Profiles and Free Fatty Acid Contents of Some Industrial and Municipal Wastes

Sample	Fatty Acid Content (% w/w)						%FFA
	C16:0	C18:0	C18:1	C18:2	C18:3	Other	
Yellow grease	23.24	12.96	44.32	6.97	0.67	2.43 (C14:0)	<15
Brown grease	22.83	12.54	42.36	12.09	0.82	1.66 (C14:0)	>15
Activated sludge	28.00	7.50	24.90	9.50	2.00	18.00 (C16:1) 3.50 (C14:0) 7.50 (UNK)	N/A
Leather industry wastes	20.59–28.40	8.36–13.23	40.50–42.06	1.80–2.97	0.16	3.05–4.20 (C14:0) 4.60–8.10 (C16:1)	12.2
Waste fish oil	27.30–28.10	11.70–12.10	42.80–44.60	9.40–10.50	0.50–0.70	4.00–4.50 (C16:1)	4.90–10.7
Tall oil	1.00	3.00	60.00	32.00	2.00	1.00 (C20:3)	
Spent coffee grounds	43.65	6.49	8.15	32.4	1.31	3.57 (C12:0) 1.99 (C14:0) 2.39 (C20:0)	3.25–6.40

Abbreviation: N/A, not available.

insects to transform the high cellulose and lipid contents into oil-rich larvae have also been introduced (Lardé 1990). On the other hand, thermochemical processes that are less sensitive to the quality of the feedstock have been used to produce lipid-based biofuel from waste fish oil (Jayasinghe and Hawboldt 2012). In conclusion, the costs of extraction, reaction, and purification should be taken into account when designing a process suitable for an individual waste.

5.4.2 Nonedible Oil Plants

To minimize the conflict between the food and fuel industries, nonedible oil plants have become potential candidates in addition to the edible oil plants. The development of nonedible oil plants, particularly in arid areas that are inappropriate for food crops, provides an alternative feedstock for biofuel manufacturing. Moreover, the advantages of nonedible oil plants include the fact that most of them have natural pest and disease resistances because of the toxic compounds in their oils. They also have the ability to capture CO_2 and restore the degraded land upon which they are grown (Atabani et al. 2013). Many nonedible oil plants have been reported; their fatty acid profiles, productivity, %oil, and %FFA contents are summarized in Table 5.12.

It should be noted that the productivities in Table 5.12 are mostly estimated values because mass production in large-scale farming is not presently available. When farming is scaled up, lowered productivity can be expected. As shown in Table 5.5, it is clear that the productivity of nonedible oil is significantly lower than that of edible oil. The development of plantation technologies, such as cultivation and selective breeding, is required to enhance productivity; this development essentially requires long-term research. For instance, the national average productivity (tons per hectare) of Malaysian palm oil in 1960 was 1.74; it increased to 3.70 in 2005 (Basiron 2007). In the same manner as oil palm, the productivity of nonedible oil could be enhanced approximately twofold in 45 years.

Regrettably, some characteristics of nonedible oil plants, such as maturity and tree height, are their drawbacks. For example, *Jatropha curcas* fruits mature slowly, one by one, on small fruit branches, whereas oil palm ripens on 15–22 kg fruit branch. The rubber seed is difficult to harvest from its 30 m height; thus, it is generally scattered on the plantation ground. This is unlike the oil palm tree, which a few farmers can harvest using metal spades and a truck because their maximum height is approximately 6 m. Without the plantation farming system, the collection of fruits, berries, and seeds from scattered locations and their subsequent transportation pose large problems for using them as feedstocks.

From the fatty acid profile in Table 5.12, even though the *Brassicaceae* family consists of up to 40% erucic acid (C22:1), the fuel properties of the biodiesel produced from rapeseed (*Brassica napus* L.) meet the international standard range. However, it has been reported that the biodiesel produced from castor oil has a high kinematic viscosity of 15 cSt, which is threefold larger than the maximum value in the international standard, because of a high content of over 80% ricinoleic acid (C18:1:1OH). The viscosity of castor oil (251.2 cSt) is higher than that

TABLE 5.12 Fatty Acid Profile, %Oil, and % Free Fatty Acid Contents of Selected Nonedible Oil Plants, Ranked by Average Reported Productivity

Sample, Productivity in kg/ha/Year	Fatty Acid Content (% w/w)						%Oil, %FFA Contents
	C16:0	C18:0	C18:1	C18:2	C18:3	Other	
Wild mustard (*Brassica juncea*), 1750–2000	2.6–3.6	0.9–1.1	7.8–13.9	14.2–21.5	13.0–13.4	0.1 (C14:0) 0.8 (C20:0) 5.3–8.7 (C20:1) 1.5 (C22:0) 33.5–45.7 (C22:1)	37.9, 0.7–1.1
Physic nut (*Jatropha curcas*), ~1500	13.6–15.6	7.4–4.2	37.6–44.7	31.4–43.2	0.2–0.3	0.1 (C12:0) 0.1–0.2 (C14:0)	30–39, 1.9–14.9
Chinese tallow tree (*Sapium sebiferum*), 740–1850	3.71	2.13	13.78	30.71	38.87	3.21 (C10:2) 3.71 (C16:1) 0.59 (C20:1)	12–29, 2.35
Caper Spurge (*Euphorbia lathyris*), 645–900	0.50	1.98	81.46	3.71	2.78	0.50 (C16:1) 0.5 (C20:1)	43.3–50.0, 12.8
Tobacco (*Nicotiana tabacum*), 252–1050	6.6	3.1	22	66	1		35–49, 1.7–6.2
Castor oil (*Ricinus communis*), 259–754	0.8–1.5	0.8–2.0	3.6	3.5–6.8	—	82.0–95.0 (C18:1:1OH)	40.8–49.8, 0.4–3.4
Karanja tree (*Pongamia pinnata*), 320–640	11	6.8	49	19	—	2.4 (C20:1) 5.3 (C22:0)	25–50, 8.3–20.0
Neem tree (*Azadirachta indica*), 320–480	18	14	46	18	—	1 (C14:0) 3 (C20:1)	20–30, >20.0
Rubber tree (*Hevea brasiliensis*), 160–640	10	8.7	25	40	16	4.1 (C20:0)	35–40, 17.0
Fennel (*Foeniculum vulgare*), N/A	4.3	1.6	80.5	12	0.4	0.2 (C16:1) 0.3 (C20:1)	18.2, N/A
Chufa sedge (*Cyperus esculentus*), N/A	14.5	2.7	63.3	17	0.6	0.1 (C16:1) 0.6 (C20:1)	17.3, N/A

Source: Sawangkeaw, R. and Ngamprasertsith, S., *Renew. Sust. Energ. Rev.*, 25, 97, 2013.
Abbreviation: N/A, not available.

(25–40 cSt) of ordinary plant oil (Karmakar et al. 2010). Thus, the high amount of special fatty acids, such as C22:1 and C18:1:1OH, somewhat affects the fuel properties of the resulting lipid-based biofuel.

5.4.3 Insects

Lipid accumulation in various insects depends on many factors, including their order, species, and stage of metamorphosis, diet, and provenance wild or bred in captivity. The lipid content was extracted from 154 samples of insects from the five major orders (Coleoptera, Hymenoptera, Orthoptera, Diptera, and Lepidoptera), as well as others. Although the data seemed heterogeneous because of the diversity of samples, conclusions could be drawn. First, the maximum accumulated lipid was observed between the larval and pupal stages because of their metabolic reserves. In other words, the fat in adult insects was lower than that in either larvae or pupae. Next, a variation in lipid content was observed, even within the same species, between wild and bred insects. Finally, the breeding environment and diet were also reported to be strong factors impacting lipid accumulation of insects (Manzano-Agugliaro et al. 2012). According to the fatty acid profiles shown in Table 5.13, it is clear that the properties of lipid-based biofuels obtained from insect oils are not significantly different from those of biofuels derived from other sources.

The main criteria for selecting insects for biodiesel production have also been presented in a conclusive work elsewhere (Manzano-Agugliaro et al. 2012). The criteria were fat content, life cycle, space requirement, and feeding cost. The selection of a suitable insect for an individual environment is quite complex. Because the high fat content of the larvae is the common target, the life cycle of the insect should be considered. After the larvae are harvested to extract lipid, the mature insect requires a certain amount of time for breeding and egg laying to produce the next generation. In general, the feeding of larvae and pupae requires a relatively smaller space than that required for adults. An isolated feeding space is necessary for both agricultural pests and insects that are carriers of disease, particularly for flying and underground insects. With respect to feeding costs, the omnivorous decomposer is an important potential candidate because of its ability to consume various inexpensive feeds, such as cellulose (except wood), spent vegetables and food, organic leachates, and fish offal. Nevertheless, it should be considered that the insect diet influences the body fat in larvae.

An example of research on the conversion of restaurant wastes into biodiesel via larvae of the oriental fly (*Chrysomya megacephala*) is briefly described herein (Li et al. 2012). One thousand oriental fly eggs were laid on 45 kg of restaurant waste as a culture medium. The waste was obtained from the local canteen at Sun Yat-sen University, Guangzhou, China. After incubation at ambient conditions (25°C–30°C and 65%–70% RH) for 5 days, all developmental stages of the oriental fly could be found in the container. All samples, including eggs, larvae, pupae, and adults that had died, were dehydrated at 60°C for 24 h before being put in a Soxhlet extractor. The eggs, larvae, pupae, and adults were found to consist of 12.6%, 24.4%–26.3%, 23.4%,

TABLE 5.13　Fatty Acid Profiles and Lipid Contents of Selected Insect Oils

Sample	Fatty Acid Content (% w/w)						% Lipid Content
	C16:0	C18:0	C18:1	C18:2	C18:3	Other	
Watermelon bug (*Aspongopus viduatus*)	30.87–30.93	3.49–3.51	46.57–46.63	3.89–3.91	N/A	30.87–30.93	45.00
Sorghum bug (*Agonoscelis pubescens*)	12.18–12.22	7.28–7.32	40.88–40.92	34.48–34.52	N/A	12.18–12.22	60.00
Black soldier fly larva (*Hermetia illucens*)	18.20–14.80	3.60–5.10	23.60–27.10	5.80–7.50	N/A	18.20–14.80	15.00–23.00
Oriental latrine fly larva (*Chrysomya megacephala*)	35.48	2.77	24.38	15.26	1.25	35.48	24.40–26.29
Darkling beetle larva (*Zophobas morio*)	32.74	9.36	29.43	22.53	0.85	32.74	33.80

Source: Sawangkeaw, R. and Ngamprasertsith, S., *Renew. Sust. Energ. Rev.*, 25, 97, 2013.

Abbreviation: N/A, not available.

and 6.3% lipid on a dry basis, respectively. From this experiment, it was estimated that 73.5 g of lipid could be harvested from a kilogram of fresh larvae. The extracted lipid had an AV of 1.1 mg KOH/g oil and was suitable for use in the conventional production of biodiesel. Finally, the resultant biodiesel met the criteria for 11 major properties among the 24 specified by international standard EN14214.

5.4.4 Oleaginous Microorganisms

The advantages of oleaginous microorganisms over other sources are high productivity and simplicity of genetic engineering. It has been estimated that the productivity of microorganisms containing 20%–70% lipid in a 1000 m^3 fermenter could be higher than that of 0.5 ha of oil palm. In terms of genetic engineering, it is more easily performed on yeast and bacteria than on either insects or plants. Genetically engineered microbes also can be quarantined inside a fermenter. The estimated oil productivity of selected oleaginous microorganisms is revealed in Table 5.14.

Since cultivation of microbes and plants is different, the productivity of microbes was estimated from the cell density/concentration reported in research articles, based on 300 days of operation/year (Sawangkeaw and Ngamprasertsith 2013). Thus, it could be revealed that the land usage for microbial cultivation is more effective than that for plant farming, except for the cultivation of microalgae in open ponds.

5.4.4.1 Microalgae

Cultivation of microalgae can be classified into autotrophic, heterotrophic, and mixotrophic cultures based on light exposure and nutrient feeding. First of all, autotrophic or photoautotrophic culture is microalgal growth in light without additional nutrients or with small amounts of additional nutrients. There are two ways to grow autotrophic microalgae: in open ponds and closed photobioreactors. Raceway ponds that are 10–35 cm in depth and of various lengths have been employed, using paddle wheels to mix the water. On the other hand, autotrophic microalgae have been fed in photobioreactors that can be categorized into three main types: (1) flat panel, (2) horizontal tubular, and (3) vertical column. The light source for a photobioreactor can be sunlight, fluorescent lamps, or LEDs. The open raceways are usually found in the form of closed-loop ponds, whereas photobioreactors can be found with or without circulating loops (Reijnders 2013). Further details on the design of photobioreactors for microalgal cultivation are described elsewhere (Eriksen 2008). Second, heterotrophic culture is microalgal growth in darkness with the feeding of nutrients, performed in an ordinary stirred-tank bioreactor. The assimilation nutrients are usually glucose, glycerol, and carboxylate salts as carbon sources, whereas ammonia, nitrite, nitrate, and urea are added as nitrogen sources (Perez-Garcia et al. 2011). Finally, a mixotrophic culture is microalgal growth in either light or dark with added nutrients, conducted in the photobioreactor that has a controllable light source.

The advantages and limitations among microalgal cultivation systems have been debated. The open pond is the simplest method with low operating costs, whereas it has the lowest relative cell concentration. Local weather dependence and the

TABLE 5.14   Lipid Content and Productivity in Terms of Biomass and Lipid from Selected Microbes

Microalgae or Yeast or Fungi of Bacteria Strain	Lipid Content (% w/w)	Productivity (kg/m³ Year)	
		Biomass	Lipid
Microalgae			
Crypthecodinium cohnii [HT]	19.9	672	133.7
Nannochloropsis oculata [AT]	22.8–23	870	199.2
Tetraselmis tetrathele [AT]	29.2–30.3	1125	334.7
Chaetoceros gracilis [AT]	15.5–60.3	1065	403.6
Schizochytrium limacinum [HT]	50.3[a]	1044	525.1
Yeasts			
Candida curvata	29.2–58.0	690.6	315.0
Cryptococcus albidus	33.0–43.8	251.7	146.0
Cryptococcus curvatus	25.0–45.8	1989.7	1154
Lipomyces starkeyi	61.5–68.0	635.7	410.0
Rhodosporidium toruloides	58.0–8.1	3362.4	2120
Fungi			
Mortierella isabellina	53.2	N/A	N/A
Mucor mucedo	62.0	N/A	N/A
Aspergillus oryzae	18.0–57.0	377.2	215.0
Cunninghamella echinulata	35.0–57.7	232.2	134.0
Mortierella isabellina	50.0–5.0	1275.8	678.8
Bacteria			
Arthrobacter sp.	>40	N/A	N/A
Acinetobacter calcoaceticus	27–38	N/A	N/A
Rhodococcus opacus	24–25	N/A	N/A
Bacillus alcalophilus	18–24	N/A	N/A

Source: Sawangkeaw, R. and Ngamprasertsith, S., *Renew. Sust. Energ. Rev.*, 25, 97, 2013.

Abbreviations: AT and HT in brackets behind microalgal strains are autotrophic and heterotrophic cultivations, respectively. N/A, not available.

[a] As total fatty acid content.

potential for contamination by other species are other disadvantages of the open raceway. Bioreactors have higher cell concentrations, as well as lower water evaporation rates and land usage than open ponds, but the number of microalgae that can be grown in photobioreactors and fermenters is restricted to a few species. The energy and nutrient expenses of bioreactors are somewhat higher than those of open ponds. Growing microalgae in the dark eliminates the dependence of metabolism on a light source and increases the growth rate. However, cell growth is inhibited by excess organic substances. On the other hand, the scale-up of heterotrophic cultivation is much simpler than the scale-up of other methods because the effect of light exposure area is negligible.

Besides the cultivation of microalgae, harvesting lipids from microalgae is an interesting topic of research. Harvesting costs contribute around 20%–30% to the total cost of dried microalgal production. The harvesting of microalgae consists of thickening and dewatering steps that concentrate the cells into 2%–7% and then 15%–25% of total suspended solids, excluding the drying process. Chemical coagulation/flocculation is the most intensive method for thickening microalgae, but it limits the recycling of culture medium and cell rupture. Polymers, either ionic or nonionic, iron, and aluminum salts were tested as excellent microalgae coagulants/flocculants. Exposure of autotrophic microalgae to sunlight, limiting the CO_2 concentration, and adjusting the pH in the range of 8.6–10.5 are general procedures for harvesting microalgae called autocoagulation or biocoagulation. This phenomenon occurs when excess O_2 is produced during photosynthesis. However, the reliability of flocculation control and cell composition during autocoagulation is the main obstacles for applying these methods on an industrial scale. Flotation and electrical-based techniques are applicable. Nonetheless, those techniques are still in the development stage with respect to harvesting microalgae. For dewatering processes, concentrated microalgae can be mechanically filtered or centrifuged to obtain microalgal slurry. For cells with diameters over 70 μm, dead-end filtration techniques are applicable, but cross-flow filtration techniques are more appropriate for smaller cells. Microfiltration is cost effective when applied to volumes less than 2 m^3/day, whereas centrifugation is economic for volumes larger than 20 m^3/day (Barros et al. 2015).

Despite their status as single-cell organisms, the fatty acid profiles of microalgae are more diverse than those of plant or insect oils, as illustrated in Table 5.15. The various PUFAs found in microalgal lipids influence the properties of lipid-based biofuels, particularly biodiesel, in terms of viscosity, density, cold flow properties, and oxidative stability. Therefore, posttreatment may be necessary to upgrade the lipid-based biofuels obtained from microalgae.

Recently, the energetic performance of lipid-based liquid biofuels produced from autotrophic microalgae has been definitively reviewed (Reijnders 2013). Since the 1950s, microalgal cultivation, for example, *Spirulina* and *Chlorella*, has been commercialized in open ponds to sell the product as a dietary supplement. Substantial research on biofuel production from microalgae was launched because of the oil crisis of the 1970s; however, biofuel production from microalgae was not commercialized. The energy return on investment (EROI) was used to indicate the energetic performance. It has been reported that the EROIs of current transport fuels (gasoline and diesel) are more than 5. Despite the high productivity of microalgae, the EROIs of autotrophic microalgae were reported to be around 2–3, which is lower than that of ethanol and lipid-based biofuels produced from starch and edible oils. Additionally, the microalgal lipid price was estimated to be approximately $3.05 ± 0.31 L^{-1} in 2011, which is higher than the prices of common plant oils (Sun et al. 2011). Therefore, the energetic performance of microalgal lipids will need to improve substantially to serve as an alternative energy source.

TABLE 5.15 Fatty Acid Profiles and Lipid Contents of Selected Microalgal Strains

Strain	Fatty Acid Content (% w/w)						% Lipid Content
	C16:0	C18:0	C18:1	C18:2	C18:3	Other	
Microalgae							
Chlorella sp.	6.7–19.1	1.1–4.4	8.5–9.1	2.4–14.4	15.5–18.8	5.24 (C15:0) 10.90 (C16:1) 0.3–13.8 (C16:2)	22.4–33.9
Scenedesmus obliquus	21.8	0.45	17.9	21.7	3.76	1.48 (C14:0) 5.95 (C16:1) 3.96 (C16:2) 0.68 (C16:3) 0.43 (C16:4)	12.6–58.3
Chaetoceros muelleri	22.6	2.09	35.7	18.5	7.75	1.97 (C16:1) 7.38 (C16:2) 1.94 (C16:3)	11.7–25.3
Crypthecodinium cohnii	22.9	2.6	7.9	—	—	2.9 (C12:0) 13.4 (C14:0) 49.5 (C22:6)	19.9
Chaetoceros gracilis	5.0–40.0	0.2–25.0	0.1–4.2	0.2–5.0	0.1–4.8	6.0–20.5 (C12:0) 18.5–40 (C14:0) 0.1–8.0 (C16:2)	15.5–60.3
Schizochytrium mangrovei	47.9–52.9	1.0–1.1	0.1	—	0.14	7.7–8.4 (C22:5) 28.7–30.7 (C22:6)	68[a]
Schizochytrium limacinum	53.6–60.5	1.34–3.9	—	—	—	3.3–4.0 (C14:0) 4.5–6.5 (C22:5) 28.6–35.1 (C22:6)	50.3[a]

[a] As total fatty acid content.

5.4.4.2 Yeasts and Fungi

The term "oleaginous" yeasts and fungi refer to microorganisms that have the ability to accumulate lipid to at least 20% of dry cell weight. Over 100 years of investigation, it has shed light on oleaginous yeasts as high potential candidates for lipid-based biofuel feedstocks. The first interest in oleaginous yeasts and molds came from the production of PFUAs that are rarely found in plants or animals. More than 70 yeast species known to be oleaginous have been intensively reviewed. The major strains utilized as oleaginous yeasts in recent publications are *Rhodosporidium toruloides*, *Rhodotorula glutinis*, *Lipomyces lipofer*, *Lipomyces starkeyi*, *Cryptococcus curvatus*, *Trichosporon pullulans*, and *Yarrowia lipolytica*. The deep information on yeast strain development, process development, and cultivation conditions are described elsewhere (Donot et al. 2014; Sitepu et al. 2014).

The cultivation modes generally used to grow yeasts and fungi are batch, repeated batch, fed batch, and continuous modes. Among those modes, the fed batch mode was reported to be the most effective in terms of cell density and lipid content. The culture temperatures and pH are typically 25°C–30°C and 5.0–6.0, respectively, with culture duration of 24–270 h. The fatty acid profiles of some yeast and fungal lipids are shown in Table 5.16. Unlike those from microalgae, the lipid–biomasses derived from yeast and fungi are simple and similar to plant oils.

The advantages of yeast and fungal cultivations are the same as those of the heterotrophic cultivation of microalgae. Since the carbon source for yeast and fungal

TABLE 5.16 Fatty Acid Profiles of Selected Yeast, Fungal, and Bacterial Lipids

Yeast, Fungal, and Bacterial Strain	Fatty Acid Composition % (w/w)					
	C16:0	C16:1	C18:0	C18:1	C18:2	C18:3
Yeast						
Candida curvata	25.0–28.0	—	5.0–44.0	9.0–33.0	15.0–27.0	—
Cryptococcus albidus	16.1–7.9	—	5.1–0.5	2.9–17.7	19.6–61.1	0.5–59.1
Cryptococcus curvatus	30.0–33.0	—	11.0–13.0	43.0–45.0	8.0–10.0	0.1–0.5
Lipomyces lipofer	37.0	4.0	7.0	48.0	3.0	—
Lipomyces starkeyi	36.2–37.1	—	4.5–5.5	45.1–46.3	3.4–4.9	—
Rhodosporidium toruloides	13.0–16.0	—	4.0–41.0	18.0–42.0	15.0–29.0	—
Rhodotorula glutinis	18.0	1.0	6.0	60.0	12.0	2.0
Trichosporon pullulans	15.0	—	2.0	57.0	24.0	1.0
Yarrowia lipolytica	11.0	6.0	1.0	28.0	51.0	1.0
Fungi						
Cunninghamella echinulata	17.7–30.1	—	5.8–11.3	52.1–56.3	4.2–12.4	—
Aspergillus oryzae	32.9	—	9.96	22.6	27.7	—
Mortierella isabellina	19.0–22.2	3.0	2.0–4.0	47.0–53.2	17.0–19.1	3.1–5.0
Bacteria						
Rhodococcus opacus	16.8–25.7	—	3.5–18.8	6.4–73.8	N/A	N/A

Sources: Sawangkeaw, R. and Ngamprasertsith, S., *Renew. Sust. Energ. Rev.*, 25, 97, 2013; Ageitos, J.M. et al., *Appl. Microbiol. Biotechnol.*, 90, 1219, 2011; Li, Q. et al., *Appl. Microbiol. Biotechnol.*, 80, 749, 2008.
Abbreviation: N/A, not available.

cultivation are usually derived from starch and sugar, the lipid-based biomasses obtained from yeasts and fungi are somewhat involved in the food versus fuel dilemma. Alternative carbon sources such as glycerol (biodiesel by-product) and lignocellulosic hydrolysates have been investigated as carbon sources for the cultivation of yeast and fungi. Unlike yeasts, which lack cellulolytic activity, some fungal species simultaneously accumulate lipids and produce cellulase (Peng and Chen 2007). Because of the presence of cellulase, these fungi are able to digest amorphous cellulose then metabolize it to lipid-based biomass in the fermenter. Therefore, the utilization of yeasts and fungi could be a link between biofuel production from lipid-based and sugar-based biomasses.

Because of their thick cell wall, yeasts can resist the solvents applied to extract intracellular lipids. Thus, the extraction of lipid-based biomass from oleaginous yeasts begins with cell breakage using heat, chemicals, and enzymes. After the dewatering and drying processes, extraction by organic solvents such as petroleum ether, chloroform, and methanol can be performed. Because of the high energy consumption of the drying process, wet extraction using ethanol–hexane and methanol–benzene has been demonstrated on the pilot scale (Ageitos et al. 2011). On the other hand, the reactive extraction of *Y. lipolytica* to obtain lipid-based biofuel was introduced to merge the reaction and extraction together (Tsigie et al. 2013).

5.4.4.3 Bacteria
Bacteria are drawing increasing interest as potential lipid-based biomasses producers; however, only a few bacteria can produce oils that can be used as feedstocks for lipid-based biofuels. The strong points of bacteria are very fast growth rate and simplicity. Because the lipids derived from bacteria are more complex than those from other microbes, they cannot be extracted and converted to biofuel using traditional methods. The lipid contents and fatty acid profiles of some bacteria are illustrated in Tables 5.14 and 5.16, respectively. The fatty acid profiles of bacteria depend strongly on carbon source, as they do for yeast and fungi. Because of the complexity of bacterial lipids, thermochemical methods such as pyrolysis and supercritical fluids are more suitable for the production of lipid-based biofuels than the transesterification process.

5.5 Conclusions

To utilize lipid-based biomasses as feedstock for biofuel production, the topics of practical importance are summarized in Table 5.17.

Lipid-based biofuels, mainly biodiesel, are produced worldwide from edible plant oils. Alternative feedstocks have been discovered to manufacture inexpensive biofuels and to avoid the food versus fuel dilemma in the future. Nonetheless, the prices of alternative feedstocks are presently higher than those of the edible plant oils, except lipid-based biomasses derived from wastes and by-products. However, the downstream processes such as collecting, handling, and pretreatment could increase the price of those wastes and by-products. The quality of the alternative feedstocks can be

TABLE 5.17 Productivity, Quality, Price, and Source of the Lipid-Based Biomasses

Lipid-Based Biomass	Productivity	Quality—Stability	Price	Source—Weather Dependent
Edible plant oil	Medium—high	High—stable	Standard	Specific area—strong
Nonedible plant oil	Low—medium	Low—stable	Higher	Specific area—medium
Waste cooking oil	Uncertain	Low—unstable	Lower	Specific area—low
Animal fat	Uncertain	Low—unstable	Lower	Specific area—low
Industrial waste and by-products	Uncertain	Low—unstable	Lower	Specific area—low
Insect oil	Low—medium	Medium—stable	Higher	All area—low
Phototrophic microalgal oil	Low—medium	Low—stable	Higher	Specific area—medium
Heterotrophic microalgal oil	High	Low—stable	Higher	All area—low
Yeasty and fungal oil	High	Low—stable	Higher	All area—low
Bacterial oil	Medium—high	Low—stable	Higher	All area—low

monitored by using the AOCS standards, referral code ACOS Ck1-07. The 23 AOCS standard testing methods comprise the testing of acidity, purity, water content, and other essential properties of feedstocks for traditional biodiesel production. Although considerable research has been presented in this chapter, it cannot be clearly concluded what is the best feedstock for each community. Because the economic compatibility of lipid-based biofuels is strongly dictated by their price, the final price of alternative feedstocks needs to be minimized.

References

Ageitos, J.M., Vallejo, J.A., Veiga-Crespo, P., and T.G. Villa. 2011. Oily yeasts as oleaginous cell factories. *Appl Microbiol Biotechnol* 90:1219–1227.

Aguerrebere, I.A., Molina, A.R., Oomah, B.D., and J.C.G. Drover. 2011. Characteristics of Prunus serotina seed oil. *Food Chem* 124:983–990.

Ahmad, A.L., Yasin, N.H.M., Derek, C.J.C., and J.K. Lim. 2011. Microalgae as a sustainable energy source for biodiesel production: A review. *Renew Sust Energ Rev* 15:584–593.

Al-Hamamre, Z., Foerster, S., Hartmann, F., Kroger, M., and M. Kaltschmitt. 2012. Oil extracted from spent coffee grounds as a renewable source for fatty acid methyl ester manufacturing. *Fuel* 96:70–76.

Andrikopoulos, N.K. 2002. Triglyceride species compositions of common edible vegetable oils and methods used for their identification and quantification. *Food Rev Int* 18:71–102.

Atabani, A.E., Silitonga, A.S., Ong, H.C. et al. 2013. Non-edible vegetable oils: A critical evaluation of oil extraction, fatty acid compositions, biodiesel production, characteristics, engine performance and emissions production. *Renew Sust Energ Rev* 18:211–245.

Banerjee, A. and R. Chakraborty. 2009. Parametric sensitivity in transesterification of waste cooking oil for biodiesel production—A review. *Resour Conserv Recycl* 53:490–497.

Barros, A.I., Goncalves, A.L., Simoes, M., and J.C.M. Pires. 2015. Harvesting techniques applied to microalgae: A review. *Renew Sust Energ Rev* 41:1489–1500.

Basiron, Y. 2007. Palm oil production through sustainable plantations. *Eur J Lipid Sci Tech* 109:289–295.

Bunyakiat, K., Makmee, S., Sawangkeaw, R., and S. Ngamprasertsith. 2006. Continuous production of biodiesel via transesterification from vegetable oils in supercritical methanol. *Energ Fuel* 20:812–817.

Burton, R., Fan, X.H., and G. Austic. 2010. Evaluation of two-step reaction and enzyme catalysis approaches for biodiesel production from spent coffee grounds. *Int J Green Energ* 7:530–536.

Calixto, F., Fernandes, J., Couto, R., Hernandez, E.J., Najdanovic-Visaka, V., and P.C. Simoes. 2011. Synthesis of fatty acid methyl esters via direct transesterification with methanol/carbon dioxide mixtures from spent coffee grounds feedstock. *Green Chem* 13:1196–1202.

Canakci, M. 2007. The potential of restaurant waste lipids as biodiesel feedstocks. *Bioresour Technol* 98:183–190.

Canakci, M. and H. Sanli. 2008. Biodiesel production from various feedstocks and their effects on the fuel properties. *J Ind Microbiol Biotechnol* 35:431–441.

Canakci, M. and J. Van Gerpen. 2001. Biodiesel production from oils and fats with high free fatty acids. *Trans ASAE* 44:1429–1436.

Catchpole, O.J., Tallon, S.J., Eltringham, W.E. et al. 2009. The extraction and fractionation of specialty lipids using near critical fluids. *J Supercrit Fluids* 47:591–597.

Chakraborty, R., Gupta, A.K., and R. Chowdhury. 2014. Conversion of slaughterhouse and poultry farm animal fats and wastes to biodiesel: Parametric sensitivity and fuel quality assessment. *Renew Sust Energ Rev* 29:120–134.

Chhetri, A.B., Watts, K.C., and M.R. Islam. 2008. Waste cooking oil as an alternate feedstock for biodiesel production. *Energies* 1:3–18.

Christie, W.W., Dijkstra, A.J., and G. Knothe. 2007. Analysis. In *The Lipid Handbook with CD-ROM* (F.D. Gunstone, J.L. Harwood, and A.J. Dijkstra, eds.), 3rd edn., pp. 415–469. Boca Raton, FL: CRC Press.

da Silva, A.C. and N. Jorge. 2014. Bioactive compounds of the lipid fractions of agro-industrial waste. *Food Res Int* 66:493–500.

Dias, J.M., Alvim-Ferraz, M.C.M., and M.F. Almeida. 2008. Mixtures of vegetable oils and animal fat for biodiesel production: Influence on product composition and quality. *Energ Fuel* 22:3889–3893.

Donot, F., Fontana, A., Baccou, J.C., Strub, C., and S. Schorr-Galindo. 2014. Single cell oils (SCOs) from oleaginous yeasts and moulds: Production and genetics. *Biomass Bioenerg* 68:135–150.

Dwivedi, G. and M.P. Sharma. 2014. Impact of cold flow properties of biodiesel on engine performance. *Renew Sust Energ Rev* 31:650–656.

Eriksen, N.T. 2008. The technology of microalgal culturing. *Biotechnol Lett* 30:1525–1536.

Espinosa, S., Fornari, T., Bottini, S.B., and E.A. Brignole. 2002. Phase equilibria in mixtures of fatty oils and derivatives with near critical fluids using the GC-EOS model. *J Supercrit Fluids* 23:91–102.

FitzPatrick, M., Champagne, P., Cunningham, M.F., and R.A. Whitney. 2010. A biorefinery processing perspective: Treatment of lignocellulosic materials for the production of value-added products. *Bioresour Technol* 101:8915–8922.

Giannelos, P.N., Sxizas, S., Lois, E., Zannikos, F., and G. Anastopoulos. 2005. Physical, chemical and fuel related properties of tomato seed oil for evaluating its direct use in diesel engines. *Ind Crop Prod* 22:193–199.

Gunstone, F. 2011. *Vegetable Oils in Food Technology: Composition, Properties and Uses.* Pondicherry, India: John Wiley & Sons.

Harwood, J.L. and C.M. Scrimgeour. 2007. Fatty acid and lipid structure. In *The Lipid Handbook with CD-ROM* (F.D. Gunstone, J.L. Harwood, and A.J. Dijkstra, eds.), pp. 1–36. Boca Raton, FL: CRC Press.

Hilten, R., Speir, R., Kastner, J., and K.C. Das. 2011. Production of aromatic green gasoline additives via catalytic pyrolysis of acidulated peanut oil soap stock. *Bioresour Technol* 102:8288–8294.

Hoekman, S.K., Broch, A., Robbins, C., Ceniceros, E., and M. Natarajan. 2012. Review of biodiesel composition, properties, and specifications. *Renew Sust Energ Rev* 16:143–169.

Ito, T., Sakurai, Y., Kakuta, Y., Sugano, M., and K. Hirano. 2012. Biodiesel production from waste animal fats using pyrolysis method. *Fuel Process Technol* 94:47–52.

Jayasinghe, P. and K. Hawboldt. 2012. A review of biooils from waste biomass: Focus on fish processing waste. *Renew Sust Energ Rev* 16:798–821.

Karmakar, A., Karmakar, S., and S. Mukherjee. 2010. Properties of various plants and animals feedstocks for biodiesel production. *Bioresour Technol* 101:7201–7210.

Kaur, D., Sogi, D.S., Garg, S.K., and A.S. Bawa. 2005. Flotation-cum-sedimentation system for skin and seed separation from tomato pomace. *J Food Eng* 71:341–344.

Keskin, A., Yasar, A., Guru, M., and D. Altiparmak. 2010. Usage of methyl ester of tall oil fatty acids and resinic acids as alternative diesel fuel. *Energ Convers Manage* 51:2863–2868.

Khoddami, A. and T.H. Roberts. 2015. Pomegranate oil as a valuable pharmaceutical and nutraceutical. *Lipid Technol* 27:40–42.

Knothe, G., Matheaus, A.C., and T.W. Ryan. 2003. Cetane numbers of branched and straight-chain fatty esters determined in an ignition quality tester. *Fuel* 82:971–975.

Kulkarni, M.G. and A.K. Dalai. 2006. Waste cooking oil-an economical source for biodiesel: A review. *Indus Eng Chem Res* 45:2901–2913.

Lam, M.K., Lee, K.T., and A.R. Mohamed. 2010. Homogeneous, heterogeneous and enzymatic catalysis for transesterification of high free fatty acid oil (waste cooking oil) to biodiesel: A review. *Biotechnol Adv* 28:500–518.

Lappi, H. and R. Alen. 2011. Pyrolysis of vegetable oil soaps-Palm, olive, rapeseed and castor oils. *J Anal Appl Pyrol* 91:154–158.

Lardé, G. 1990. Recycling of coffee pulp by *Hermetia illucens* (Diptera: Stratiomyidae) larvae. *Biol Wastes* 33:307–310.

Lee, J.S. and S. Saka. 2010. Biodiesel production by heterogeneous catalysts and supercritical technologies. *Bioresour Technol* 101:7191–7200.

Lee, K. and C. Ofori-Boateng. 2013. Production of palm biofuels toward sustainable development. In *Sustainability of Biofuel Production from Oil Palm Biomass*, pp. 107–146. Singapore: Springer.

Li, Q., Du, W., and D. Liu. 2008. Perspectives of microbial oils for biodiesel production. *Appl Microbiol Biotechnol* 80:749–756.

Li, Z.X., Yang, D.P., and M.L. Huang. 2012. Chrysomya megacephala (Fabricius) larvae: A new biodiesel resource. *Appl Energ* 94:349–354.

Lin, C.Y. and R.J. Li. 2009. Fuel properties of biodiesel produced from the crude fish oil from the soapstock of marine fish. *Fuel Process Technol* 90:130–136.

Lin, C.Y. and Y.W. Lin. 2012. Fuel characteristics of biodiesel produced from a high-acid oil from soybean soapstock by supercritical-methanol transesterification. *Energies* 5:2370–2380.

Manzano-Agugliaro, F., Sanchez-Muros, M.J., Barroso, F.G., Martinez-Sanchez, A., Rojo, S., and C. Perez-Banon. 2012. Insects for biodiesel production. *Renew Sust Energ Rev* 16:3744–3753.

Marulanda, V.F., Anitescu, G., and L.L. Tavlarides. 2010. Investigations on supercritical transesterification of chicken fat for biodiesel production from low-cost lipid feedstocks. *J Supercrit Fluid* 54:53–60.

Mata, T.M., Cardoso, N., Ornelas, M., Neves, S., and N.S. Caetano. 2011. Evaluation of two purification methods of biodiesel from beef tallow, pork lard, and chicken fat. *Energ Fuel* 25:4756–4762.

Mazubert, A., Poux, M., and J. Aubin. 2013. Intensified processes for FAME production from waste cooking oil: A technological review. *Chem Eng J* 233:201–223.

Meng, X., Yang, J.M., Xu, X., Zhang, L., Nie, Q.J., and M. Xian. 2009. Biodiesel production from oleaginous microorganisms. *Renew Energ* 34:1–5.

Muller, E.E., Sheik, A.R., and P. Wilmes. 2014. Lipid-based biofuel production from wastewater. *Curr Opin Biotechnol* 30:9–16.

O'Brien, R.D. 2008. *Fats and Oils: Formulating and Processing for Applications*, 3rd edn. Boca Raton, FL: CRC Press.

Ofori-Boateng, C. and K.T. Lee. 2013. Sustainable utilization of oil palm wastes for bioactive phytochemicals for the benefit of the oil palm and nutraceutical industries. *Phytochem Rev* 12:173–190.

Özbay, N., Oktar, N., and N.A. Tapan. 2008. Esterification of free fatty acids in waste cooking oils (WCO): Role of ion-exchange resins. *Fuel* 87:1789–1798.

Peng, X.W. and H.Z. Chen. 2007. Microbial oil accumulation and cellulase secretion of the endophytic fungi from oleaginous plants. *Ann Microbiol* 57:239–242.

Perez-Garcia, O., Escalante, F.M., de-Bashan, L.E., and Y. Bashan. 2011. Heterotrophic cultures of microalgae: Metabolism and potential products. *Water Res* 45:11–36.

Piloto-Rodríguez, R., Melo, E.A., Goyos-Pérez, L., and S. Verhelst. 2014. Conversion of byproducts from the vegetable oil industry into biodiesel and its use in internal combustion engines: A review. *Braz J Chem Eng* 31:287–301.

Rashid, U., Ibrahim, M., Yasin, S., Yunus, R., Taufiq-Yap, Y.H., and G. Knothe. 2013. Biodiesel from *Citrus reticulata* (mandarin orange) seed oil, a potential non-food feedstock. *Ind Crop Prod* 45:355–359.

Reijnders, L. 2013. Lipid-based liquid biofuels from autotrophic microalgae: Energetic and environmental performance. *Wires Energ Environ* 2:73–85.

Savikin, K.P., Ethordevic, B.S., Ristic, M.S., Krivokuca-Ethokic, D., Pljevljakusic, D.S., and T. Vulic. 2013. Variation in the fatty-acid content in seeds of various black, red, and white currant varieties. *Chem Biodivers* 10:157–165.

Sawangkeaw, R. and S. Ngamprasertsith. 2013. A review of lipid-based biomasses as feedstocks for biofuels production. *Renew Sust Energ Rev* 25:97–108.

Shao, P., He, J.Z, Sun, P.L., and S.T. Jiang. 2009. Process optimisation for the production of biodiesel from rapeseed soapstock by a novel method of short path distillation. *Biosyst Eng* 102:285–290.

Shin, H.Y., Lee, S.H., Ryu, J.H., and S.Y. Bae. 2012. Biodiesel production from waste lard using supercritical methanol. *J Supercrit Fluid* 61:134–138.

Sitepu, I.R., Garay, L.A., Sestric, R. et al. 2014. Oleaginous yeasts for biodiesel: Current and future trends in biology and production. *Biotechnol Adv* 32:1336–1360.

Subramaniam, R., Dufreche, S., Zappi, M., and R. Bajpai. 2010. Microbial lipids from renewable resources: Production and characterization. *J Ind Microbiol Biotechnol* 37:1271–1287.

Sun, A., Davis, R., Starbuck, M., Ben-Amotz, A., Pate, R., and P.T. Pienkos. 2011. Comparative cost analysis of algal oil production for biofuels. *Energy* 36:5169–5179.

Tickell, J. and K. Tickell. 2003. *From the Fryer to the Fuel Tank: The Complete Guide to Using Vegetable Oil as an Alternative Fuel.* Hollywood, CA: Tickell Energy.

Tsigie, Y.A., Huynh, L.H., Nguyen, P.L.T., and Y.H. Ju. 2013. Catalyst-free biodiesel preparation from wet *Yarrowia lipolytica* Po1g biomass under subcritical condition. *Fuel Process Technol* 115:50–56.

White, K., Lorenz, N., Potts, T. et al. 2011. Production of biodiesel fuel from tall oil fatty acids via high temperature methanol reaction. *Fuel* 90:3193–3199.

Wisniewski Jr., A., Wiggers, V.R., Simionatto, E.L., Meier, H.F., Barros, A.A.C., and L.A.S. Madureira. 2010. Biofuels from waste fish oil pyrolysis: Chemical composition. *Fuel* 89:563–568.

Zhang, H.M., Ozturk, U.A., Wang, Q.W., and Z.Y. Zhao. 2014. Biodiesel produced by waste cooking oil: Review of recycling modes in China, the US and Japan. *Renew Sust Energ Rev* 38:677–685.

6

Solid Acid–Mediated Hydrolysis of Biomass for Producing Biofuels

Rajeev Kumar Sukumaran, Anil K. Mathew, and Meena Sankar

Contents

Abstract

Hydrolysis of biomass is one of the most challenging steps in its conversion to sugars and hence to biofuels. Conventional catalysts are expensive and difficult to be recycled. Solid acid catalysts (SACs) have been proven recently as effective in lignocellulose hydrolysis with the advantage of easy recovery and recycling and are environmentally benign compared to the homogeneous mineral acids. The review introduces SACs in the context of biomass hydrolysis and discusses the various aspects of SAC-mediated hydrolysis of cellulosic materials, including the catalysts themselves, the conditions of hydrolysis, the recent advances in the use of SACs for biomass hydrolysis, the recovery and reuse of SACs, and the limitations in the use of SACs.

6.1 Introduction

Plant biomass inarguably is the prominent renewable resource for future energy requirements and the only sustainable source of organic carbon on earth (Huang and

Fu 2013). Lignocellulose, which makes up most of the plant biomass, has considerable amount of carbohydrate polymers—cellulose (35%–55%) and hemicellulose (20%–40%) in addition to lignin (10%–25%) (Ghosh and Ghose 2003, Wyman et al. 2005). Cellulose is a linear polymer of β-1,4-linked glucose units, whereas hemicellulose is a branched-chain heteropolysaccharide containing several sugars like xylose, arabinose, mannose, glucose, galactose, and rhamnose with variable structure depending on the source. Hemicelluloses include xyloglucans, xylans, mannans and glucomannans, and β-(1 → 3, 1 → 4)-glucans with xylose often being the sugar moiety present in largest quantities (Scheller and Ulvskov 2010). Lignin, on the other hand, is a heterogeneous nonsugar polymer comprising phenylpropanoid units—cross-linked with a variety of different bonds and without any clearly defined repeating structural units (Ralph et al. 2004). The main building blocks of lignin are the hydroxyl cinnamyl alcohols—coniferyl and sinapyl alcohol with minor amounts of p-coumaryl alcohol. Both cellulose and hemicellulose can be hydrolyzed into their component sugars and may be fermented to bioethanol/biobutanol or may be converted either chemically or fermentatively to several chemical building blocks (e.g., succinic acid, levulinic acid) or chemicals. Lignocellulose-derived bioethanol is considered as one of the most promising future alternative renewable fuels, and worldwide there are an increasing number of efforts to commercialize lignocellulosic ethanol with a limited number of success stories. Hydrolysis of the sugar polymers that make up the lignocellulosic biomass is essential for the generation of sugars and hence for bioethanol production. Since *Saccharomyces cerevisiae*, the yeast strain commonly employed for alcohol fermentation, can utilize only hexose sugars (e.g., glucose), the hydrolysis of cellulose that yields glucose has received much attention compared to the hydrolysis of hemicellulose. Moreover, the hydrolysis of hemicellulose with homogeneous catalysts (e.g., mineral acids) is relatively easy.

Hydrolysis of lignocellulosic biomass is typically accomplished through cellulose hydrolyzing enzymes (cellulases), which is considered as the most efficient method for depolymerization of the sugar polymer (Zhang and Lynd 2004). Nevertheless, enzymatic hydrolysis does have several limitations, the most important ones being the cost of cellulases, the inability of the enzymes to efficiently attack the cellulose in raw biomass, the low reaction rates, and the long duration required to achieve near-complete hydrolysis, in addition to product inhibition of the enzymes (Lynd et al. 2008, Binder and Raines 2010, Arantes and Saddler 2011). One of the major barriers in enzymatic hydrolysis of lignocellulosic biomass is the accessibility of enzymes to cellulose, and this makes the pretreatment of biomass an absolute necessity for efficient hydrolysis (Arantes and Saddler 2010). Acid-catalyzed hydrolysis of biomass is the major alternative to enzymatic hydrolysis, and this method dates back to the early nineteenth century (Huang and Fu 2013). Sulfuric acid–catalyzed hydrolysis of biomass had been established, and large-scale processes were operational even several decades back (Bergius 1937, Faith 1945, Sherrard and Kressman 1945, Harris and Beglinger 1946). Cellulose is completely solubilized at H_2SO_4 concentrations above 62% and promoted the breakage of glycosidic bonds through the action of water to liberate free sugars and oligosaccharides (Camacho et al. 1996). Dilute acids can also perform cellulose hydrolysis but need higher temperatures (>180°C) and pressures

(1.2–1.3 MPa) and have low selectivity for glucose (Guo et al. 2012). Besides, the hydrolysates generated using high-temperature treatments are often contaminated with degradation products of lignin and hemicelluloses, products of secondary transformations of monosaccharides and other intermediates of hydrolysis (Kuznetzov et al. 2013). While sulfuric acid is highly active as a catalyst and is inexpensive, usage of this on a large scale poses several difficulties and technical limitations (Hara 2010). Concentrated sulfuric acid is highly corrosive and its usage for hydrolysis will require special reactors, besides the necessity to separate and neutralize the acidic residues and which generate large quantities of waste (Van de Vyver et al. 2011). Recycling of the catalyst is difficult since it forms a homogeneous mixture with the reactants and products that are soluble. Nevertheless, there is a renewed interest in such acid-catalyzed processes due to the possibility of generating several fine chemicals and platform chemicals from further catalytic conversion of the reducing sugars generated from biomass, in addition to bioethanol or biobutanol (Zhou et al. 2011).

While the use of homogeneous catalysts is attractive with respect to higher reaction efficiency and mass transfer, heterogeneous solid acid catalysts (SACs) that can be recovered and reused have considerable advantages and have received greater attention in recent years (Guo et al. 2012). The main advantages offered by the SACs are the ability to separate them easily from the reaction mixture allowing the reuse of catalyst, lesser corrosion, and lesser contamination of sugars with degradation products from lignin and hemicelluloses. They also offer higher selectivity and extended life allowing multiple reuses. There have been a large number of reports on the use of SACs for cellulose hydrolysis, which have addressed SACs in various contexts including design criteria for SACs (Rinaldi and Schuth 2009), types and mode of action of different SACs (Hara 2010, Guo et al. 2012, Huang and Fu 2013, Hara et al. 2015), preparation of SACs (Liu et al. 2010, Ormsby et al. 2012, Ma et al. 2014), methods for heterogeneous catalysis (Kuznetzov et al. 2013, Bhaumik et al. 2014; Hara et al. 2015), catalytic materials (Dutta et al. 2012, Nakajima and Hara 2012, Lam and Luong 2014), cellulose pretreatment for SAC-mediated hydrolysis (Kim et al. 2010), magnetic SACs allowing rapid recovery and recycle (Lai et al. 2011a,b), and improvement in catalytic efficiency through cellulose binding sites (Kitano et al. 2009, Hu et al. 2014). Most of these reports have dealt with hydrolysis of pure cellulose at large, while some have addressed the hydrolysis of the native or pretreated plant biomass (e.g., Jiang et al. 2012, Li et al. 2012, Rinaldi and Dwiatmoko 2012, Wu et al. 2012). This chapter therefore tries to address solid acid–mediated hydrolysis of lignocellulosic biomass for biofuels—especially bioethanol. It would cover the studies conducted on hydrolysis of cellulose and plant biomass using different types of catalysts and the current state of the art in the subject with discussions on the directions and future possibilities. Since the majority of the studies in the field are restricted to hydrolysis of biomass to generate sugars and also since alcohol fermentation is an easy task once the sugars are generated, most of the discussions are centered on the hydrolysis step. Also, it is not the objective of this chapter to cover the chemistries of SAC-mediated hydrolysis or the synthesis of the catalysts in detail, though some references to these are presented. Some of the major types of SACs currently being evaluated would be covered with some details on the experiments to hydrolyze biomass and the efficiencies achieved.

6.2 Why Solid Acid Catalysts for Biomass Hydrolysis?

Hydrolysis of cellulose is accomplished by the breakage of the β-1,4-glycosidic bond mediated by the attack of H^+ ion (resulting from splitting of water molecule) on the oxygen atom in the 1,4-β-glycosidic linkage (Figure 6.1). This breaks the bond resulting in a cyclic carbonium ion and is considered as the rate-limiting step in the hydrolysis (Xiang et al. 2003, Zhou et al. 2011). Glucose is now formed by a rapid transfer of OH^- from the dissociation of water to the carbonium ion. Liquid acid catalysts commonly employed for cellulose hydrolysis like H_2SO_4, HCl, or H_3PO_4 are excellent proton donors, which explain their relatively high catalytic activity in the hydrolysis of cellulose.

In general, the yield of reducing sugars increases with an increase in acid concentration. Use of dilute acid requires higher temperatures and longer reaction times, but similar hydrolytic efficiencies may be achieved by concentrated acids at lower temperatures and in a shorter duration. However, with concentrated acids, prolonged exposure of the sugars generated can result in their further degradation to other compounds like 5-hydroxymethyl-2-furfural (HMF). Commonly employed liquid acid catalysts are highly corrosive in their concentrated form, and due to concerns regarding safety of usage, equipment life, operational and maintenance costs, and environmental issues, dilute acids are preferred for use in biomass hydrolysis (Wyman 1996). An impregnation pretreatment of biomass with very dilute acid can also facilitate further hydrolysis as was demonstrated by Emmel et al. (2003). Despite several advantages that the liquid acid catalysts possess, the major limitation lies in the difficulty to separate the products from the bulk phase that contains the catalysts themselves. The separation of sugars from the acid is costly and leaving the sugars in the acid can lead to their further degradation, resulting in a lower effective yield.

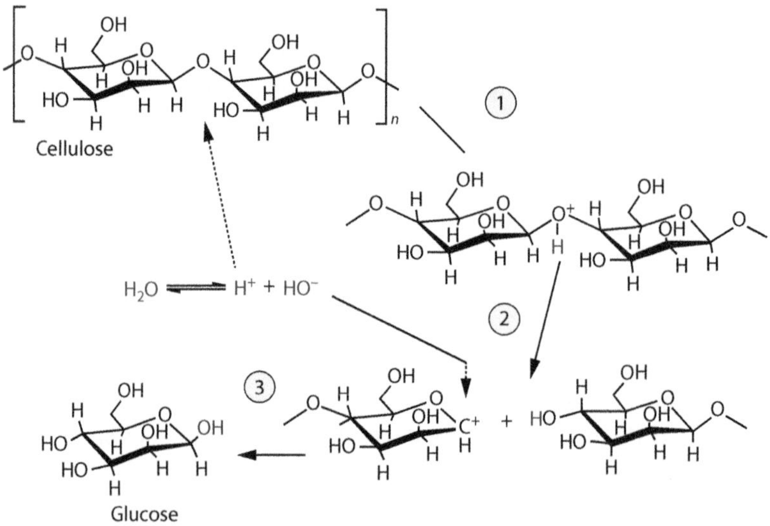

Figure 6.1 Scheme for hydrolysis of cellulose.

Moreover, neutralization of the acid can lead to accumulation of a huge amount of waste in the form of gypsum considering the implementation of the operation at commercial scales. Based on the earlier considerations, it is advisable to have catalysts that act rapidly but at the same time are less hazardous, are environmentally benign/green, offer better selectivity, and are recyclable. SACs fit most of these criteria and the major advantage is the ability to be recycled, and since no mineral acids or buffers are used to adjust the pH of the reaction mixture, the whole system is corrosion-free and simple to handle (Dhepe and Fukuoka 2008). Solid acids have the potential to be used in fixed bed kind of reactors to be adapted for continuous processes. They also have some unique physical or chemical properties such as large surface areas, specific surface structures, unique channels, and substrate adsorption capabilities (Wang et al. 2015).

While a large variety of solid catalysts ranging from commercial ion-exchange resins to lignocellulose-derived carbonaceous catalysts are now being tried for depolymerization of cellulose, very less is known about the precise mechanisms of action and the catalyst properties and reaction conditions that favor efficient and selective hydrolysis of cellulose without the generation of sugar degradation products. Based on the mechanism outlined in scheme 1, it may be expected that a solid catalyst and reaction conditions that favor the dissociation of water shall favor cellulose hydrolysis. Water dissociation into H^+ and OH^- ions is considered as a deprotonation. It is also demonstrated that the catalytic activity of a catalyst for cellulose hydrolysis increases with a decrease in deprotonation enthalpy of water on the surface of a SAC (Shimizu et al. 2009). Consequently, a stronger Brønsted acidity is more favorable for catalytic hydrolysis of cellulose. Also the amount of water has a positive effect on the breakage of β-1,4-linkages and intramolecular hydrogen bonds in cellulose. On the one hand, the depolymerization of cellulose requires conditions that favor water splitting, while on the other hand, the same conditions are not conductive for preventing further breakdown of sugars (Zhou et al. 2011). Consequently, recent trends are toward developing catalysts that have better surface properties to allow hydrolysis at milder conditions with a higher turnover frequency and at the same time to allow higher specificity.

6.3 Types of Solid Acid Catalysts

Different types of SACs, their structures, chemical properties, mode of action, and case studies on their use in hydrolysis of cellulose, hemicellulose, or plant biomass have been the subject of several recent reviews (Dhepe and Fukuoka 2008, van de Vyver et al. 2010, Guo et al. 2012, Huang and Fu 2013, Lam and Luong 2014, Hara et al. 2015). The reader is referred to these for an extensive coverage of the different types of SACs. The attempt here is to introduce the major types of SACs currently being evaluated worldwide for biomass hydrolysis, mostly in the context of this application. While there is no universally accepted classification of SACs, several types of catalysts are recognized based on their compositions, chemistries, structures, and active groups. The descriptions mentioned here would follow the categories described by the previous reports (Guo et al. 2012, Huang and Fu 2013, Hara et al. 2015). While solubilization of cellulose in ionic liquids can greatly aid in the catalytic hydrolysis, the difficulties in separation of sugars from the reaction mixtures, challenges in recycling of the ionic liquids, toxicity of the ionic liquids in several cases, and finally their prohibitive cost that impedes large-scale applications seriously limit their use as solvents for cellulose hydrolysis while using SACs. So the following descriptions have not discussed the use of SACs with cellulose/biomass dissolved in ionic liquids.

6.3.1 Metal Oxides

Several metal oxides and phosphates have both Brønsted and Lewis acid sites on the surfaces, which are catalytically active (Hara et al. 2015). Some of the examples include the metal oxides and phosphates of group IV and V elements. Broadly, they can be classified as single metal oxides and mixed metal oxides (Guo et al. 2012). Metal oxides are prepared with high specific surface and pore sizes to allow easy access of reactants to the active sites inside the pores (Huang and Fu 2013). Mesoporous transition metal oxides such as niobium oxide (Nb_2O_5), tungsten oxide (WO_3), titanium oxide (TiO_2), mixed metal oxide of tantalum oxide and tungsten oxide (Ta_2O_5–WO_5), and zirconium transition metal oxide mesoporous molecular sieves (Zr-TMS) have been tried as SACs in various reactions (Chidambaram et al. 2003, Kondo et al. 2005, Tagusagawa et al. 2010a,b). Mesoporous metal oxides have high specific surface areas, adjustable pore sizes, and enhanced stability (Guo et al. 2012). Major advantages of such catalysts are the high catalytic efficiencies and the easiness of separation and reuse. Mixed metal oxides like that of Nb and W have been tried for hydrolysis of cellulose. While the mesoporous Nb–W oxide exhibited higher turnover frequencies compared to nonporous Nb–W oxide, the glucose yields were very low indicating a lower activity for the catalyst (Tagusagawa et al. 2010b). In contrast, a layered transition metal oxide $HNbMoO_6$ showed a higher turnover frequency than Amberlyst® 15 (a strongly acidic polymer-based SAC) and a higher glucose yield of ~41%. This was projected as partly due to the facile intercalation of saccharides in the strongly acidic interlayer gallery of the catalyst (Figure 6.2). Other features that contributed to the improved efficiency were speculated to be the strong acidity and water tolerance (Takagaki et al. 2008). Nevertheless, the efficiency of cellulose hydrolysis was low with the total yield of glucose and cellobiose from it being only 8.5%. Lower efficiency of the layered transition metal catalyst was attributed to the limited accessibility of the larger polymers like cellulose or starch to the protons of the catalyst. It was therefore concluded that a large number of protons need to be generated to compensate for the limited accessibility of the substrate (Tagusagawa et al. 2010b). Since the hydrolysis efficiency is highly dependent on the cocatalysis of Lewis acid and Brønsted base sites, a high reaction rate and acid strength may be achieved by adjusting the relative proportions of the transition metal oxides.

Nanoscale mixed metal oxide catalysts are another important catalyst type being tried with impressive results. Zhang et al. (2011) had tested a nano-Zn–Ca–Fe oxide

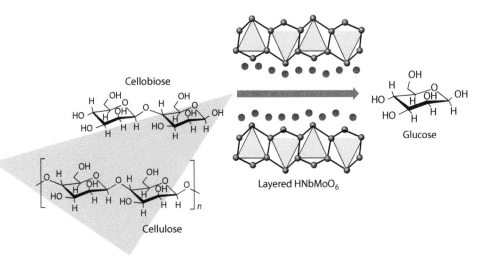

Figure 6.2 Hydrolysis of cellulose by layered transition metal catalyst. (From Takagaki, A., Tagusagawa, C., and Domen, K., Glucose production from saccharides using layered transition metal oxide and exfoliated nanosheets as a water-tolerant solid acid catalyst, *Chem Commun.*, 42, 5363–5365. 2008. Reproduced by permission of The Royal Society of Chemistry.)

catalyst for hydrolysis of crystalline cellulose and obtained a cellulose conversion rate of 42.6%. The nanoscale catalyst suspension can provide more active sites per gram and also possess several characteristics of a fluid. Moreover, due to the presence of iron, the material is paramagnetic allowing easy separation from reaction mixture by the use of a magnetic field. Other concepts being discussed in the mixed metal oxide catalysts include the development of 1D nanomaterials and strategies to improve the surface area and reactant accessibility by engineering mesoporous structures (Guo et al. 2012).

6.3.2 Supported Metal Catalysts

Supported metal catalysts are known for their catalytic abilities for hydrogenation reactions and have been used for hydrolytic hydrogen transfer reactions to produce sugar alcohols (Palkovits et al. 2010, Kobayashi et al. 2012). Supported ruthenium (Ru) catalysts have been tried for the hydrolysis of cellulose (Kobayashi et al. 2010). In this study, Ru supported on CMK (mesoporous carbon material)—Ru/CMK-3 hydrolyzed cellulose to glucose with 24% yield. The reaction had a high turnover and short reaction period. Other major product was cellooligosaccharides, which formed about 16%, while low percentages of glucose derivatives were also formed. The reaction scheme is represented in Figure 6.3.

Studies conducted with different concentrations of Ru on the supports indicated that the increase in Ru concentration resulted in an increased yield of glucose, which went up to 31% with 10% Ru loading on the support. Correspondingly, there was a decrease in concentration of oligosaccharides indicating that Ru hydrolyzes

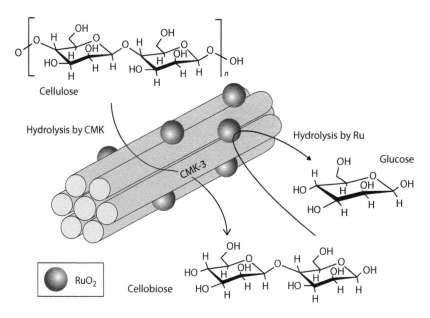

Figure 6.3 Hydrolysis of cellulose by layered transition metal catalyst. (From Huang, Y.B. and Fu, Y., Hydrolysis of cellulose to glucose by solid acid catalysts, *Green Chem.*, 15, 1095–1111, 2013. Reproduced by permission of The Royal Society of Chemistry.)

both the cellulose and oligosaccharides, and a higher activity is shown for the latter (Kobayashi et al. 2010). It was later determined that the Ru species in the catalyst is $RuO_2 \cdot 2H_2O$, which was speculated to desorb the hydrated water to give a Lewis acid site that can depolymerize cellulose (Kobayashi et al. 2012). Metal oxides are also widely used as catalyst supports because of their thermal and mechanical stability, high specific surface areas, and large pore size and pore volume. Sulfonated metal oxides such as SO_4^{2-}/Al_2O_3, SO_4^{2-}/TiO_2, SO_4^{2-}/ZrO_2, SO_4^{2-}/SnO_2, and SO_4^{2-}/V_2O_5 can supply the acidic groups and may serve as good catalysts (Guo et al. 2012). Fukuoka and Dhepe (2006) described the hydrolysis of cellulose over Pt or Ru supported on metal oxides. Pt supported on γ-Al_2O_3 gave the highest yield and was recyclable with no significant reduction in activity for three cycles. One of the proposed limitations in such systems is the leaching of acidic sites from the support under conditions of hydrolysis (and hence the thermal stability of catalyst becomes important [Guo et al. 2012]).

6.3.3 Heteropolyacids

Heteropolyacids (HPAs) are solid acids consisting of early transition metal-oxygen anion clusters, with the most common and widely used ones being the Keggin-type acids with the formula $[XY_xM_{(12-x)}O_{40}]^{n-}$ where X is the heteroatom and M and Y are the addendum atoms (Huang and Fu 2013). They have high Brønsted acidity, stability, and high proton mobility (Hill 2007). HPAs are soluble in polar solvents and release protons whose acidic strength is stronger than the mineral acids like sulfuric acid (Macht et al. 2007). The HPAs $H_3PW_{12}O_{40}$ and $H_4SiW_{12}O_{40}$ were evaluated by Shimizu et al. (2009) as homogeneous catalysts for the hydrolysis of cellulose and Japanese cedar wood fiber. The catalytic efficiencies correlated with the deprotonation enthalpy indicating that the stronger Brønsted acid was a better catalyst. The maximum reducing sugar yield attained was ~18%. Recovery of the catalyst was attempted through concentration and precipitation, which resulted in about 50% of the catalyst being recovered. Water solubility of the HPAs can be an issue while projecting them as easily separable heterogeneous catalysts. Solvent extraction of the catalyst is another strategy explored for the recovery of the HPA catalyst. The HPA catalyst $H_3PW_{12}O_{40}$ tested for catalyzing hydrolysis of microcrystalline cellulose by Tian et al. (2010) gave a glucose yield of 50.5%, with >90% selectivity at 453 K in 2 h. The soluble catalyst was recovered from the reaction mixture with diethyl ether and was recycled. The catalyst loss was 8.8% after six cycles of hydrolysis. The issue of water solubility of the HPAs may be addressed by substitution of the protons with larger monovalent cations like Cs^+. This makes the catalyst insoluble in water and other polar solvents (Deng et al. 2012). A series of water-insoluble heteropolytungstate salts $Cs_xH_{3-x}PW_{12}O_{40}$ (x =1–3) were prepared and evaluated for cellulose hydrolysis by Tian et al. (2011). Among them, $Cs_1H_2PW_{12}O_{40}$ exhibited the highest conversion of cellulose and yields of total reducing sugar and glucose. A sugar yield of ~27% and the ability to separate the catalyst from the reaction phase by simple filtration make these catalysts a serious candidate for cellulose conversion. Another major limitation of using the HPAs and SACs, in general, is the limited and poor contact between the catalyst and the substrate. This often necessitates a higher catalyst to biomass ratio, longer reaction time, and higher temperature to complete the reaction. Better conversion and glucose yields have been obtained by the use of microwave irradiation (Li et al. 2012). The study, which used about 2% cellulose or lignocellulosic biomass as substrates in concentrated (50%–88%) $H_3PW_{12}O_{40}$ catalyst and reaction conducted at 90°C for 3 h, resulted in 75.6% glucose yield from cellulose. Similar conditions gave glucose yields of ~28%, 37%, and 43% for bagasse, corncob, and corn stover, respectively. While the catalyst could be recovered through extraction with solvent and reused without the loss of activity at least six times in the case of cellulose, there was a drastic decline in activity when hydrolyzing biomass that the authors attributed to the damage of catalytic sites by the by-products of hydrolysis reaction.

6.3.4 H-Form Zeolites

Zeolites are a large group of natural and synthetic crystalline aluminosilicates characterized by complex 3D structures, made of corner-sharing AlO_4 and SiO_4 tetrahedra joined into 3D frameworks with pores of molecular dimensions (Figure 6.4). Clusters of tetrahedra form boxlike polyhedral units that are further linked to build up the entire framework (Kaduk and Faber 1995, Chang 2002). In different zeolites, the polyhedral units may be equidimensional, sheetlike, or chain-like. The aluminosilicate framework of a zeolite has a negative charge, which is balanced by the cations housed in the cagelike cavities. Ions and molecules in the pores of the zeolites can be removed

Figure 6.4 Representative molecular structure of zeolite showing distinctive porous lattice. (Image from ChemTube3D by Nick Greeves, http://www.chemtube3d.com.)

or exchanged without destroying the structural framework. The number of Brønsted acid sites in H-form zeolites and hence their acidity are related to the atomic ratio of Si/Al, and in general, a higher Al content can give higher acidity (Salman et al. 2006).

Zeolites can be synthesized with a range of acidic or textural properties, and the hydrophilic or hydrophobic properties can be modulated without compromising the functionalized acidic sites (Guo et al. 2012). Since they are nontoxic, noncorrosive, and easily recovered from the reaction mixtures, they are projected as potential SACs for cellulose hydrolysis. The study by Onda et al. (2008) demonstrated that the catalytic efficiencies of H-form zeolites are lower compared to other SACs and the best zeolite tested (H-beta [75]) in the study could give a sugar yield of only 12%. Major limitation in this case was proposed as the limited accessibility of cellulose to catalytic sites due to the small pore sizes of the zeolites. In order to improve the efficiency of H-form zeolites, either the cellulosic materials must be dissolved in a solvent or the catalysts should be synthesized to have larger pore sizes that allow substrate access to their catalytic sites. The former approach was tried by Zhang and Zhao (2009) for the hydrolysis of cellulose from different sources, dissolved in ionic liquid, and the glucose yields ranged from 32.5% to 36.9% when the reaction was conducted using a microwave reactor. However, with conventional heating, the glucose yield was only 12% from milled cellulose. Guo et al. (2012) had discussed some of the proposed approaches to improve the efficiency of zeolites for biomass hydrolysis. These include the loading of alternate cations like lanthanum and cerium to increase the weak acid sites on the zeolite, synthesis of zeolites with larger pore sizes accomplished through change of synthesis techniques, loading of superacids or heteropolyacids by direct impregnation/through incorporation during synthesis, and synthesis of composite zeolites. Nevertheless, the efficiency of such methods in actual hydrolysis of cellulose and lignocellulosic biomass is yet to be demonstrated.

6.3.5 Polymer-Based Solid Acid Catalysts

Polymer-based SACs, especially the strong cation-exchange resins, are used as catalysts in several commercial processes of organic synthesis (Altava et al. 2008). Macroreticulated styrene divinylbenzenes with sulfonic groups commercially known as Amberlyst have been described for the hydrolysis of cellulose and even lignocellulosic biomass (Rinaldi et al. 2008, Meena et al. 2015). Other styrene-based SACs have also been tried with varying yields. Zhang and Zhao (2009) evaluated the styrene-based sulfonic acid resin NKC-9 that gave glucose and total reducing sugar yields of 26.9% and 38.4%, respectively, for the hydrolysis of Avicel. Another major polymer-based SAC being tested for cellulose conversion is Nafion (sulfonated tetrafluoroethylene–based fluoropolymer–copolymer). Hydrolysis of cellulose by Nafion supported on amorphous silica was reported by Hegner et al. (2010). The study reported a glucose yield of 11% with a reaction temperature of 190°C, and though the catalyst could be recycled, the efficiency dropped by 28% and 37% for the second and third cycles. Similar to zeolites, the efficiency of hydrolysis is greatly influenced by the accessibility of the inner acidic sites of the polymer-based resins by cellulose, and hence, solubilization of cellulose in appropriate solvents like ionic liquids may improve the catalytic efficiencies of such catalysts. Hydrolysis of cellulose and palm stem on Amberlite IR 120—another sulfonated polystyrene–type cation-exchange resin with the use of 1-n-butyl-3-methylimidazolium chloride ([bmim]Cl) as the medium at 130°C for 4 h—gave a sugar yield of 13.4% that was more than three times the yield obtained with the ionic liquid alone. The catalyst also supported a total sugar yield of 19.9% from palm stem biomass (Rinaldi and Dwiatmoko 2012). Ishida et al. (2014) studied the effect of multistep water addition for improving the catalytic hydrolysis of cellulose using different catalysts including the polymeric catalysts Amberlyst 15 and Dowex 50wx8-100. The reaction was conducted in [bmim]Cl with microwave heating and high glucose yields of 68.5% and 75% were obtained by two-step and three-step water addition, respectively. The study by Rinaldi et al. (2008) showed that the reaction proceeded by initial conversion of the cellulose to oligosaccharides that could also be an effective method for cellooligomer production. The separation of cellooligosaccharides was possible by precipitation that gave 90% yield. Prolonging the reaction resulted in the production of sugars, but the separation of sugars from reaction mixture was a problem due to the high solubility of sugars in ionic liquids. Our studies with Amberlyst 15 as catalyst for the hydrolysis of alkali-pretreated rice straw yielded 34% total reducing sugars of which the major part was pentose sugars from hemicellulose hydrolysis (Meena et al. 2015). There was also a significant loss of catalytic activity on reuse of the catalyst. One of the major limitations here was the difficulty in the separation of the catalyst from the unhydrolyzed residues, which could be a serious issue if operating in a larger scale. Nevertheless, the advantages included a lower reaction temperature (140°C), relatively

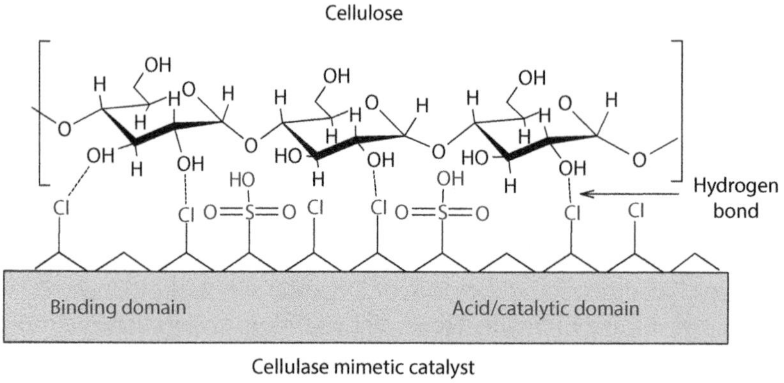

Figure 6.5 Cellulase-mimetic polymer-based solid acid catalyst. (From Huang, Y.B. and Fu, Y., Hydrolysis of cellulose to glucose by solid acid catalysts, *Green Chem.*, 15, 1095–1111, 2013. Reproduced by permission of The Royal Society of Chemistry.)

higher biomass loadings (5.4%), and shorter residence time (~4 h). Another novel concept in biomass hydrolysis using polymer-based SACs includes the development of cellulase-mimetic catalysts, which consists of a cellulose binding domain and a catalytic domain (Shuai and Pan 2012). Sulfonated chloromethyl polystyrene resin (CP–SO$_3$H), containing cellulose binding sites (–Cl) and catalytic sites (–SO$_3$H), was synthesized for hydrolyzing cellulose (Figure 6.5). Avicel was hydrolyzed by the catalyst into glucose with a yield of 93% within 10 h at moderate temperature of 120°C. Apparently, the method offers several advantages and could be considered as a possible future method for biomass hydrolysis on larger scales.

6.3.6 Carbonaceous Solid Acid Catalysts

Carbonaceous solid acid catalysts (CSACs) are considered superior to all of the solid catalysts tested so far, due to the high catalytic activity and recyclability (Guo et al. 2012). Carbon-based solid acids may be prepared by the sulfonation of incompletely carbonized natural polymers such as sugars and cellulose starch or by the incomplete carbonization of sulfopolycyclic aromatic compounds in concentrated sulfuric acid (Nakajima and Hara 2012). While the latter method can yield CSACs with high density of –SO$_3$H groups, there are challenges involved due to the use of high temperatures and concentrated sulfuric acid. Hence, most of the work on amorphous carbon bearing –SO$_3$H has been done using the former method. Pioneering work on carbonaceous acid catalysts was done by Hara's group when they synthesized CSACs from D-glucose/sucrose (Toda et al. 2005). They did incomplete carbonization of these sugars at low temperature to form small polycyclic aromatic carbon rings that were then sulfonated using sulfuric acid to introduce the sulfonic (–SO$_3$H) groups. Several studies using such amorphous catalysts had followed later, which evaluated their use in cellulose hydrolysis. For the catalyst preparation, the carbonization step is typically done at 400°C in a nitrogen atmosphere and the sulfonation at 150°C. Hara's group and others had conducted several studies in such catalysts using different starting materials to prepare the catalysts, which included glucose (Toda et al. 2005), glycerol (Prabhavathi et al. 2009, Goswami et al. 2015), activated carbon (Onda et al. 2008, Liu et al. 2010), starch (Liang et al. 2011), cellulose (Suganuma et al. 2010, Fukuhara et al. 2011), and even lignocellulosic biomass of various types (Jiang et al. 2012, Wu et al. 2012, Namchot et al. 2014), and several of them were evaluated for cellulose/biomass hydrolysis.

The carbon catalysts possess a graphene-like structure containing 1.2–1.3 nm aromatic groups with a surface area <5 m²/g (Nakajima and Hara 2012). Several functionalized graphene sheets accumulate to form flexible domains that are linked together to form the catalyst particles. The acid strengths of these catalysts are comparable to that of concentrated sulfuric acid (Lam and Luong 2014). Hara and coworkers had characterized the sulfonated CSAC to show that the flexible polycyclic carbon sheets of the catalysts bear SO_3H, COOH, and phenolic hydroxyl group (OH) (Figure 6.6). These studies had also demonstrated that larger amount of water was also incorporated into the bulk of the catalyst that facilitated the cellulose chain in solution to be in contact with the SO_3H group of the catalyst, which in turn helps to achieve higher catalytic efficiencies (Huang and Fu 2013, Wang et al. 2015).

The high density of SO_3H, COOH, and OH functional groups bonded to the flexible carbon sheets makes the accessibility of the active sites to reactants easier in solution leading to the extraordinary performance of these catalysts in cellulose hydrolysis. Microcrystalline cellulose was converted with 68% efficiency at a low temperature (100°C) and short time (3 h). Glucose yield was 4% and soluble β-1,4-glucan yield was 64%, which was higher than that catalyzed by sulfuric acid (Suganuma et al. 2008). Absorption experiments conducted using cellobiose indicated that the enhanced catalytic activities could be the outcome of strong interactions between the phenolic OH groups and the glycosidic bonds in cellulose—that is, the hydrogen bonds between the phenolic OH and oxygen atoms in the glycosidic bonds. Such hydrogen bonds were speculated to bind cellulose to the catalyst surface making it easy for the reaction to proceed. The study also revealed that very little SO_3H groups were leached into the reaction mixture.

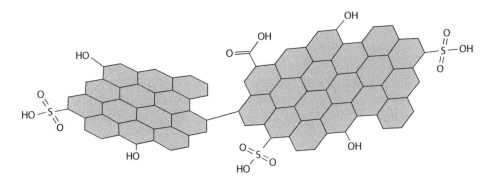

Figure 6.6 Sulfonated carbonaceous solid acid catalyst. (From Hara, M., Biomass conversion by a solid acid catalyst, *Energy Environ. Sci.*, 3, 601–607, 2010; Wang, J., Xi, J., and Wang, Y., Recent advances in the catalytic production of glucose from lignocellulosic biomass, *Green Chem.*, 17, 737–751, 2015. Reproduced by permission of The Royal Society of Chemistry.)

Natural lignocellulosic biomass was also used effectively as carbon source for the preparation of CSACs. A biomass char sulfonic acid (BC-SO$_3$H) prepared from bamboo could catalyze 32.4% reducing sugar yield from treated bamboo powder (Wu et al. 2012). The study used lower temperatures (70°C–100°C) and microwave heating to accelerate the rate of reaction. High efficiencies were attributed to the strong affinity of catalyst to the glycosidic bonds of cellulose that is not shown by other SACs. In another similar attempt, Jiang et al. (2012) used CSACs prepared from the residues of corncob hydrolysis to hydrolyze the same biomass treated by ball milling and using microwave heating. The maximum glucose, xylose, and arabinose yields in this case were 34.6%, 77.3%, and 100%, respectively. It has also been demonstrated that the acid density and hence the catalytic efficiency of CSACs may be tuned by changing the sulfonation temperature (Pang et al. 2010). A sulfonated activated carbon catalyst—AC-N-SO$_3$H sulfonated at 250°C—gave one of the highest glucose yields (62.6%) reported using CSACs. Novel concepts being tried in CSACs include the preparation of sulfonated silica–carbon nanocomposites (Van de Vyver et al. 2010), the use of mesoporous materials like CMK-3 (Pang et al. 2010, Chung et al. 2012), and the development of cellulase-mimetic catalysts (Hu et al. 2014). Specific advantages of the sulfonated carbon acid catalysts include the high adsorption affinity toward the substrates, excellent activity, and great reusability (Wang et al. 2015). While hydrolysis efficiencies as high as 74% have been achieved (Pang et al. 2010), in general the glucose yields are low, and the major limit has been proposed as the solid diffusion barrier as the reaction involves solid–solid contact. Though mechanocatalytic conversion that involves milling of catalyst and substrate together has shown some promise in improving this, the method is highly energy consuming and may not be feasible in larger scales.

6.3.7 Nanosolid Acid Catalyst

Nanocatalysis is performed by mixing catalyst nanoparticles with solvent and lignocellulosic biomass. These nanoparticles are more accessible to the oxygen atom in the ether linkage of cellulose. Catalytic species such as heavy metal complexes, enzymes, and organic catalysts are able to form bonds to the nanoparticles via monolayer of long-chain alkanethiols with an SH or SS terminus (Gill et al. 2007, Guo et al. 2012). The reaction efficiency may be related to the position and degree of substitution of catalytically active sites in the framework structure, where the active site can be an acidic group that might function by hydrolysis (Gill et al. 2007). A problem faced by nano-SAC is the recycling. Recycling of nano-SAC may be difficult due to adsorption, agglomeration, and viscous effects in the reaction mixture (Nanda 2009). With the addition of paramagnetic compounds into the nanoparticle catalysts, it may be possible to separate (by applying an external magnetic field) and recycle nano-SAC.

6.4 Hydrolysis of Lignocellulosic Biomass for Biofuels Using Solid Acid Catalysts

Most of the studies on the use of SACs have concentrated on hydrolysis of pure cellulose and very few have attempted to hydrolyze the natural lignocellulosic biomass either in the untreated or pretreated form that is eventually the end use targeted for all these studies. Li et al. (2012) studied the catalytic hydrothermal saccharification of rice straw using sulfonated mesoporous silica-based catalyst (SBA-15). The monomeric sugar yield was directly proportional to the weight ratio of SAC to rice straw up to 25% and thereafter remained constant. A maximum monomeric sugar yield of 35% was obtained at 180°C for 1 h. Based on the compositional analysis, authors concluded that hemicellulose was decomposed completely and only a part of glucan from cellulose was decomposed even at above 200°C. Our group had evaluated the strongly acidic cation-exchange resin Amberlyst 15 for the hydrolysis of alkali-treated rice straw and subsequent fermentation to ethanol (Meena et al. 2015). Under the optimum conditions, a maximum sugar yield of 255 mg/g pretreated biomass was obtained in 4 h of the reaction. The study was further extended by comparing the hydrolysis performance of Amberlyst 15 against mineral sulfuric acid. A higher sugar yield was obtained for mineral acid–catalyzed hydrolysis that is accounted as a result of the easily available H^+ ions in the solution. In another study, we had evaluated a glycerol-based SAC for hydrolysis of alkali-treated rice straw. A maximum sugar yield of 262 mg/g pretreated biomass was obtained corresponding to 31% conversion efficiency. Moreover, this catalyst had the potential to convert untreated biomass into monomeric sugars through at a lower efficiency (Goswami et al. 2015).

Carbonaceous solid acids have given some of the highest sugar yields on the hydrolysis of lignocellulosic biomass. Jiang et al. (2012) reported a 34.6% yield of glucose from the hydrolysis of corncob using CSAC prepared from the unhydrolyzed corncob residues coming from xylose production. A total reducing sugar yield of 32.4% from bamboo hydrolysis was reported using bamboo biomass char–derived CSACs by Wu et al. (2012). More recently, CSACs prepared using sugarcane bagasse and lignin could yield 41% and 65% sugars from milled sugarcane bagasse (Namchot et al. 2014). A compilation of the different studies that have demonstrated the hydrolysis of lignocellulosic biomass is given in Table 6.1. While there are a considerable number of such studies describing the hydrolysis of biomass using SACs, the efficiencies attained have been very less except probably in cases where CSACs have been used. Even in these cases, the scales of operation have been very small, reaction systems limiting, and quite often employing conditions that may not be scalable to propose any future large-scale applications. Nevertheless, the progress made in this field has been considerable, especially in CSACs. With improvements in reaction conditions and reactors themselves and in the recovery and reuse of catalysts, realization of the use of SACs for lignocellulosic biomass hydrolysis in practical scales may not be too far.

While limited, there have been some attempts at developing reaction systems that include the study by Avram et al. (2013) who developed a novel membrane reactor based on polymeric SAC for combined biomass hydrolysis and sugar recovery. The cellulose and hemicellulose in water/ionic liquid mixture are pumped from one side of the reactor, the polyionic liquids solubilize the polysaccharides, and the polysulfonic acids catalyze the hydrolysis of polysaccharides. Another very interesting concept, which was demonstrated recently, was to selectively convert lignin from biomass leaving behind the carbohydrate part (Parsell et al. 2015). The authors reported a bimetallic Zn/Pd/C catalyst that converts lignin into two methoxyphenol products,

TABLE 6.1   Hydrolysis of Lignocellulosic Biomass Using Solid Acid Catalysts

Biomass	Pretreatment	Catalyst	Process Conditions	Efficiency (%)	Reference
Rice straw (RS)	Chopping (1.8 mm)	Sulfonated mesoporous silica (SBA-15)	Weight ratio of SAC to RS 8.3%–25%, DW: 30 g, time 0.5–2 h, temp 150°C–220°C	38	Li et al. (2012)
Sugarcane bagasse	DMA containing LiCl at 80°C for 5 min	$Zr(O)Cl_2/CrCl_3$ Additive: [BMIM]Cl	SCB 1 g, ethanol 20 mL, 20 mol% of $Zr(O)Cl_2/CrCl_3$ catalyst in [BMIM]Cl	42	Dutta et al. (2012)
Rice straw	Dilute alkali pretreatment	Amberlyst® 15	RS 5.41%, catalyst to RS ratio 0.7:1, temp 140°C, time 3.7 h	34	Meena et al. (2015)
Rice straw	Dilute alkali pretreatment	Glycerol-based CSAC	RS 3%, catalyst to RS ratio 1:1, temp 140°C, time 4 h	31	Goswami et al. (2015)
Rice straw	Milling to <2 mm	Glycerol-based CSAC	RS 3%, catalyst to RS ratio 1:1, temp 140°C, time 4 h	17	Goswami et al. (2015)
Corn stover (CS)	None	$RhCl_3/H_1/H_2$	CS 0.18 g, $RhCl_3 \times H_2O$ 10 mg, DW 1.8 mL, HCl 70 mL, NaI 300 mg, and organic solvent 2 mL	41	Yang and Sen (2010)
Sugarcane bagasse (SCB)	None	MTiP-1 catalyst Additive: DMA–LiCl	SCB 50 mg, MTiP-1 catalyst 10 mg, DMA–LiCl solvent 0.5 g, temp 140°C	27	Dutta et al. (2012)
Corncob (CC)	None	Fe_3O_4—SBA—SO_3H	CC 1 g, catalyst 1.5 g, DW 15 mL, temp 150°C, time 3 h	45	Lai et al. (2011a)
Corn stover (CS)	Ball milling to 40–80 mesh	$H_3PW_{12}O_{40}$ (HPW)	CS 60 mg, HPW: 3 mL, temp 90°C, time 180 min	30	Li et al. (2012)
Corncob (CC)	Ball milling to 40–80 mesh	$H_3PW_{12}O_{40}$ (HPW)	CS 60 mg, HPW: 3 mL, temp 90°C, time 180 min	24.6	Li et al. (2012)

(Continued)

TABLE 6.1 (*Continued*)   Hydrolysis of Lignocellulosic Biomass Using Solid Acid Catalysts

Biomass	Pretreatment	Catalyst	Process Conditions	Efficiency (%)	Reference
Bagasse (SCB)	Ball milling to 40–80 mesh	$H_3PW_{12}O_{40}$ (HPW)	SCB 60 mg, HPW: 3 mL, temp 90°C, time 180 min	19	Li et al. (2012)
Corncob (CC)	Milling	CSAC from unhydrolyzed corncob residues	CC 0.2 g, CSA 0.2 g, DW 2 mL, temp 130°C, time 1 h	34.6	Jiang et al. (2012)
Palm stem	Not specified	Amberlite IR 120	Palm stem 0.2 g, [bmim]Cl 5 g, Amberlite 0.1 g, temp 160°C, time 3 h	19.9	Rinaldi and Dwiatmoko (2012)
Birch kraft pulp	Bleached and dried	Mesoporous material H-MCM-48	Birch kraft pulp 0.64 g, catalyst 0.3 g, DW 150 mL, temp 458 K, 20 bar hydrogen pressure	14	Kaldstrom et al. (2012)
Bamboo (BM)	Milled to 200 mesh particles	CSAC based on biomass char (from bamboo) BC–SO$_3$H	BM 0.1 g, catalyst 0.1 g, DMA with 9% LiCl (DMA–Li) 10 mL, DW 5 mL, microwave irradiation 75 W, 20 min	32.4	Wu et al. (2012)
Sugarcane bagasse (SCB)	Milled to 70 mesh particles	CSAC prepared from sugarcane bagasse 100 mesh	SCB 0.05 g, catalyst 0.03 g, DW 5 mL, temp 140°C, time 3 h	41.6	Namchot et al. (2014)
Sugarcane bagasse	Milled to 70 mesh particles	CSAC prepared from lignin (100 mesh)	SCB 0.05 g, catalyst 0.03 g, DW 5 mL, temp 140°C, time 3 h	65.0	Namchot et al. (2014)

Abbreviations: DMA, dimethylacetamide; DW, distilled water/water; CSAC, carbonaceous solid acid catalyst.

namely, 2-methoxy-4-propylphenol (dihydroeugenol) and 2,6 dimethoxy-4-propyl-phenol, leaving carbohydrates as a solid residue. The carbohydrate-rich residue was then hydrolyzed by cellulase and a higher glucose yield of 95%, which is comparable to lignin-free cellulose.

6.5 Recovery and Reuse of Solid Acid Catalysts

The greatest and most important advantage proposed for SACs is the easiness in their recovery and reuse. The catalyst being a solid is expected to be easily separable, though in reality it is seldom so, because the substrate for the reaction—biomass—is also insoluble and in a solid phase. A complete hydrolysis of the lignocellulosic biomass will solubilize the sugar polymers but still may leave lignin and sometimes humins as solid residues, which makes it difficult to separate the catalyst (Huang and Fu 2013). Typical methods to separate the catalysts from unhydrolyzed biomass include resuspending the mixture of unhydrolyzed biomass and catalyst in water followed by gravity separation. This method may be feasible for cases where the catalyst is having a higher density than the biomass as is the case with metal catalysts, zeolites, etc., but is not feasible for catalysts that do not have appreciable density differences with the biomass. Reuse of CSACs by adding the fresh biomass to the mixture of catalysts and unhydrolyzed biomass was reported by Jiang et al. (2012). While the catalyst retained its catalytic ability, the efficiency was reduced indicated by the reduction in sugar yield from 55.6% to 26.7% for the fresh and recycled catalyst. Regeneration of catalyst by carbonization and sulfonation increased the catalytic activity to the original levels. We had attempted a similar strategy for catalyst reuse with Amberlyst 15, where fresh biomass was added to the mixture containing catalysts and unhydrolyzed biomass to start a fresh reaction. Here also the catalytic efficiency was decreased with every cycle. While the fresh catalyst gave 34% total sugar yield, the sugar yields for second and third cycles were 19.1% and 12.6%, respectively (Meena et al. 2015). In another study by us where a glycerol-based carbon acid catalyst was used for the hydrolysis of pretreated rice straw, the recovery of the catalyst was accomplished by density-based sedimentation where the mixture of catalyst and residual biomass was vortexed and allowed to settle for shorter duration when the denser catalyst particles settled down faster and was recovered by aspirating the suspension of unhydrolyzed biomass. Here, the reuse of catalyst after a wash gave a sugar yield that was 68.45% of the yield obtained with fresh catalyst (Goswamy et al. 2015). While these methods conceptually prove that the catalysts may be separated and reused effectively, practical limitations do exist, especially on a larger scale.

Considering the earlier discussion, the best strategy to recover the catalyst from the reaction mix seems to be the use of catalysts with magnetic properties. Acid-functionalized paramagnetic nanoparticle catalysts, core–shell-type magnetic particles where there is magnetic core and acid-functionalized shell, and mesoporous catalysts synthesized incorporating magnetic particles have been tried successfully for the purpose (Guo et al. 2012, Huang and Fu 2013). In the work by Gill et al. (2007), two types of acid-functionalized magnetic nanoparticles were generated, which differed in the acid group. Silica-coated nanoparticles were functionalized with alkyl sulfonic acid or perfluoroalkyl sulfonic acid and were used for hydrolysis of cellulose with separation of catalyst effected after hydrolysis using a magnetic field. Magnetic sulfonated mesoporous silica was prepared by Lai et al. (2011a,b) by incorporating Fe_3O_4 magnetic nanoparticles during the synthesis. The catalyst supported 50% yield of glucose from cellulose dissolved in ionic liquid and a total reducing sugar yield of 45% from corncob. The catalyst could be separated easily using a magnetic field and reused without deactivation. Takagaki et al. (2011) reported the use of a magnetic solid acid with $CoFe_2O_4$ as magnetic core. The catalyst converted cellulose with a 7% yield of glucose and 30% yield of total reducing sugars. The catalyst was recovered from the solution using a magnet after the reaction. It is expected that further research in this area would yield better catalysts with higher efficiencies and stability and that are easily recovered and recycled.

6.6 Concluding Remarks and Outlook

Current strategies for hydrolysis of cellulose are either expensive or not environment friendly, the void that is proposed to be filled by SACs. The major features proposed in an ideal catalyst for biomass conversion include high efficiency, low cost, easy recoverability and reusability after hydrolysis, ambient to moderate conditions for reaction, easiness in synthesis and storage, and environment friendly and nontoxic nature. The optimal SAC may fit these criteria, while the currently available ones are far from optimal. Nevertheless, compared to homogeneous acid catalysts or even enzymes, SACs also have several advantages of which the major one is the easiness in separation and recycling. While the current strategies for the recovery of SACs from the reaction mix after hydrolysis are far from perfect or economic, it still seems to be better than those for separation of enzyme catalysts. Different types of materials have been employed as SACs of which the CSACs seem to be the most promising with efficiencies more than 60% achieved for hydrolysis of lignocellulosic biomass.

The major concern in the use of SACs seems to be the low efficiencies of hydrolysis that is primarily due to the limitations of solid–solid contact since both the substrate and the catalyst are in insoluble form. While studies have addressed this by dissolving the biomass in ionic liquid, it may not be a feasible option due to the prohibitive cost of these solvents. Alternative strategies include the mechanical methods where the catalyst and biomass are milled together or vigorously blended under appropriate reaction conditions. However, the use of such systems is also not efficient due to the high energy requirement. Nevertheless, it is apparent that mixing of the biomass and catalyst plays an important role in determining the hydrolysis efficiency, and hence, the development of appropriate reactor systems is of top priority in biomass conversion using SACs. Another important concern is regarding the specificity of hydrolysis that is now being addressed by the development of cellulase-mimetic catalysts, which have both cellulose binding and hydrolyzing functional groups.

The development of magnetic SACs can go a long way toward effective recovery and reuse of the catalysts. Research in this field looks highly promising with the development of acid-functionalized paramagnetic nanoparticles and other types of magnetic catalysts. Combining the efficiency of CSACs and the easy recovery and reusability of magnetic nanoparticles could be an interesting strategy to look forward. It may be safely concluded that the developments in the area of SACs for biomass hydrolysis are progressing rapidly and practical solutions to the efficient and economic hydrolysis of lignocellulosic biomass using reusable catalysts may not be too far.

References

Altava, B., Burguete, M.S., and S.V. Luis. 2008. Polymer supported organocatalysts. In *The Power of Functional Resins in Organic Synthesis*, eds. J. Tulla-Puche and F Albericio, pp. 247–298. Weinheim, Germany: Wiley-VCH Verlag GmbH Co.

Arantes, V. and J.N. Saddler. 2010. Access to cellulose limits the efficiency of enzymatic hydrolysis: The role of amorphogenesis. *Biotechnol Biofuel* 3:1–11.

Arantes, V. and J.N. Saddler. 2011. Cellulose accessibility limits the effectiveness of minimum cellulase loading on the efficient hydrolysis of pretreated lignocellulosic substrates. *Biotechnol Biofuel* 4:1–17.

Avram, A., Lei, J., Qian, X., Wickramasinghe, S.R., and A. Beier. 2013. Novel catalytic membranes for combined biomass hydrolysis and sugar recovery. In *Proceedings of the 2013 International Congress on Energy (ICE)*, American Institute of Chemical Engineers, 330e.

Bergius, F. 1937. Conversion of wood to carbohydrates. *Ind Eng Chem* 29:247–253.

Bhaumik, P., Deepa, A.K., Kane, T., and P.L. Dhepe. 2014. Value addition to lignocellulosics and biomass-derived sugars: An insight into solid acid-based catalytic methods. *J Chem Sci* 126:373–385.

Binder, J.B. and R.T. Raines. 2010. Fermentable sugars by chemical hydrolysis of biomass. *Proc Natl Acad Sci USA* 107:4516–4521.

Camacho, F., González-Tello, P., Jurado, E., and A. Robles. 1996. Microcrystalline-cellulose hydrolysis with concentrated sulphuric acid. *J Chem Technol Biotechnol* 6:350–356.

Chang, L.L.Y. 2002. *Industrial Mineralogy: Materials, Processes, and Uses*. Upper Saddle River, NJ: Prentice Hall.

Chidambaram, M., Curulla-Ferre, D., Singh, A.P., and B.G. Anderson. 2003. Synthesis and characterization of triflic acid-functionalized mesoporous Zr-TMS catalysts: Heterogenization of CF_3SO_3H over Zr-TMS and its catalytic activity. *J Catal* 220:442–456.

Chung, P.W., Charmot, A., Gazit, O.M., and A. Katz. 2012. Glucan adsorption on mesoporous carbon nanoparticles: Effect of chain length and internal surface. *Langmuir* 28:15222–15232.

Deng, W., Zhang, Q., and Y. Wang. 2012. Polyoxometalates as efficient catalysts for transformations of cellulose into platform chemicals. *Dalton Trans* 41:9817–9831.

Dhepe, P.L. and A. Fukuoka. 2008. Cellulose conversion under heterogeneous catalysis. *ChemSusChem* 1:969–975.

Dutta, A., Patra, A.K., Dutta, S., Saha, B., and A. Bhaumik. 2012. Hierarchically porous titanium phosphate nanoparticles: An efficient solid acid catalyst for microwave assisted conversion of biomass and carbohydrates into 5-hydroxymethylfurfural. *J Mater Chem* 22:14094–14100.

Dutta, S., De, S., Alam, I., Abu-Omar, M., and B. Saha. 2012. Direct conversion of cellulose and lignocellulosic biomass into chemicals and biofuel with metal chloride catalysts. *J Catal* 288:8–15.

Emmel, A., Mathias, A.L., Wypych, F., and L.P. Ramos. 2003. Fractionation of Eucalyptus grandis chips by dilute acid-catalysed steam explosion. *Bioresour Technol* 86:105–115.

Faith, W.L. 1945. Development of the scholler process in the United States. *Ind Eng Chem* 37:9–11.

Fukuhara, K., Nakajima, K., Kitano, M., Kato, H., Hayashi, S., and M. Hara. 2011. Structure and catalysis of cellulose-derived amorphous carbon bearing SO$_3$H groups. *ChemSusChem* 4:778–84.

Fukuoka, A. and P.L. Dhepe. 2006. Catalytic conversion of cellulose into sugar alcohols. *Angew Chem Int Ed* 45:5161–5163.

Ghosh, P. and T.K. Ghose. 2003. Bioethanol in India: Recent past and emerging future. *Adv Biochem Eng Biotechnol* 85:1–27.

Gill, C.S., Price, B.A., and C.W. Jones. 2007. Sulfonic acid-functionalized silica-coated magnetic nanoparticle catalysts. *J Catal* 251:145–152.

Goswami, M., Meena, S., Navatha, S. et al. 2015. Hydrolysis of biomass using a reusable solid carbon acid catalyst and fermentation of the catalytic hydrolysate to ethanol. *Bioresour Technol* 188:99–102.

Guo, F., Fang, Z., Xu, C.C., and R.L. Smith. 2012. Solid acid mediated hydrolysis of biomass for producing biofuels. *Prog Energ Combust Sci* 38:672–690.

Hara, M. 2010. Biomass conversion by a solid acid catalyst. *Energ Environ Sci* 3:601–607.

Hara, M., Nakajima, K., and K. Kamata. 2015. Recent progress in the development of solid catalysts for biomass conversion into high value-added chemicals. *Sci Technol Adv Mater* 16:034903–034924.

Harris, E.E. and E. Beglinger. 1946. Madison wood sugar process. *Ind Eng Chem* 38:890–895.

Hegner, J., Pereira, K.C., DeBoef, B., and B.L. Lucht. 2010. Conversion of cellulose to glucose and levulinic acid via solid-supported acid catalysis. *Tetrahedron Lett* 51:2356–2358.

Hill, C.L. 2007. Progress and challenges in polyoxometalate-based catalysis and catalytic materials chemistry. *J Mol Catal A Chem* 262:2–6.

Hu, S., Smith, T.J., Lou, W., and M. Zong. 2014. Efficient hydrolysis of cellulose over a novel sucralose-derived solid acid with cellulose-binding and catalytic sites. *J Agric Food Chem* 62:1905–1911.

Huang, Y.B. and Y. Fu. 2013. Hydrolysis of cellulose to glucose by solid acid catalysts. *Green Chem* 15:1095–1111.

Ishida, K., Matsuda, S., Watanabe, M. et al. 2014. Hydrolysis of cellulose to produce glucose with solid acid catalysts in 1-butyl-3-methyl-imidazolium chloride ([bmIm][Cl]) with sequential water addition. *Biomass Conv Bioref* 4:323–331.

Jiang, Y., Li, X., Wang, X. et al. 2012. Effective saccharification of lignocellulosic biomass over hydrolysis residue derived solid acid under microwave irradiation. *Green Chem* 14:2162–2167.

Kaduk, J.A. and J. Faber. 1995. Crystal structure of Zeolite Y as a function of ion exchange. *Rigaku J* 12:14–34.

Kaldstrom, M., Kumar, N., Tenho, M., Mokeev, M.V., Moskalenko, Y.E., and D.Y. Murzin. 2012. Catalytic transformations of birch kraft pulp. *ACS Catal* 2:1381–1393.

Kim, S.J., Dwiatmoko, A.A., Choi, J.W., Suh, Y.W., Suh, D.J., and M. Oh. 2010. Cellulose pretreatment with 1-n-butyl-3-methylimidazolium chloride for solid acid catalyzed hydrolysis. *Bioresour Technol* 101:8273–8279.

Kitano, M., Yamaguchi, D., Suganama, S. et al. 2009. Adsorption-enhanced hydrolysis of β-1,4-glucan on graphene-based amorphous carbon bearing SO_3H, COOH, and OH groups. *Langmuir* 25:5068–5075.

Kobayashi, H., Komanoya, T., Hara, K., and A. Fukuoka. 2010. Water-tolerant mesoporous-carbon-supported ruthenium catalysts for the hydrolysis of cellulose to glucose. *ChemSusChem* 3:440–443.

Kobayashi, H., Ohta, H., and A. Fukuoka. 2012. Conversion of lignocellulose into renewable chemicals by heterogeneous catalysis. *Catal Sci Technol* 2:869–883.

Kondo, J.N., Yamashita, T., Nakajima, K., Lu, D., Hara, M., and K. Domen. 2005. Preparation and crystallization characteristics of mesoporous TiO_2 and mixed oxides. *J Mater Chem* 15:2035–2040.

Kuznetzov, B.N., Chesnokov, N.V., Yatsenkova, O.V., and V.I. Sharypov. 2013. New methods of heterogeneous catalysts for lignocellulosic biomass conversion to chemicals. *Russ Chem Bull Int Ed* 62:1493–1502.

Lai, D., Deng, L., Guo, Q., and Y. Fu. 2011a. Hydrolysis of biomass by magnetic solid acid. *Energ Environ Sci* 4:3552–3557.

Lai, D.M., Deng, L., Li, J., Liao, B., Guo, Q.X., and Y. Fu. 2011b. Hydrolysis of cellulose into glucose by magnetic solid acid. *ChemSusChem* 4:55–58.

Lam, E. and J.H.T. Luong. 2014. Carbon materials as catalyst supports and catalysts in the transformation of biomass to fuels and chemicals. *ACS Catal* 4:3393–3410.

Li, S., Qian, E.W., Shibata, T., and M. Hosom. 2012. Catalytic hydrothermal saccharification of rice straw using mesoporous silica based solid acid catalyst. *J Jpn Pet Inst* 55:250–260.

Li, X., Jiang, Y., Wang, L., Menq, L., Wang, W., and X. Mu. 2012. Effective low temperature hydrolysis of cellulose catalyzed by concentrated $H_3PW_{12}O_{40}$ under microwave irradiation. *RSC Adv* 2:6921–6925.

Liang, X., Li, C., and C. Qi. 2011. Novel carbon-based strong acid catalyst from starch and its catalytic activities for acetalization. *J Mater Sci* 46:5345–5349.

Liu, X.Y., Huang, M., Ma, H.L. et al. 2010. Preparation of a carbon-based solid acid catalyst by sulfonating activated carbon in a chemical reduction process. *Molecules* 15:7188–7196.

Lynd, L.R., Laser, M.S., Bransby, D. et al. 2008. How biotech can transform biofuels. *Nat Biotechnol* 26:169–172.

Ma, H., Li, J., Liu, W. et al. 2014. Hydrothermal preparation and characterization of novel corncob-derived solid acid catalysts. *J Food Agric Chem* 62:5345–5353.

Macht, J., Janik, M.J., Neurock, M., and E. Iglesia. 2007. Catalytic consequences of composition in polyoxometalate clusters with keggin structure. *Angew Chem Int Ed* 46:7864–7868.

Meena, S., Navatha, S., Devi, B.L.A.P., Prasad, R.B.N., Pandey, A., and R.K. Sukumaran. 2015. Evaluation of Amberlyst 15 for hydrolysis of alkali pretreated rice straw and fermentation to ethanol. *Biochem Eng J* 102:49

Nakajima, K. and M. Hara. 2012. Amorphous carbon with SO_3H groups as a solid Brønsted acid catalyst. *ACS Catal* 2:1296–1304.

Namchot, W., Panyacharay,N., Jonglertjunya,W., and C. Sakdaronnarong. 2014. Hydrolysis of delignified sugarcane bagasse using hydrothermal technique catalyzed by carbonaceous acid catalysts. *Fuel* 116:608–616.

Nanda, K.K. 2009. Size-dependent melting of nanoparticles: Hundred years of thermodynamic model. *Pramana* 72:617–628.

Onda, A., Ochi, T., and K. Yanagisawa. 2008. Selective hydrolysis of cellulose into glucose over solid acid catalysts. *Green Chem* 10:1033–1037.

Ormsby, R., Kastner, J.R., and J. Miller. 2012. Hemicellulose hydrolysis using solid acid catalysts generated from biochar. *Catal Today* 190:89–97.

Palkovits, R., Tajvidi, K., Procelewska, J., Rinaldi, R., and A. Ruppert. 2010. Hydrogenolysis of cellulose combining mineral acids and hydrogenation catalysts. *Green Chem* 12:972–978.

Pang, J., Wang, A., Zheng, M., and T. Zhang. 2010. Hydrolysis of cellulose into glucose over carbons sulfonated at elevated temperatures. *Chem Comm* 46:6935–6937.

Parsell, T., Yohe, S., Degenstein, J. et al. 2015. A synergistic biorefinery based on catalytic conversion of lignin prior to cellulose starting from lignocellulosic biomass. *Green Chem* 17:1492–1499.

Prabhavathi, D., Gangadhar, B.L.A., Sai Prasad, K.N., Jagannadh, P.S., and R.B.N. Prasad. 2009. Glycerol-based carbon catalyst for the preparation of biodiesel. *ChemSusChem* 2:617–620.

Ralph, J., Lundquist, K., Brunow, G. et al. 2004. Lignins: Natural polymers from oxidative coupling of 4-hydroxyphenylpropanoids. *Phytochem Rev* 3:29–60.

Rinaldi, N. and A.A. Dwiatmoko. 2012. Hydrolysis process of cellulose and palm stem into total reducing sugars (TRS) over solid acid in 1-*n*-butyl-3-methylimidazoliu chloride: A preliminary study. *Int J Eng Technol* 12:26–29.

Rinaldi, R., Palkovits, R., and F. Schüth. 2008. Depolymerization of cellulose using solid catalysts in ionic liquids. *Angew Chem Int Ed* 47:8047–8050.

Rinaldi, R. and F. Schüth. 2009. Design of solid catalysts for the conversion of biomass. *Energy Environ Sci* 2:610–626.

Salman, N., Rüscher, C.H., Buhl, J.C., Lutz, W., Toufar, H., and M. Stöcker. 2006. Effect of temperature and time in the hydrothermal treatment of HY zeolite. *Micropor Mesopor Mat* 90:339–346.

Scheller, H.V. and P. Ulvskov, P. 2010. Hemicelluloses. *Annu Rev Plant Biol* 61:263–289.

Sherrard, E.C. and F.W. Kressman. 1945. Review of processes in the United States prior to world war II. *Ind Eng Chem* 37:5–8.

Shimizu, K.I., Uozumi, R., and A. Satsuma. 2009. Enhanced production of hydroxymethylfurfural from fructose with solid acid catalysts by simple water removal methods. *Catal Comm* 10:1849–1853.

Shuai, L. and X. Pan. 2012. Hydrolysis of cellulose by cellulase-mimetic solid catalyst. *Energy Environ Sci* 5:6889–6894.

Suganuma, S., Nakajima, K., Kitano, M. et al. 2008. Hydrolysis of cellulose by amorphous carbon bearing SO_3H, COOH, and OH groups. *J Am Chem Soc* 130:12787–12793.

Suganuma, S., Nakajima, K., Kitano, M. et al. 2010. Synthesis and acid catalysis of cellulose-derived carbon-based solid acid. *Solid State Sci* 12:1029–1034.

Tagusagawa, C., Takagaki, A., Iguchi, A. et al. 2010a. Highly active mesoporous Nb-W oxide solid-acid catalyst. *Angew Chem Int Ed* 122:1146–1150.

Tagusagawa, C., Takagaki, A., Iguchi, A. et al. 2010b. Synthesis and characterization of mesoporous Ta–W oxides as strong solid acid catalysts. *Chem Mater* 22:3072–3078.

Takagaki, A., Nishimura, M., Nishimura, S., and K. Ebitani. 2011. Hydrolysis of sugars using magnetic silica nanoparticles with sulfonic acid groups. *Chem Lett* 40:1195–1197.

Takagaki, A., Tagusagawa, C., and K. Domen. 2008. Glucose production from saccharides using layered transition metal oxide and exfoliated nanosheets as a water-tolerant solid acid catalyst. *Chem Commun* 42:5363–5365.

Tian, J., Fan, C., Cheng, M., and X. Wang. 2011. Hydrolysis of cellulose over $Cs_xH_{3-x}PW_{12}O_{40}$ (x =1–3) heteropoly acid catalysts. *Chem Eng Technol* 34:482–486.

Tian, J., Wang, J., Zhao, S., Jiang, C., Zhang, X., and X. Wang. 2010. Hydrolysis of cellulose by the heteropoly acid $H_3PW_{12}O_{40}$. *Cellulose* 17(3):17587–17594.

Toda, M., Takagaki, A., Okamura, M. et al. 2005. Biodiesel made with sugar catalyst. *Nature* 438:178.

Van de Vyver, S., Geboers, J., Jacobs, P.A., and B.F. Sels. 2011. Recent advances in the catalytic conversion of cellulose. *ChemCatChem* 3:82–94.

Van de Vyver, S., Peng, L., Geboers, J. et al. 2010. Sulfonated silica/carbon nanocomposites as novel catalysts for hydrolysis of cellulose to glucose. *Green Chem* 12:1560–1563.

Wang, J., Xi, J., and Y. Wang. 2015. Recent advances in the catalytic production of glucose from lignocellulosic biomass. *Green Chem* 17:737–751.

Wu, Y., Zhang, C., Liu, Y., Fu, Z., Dai, B., and D. Yin. 2012. Biomass char sulfonic acids (BC-SO$_3$H)-catalyzed hydrolysis of bamboo under microwave irradiation. *BioResources* 7:5950–5959.

Wyman, C.E. 1996. *Handbook on Bioethanol Production and Utilization.* Washington, DC: Taylor & Francis.

Wyman, C.E., Dale, B.E., Elander, R.T., Holtzapple, M., Ladisch, M.R., and Y.Y. Lee. 2005. Coordinated development of leading biomass pretreatment technologies. *Bioresour Technol* 96:1959–1966.

Xiang, Q., Kim, J.S., and Y.Y. Lee. 2003. A comprehensive kinetic model for dilute-acid hydrolysis of cellulose. *Appl Biochem Biotechnol* 106:337–352.

Yang, W. and A. Sen. 2010. One step catalytic transformation of carbohydrates and cellulosic biomass to 2,5-dimethyltetrahydrofuran for liquid fuels. *ChemSusChem* 3:597–603.

Zhang, F., Deng, X., Fang, Z., Zeng, H.Y., Tian, X.F., and J.A. Kozinski. 2011. Hydrolysis of crystalline cellulose over Zn–Ca–Fe oxide catalyst. *Petrochem Technol* 40:43–48.

Zhang, Y.H.P. and L.R. Lynd. 2004. Toward an understanding of enzymatic hydrolysis of cellulose: Noncomplexed cellulase systems. *Biotechnol Bioeng* 88:797–824.

Zhang, Z. and Z.K. Zhao. 2009. Solid acid and microwave-assisted hydrolysis of cellulose in ionic liquid. *Carbohyd Res* 344:2069–2072.

Zhou, C.H., Xia, X., Liu, C.X., Tong, D.S., and J. Beltramini. 2011. Catalytic conversion of lignocellulosic biomass to fine chemicals and fuels. *Chem Soc Rev* 40:5588–5617.

7

Microwave-Assisted Pyrolysis of Biomass for Liquid Biofuels Production

Pallavi Yadav, Gaurav Mundada, Bijoy Biswas,
Vartika Srivastava, Rawel Singh, Bhavya B. Krishna,
Jitendra Kumar, and Thallada Bhaskar

Contents

Abstract

Biomass is the only sustainable, renewable, and environment-friendly source of organic carbon that we now obtain from fossil resources. Biomass can be converted to various fuels and chemicals by different biochemical and thermochemical processes. The conventional methods of conversion have several disadvantages of low energy efficiency, poor heat transfer, low carbon utilization, etc. Increasing the efficiency of biomass conversion to obtain more value-added hydrocarbons is a challenging task. In this regard, microwave energy can efficiently be used for different biomass pretreatment and conversion processes with higher energy efficiency compared to the conventional processes. The selective cleavage by molecular interactions would enable the complete organic content utilization thereby producing bulk/specialty chemicals/fuels. Microwave pyrolysis can be used for diversified feedstocks without any energy-intensive pretreatment steps like drying and crushing. Compared to conventional pyrolysis, microwave pyrolysis produces better-quality bio-oil, gas, and biochar. The chapter aims to provide an updated comprehensive review on the usage of microwave energy for the biomass pyrolysis.

7.1 Introduction

Energy is one of several essential inputs to economic and social development, and its demand is increasing day by day. Some commonly used sources of energy around the world are fossil fuels, hydroelectric, solar energy, wind, geothermal, biomass, etc. Fossil-based energy resources, such as petroleum, coal, and natural gas, are responsible for about three quarters of the world's primary energy consumption. As the crude-oil reserves are decreasing, there is an enhanced demand for fuels worldwide along with which are increased climate concerns about the use of fossil-based energy carriers. In this scenario, the focus has recently turned toward improved utilization of renewable energy resources. Alternatives to fossil-based energy resources are nuclear power, hydropower, and biomass.

Biomass is photosynthesized from CO_2 and H_2O, and its storage of solar energy in the form of various polymerized organic compounds is characterized by its abundance, renewability, and CO_2 neutrality. Biomass is probably our oldest source of energy after the sun. Biomass is an abundant and carbon-neutral renewable energy resource for the production of biofuels and valuable chemicals. Energy production from biomass has the advantage of forming smaller amounts of greenhouse gases compared to the conversion of fossil fuels, as the carbon dioxide generated during the energy conversion is consumed during subsequent biomass regrowth (Stocker 2008).

Biomass has become one of the most commonly used renewable sources of energy in the last two decades due to its low cost and indigenous nature, and in that it accounts for almost 15% of the world's total energy supply and as much as 35% in developing countries, mostly for cooking and heating.

The energy policy of India is largely defined by the country's burgeoning energy deficit and increased focus on developing alternative sources of energy. Due to rapid economic expansion, India has one of the world's fastest growing energy markets and is expected to be the second-largest contributor to the increase in global energy demand by 2035, accounting for 18% of the rise in global energy consumption. Given India's growing energy demands and limited domestic fossil fuel reserves, the country has ambitious plans to expand its renewable and nuclear power industries (https:// en.wikipedia.org/wiki/Energy_policy_of_India). Biomass energy constitutes <13% of renewable energy in India. India is an agri-based economy. Every year tons of agricultural and horticulture wastes are produced. Some statistical data about biomass production are given in Table 7.1 (http://powermin.nic.in/upload/pdf/ddg_based_ renewable.pdf). Crop residues that are not used as animal fodder like cane trash, paddy straw, and coconut fronds can be used as feedstock for fuel and chemical production.

Abundant and inexpensive lignocellulosic biomass does not compete with the production of food crops. Cellulose is a polydisperse linear homopolymer consisting of regio- and stereoselectively β-1,4-glycosidic linked D-glucopyranose units. Unlike the β-1,4-glycosidic bonding present in starch, β-linkages between the equatorial hydroxyl groups on the C1 and C4 carbon atoms in cellulose lead to a straight chain polymer with no coils or branches, and these molecules adopt an extended and rather stiff rodlike conformation (Stocker 2008; Yang et al. 2011).

Hemicellulose, the second most abundant component of lignocellulose, is composed of various 5- and 6-carbon sugars such as arabinose, galactose, glucose, mannose, and xylose. Hemicelluloses (copolymer of any of the monomers glucose, galactose, mannose, xylose, arabinose, and glucuronic acid) are plant cell wall polysaccharides that are not solubilized by water but are solubilized by aqueous alkali. Hemicellulose surrounds the cellulose fibers and is a linkage between cellulose and lignin (Stocker 2008). Major hemicelluloses in coniferous wood (softwood) are glucomannan, galactoglucomannan, and arabinoglucuronoxylan. Other softwood hemicelluloses are arabinogalactan, xyloglucan, and other glucans. Glucomannans are the principal hemicelluloses in softwood. The main hemicellulose in hardwood is a xylan, more specifically an O-acetyl-4-O-methylglucurono-β-D-xylan. Hardwood also contains glucomannan with a backbone of β-(1,4)-linked D-mannopyranose and D-glucopyranose.

Lignin is a three-dimensional amorphous polymer consisting of methoxylated phenyl propane structures. In plant cell walls, lignin fills the spaces between cellulose and hemicelluloses, and it acts like a resin that holds the lignocellulose matrix together. Cross-linking with the carbohydrate polymers then confers strength and rigidity to

TABLE 7.1 Biomass Availability in India

Biomass Residue Type	Amount Produced (Mill MT/Year)
Branches/toppings of eucalyptus, cotton/pulses stalks, mustard waste, etc.	75
Mill residues (e.g., bagasse, rice/groundnut husk, corn cobs, saw mill waste, de-oiled cake, etc.)	145
Animal waste (cow dung, poultry waste, etc.)	1200
Horticultural waste (farms and wholesale markets)	75

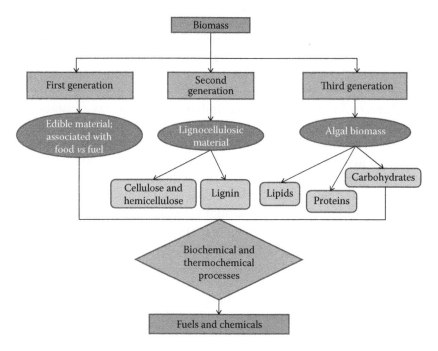

Figure 7.1 Various generations of biofuels.

the system. The three primary monolignols p-coumaryl alcohol, coniferyl alcohol, and sinapyl alcohol are the building blocks of lignin; these monolignols are linked by β-*O*-4,5-5, β-5, 4-*O*-5, β-1, dibenzodioxocin, and β-β linkages, of which the β-*O*-4 linkage is dominant, consisting of more than half of the linkage structures of lignin.

Algae can be referred to as plantlike organisms that are usually photosynthetic and aquatic, but do not have true roots, stems, leaves, and vascular tissue and do have simple reproductive structures. They are distributed worldwide in the sea, in freshwater, and in wastewater. Algae are classified as Bacillariophyta (diatoms), Charophyta (stoneworts), Chlorophyta (green algae), Chrysophyta (golden algae), Cyanobacteria (blue-green algae), Dinophyta (dinoflagellates), Phaeophyta (brown algae), and Rhodophyta (red algae). Based on the type of biomass, various generations of biofuels can be proposed that is presented in Figure 7.1.

7.2 Biomass Conversion Methods

Many biochemical and thermochemical processes have been used for the conversion of biomass to energy/value-added hydrocarbons (Faaij 2006; Fernández et al. 2011) While both methods of processing can be used to produce fuels and chemicals, thermochemical processing can be seen as being the easiest to adapt to current energy infrastructures and to deal with the inherent diversity in biomass. Biochemical routes involve several steps such as pretreatment, fermentation, and separation of products by various unit processes. The base of thermochemical conversion is the pyrolysis

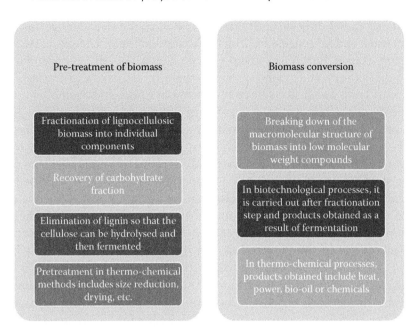

Figure 7.2   Comparison of biomass pretreatment and biomass conversion methods.

process in most cases. The products of conversion include water, charcoal (carbonaceous solid), biocrude, tars, and permanent gases including methane, hydrogen, carbon monoxide, and carbon dioxide depending upon the reaction parameters such as environment, reactors used, final temperature, rate of heating, and source of heat (Bhaskar et al. 2011). A comparison of biomass pretreatment and biomass conversion processes is shown in Figure 7.2.

Basically there are three different thermochemical conversion routes used for biomass: combustion (complete oxidation), gasification (partial oxidation) and pyrolysis (thermal decomposition without oxygen). Pyrolysis offers a flexible and attractive way of converting solid biomass into an easily stored and transportable fuel, which can be successfully used for the production of heat, power, and chemicals. In pyrolysis processes, the feed material is heated in the absence of oxygen across a range of temperatures from 200°C to 600°C, depending on the exact composition of the feedstock, which leads to thermal decomposition of the solid and the evolution of volatile and semivolatile compounds. Gasification process is a thermal process that utilizes controlled air to support combustion that results in a N_2-rich, low-Btu fuel gas. If gasification is conducted using pure oxygen, then higher-Btu fuel is produced, and if the gasification process uses steam to support combustion, then the output is a synthetic gas. Among them, combustion (also called incineration) is the most established route in industry, but this is also associated with the generation of carbon oxides, sulfur, nitrogen, chlorine products (dioxins and furans), volatile organic compounds, polycyclic aromatic hydrocarbons, dust, etc. On the contrary, gasification and pyrolysis offer the potential for greater efficiencies in energy production and less pollution (Elías 2005). Although pyrolysis is still under development, this process has received special attention, not only as a primary process of combustion and gasification but also as an independent process leading to the production of energy-dense products with numerous uses. Pyrolysis converts a raw material into different products: solid (char), liquid (bio-oil), and gaseous products (light-molecular-weight gases). The heat of reaction is from an ex situ source producing bio-oil, which contains up to 70% of the energy of the biomass feed in catalytic process like fast pyrolysis. However, the bio-oil properties such as its low heating value, incomplete volatility, acidity, instability, and incompatibility with standard petroleum fuels significantly restrict its application. The undesirable properties of pyrolysis oil result from the chemical composition of bio-oil that mostly consists of different classes of oxygenated organic compounds. Thermochemical routes also require feedstock pretreatment steps such as moisture reduction and size reduction. Thermal energy required for the conventional pyrolysis is derived from fossil carbon–based resources, and the efficiency of thermal energy production is very less. On the other hand, biomass is a bad heat conductor, so during conventional pyrolysis, there are more heat losses and very less energy efficiency. To overcome such issues, microwave heating as substitute to conventional heating during biomass conversion processes has been proposed in the literature. The efficiency during microwave processing is high compared to conventional heating as heat is directly generated due to interaction of electromagnetic radiations with biomass, and less time is required to heat biomass. Microwave pyrolysis of various biomass feedstocks has also been reported in the literature and has been reviewed by various authors (Luque et al. 2012; Macquarrie et al. 2012). Microwave pyrolysis of biomass is discussed in detail in the following sections.

7.3 Advantages of Microwave Pyrolysis over Conventional Pyrolysis

Microwave generates thermal energy through dielectric heating, and the energy is introduced into the reactor remotely without any contact between the energy source and the reaction mixture (http://www.biotage.com). The interaction of microwaves with any material depends on its dielectric properties such as dielectric constant and dielectric loss factor. The dielectric constant is a measure of the ability of a material to store electromagnetic energy, and the dielectric loss factor is a measure of the ability of a material to convert electromagnetic energy into heat (Kumar et al. 2007). Loss tangent, which is a ratio of the dielectric loss factor to the dielectric constant, is a parameter used to describe the overall efficiency of a material to utilize energy from microwave radiation (Nelson and Datta 2001). During microwave heating process, energy transfer occurs through the interaction of molecules or atoms. Compared with conventional heating methods, more uniform temperature distribution can be achieved, and the undesired secondary reactions may be avoided. As a result, better control of the process and more desired products will be obtained (Yu et al. 2006).

Microwave heating reveals higher heating rates due to the fact that microwave energy is delivered directly into the material through molecular interaction with the electromagnetic field, and no time is wasted in heating the surrounding area. Therefore, significant savings of time and energy are achieved in microwave pyrolysis, although other effects can also be deduced affecting the volatiles' profiles. Generally, high heating rates improve the devolatilization of the material reducing the conversion times. The heating rate also has an influence on the residence time of volatiles, whose flow occurs from the more internal hot zones toward the external cold regions of the sample. The higher heating rate, the shorter residence time, and the faster the volatiles arrive to the external cold regions, which, in turn, reduces the activity of secondary reactions of vapor phase products. This results in high yields of liquid and reduced deposition of refractory condensable material on the char's internal surface.

During microwave processing of biomass, the enthalpy of the reaction is provided by the microwave energy. The comparison of microwave heating with conventional heating methods is given in Table 7.2 (http://www.milestonesci.com; http://www.maos.net; http://www.cem.com; Abramovitch 1991; Caddick 1995; Saillard et al. 1995; Strauss and Trainor 1995; Westaway and Gedye 1995; Stuerga and Gaillard 1996; Bose et al. 1997; Langa et al. 1997; Gabriel et al. 1998; Lidstrom et al. 2001; Kuhnert 2002; Nuchter et al. 2003, 2004; Leadbeater 2004; Adnadjevic and Jovanovic 2011).

In contrast with conventional heating mechanisms, where energy is first converted to heat then transferred along temperature gradients from the surface to the core of the material, microwaves induce heat at the molecular level by direct conversion of the

TABLE 7.2 Comparison of Microwave Heating with Conventional Heating

Microwave Heating	Conventional Heating
Heating of reaction mixture starts from inside core to outside surface.	Heating of reaction mixture starts from outside surface to inside core.
No need of physical contact of reaction mixture with the higher-temperature source.	The vessel should be in physical contact with higher-temperature source (e.g., oil bath, steam bath).
Type of heating is electromagnetic wave heating.	Type of heating is thermal or electric source heating.
Heating takes place by dielectric polarization and conduction.	Heating takes place by conduction mechanism.
In microwave, the temperature of mixture can be raised more than its boiling point, i.e., superheating can take place.	In conventional heating, the highest temperature (for open vessels) that can be achieved is limited by boiling point of particular mixture.
Heating rate is rapid.	Heating rate is slow.
Heating is volumetric and selective.	Heating is superficial and nonselective.
Heating is uniform throughout as electromagnetic energy is directly converted into heat energy.	Heating is nonuniform.
Dependent on the properties of the materials.	Less dependent on the properties of the materials.
Heating efficiency is high as heating is in situ.	Heating efficiency is low as heat is transferred from an external source.
The core mixture is heated directly so energy loss is minimal.	Transfer of energy occurs from the wall or surface of vessel, to the mixture and eventually to reacting species so energy loss is high.
Superior product quality and temp control is easier.	Product quality and temperature control is difficult.

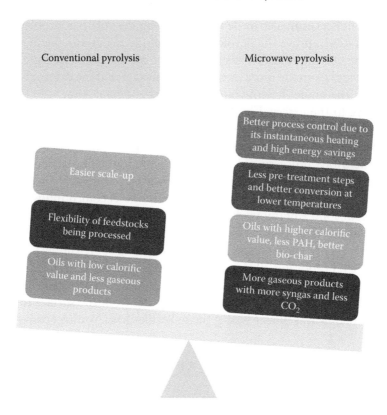

Figure 7.3   Comparison of microwave pyrolysis and conventional pyrolysis.

electromagnetic field into heat. The electromagnetic field enters the solid and induces heating throughout the penetration depth through interaction with polarizable dipoles present in the target material. Two main factors affect the penetration depth and level of heating with microwaves: the nature of the electromagnetic field (strength and frequency) and dielectric properties of the material being heated (polarizability of the dipoles and capacity for interacting with microwave energy) (Veggi et al. 2013).

Microwave irradiation could change the ultra structure of cellulose and degrade lignin and hemicellulose. Microwave causes cellulosic breakdown mainly through molecular collision due to dielectric polarization, and during microwave heating process, energy transfer occurs through the interaction of molecules or atoms. A general comparison of microwave and conventional pyrolysis is shown in Figure 7.3 (Luque et al. 2012; Macquarrie et al. 2012).

7.4 Basic Properties of Microwaves

Microwave is electromagnetic wave lying in electromagnetic spectrum between infrared waves and radio waves. Microwaves have frequency range of 0.3–300 GHz having wavelength of 0.01 and 1 m. During World War II, two scientists, John Randall and Henry H.A. Boot, invented the magnetron, a tube that produced microwave.

Microwave energy quantum is given by the well-known equation

$$\text{Energy, } E = h\nu,$$

where
   h is the Plank's constant with value 6.634×10^{-34} Js
   ν is the frequency

The energy associated with frequency $0.3 - 300$ GHz is 1.24×10^{-6} to 1.24×10^{-3} eV, respectively. Microwave energy is lower than Brownian motion, so it is not enough strong to break chemical bond as such microwave cannot induce chemical reaction. The energy associated with microwave are much lower than ionization energy of biological compound 13.6 eV, covalent bond 5 eV, hydrogen bond 2 eV, van der Waals intermolecular interaction less than 2 eV, and Brownian motion 2.7×10^{-3} eV (Metaxas and Meredith 1993; Chemat-Djenni et al. 2007). The influence of microwave energy in both chemical and biochemical method is thermal and nonthermal.

Microwave will reach the object to be heated at the same speed of light. Then it enters into the object as a wave, and by getting absorbed, the object generates heat (Surati et al. 2012).

In conventional heating, the object's temperature rises by spreading heat energy from the surface to the inside (external heating). On the other hand, by microwave heating, the object will generate heat on its own by the penetration of the microwave. This is not necessary to consider about the heat conduction. That is why rapid heating is possible by microwave. Although the object has to be large enough for the microwave to penetrate, the smaller objects will also be heated from the inside as the depth of microwave penetrates.

As we know, microwave heating is dependent on the microwave energy absorbed by the material. For example, a borosilicate glass is sold as microwavable glass container. When this glass is heated with water in it, only the water gets heated. That's because microwave power absorption of the glass is ignored since it is only 3000/1th of the water. When a good material of container is selected, microwave can only heat the object and heat efficiency improves substantially. Due to internal and high heating efficiency, microwave heating allows rapid response. For example, heating can be started and stopped instantly. In addition, by the adjustment of microwave output, the amount of heat energy generated inside the heated object can be controlled.

During microwave heating, objects with complicated shape can be heated relatively uniform. Microwave doesn't require a medium, because it propagates only by changes of electric fields and magnetic fields. It can propagate in a vacuum. It reaches the object and penetrates without heating the air. The heated object generates heat by absorbing microwave energy to convert it to heat energy. Conventional heating requires a heat source, and the temperature rises not only of heated object but also of heat source and the heating furnace. So the temperature of room equipped with heating furnace goes high because of radiant heat. During microwave heating, only the temperature of the object rises. (http://www.microdenshi.co.jp/en/microwave/).

7.5 Interaction of Microwaves with Biomass

The dielectric properties of biomass will directly affect the heating characteristics of a reaction system under microwave irradiation. The most important dielectric property is the loss tangent (tan δ), which is the ability of a specific substance to convert electromagnetic energy into heat at a given temperature and frequency and can be obtained as a quotient of dielectric loss (ε'') and the dielectric constant (ε'):

$$\frac{\varepsilon''}{\varepsilon'} = \tan \delta$$

High–tan δ reaction media have high absorption of microwaves and consequently give fast and efficient heating. The loss tangent is not directly related to dielectric constant, and a solvent such as water, which has a high dielectric constant value, does not necessarily have a high tan δ. It is important to note that microwave heating also depends on the penetration depth, relaxation time, etc.

The loss tangent is dependent on both temperature and frequency applied to the reaction media. For water, for a frequency of 2.45 GHz, the dielectric constant and dielectric loss both decrease with increasing temperatures being the solvent almost transparent at high frequency medium. It is very difficult to heat water at high temperatures, above 100°C, with microwave irradiation (2.45 GHz) since the dielectric loss (ε'') is close to zero (Kappe et al. 2008).

7.5.1 Microwave Heating Mechanism

7.5.1.1 Dipolar Polarization Heating Mechanism

Dipolar polarization is a process by which heat is generated in polar molecules. On exposure to an oscillating electromagnetic field of appropriate frequency, polar molecules try to follow the field and align themselves in phase with the field. However, owing to intermolecular forces, polar molecules experience inertia and are unable to follow the field. This results in the random motion of particles, and this random interaction generates heat. Dipolar polarization can generate heat by either one or both the following mechanisms:

1. Interaction between polar solvent molecules such as water, methanol, and ethanol
2. Interaction between polar solute molecules such as ammonia and formic acid

The key requirement for dipolar polarization is that the frequency range of the oscillating field should be appropriate to enable adequate interparticle interaction. If the frequency range is very high, intermolecular forces will stop the motion of a polar molecule before it tries to follow the field, resulting in inadequate interparticle interaction. On the other hand, if the frequency range is low, the polar molecule gets sufficient time to align itself in phase with the field. Hence, no random interaction takes place between the adjoining particles. Microwave radiation has the appropriate frequency (0.3–30 GHz) to oscillate polar particles and enable enough interparticle interaction. This makes it an ideal choice for heating polar solutions (De Souza 2014). In addition, the energy in a microwave photon (0.037 kcal/mol) is very low, relative to the typical energy required to break a molecular bond (80–120 kcal/mol). Therefore, microwave excitation of molecules does not affect the structure of an organic molecule, and the interaction is purely kinetic (Boye 2005).

7.5.1.2 Interfacial Polarization Heating Mechanism

Interfacial polarization is an effect, which is very difficult to treat in a simple manner, and easily viewed as a combination of the conduction and dipolar polarization effects. This mechanism is important for a system where a dielectric material is not homogenous but consists of conducting inclusion of one dielectric in other.

7.5.1.3 Conduction Heating Mechanism

The conduction mechanism generates heat through resistance to an electric current. The oscillating electromagnetic field generates an oscillation of electrons or ions in a conductor, resulting in an electric current. This current faces internal resistance, which heats the conductor. The main limitation of this method is that it is not applicable for materials that have high conductivity, since such materials reflect most of the energy that falls on them (Bogdal 2005).

7.6 Microwave Furnaces and Reactors

Microwave furnace consists of three major components, namely, source, transmission lines, and applicator. The source generates electromagnetic radiation, and the transmission lines deliver the electromagnetic energy from the source to the applicator while the energy is either absorbed or reflected by the material in the applicator. Generally, the processing is performed within a metallic applicator (e.g., single-mode applicator, travelling wave applicator, and multimode applicator). The type of applicator used depends on the materials to be processed. The single-mode applicator and the travelling wave applicator are successful in processing materials of simple geometries. However, the multimode applicator has the capability to produce large and complex components (Sutton 1989; Thostenson and Chou 1999; Das et al. 2009).

In the monomode microwave oven, the dimensions of wave belt (wave guide) and excitations are specially calculated so to allow only one mode of propagation or resonance. They are able to obtain a homogeneous distribution of the electric field in the wave belt (focalized fasciculus) and hence in the heated reaction mixtures. They are used with less power emitted with a high return of energy, and thus, the utilization of monomode reactor is energy efficient and leads to better yields in organic synthesis, while preserving the thermally unstable products (Zhang and Li 1999).

Due to different technical difficulties, specifically designed microwave reactors are required. Some of the microwave reactors used in organic synthesis are rotative solid-phase microwave reactor, continuous microwave reactor, and microwave batch reactor.

Multimode ovens (with limited power 800–1000 W) are characterized by a nonhomogeneous distribution of electric field, and their use in synthetic purposes requires mapping of the field that involves determination of hot spots of high energy (Loupy and Petit 1997). The use of multimode reactors has, however, the following limitations: (1) the distribution of electric field inside the cavity results from multiple reflections off the walls and reaction vessel and is consequently heterogeneous, (2) the temperature cannot be simply and accurately measured, and (3) the power is not tunable.

Microwave heating has also been used in stirred-tank reactors; some of the examples include esterification of benzoic acid with ethanol by conventional and microwave heating (Plazl et al. 1995), hydrolysis of sucrose by conventional and microwave heating (Plazl 2002), and esterification of benzoic acid and with 2-ethylhexanol (Plazl et al. 1997; Surati et al. 2012).

7.7 Microwave-Assisted Conversion of Biomass

7.7.1 Microwave Pyrolysis of Biomass

7.7.1.1 Role of Microwave Heating on the Pyrolysis Conditions

During microwave pyrolysis, different temperature distributions, the heating rate, and the residence time of volatiles are observed when compared with those during conventional pyrolysis, which leads to the different final pyrolysis products portfolio. During conventional pyrolysis, heat is transferred from the surface toward the center of the material by convection, conduction, and radiation. On the other hand, during microwave pyrolysis, heat can be generated throughout the volume of the material, rather than from an external because microwaves can penetrate materials and deposit energy. Microwave heating is energy conversion rather than heat transfer where the electromagnetic energy of microwave interacts with materials with different properties generating heat energy. Opposite thermal gradients, that is, temperature distribution, are found in both heating systems. In microwave heating, the material is at higher temperature than the surrounding area, unlike conventional heating where conventional furnace cavity reaches the operating temperature, to begin heating the material. Consequently, microwave heating favors the reactions involving the solid material, that is, devolatilization or heterogeneous reactions (Zhang and Hayward 2006), and conventional heating improves the reactions that take place in surroundings, such as homogeneous reactions in the gas phase. Additionally, the lower temperatures in the microwave cavity can be useful to avoid undesirable reactions during conventional pyrolysis.

The conversions in microwave are always higher than those observed in conventional heating at any temperature. Additionally, the differences between both heating methods seem to be reduced with the temperature increase, which points to the higher efficiency of microwave heating at lower temperatures. Other heterogeneous reactions that can be found in a pyrolysis process are those in which the solid char acts not only as a reactive but also as a catalyst owing to the carbonaceous or metallic active center located in the surface. These provide a catalytic effect for specific reactions, such as the methane decomposition reaction. This reaction has also been proved separately to give better conversion under microwave heating (Domínguez et al. 2008).

The differing performance between conventional and microwave heating is also translational differential heating rates of the material. During microwave heating, no time is wasted in heating the surrounding area. Therefore, significant savings of time and energy are achieved in microwave pyrolysis. The heating rate also influences the residence time of volatiles, whose flow occurs from the more internal hot zones toward the external cold regions of biomass. The higher heating rate, the shorter residence time, and the faster the volatiles arrive to the external cold regions, which reduces the secondary reactions of vapor phase products that produce high amount of liquid and reduce the deposition of refractory condensable material on the char's internal surface (Allan et al. 1980).

In the case of materials with significant moisture content, like most biomass feedstocks, microwave pyrolysis can effectively be utilized. Since water molecules have high affinity for microwaves, moisture content within a given biomass particle is selectively targeted by incidental microwaves. Microwaves vaporize moisture in the depth of the particle, prior to volatilizing organic content. The steam generated is rapidly released into the surrounding area, not only sweeping volatiles but also creating preferential channels in the carbonaceous solid that increase its porosity. This, in turn, favors the release of volatiles at low temperatures, and hence, its reaction with the steam produced leads to partial oxidation and formation of permanent gases (H_2, CO, CH_4, CO_2) (Fernández et al. 2011).

7.7.1.2 Microwave-Assisted Pyrolysis of Biomass and Model Compounds
The yield and quality of bio-oil, biochar, and gases during microwave pyrolysis are affected by some critical parameters that should be optimized to get the desired products. These variables include type and size of biomass; moisture content of biomass; reaction temperature; reaction time; microwave output power; microwave reactor type (multimode or single mode); reactor design; microwave receptor type, size, and amount; catalyst type and concentration; stirring; and type and flow rate of carrier gas (Motasemi and Afzal 2013). Effect of these variables on microwave pyrolysis of various biomass and model compounds studied by various authors is presented in the following text.

Direct conversion cellulose and sugarcane bagasse into 5-hydroxymethylfurfural (HMF) using single or combined metal chloride catalysts in DMA-LiCl solvent under microwave-assisted heating was carried out. $Zr(O)Cl_2/CrCl_3$ combined catalyst was most effective enabling 57% HMF from cellulose fiber. $Zr(O)Cl_2/CrCl_3$ catalyst was also effective for the conversion of sugarcane bagasse to HMF and 5-ethoxymethyl-2-furfural (Dutta et al. 2012).

The decomposition of cellulose by microwave plasma (MWP) and radio-frequency plasma (RFP) was examined. The conversion of cellulose by MWP (XMWP) was higher than that by RFP (XRFP) at all residence times. It was found that XMWP and XRFP reach 92.8 wt.% at 10 min and 68.1 wt.% at 30 min (Konno et al. 2011).

Pretreatment of cellulose in ionic liquids (ILs) with microwave heating could cause a significant decrease in the degree of polymerization of cellulose dissolved in ILs that enhance the cellulase-catalyzed cellulose hydrolysis (Koo and Ha 2010).

Xie et al. (2010) demonstrated that cellulose and lignocellulose can be hydrolyzed much more efficiently in ILs by mineral acid or acidic ILs under mild conditions, and use of microwave irradiation could accelerate the conversion of cellulose to hydroxymethyl furfural or furan dramatically with high yield and selectivity (Xie et al. 2010).

Catalytic degradation of lignin model compounds (anisole, diphenyl ether, and phenethyl phenyl ether) under microwave was carried out by Xiong using tetralin as solvent and *p*-toluenesulfonic acid as catalyst (Xiong 2012). Selective cleavage of ether bonds in lignin molecule were observed under microwave. In addition, activation energies in the conversion of anisole (62.91 kJ/mol) and phenethyl phenyl ether (97.77 kJ/mol) in microwave reactor were lower than that in batch reactor (81.97 kJ/mol and 113.37 kJ/mol) indicating that microwave can accelerate degradation of lignin model compounds. Badamali et al. (2009) investigated microwave-assisted H_2O_2 oxidation of lignin model monomer apocynol, 1-(4-hydroxy-3-methoxyphenoxy)-ethanol with mesoporous SBA-15 as catalyst to produce acetovanillone, vanillin, and 2-methoxybenzoquinone. The oxidative degradation of a lignin model phenolic dimer (1-(4-hydroxy-3-methoxyphenoxy)-2-(2-methoxyphenoxy)-propane-1,3-diol) catalyzed by Co (salen)/SBA-15 with the assistance of microwave irradiation was also studied by Badamali et al. (2011). The phenolic dimer was found to be selectively oxidized to 2-methoxy phenol. Comparatively, reactions with conventional heating led to oligomerization of the dimer and resulted in a mixture of products. These authors used MCM-41, HSM, SBA-15, and amorphous silica as catalyst to convert lignin model compounds, apocynol, under microwave irradiation (Badamali et al. 2013; Li et al. 2014).

The microwave and conventional pyrolysis of sewage sludge using graphite and char as microwave absorbers was investigated (Dominguez et al. 2006). Both H_2 and CO were produced in a higher proportion by microwave pyrolysis than by conventional pyrolysis.

Microwave pyrolysis of coffee hulls was studied (Dominguez et al. 2007; Menendez et al. 2007). Different mechanisms of heating that take place in the microwave, in comparison to conventional heating, give rise to the formation of "micro plasmas," which induce self-gasification of the char that is being formed. Microwave treatment produces more gas and less oil than conventional pyrolysis. Also gas from the microwave pyrolysis has much higher H_2 and syngas (H_2 + CO) contents (up to 40 and 72 vol.%, respectively) than those obtained by conventional pyrolysis (up to 30 and 53 vol.%, respectively), in an electric furnace, at similar temperatures. Compared with the conventional pyrolysis conducted by electric furnace, the microwave pyrolysis produces more content of H_2 and CO (Menendez et al. 2004). In addition to microwave pyrolysis generating less polycyclic aromatic hydrocarbons (PAHs), it provides less hazardous compounds (Dominguez et al. 2003).

The microwave pyrolysis of rice straw has been reported (Huang et al. 2008). Under 400–500 W microwave power, the reduction of fixed carbon in the biomass was significant. The major compositions in gaseous product were H_2, CO_2, CO, CH_4 of 55, 17, 13, 10 vol.%, respectively. The high H_2 content might imply that microwave-induced pyrolysis of biomass waste has the potential to produce the H_2-rich fuel gas. The condensable part of product was highly alkylated and oxygenated, and the harmful PAH content was quite less.

Microwave-assisted pyrolysis was used to fabricate porous carbon nanostructures from biomass (wood, cotton, filter paper) precursors filled with a conducting polymer and Fe catalyst species (Wang et al. 2008). The morphology and porosity of the biomass precursors were retained, but their infrastructure became highly graphitic after the microwave treatment. Various graphitic nanostructures (viz., nanofoams, nanoflakes, nanoribbons, and spongelike nanosheets) were identified as the building blocks in the porous graphitic carbon materials produced.

Effect of various parameters like particle size, temperature, and reaction time on microwave pyrolysis of corn stover was studied (Lei et al. 2009). Corn stover with different particle sizes was found to be similarly pyrolyzed by microwave heating; the maximum volatile matter of corn stover was 76%, with bio-oil yield of 34% and biogas yield of 42% at a reaction temperature of 650°C, residence time of 8 min, and corn stover particle size of 4 mm. The results showed that very fine feedstock grinding required by conventional pyrolysis is not necessary for microwave pyrolysis process, resulting in substantial energy savings.

Microwave and conventional pyrolysis of cornstalk bale (Zhao et al. 2010) was performed. The content of H_2 attained the highest value of 35 vol.% and syngas (H_2 and CO) was greater than 50 vol.% during microwave pyrolysis. More liquids (bio-oil) were obtained during microwave pyrolysis due to the high heating rate.

Microwave pyrolysis of Douglas fir sawdust pellet was studied. The yields of bio-oil and syngas were increased with the reaction temperature and time. The highest yield of bio-oils was 57.8% obtained at 471°C and 15 min. A third-order reaction mechanism fits well the microwave pyrolysis of Douglas fir pellet with activation energy of 33.5 kJ/mol and a frequency factor of 3.03 s^{-1} (Ren et al. 2012).

7.7.1.3 Catalytic Pyrolysis of Biomass under Microwave
Bu et al. (2011) used activated carbon as catalyst during microwave-assisted biomass pyrolysis to produce bio-oil. Activated carbon increased the production of phenol (38.9%) and phenolic compounds (66.9%). The catalyst can be reused at least twice with stable catalytic efficiency. Wan et al. (2009) studied the effect of metal oxides, salts, and acids added on products selectivity of corn stover and aspen wood in microwave-assisted pyrolysis process. $K_2Cr_2O_7$, Al_2O_3, KAc, H_3BO_3, Na_2HPO_4, $MgCl_2$, $AlCl_3$, $CoCl_2$, and $ZnCl_2$ were premixed with corn stover powder and aspen wood pellets prior to pyrolysis by microwave. KAc, Al_2O_3, $MgCl_2$, H_3BO_3, and Na_2HPO_4 increased the bio-oil yield and suppressed gas yield. These catalysts may be used as a microwave absorbent to accelerate the heating process or involved in "in situ upgrading" of pyrolytic volatile compounds obtained from microwave pyrolysis process.

The catalytic microwave pyrolysis of pine wood sawdust using inorganic additives (NaOH, Na_2CO_3, Na_2SiO_3, NaCl, TiO_2, HZSM-5, H_3PO_4, $Fe_2(SO_4)_3$) was carried out under N_2. All of the eight additives increased the yields of solid products greatly and decreased the yields of gaseous products. There was not much effect on yields of liquid products. The incondensable gases produced from pyrolysis consist mainly of H_2, CH_4, CO, and CO_2. The amount of CH_4 and CO_2 decreased using the catalyst, while all of them except NaCl, TiO_2 and $Fe_2(SO_4)_3$ increased the amount of H_2. CO decreases under all catalyst except Na_2SiO_3 and HZSM-5. Acetol was the main product obtained in liquid products, and a possible pathway for acetol formation has been tentatively proposed (Chen et al. 2008).

Microwave pyrolysis of kraft lignin was carried at different conditions of weight% of a microwave absorber (char., 20%–40%) and power setting (1.5–2.7 kW). It was observed that char wt.% has more pronounced effects than power setting on the heating rate during microwave pyrolysis. High liquid yield was observed at high temperatures. It was observed from NMR that 80% of the carbon atoms in the oil phase were aromatic carbons (Faraga et al. 2014).

7.7.2 Microwave-Assisted Gasification

Microwave-assisted gasification of biomass under different conditions using various catalysts, namely, Fe, Co and Ni with Al_2O_3 was studied, and the effects of various catalysts on syngas production and tar removal were examined. Biomass gasification under microwave produced gas yield of more than 65% with catalyst. When catalyst/biomass ratio was changed from 1:10 to 1:5, gas yield was increased and tar produced was decreased. Also it was observed that H_2 production was also increased and amount of CO was decreased, indicating that syngas of higher quality was obtained (Xie et al. 2014).

7.7.2.1 Self-Gasification during Microwave Pyrolysis

The microwave pyrolysis of coffee hulls pellets produces large amounts of syngas ($H_2 + CO$), whereas under conventional pyrolysis the main gaseous product was CO_2. The different gas composition under microwave and conventional pyrolysis can be explained due to self-gasification of the char by the CO_2 that is produced during microwave pyrolysis. These differences are especially significant at low temperatures. Self-gasification would be favored by the relatively high amount of K present in the inorganic matter of the raw material, which would act as a catalyst. Moreover, at the relatively low temperature (500°C), the CO/CO_2 ratio is much higher for microwave pyrolysis than for conventional pyrolysis.

During microwave pyrolysis, the char is at a much higher temperature than the surrounding atmosphere, while in conventional heating the temperature gradient is the opposite, especially during the first stages of the process. Also during microwave heating, numerous small sparks are observed practically all the time. These sparks in a certain way could be considered as "microplasmas," both from the point of view of time (since they only last for a fraction of a second) and space (since they are located in a very tiny spot). Thus, while the large-scale temperature corresponds to the one measured by the optical pyrometer, the temperature in these microplasmas must be extremely high. Such a situation can be expected to favor heterogeneous reactions between the solid and the gases produced and, in consequence, the reaction of CO_2, with C, even at the low global temperature of 500°C. The char gasification mechanism under conventional heating is different than that of microwave heating. During conventional heating, the transition in the reaction mechanism that controls the Boudouard reaction (i.e., chemical or diffusion control) occurs at about 800°C.

However, in the case of microwave heating, the temperature is much lower and the reaction never proceeds under pure chemical control (Menendez et al. 2007).

7.7.3 Microwave-Assisted Liquefaction

Li carried out wheat straw alkali lignin liquefaction under microwave irradiation. Maximum yield (13.47 wt.%) was obtained using glycol as solvent and sulfuric acid as catalyst. G-type phenolic compounds were predominant with 2-methoxy-4-methyl phenol contributing 50% of the total monophenolic compounds (Li 2013). Xu et al. (2012) studied lignocellulosic biomass liquefaction at 180°C for 15 min. Glycerol–methanol mixture produced lower bio-residue yields compared to when liquefaction was carried out in methanol. The total yield of phenolics and poly-hydroxy compounds (including glycerol and sugar derivatives) was 65.9% and 84.9%, respectively. Sequeiros et al. (2013) studied the liquefaction of organosolv lignin from olive tree pruning under microwave heating. The highest liquefaction yield of 99.07% was obtained at 155°C in 5 min with 1% of sulfuric acid. Xie et al. (2013) studied the liquefaction of bamboo lignin-enriched residues under microwave irradiation for the production of crude biopolyols (CBP) (Xiao et al. 2011).

7.7.4 Microwave-Assisted Hydrothermal Carbonization

Microwave-assisted hydrothermal carbonization (MAHC) of pine sawdust (Pinus sp.) and a-cellulose (SolucellW) at three different reaction times using microwave heating at 200°C in acidic aqueous media was investigated. Elemental analysis showed that the lignocellulosic samples subjected to MAHC yielded carbon-enriched material 50% higher than raw materials. These results showed that microwave-assisted HTC is an innovative approach to obtain carbonized lignocellulosic materials (Guiotoku et al. 2009).

7.7.5 Microwave-Assisted Digestion

Microwave-assisted acid digestion of plants using nitric acid with hydrogen peroxide was studied. Under all experimental conditions, the residual carbon content values were always lower than 13% w/v. It was observed that for plant materials, microwave-assisted acid digestion can be carried out under mild conditions. An additional advantage is the lower amount of residue generated when working with less concentrated acid solutions (Araujoa et al. 2002).

7.7.6 Transesterification and Biodiesel Production from Algae

Transesterification or alcoholysis is the displacement of alcohol from an ester by another in a process similar to hydrolysis, except that alcohol is used instead of water. This process has been widely used to reduce the high viscosity of triglycerides. If methanol is used in this process, it is called methanolysis. Transesterification is one of the reversible reactions and proceeds essentially by mixing the reactants. However, the presence of a catalyst (a strong acid or base) accelerates the conversion. Conventional transesterification process is used for biodiesel production. The major challenges for industrial commercialized biodiesel production from microalgae are the high cost of downstream processing such as dewatering and drying, utilization of large volumes of solvent, and laborious extraction processes. Microwave irradiation method was used to produce biodiesel directly from wet microalgae biomass. The microwave irradiation extracted more lipids and high biodiesel conversion was obtained compared to the water bath–assisted extraction method due to the high cell disruption achieved and rapid transesterification. The total content of lipid extracted from microwave irradiation and water bath–assisted extraction were 38.31% and 23.01%, respectively. The biodiesel produced using microwave irradiation was higher (86.41%) compared to the conventional method. The FAMEs composition obtained via microwave irradiation has higher content in C16:0 and C18:0 but low in C20:0, which has the advantage of lower viscosity (Wahidin et al. 2014).

7.8 Microwave-Assisted Conversion of Biomass to Platform Fuel Chemicals

7.8.1 Microwave Reaction in Ionic Liquids

The dielectric properties of ILs make them highly suitable for use as solvents or additives in microwave-assisted organic synthesis. ILs consist entirely of ions and therefore absorb microwave irradiation extremely efficiently. ILs dissolve in a wide range of organic solvents so they can be used to increase the microwave absorption of low absorbing reaction mixtures (Yinghuai et al. 2011).

Direct conversion of cellulose into HMF in ILs under microwave irradiation in the presence of $CrCl_3$, with a yield of 62%, has been reported to occur within few minutes (Li et al. 2009). Zhang and Zhao (2011) reported microwave-assisted direct conversion of biomass into HMF and furfural in ILs in the presence of $CrCl_3$. When xylan was treated under identical condition, furfural yield of 63% was observed. When cornstalk was used, HMF and furfural yields were 45% and 23%, respectively. Microwave irradiation improved the yields for furanic compounds and reduced the reaction time (Nuchter et al. 2004; Zhang and Zhao 2010).

ILs containing different cations and anions were employed during microwave conversion of microcrystalline cellulose (MCC) to 5-HMF. The highest yields of 5-HMF (28.16% and 10.45%, respectively) were obtained with $[TMG]BF_4$ as the catalyst. Higher yields of HMF were produced from cellulose under microwave irradiation at lower reaction times. This might be explained due to the fact that ILs are good absorber of microwaves and also have nonthermal effects of microwave irradiation such as electric, magnetic, and chemical effects that accelerate deformation and vibration of ions and molecules, which made the transformation of cellulose into 5-HMF more efficient than was done under conventional heat sources (Qu et al. 2014).

7.8.2 Microwave-Assisted Production of Chemicals from Carbohydrates

The acid-catalyzed conversion of xylose, xylan, and straw biomass to furfural by microwave energy was carried out with different acids under different reaction conditions. Major compounds of the acid catalyzed conversion of straw were furfural, HMF, glucose, xylose, galactose, arabinose, mannose, etc. The major furanic product obtained was furfural. The furfural yields obtained from triticale straw, wheat straw, and flax shives were 10.92, 10.36, and 10.16 g/100 g of straw on dry weight basis, respectively. Although the furfural yields on a dry weight basis of straw were very similar, the furfural yields based on the straw biomass composition show differences. Quantitatively, the furfural yield from flax shives was much higher than the furfural yields from wheat and triticale straw calculated on the basis of pentose content. The high levels of glucose in the reaction media showed that the cellulose in the straw biomass samples was partially hydrolyzed under the process conditions. Consumption of mannose and arabinose that were present in very low amounts in triticale straw and flax shives was observed during the microwave process (Yemis and Mazza 2011).

Yields in 5-HMF reach 35% for a fructose conversion of 78% after only 5 min heating at 200°C in closed vessels conditions MW ("microwave hot compressed water process"), while 5-HMF yield and fructose conversion are, respectively, of 12% and 27% after an identical runtime using a sand bath heating. Zirconium oxide demonstrates great performances for the catalytic dehydration of glucose, favoring isomerization of glucose to fructose prior to dehydration. When using an acid ion-exchange resin in an acetone/dimethyl sulfoxide system, improvement in 5-HMF selectivity is encountered together with high yields (98% fructose conversion with 5-HMF selectivity of 92% for 10 min MW-assisted reaction at 150°C in closed vessel conditions).With microwave heating at 140°C for 30 s, a 5-HMF yield of 71% is obtained for 96% glucose in 1-butyl-3-methyl imidazolium chloride as the solvent (Qi et al. 2008; Richel and Paquot 2012).

7.9 Conclusions and Recommendations

Microwave pyrolysis is an effective and energy-efficient route to valorize biomass to produce fuel/chemicals. Microwave pyrolysis can be used for the pyrolysis of diversified feedstocks like sewage sludge, agricultural residues, forest residues, and algal biomass. Microwave processing can also be utilized for converting biomass to useful platform molecules. Microwave processing of biomass does not require energy-intensive pretreatment steps like biomass crushing to submillimeter particle and feedstock drying. Microwave pyrolysis not only avoids various disadvantages of conventional pyrolysis like slow heating and the necessity of feedstock shredding but also produces bio-oil, biochar, and gas of better quality and reduces the processing time than conventional pyrolysis. Irrespective of the various advantages associated with microwave processing, it has been observed from the literature that microwave processes have not been widely exploited for biomass processing. These factors include the absence of sufficient data to quantify the dielectric properties of the biomass; extensive research needs to be carried out to understand the various biomass decomposition mechanisms and reaction pathways during microwave processing. During microwave pyrolysis, other phenomenon like heat, mass transportation mechanism, and chemical reaction has to be studied in detail. Development of continuous microwave pyrolysis reactors will help to study in detail the mechanisms of activation and interaction of microwaves with biomass. The role of reaction media, the development of new catalyst supports/catalysts for microwave processing, synergistic interactions due to microwave energy processes, and the generation of database using various biomass feedstocks during microwave pyrolysis are also very essential.

Acknowledgments

The authors thank the director of CSIR-Indian Institute of Petroleum, Dehradun, for his constant encouragement and support. RS thanks Council of Scientific and Industrial Research (CSIR), New Delhi, India, for providing Senior Research Fellowship (SRF). The authors also thank CSIR for the XII Five Year Plan project (CSC0116/BioEn) and the Ministry of New and Renewable Energy for providing financial support.

References

Abramovitch, R.A. 1991. Applications microwave-energy in organic-chemistry—A review. *Org Prep Proced Int* 23:685–711.

Adnadjevic, B. and J. Jovanovic. 2011. The effects of microwave heating on the isothermal kinetics of chemicals reactions and physicochemical processes. In *Advances in Induction and Microwave Heating of Mineral and Organic Materials*, ed. S. Grundas, pp. 391–422. Rijeka, Croatia: InTech.

Allan, G.G., Krieger-Brockett, B., and D.W. Work. 1980. Dielectric loss microwave degradation of polymers: Cellulose. *J Appl Poly Sci* 25:1839–1859.

Araujo, G.C.L., Gonzalez, M.H., Ferreira A.G., Nogueira, A.A., and A.J. Nobrega. 2002. Effect of acid concentration on closed-vessel microwave-assisted digestion of plant materials. *Spectrochim Acta Part B* 57:2121–2132.

Badamali, S.K., Luque, R., Clark, J.H., and S.W. Breeden. 2009. Microwave assisted oxidation of a lignin model phenolic monomer using Co (salen)/SBA-15. *Catal Comm* 10:1010–1013.

Badamali, S.K., Luque, R., Clark, J.H., and S.W. Breeden. 2011. Co (salen)/SBA-15 catalysed oxidation of a β-O-4 ic dimer under. *Catal Comm* 12:993–995.

Badamali, S.K., Luque, R., Clark, J.H., and S.W. Breeden. 2013. Unprecedented oxidative properties of mesoporous silica materials: Towards microwave-assisted oxidation of lignin model compounds. *Catal Comm* 31:1–4.

Bhaskar, T., Bhavya, B., Singh, R., Naik, D.V., Kumar, A., and H.B. Goyal. 2011. Thermochemical conversion of biomass to biofuels. In *Biofuels*, eds. A. Pandey, C. Larroche, S.C. Ricke, C.G. Dussap, and E. Gnansounou, pp. 51–77. Burlington, MA: Academic Press.

Bogdal, D. 2005. *Microwave Assisted Organic Synthesis*. U.K.: Elsevier.

Bose, A.K., Banik, B.K., Lavlinskaia, N., Jayaraman, M., and M.S. Manhas. 1997. More chemistry in a microwave. *Chemtech* 27:18–24.

Boye, A.C. 2005. Microwaves in inorganic chemistry: Is it just a bunch of hot air? http://www.chemistry.illinois.edu/research/organic/seminar_extracts/2004_2005/7_Boye_Abstract_SP05.pdf. Accessed on March 3, 2005.

Bu, Q., Lei, H., Ren, S. et al. 2011. Phenol and phenolics from lignocellulosic biomass by catalytic microwave pyrolysis. *Bioresour Technol* 102:7004–7007.

Caddick, S. 1995. Microwave assisted organic reactions. *Tetrahedron* 51:10403–10432.

Chemat-Djenni, Z., Hamada, B., and F. Chemat. 2007. Atmospheric pressure microwave assisted heterogeneous catalytic reactions. *Molecules* 12:1399–1409.

Chen, M.-Q., Wang, J., Zang, M. et al. 2008. Catalytic effects of eight inorganic additives on pyrolysis of pine wood sawdust by microwave heating. *J Anal Appl Pyrol* 82:145–150.

Das, S., Mukhoupdhayay, A.K., Datta, S., and D. Basu. 2009. Prospects of microwave processing: An overview. *Bull Mater Sci* 32:1–13.

De Souza, R.O.M.A. 2014. Theoretical aspects of microwave irradiation practices. In *Biofuels and Biorefineries*, eds. Z. Fang, R.L. Smith Jr., X. Qi et al. pp. 3–16. Netherlands: Springer.

Dominguez, A., Fernández, Y., Fidalgo, B., Pis, J.J., and J.A. Menéndez. 2008. Biogas production with low concentration of CO_2 and CH_4 from microwave-induced pyrolysis of wet and dried sewage sludge. *Chemosphere* 70:397–403.

Dominguez, A., Menéndez, J.A., Fernández, Y. et al. 2007. Conventional and microwave induced pyrolysis of coffee hulls for the production of a hydrogen rich fuel gas. *J Anal Appl Pyrol* 79:128–135.

Dominguez, A., Menendez, J.A., Inguanzo, M., Bernad, P.L., and J.J. Pis. 2003. Gas chromatographic–mass spectrometric study of the oil fractions produced by microwave-assisted pyrolysis of different sewage sludges. *J Chromatogr A* 1012:193–206.

Dominguez, A., Menéndez, J.A., Inguanzo, M., and J.J. Pis. 2006. Production of biofuels by high temperature pyrolysis of sewage sludge using conventional and microwave heating. *Bioresour Technol* 97:1185–1193.

Dutta, S., De, S., Alam, Md.I., Abu-Omar, M.M., and B. Saha. 2012. Direct conversion of cellulose and lignocellulosic biomass into chemicals and biofuel with metal chloride catalysts. *J Catal* 288:8–15.

Elías, X. 2005. *Tratamiento y valorización energética de residuos*, ed. D. Santos. Barcelona, Spain: Fundación Universitaria Iberoamericana.

Faaij, A.P.C. 2006. Bioenergy in Europe: Changing technology choices. *Energ Pol* 34:322–343.

Faraga, S., Fu, D., Jessop, P.G., and J. Chaoukia. 2014. Detailed compositional analysis and structural investigation of a biooil from microwave pyrolysis of kraft lignin. *J Anal Appl Pyrol* 109:249–257.

Fernández, Y., Arenillas, A., and J.A. Menéndez. 2011. Microwave heating applied to pyrolysis. In *Advances in Induction and Microwave Heating of Mineral and Organic Materials*, ed. S. Grundas, pp. 723–751. Rijeka, Croatia: InTech.

Gabriel, C., Gabriel, S., Grant, E.H., Holstead, B.S.T., and D.M.P. Mingos. 1998. Dielectric parameters relevant to microwave dielectric heating. *Chem Soc Rev* 27:213–223.

Guiotoku, M., Rambo, C.R., Hansel, F.A., Magalhaes, W.L.E., and D. Hotza. 2009. Microwave-assisted hydrothermal carbonization of lignocellulosic materials. *Mater Lett* 63:2707–2709.

https://iosrjournals.org/iosr-jeee/paper/vol10-issue1/version-3/I010136772.pdf

Huang, Y.F., Kuan, W.H., Lo, S.L., and C.F. Lin. 2008. Total recovery of resources and energy from rice straw using microwave-induced pyrolysis. *Bioresour Technol* 99:8252–8258.

Kappe, C.O., Dallinger, D., and D.D. Murphree. 2008. Microwave synthesis—An introduction. *Practical Microwave Synthesis for Organic Chemists: Strategies, Instruments, and Protocols*, Wiley-VcH Verlag GmbH and Co.KGaA, Germany.

Konno, K., Onodera, H., Murata, K., Onoe, K., and T. Yamaguchi. 2011. Comparison of cellulose decomposition by microwave plasma and radio frequency plasma. *Green Sustain Chem* 1:85–91.

Koo, Y. and S. Ha. 2010. Efficient enzymatic hydrolysis of cellulose in ionic liquids under microwave irradiation. Paper presented at *Pacifichem, International Chemical Congress of Pacific Basin Societies*, Honolulu, HI.

Kuhnert, N. 2002. Microwave-assisted reactions in organic synthesis-are there any nonthermal microwave effects? *Angrew Chem Int* 41:1863–1866.

Kumar, P., Coronel, P., Simunovic, J., Truong, V.D., and K.P. Sandeep. 2007. Measurement of dielectric properties of pumpable food materials under static and continuous flow conditions. *J Food Sci* 72:77–83.

Langa, F., DelaCruz, P., DelaHoz, A., Diaz-Ortiz, A., and E. Diez-Barra. 1997. Microwave irradiation: More than just a method for accelerating reactions. *Contemp Org Synth* 4:373–386.

Leadbeater, N.E. 2004. Making microwaves. *Chem World* 1:38–41.

Lei, H., Ren, S., and J. Julson. 2009. The effects of reaction temperature and time and particle size of corn stover on microwave pyrolysis. *Energ Fuel* 23:3254–3261.

Li, B. 2013. Study on liquefaction and degradation of wheat straw alkali lignin assisted by microwave irradiation. Dissertation ID:1685374. South China University of Technology, Guangzhou, China.

Li, C., Zhang, Z., and Z.K. Zhao. 2009. Direct conversion of glucose and cellulose to 5-hydroxymethylfurfural in ionic liquid under microwave irradiation. *Tetrahedron Lett* 50:5403–5405.

Li, H., Qu, Y., and J. Xu. 2014. Production of biofuels and chemicals with microwave. In *Biofuels and Biorefineries*, eds. Z. Fang, R.L. Smith Jr., X. Qi, L. Fan, J.R. Grace, Y. Ni, and N.R. Scott, pp. 61–82. Netherlands Springer.

Lidstrom, P., Tierney, J., Wathey, B., and J. Westman. 2001. Microwave assisted organic synthesis—A review. *Tetrahedron* 57:9225–9283.

Loupy, A. and A. Petit. 1997. Chime sous micro-ondes, Vers des technologies propres en synthèse organique. Pelin-Sud, Special researches. Universite Paris-Sud, Orsay, France.

Luque, R., Menendez, J.A., Arenillas, A., and J. Cot. 2012. Microwave-assisted pyrolysis of biomass feedstocks: The way forward? *Energ Environ Sci* 5:5481–5488.

Macquarrie, D.J., Clark, J.H., and E. Fitzpatrick. 2012. The microwave pyrolysis of biomass. *Biofuels Bioprod Bioref* 6:549–560.

Menendez, J.A., Dominguez, A., Fernandez, Y., and J.J. Pis. 2007. Evidence of self-gasification during the microwave-induced pyrolysis of coffee hulls. *Energ Fuels* 21:373–378.

Menendez, J.A., Dominguez, A., Inguanzo, M., and J.J. Pis. 2004. Microwave pyrolysis of sewage sludge: Analysis of the gas fraction. *J Anal Appl Pyrol* 71:657–667.

Metaxas, A.C. and R.J. Meredith. 1993. *Industrial Microwave Heating*. IEEE Power Engineering Series 4. London, U.K.: Peter Peregrinus Ltd.

Motasemi, F. and M.T. Afzal. 2013. A review on the microwave-assisted pyrolysis technique. *Renew Sustain Energ Rev* 28:317–330.

Nelson, S.O. and C.J. Datta. 2001. Dielectric properties of food materials and electric field interactions. In *Handbook of Microwave Technology for Food Applications*, eds. A.K. Datta and R.C. Anantheswaran, pp. 69–107. New York: Marcel Dekker, Inc.

Nuchter, M., Ondruschka, B., Bonrath, W., and A. Gum. 2004. Microwave assisted synthesis—A critical technology overview. *Green Chem* 6:128–141.

Nuchter, M., Ondruschka, B., and W. Lautenschlager. 2003. Microwave-assisted chemical reactions. *Chem Eng Technol* 26:1208–1216.

Plazl, I. 2002. Esterification of benzoic acid with 2-ethylhexanol in a microwave-stirred tank reactor. *Ind Eng Chem Res* 41:1129–1134.

Plazl, I., Leskovek, S., and T. Koloini. 1995. Hydrolysis of sucrose by conventional and microwave heating in stirred tank reactor. *Chem Eng J* 59:253–257.

Plazl, I., Pipus, G., and T. Koloini. 1997. Microwave heating of the continuous flow catalytic reactor in a nonuniform electric field. *AIChE J* 43:754–760.

Qi, X., Watanabe, M., Aida T.M., and R.L. Smith Jr. 2008. Catalytical conversion of fructose and glucose into 5-hydroxymethylfurfural in hot compressed water by microwave heating. *Catal Comm* 9:2244–2249.

Qu, Y., Wei, Q., Li, H., Oleskowicz-Popiel, P., and P.C. Huan. 2014. Microwave-assisted conversion of microcrystalline cellulose to 5-hydroxymethylfurfural catalyzed by ionic liquids. *Bioresour Technol* 162:358–364.

Rana, K.K. and Rana, S. 2014. Microwave reactors: A brief review on its fundamentals aspects and applications. *Open Access Library Journal*, 1. DOI:10.4236/oalib.1100686.

Ren, S., Lei, H., Wang, L. et al. 2012. Biofuel production and kinetics analysis for microwave pyrolysis of Douglas fir sawdust pellet. *J Anal Appl Pyrol* 94:163–169.

Richel, A. and M. Paquot. 2012. Conversion of carbohydrates under microwave heating. In *Carbohydrates—Comprehensive Studies on Glycobiology and Glycotechnology*, ed. C.F. Chang, pp. 21–36. Rijeka, Croatia: Intech.

Saillard, R., Poux, M., Berlan, J., and M. Audhvy-peaudecert. 1995. Microwave heating of organic solvents: Thermal effects and field modelling. *Tetrahedron* 51:4033–4042.

Sequeiros, A., Serrano, L., Briones, R., and J. Labidi. 2013. Lignin under microwave heating. *J Appl Polym Sci* 130:3292–3298.

Stocker, M. 2008. Biofuels and biomass-to-liquid fuels in the biorefinery: Catalytic conversion of lignocellulosic biomass using porous materials. *Angew Chem Int* 47:9200–9211.

Strauss, C.R. and R.W. Trainor. 1995. Developments in microwave-assisted organic chemistry. *Aust J Chem* 48:1665–1692.

Stuerga, D. and P. Gaillard. 1996. Microwave heating as a new way to induce localized enhancements of reaction rate: Non-isothermal and heterogeneous kinetics. *Tetrahedron* 52:5505–5510.

Surati, M.A., Jauhari, S., and K.R. Desai. 2012. A brief review: Microwave assisted organic reaction. *Arch Appl Sci Res* 4:645–661.

Sutton, W.H. 1989. Microwave processing of ceramic materials. *Am Ceram Soc Bull* 68:376–386.

Thostenson, E.T. and T.W. Chou. 1999. Microwave processing: Fundamentals and applications. *Composit A* 30:1055–1071.

Veggi, P.C., Martinez J., and M.A.A. Meireles. 2013. Fundamentals of microwave extraction. In *Microwave Assisted Extraction for Bioactive Compounds: Theory and Practice*, Food Engineering Series 4, eds. F. Chemat and G. Cravotto, pp. 15–52. New York: Springer Science.

Wahidin, S., Idris, A., and S.R.M. Shaleh. 2014. Rapid biodiesel production using wet microalgae via microwave irradiation. *Energ Convers Manage* 84:227–233.

Wan, Y., Chen, P., Zhang, B. et al. 2009. Microwave-assisted pyrolysis of biomass: Catalysts to improve product selectivity. *J Anal Appl Pyrol* 86:161–167.

Wang, C., Ma, D., and X. Bao. 2008. Transformation of biomass into porous graphitic carbon nanostructures by microwave irradiation. *J Phys Chem C* 112:17596–17602.

Westaway, K.C. and R.J. Gedye. 1995. The question of specific activation of organic reactions by microwaves. *J Microw Power Electromagn Energ* 30:219–230.

Xiao, W., Han, L., and Y. Zhao. 2011. Comparative study of conventional and microwave-assisted liquefaction of corn stover in ethylene glycol. *Ind Crops Prod* 34:1602–1606.

Xie, H., Zhang, Z., and Z.K. Zhao. 2010. Microwave assisted conversion of cellulose and lignocellulose into value-added chemicals in ionic liquids. Paper presented at *Pacifichem, International Chemical Congress of Pacific Basin Societies*, Honolulu, HI.

Xie, J., Qi, J., Hse, C., and T. Shupe. 2013. Effect of lignin derivatives in the biopolyols from microwave liquefied bamboo on the properties of polyurethane foams. *BioResources* 9:578–588.

Xie, Q., Borges, F.C., Cheng, Y. et al. 2014. Fast microwave-assisted catalytic gasification of biomass for syngas production and tar removal. *Bioresour Technol* 156:291–296.

Xiong, J. 2012. Studies on catalytic degradation of lignin model compounds with the assistance of microwave and hydrogen-donor solvents. Master thesis, Zhejiang University, Hangzhou, China.

Xu, J., Jiang, J., Hse, C., and T.F. Shupe. 2012. Renewable chemicals from integrated processing of lignocellulosic materials using microwave energy. *Green Chem* 14:2821–2830.

Yang, P., Kobayashi, H., and A. Fukuoka. 2011. Recent developments in the catalytic conversion of cellulose into valuable chemicals. *Chin J Catal* 32:716–722.

Yemis, O. and G. Mazza. 2011. Acid-catalyzed conversion of xylose, xylan and straw into furfural by microwave-assisted reaction. *Bioresour Technol* 102:7371–7378.

Yinghuai, Z., Biying, A.O., Siwei, X., Hosmane, N.S., and J.A. Maguire. 2011. Ionic liquids in catalytic biomass transformation. In *Applications of Ionic Liquids in Science and Technology*, ed. S. Handy, pp. 3–26. Rijeka, Croatia: InTech.

Yu, F., Ruan, R., Deng, S.B., Chen, P., and X.Y. Lin. 2006. Microwave pyrolysis of biomass. Paper presented at an *ASABE Meeting Presentation*, Portland, OR, ASABE Number P: 066051.

Zhang, X. and D.O. Hayward. 2006. Applications of microwave dielectric heating in environment-related heterogeneous gas-phase catalytic systems. *Inorg Chim Acta* 359:3421–3433.

Zhang, Z. and X. Li. 1999. Microwave-assisted organic chemistry. *Shenyang Yaoke Daxue Xuebao* 16:304–309.

Zhang, Z. and Z.B. Zhao. 2010. Microwave-assisted conversion of lignocellulosic biomass into furans in ionic liquid. *Bioresour Technol* 101:1111–1114.

Zhang, Z.H. and Z.B. Zhao. 2011. Production of 5-hydroxymethylfurfural from glucose catalyzed by hydroxyapatite supported chromium chloride. *Bioresour Technol* 102:3970–3972.

Zhao, X., Song, Z., Liu, H., Li, Z., Li, L., and C. Ma. 2010. Microwave pyrolysis of corn stalk bale: A promising method for direct utilization of large-sized biomass and syngas production. *J Anal Appl Pyrol* 89:87–94.

8

Hydroprocessing Challenges in Biofuel Production

Vartika Srivastava, Bijoy Biswas, Pallavi Yadav, Bhavya B. Krishna,
Rawel Singh, Jitendra Kumar, and Thallada Bhaskar

Contents

Abstract

The increasing concerns of availability of fossil resources have fueled research in the area of biomass conversion to hydrocarbons. Biooil produced by thermochemical conversion of biomass cannot be used directly for fuel applications and an upgradation step is required for the same. Hydroprocessing helps in reducing the oxygen content of the biooil, thereby improving its properties such as total acid number, viscosity, and increased thermal stability and eliminating repolymerization. Catalysts play a major role in the hydroprocessing of biooil, and several sulfided and nonsulfided or noble metal and nonnoble metal catalysts have been tested for the biooil model compounds and its simulated mixtures. Higher hydrogen consumption, catalyst deactivation by coking, and poor heat and mass transfer in the reactor are some of the challenges in this process. Several opportunities exist in the development of catalysts and processes for the hydroprocessing of biooil, which will be addressed in this chapter.

8.1 Introduction

The transportation sector over the past several decades has been heavily dependent on fossil resources. However, the existing environmental issues of climate change, growing demand for energy, and geopolitical issues related to fossil resources have compelled humans to look out for alternative and sustainable sources of energy. It is highly essential for the development of sustainable processes based on renewable raw materials. In this scenario, biomass is the only renewable source of organic carbon, which is very much required for the production of hydrocarbons. The term "biofuel" refers to the fuel derived from biomass and has been gaining increased attention to supplement the growing demand for fuels. Biomass, being a natural polymer, is also a source of valuable chemicals/petrochemical feedstocks, which are now produced after several steps of functionalization/rearrangements of hydrocarbons present in crude oil. The utilization of biomass offers the advantage of utilizing the domestic resources for the production of fuels that can be used locally. Some of the other advantages it provides are the additional income/self-sustenance to farmers who can utilize the residues in the farm to produce fuels. This provides a way to utilize biomass in decentralized units.

Biomass is available in almost all parts of the world with variations in quality and quantity throughout the year. It consists of any organic matter produced by the process of photosynthesis. Biofuels can be classified into first-generation, second-generation, third-generation, or fourth-generation biofuels based on the source and kind of feedstock used. A first-generation biofuel is made from edible portions of a plant, for example, sugar or starch, vegetable oil, and also animal fats. First-generation biofuels have limitations in terms of food vs. fuel issue, which has led to the search for alternative feedstocks for biofuel production. The second-generation biofuels are made from lignocellulosic biomass, which are not edible for humans and therefore do not compete directly with food production. Some of the sources of lignocellulosic biomass are agricultural and forest residues. In addition, energy crops can be bred specifically for the purpose of biofuel production and some of the examples are miscanthus and switchgrass, which enables higher production per unit land area.

Lignocellulosic biomass is composed of cellulose (~50 wt.% dry base [d.b.]), hemicellulose (~25 wt.% d.b.), and lignin (~25 wt.% d.b.). These polymers are linked together to obtain structural strength in combination with flexibility. In addition, biomass also contains some amounts of extractives and inorganic compounds termed as ash. Cellulose, whose monomer is glucose, is a long-chain linear polymer arranged in microfibrils that are organized in fibrils, which combine to form cellulose fibers. Hemicelluloses are shorter or branched amorphous polymers, of five- or six-carbon sugars. Together with lignin, hemicellulose forms the matrix in which the cellulose fibrils are embedded (Fengel and Wegener 2003). Finally, lignin, the third cell wall component, is a three-dimensional polymer formed from phenyl propane units with many different types of linkages between the building blocks (Alonso et al. 2010).

Third-generation biofuels are derived from algae (micro-/macroalgae), which addresses some of the drawbacks associated with first- and second-generation biofuels. Algae are able to produce many times more oil than traditional crops on an area basis. In contrast to conventional crop plants, microalgae can be harvested after every cycle, which normally lasts up to 10 days (depending on process), while the former can be harvested only once or twice a year (Dragone et al. 2010). Fourth-generation biofuels are obtained from engineered plants or genetically modified microorganisms, which have one or more required qualities enhanced with the help of biological tools.

8.2 Biomass Conversion Technologies

Biomass can be converted to various hydrocarbons through several processes. Some of the main factors influencing the choice of conversion process are the quality and quantity of biomass feedstock and the end-use requirement. In addition, environmental standards have to be met and the technoeconomic feasibility of the process has to be considered. There are several methods of biomass conversion such as mechanical, biological, and thermochemical processes. Thermochemical processes have advantages over the other processes in terms of feedstock flexibility, complete utilization of biomass, faster processing times, minimal pretreatment techniques, etc.

Under the umbrella of thermochemical conversion, the four main routes possible are combustion, gasification, pyrolysis, and carbonization. Combustion is the complete oxidation of biomass in the presence of oxygen mainly producing heat. Gasification is partial oxidation of biomass mainly producing syngas (a mixture of

CO and H_2), which can be catalytically transformed to fuels/chemicals using the Fischer–Tropsch process. Electricity can also be generated using the integrated gasification combined cycle. Carbonization is used mainly for the production of solid biochar, which has several applications of its own.

Pyrolysis of biomass takes place in the absence of oxygen mostly under the presence of inert atmosphere though at times even hydrogen can be used for the process. It is the decomposition of biomass material for the production of biooil, noncondensable gases, and biochar. The products are obtained in different ratios with variations in quality and the process parameters used. There are various kinds of pyrolysis processes depending on the residence time inside the reactor, the reactor used, and the atmosphere in which pyrolysis is conducted. Some of the types of pyrolysis are fast, slow, auger, vacuum, rotating cone, bubbling fluidized-bed, circulating fluidized-bed pyrolysis, etc.

8.3 Composition and Properties of Biooil

Biooils are dark-brown-colored liquid having a smoky order and composed of polar organics (ca. 75–80 wt.%) and water (ca. 20–25 wt.%) (Bridgwater 1996, 1999; Bridgwater et al. 1999; Czernik and Bridgwater 2004). Biooil is composed of a complex mixture of oxygenated organic compounds and its elemental composition approximates as that of the biomass feedstock (Bridgwater 1994). It consists of two phases: an aqueous phase containing a wide variety of oxygenated organic compounds of low molecular weight (such as acetic acid, methanol, and acetone) and a nonaqueous phase containing oxygenated compounds (such as aliphatic alcohols, carbonyls, acids, phenols, cresols, benzenedioles, guaiacol, and their alkylated derivatives) and aromatic hydrocarbons, which are single ring aromatic compounds (such as benzene, toluene, indene, and their alkylated derivatives), and polycyclic aromatic hydrocarbons (such as naphthalene, fluorene, phenanthrene, and their alkylated derivatives) (Bridgwater 1994; Meier and Faix 1999; Williams and Nugranad 2000).

Biooil has a negligible content of sulfur and nitrogen when derived from lignocellulosic feedstocks. In cases where algal biomass is used, biooil has some amount of nitrogen content due to its protein content. In most cases, the oxygen content in biooil is very high up to 30%, which leads to several disadvantages such as low heating value, thermal and chemical instabilities, and tendency to polymerize. In addition, it also has high total acid number and is immiscible with crude oil or its fractions (Furimsky 2000; Marker et al. 2012). The oxygenated compounds present in raw biooils impart a number of unwanted characteristics such as thermal instability (reflected in increasing viscosity upon storage), corrosiveness, and low heating value (Wang et al. 2011a). This instability is associated with the presence of reactive chemical species, notably aldehydes, ketones, carboxylic acids, alkenes, and guaiacol-type molecules (Diebold 2000). Upon prolonged storage, condensation reactions involving these functional groups result in the formation of heavier compounds. High total acid number also means that it cannot be transported via pipelines. Typical composition and properties of biooil are available in the open literature (Oasmaa et al. 2010, 2012; Solantausta et al. 2012). In short, biooils have poor volatility, cold flow problems, low heating value (16–19 MJ kg^{-1}; less than half that of petroleum-derived fuels), strong corrosiveness (pH of 2–3), high viscosity (35–1000 cp at 40°C), and poor chemical stability (viscosity and phase change with time) (Czernik and Bridgwater 2004). This leads to the restrictions in usage or application of biofuel as liquid transportation fuel. Therefore, it is necessary to remove the oxygen from biooil if it is to be used as a substitute for fossil-derived transportation fuels.

8.4 Upgradation of Biooil

In order to make biooils usable, some chemical/physical transformations are required to increase volatility and thermal stability, reduce viscosity and the tendency to repolymerize, reduce the oxygen content of the biooil and total acid number, and improve the miscibility with crude oil fractions.

8.4.1 Physical Methods for Biooil Upgradation

Some of the physical methods of upgradation techniques are filtration, solvent addition, and emulsification, which are explained in the following.

8.4.1.1 Filtration
Hot-vapor filtration reduces the ash content of biooil to less than 0.01% and the alkali content to less than 10 ppm, which is much lower than those reported for biooils produced in units using only cyclones. This reduces viscosity and lowers the average molecular weight of the liquid product. Hot gas filtration has not yet been demonstrated over a long-term process operation. Filtration has been used by NREL (Zhang et al. 2007), VTT, and Aston University (Sitzmann and Bridgwater 2007).

8.4.1.2 Solvent Addition
Polar solvents have been used for many years to homogenize and reduce the viscosity of biomass oils. The addition of solvents, especially methanol, has shown a significant effect on the biooil stability.

8.4.1.3 Emulsions
Biooil can be emulsified with diesel using surfactants. A process for producing stable microemulsions with 5%–30% of biooil in diesel has been developed at CANMET (Ikura et al. 1998). The University of Florence, Italy, has been working on emulsions of 5%–95% biooil in diesel (Baglioni et al. 2001, 2003), for producing usable fuels or power from generator that does not require engine modification. Some of the disadvantages are the observed high levels of corrosion/erosion

in engine applications. In addition, the cost of surfactants and energy required for emulsification is also high.

8.4.2 Catalytic Routes for Biooil Upgradation

The quality of biooils can be improved by the partial or total elimination of the oxygenated functionalities present. There are several methods for the purpose of deoxygenation (Czernik et al. 2002; Elliott 2007; Qi et al. 2006). Some of the catalytic routes to improve the quality of biooil (considering $C_6H_8O_4$ as the stoichiometric composition of biooil) are given as follows.

8.4.2.1 Fluidized Catalytic Cracking

$$C_6H_8O_4 \rightarrow C_{4.5}H_6 + H_2O + 1.5CO_2$$

The reactions that occur in the fluidized catalytic cracking (FCC) process include cracking reactions (cracking of alkanes, alkenes, naphthenes, and alkyl aromatics to form lighter products), hydrogen transfer, isomerization, and coking reactions (Corma et al. 1994). Catalytic cracking catalysts can be solid acid catalysts (typically Y-zeolite), binder (kaolin) catalysts, or alumina or silica–alumina catalysts. ZSM-5 is a common additive to FCC catalysts. Zeolites, and in general solid acids, are the most widely used industrial catalyst for oil refining, petrochemistry, and the production of fine and specialty chemicals (Chica and Corma 1999; Corma 1995, 2003; Corma et al. 1985). Biooil can also be mixed with vacuum gas oil and then treated in a fluid catalytic cracking unit to obtain fungible hydrocarbons directly.

8.4.2.2 Steam Reforming of Biooil

Biooil can also be used for the production of hydrogen by the steam reforming process. It is a complicated process as some of the biooil components are thermally unstable and decompose upon heating. Deactivation of the catalysts due to coking is one of the major problems, and biooils cause deactivation faster than petroleum-derived feedstocks. It has been observed that the catalysts used in the steam reforming of biooils in fixed-bed reactors require a catalyst regeneration step after 3–4 h of time on stream. High temperature and high ratios of steam to carbon are necessary to avoid catalyst deactivation by coking. Czernik et al. have developed a fluidized-bed reactor for steam reforming of biooils and observed that the catalysts were more stable in the fluidized-bed reactor than in the fixed-bed reactor due to better contact of the catalyst particle with steam (Czernik et al. 2002). When water is added to biooil, phase separation into aqueous and organic fraction occurs. The organic phase can be used to make specialty chemicals and the aqueous phase can be reformed to produce hydrogen or synthesis gas, which has several applications (Czernik et al. 2002; Garcia et al. 2000). Partial hydrogenation of biooils before steam reforming would lead to increased thermal stability and reduced coke formation on the catalyst.

The hydrogen required for the process could be obtained from the product stream (Huber et al. 2006).

8.4.2.3 Gasification of Biooil

Biooil can also be gasified to produce syngas ($CO + H_2$), which can be catalytically converted to fuels/petrochemical feedstocks/chemicals by the Fischer–Tropsch process (Van Rossum et al. 2007).

8.4.2.4 Decarboxylation

Decarboxylation is the removal of oxygen present in the biooil in the form of CO_2, and hydrogen is not required in this process for the production of hydrocarbons (Kersten et al. 2007). Decarboxylation of the organic acids leads to reduced acid content, thereby making the biooil less corrosive but more stable with higher energy content. Biooil can also be reacted over solid acid catalysts at atmospheric pressure, resulting in simultaneous dehydration and decarboxylation.

$$C_6H_8O_4 \rightarrow C_4H_8 + 2CO_2$$

8.4.2.5 Direct Oxygenation

In the direct deoxygenation method carried out at atmospheric pressure, C–O bonds are broken without the assistance of a reducing gas such as hydrogen (Filley and Roth 1999). This has been reported on tungsten (IV) compounds (Sharpless and Flood 1972) and acidic zeolites such as HZSM-5 (Prasad et al. 1986). In case of hydrogenation, the aromatic rings are hydrogenated before removal of oxygen (Grange et al. 1996) and carried out at high pressure and temperature. Thus, this process can be integrated with the existing petroleum refining infrastructure for processing (Huber et al. 2006).

8.4.2.6 Hydroprocessing

Hydrodeoxygenation (HDO)

$$C_6H_8O_4 + 4H_2 \rightarrow C_6H_8 + 4H_2O$$

Hydrotreating

$$C_6H_8O_4 + 7H_2 \rightarrow C_6H_{14} + 4H_2O$$

Typical hydrotreating conditions such as high hydrogen pressures in combination with conventional hydrodesulfurization catalysts are used for the hydrogenation of unsaturated groups (with elimination of oxygen as water) and hydrogenation-hydrocracking of large molecules. The technoeconomic studies indicate that the process may not be feasible at times due to the very high requirement of hydrogen in the process (Bridgwater 1996). The process might not be suitable in decentralized units, but in centralized units where hydrogen can be supplied easily, this process is very attractive, for example, in cases where excess hydrogen from refinery is available.

Acid-catalyzed cracking can also be carried out but the tendency of catalysts to undergo rapid deactivation due to coking, and the production of relatively high yields of low-value light hydrocarbons is one of its disadvantages (Diebold and Scahill 1988; Fisk et al. 2009).

8.5 Hydroprocessing of Biooil

HDO without saturation of the aromatic rings is carried out at moderate temperatures (300°C–600°C) with high pressure of H_2 in the presence of solid catalysts. During HDO, the oxygen in the biooil reacts with H_2 to form water and compounds with saturated C–C bonds. The main HDO reaction is

$$-(CH_2O)-+H_2 \rightarrow -(CH_2)-+H_2O$$

The conventional process used for petroleum fractions cannot be directly used for the biooil HDO as several modifications in the process are required (Elliott and Baker 1987). A low-temperature stabilization step was found to be essential before HDO at higher temperatures (Elliott and Baker 1989). Hydroprocessing of petroleum or coal liquids is mainly for nitrogen or sulfur removal, but in the case of biooils, oxygen removal is the most important step (Elliott 2007). In conventional crudes, the content of oxygen is less than 2 wt.%, hence reducing emphasis on oxygen removal. In the case of synthetic crudes derived from coal, the oxygen content may be around 10 wt.%. In case of biomass-derived liquids, the oxygen content may approach 50 wt.% (Elliott et al. 1991; Furimsky 2000).

Delmon and coworkers studied the HDO of model biooil compounds, 4-methyl-acetophenone, ethyldecanoate, and guaiacol, with sulfided CoMo and NiMo catalysts (Ferrari et al. 2001; Laurent and Delmon 1994a,b). The ketone group was seen to easily and selectively hydrogenate into methylene group above 200°C (Laurent and Delmon 1994a). Guaiacol was hydrogenated to catechol and then to phenol. Guaiacol caused catalyst deactivation due to coking reactions.

Biooil upgradation work at Pacific Northwest National Laboratory (PNNL) has focused on heterogeneous catalytic hydroprocessing. Initially, batch reactor studies of model phenolic compounds (Elliott 1983) with various catalysts were carried out. Commercial samples of catalysts were tested such as CoMo, NiMo, NiW, Ni, Co, Pd, and CuCrO to hydrogenate phenol at 300°C or 400°C (1 h at temperature). p-Cresol, ethylphenol, dimethylphenol, trimethylphenol, naphthol, and guaiacol (methoxy-phenol) were also tested with CoMo catalyst at 400°C. Of the catalysts tested, it was observed that sulfided form of CoMo was the most active as it gave a product containing 33.8% benzene and 3.6% cyclohexane at 400°C (Elliott 2007).

Hydrotreating process removes the oxygen present in the biooil through the reaction that can be expressed as

$$C_6H_8O_4 +6H_2 \rightarrow 6CH_2 +4H_2O$$

This is a carbon limited system that gives a maximum stoichiometric yield of 58% by weight on liquid biooil or a maximum energetic yield of about 50 wt.%. Hydrotreating can be used to convert biooils into a more stable fuel with a higher energy density so that it can be blended with petroleum-derived feedstocks. In a petroleum refinery, hydrotreating is typically carried out at temperatures of 300°C–600°C and H_2 pressures of 35–170 atm with sulfided CoMo- and NiMo-based catalysts. Most HDO of biooils has focused on sulfided CoMo- and NiMo-based catalysts, which are used for hydrotreating industrial feedstocks. When sulfided catalysts are used, additional sulfur must be added to the catalyst to avoid deactivation as the sulfur content of biooil is very minimal. Nonsulfided catalysts such as $Pt/SiO_2–Al_2O_3$ (Sheu et al. 1988), vanadium nitride (Ramanathan and Oyama 1995), and Ru have also been used for HDO. Deoxygenation to less than 5 wt.% oxygen is required for fuel applications (Elliott and Neuenschwander 1996).

Elliot and coworkers developed a two-step hydrotreating process for the upgradation of biooils derived from pyrolysis (Elliott and Neuenschwander 1996; Elliott and Oasmaa 1991; Elliott et al. 1988; Joshi and Lawal 2012). The first step involved a low-temperature (270°C, 136 atm) catalytic treatment that hydrogenated the thermally unstable biooil compounds, which thermally decompose to form coke and plugged the reactor. The second step involved catalytic hydrogenation at higher temperature (400°C, 136 atm). Sulfided $CoMo/Al_2O_3$ or sulfided $NiMo/Al_2O_3$ was used for both steps.

8.6 Reactions Involved in Hydroprocessing

Several reactions are known to take place during catalytic hydrotreatment of biooil, and they depend on the type of biomass feedstock, operating conditions, and catalyst employed. Some of the major reactions taking place during the hydroprocessing process are cracking, saturation, heteroatom removal, and isomerization.

8.6.1 Cracking

High-molecular-weight biomass is converted to biooil, which is relatively of low molecular weight, but it is still required to carry out cracking reactions to obtain fuels such as gasoline and diesel. Such reactions are known to occur during catalytic hydrotreating of oils/fats to fatty acids (carboxylic acids) and propane (Birchem 2010; Donnis et al. 2009). They are also known to occur during catalytic hydrotreatment of pyrolysis biooil and upgradation of Fischer–Tropsch wax.

8.6.2 Saturation

When saturation reactions occur, excess hydrogen allows the breakage of C=C bonds and enables their conversion to single bonds. Unsaturated cyclic compounds and aromatic compounds are converted to naphthenes during upgradation of pyrolysis biooils. Saturated molecules are less active and less prone to polymerization and oxidation reactions, mitigating the sediment formation and corrosion phenomena.

8.6.3 Heteroatom Removal

Heteroatoms are atoms other than carbon (C) and hydrogen (H) such as sulfur (S), nitrogen (N), and oxygen (O). With regard to biooil upgradation, oxygen removal is of utmost importance as the presence of oxygen is the main hindrance in direct utilization of biooil. Deoxygenation reactions mainly produce n-paraffins along with the release of oxygen in the form of H_2O, CO_2, and CO. Heteroatoms such as S and N are removed in the form of H_2S and NH_3, respectively, similar to crude fraction hydrotreatment. The presence of paraffinic hydrocarbons increases the cetane number of the fuel produced, and the hydroprocessed biooil as such has increased heating value and better oxidation stability.

8.6.4 Isomerization

The presence of paraffinic compounds degrades the cold flow properties of the biooil. Isomerization reactions produce compounds that improve the cold flow properties, but they generally occur on a different catalyst in a second step after all the reactions mentioned earlier occur (Bezergianni 2013).

8.7 Operating Parameters for Hydroprocessing

Variables such as reactor temperature, hydrogen partial pressure, liquid hourly space velocity (LHSV), hydrogen feed rate, and catalyst used are the most important operating parameters that have to be optimized for the hydroprocessing of biooil.

8.7.1 Temperature

Temperatures in the range of 290°C–450°C are generally used for the hydrotreating and hydroprocessing reactions. The temperature is generally selected depending on the catalyst and feed treated in the process. The temperature is low in the initial stages due to the high catalytic activity of the fresh catalyst used. Once catalyst deactivation due to coke formation is observed, the temperature used in the process is increased to compensate for the drop in catalytic activity as the desired product has to be produced.

8.7.2 Liquid Hourly Space Velocity

LHSV is defined as the ratio of the liquid mass feed rate (g h^{-1}) over the catalyst mass (g) and as a result is expressed in h^{-1}. The inverse of LHSV is proportional to the residence time of the liquid feed in the reactor. Higher LHSV reduces the contact time between the feed and catalyst leading to reduced conversions, but in contrast, reduced LHSV leads to faster loss in catalytic activity by degradation.

8.7.3 Hydrogen Partial Pressure

The next important process parameter is the hydrogen partial pressure used in the process as the hydrotreating reactions and catalyst deactivation are dependent on the presence of hydrogen. Catalyst deactivation reduces with increase in hydrogen partial pressure and hydrogen feed rate. The drawback is the increased operating expenses of the process due to high hydrogen consumption, which is higher in case of olefin-rich feedstocks.

8.7.4 Hydrogen Feed Rate

Hydrogen feed rate dictates the hydrogen partial pressure in the reactor depending on the hydrogen consumption of the process. It is used for heteroatom removal and saturation reactions. Similar to the hydrogen partial pressure, the hydrogen feed rate is also optimized after several technical and economical considerations due to the high cost of hydrogen. If cheaper or renewable sources of hydrogen are available, then the process would become even more favorable (Bezergianni 2013).

8.7.5 Catalyst Used

The catalyst used in hydroprocessing is the heart of the process as it dictates the quality and quantity of the desired product obtained. The reactions taking place during the process have already been discussed in the earlier sections. Acid sites are required for cracking reactions and metal functionalities are required for the hydrogenation reactions occurring in the process. The catalyst used, its stability, lifetime, deactivation, recyclability, etc., dictate the economics of the process. Various types of catalysts have been tested so far for the process of hydroprocessing, which will be discussed in the following sections.

8.8 Hydroprocessing Catalysts

Thermal processes have also been tested for removing the oxygen from biooil, but it has been noticed that oxygen removal above 10% is difficult without the use of catalysts.

8.8.1 Conventional Sulfide Catalysts

Conventional sulfide catalysts have been used for the past several decades in petroleum hydroprocessing (Laurent and Delmon 1994a). Oxygenated groups in biooil such as ketones, aldehydes, and organic acids require lower temperatures for elimination of the reactive functionalities but guaiacol-type molecules and other phenolic species require higher temperatures for oxygen removal (Centeno et al. 1995). Phenyl–oxygen bonds are cleaved at around 225°C–375°C using hydroprocessing catalysts under hydrogen pressure where oxygen is removed as water. Typical hydrodesulfurization catalysts such as $NiMoS/Al_2O_3$ and $CoMoS/Al_2O_3$ were found to quickly deactivate due to coke deposition in model HDO reactions because of the acidity of the reactant (Hong et al. 2010; Laurent et al. 1994). Sulfides on neutral supports including carbon, silica, and alumina modified by K for HDO reactions have also been reported (Centeno et al. 1995; Ferrari et al. 2001, 2002; Laurent et al. 1994). When molybdenum sulfide supported on activated carbon was used for HDO of guaiacol, coking reactions were found to be negligible with good catalyst stability (Puente et al. 1999). Noble metal Pt as an active component was added to a conventional CoMoS catalyst and showed no significant improvement (Centeno et al. 1995). Monometallic and bimetallic noble metal catalysts supported on zirconia have lower coke formation than $CoMoS/Al_2O_3$ catalyst, and in particular, Rh-containing catalysts demonstrate potential in biofuel upgradation (Zhao et al. 2011b).

8.8.2 Zeolite Catalysts

Zeolites are crystalline microporous silica-based materials that are extensively used as heterogeneous catalysts in industry. They offer a well-defined and ordered pore structure on a molecular level. By choosing the appropriate organic template and synthesis conditions, the pore size and pore shape (dimensionality, intersections, and cages) can be directly influenced. The acidity of the zeolite may also be controlled through various methods such as silicon/aluminum (Si/Al) ratio, ion exchange, and calcination conditions. Their high thermal stability permits them to be used at high temperatures that often result in higher yields and easier heat recovery. All of these characteristics, and especially the control of structure and acidity, make zeolites well suited as catalysts for organic reactions (Corma and Garcia 1997; Ram Reddy et al. 1998; Sheldon and Downing 1999). Traditionally, zeolites (aluminosilicates) are defined as crystalline materials in which Si and Al are tetrahedrally coordinated by oxygen atoms in a three-dimensional network creating uniform sized pores of molecular dimensions.

Zeolites, and in general molecular sieve inorganic materials, are the most widely used industrial catalysts for petroleum refining (Corma 1995, 1997, 2003). These reactions have been postulated to proceed through intermediate compounds called carbenium ions, which are created on the surface of zeolites through interaction of reaction molecules with active sites. Zeolites then act as solid acids, which are capable of converting the adsorbed molecule into various forms. This is accomplished either by transforming a proton from the solid acid to the adsorbed molecule (Bronsted acid site) or by transferring electron pairs from the adsorbed molecule to the solid surface (Lewis acid site). Zeolite acidity arises from the imbalance in charge between silicon and aluminum atoms within the framework so that each aluminum atom is capable of inducing a potential active site.

Mono- and bifunctional catalysts such as zeolites, mesoporous materials with uniform pore size distribution (MCM-41, MSU, SBA-15), micro-/mesoporous hybrid materials doped with noble and transition metals, and base catalysts have been explored for the hydroprocessing reactions. The catalysts should be able to selectively favor the decarboxylation reactions producing high-quality biooils with low amounts of oxygen (Stöcker 2008).

Temperatures of 350°C–500°C, atmospheric pressure, and gas hourly space velocities of around 2 are generally used for biooil upgradation reactions using zeolite catalysts. The products from this reaction include hydrocarbons (aromatic, aliphatic), water-soluble organics, water, oil-soluble organics, gases (CO_2, CO, light alkanes), and coke (Bridgwater 1994). However, poor hydrocarbon yields and high yields of coke generally occur under reaction conditions limiting the usefulness of zeolite upgrading (Huber et al. 2006).

Zeolite upgradation of fast-pyrolysis biooils using wood as feedstock was studied by Bakhshi and coworkers with different catalysts (Adjaye et al. 1996; Katikaneni et al. 1995; Sharma and Bakhshi 1993). Between 30 and 40 wt.% of the biooil was deposited on the catalyst as coke or in the reactor as char. The ZSM-5 catalyst produced the highest amount (34 wt.% of feed) of liquid organic products of any catalyst tested. The products in the organic carbon were mostly aromatics for ZSM-5 and aliphatics for SiO_2–Al_2O_3. Gaseous products include CO_2, CO, light alkanes, and light olefins. Biooils are thermally unstable and thermal cracking reactions occur during zeolite upgrading. Bakhshi and coworkers developed a two-reactor process, where only thermal reactions occur in the first empty reactor and catalytic reactions occur in the second reactor that contains the catalyst (Srinivas et al. 2000). The advantage of the two-reactor system is that it improved catalyst life by reducing the amount of coke deposited on the catalyst.

Adam et al. (2005) illustrated the effects and catalytic properties of Al-MCM-41, Cu/Al-MCM-41, and Al-MCM-41 with pores enlarged on biooil upgrading. The resulting compositions of vapors were changed through the catalyst layer. Adjaye and Bakhshi (1995a,b) studied the catalytic performance of the different catalysts for the upgradation of biooil. Among the five catalysts studied, HZSM-5 was the most effective catalyst for producing the organic distillate fraction, overall hydrocarbons, and aromatic hydrocarbons and the least coke formation. Also, coking with HZSM-5 was low due to its shape-selective properties and intracrystalline network (Chang et al. 1979).

In order to study the chemical mechanisms of catalytic hydroprocessing of biooil, different model compounds have been chosen to represent biooil components by various researchers. Table 8.1 lists the different model compounds tested for hydroprocessing using various catalysts.

TABLE 8.1 Catalytic Hydroprocessing of Different Model Compounds

Model Compounds	Process Conditions	Remarks
Methanol, acetic acid, aqueous solution of hydroxyl acetaldehyde, methanol solution of 4-allyl-2,6 dimethoxyphenol	Microreactor coupled to a molecular beam mass spectrometer (MBMS) and a bench-scale, fixed-bed unit; with the catalyst G-90C, ICI 46-1/4 used	Biooil or its aqueous fraction can be efficiently reformed to generate hydrogen by a thermocatalytic process using commercial, nickel-based catalysts. The hydrogen yield is as high as 85% of the stoichiometric value (Wang et al. 1998).
Acetic acid; xylose, glucose, and sucrose (saccharides with 20% in water); m-cresol and dibenzyl aromatic ether group	Nickel-based catalysts (UC G-90C, ICI 46-1) tested in two fixed-bed reactors at a temperature range of 550°C–810°C	Acetic acid is almost completely converted to hydrogen and carbon oxides provided that the catalyst temperature is above 650°C. Carbonaceous deposits are formed on the catalyst particles when temperatures are lower than this (Marquevich et al. 1999).
1-Propanol, 2-propanol, 1-butanol, 2-butanol, phenol, 2-methoxyphenol	HZSM-5 zeolite catalyst tested in fixed-bed catalytic reactors used at an atmospheric pressure of 200°C–450°C	Alcohols dehydrate to olefins rapidly at low temperatures. Iso-alcohols dehydrate more rapidly than linear alcohols. Phenol has low reactivity on HZSM-5 and produces small amount of propylene and butenes. 2-Methoxyphenol has a low reactivity to hydrocarbons (Gayubo et al. 2004).
p-Cresol, ethylphenol, dimethylphenol, trimethylphenol, naphthol, guaiacol	CoMo, NiMo, NiW, Ni, Co, Pd, and CuCrO catalysts used in a batch and continuous flow reactor at a temperature of 300°C or 400°C	Sulfided form of CoMo was more active, producing 33.8% benzene and 3.6% cyclohexane at 400°C. Ni catalyst was active with 16.9% benzene and 7.6% cyclohexane (Elliott 2007).
Guaiacol, acetic acid, furfural	Palladium or ruthenium catalyst tested in a parr reactor (450 mL) or Hastelloy C pressure vessel at a temperature range of 150°C–300°C and a pressure of 4.2 MPa (H_2) for 4 h	Acetic acid effectively hydrogenated using Pd, guaiacol with Ru, and furfural with both. Ruthenium appears to be an active catalyst for hydrogenations at lower temperatures. Palladium can be used at higher temperatures for hydrogenation (Elliott and Hart 2009).
Glacial acetic acid	200°C and 3 MPa H_2 pressure, Mo–10Ni/γ–Al$_2$O$_3$	Reduced Mo–10Ni/γ–Al$_2$O$_3$ catalyst exhibited the highest activity in the reaction of model compound. The conversion of acetic acid reached the highest point (Xu et al. 2009).
2-Propanol, cyclopentanone, anisole, guaiacol, propanoic acid, ethyldecanoate	Hydrotreating process carried out using a pilot-scale fixed-bed reactor at a temperature of 330°C and a pressure of 3–5 MPa with CoMo/γ–Al$_2$O$_3$ catalyst	2-Propanol, cyclopentanone, anisole, and guaiacol were not found to be inhibitors of catalytic performances. Propanoic acid and ethyldecanoate had an inhibiting effect on HDS, HDN, and HDCA reactions (Pinheiro et al. 2009).

(Continued)

TABLE 8.1 (*Continued*)   Catalytic Hydroprocessing of Different Model Compounds

Model Compounds	Process Conditions	Remarks
Glucose, cellobiose	Batch autoclave HDO process performed at a temperature of 250°C and a pressure of 100 bar with the use of Ru/C and Pd/C catalysts	Catalytic hydrotreatment was preferred over thermal decomposition. Main products were polyols and gas products (mostly methane) (Wildschut et al. 2009).
Mixtures of representative compounds of biooil	HZSM-5 catalyst tested for Mixture 1 (acetone, acetic acid, methanol, 2-butanol, phenol, and acetaldehyde) and Mixture 2 (acetone, acetic acid, 2-methoxy phenol, 2- propanol, and furfural) at a temperature range of 400°C–450°C	Such transformation requires previous separation of components like aldehydes, oxyphenol, and furfural as they undergo severe thermal degradation and deactivate the catalyst (Gayubo et al. 2009).
Phenol, *p*-methoxyphenol furfural	CeO$_2$ or alumina (Degussa type C), silica (Aerosil 200), and titania (P25) supports used in a tubular reactor	CeO$_2$ was chosen as active phase due to its superior properties: high catalytic activity and tolerance to water. Al$_2$O$_3$ and TiO$_2$ supports or doping with strong base can improve the catalytic activity to a certain degree (Deng et al. 2009).
Phenol, guaiacols and syringols	Carbon-supported palladium catalysts tested in a parr reactor at a pressure of 5 MPa (H$_2$) and temperature of 200°C or 250°C for 0.5 h at 1000 rpm	Carbon-supported noble metal catalyst in combination with the mineral acid H$_3$PO$_4$ act as bifunctional catalysts for the one-pot HDO conversion of biooil through the multistep reactions involving hydrogenation, hydrolysis, and dehydration (Zhao et al. 2009).
Mixture of methanol (5 wt.%), acetaldehyde (12 wt.%), acetic acid (14 wt.%), glyoxal (4 wt.%), acetol (8 wt.%), glucose (8 wt.%), guaiacol (17 wt.%), furfural (4 wt.%), vanillin (8 wt.%), and deionized water (20 wt.%)	Supported platinum catalyst tested in a 300 mL autoclave at a pressure of 100 psi of N$_2$ and a temperature of 350°C	Pt/Al$_2$O$_3$ showed the highest activity for oxygen removal and treatment of synthetic biooil at 350°C. Upgraded product rich in alkylbenzenes and alkylcyclohexanes and alkylsubstituted phenols constituted the main residual oxygen-containing compounds (Fisk et al. 2009).
Acetic acid	MoNi/γ–Al$_2$O$_3$ catalyst mixed in a 250 mL autoclave at a temperature of 200°C and a pressure of 3 MPa (H$_2$) for 2 h	Reduced Mo–10Ni/γ–Al$_2$O$_3$ catalyst exhibited the highest activity in the reaction of model compound (Xu et al. 2010).

(*Continued*)

TABLE 8.1 (Continued) Catalytic Hydroprocessing of Different Model Compounds

Model Compounds	Process Conditions	Remarks
Acetaldehyde, furfural, vanillin	Catalytic hydrogenation carried out through a 60 mL autoclave at a temperature range of 55°C–90°C and a pressure range of 1.3–3.3 MPa ($RuCl_2(PPh_3)_3$) for 3 h	Furfural and acetaldehyde were singly converted to furfuryl alcohol and ethanol after hydrogenation. Vanillin was mainly converted to vanillin alcohol, together with small amounts of 2-methoxy-4 methylphenol and 2-methoxyphenol (Huang et al. 2010).
Phenol, anisole, guaiacol	Carried out in the presence of silica, alumina (pure or doped with K or F), and silica–alumina catalysts	Phenol, anisole, and guaiacol mainly interact via H-bonding with silica. By contrast, chemisorption is their main adsorption mode on alumina. Anisole requires temperature greater than 200°C to transform coordinated species into phenate species. By contrast, silica addition to alumina markedly decreases the presence of phenate species (Popov et al. 2010).
Phenol	Pt catalyst supported on HY zeolite tested in a fixed-bed flow reactor at a temperature range of 200°C–250°C and a pressure of 4 MPa	Zeolite-supported Pt catalyst acts as a bifunctional catalyst, with hydrogenation and acid functions yielding high activity and selectivity to monocyclics as well as production of useful bicyclics (Hong et al. 2010).
Phenols, guaiacols syringols, an equimolar mixture of phenolic monomers including 4-*n*-propylphenol, 2-methoxy-4-*n*-propylphenol, and 4-allyl-2 methoxyphenol	Raney Ni and Nafion/SiO_2 catalysts tested in a parr reactor (300 mL) at a pressure of 4 MPa (H_2, room temperature) and a temperature of 200°C for 0.5 h or 300°C for 2 h (1000 rpm)	Raney Ni acted as hydrogenation catalyst and Nafion/SiO_2 acted as Brønsted solid acid for hydrolysis and dehydration. No product from polymerization among phenolic compounds was observed (Zhao et al. 2010).
Acid-catalyzed 1-octene reactions with phenol and mixtures of phenol with water, acetic acid, and 1-butanol that were studied as partial biooil upgrading models	Amberlyst 15, Dowex50WX2, and Dowex50WX4 put in 25 mL two-necked round-bottom flasks at atmospheric pressure and a temperature range of 65°C–120°C	Additions across olefins offer a route to simultaneously lower water content and acidity while increasing hydrophobicity, stability, and heating value. Amberlyst 15, Dowex50WX2, and Dowex50WX4 effectively catalyzed phenol *O*- and *C*-alkylation at a temperature of 65°C–120°C, giving high *O*-alkylation selectivity in the presence of water, acetic acid, and 1-butanol. Dowex50WX2 and Dowex50WX4 were more stable in the presence of water than Amberlyst 15 and were successfully recycled (Zhang et al. 2010).

(Continued)

TABLE 8.1 (Continued) Catalytic Hydroprocessing of Different Model Compounds

Model Compounds	Process Conditions	Remarks
Phenol, water, acetic acid, methanol, and 2-hydroxy methylfuran with olefins	$40°C–120°C$ 30 wt.% acidic salt $Cs_{2.5}H_{0.5}PW_{12}O_{40}$ supported on K-10 clay ($30\%Cs_{2.5}/K-10$), Nafion (NR50) and Amberlyst 15	Both catalysts had a high activity and selectivity for O-alkylation of phenol with 1-octene but not with 2,4,4-trimethylpentene. 30% $Cs_{2.5}/K-10$ is an excellent water-tolerant catalyst, while Amberlyst 15 decomposed at higher temperature and water concentration (Yang et al. 2010).
Anisole	Zeolite catalyst tested on a continuous flow and pulse reactors at a temperature of $400°C$ and at atmospheric pressure	Transalkylation produces phenol, cresols, xylenols, etc. (Prasomsri et al. 2011).
Guaiacol	Use of batch-type mono- and bimetallic Rh-based catalysts (Rh, PtRh, and PdRh on ZrO_2) and classical sulfided CoMo and NiMo on Al_2O_3 catalysts	Substituting some of the Rh ions with another noble metal (i.e., Pd or Pt) produced no positive effect. Rh/ZrO_2 was the most effective catalyst (Lin et al. 2011).
Phenol, benzaldehyde, acetophenone	Amorphous Co-Mo-B catalyst tested on a 300 mL sealed autoclave at a temperature of $10°C$ min^{-1} and pressure of 4 MPa (700 rpm)	Catalyst activity increased with increasing Co/Mo ratio (Wang et al. 2011a).
Dibenzofuran	Noble metal (Pt) supported on mesoporous zeolite	Pt/mesoporous ZSM-5 is better than Pt/ZSM-5 and Pt/Al_2O_3 (Wang et al. 2011b).
Anisole, guaiacol	Inconel monoliths coated with in situ grown carbon nanofibers (CNFs), which were subsequently impregnated with catalytic species (Pt, Sn, and bimetallic Pt-Sn)	The main products obtained from these feeds on the monolithic catalysts were phenol and benzene. Bimetallic Pt–Sn catalysts showed higher activity and stability than monometallic Pt and Sn catalysts (Gonzalez-Borja and Resasco 2011).
Acetic acid	$Ru/\gamma-Al_2O_3$ catalyst tested on a 250 mL autoclave at a pressure of 3 MPa (H_2) for 2 h	The $0.5Ru/\gamma-Al_2O_3$ catalyst with 0.5% Co addition exhibited the highest activity, giving the highest conversion of acetic acid (30.98%) (Ying et al. 2012).
4-Methyl anisole	Pt/Al_2O_3, $Pt/SiO_2-Al_2O_3$, or HY zeolite tested out at a temperature of $300°C$ and at atmospheric pressure	The most abundant products are 4-methylphenol, 2,4-dimethylphenol, and 2,4,6-trimethylphenol; toluene was also a major product when the catalyst was supported platinum with H_2 as a coreactant (Runnebaum et al. 2012).
Phenol	Use of Ru/SBA-15 catalyst	This is converted to C_3 to C_{10} alcohols at mild conditions. Ru/SBA-15 is stable with phenolic compounds in the simulated biooil, but not in a real one (Guo et al. 2012).

(Continued)

TABLE 8.1 (*Continued*)   **Catalytic Hydroprocessing of Different Model Compounds**

Model Compounds	Process Conditions	Remarks
2-Methoxy phenol	Ni/kieselguhr catalysts used in a batch reactor at a temperature of 100°C and a pressure of 13.6 bar	Ni (40 wt.%)–MgO (1.8 wt.%)/kieselguhr showed much higher activity due to some factors such as optimal nickel particle size, surface area, and reducibility of the catalyst (Jeon et al. 2013).
Nickel-based zeolite catalyst	NiO loading at 8 wt.% and the calcination temperature at 350°C–650°C	At calcination temperature of 550°C, the conversion rates of toluene and guaiacol are 83% and 80%, respectively, but it decreased to 40% and 50%, respectively, after 6 h of continuous catalytic reaction (Qin et al. 2014).
Vanillin	Au/CNTs assembled at the interfaces of a Pickering emulsion	A good catalytic activity and 100% selectivity for the HDO of vanillin to *p*-creosol under mild reaction conditions is achieved (Yang et al. 2014a).
Guaiacol	Ni-based catalysts using mixed oxides of Al_2O_3–SiO_2, Al_2O_3–TiO_2, TiO_2–SiO_2, and TiO_2–ZrO_2 as supports	Guaiacol conversion of 100% with cyclohexane selectivity of 86.4% was obtained over the Ni/TiO_2–ZrO_2 catalyst at a temperature of 300°C and a pressure of 40 bar (H_2) when a decalin solvent is utilized (Zhang et al. 2014).
Anisole	Ni-containing (20% loading) catalysts supported on SBA-15, Al-SBA-15, γ-Al_2O_3, microporous C, TiO_2, and CeO_2	Under low H pressure (three bar) and moderate temperatures (290°C–310°C) and space velocity (20.4 and 81.6 h^{-1}), these catalysts showed high HDO activity. Maximum yield of 64% of benzene was obtained over Ni/C catalysts (Yang et al. 2014b).
Eugenol	Hydrodeoxygenated in aqueous phase with Pd/C as hydrogenation catalyst and ZSM-5 zeolite as hydrolysis and dehydration catalysts	Alkali-treated ZSM-5 zeolites showed higher conversion and hydrocarbon selectivity compared to the parent ZSM-5 zeolite. At 65°C, conversion and hydrocarbon selectivity increased by 21% and 24%, respectively (Xing et al. 2015).
Propanol	Mesoporous aluminosilicate MFI nanosheets (Al-MFI-ns) of single-unit-cell thickness and conventional Al-MFI zeolite (Al-MFI) at 400°C and atmospheric pressure	Overall ratio of olefin to aromatic products is similar for Al-MFI-ns and Al-MFI at all conversion levels. Al-MFI-ns generated a fivefold increase in selectivity to C_{6-8} olefins and a twofold increase in selectivity to C_{9-10} aromatics compared to Al-MFI (Luo et al. 2015).

(Continued)

TABLE 8.1 (*Continued*) Catalytic Hydroprocessing of Different Model Compounds

Model Compounds	Process Conditions	Remarks
4-Methyl anisole	Pt/γ–Al₂O₃, in the presence of hydrogen with a fixed-bed tubular flow reactor at a temperature range of 300°C–400°C, a pressure of 8–20 bar, and space velocities in the range of 3–120 (g of 4-methylanisole)/(g of catalyst × h)	4-Methylphenol formed as a primary intermediate product via scission of the C methyl-O bond and then hydrogenolysis of the C aromatic-O occurred giving toluene. Major products were 4-methylphenol, 2,4-dimethylphenol, and 2,4,6-trimethylphenol (Saidi et al. 2015).
Guaiacol	Pt/Al-SBA-15 (with the Si/Al ratios of 20, 40, and 80) and Pt/HZSM-5 catalysts used in a batch reactor at a pressure of 40 bar and a temperature of 250°C	The order of cyclohexane yield was Pt/Al-SBA-15 (Si/Al = 20) > Pt/Al-SBA-15(40) > Pt/Al-SBA-15 (80). The quantity of acid sites plays an important role in the HDO reaction. Pt/HZSM-5 led to a very low cyclohexane yield, in spite of its abundant strong acid sites, due to its small pore size (Yu et al. 2015).
Phenol	Supported palladium nanoparticles on carbon nitrogen composites	Conversion of phenol was 91.7% within 1 h at 80°C (Feng et al. 2015).
Acetone and phenol	Methanol as hydrogenation liquid donor, instead of hydrogen gas over Raney Ni catalyst	Conversions of acetone and phenol reached the highest point with 55.76% and 64.65%, respectively, at a temperature of 220°C and a pressure of 30 bar (N₂) (Ying et al. 2015).

8.8.3 Noble Metal Catalysts

In cases where biooil contains fewer amounts of sulfur and nitrogen, catalyst supports such as Al_2O_3 or other novel supports impregnated with noble metals (including bifunctional catalysts) have also been studied. Coke deposition was found to be the most important reason for catalyst deactivation.

8.8.4 Nonconventional Catalysts

Several other kinds of catalysts have been tested for the hydroprocessing reactions that do not require the presulfidation step and are also stable to some extent in the presence of water. Catalysts such as metal carbides, nitrides, and phosphides have been tested for the process. Carbides can also be prepared with a high surface area.

The activity of metal carbides and nitrides for hydroprocessing follows a peculiar behavior as the catalytic activity is seen to increase with increasing particle size or decreasing surface area (Furimsky 2003). This is contradictory to the generally observed trends involving conventional hydroprocessing catalysts (Lostaglio and Carruthers 1986; Topsoe et al. 1996; Trimm 1980; Trimm and Stanislaus 1986). This might be due to the structural features of the crystals of metal carbides and nitrides that allow diffusion of hydrogen to subsurfaces followed by its activation (Furimsky 2003). The active hydrogen may then migrate to the surface to participate during hydroprocessing. The possible reaction mechanism as mentioned earlier is yet to be proved, and the information regarding the nature of active sites on metal carbides and nitrides is rather limited. The three types of sites may include acidic, metallic, and dual sites. It is known that nitrogen, carbon, and oxygen atoms can be inserted into the octahedral sites located in either the center or edges of the face-centered cubic lattice of Mo(W) atoms, which facilitates the formation of oxycarbonitride-like structure. This means that if metal nitrides and carbides are used for hydroprocessing of biooil, gradual change to corresponding oxynitrides and oxycarbides caused by the presence of H_2O may not be avoided (Furimsky 2003). Recent attempts made by Tominaga and Nagai (Tominaga and Nagai 2010) have focused on potential active sites on Mo nitrides.

Zhao et al. (2011b) studied the HDO of guaiacol over the Ni_2P, Co_2P, Fe_2P, WP, and MoP catalysts. The absence of catechol among the products obtained at lesser contact time over Ni_2P compared with the other metal phosphides suggested a different involvement of metals either during hydrogen activation or the interaction of reactants with catalyst surface. This is supported by the difference in activation energies estimated for these phosphides between 200°C and 300°C, that is, from 23 to 63 kJ mol^{-1}. Turnover frequency determined by titration of active sites using the chemisorption of CO followed the order $Ni_2P > Co_2P > Fe_2P \approx WP \approx MoP$. This order coincided with the activity order obtained during the HDO of guaiacol (Furimsky 2013; Zhao et al. 2011).

The use of carbon supports or nitrides or borides or phosphides, which do not require presulfidation step, must be developed since biooil has negligible amount of sulfur. The sulfiding agent leads to emanation of poisonous H_2S gas and continuous supply of sulfiding agent is required to maintain the sulfur level in the feed stream, which is necessary for good catalytic activity.

The products obtained from the hydroprocessing using low-cost catalysts will differ from that using standard Co(Ni)/Mo(W) hydroprocessing catalyst. Therefore, several criteria have to be considered to make a choice between the low-cost catalyst systems and the systems employing conventional hydroprocessing catalysts. A number of processes employing a disposable catalyst are in a developmental stage, and few others are in a near commercial stage (Furimsky 1998; Graeser and Niemann 1982; Guitian et al. 1988; Niemann 1991; Silva et al. 1983).

8.8.5 Hydroprocessing of Aqueous Phase of Biooil

Biooil obtained by the process of hydrothermal liquefaction is present in an aqueous medium in a significant amount, and hence, separation of water-soluble components from the aqueous phase becomes inefficient. Similar situation also exists with pyrolysis biooils though the aqueous fraction obtained may not be as high as that of liquefaction biooils. In this scenario, conversion of polar compounds to hydrocarbons directly by hydroprocessing in the same environment may have some advantages as product separation may become easier (Amin 2009; Chen et al. 2011; Hong et al. 2010; Huber and Dumesic 2006; Kanie et al. 2011; Ohta et al. 2011; Peng et al. 2012; Zhao et al. 2009, 2010, 2011a). The temperatures used for aqueous hydroprocessing cover the subcritical and super critical water temperature regions ranging from 200°C to 450°C (Peterson et al. 2008). As the reactions occur in aqueous media, the supports used as catalysts in this process must be hydrophobic in nature (Furimsky 2013; Olarte et al. 2012).

8.9 Reactors for Biooil Upgradation

Reactors play a major role in the hydroprocessing process and are of great concern as the product yield and selectivity depend on the type of reactor used. Different reactors have different heat and mass transfer effects between the feed, catalyst, and products, and hence, choosing the right reactor for the process is very important.

It was observed that when HDO of pyrolysis oil is performed in a conventional macroreactor system, severe heat and mass transfer limitations as well as concerns of high operating cost and safety existed. Biomass is a low-density material and, when converted into biooil, is not available in huge quantities like crude oil, and hence, large-scale processing is not possible (Rotman 2008). In the case of distributed units where less biooil has to be processed, microreactor systems can be utilized where diffusion time is short and influence of mass transfer on rate of reaction is greatly reduced. The other advantages with microreactor are the improved heat and mass transfer and reduced risk with high-temperature and high-pressure options in addition to high yield, improved product quality, and better selectivity

(Besser et al. 2003; Halder et al. 2007; Hendershot 2003; Joshi and Lawal 2012; Okafor et al. 2010; Stankiewicz and Drinkenburg 2003; Tadepalli et al. 2007; Voloshin and Lawal 2010).

8.10 Conclusions

There are several challenges in the hydroprocessing of biooil and the options to overcome the same become the opportunities available in this area of research. The size of the reactor plays a major role in the throughput of the plant and its design has to be in such a way that effective heat and mass transfer occurs in the reactor. Hotspots should not be formed in the catalyst bed that will lead to several other unwanted products to be formed, thus reducing the selectivity of the desired products. The other operating parameters for hydroprocessing such as hydrogen feed rate, hydrogen partial pressure, temperature, and LHSV follow their own individual patterns and have different relationship between variables (i.e., they are directly or indirectly proportional to some variables) when varied separately. All these parameters have to be set in such a way that an optimum balance is maintained between the technical and economical aspects of the process. Several challenges exist in designing the catalyst for the process of hydroprocessing. First, the supports required for the process must be robust, and since cracking activity is required for the process, the acidity of the catalysts must be well maintained. Second, the hydrogenation functionality is also required for the process, and hence, noble metals are used for the process. Opportunities to reduce the loading of noble metals must be explored as their costs are very high, and similarly, the use of nonnoble metals must be encouraged. The catalysts must be hydrothermally stable over a wide range of temperatures and this becomes even more important when hydroprocessing occurs in aqueous media. In case where algal biooils are upgraded, the effect of high nitrogen content in the biooil must be taken into consideration before designing the catalyst for the hydroprocessing of the same. The main reason for loss in catalytic activity is the coke deposition that occurs on the catalyst due to the polymerization reactions of unstable species formed during the reaction. Hence, efforts have to be taken to develop catalyst that has a high on-stream time, followed by an easy regeneration step. The catalyst must also be usable for several cycles of operation, which defines the recyclability of the catalyst. In cases where fluidized-bed reactors are used, attrition-resistant catalysts have to be developed to enable longer on-stream catalytic activity. In addition to the points mentioned earlier, the steps used to recover the metals from spent catalysts must be identified and options for reuse of supports must be looked into. This is very essential as the operating cost of the process is also dependent on the catalyst inventory. On a holistic level, several efforts have to be taken to understand the detailed reaction mechanism of the process using advanced analytical tools that are available. This will help in identifying the intermediates of the process and help in designing catalysts that has a higher selectivity to the desired products. The kinetics and thermodynamic parameters of the process have to be identified to completely comment on the possible products that can be obtained from the process. The elucidation of the reaction mechanism will help the researchers throw a light on the structure–activity relationships involved in the process. Thus, there exists several opportunities for young and experienced researchers to carry out research in the area of hydroprocessing of biofuels to produce fuels or chemicals from a renewable source of energy such as biomass and contribute in creating a better world for the future generations to come.

Acknowledgments

The authors thank the director of CSIR-Indian Institute of Petroleum, Dehradun, for his constant encouragement and support. RS thanks the Council for Scientific and Industrial Research (CSIR), New Delhi, India, for providing senior research fellowship (SRF). The authors also thank CSIR for initiating the XII Five Year Plan Project (CSC0116/BioEn) and the Ministry of New and Renewable Energy for providing financial support.

References

Adam, J., Blazso, M., and E. Mészáros. 2005. Pyrolysis of biomass in the presence of Al-MCM-41 type catalysts. *Fuel* 84:1494–1502.

Adjaye, J.D. and N.N. Bakhshi. 1995a. Production of hydrocarbons by catalytic upgrading of a fast pyrolysis biooil. Part I: Conversion over various catalysts. *Fuel Process Technol* 45:161–183.

Adjaye, J.D. and N.N. Bakhshi. 1995b. Production of hydrocarbons by catalytic upgrading of a fast pyrolysis biooil. Part II: Comparative catalyst performance and reaction pathways. *Fuel Process Technol* 45:185–202.

Adjaye, J.D., Katikaneni, S.P.R., and N.N. Bakhshi. 1996. Catalytic conversion of a biofuel to hydrocarbons: Effect of mixtures of HZSM-5 and silica-alumina catalysts on product distribution. *Fuel Process Technol* 48:115–143.

Alonso, D.M., Bond, J.Q., and J.A. Dumesic. 2010. Catalytic conversion of biomass to biofuels. *Green Chem* 12:1493–1513.

Amin, S. 2009. Review on biofuel oil and gas production processes from microalgae. *Energ Convers Manage* 50:1834–1840.

Baglioni, P., Chiaramonti, D., Bonini, M., Soldaini, I., and G. Tondi. 2001. BCO/diesel oil emulsification: Main achievements of the emulsification process and preliminary results of tests on diesel engine. In *Progress in Thermochemical Biomass Conversion*, ed. A.V. Bridgwater, pp. 1525–1539. Blackwell Science Limited, U.K.

Baglioni, P., Chiaramonti, D., Gartner, K., Grimm, H.P., Soldaini, I., and G. Tondi. 2003. Development of bio crude oil/diesel oil emulsions and use in diesel engines part 1: Emulsion production. *Biomass Bioenerg* 25:85–99.

Besser, R.S., Ouyang, X., and H. Surangalikar. 2003. Hydrocarbon hydrogenation and hydrogenation reactions in microfabricated catalytic reactors. *Chem Eng* 58:19–26.

Bezergianni, S. 2013. Catalytic hydroprocessing of liquid biomass for biofuels production. In *Liquid, Gaseous and Solid Biofuels—Conversion Techniques*, ed. Z. Fang, Chemical Processes & Energy Resources Institute (CPERI), Centre for Research & Technology Hellas (CERTH), Thermi-Thessaloniki, Greece. pp. 299–326.

Birchem, T. 2010. Latest improvements in ACETM catalysts technology for ULSD production and deep cetane increase. In *Fifth ERTC Annual Meeting*, Istanbul, Turkey.

Bridgwater, A.V. 1994. Catalysis in thermal biomass conversion. *Appl Catal A Gen* 116:5–47.

Bridgwater, A.V. 1996. Production of high grade fuels and chemicals from catalytic pyrolysis of biomass. *Catal Today* 29:285–295.

Bridgwater, A.V. 1999. Principles and practice of biomass fast pyrolysis processes for liquids. *J Anal Appl Pyrolysis* 51:3–22.

Bridgwater, A.V., Meierand, D., and D. Radlein. 1999. An overview of fast pyrolysis of biomass. *Org Geochem* 30:1479–1493.

Centeno, A., Laurent, E., and B. Delmon. 1995. Influence of the support of como sulfide catalysts and of the addition of potassium and platinum on the catalytic performances for the hydrodeoxygenation of carbonyl, carboxyl, and guaiacol-type molecules. *J Catal* 154:288–298.

Chang, C.D., Lange, W.H., and R.L. Smith. 1979. The conversion of methanol and other O-containing compounds to hydrocarbons over zeolite catalyst. *J Catal* 56:169–173.

Chen, L., Zhu, Y., Zheng, H., Zhang, C., and Y. Li. 2011. Aqueous phase hydrodeoxygenation of propanoic acid over the Ru/ZrO_2 and $Ru-Mo/ZrO_2$ catalysts. *Appl Catal A Gen* 411–412:95–104.

Chica, A. and A. Corma. 1999. Hydroisomerisation of pentane, hexane and heptanes for improving the octane number of gasoline. *J Catal* 187:167–176.

Corma, A. 1995. Inorganic solid acids and their use in acid-catalyzed hydrocarbon reactions. *Chem Rev* 95:559–614.

Corma, A. 1997. From microporous to mesoporous molecular sieve materials and their use in catalysis. *Chem Rev* 97:2373–2420.

Corma, A. 2003. State of the art and future challenges of zeolites as catalysts. *J Catal* 216:298–312.

Corma, A., Fornes, V., and E. Ortega. 1985. The nature of acid sites on fluorinated γ-Al_2O_3. *J Catal* 92:284–290.

Corma, A. and H. Garcia. 1997. Organic reactions catalyzed over solid acids. *Catal Today* 38:257–308.

Corma, A., Miguel, P.J., and A.V. Orchilles. 1994. The role of reaction temperature and cracking catalyst characteristics in determining the relative rates of protolytic cracking, chain propagation, and hydrogen transfer. *J Catal* 145:171–180.

Czernik, S. and A.V. Bridgwater. 2004. Overview of applications of biomass fast pyrolysis oil. *Energ Fuels* 18:590–598.

Czernik, S., French, R., Feik, C., and E. Chornet. 2002. Hydrogen by catalytic steam reforming of liquid byproducts from biomass thermoconversion processes. *Ind Eng Chem Res* 41:4209–4215.

Czernik, S., Maggi, R., and G.V.C. Peacocke. 2002. Review of methods for upgrading biomass-derived fast pyrolysis oils. In *Fast Pyrolysis of Biomass: A Handbook*, ed. A.V. Bridgwater, Vol. 2. Newbury, U.K.: CPL Press.

Deng, L., Fu, Y., and Q.X. Guo. 2009. Upgraded acidic components of biooil through catalytic ketonic condensation. *Energ Fuels* 23:564–568.

Diebold, J.P. 2000. A review of the chemical and physical mechanisms of the storage stability of fast pyrolysis biooils. NREL/SR-570-27613. http://www.nrel.gov/docs/fy00osti/27613.pdf. Accessed on January 2000.

Diebold, J.P. and J.W. Scahill. 1988. *Upgrading Pyrolysis Vapors to Aromatic Gasoline with Zeolite Catalysis at Atmospheric Pressure*. ACS Symposium Series, Vol. 376, pp. 264–276. Washington, DC: American Chemical Society.

Donnis, B., Egeberg, R.G., Blom, P., and K.G. Knudsen. 2009. Hydroprocessing of bioo-ils and oxygenates to hydrocarbons. Understanding the reaction routes. *Top Catal* 52:229–240.

Dragone, G., Fernandes, B.,Vicente, A.A., and J.A. Teixeira. 2010. Third generation bio-fuels from microalgae. In *Current Research, Technology and Education Topics in Applied Microbiology and Microbial Biotechnology*, ed. A. Mendez-Vilas, pp. 1355–1366. Formatex, U.K.

Elliott, D.C. 1983. Hydrodeoxygenation of phenolic components of wood-derived oil. *Am Chem Soc Div Petrol Chem Prepr* 28:667–674.

Elliott, D.C. 2007. Historical developments in hydroprocessing biooils. *Energ Fuels* 21:1792–1815.

Elliott, D.C. and E.G. Baker. 1987. Hydrotreating biomass liquid to produce hydrocarbon fuels, in *Energy from Biomass and Waste*, ed. D.L. Klass, pp. 765–784. Chicago, IL: Institute of Gas Technology.

Elliott, D.C. and E.G. Baker. 1989. Process for upgrading biomass pyrolyzates. US Patent 4,795,841.

Elliott, D.C., Baker, E.G., Piskorz, J., Scott, D.S., and Y. Solantausta. 1988. Production of liq-uid hydrocarbon fuels from peat. *Energ Fuels* 2:234–235.

Elliott, D.C., Beckman, D., Bridgwater, A.V., Diebold, J.P., Gevert, S.B., and Y. Solantausta. 1991. Developments in direct thermochemical liquefaction of biomass. *Energ Fuels* 5:399–410.

Elliott, D.C. and T.R. Hart. 2009. Catalytic hydroprocessing of chemical models for biooil. *Energ Fuels* 23:631–637.

Elliott, D.C. and G.G. Neuenschwander. 1996. Liquid fuels by low-severity hydrotreating of biocrude. In *Developments in Thermochemical Biomass Conversion*, eds. A.V. Bridgwater and D.G.B. Boocock, pp. 611–621. London, U.K.: Blackie Academic and Professional.

Elliott, D.C. and A. Oasmaa. 1991. Catalytic hydrotreating of black liquor oils. *Energ Fuels* 5:102–109.

Feng, G., Chen, P., and H. Lou. 2015. Palladium catalysts supported on carbon-nitrogen com-posites for aqueous-phase hydrogenation of phenol. *Catal Sci Technol* 5:2300–2304.

Fengel, D. and G. Wegener. 2003. *Wood: Chemistry, Ultrastructure, Reactions*. Verlag Kessel. Remagen-Oberwinter, Germany.

Ferrari, M., Bosmans, S., Maggi, R., Delomon, B., and P. Grange. 2001a. CoMo/carbon hydro-deoxygenation catalysts: Influence of the hydrogen sulfide partial pressure and of the sulfidation temperature. *Catal Today* 65:257–264.

Ferrari, M., Delmon, B., and P. Grange. 2002. Influence of the impregnation order of molyb-denum and cobalt in carbon-supported catalysts for hydrodeoxygenation reactions. *Carbon* 40:497–511.

Ferrari, M., Maggi, R., Delmon, B., and P. Grange. 2001b. Influences of the hydrogen sulfide partial pressure and of a nitrogen compound on the hydrodeoxygenation activity of a CoMo/carbon catalyst. *J Catal* 198:47–55.

Filley, J. and C. Roth. 1999. Vanadium catalysed guaiacol deoxygenation. *J Mol Catal A Catal* 139:245–252.

Fisk, C.A., Morgan, T., Ji, Y., Crocker, M., Crofcheck, C., and S.A. Lewis. 2009. Biooil upgrading over platinum catalysts using in situ generated hydrogen. *Appl Catal A Gen* 358:150–156.

Furimsky, E. 1998. Selection of catalysts and reactors for hydroprocessing. *Appl Catal A Gen* 171:177–206.

Furimsky, E. 2000. Catalytic hydrodeoxygenation. *Appl Catal A Gen* 199:147–190.

Furimsky, E. 2003. Metal carbides and nitrides as potential catalysts for hydroprocessing. *Appl Catal A Gen* 240:1–28.

Furimsky, E. 2013. Hydroprocessing challenges in biofuels production. *Catal Today* 217:13–56.

Garcia, L., French, R., Czernik, S., and E. Chornet. 2000. Catalytic steam reforming of biooils for the production of hydrogen. *Appl Catal A Gen* 201:225–239.

Gayubo, A.G., Aguayo, A.T., Atutxa, A., Aguado, R., and J. Bilbao. 2004. Transformation of oxygenate components of biomass pyrolysis oil on a HZSM-5 zeolite. I. Alcohols and phenols. *Ind Eng Chem Res* 43:2610–2618.

Gayubo, A.G., Aguayo, A.T., Atutxa, A., Valle, B., and J. Bilbao. 2009. Undesired components in the transformation of biomass pyrolysis oil into hydrocarbons on an HZSM-5 zeolite catalyst. *J Chem Technol Biotechnol* 80:1224–1251.

Gonzalez-Borja, M.A. and D.E. Resasco. 2011. Anisole and guaiacol hydrodeoxygenation over monolithic Pt-Sn catalysts. *Energ Fuels* 25:4155–4162.

Graeser, U. and K. Niemann. 1982. Proven hydrogenation processes for upgradation residueal being revived in Germany. *Oil Gas J* 12:121–124.

Grange, P., Laurent, E., Maggi, R., Centeno, A., and B. Delmon. 1996. Hydrotreatment of pyrolysis oils from biomass: Reactivity of the various categories of oxygenated compounds and preliminary techno-economical study. *Catal Today* 29:297–301.

Guitian, J., Souto, A., Ramirez, L., Marzin, R., and B. Solari. 1988. Commercial design of new HDH process. In *Fourth UNITAR/UNDP International Conference on Heavy Crudes and Tar Sands*, Edmonton, Alberta, Canada, Vol. 4, p. 237.

Guo, J., Ruan, R., and Y. Zhang. 2012. Hydrotreating of phenolic compounds separated from biooil to alcohols. *Ind Eng Chem Res* 51:6599–6604.

Halder, R., Lawal, A., and R. Damavarapu. 2007. Nitration of toluene in a microreactor. *Catal Today* 125:74–80.

Hendershot, D.C. 2003. Process intensification for safety. In *Re-Engineering the Chemical Processing Plant*, eds. A. Stankiewicz and J.A. Moulijn, pp. 441–463. New York: CRC Press, Marcel Dekker.

Hong, D.Y., Miller, S.J., Agrawal, P.K., and C.W. Jones. 2010. Hydrodeoxygenation and coupling of aqueous phenolics over bifunctional zeolite-supported metal catalysts. *Chem Comm* 46:1038–1040.

Huang, F., Li, W., Lu, Q., and X. Zhu. 2010. Homogeneous catalytic hydrogenation of biooil and related model aldehydes with $RuCl_2(PPh_3)_3$. *Chem Eng Technol* 33:2082–2088.

Huber, G.W. and J.A. Dumesic. 2006. An overview of aqueous phase catalytic processes for production of hydrogen and alkanes in a biorefinery. *Catal Today* 111:119–132.

Huber, G.W., Iborra, S., and A. Corma. 2006. Synthesis of transportation fuels from biomass: Chemistry, catalysts, and engineering. *Chem Rev* 106:4044–4098.

Ikura, M., Slamak, M., and H. Sawatzky. 1998. Pyrolysis liquid-in-diesel oil microemulsions. US Patent 5,820,640.

Jeon, J.K., Huh, B., Lee, C.H. et al. 2013. The hydrodeoxygenation of 2-methoxyphenol over Ni/kieselguhr catalysts as a model reaction for biooil upgrading. *Energ Sour A Recov Util Environ Eff* 35:271–277.

Joshi, N. and A. Lawal. 2012. Hydrodeoxygenation of pyrolysis oil in a microreactor. *Chem Eng Sci* 74:1–8.

Kanie, Y., Akiyama, K., and M. Iwamoto. 2011. Reaction pathways of glucose and fructose on Pt nanoparticles in subcritical water under a hydrogen atmosphere. *Catal Today* 178:58–63.

Katikaneni, S.P.R., Adjaye, J.D., and N.N. Bakhshi. 1995. Catalytic conversion of a biofuel to hydrocarbons: Effect of mixtures of HZSM-5 and silica-alumina catalysts on product distribution. *Energ Fuels* 9:1065–1078.

Kersten, S.R.A.,Van Swaaij, W.P.M., Lefferts, L., and K. Seshan. 2007. Options for catalysis in the thermochemical conversion of biomass into fuels, In *Catalysis for Renewables: From Feedstock to Energy Production*, eds. G. Centi and R.A. van Santen, pp. 119–145. Weinheim, Germany: Wiley-VCH.

Laurent, E., Centeno, A., and B. Delmon. 1994. Coke formation during the hydrotreating of biomass pyrolysis oils: Influence Guaiacol type compounds. *Proceedings of the Sixth International Symposium on Studies in Surface Science and Catalysis*, Vol. 88, pp. 573–578. Belgium, U.K.

Laurent, E. and B. Delmon. 1994a. Study of the hydrodeoxygenation of carbonyl, carboxylic and guaiacyl groups over sulfided CoMo/γ-Al$_2$O$_3$ and NiMo/γ-Al$_2$O$_3$ catalysts: I. Catalytic reaction schemes. *Appl Catal A Gen* 109:77–96.

Laurent, E. and B. Delmon. 1994b. Influence of water in the deactivation of a sulfided NiMo/ γ-Al$_2$O$_3$ catalyst during hydrodeoxygenation. *J Catal* 146:281–285.

Lin, Y.C., Li, C.L., Wan, H.P., Lee, H.T., and C.F. Liu. 2011. Catalytic hydrodeoxygenation of guaiacol on Rh-based and sulfided CoMo and NiMo catalysts. *Energ Fuels* 25:890–896.

Lostaglio, V.J. and J.D. Carruthers. 1986. New approaches in catalyst manufacture. *Chem Eng Prog* 82:46–51.

Luo, H., Prasomsri, T., and Y. Román-Leshkov. 2015. Al-MFI nanosheets as highly active and stable catalysts for the conversion of propanal to hydrocarbons. *Top Catal* 58:529–536.

Marker, T.L., Felix, L.G., Linck, M.B., and M.J. Roberts. 2012. Integrated hydropyrolysis and hydroconversion (IH2) for the direct production of gasoline and diesel fuels or blending components from biomass, Part 1: Proof of principle testing. *Environ Prog Sustain Energ* 31:191–199.

Marquevich, M., Czernik, S., Chornet, E., and D. Montane. 1999. Hydrogen from biomass: Steam reforming of the model compounds of the fast pyrolysis oil. *Energ Fuels* 13:1160–1166.

Meier, D. and O. Faix. 1999. State of the art of applied fast pyrolysis of lignocellulosic materials—A review. *Bioresour Technol* 68:71–77.

Niemann, K. 1991. Status of the VCC technology – An updates. In Paper presented at *UNITAR/UNDP Fifth Conference on Heavy Oils and Tar Sands*, Caracas, Venezuela, p. 4.

Oasmaa, A., Kalli, A., Lindfors, C. et al. 2012. Guidelines for transportation, handling, and use of fast pyrolysis biooil. 1. Flammability and toxicity. *Energ Fuels* 26:3864–3873.

Oasmaa, A., Solantausta, Y., Arpiainen, V., Kuoppala, E., and K. Sipila. 2010. Fast pyrolysis biooils from wood and agricultural residues. *Energ Fuels* 24:1380–1388.

Ohta, H., Kobayashi, H., Hara, K., and A. Fukuoka. 2011. Hydrodeoxygenation of phenols as lignin models under acid-free conditions with carbon-supported platinum catalysts. *Chem Comm* 47:12209–12211

Okafor, O.C., Tadepalli, S., Tampy, G., and A. Lawal. 2010. Cyclo addition of iso-amylene and alfa-methyl styrene in a micro reactor using Filtrol-24 catalyst: Microreactor performance study and comparison with semi-batch reactor performance. *Int J Chem React Eng* 8, Article A71.

Olarte, M.V., Lebarbier, V.M., Brown, H.M., Swita, M., Lemmon, T., and D.C. Elliott. 2012. *Am Chem Soc Div Fuel Chem Prepr* 57:754.

Peng, B., Zhao, C., Mejía-Centeno, I., Fuentes, G.A., Jentys, A., and J.A. Lercher. 2012. Comparison of kinetics and reaction pathways for hydrodeoxygenation of C_3 alcohols on Pt/Al_2O_3. *Catal Today* 183:3–9.

Peterson, A.A., Vogel, F., Lachance, R.P., Froling, M., Antal Jr, M.J., and J.W. Tester. 2008. Thermochemical biofuel production in hydrothermal media: A review of sub-and supercritical water technologies. *Energ Environ Sci* 1:32–65.

Pinheiro, A., Hudebine, D., Dupassieux, N., and C. Geantet. 2009. Impact of oxygenated compounds from lignocellulosic biomass pyrolysis oils on gas oil hydrotreatment. *Energ Fuels* 23:1007–1014.

Popov, A., Kondratieva, E., Goupil, J.M. et al. 2010. Biooils hydrodeoxygenation: Adsorption of phenolic molecules on oxidic catalyst supports. *J Phys Chem C* 114:15661–15670.

Prasad, Y.S., Bakhshi, N.N., Mathews, J.F., and R.L. Eager. 1986. Catalytic conversion of canola oil to fuels and chemical feedstocks Part I. Effect of process conditions on the performance of HZSM-5 catalyst. *Can J Chem Eng* 64:278–284.

Prasomsri, T., To, A.T., Crossley S., Alvarez, W.E., and D.E. Resasco. 2011. Catalytic conversion of anisole over HY and HZSM-5 zeolites in the presence of different hydrocarbon mixtures. *Appl Catal B Environ* 106:204–211.

Puente, G.D., Gil, A., Pis, J.J., and P. Grange. 1999. Effects of support surface chemistry in hydrodeoxygenation reactions over CoMo/activated carbon sulfided catalysts. *Langmuir* 15:5800–5806.

Qi, Z., Chang, J., Wang, T., and Y. Xu. 2006. Review of biomass pyrolysis oil properties and upgrading research. *Energ Conver Manage* 48:87–92.

Qin, L.Y., Jiang, E.C., Sun, Y., and S. Li. 2014. Catalytic and regenerative properties of nickel based zeolite catalysts. *Appl Mech Mater* 633–634:491–494.

Ram Reddy, P., Subba Rao, K.V., and M. Subrahmanyam. 1998. Selective synthesis of 2-ethylquinoline over zeolites. *Catal Lett* 56:155–158.

Ramanathan, S. and S.T. Oyama. 1995. New catalysts for hydroprocessing: Transition metal carbides and nitrides. *J Phys Chem* 99:163–165.

Rotman, D. 2008. The price of biofuels. *Tech Rev* 111:42–51.

Runnebaum, R.C., Tarit, N., Limbo R.R., Block, D.E., and C.B. Gates. 2012. Conversion of 4-methylanisole catalyzed by Pt/γ – Al_2O_3 and by Pt/SiO_2-Al_2O_3: Reaction networks and evidence of oxygen removal. *Catal Lett* 142:7–15.

Saidi, M., Rahimpour, M.R., and S. Raeissi. 2015. Upgrading process of 4-methylanisole as a lignin-derived biooil catalyzed by Pt/γ-Al_2O_3: Kinetic investigation and reaction network development. *Energ Fuels* 29:3335–3344.

Sharma, R.K. and N.N. Bakhshi. 1993. Catalytic upgrading of pyrolysis oil. *Energ Fuels* 7:306–314.

Sharpless, B.K. and T.B. Flood. 1972. Direct deoxygenation of vicinal diols with tungsten (IV): A new olefin synthesis. *J Chem Soc Chem Comm*, Issue 7, 370–371.

Sheldon, R.A. and R.S. Downing. 1999. Heterogeneous catalytic transformations for environmentally friendly production. *Appl Catal A Gen* 189:163–183.

Sheu, Y.H.E., Anthony, R.G., and E.J. Soltes. 1988. Kinetic studies of upgrading pine pyrolytic oil by hydrotreatment. *Fuel Process Technol* 19:31–50.

Silva, F., Guitian, G., Galiasso, R., and J. Krasuk. 1983. The HDH process for upgrading of heavy Venezuelan crudes. In *11th World Petroleum Congress*, London, U.K., Vol. 4, pp. 199.

Sitzmann, J. and A.V. Bridgwater. 2007. Upgrading fast pyrolysis oils by hot vapour filtration. In Paper presented at *Proceeding of 15th European Energy from Biomass Conference*, Berlin, Germany.

Solantausta, Y., Oasmaa, A., Sipila, K. et al. 2012. Biooil production from biomass: Steps toward demonstration. *Energ Fuels* 26:233–240.

Srinivas, S.T., Dalai, A.K., and N.N. Bakhshi. 2000. Thermal and catalytic upgrading of a biomass-derived oil in a dual reaction system. *Can J Chem Eng* 78:343–354.

Stankiewicz, A. and A.A.H. Drinkenburg. 2003. Process intensification: History, philosophy, principles. In *Re-Engineering the Chemical Processing Plant*, eds. A. Stankiewicz and J.A. Moulijn, pp. 1–28. New York: CRC Press, Marcel Dekker.

Stöcker, M. 2008. Biofuels and biomass-to-liquid fuels in the biorefinery: Catalytic conversion of lignocellulosic biomass using porous materials. *Angew Chem Int Ed* 47:9200–9211.

Tadepalli, S., Qian, D., and A. Lawal. 2007. Comparison of performance of microreactor and semi batch reactor for catalytic hydrogenation of o-nitroanisole. *Catal Today* 125:64–73.

Tominaga, H. and M. Nagai. 2010. Reaction mechanism for hydrodenitrogenation of carbazole on molybdenum nitride based on DFT study. *Appl Catal A Gen* 389:195–204.

Topsoe, H., Clausen, B.S., and F.E. Massoth. 1996. Hydrotreating catalysis. In *Catalysis-Science and Technology*, eds. J.R. Anderson and M. Boudart, pp. 1–269. Berlin, Germany: Springer.

Trimm, D.L. 1980. *Design of Industrial Catalysts*. Amsterdam, the Netherlands: Elsevier Scientific Publishing Company.

Trimm, D.L. and A. Stanislaus. 1986. The control of pore size in alumina catalyst supports: A review. *App Cat* 21:215–238.

Van Rossum, G., Kersten, S.R.A., and W.P.M. Van Swaaij. 2007. Catalytic and non catalytic gasification of pyrolysis oil. *Ind Eng Chem Res* 46:3959–3967.

Voloshin, Y. and A. Lawal. 2010. Overall kinetics of hydrogen peroxide formation by direct combination of H_2 and O_2 in a micro reactor. *Chem Eng Sci* 65:1028–1036.

Wang, D., Czernik, S., and E. Chornet. 1998. Production of hydrogen from biomass by catalytic steam reforming of the biooil. *Energ Fuels* 12:19–24.

Wang, W., Yang, Y., Luo, H., Hu, T., and W. Liu. 2011a. Amorphous Co-Mo-B catalyst with high activity for the hydrodeoxygenation of biooil. *Catal Comm* 12:436–440.

Wang, Y., Fang, Y., He, T., Hu, H., and J. Wu. 2011b. Hydrodeoxygenation of dibenzofuran over noble metal supported on mesoporous zeolite. *Catal Comm* 12:1201–1205.

Wildschut, J., Arentz J., Rasrendra, C.B., Venderbosch, R.H., and H.J. Heeres. 2009. Catalytic hydrotreatment of fast pyrolysis oil: Model studies on reaction pathways for carbohydrate fraction. *Environ Prog Sustain Energ* 28:450–460.

Williams, P.T. and N. Nugranad. 2000. Comparison of products from the pyrolysis and catalytic pyrolysis of rice husks. *Energ Fuels* 25:493–513.

Xing, J., Song, L., Zhang, C., Zhou, M., Yue, L., and X. Li. 2015. Effect of acidity and porosity of alkali treated ZSM-5 zeolite on eugenol hydrodeoxygenation. *Catal Today*. 258:90–95.

Xu, Y., Wang, T., Ma, L., Zhang, Q., and G.W. Lian. 2010. Upgrading of the liquid fuel from fast pyrolysis of biomass over MoNi/γ-Al_2O_3 catalysts. *Appl Energ* 87:2886–2891.

Xu, Y., Wang, T., Ma, L., Zhang, Q., and L. Wang. 2009. Upgrading of liquid fuel from the vacuum pyrolysis of biomass over the Mo-Ni/γ-Al_2O_3 catalysts. *Biomass Bioenerg* 33:1030–1036.

Yang, X., Chatterjee, S., Zhang, Z., Zhu, X., and C.U. Pittman. 2010. Reactions of phenol, water, acetic acid, methanol, and 2-hydroxy methylfuran with olefins as models for biooil upgrading. *Ind Eng Chem Res* 49:2003–2013.

Yang, X., Liang, Y., Zhao, X. et al. 2014a. Au/CNTs catalyst for highly selective hydrodeoxygenation of vanillin at the water/oil interface. *RSC Adv* 4:31932–31936.

Yang, Y., Ochoa-Hernández, C., O'Shea, V.A.d.P., Pizarro, P., Coronado, J.M., and D.P. Serrano. 2014b. Effect of metal support interaction on the selective hydrodeoxygenation of anisole to aromatics over Ni-based catalysts. *Appl Catal B* 145:91–100.

Ying, X., Jinxing, L., Qiying, L. et al. 2015. In situ hydrogenation of model compounds and raw biooil over Raney Ni catalyst. *Energ Convers Manage* 89:188–196.

Ying, X., Tiejun, W., Longlong, M., and C. Guanyi. 2012. Upgrading of fast pyrolysis liquid fuel from biomass over Ru/γ-Al$_2$O$_3$ catalyst. *Energ Convers Manage* 55:172–177.

Yu, M.J., Park, S.H., Jeon, J.K. et al. 2015. Hydrodeoxygenation of guaiacol over Pt/Al-SBA-15 catalysts. *J Nanosci Nanotechnol* 15:527–531.

Zhang, Q., Chang, J., Wang, T., and Y. Xu. 2007. Review of biomass pyrolysis oil properties and upgrading research. *Energ Convers Manage* 48:87–92.

Zhang, X., Long, J., Kong, W. et al. 2014. Catalytic upgrading of biooil over ni-based catalysts supported on mixed oxides. *Energ Fuels* 28:2562–2570.

Zhang, Z.J., Wang, Q.W., Yang, X.L., Chatterjee, S., and C.U. Pittman Jr. 2010. Sulfonic acid resin catalyzed addition of phenols, carboxylic acids, and water to olefins: Model reactions for catalytic upgrading of biooil. *Bioresour Technol* 101:3685–3695.

Zhao, C., He, J., Lemonidou, A.A., Li, X., and J.A. Lercher. 2011a. Aqueous phase hydrodeoxygenation of bioderived phenols to cycloalkanes. *J Catal* 280:8–16.

Zhao, C., Kou, Y., Lemonidou, A.A., Li, X., and J.A. Lercher. 2009. Highly selective catalytic conversion of phenolic biooil to alkanes. *Angew Chem Int Ed* 48:3987–3990.

Zhao, C., Kou, Y., Lemonidou, A.A., Li, X., and J.A. Lercher. 2010. Hydrodeoxygenation of bioderived phenols to hydrocarbons using Raney® Ni and Nafion/SiO$_2$ catalysts. *Chem Comm* 46:412–414.

Zhao, H.Y., Li, D., Bui, P., and S.T. Oyama. 2011b. Hydrodeoxygenation of guaiacol as model compound for pyrolysis oil on transition metal phosphide hydroprocessing catalysts. *Appl Catal A Gen* 391:305–310.

9

Production of Biodiesel from Renewable Resources

Ali Shemsedin Reshad, Pankaj Tiwari, and Vaibhav V. Goud

Contents

Abstract

Worldwide biodiesel is being produced mainly from edible and nonedible vegetable oils. The choice of vegetable oils depends primarily on availability, price, and the policy adopted by governing agencies. Many countries are making their policies to produce nonedible vegetable oils on large scale. The economic aspects of biodiesel production are a barrier for its commercial success. Presently, the current prices of fossil fuels are lower than the cost of biodiesel. However, research on biodiesel production processes will result in multidirectional benefits, namely, employment generation (particularly in rural areas), efficient usage of waste and fallow land, drought proofing, energy security for country and promotion of organic farming, and less dependency on foreign exchange. The production of biodiesel from nonedible vegetables oils may offer solutions for various issues involved in the assessment of biomass-based industry for energy generation. Transesterification of nonedible vegetable oils using heterogeneous catalyst in an ultrasonic-assisted reactor could be an effective approach to produce a clean and renewable source of energy. Biodiesel can be used as a blend with mineral diesel in the existing engine to mitigate environment emissions.

9.1 Introduction

Excessive demand of fossil fuels and their limited availability have imposed scarcity of supplying energy for various needs, especially in the developing countries. Growth of population, revolution in industrialization, improved living standard, and environmental concerns have made special attention to search for an alternative source of renewable energy (Balat 2011, Bokhari et al. 2012, Jull et al. 2007, Kole et al. 2012). Production of biofuels from different renewable feedstocks is one way of partially substituting dependence on fossil fuel as well as significant reduction of emission of greenhouse gases to the environment. Biofuels can be produced from corn, soybean seed, palm seed, rapeseed, flaxseed, castor seed, rubber seed, sugar beet, raw sewage, animal fats, etc. (Speight 2011). Biodiesel, an alternative fuel for diesel engine, is a promising, nontoxic, biodegradable fuel that can be produced by chemical transformation of triglyceride present in the seed oil and/or animal fats with short chain of alcohol in the presence of a catalyst and/or absence of a catalyst (Bart et al. 2010, Freedman et al. 1984, Geuens et al. 2014, Liu 2013). Chemically, it is a mixture of long-chain fatty acid alkyl esters (FAAEs) of feedstock esters, which fall in the carbon range of C_{12}–C_{22} (Rakopoulos et al. 2006). According to the U.S. Energy Information Administration (EIA 2012), worldwide consumption of biodiesel has exponentially been increased from 605,000 to 17,606,000 gallons per day in the last decade with a projected value of 71,230,000 gallons per day by 2015 (Khan et al. 2014). Currently, more than 95% biodiesel is being produced from edible feedstocks, which may affect the sustainability of biodiesel industries (Abdelfatah et al. 2012, Khan et al. 2014). The use of edible vegetable oils and animal fats for biodiesel production has been a great concern because they compete with the food materials and leads to food versus fuel crisis (Atabani et al. 2013, Chhetri et al. 2008, No 2011).

The history of biodiesel synthesis using various feedstock materials can be classified as (1) first generation, (2) second generation, and (3) third generation of biodiesel. The detail of various feedstocks is listed in Table 9.1. First-generation biodiesel is produced from vegetable oils, which are usually part of the food chain. The fuels, produced from transesterification process, such as soybean oil methyl esters (Keera et al. 2011, Kouzu et al. 2008, Rahimi et al. 2014, Silva et al. 2011), palm oil methyl esters (Chen et al. 2014, Manickam et al. 2014, Salamatinia et al. 2010), coconut oil methyl esters, corn oil methyl esters, sunflower methyl esters, rapeseed oil methyl esters, and olive oil methyl esters (No 2011) are a few examples of alternative for diesel fuel. Despite having several benefits of using first-generation biodiesel for combustion in compression ignition (CI) engines, high feedstock cost, lower calorific value, and higher NO_x emission compared to conventional diesel are major concerns (Chhetri et al. 2008). The cost of first-generation biodiesel feedstock accounts around 60%–80% of the total production depending on the availability (Atabani et al. 2013, Chhetri et al. 2008). The problem of higher cost of feedstock for first-generation biodiesel can be minimized by the selection of nonedible materials. Therefore, second-generation renewable feedstocks such as oilseed plants of

TABLE 9.1   Renewable Sources for Biodiesel Production

Common Name	Botanical Name	Plant Type	Plant Part	Oil Content Seed (wt%)
Edible feedstock (Ardebili et al. 2011, Issariyakul and Dalai 2014, Kole et al. 2012, Kumar and Sharma 2015, Salamatinia et al. 2010, Santori et al. 2012)				
Soybean	*Glycine max*	Oilseed	Seed	15–22
Cotton	*Gossypium hirsutum*		Seed	20
Sesame	*Sesamum indicum*	Flowering plant	Seed	50
Canola	*Brassica campestris*	Flowering plant	Seed	40
Olive	*Olea europaea*	Small tree	Seed	20
Sunflower	*Helianthus annuus*	Flowering		44–50
Almond	*Prunus dulcis*	Tree	Seed	55
Corn	*Zea mays subsp. Mays*	Flowering plant	Seed	10
Walnut	*Juglans regia*	Tree	Seed	60
Hazelnut		Tree	Seed	55
Rapeseed	*Brassica napus*	Flowering plant	Seed	35
Peanut	*Arachis hypogaea*	Herbaceous annual	Seed	36–56
Safflower	*Carthamus tinctorius*	Herbaceous, thistle-like annual plant	Seed	32
Coconut	*Cocos nucifera*	Tree	Seed	60–70

Common Name	Botanical Name	Plant Type	Plant Part	Seed (wt%)	Kernel (wt%)
Nonedible feedstock (Ahmad et al. 2014, Atabani et al. 2013, Bankovic-Ilic et al. 2012, Bokhari et al. 2012, Chhetri et al. 2008, Kamalakar et al. 2013, Kole et al. 2012, No 2011)					
Rubber	*Hevea brasiliensis*	Tree	Seed, kernel	40–60	40–50
Neem	*Azadirachta indica*	Tree	Seed, kernel	20–30	25–45
Milk weed	*Asclepias* L.	Herbaceous perennial	Seeds	20–25	—
Ethiopian mustard	*Brassica carinata*	Herbaceous annual	Seed, kernel	42	2.2–10.8
Desert date	*Balanites aegyptiaca*	Tree	Kernel	36–47	
Annona squamosa	*Annona squamosa*	Tree	Seed	15–20	—
Purging Croton	*Croton tiglium*	Herbaceous perennial	Seed, kernel	30–45	50–60

(Continued)

TABLE 9.1 (Continued) Renewable Sources for Biodiesel Production

Common Name	Botanical Name	Plant Type	Plant Part	Seed (wt%)	Kernel (wt%)
Cuphea	—	Herbaceous annual	Seed	20–38	—
Garcinia indica	*Vrikshamla*	Tree	Seed	45.5	—
Jatropha	*Jatropha curcas*	Tree	Seed, kernel	20–60	40–60
Sapindus	*Sapindus saponaria*	Tree	Seed	51.8	—
Ximenia	*Ximenia americana*	Tree	Kernel	—	49–61
Terminalia catappa	*Terminalia catappa*	Tree	Seed	49	—
Tung	*Vernicia fordii*	Tree	Seed	35–40	—
Linseed	*Linum usitatissimum*	Herbaceous annual	Seed	35–45	—
Karanja	*Pongamia pinnata*	Tree	Seed, kernel	25–50	30–50
Castor	*Ricinus communis* L.	Tree	Seed	45–50	—
Kusum	*Schleichera oleosa*	Tree	Kernel	—	55–70
Mahua	*Madhuca longifolia*	Tree	Seed, kernel	35–50	50
Nahor	*Mesua ferrea*	Tree	Seed, kernel	58–75	—

				Oil Content (dry wt%)
Microalgae Feedstock (Borugadda and Goud 2012, Chisti 2007, Menetrez 2012, Vyas et al. 2010)				
Botryococcus braunii				25–75
Chlorella sp.				28–32
Dunaliella primolecta				23
Monallanthus salina				>20
Nannochloropsis sp.				31–68
Neochloris oleoabundans				35–54
Nitzschia sp.				45–47
Phaeodactylum tricornutum				20–30
Oedogonium, Spirogyra sp.				—
Tetraselmis suecica				15–23
Schizochytrium sp.				50–77

Jatropha curcas L. (jatropha), *Ricinus communis* (castor), *Hevea brasiliensis* (rubber tree), *Pongamia pinnata* L. (karanja), *Moringa oleifera, Calophyllum inophyllum* L. (polanga), and *Mesua ferrea* (nahor) are sustainable raw materials for biodiesel synthesis (Atabani et al. 2013, Bokhari et al. 2012, Chhetri et al. 2008, No 2011, Speight 2011). Biofuel derived from algae, either unicellular or multicellular autotrophic organisms, is considered as third-generation feedstock material for renewable energy (Menetrez 2012). More than 25,000 microalgae species are identified, out of which only 15 are potential candidates for biodiesel production (Hossain et al. 2008). Chisti (2007) stated that in a hectare land, possible yield of algal oil can be around 200 barrels. Algae can produce ~250 and ~31 times more oil per acre compared to soybean and palm oil, respectively (Hossain et al. 2008). Oil yield from microalgae is much more than macroalgae. The oil content in some of the microalgae can be found around 80% of its dry weight, and 20%–50% is most common for any microalgae. The most common microalgae species for production of biodiesel are *Botryococcus braunii, Chlorella* sp., *Dunaliella primolecta, Monallanthus salina, Nannochloropsis* sp., *Neochloris oleoabundans, Nitzschia* sp., *Phaeodactylum tricornutum, Oedogonium, Spirogyra* sp., *Tetraselmis suecica*, and *Schizochytrium* sp. (Vyas et al. 2010). Algae-based feedstock for biodiesel production has a number of advantages: (1) a large variety of algae can be used, (2) higher growth rate (1–3 doubling per day), (3) barren land can be used for cultivation, (4) growth nutrients such as nitrogen, phosphorus, and silicon sulfate from waste can be utilized, (5) efficient use of CO_2, (6) and responsible for more than 40% of the global carbon fixation (Shahzad et al. 2010). However, strain isolation, nutrient sourcing and utilization, harvesting, residual biomass utilization, water, etc., are major challenges in the process of using algae for biodiesel synthesis (Hannon et al. 2010). Therefore, in this report, an attempt has been made to give an overview on the synthesis of biodiesel using different processes, and their importance is also discussed here.

9.2 Modification of Feedstock

Oil extracted by any means cannot be used directly in existing diesel engines due to high free fatty acid (FFA) content, low volatility, high viscosity, presence of moisture, and other impurities. The injection, atomization, and combustion characteristics of vegetable oils in diesel engines are significantly different from those of mineral diesel. This is attributed to the fact that the fuel properties of vegetable oils and mineral diesel are different with respect to viscosity, molecular weight, density, flash point, and cetane number. This leads to a number of problems specific to the feedstock oil being used. Direct use in engine often results in short- and long-term operational problems. Plugging and gumming of filters, injectors, and engine knocking are short-term engine problems. Coking of injectors, carbon deposits, and failure of engine are the most probable long-term problems (Ma and Hanna 1999, Sharma and Singh 2009). Therefore, modification of feedstock oil, physical and chemical transformation, is required through blending with diesel (dilution), pyrolysis (cracking), emulsification, and transesterification. Transesterification has been demonstrated as the

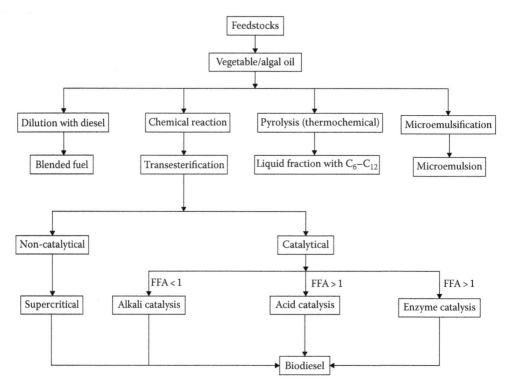

Figure 9.1   A process chart for utilization of vegetable/algal oil as renewable energy source.

simplest and most efficient route for biodiesel production at large scale (Leung et al. 2010, Ma and Hanna 1999, Salamatinia et al. 2010, Yin et al. 2012). The summary of feedstock utilization processes is illustrated in Figure 9.1.

9.2.1 Blending of Feedstock Oil with Diesel

Several studies are reported on the effect of blending of vegetable oils with diesel. Adams et al. (1983) investigated the performance of John Deere 6-cylinder engine for a total of 600 h operated with blended fuel of soybean oil and diesel fuel. The results indicated lubricating oil thickening and potential gelling with 1:1 blend ratio, while a blend of one-third soybean oil and two-thirds diesel (ratio of 1:2) suggested being suitable fuel for agricultural machines during shortage of diesel fuel. Comparative test of agricultural engine fuelled by refined palm oil and diesel fuel showed that the specific fuel consumption of engine was increased 15%–20% due to lower calorific value and incomplete combustion of palm oil (Prateepchaikul and Apichato 2003). Use of vegetable oil directly or its blend with diesel fuel has generally been considered to be unsatisfactory and impractical for diesel engine. The high viscosity, FFA content, and gum formation can be reduced by other processes. A long-term evaluation of the engine when using 100% crude vegetable oil was prematurely terminated as severe injector coking led to a decrease in power output and thermal efficiency.

9.2.2 Pyrolysis

Pyrolysis or thermal cracking is a process of conversion of feedstocks (high molecular weight compounds) into products (smaller compounds) by means of thermal energy (heat) in the absence of air or oxygen with the aid of a catalyst. Thermochemical conversion of feedstock has become an alternative method for the synthesis of high-grade biofuels (gasoline fraction with carbon chain C_6–C_{12} (Xu et al. 2013). The long chain of the feedstock is cleavages of chemical bond to yield small molecules (Ma and Hanna 1999, No 2011). The mechanisms for thermal decomposition of triglycerides to several products are likely to be complex in nature because of many structures and multiplicity of possible reaction pathways of mixed triglycerides.

During catalytic cracking of feedstock, the catalysts have significant role on molecular distribution of the product (Biswas et al. 2012, Ooi et al. 2004, Siswanto et al. 2008, Xu et al. 2013). The pyrolytic products are primarily olefins, paraffins, carboxylic acids, and aldehydes. The large variations in the compounds are directly related to the type of vegetable oils and its origin. Catalytic upgrading of the soybean pyrolytic product by HZSM-5 zeolite at 400°C supported partial deoxygenation (Lima et al. 2004). Xu et al. (2013) investigated thermal and catalytic cracking of soybean and high acid value waste oil using basic catalyst. About 70% biooil yield was obtained at a reaction temperature of 450°C with a heating rate of 5°C/min. Furthermore, the undesired product in pyrolysis process, that is, carboxylic acid, is converted into its corresponding esters by esterification process. In addition to liquid and gaseous products (i.e., CO, CO_2), considerable amounts of propane and butane are also produced.

9.2.3 Microemulsions

A high viscosity of the feedstock for diesel engine application can be solved by microemulsions with solvent or cosurfactants such as methanol, ethanol, 1-butanol, hexanol, and octanol (Ma and Hanna 1999). Microemulsion is a colloidal equilibrium dispersion of two or more immiscible liquids, of which at least one is immiscible and dispersed into the other. Microemulsions are spontaneously formed system and have special features, that is, thermodynamically stable. Formation of emulsified vegetable oil fuel is a promising technique for solving high surface tension and high viscosity and for reducing smoke, particulate matter, nitric oxide emission. Emulsification is a cheaper method for use of vegetable oil for diesel engine (Melo-Espinosa et al. 2015, Mubarak and Kumar 2012). Unlike thermal cracking and transesterification processes, no complex chemical reaction and by-product formation were involved during an emulsification of animal fats and vegetable/algal oil (Melo-Espinosa et al. 2015).

Ternary phase equilibrium diagram and the plot of viscosity versus solvent fraction are often used to determine the emulsified fuel formulations. Water is one of the essential components used to prepare emulsified fuels. The presence of water in fuel can lead to corrosion of engine part. Hence, special attention is needed on water issue, mainly salt content (Badran et al. 2011, Fayyad et al. 2010, Graboski and McCormick 1998).

9.2.4 Transesterification

Some of the problems encountered in diesel engines on direct use of vegetable oils are associated with the polyunsaturation of large triglyceride molecules and its higher molecular mass. These problems cannot be solved with the physical means discussed earlier. Appropriate chemical means need to be employed to get vegetable oil derivatives that are saturated and possess properties similar to mineral diesel oil. Transesterification process is one of the most successful and promising processes to convert vegetable oils into cleaner and environmentally safe diesel fuel-like liquid. Chemically, natural oils and fats are made up mainly of triglycerides. These triglycerides when reacted chemically with lower chain of alcohols in the presence of a catalyst result in alkyl esters. Methanol and ethanol are the most widely used alcohols, because of their lower cost, higher polarity, and short-chain over other alcohols. If methanol is used in this process, it is called methanolysis. Transesterification of triglycerides produces FAAEs and glycerol. Diglycerides and monoglycerides are the intermediates in this process. Overall three moles of monoalkyl ester are obtained from one mole of triglyceride. Due to reversible nature of reactions involved, excess alcohol is used to shift the equilibrium reaction to forward direction. The equilibrium conversions of triglycerides during transesterification are affected by various factors, namely, feedstock quality (like FFA content, water content), type of alcohol used, molar ratio of alcohol to triglyceride, type of catalyst, amount of catalyst, reaction temperature, stirring rate, and reaction time (Bart et al. 2010, Canakci and Gerpen 1999, Freedman, et al. 1984, Kole et al. 2012, Leung et al. 2010, Ma and Hanna 1999, Manickam et al. 2014, Sharma and Singh 2009). Transesterification significantly reduces the viscosity of triglyceride without affecting heating value of original feedstock. Therefore, engine performance and emission characteristics of transesterified product give better results compared to original feedstock (Barnwal and Sharma 2005, Canakci 2007, Ma and Hanna 1999, Sharma and Singh 2009). FFA content is the most important parameter for path selection of FAAE synthesis (Figure 9.1). Higher conversion of feedstock is not achieved if the oil contains higher amount of FFA > 1% (Freedman et al. 1984, Georgogianni et al. 2008, Ma and Hanna 1999, Tiwari et al. 2007). The higher content of FFA leads to increased acid value of the feedstock, which causes soap formation with alkaline (base) catalyst, thereby preventing the separation of the biodiesel from the glycerol fraction and reducing the yield of biodiesel. Therefore, a two-step process is required for these kinds of feedstocks. Initially, the FFA of these oils can be converted to fatty acid methyl esters (FAMEs) by an acid-catalyzed pretreatment via esterification process (9.1) followed by transesterification process (9.2) to achieve higher conversion.

$$R' - \overset{\overset{\text{O}}{\|}}{C} - OH \quad + \quad R - OH \quad \underset{\text{Acid catalyst}}{\rightleftharpoons} \quad R' - \overset{\overset{\text{O}}{\|}}{C} - O - R \quad + \quad H \overset{O}{\diagdown} H$$

Free fatty acid (FFA)   (Alcohol)   (Fatty acid alkyl esters (FAAE))   (Water)

$$(9.1)$$

(Triglyceride)   (Alcohol)   (Fatty acid alkyl esters (FAAE))   (Glycerol)

$$(9.2)$$

Esterification (9.1) and transesterification (9.2) reaction for high FFA vegetable/algal oil, R′: free fatty acid alkyl group, R_1, R_2, and R_3 are triglyceride alkyl group and R: alcohol alkyl group.

9.3 Biodiesel Synthesis

Vegetable oil and alcohol are immiscible in nature (Mostafaei et al. 2013, Wan Omar and Amin 2011). So, both esterification and transesterification reactions are limited due to high mass transfer resistance present primarily because of poor diffusion process (physical step) (He and Van Gerpen 2012). This can be overcome by employing an external force such as an intensive agitation (Yin et al. 2012). Mechanical stirring is the most common technique to increase the rate of transesterification reaction (Kouzu et al. 2008, Leung et al. 2010, Liu 2013, Ma and Hanna 1999). Constant vigorous agitation and high temperatures are required because the reaction can take place only in the interfacial region between the reactant mixtures (Mootabadi et al. 2010). Correia et al. (2014) studied transesterification of sunflower oil via mechanical stirring methods, and around 97.75% yield was obtained with 1000 rpm stirring speed for 4 h at a reaction temperature of 65°C. A 94% conversion of nahor oil was reported with 900 rpm for 4 h (Boro et al. 2014). Mustata and Bicu (2014) also studied optimization of corn oil methyl ester production and summarized that around 99% conversion can be achieved at 500 rpm in a 2 h duration using 3.6 wt% Ba(OH)$_2$. Waste cooking oil at lower stirring rate (400 rpm) for 6 h resulted in 90.6% conversion (Jain et al. 2011).

In conventional transesterification process, longer reaction time causes glycerol accumulation in the polar phase (alcohol), which makes catalyst separation difficult

(He and Van Gerpen 2012, Mootabadi et al. 2010, Yin et al. 2012). Therefore, different alternative techniques such as application of microwave (as an alternative energy source) and an ultrasonication (using an ultrasound tools) were reported to overcome the problems related to diffusion limitation, an improvement of methanol solubility in vegetable oil or fat, and to shorten the reaction time of transesterification process (Borugadda and Goud 2012, Choedkiatsakul et al. 2014, He and Van Gerpen 2012). In conventional heating, heat transfer to reaction mixture is inefficient. This is due to transferred heat from the surface of reactor through convection, conduction, and radiation. Microwave radiations supply heat directly and efficiently to the reacting mixtures (Encinar et al. 2012, Sajjadi et al. 2014). Several studies on transesterification intensification via microwave and an ultrasonication have been reported for a variety of feedstocks using various catalysts (Manickam et al. 2014, Mootabadi et al. 2010, Pukale et al. 2015, Salamatinia et al. 2010, Wan Omar and Amin 2011, Yin et al. 2012). A comparative study on transesterification reaction of cotton seed oil using KOH catalyst (1.5 wt%) with 6:1 molar ratio of methanol to oil showed improved yield in a lesser duration in the case of microwave-assisted process (92.4% in 7 min reaction time) than conventional process (91.4% in 30 min reaction time) (Nezinhe and Danisman 2007). Encinar et al. (2012) reported that the use of microwave irradiation-assisted transesterification of soybean oil yields highest conversion (99%).

Due to the immiscible nature of the reactant compounds, alcohol and triglyceride of vegetable/algal oil, contact surface area between the reactants is only at the interface, which and causes lower reaction rate. An ultrasonic phenomenon in a reactive solution is one way to increase the liquid–liquid interfacial area, which is important to enhance the rate of transesterification reaction. During ultrasonic irradiations vapor bubble in methanol and cavitation in vegetable oil are generated ultrasonically (Gole and Gogate 2012, He and Van Gerpen 2012). Even though an ultrasound can be extended to a wider range of frequencies, 20–60 kHz is the most commonly used ultrasound frequency range (Chen et al. 2014, Pukale et al. 2015, Salamatinia et al. 2010). Deng et al. (2010) studied an ultrasonic-assisted transesterification of *Jatropha curcas* L. oil 5.25% FFA and achieved 96.4% yield of biodiesel with a total reaction time of 1.5 h using NaOH as catalyst. Georgogianni et al. (2008) achieved 98% yield of ethyl ester from sunflower seed oil within 40 min using ultrasonic-assisted (24 kHz) transesterification, whereas stirring of reaction mixture for 4 h resulted in only 88% yield. Mostafaei et al. (2013) also reported an improvement in the conversion of waste cooking oil to methyl esters using ultrasonic approach (89% yield) compared to conventional stirrer technique (50% yield).

9.3.1 Catalytic Transesterification

Like other chemical reactions, catalyst has an important role to achieve a significant increase in the rate of transesterification reaction and the yield of FAAE (biodiesel). Catalyst used for transesterification reaction can be classified as alkali, acid, and biocatalyst (enzyme) (Freedman et al. 1984, Sivasamy et al. 2009). Homogeneous alkali and acid catalysts have been used at a large-scale production of biodiesel.

9.3.1.1 Acid-Catalyzed Transesterification

The most common homogeneous acid catalysts used for the synthesis of biodiesel from higher FFA content feedstock are sulfuric acid (H_2SO_4), hydrochloric acid (HCl), boron trifluoride (BF_3), phosphorous acid (H_3PO_3), and organic sulfonic acid. Acid-catalyzed transesterification reaction requires the use of high molar ratio of alcohol to oil in order to obtain good product yields. However, the feedstock conversion or FAAE yield does not proportionally increase with molar ratio. Transesterification of soybean oil using sulfuric acid showed an improvement in yield from 77% to 87.8% using a methanol to oil ratio of 3.3:1 to 6:1, but with much higher molar ratio (30:1), only moderate improvement in yield (98.4%) was achieved (Helwani et al. 2009). The rate of transesterification reaction using homogeneous acid catalyst is about 4000 times slower than alkali catalyst. Canakci and Gerpen (1999) reported highest conversion of 95.1% of soybean oil using acid catalyst (3%) at 65°C and reaction time of 96 h. However, Freedman et al. (1984) reported that the acid catalyst is more effective than alkali catalyst for feedstock containing higher FFA more than 1%. The overall FFA content of particular oil can be reduced by blending it with less FFA content oil. Khan et al. (2010) studied acid esterification of a blended feedstock of rubber seed oil (high FFA) and palm oil via sulfuric acid (H_2SO_4) with 0.5 wt% H_2SO_4 for 3 h reaction time and achieved 95% reduction of FFA. Sulfuric acid has been also used for esterification of high FFA *Chlorella* sp. lipid and maximum 60% yield of methyl esters was observed with 3.5 wt% H_2SO_4 and 2.5 h stirring and reaction time (Mathimani et al. 2015). A better conversion of high FFA feedstock into its methyl esters was observed by homogeneous acid catalyst compared to homogeneous alkali catalyst (Mathimani et al. 2015, Vyas et al. 2010). The homogeneous acid catalyst can perform both esterification and transesterification reaction to produce corresponding alkyl esters and prevent saponification reaction. Higher FFA amounts of *Jatropha curcas* oil (21.5%) and waste cooking oil (21.84%) were reduced to less than 1% using homogeneous acid catalyst (H_2SO_4) and about 21.2% and 21.5% methyl ester yields were obtained, respectively (Jain and Sharma 2010, Jain et al. 2011). Several research suggested that around 1% (v/v) homogeneous acid catalyst should be used because higher amount of acid may burn the feedstock and darken the products (Sharma and Singh 2009).

Heterogeneous acid catalysts are promising alternatives to replace strong homogeneous acids. The corrosion and environmental hazard caused by homogenous (liquid) acid catalyst and product purification problems can be improved with the use of solid acid catalysts (Helwani et al. 2009). Sulfated zinc oxide, zeolites, heteropoly acids, functionalized zirconia and silica, tungsten oxides, sulfonated zirconia (SZ), Amberlyst 115, Lewatit GF 101, sulfonated saccharides, Nafion 1 resins, and organosulfonic-functionalized mesoporous silica are solid acid catalysts that have been used effectively in esterification of carboxylic acid as well as in obtaining high conversion of triglycerides of feedstock (Helwani et al. 2009, Istadi et al. 2015, Lopez et al. 2005, Shokrolahi et al. 2011). An active solid catalyst (SO_4^{2-}–ZnO) for methyl ester synthesis from soybean oil under mild reaction condition resulted in 80.19% yield within 4 h (Istadi et al. 2015). However, the shortcomings of solid acid catalysts are (1) complicated catalyst preparation, (2) energy intensive, (3) leaching of the catalyst causing product contamination and side reaction, (4) low acid site, (5) low microporosity, (6) diffusion limitation, and (7) high temperature and longer time for higher yield of products.

9.3.1.2 Alkali Transesterification

The most commonly adopted route for biodiesel production at industrial scale is based on homogeneous alkali transesterification process using sodium hydroxide (NaOH) and potassium hydroxide (KOH) catalysts. Several research studies have been reported on parameter optimization of transesterification process using homogeneous alkali catalysts. Keera et al. (2011) considered (NaOH) catalyst (0.5–1.5 wt%), molar ratio of methanol to oil (3:1–9:1), reaction time (1–3 h) and a constant reaction temperature (60°C) as process parameters to optimize the yield of biodiesel. They reported that the best yields, 90% and 98.5%, from soybean and cotton seed oil, respectively, were obtained with 1 wt% NaOH 6:1 molar ratio and 1 h reaction time.

Silva et al. (2011) applied response surface methodology (RSM) considering a wide range of process parameters and obtained an optimum yield of 95% ethyl esters at molar ratio of ethanol to soybean oil (9:1) with a catalyst loading of 1.7 wt% NaOH during 80 min reaction at 40°C. Lueung and Guo (2006) studied characteristics and performance of sodium hydroxide (NaOH), potassium hydroxide (KOH), sodium methoxide (CH_3ONa), and potassium methoxide (CH_3OK) catalysts for alkaline-catalyzed transesterification of edible canola oil and used frying oil. Both CH_3ONa and CH_3OK provided higher yield of biodiesel compared to NaOH and KOH. Due to the sensitiveness of FFAs and formation of undesired products (i.e., water and soap), alkali catalyst transesterification reaction lowers the conversion of feedstock into a desired product (Lueung and Guo 2006, Sharma and Singh 2009). Formation of soap not only lowers the conversion of the feedstock but also makes purification of the final product difficult and causes emulsion formation. Separation of catalysts after transesterification reaction requires washing the final product with distilled water, which gives rise to generation of wastewater. Heterogeneous alkali-catalyzed transesterification is included under green technology, because it generates negligible wastewater, reusability of catalyst, relatively faster reaction rate than acid-catalyzed transesterification, and separation of biodiesel from glycerol is much easier (Lee and Saka 2010). The main drawbacks of heterogeneous catalysts are anhydrous condition, poisoning of the catalyst when exposed to ambient air, diffusion limitation, and low FFA requirement. Commonly used solid heterogeneous catalysts for transesterification reaction are CaO, ZnO, MgO, BrO, Na_3PO_4, CaMgO, CaZnO, tripotassium phosphate (K_3PO_4), trisodium phosphate (Na_3PO_4), disodium phosphate (Na_2HPO_4), dipotassium phosphate (K_2HPO_4), monosodium phosphate (NaH_2PO_4) and monopotassium phosphate (KH_2PO_4), magnesium–lanthanum mixed oxide (Mg/La), strontium–titanium mixed oxide (Sr-Ti), calcium–zirconium mixed oxide (Ca-Zr), ($Ca(OCH_2CH_3)_2$), KOH/NaX zerolite, calcium ethoxide, calcium methoxide, calcium glyceroxide, calcium hydroxide ($Ca(OH)_2$), barium hydroxide ($Br(OH)_2$), etc.

Biodiesel production from kapok seed oil was investigated using a two-step process, esterification using H_2SO_4 followed by transesterification with CaO catalyst.

The results of the study revealed that the high FFA (9.32%) content of the oil could be reduced to 1% FFA by esterification reaction (Putri et al. 2012). An optimum yield of 88.57% of biodiesel was obtained at a molar ratio of 15:1, a reaction time of 1 h, and a reaction temperature of 60°C. Catalyst CaO has also been used for transesterification of several feedstocks such as soybean oil, *Jatropha curcas* oil, rapeseed oil, and sunflower oil (Helwani et al. 2009, Kawashima et al. 2009, Kouzu et al. 2008, Vyas et al. 2010). Pukale et al. (2015) have compared transesterification of waste cooking oil (WCO) using 2 wt% heterogeneous solid catalysts. Catalytic performance was obtained in the following order—tripotassium phosphate (K_3PO_4) (70.5%) > trisodium phosphate (Na_3PO_4) (63.1%) > disodium phosphate (Na_2HPO_4) (14.3%)—and negligible conversion of WCO was observed in dipotassium phosphate (K_2HPO_4), monosodium phosphate (NaH_2PO_4), and monopotassium phosphate (KH_2PO_4).

Mootabadi et al. (2010) studied the catalytic performance of alkaline earth metal oxides (CaO, BaO, and SrO) for transesterification of palm oil. The result shows that, under similar reaction condition, the catalytic activity was found in the order of CaO < SrO < BaO. This is attributed to the difference in basic strength of the catalysts. Eggshell-derived CaO catalyst has also been used for transesterification reaction of vegetable oil (Boro et al. 2014, Buasri et al. 2013, Chen et al. 2014, Correia et al. 2014, Oliveira et al. 2013). The eggshell (crushed) calcined at 600°C–900°C for 2–4 h contents more than 98% CaO. A biodiesel yield of 90%–97% has been reported using activated eggshell as heterogeneous solid catalyst (Chen et al. 2014, Khemthong et al. 2012, Niju et al. 2014, Viriya-empikul et al. 2010).

9.3.1.3 Enzyme-Catalyzed Transesterification

Biocatalytic transesterification reaction is another alternative of chemical catalysts. Enzymes catalyze both esterification and transesterification reactions. Lipases from *Pseudomonas flseudomona*, *Pseudomonas cepacia*, *Candida rugosa*, *Candida antarctica*, *Rhizomucor miehei*, and *Thermomyces lanuginosa* are the most extensively studied enzymes for biodiesel production (Li et al. 2013, Vyas et al. 2010). Transesterification reaction using lipase enzyme looks attractive and encouraging due to easy product separation, minimal wastewater disposal, less energy intensity, zero side reactions, and environmental friendliness (Borugadda and Goud 2012, Khan et al. 2014, Leung et al. 2010, Vyas et al. 2010). Enzymatic transesterification process has some disadvantages: (1) contamination of product with residual enzyme, (2) longer reaction time, and (3) cost of enzyme (Vyas et al. 2010). Shimada et al. (2002) proposed an immobilization of enzyme and addition of alcohol solvent in multistep (ethanol and methanol) to overcome the issues. Rodrigues et al. (2010) found that a two-step ethanolysis was very effective to avoid the negative effect caused by ethanol and 100% conversion of soybean oil could be obtained in the presence of *n*-hexane. A stepwise addition of ethanol could enhance conversion of fish oil up to 95% (Watanabe et al. 1999). Raita et al. (2010) observed that operational stability of protein-coated microcrystals (PCMC-lipase) could be improved by introducing hydrophilic solvent *tert*-butanol and allowing recycling of the biocatalyst for at least eight consecutive batches.

9.3.2 Noncatalytic Transesterification

Supercritical alcohol transesterification provides a new way to produce biodiesel from vegetable oils. Triglyceride and fatty acids of vegetable oil and short-chain alcohols (ethanol and methanol) are immiscible in nature and result in incomplete transesterification reaction. Under supercritical conditions, short-chain alcohols such as methanol and ethanol are hydrophobic, and triglyceride dissolves well in them (Geuens et al. 2014, Li et al. 2013). Hence, there is no issue on soap formation, catalyst efficiency and consumption, as well as yield of biodiesel (Borugadda and Goud 2012, Warabi et al. 2004). Supercritical fluid has characteristics of both liquid (density) and gas (viscosity). Dissolving capacity of supercritical solvent depends on its density, which is one of the main advantages of supercritical transesterification reaction. The density of a fluid is highly adjustable properties by changing pressure and temperature (Li et al. 2013). However, the supercritical method requires high molar ratios of alcohol to oil and the use of high temperatures and pressures to afford satisfactory conversion levels, leading to high operational and investment (Borugadda and Goud 2012, Silva et al. 2011). Supercritical reaction using methanol achieved optimum conversion of palm oil (81.5%) at relatively lower reaction time compared to supercritical ethanol process (79.2%) (Tan et al. 2010). Conversion achieved by supercritical methanolysis is relatively higher than ethanolysis (Vieitez et al. 2010).

9.4 Summary

The nature of FFA and fatty acid chain in feedstocks such as degree of unsaturation and presence of other functional groups strongly affect the properties of biodiesel produced. Transesterification reaction does not change fatty acid chain and degree of unsaturation. The most common carbon number of fatty acids of vegetable oils ranges between C10 and C24 (Table 9.2). The physicochemical properties of feedstock, vegetable oils from different origin, make differences in the biodiesel produced. Even with similar feedstock material slight changes in physicochemical properties of the product (FAAE) are due to genotype of the feedstocks and performance of transesterification reaction at different conditions. Biodiesel may be blended with petroleum-based diesel fuel to be used in existing diesel engine. Considering engine performance and emission characteristics, several standard-setting organizations worldwide have adopted biodiesel specifications: D6751 (ASTM International), DIN5160611 (German authorities issued), and EN14214 (Europe's Committee for Standardization). Biodiesel prepared from various renewable resources and processes should meet the international standard, ASTM D6751 (Gerpen et al. 2004). Pure biodiesel and biodiesel blends can achieve significant reduction in reactive hydrocarbons and carbon monoxide emissions when used in an unmodified diesel engine. Most of the emissions reduce while using biodiesel with the exception of nitrogen oxides, which either remain the same or are slightly increased. Moreover, due to certain other properties, like biodegradable nature, nontoxicity, high flash point, and higher cetane number, biodiesel is being viewed as a suitable alternative to mineral diesel.

TABLE 9.2   The Most Common Fatty Acids Found in Vegetable Oil and Algal Oil

System Name	Common Name	Symbol	Chemical Formula	Number of Unsaturation and Position
Unsaturated				
Hexadecenoic	Palmitoleic	C16:1	$C_{16}H_{30}O_2$	1, 9 *cis*
Octadecenoic	Petroselinic	C18:1	$C_{18}H_{34}O_2$	1, 6 *cis*
Octadecenoic	Oleic	C18:1	$C_{18}H_{34}O_2$	1, 9 *cis*
Octadecenoic	Elaidic	C18:1	$C_{18}H_{34}O_2$	1, 9 *trans*
Octadecenoic	Vaccenic	C18:1	$C_{18}H_{34}O_2$	1, 11 *cis*
Eicosenoic	Gadoleic	C20:1	$C_{20}H_{38}O_2$	1, 9 *cis*
Eicosenoic	Gondoic	C20:1	$C_{20}H_{38}O_2$	1, 11 *cis*
Docosenoic	Erucic	C22:1	$C_{22}H_{42}O_2$	1, 13 *cis*
Hexadecadienoic	Dienoic	C16:2	$C_{16}H_{28}O_2$	2, —
Octadecadienoic	Linoleic	C18:2	$C_{18}H_{32}O_2$	2, 9 and 12 *cis*
Octadecatrienoic	α-Linolenic	C18:3	$C_{18}H_{30}O_2$	3, 9, 12 and 15 *cis*
Octadecatrienoic	γ-Linolenic	C18:3	$C_{18}H_{30}O_2$	3, 6, 9 and 12 *cis*
Octadecatrienoic	Eleostearic	C18:3	$C_{18}H_{30}O_2$	3, 9 *cis*, 11 and 13 *trans*
Octadecatrienoic	Calendic	C18:3	$C_{18}H_{30}O_2$	3, 8 and 10 *trans*, 12 *cis*
Saturated				
Decanoic	Capric	C10:0	$C_{10}H_{20}O_2$	—
Dodecanoic	Lauric	C12:0	$C_{12}H_{24}O_2$	—
Tetradecanoic	Myristic	C14:0	$C_{14}H_{28}O_2$	—
Hexadecanoic	Palmitic	C16:0	$C_{16}H_{32}O_2$	—
Octadecanoic	Stearic	C18:0	$C_{18}H_{36}O_2$	—
Eicosanoic	Arachidic	C20:0	$C_{20}H_{40}O_2$	—
Docosanoic	Behenic	C22:0	$C_{22}H_{44}O_2$	—
Tetracosanoic	Lignoceric	C24:0	$C_{24}H_{48}O_2$	—
Docosanoic	Behenic	C22:0	$C_{22}H_{44}O_2$	—
Tetracosanoic	Lignoceric	C24:0	$C_{24}H_{48}O_2$	—

Sources: Bankovic-Ilic, I.B. et al., *Renew. Sustain. Energ. Rev.*, 16, 3621, 2012; Borugadda, V.B. and Goud, V.V., *Renew. Sustain. Energ. Rev.*, 16, 4763, 2012; Chhetri, A.B. et al., *Int. J. Mol. Sci.*, 9, 169, 2008; Gimbun, J. et al., *Procedia Eng.*, 53, 13, 2013; Issariyakul, T. and Dalai, A.K., *Renew. Sustain. Energ. Rev.*, 31, 446, 2014; Khan, T.M.Y. et al., *Renew. Sustain. Energ. Rev.*, 37, 840, 2014; Kumar, M.S. and Sharma, M.P., *Renew. Sustain. Energ. Rev.*, 44, 814, 2015; Ma, F. and Hanna, M.A., *Bioresour. Technol.*, 70, 1, 1999.

References

Abdelfatah, M., Farag, H.A., and M.E. Ossman. 2012. Production of biodiesel from non-edible oil and effect of blending with diesel on fuel properties. *Eng Sci Technol Int J* 2:583–591.

Adams, C., Peters, J.F., Rand, M.C., Schroer, B., and M.C. Ziemke. 1983. Investigation of soybean oil as a diesel fuel extender: Endurance tests. *J Oil Fat Ind* 60:1574–1579.

Ahmad, J., Yusup, S., Bokhari, A., and R.N.M. Kamil. 2014. Study of fuel properties of rubber seed oil based biodiesel. *Energy Convers Manage* 78:266–275.

Ardebili, M.S., Ghobadian, B., Najafi, G., and A. Chengeni. 2011. Biodiesel production potential from edible oil seeds in Iran. *Renew Sustain Energ Rev* 15:3041–3044.

Atabani, A.E., Silitonga, A.S., Ong, H.C. et al. 2013. Non-edible vegetable oils: A critical evaluation of oil extraction, fatty acid compositions, biodiesel production, characteristics, engine performance and emissions production. *Renew Sustain Energ Rev* 18:211–245.

Badran, O., Emeish, S., Abu-Zaid, M., Abu-Rahmaa, T., Al-Hasana, M., and M. Al-Ragheb. 2011. Impact of emulsified water/diesel mixture on engine performance and environment. *Int J Therm Environ Eng* 3:1–7.

Balat, M. 2011. Potential alternatives to edible oils for biodiesel production—A review of current work. *Energy Convers Manage* 52:1479–1492.

Bankovic-Ilic, I.B., Stamenkovic, O.S., and V.B. Veljkovic. 2012. Biodiesel production from non-edible plant oils. *Renew Sustain Energ Rev* 16:3621–3647.

Barnwal, B.K. and M.P. Sharma. 2005. Prospects of biodiesel production from vegetable oils in India. *Renew Sustain Energ Rev* 9:363–378.

Bart, J.C.J., Palmeri, N., and S. Cavallaro. 2010. *Biodiesel Science and Technology from Soil to Oil*. Washington, DC: Woodhead Publishing Series in Energy.

Biswas, S., Biswas, P., and A. Kumar. 2012. Catalytic cracking of soybean oil with zirconium complex chemically bonded to alumina support without hydrogen. *Int J Chem Sci Appl* 3:306–313.

Bokhari, A., Yusup, S., and M.M. Ahmad. 2012. Optimization of the parameters that affects the solvent extraction of crude rubber seed oil using response surface methodology (RSM). In *Recent Advance in Engineering,* WSEAS, Paris, France.

Boro, J., Konwar, L.J., and D. Deka. 2014. Transesterification of non edible feedstock with lithium incorporated egg shell derived CaO for biodiesel production. *Fuel Process Technol* 122:72–78.

Borugadda, V.B. and V.V. Goud. 2012. Biodiesel production from renewable feedstocks: Status and opportunities. *Renew Sustain Energ Rev* 16:4763–4784.

Buasri, A., Chaiyut, N., Loryuenyong, V., Wongweang, C., and S. Khamsrisuk. 2013. Application of eggshell wastes as a heterogeneous catalyst for biodiesel production. *Sustain Energ* 1:7–13.

Canakci, M. 2007. The potential of restaurant waste lipids as biodiesel feedstocks. *Bioresour Technol* 98:183–190.

Canakci, M. and J.V. Gerpen. 1999. Biodiesel production via acid catalysis. *Trans ASAE* 42:1203–1210.

Chen, G., Shan, R., Shi, J., and B. Yan. 2014. Ultrasonic-assisted production of biodiesel from transesterification of palm oil over ostrich eggshell-derived CaO catalyst. *Bioresour Technol* 171:428–432.

Chhetri, A.B., Tango, M.S., Budge, S.M., Watts, K.C., and M.R. Islam. 2008. Non-edible plant oils as new sources for biodiesel production. *Int J Mol Sci* 9:169–180.

Chisti, Y. 2007. Biodiesel from microalgae. *Biotechnol Adv* 25:294–306.

Choedkiatsakul, I., Ngaosuwan, K., Cravotto, G., and S. Assabumrungrat. 2014. Biodiesel production from palm oil using combined mechanical stirred and ultrasonic reactor. *Ultrason Sonochem* 21:1585–1591.

Correia, L.M., Saboya, R.M.A., Campelo N.S. et al. 2014. Characterization of calcium oxide catalysts from natural sources and their application in the transesterification of sunflower oil. *Bioresour Technol* 151:207–213.

Deng, X., Fang, Z., and Y. Liu. 2010. Ultrasonic transesterification of *Jatropha curcas L.* oil to biodiesel by a two-step process. *Energ Convers Manage* 51:2802–2807.

Encinar, J.M., Gonzalez, J.F., Martinez, G., Sanchez, N., and A. Pardal. 2012. Soybean oil transesterification by the use of a microwave flow system. *Fuel* 95:386–393.

Fayyad, S.M., Abu-Ein, S., Al-Marahleh, G. et al. 2010. Experimental emulsified diesel and benzen investigation. *Res J Appl Sci Eng Technol* 2:268–273.

Freedman, B., Pryde, E.H., and T.L. Mounts. 1984. Variables affecting the yields of fatty esters from transesterified vegetable oils. *J Am Oil Chem Soc* 61:1638–1642.

Georgogianni, K.G., Kontominas, M.G., Pomonis, P.J., Avlonitis, D., and V. Gergis. 2008. Conventional and *in situ* transesterification of sunflower seed oil for the production of biodiesel. *Fuel Process Technol* 89:503–509.

Gerpen, J.V., Shanks, B., Pruszko, R., Clements, D., and G. Knothe. 2004. Biodiesel production technology: Subcontractor report. NREL, Golden, CO.

Geuens, J., Sergeyev, S., Maes, B.U.W., and S.M.F. Tavernier. 2014. Catalyst-free microwave-assisted conversion of free fatty acids in triglyceride feedstocks with high acid content. *Renew Energ* 68:524–528.

Gimbun, J., Ali, S., Kanwal, C.C.S.C. et al. 2013. Biodiesel production from rubber seed oil using activated cement clinker as catalyst. *Procedia Eng* 53:13–19.

Gole, V.L. and P.R. Gogate. 2012. A review on intensification of synthesis of biodiesel from sustainable feed stock using sonochemical reactors. *Chem Eng Process* 53:1–9.

Graboski, M.S. and R.L. McCormick. 1998. Combustion of fat and vegetable oil derived fuels in diesel engines. *Prog Energ Combust Sci* 24:125–164.

Hannon, M., Gimpel, J., Rasala, B., and S. Mayfield. 2010. Biofuels from algae: Challenges and potential. *Biofuels* 1:763–784.

He, B. and J.H. Van Gerpen. 2012. Application of ultrasonication in transesterification processes for biodiesel production. *Biofuels* 3:479–488.

Helwani, Z., Othman, M.R., Aziz, N., Kim, J., and W.J.N. Fernando. 2009. Solid heterogeneous catalysts for transesterification of triglycerides with methanol: A review. *Appl Catal A* 363:1–10.

Hossain, A.B.M.S., Salleh, A., Boyce, A.N., Chowdhury, P., and M. Naqiuddin. 2008. Biodiesel fuel production from algae as renewable energy. *Am J Biochem Biotechnol* 4:250–254.

International Energy statistics. https://www.eia.gov/, February 10, 2015.

Issariyakul, T. and A.K. Dalai. 2014. Biodiesel from vegetable oils. *Renew Sustain Energ Rev* 31:446–471.

Istadi, I., Anggoro, D.D., Buchori, L., Rahmawati, D.A., and D. Intaningrum. 2015. Active acid catalyst of sulphated zinc oxide for transesterification of soybean oil with methanol to biodiesel. *Procedia Environ Sci* 25:385–393.

Jain, S. and M.P. Sharma. 2010. Kinetics of acid base catalyzed transesterification of *Jatropha curcas* oil. *Bioresour Technol* 101:7701–7706.

Jain, S., Sharma, M.P., and S. Rajvanshi. 2011. Acid base catalyzed transesterification kinetics of waste cooking oil. *Fuel Process Technol* 92:32–38.

Jull, C., Renondo, P.C., Mosoti, V., and J. Vapnek. 2007. *Recent Trend in the Law and Policy of Bioenergy Production, Promotion and Use*. Rome, Italy: FAO.

Kamalakar, K., Rajak, A.K., Prasad, R.B.N., and M.S.L. Karuna. 2013. Rubber seed oil-based biolubricant base stocks: A potential source for hydraulic oils. *Ind Crops Prod* 51:249–257.

Kawashima, A., Matsubara, K., and K. Honda. 2009. Acceleration of catalytic activity of calcium oxide for biodiesel production. *Bioresour Technol* 100:696–700.

Keera, S.T., Sabagh, S.M.E., and A.R. Taman. 2011. Transesterification of vegetable oil to biodiesel fuel using alkaline catalyst. *Fuel* 90:42–47.

Khan, M.A., Yusup, S., and M.M. Ahmad. 2010. Acid esterification of a high free fatty acid crude palm oil and crude rubber seed oil blend: Optimization and parametric analysis. *Biomass Bioenerg* 34:1751–1756.

Khan, T.M.Y., Atabani, A.E., Badruddin, I.A., Badarudin, A., Khayoon, M.S., and S. Triwahyono. 2014. Recent scenario and technologies to utilize non-edible oils for biodiesel production. *Renew Sustain Energ Rev* 37:840–851.

Khemthong, P., Luadthong, C., Nualpaeng, W. et al. 2012. Industrial eggshell wastes as the heterogeneous catalysts for microwave-assisted biodiesel production. *Catal Today* 190:112–116.

Kole, C., Joshi, C.P., and D.R. Shonnard. 2012. *Handbook of Bioenergy Crop Plants*. New York: CRC Press.

Kouzu, M., Kasuno, T., Tajika, M., Yamanaka, S., and J. Hidaka. 2008. Active phase of calcium oxide used as solid base catalyst for transesterification of soybean oil with refluxing methanol. *Appl Catal A* 334:357–365.

Kumar, M.S. and M.P. Sharma. 2015. Assessment of potential of oils for biodiesel production. *Renew Sustain Energ Rev* 44:814–823.

Lee, J.-S. and S. Saka. 2010. Biodiesel production by heterogeneous catalysts and supercritical technologies. *Bioresour Technol* 101:7191–7200.

Leung, D.Y.C., Wu, X., and M.K.H. Leung. 2010. A review on biodiesel production using catalyzed transesterification. *Appl Energ* 87:1083–1095.

Li, Q., Xu, J., Du, W., Li, Y., and D. Liu. 2013. Ethanol as the acyl acceptor for biodiesel production. *Renew Sustain Energ Rev* 25:742–748.

Lima, D.G., Soares, V.C.D., Ribeiro, E.B. et al. 2004. Diesel-like fuel obtained by pyrolysis of vegetable oils. *J Anal Appl Pyrolysis* 71:987–996.

Liu, J. 2013. Biodiesel synthesis via transesterification reaction in supercritical methanol: a) a kinetic study, b) biodiesel synthesis using microalgae oil. Master of Science, Biomedical and Chemical Engineering, Syracuse, NY.

Lopez, D.E., Goodwin Jr., G., Bruce, D.A., and E. Lotero. 2005. Transesterification of triacetin with methanol on solid acid and base catalyst. *Appl Catal A* 295:97–105.

Lueung, D.Y.C. and Y. Guo. 2006. Transesterification of neat and used frying oil: Optimization for biodiesel production. *Fuel Process Technol* 87:883–890.

Ma, F. and M.A. Hanna. 1999. Biodiesel production: A review. *Bioresour Technol* 70:1–15.

Manickam, S., Arigela, V.N.D., and P.R. Gogate. 2014. Intensification of synthesis of biodiesel from palm oil using multiple frequency ultrasonic flow cell. *Fuel Process Technol* 128:388–393.

Mathimani, T., Uma, L., and D. Prabaharan. 2015. Homogeneous acid catalysed transesterification of marine microalga *Chlorella* sp. BDUG91771 lipid—An efficient biodiesel yield and its characterization. *Renew Energ* 81:523–533.

Melo-Espinosa, E.A., Piloto-Rodriguez, R., Goyos-Perez, L., Sierens, R., and S. Verhelst. 2015. Emulsification of animal fats and vegetable oils for their use as a diesel engine fuel: A overview. *Renew Sustain Energ Rev* 47:623–633.

Menetrez, M.Y. 2012. An overview of algae biofuel production and potential environmental impact. *Environ Sci Technol* 46:7073–7085.

Mootabadi, H., Salamatinia, B., Bhatia, S., and A.Z. Abdullah. 2010. Ultrasonic-assisted biodiesel production process from palm oil using alkaline earth metal oxides as the heterogeneous catalysts. *Fuel* 89:1818–1825.

Mostafaei, B., Ghobadian, B., Barzegar, M., and A. Banakar. 2013. Optimization of ultrasonic reactor geometry for biodiesel production using response surface methodology. *J Agric Sci Technol* 15:697–708.

Mubarak, M. and M.S. Kumar. 2012. An experimental study on waste cooking oil and its emulsions as diesel engine fuel. In *Advances in Engineering, Science and Management (ICAESM)*, Nagapattinam, Tamil Nadu, India.

Mustata, F. and I. Bicu. 2014. The optimization of the production of methyl esters from corn oil using barium hydroxide as a heterogeneous catalyst. *J Am Oil Chem Soc* 91:839–847.

Nezinhe, A. and A. Danisman. 2007. Alkali catalyzed transesterification of cottonseed oil by microwave irradiation. *Fuel* 86:2639–2644.

Niju, S., Begum, K.M.M.S., and N. Anantharaman. 2014. Modification of egg shell and its application in biodiesel production. *J Saudi Chem Soc* 18:702–706.

No, S.-Y. 2011. Inedible vegetable oils and their derivatives for alternative diesel fuels in CI engines: A review. *Renew Sustain Energ Rev* 15:131–149.

Oliveira, D.A., Benelli, P., and E.R. Amante. 2013. A literature review on adding value to solid residues: Egg shells. *J Clean Prod* 46:42–47.

Ooi, Y.-S., Zakaria, R., Mohamed, A.R., and S. Bhatia. 2004. Catalytic cracking of used palm oil and palm oil fatty acid mixture for the production of liquid fuel: Kinetic modeling. *Energ Fuels* 18:1555–1561.

Prateepchaikul, G. and T. Apichato. 2003. Palm oil as a fuel for agricultural diesel engines: Comparative testing against diesel oil. *Songklanakarin J Sci Technol* 25:317–326.

Pukale, D.D., Maddikeri, G.L., Gogate, P.R., Pandit, A.B., and A.P. Pratap. 2015. Ultrasound assisted transesterification of waste cooking oil using heterogeneous catalyst. *Ultrason Sonochem* 22:278–286.

Putri, E.M.M., Rachimoellah, M., Santoso, N., and F. Pradana. 2012. Biodiesel production from Kapok seed oil (*Ceiba Pentandra*) through the transesterification process by using CaO as catalyst. *Glob J Res Eng* 12:6–12.

Rahimi, M., Aghel, B., Alitabar, M., Sepahvand, A., and H.R. Ghasempour. 2014. Optimization of biodiesel production from soybean oil in a microreactor. *Energ Convers Manage* 79:599–605.

Raita, M., Champreda, V., and N. Laosiripojana. 2010. Biocatalytic ethanolysis of palm oil for biodiesel production using microcrystalline lipase in *tert*-butanol system. *Process Biochem* 45:829–834.

Rakopoulos, C.D., Antonopoulos, K.A., Rakopoulos, D.C., Hountalas, D.T., and E.G. Giakoumis. 2006. Comparative performance and emissions study of a direct injection diesel engine using blends of diesel fuel with vegetable oils or biodiesels of various origins. *Energ Convers Manage* 47:3272–3287.

Rodrigues, R.C., Passela, B.C.C., Volpato, G., Fernandez-Lafuente, R., Guisan, J.M., and M.A.Z. Ayub. 2010. Two step ethanolysis: A simple and efficient way to improve the enzymatic biodiesel synthesis catalyzed by an immobilized-stabilized lipase from *Thermomyces lanuginosus*. *Process Biochem* 45:1268–1273.

Sajjadi, B., Abdul Aziz, A.R., and S. Ibrahim. 2014. Investigation, modelling and reviewing the effective parameters in microwave-assisted transesterification. *Renew Sustain Energ Rev* 37:762–777.

Salamatinia, B., Mootabadi, H., Bhatia, S., and A.Z. Abdullah. 2010. Optimization of ultrasonic-assisted heterogeneous biodiesel production from palm oil: A response surface methodology approach. *Fuel Process Technol* 91:441–448.

Santori, G., Nicola, G.D., Moglie, M., and F. Polonara. 2012. A review analyzing the industrial biodiesel production practice starting from vegetable oil refining. *Appl Energ* 92:109–132.

Shahzad, I., Hussain, K., Nawaz, K., Nisar, M.F., and K.H. Bhatti. 2010. Algae as an alternative and renewable resource for biofuel production. *BIOL* 1:16–23.

Sharma, Y.C. and B. Singh. 2009. Development of biodiesel: Current scenario. *Renew Sustain Energ Rev* 13:1646–1651.

Shimada, Y., Watanabe, Y., Sugihara, A., and Y. Tominaga. 2002. Enzymatic alcoholysis for biodiesel fuel production and application of the reaction to oil processing. *J Mol Catal B Enzym* 17:133–142.

Shokrolahi, A., Zali, A., and H.R. Pouretedal. 2011. Sulfonated porous carbon catalyzed esterification of free fatty acids. *Iran J Catal* 1:37–40.

Silva, C., Lima, A.P., Castilhas, F., Filho, L.C., and J.V. Oliveira. 2011. Non-catalytic production of fatty acid ethyl esters from soybean oil with supercritical ethanol in a two-step process using a microtube reactor. *Biomass Bioenerg* 35:526–532.

Siswanto, D.Y., Salim, G.W., Wibisono, N., Hindarso, H., Sudaryanto, Y., and S. Ismadji. 2008. Gasoline production from palm oil via catalytic cracking using MCM-41: Determination of optimum condition. *ARPN J Eng Appl Sci* 3:42–46.

Sivasamy, A., Cheah, K.Y., Fornasiero P., Kemausuor, F., Zinoviev, S., and S. Miertus. 2009. Catalytic applications in the production of biodiesel from vegetable oils. *ChemSusChem* 2:278–300.

Speight, J.G. 2011. *The Biofuels Handbook*. Cambridge, UK: RSC Publishing.

Tan, K.T., Gui, M.M., Lee, K.T., and A.R. Mohamad. 2010. An optimized study of methanol and ethanol in supercritical alcohol technology for biodiesel production. *J Supercrit Fluids* 53:82–87.

Tiwari, A.K., Kumar, A., and H. Raheman. 2007. Biodiesel production from jatropha oil (*Jatropha curcas*) with high free fatty acids: An optimized process. *Biomass Bioenerg* 31:569–575.

Vieitez, I., Silva, C., Alcmin I. et al. 2010. Continuous catalyst-free methanolysis and ethanolysis of soybean oil under supercritical alcohol/water mixtures. *Renew Energ* 35:1976–1981.

Viriya-empikul, N., Krasae, P., Puttasawat, B., Yoosuk, B., Chollacoop, N., and K. Faungnawakij. 2010. Waste shells of mollusk and egg as biodiesel production catalysts. *Bioresour Technol* 101:3765–3767.

Vyas, A.P., Verman, J.L., and N. Subrahmanyam. 2010. A review on FAME production processes. *Fuel* 89:1–9.

Wan Omar, W.N. and A.S. Amin. 2011. Optimization of heterogeneous biodiesel production from waste cooking palm oil via response surface methodology. *Biomass Bioenerg* 35:1329–1338.

Warabi, Y., Kusdiana, D., and S. Saka. 2004. Reactivity of triglycerides and fatty acids of rapeseed oil in supercritical alcohols. *Bioresour Technol* 91:283–287.

Watanabe, Y., Shimada, Y., Sugihara, A., and Y. Tominaga. 1999. Stepwise ethanolysis of tuna oil using immobilized *Candida antarctica* lipase. *J Biosci Bioeng* 88:622–626.

Xu, J., Jiang, J., Zhang, T., and W. Dai. 2013. Biofuel production from catalytic cracking of triglyceride materials followed by an esterification reaction in a scale-up reactor. *Energ Fuel* 27:255–261.

Yin, X., Ma, H., You, Q., Wang, Z., and J. Chang. 2012. Comparison of four different enhancing methods for preparing biodiesel through transesterification of sunflower oil. *Appl Energ* 91:320–325.

10

Bioethanol Production from Lignocellulosics

Raveendran Sindhu, Ashok Pandey, and Parameswaran Binod

Contents

Abstract

Increasing environmental concerns and depletion of fossil fuels lead to a search for alternative sources of energy. Lignocellulosic biomass is abundantly available, does not compete with food, and serves as a potential candidate for the production of second-generation biofuels. Several physical, chemical, structural, and compositional factors limit enzymatic saccharification of cellulose present in the lignocellulosic biomass. In order to develop an economically viable ethanol conversion process, fine-tuning of pretreatment, enzymatic saccharification, and fermentation is required. Several research and developmental activities are going on in this direction. Since there is a wide diversity in the composition of lignocellulosic biomass, tailor-made pretreatment strategies are to be developed for each biomass to obtain a better cellulose conversion rate.

10.1 Introduction

Transportation sector greatly depends on petroleum-based fuels, constitutes about 60% of world's total oil consumption, and contributes to 70% of global CO emission (Balat 2011). Lignocellulosic biomass constitutes a renewable substrate for bioethanol production and does not compete with food production and animal feed. These materials also contribute to environmental sustainability (Demirbas 2003). Currently the most abundant lignocellulosic feedstocks available are sugarcane tops, sugarcane bagasse, rice straw, wheat straw, corn stover, and so on.

The presence of high level of cellulose and hemicelluloses, their surplus availability, their relatively low cost, and their renewability are the major reasons behind their use in the production of bioethanol (Cheng et al. 2008).

Fuel ethanol production from lignocellulosic biomass has several disadvantages such as high production cost, special equipment requirements, large water consumption, and complex production technology (Sun and Cheng 2002) and, hence, is currently not economically viable.

Biofuels offer economic, energy, and environmental benefits. Their major economic benefits include sustainability, fuel diversity, and reduced dependency on imported petroleum. Their environmental impacts include greenhouse gas reduction, reduced air pollution, biodegradability, higher combustion efficiency, improved land and water use, and carbon sequestration. They offer energy security, including domestic targets, supply reliability, reduction in the use of fossil fuels, ready availability, and domestic distribution (Balat 2011). Cost of bioethanol production depends on several factors such as cost of feedstocks, by-product revenues, energy cost, plant location, investment cost, transportation cost, and financing costs (Fraiture et al. 2008). Bioethanol-blended fuel can significantly reduce petroleum use and greenhouse gas emission.

This chapter discusses the importance of bioethanol production from lignocellulosic biomass, biomass composition and availability, bioethanol conversion processes, and current bottlenecks in the economical production of bioethanol.

10.2 Why Lignocellulosic Biomass?

Lignocellulosic biomass is an important renewable source for the production of second-generation biofuels. It is mainly composed of three types of polymers— cellulose, hemicelluloses, and lignin—that are bonded by covalent cross-linkages and noncovalent forces. Many lignocellulosic biomasses like rice straw, sugarcane bagasse, wheat straw, cotton stalk, bamboo, and sugarcane tops are some of the abundantly available agro-residues. Many agro-residues are well known for ethanol production. Lignocellulosic materials constitute a substantial renewable substrate for the production of bioethanol, do not compete with food or feed, and thus contribute to environmental sustainability.

10.3 Biomass Composition

Composition of the biomass is one of the key factors affecting the efficiency of biofuel production during conversion process. The structural and chemical compositions of lignocellulosic biomass are highly variable because of genetic and environmental factors and their interactions (Lee et al. 2007). The enzymatic digestibility of native biomass is very low unless a very large amount of enzyme is used because of the structural characteristics of the biomass. Lignocellulosic biomass is a heterogeneous complex of carbohydrate polymers and lignin that typically contains 55%–75% carbohydrates by dry weight. Lignocellulosic biomass is composed of three main components—cellulose, hemicelluloses, and lignin and nonorganic components like proteins, lipids, and extractives. Composition of the feedstock varies among species and variety. It also shows variation, depends on growth conditions and maturation, within the species itself. Normally, the cellulose constitutes about 30–50%, hemicelluloses about 15–35%, and lignin around 10–20%. Cellulose and hemicelluloses constitute about 70% of the total biomass and are tightly linked to lignin through covalent and hydrogenic bonds, which are resistant to any pretreatment (Pettersen 1984). Cellulose is a polymer of glucose, and the specific structure of cellulose favors the ordering of polymer chains into tightly packed, highly crystalline structures that are water-insoluble and resistant to depolymerization (Mosier et al. 2005). Other components of lignocellulosic biomass are hemicelluloses that are branched polymers of glucose or xylose substituted with arabinose, xylose, galactose, fucose, mannose, glucose, or glucuronic acid or with some side chains containing acetyl groups of ferulate (Carpita and Gibeaut 1993). Hemicellulose forms hydrogen bonds with cellulose microfibrils that provide the structural backbone to plant cell wall. Lignin present in the cell wall imparts further strength. Recalcitrance of lignocellulosic biomass for hydrolysis is due to the crystallinity of cellulose, accessible surface area, heterogeneous nature of biomass, protection of cellulose by lignin, and so on (Chang and Holtzapple 2000). The composition of the feedstock greatly affects the type of pretreatment to be used for the removal of hemicelluloses and lignin. The relationship between structural and compositional factors reflects the complexity of lignocellulosic materials, and the variability in these characteristics accounts for the variability in the digestibility between different sources of lignocellulosic biomass (Mosier et al. 2005).

10.4 Biomass Availability

Lignocellulosic biomass serves as cheap and abundant feedstock. For the production of bioethanol, feedstock availability, its variability, and sustainability are the key issues to be considered (Sukumaran et al. 2010). Though there is a huge generation of biomass residues, a major portion is used for other purposes. Hence, the actual availability of biomass is questionable. A significant part of the agro-residues generated is consumed as fodder and has other applications. The residue to crop ratios derived from the studies are useful in calculating the amount of residues obtained for each crop (Ravindranath et al. 2005). Apart from the issue of biomass availability, other problems to be addressed include sustainability and logistics. Since farming and residue generation are concentrated in distributed pockets, collection and transportation face serious limitations (Sukumaran et al. 2010). Based on the source, lignocellulosic biomass can be classified into four groups based on the type of resource as forest residues, municipal solid waste, waste paper, and crop residues. One of the main problems associated with the bioethanol production is the availability of raw materials for production. Feedstock availability can vary considerably from season to season and depends on geographic location (Balat 2011).

10.5 Bioethanol Conversion Process

The production of lignocellulosic ethanol from biomass generally involves four major steps: pretreatment of biomass and hydrolysis of the pretreated biomass followed by fermentation of released sugars and ethanol separation.

10.5.1 Pretreatment

Pretreatment of biomass is the first step in bioethanol process. An effective pretreatment is essential for successful hydrolysis and downstream operation. Pretreatment alters biomass macroscopic and microscopic structure as well as submicroscopic chemical composition and structure so that hydrolysis of carbohydrate to monomeric sugars takes place more rapidly with better sugar yields. The first and foremost critical step in lignocellulosic biorefinery is pretreatment since it affects downstream costs involved in detoxification, enzyme loading, waste treatment demands, and other variables (Zhang 2008). Pretreatment makes up for more than 40% of the total processing cost. Hydrolysis of lignocellulosic biomass without any pretreatment can yield less than 20% of total sugars, while after pretreatment it can reach up to 0% to 90% with some pretreatment methods (Alizadeh et al. 2005). Pretreatment of woody biomass differs substantially from the agricultural biomass due to differences in their chemical composition and physical properties (Zhu and Pan 2010).

Though several pretreatment regimens are currently developed to overcome biomass recalcitrance, only very few seem to be promising as an industrial process; each one has its own merits and demerits. Pretreatment is probably the most energy-intensive operation in biomass conversion to fuels or chemicals. Conventional pretreatment techniques are focused on removing either lignin or hemicellulose from the biomass structure, thereby decompacting it and making it more susceptible to enzyme attack. While these methods are widely accepted and practiced, even minor tweaks to these established technologies are often sufficient to improve the pretreatment efficiency.

The most commonly used treatments include acid hydrolysis, dilute acid treatment, hot water, and lime. Among the different pretreatment methods, acidic pretreatment is extensively studied and is close to commercialization. One of the major

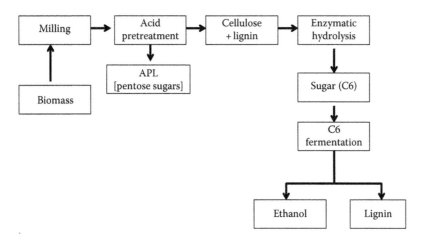

Figure 10.1 Biomass to ethanol—general scheme of acid pretreatment.

advantages of acidic pretreatment is the generation of a separate pentose and hexose stream. Dilute acid treatments are always accompanied by formation of inhibitors like furfural, hydroxymethylfurfural, phenolics, and organic acids. The schematic diagram for acid pretreatment is depicted in Figure 10.1.

Other pretreatment strategies include physical, chemical, and biological pretreatments.

These include hydrothermal pretreatment, lime pretreatment, flow-through acid pretreatment, ammonia pretreatment, microwave pretreatment, ultrasound pretreatment, and combination of methods or hybrid methods. Table 10.1 shows different pretreatments currently employed and their mode of action.

TABLE 10.1 Different Pretreatment Strategies and Their Mode of Action on Lignocellulosic Biomass

Mode of Action	Pretreatment Agent	Reference
Hemicellulose removal	Acid	Sindhu et al. (2014a)
Delignification	Alkali	Sindhu et al. (2014b)
Delignification and hemicellulose removal	Ammonia fiber explosion	Holtzapple et al. (1994)
Hemicellulose and lignin removal	Ultrasound	Chen et al. (2011)
Delignification	Surfactant	Tu et al. (2007)
Delignification	Biological	Wan and Li (2011)
Hemicellulose removal	Hot water	Hu et al. (2008)
Delignification	Ozone	Sun and Cheng (2002)
Delignification	Hydrogen peroxide	Dias et al. (2013)
Delignification	Organosolvent	Amiri et al. (2014)
Delignification	Crude glycerol	Guragain et al. (2011)
Hemicelluloses removal	Steam explosion with catalyst	Lloyd and Wyman 2005
Delignification	Alkaline wet oxidation	Klinke et al. (2004)

Developments of advanced pretreatment technologies that control mechanism are tuned to unique characteristics of different biomass types, and minimizations of cost are still needed. Understanding of the fundamental chemical and physical mechanisms that occur during pretreatment as well as improved understanding of the relationship between the chemical composition and physicochemical structure of lignocellulose on the enzymatic digestibility of cellulose and hemicellulose is required for the generation of effective pretreatment models. Predictive pretreatment models enable the selection, design, optimization, and process control pretreatment technologies that match biomass feedstock with the appropriate method and process configuration (Mosier et al. 2005). Development of an integrated process for full fractionation is essential to make the process economically viable.

10.5.2 Enzymatic Hydrolysis

One of the major factors that limit commercialization of bioethanol production from lignocellulosic biomass is the cost as well as hydrolytic efficiency of the enzymes. The success of hydrolysis depends on the effectiveness of pretreatment. During pretreatment, the enzymatic hydrolysis rate of lignocellulosic biomass is improved with an increase in porosity of the substrate and cellulose accessibility to cellulases. Accessory enzymes are those enzymes that act on less abundant linkages found in plant cell walls. These include arabinases, lyases, pectinases, galactanases, and several types of esterases.

Hydrolysis efficiency can be improved by supplementation of accessory enzymes. Several research and developmental (R&D) activities are on to improve the economic viability of enzymatic hydrolysis. Development of hydrolysis kinetics helps in the design and operation of hydrolysis reactors that help in improving specific activities, thereby improving the economic viability of enzymatic hydrolysis for bioethanol production.

One of the most efficient cellulase-producing fungi is *Trichoderma reesei*. Cellulolytic enzymes include endoglucanase, exoglucanase, and β-glucosidase. *Trichoderma reesei* catalyzes the hydrolysis of β-1,4 linkages in cellulose. The conversion of cellulose to glucose monomers takes place with the help of the combined action of three enzymes—endoglucanase, exoglucanase, and β-glucosidase. Endoglucanases hydrolyze β-1,4 glycosidic linkages in the cellulose chain; cellobiohydrolase cleaves off cellobiose units from the end of the chain, and β-glucosidase converts cellobiose to glucose (Himmel et al. 1996). Phenolic compounds derived from lignin are known to inhibit cellulases. Another drawback of lignin is that it causes nonproductive adsorption that limits the accessibility of cellulose to cellulases.

Several reports are available on the role of additives in enzymatic hydrolysis. A study conducted by Erickson et al. (2002) revealed that addition of surfactant bovine serum albumin helps in blocking lignin's interaction with cellulases. Though there are several advantages of enzymatic hydrolysis over acid hydrolysis, the main inhibiting factor in ethanol production at present is the cost of enzymes. Proteins like swollenin play an important role in nonhydrolytic loosening of the cellulosic fibril network and do not act on β-1,4 glycosidic bonds in cellulose. Swollenin increases the access of cellulases to cellulose chains by dispersion of cellulose aggregations, thereby exposing individual cellulose chains to the enzyme. Enzyme-related factors that affect hydrolysis include enzyme concentration, enzyme adsorption, end-product inhibition, thermal inactivation, and unproductive binding to lignin. The rate of enzymatic hydrolysis is mainly affected by the structural features of cellulose that include cellulose crystallinity, degree of polymerization, accessible surface area, particle size, and the presence of associated materials like hemicelluloses and lignin (Binod et al. 2011).

The accessibility of cellulose by cellulase is more affected by xylan than lignin. Though removal of lignin and xylan enhances the saccharification rate, xylan removal directly impacts glucan chain accessibility. Hence, the removal of xylan is more advantageous than the removal of lignin. It also helps in reduction of enzyme inhibition by xylooligomers as well as reduction of requirements of accessory enzymes.

Recycling of cellulases is an effective way to reduce the cost of enzymatic hydrolysis for the production of bioethanol. Studies revealed that lignin content plays a significant role in the distribution of cellulases between solid and liquid phases for the nonproductive adsorption of cellulases. The recycling efficiency is highly dependent on the adsorption characteristics of cellulases to lignocellulosic biomass. pH-triggered adsorption–desorption is a simple, feasible, and efficient strategy for the recycling of cellulases during the hydrolysis of pretreated lignocellulosic biomass (Shang et al. 2014).

The amount of protein/enzyme required for effective cellulose hydrolysis is too high. This can be reduced by formulating enzyme cocktail by adding accessory enzymes like xylanase and lytic polysaccharide monooxygenase to the cellulase mixture (Hu et al. 2013). Studies revealed that xylanase enhances the hydrolytic potential of *cellulase mixture* by helping in the release of cellulase adsorbed or stuck on the substrate while hydrolyzing the noncellulose components of the substrate that restrict access of cellulases to the cellulose. One of the main challenges in achieving effective cellulose hydrolysis is overcoming a decrease in hydrolysis rate at later stages of hydrolysis. Different mechanisms like inactivation of cellulase by denaturation, product inhibition, unproductive binding, and an addition of recalcitrance of the residual substrate play an important role (Erikson et al. 2002; Yang et al. 2006). Crystalline structures and type of chemical bonding have significant impact on the physical properties and chemical reactivity.

Several challenges need to be addressed to improve hydrolysis efficiency. Enzymatic hydrolysis is an attractive and eco-friendly process, and several R&D activities need to be carried out to make it economically viable. Due to the heterogeneous nature of lignocellulosic biomass, it is difficult to develop a specific cocktail for a specific biomass. Several R&D studies are ongoing in this direction to improve the enzymatic conversion efficiency of lignocellulosic biomass to fermentable sugars by protein engineering approaches.

10.5.3 Fermentation

Ethanol fermentation from enzymatic hydrolyzate is the last step in lignocellulosic bioethanol production process. Most ethanol production is accomplished using the yeast *Saccharomyces cerevisiae*. Besides *S. cerevisiae*, other yeasts used for ethanol production are *Schizosaccharomyces pombe*, *Kluyveromyces lactis*, *Candida* sp., and *Pichia* sp. that are able to ferment even pentose sugars. In addition to yeasts, several bacteria and fungi can also produce ethanol by fermentation.

The enzymatic hydrolyzate will be used for fermentation. Yeast fermentation is always accompanied by the formation of CO_2 by-products. The optimum temperature for fermentation of conventional yeast strain is 30°C, and this strain resists high osmotic pressure. *S. cerevisiae* can produce a high yield of ethanol (90% of theoretical) from hexose sugars (Limayem and Ricke 2012).

One of the major challenges in fermentation of biomass hydrolyzate to produce ethanol is to overcome the effect of inhibitors by fermenting microorganisms. Several strategies are currently adopted to overcome inhibition. These include microbial and enzymatic treatment, adaptation of fermenting microbes in lignocellulosic hydrolyzate prior to fermentation, and overexpression of genes encoding enzymes for resistance against specific inhibitors (Parawira and Tekere 2011). These strategies seem to improve the ability of *S. cerevisiae* to more efficiently produce bioethanol. The advantages of biological detoxification over chemical methods are mild reaction conditions, fewer by-products, and less energy demand. Adaptation as well as targeted engineering may lead to improvement of tolerance to inhibitors.

Fermentative microorganism should be tolerant to high ethanol concentration and to chemical inhibitors formed during pretreatment. Several genetically modified organisms have been developed for the fermentation of C5 and C6 sugars simultaneously. Though several eukaryotic and prokaryotic microorganisms can produce ethanol, most of them remain limited due to sugar co-fermentation, ethanol yield, and tolerance to inhibitors, high temperature, and ethanol.

The concentration of ethanol produced from lignocellulosic biomass may be at least 40 g/L (5% v/v) to minimize the cost of distillation (Park et al. 2013). Conditions for the simultaneous saccharification and fermentation (SSF) at fed-batch mode for the production of ethanol from empty palm oil fruit bunches (EFBs) by *S. cerevisiae* were investigated. An economic concentration of ethanol (5%) was achieved within the first 17 h and was equivalent to 94.1% of the theoretical yield. This is the first report on ethanol production from EFB in fed-batch SSF using *S. cerevisiae*.

Sankh et al. (2011) reported improvement of ethanol production using *S. cerevisiae* by enhancement of biomass and nutrient supplementation. Oil seed meal extract was used as protein lipid supplement, and rice husk was used as substratum. Different oil seed meal extracts from peanut, safflower, and sunflower were evaluated. The study revealed that oil seed meal extract of 4% was found to enhance ethanol production by 50% and sugar tolerance was enhanced to 16% from 8%. The addition of substratum up to 4% oil seed meal extract leads to an increase of ethanol production, but further addition of rice husk does not contribute in any way as nutrient source.

The study revealed that addition of a substratum, protein, and lipid supplements enhances the ethanol yield and improves both sugar tolerance and ethanol tolerance of the yeast strain.

López-Abelairas et al. (2013) studied the fermentation of biologically pretreated wheat straw for ethanol production. The pretreatment was carried out using a white-rot fungus to produce bioethanol in an eco-friendly process alternative to physico-chemical process. Six different microbes like the fungus *Fusarium oxysporum* CECT 2159 and yeasts like *S. cerevisiae* CECT 1332, *Pachysolen tannophilus* CECT 1426, *Pichia stipitis* CECT 1922, *Kluyveromyces marxianus* CECT 10585, and *Candida shehatae* CECT 10810 were evaluated for ethanol production from the hydrolyzate obtained from wheat straw pretreated with the white-rot fungus *Irpex lacteus*. All these microbes are able to utilize pentoses in addition to glucose. The highest ethanol yield was observed with *P. tannophilus* CECT 1426. Its application in combination with the best process configuration yielded 163 mg ethanol per gram of raw wheat straw, which was between 23% and 35% greater than the yields typically obtained with a conventional bioethanol process, in which wheat straw is pretreated using steam explosion and fermented with the yeast *S. cerevisiae*.

Desirable characteristics for ethanol production by microorganisms include high ethanol yield, low by-products, and metabolization of a wide variety of sugars (Ingram et al. 1987). Promising engineered ethanologenic bacteria include *Escherichia coli*, *Klebsiella oxytoca*, *Lactobacillus casei*, *Zymomonas mobilis*, and *Clostridium cellulolyticum*. The construction of *E. coli* strains to selectively produce ethanol is the first successful application of metabolic engineering. The parameters used for screening for xylose fermentation included ethanol tolerance, plasmid stability, and ethanol yields. Under optimum conditions, the xylose-fermenting strains showed maximum ethanol tolerance and productivity of 53–56 g/L and productivity of 0.72 g/L/h (Alterthum and Ingram 1989).

Comparison of glucose/xylose co-fermentation by different recombinant *Z. mobilis* under different genetic and environmental conditions was reported by Ma et al. (2012). They engineered and developed three xylose-fermenting recombinant *Z. mobilis* strains containing different peno-talB/tktA operon terminators. All these strains exhibited similar profile in foreign protein expression and xylose fermentation performance. The study revealed that optimum co-fermentation was achieved at 30°C–34°C and pH 5.5 with an ethanol productivity and yield of 20.5 g/L and 0.43 g/L/h, respectively. The key factors affecting co-fermentation at molecular level were also explored and provide valuable insights for the effective harnessing of biomass resources.

Various process parameters affecting ethanol production from pretreated wheat straw hydrolyzate by *S. cerevisiae* were tested by Singh and Bishnoi (2013). Process parameters affecting ethanol production were first screened using the Plackett–Burman design, and significant parameters like pH, incubation temperature, initial total reducing sugar concentration, and inoculums level were optimized using the Box–Behnken design. Maximum ethanol productivity and yield were 5.6 g/L and 0.43 g/L/h, respectively.

After fermentation, bioethanol has to be recovered from fermentation broth. Since fermentation products are volatile than water, recovery by distillation is the common technology used. Distillation allows economic recovery of volatile products containing high impurities. Distillation system separates the bioethanol from water in the liquid mixture. The product contains approximately 37% ethanol that can be concentrated in a rectifying column to a column just below the azeotrope (95%). It is then passed through stripping column to remove additional water. The recovery of bioethanol from the distillation column in the plant is fixed to 99.6% to reduce bioethanol loss (Balat 2011). Fermentation and distillation are very important steps in bioethanol production.

10.5.3.1 Dehydration

Ethanol is recovered from the fermentation broth by distillation or distillation combined with adsorption. The residue left out is composed of residual lignin, unhydrolyzed cellulose and hemicelluloses, ash, enzyme, organisms, and other components. These residues can be concentrated and burnt for power generation.

After fermentation, distillation is carried out to get 95% ethanol that is known as rectified spirit. The remaining water cannot be removed by distillation since ethanol forms a constant boiling mixture with water known as azeotrope. In a small-scale process, this water can be removed by adding some dehydrants like lime that are capable of separating water from ethanol. But this is mainly used in small-scale processes. Dehydration using molecular sieve is used in industry, and the main advantage of this technology is that the process is simple and very easy to automate, thereby reducing labor. The process is inert and there is no usage of chemicals. A properly designed molecular sieve can dehydrate ethanol near theoretical recovery.

10.6 Various Strategies for Biomass to Bioethanol Production

Several strategies have been developed for enzymatic hydrolysis and fermentation. These include separate hydrolysis and fermentation (SHF), SSF, simultaneous saccharification and co-fermentation (SSCF), and consolidated bioprocessing (CBP). Figure 10.2 describes the various strategies adopted for biomass to bioethanol production.

10.6.1 Separate Hydrolysis and Fermentation

In SHF, the pretreated biomass is hydrolyzed to glucose and subsequently fermented to ethanol in two separate reactors. One of the most important benefits of this strategy is that both hydrolysis and fermentation can be carried out at their optimum conditions. The hydrolysis and fermentation were carried out at their optimum temperature—50°C and 37°C, respectively. One of the main drawbacks of this method

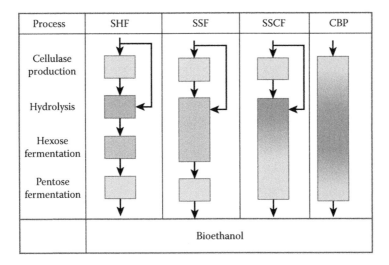

Figure 10.2 Various strategies for biomass to ethanol production.

is feedback inhibition of the cellulase by the end products (Taherzadeh and Karimi 2007). This can be minimized to a certain extent by supplying glucose-tolerant β-glucosidase. Another problem associated with SHF is contamination, and this can be avoided to an extent by the addition of antibiotics. Use of antibiotics will make the process economically unviable.

10.6.2 Simultaneous Saccharification and Fermentation

One of the most successful methods for production of ethanol from lignocellulosic biomass is SSF. In SSF, both hydrolysis and fermentation are carried out in a single reactor where the glucose released after enzymatic hydrolysis is consumed immediately by the fermenting microorganism present in the culture maintaining a low concentration of sugar in the media, thereby reducing the end-product inhibition of cellulase (Binod et al. 2011). The optimum temperature for SSF is maintained at 38°C. Other advantages of SSF include low enzyme requirement, high ethanol yield, lower requirement of sterile conditions, and shorter process time. Though SSF has several benefits, the major drawback of this process is the inhibition of cellulase enzyme by the ethanol produced after fermentation. Another drawback is the partial hydrolysis of pretreated biomass at the end of hydrolysis that leads to adsorption of cellulases with the recalcitrant residue. Another benefit is that the risk of contamination in SSF is lower than in SHF since the presence of ethanol reduces the chances of contamination. The most serious drawback of this strategy is the difference in the optimum temperature between the hydrolyzing enzymes and fermenting microorganism. Inhibition of cellulase by ethanol is another drawback and may be a limiting factor in producing high ethanol concentration.

10.6.3 Simultaneous Saccharification and Co-fermentation

SSCF involves co-fermentation of pentoses and hexoses that is carried out simultaneously. In this process, both hexoses and pentoses are fermented in a single bioreactor with a single microorganism. Hence, a single fermentation step is required to process hydrolyzed and solid fractions of the pretreated lignocellulosic biomass (Taherzadeh and Karimi 2007). Recombinant strains of *Z. mobilis* and *E. coli* are capable of producing ethanol by hydrolyzed SSCF.

10.6.4 Consolidated Bioprocessing

In CBP, ethanol together with all required enzymes is produced in a single bioreactor by a single microorganism community (Taherzadeh and Karimi 2007). One of the most important benefits of this process is that it requires no operational or capital costs for purchasing enzymes or their production. Two strategies are commonly adopted for developing strains for CBP. The first strategy involves modification of efficient ethanol producers so that they become excellent cellulase producers, and the second strategy involves modification of excellent cellulase producers so that they become efficient ethanol producers (Lynd et al. 2005).

10.7 Current Bottlenecks in Economical Production

Lignocellulosic biomass is highly recalcitrant and requires intensive labor and high capital cost for processing. Though several strategies are currently available, none of them seems to be economically viable. Lignocellulosic biomass is a promising feedstock based on its availability and low cost, but a large-scale commercial production of bioethanol from lignocellulosic biomass has yet to materialize.

Effective pretreatment is fundamental for the successful hydrolysis and downstream operations. Pretreatment is the most expensive step due to recalcitrance of the lignocellulosic biomass. Integrated or hybrid strategies need to be carried out for the better removal of hemicelluloses and lignin from the biomass. Strategies with minimum inhibitor generation, more sugar yield, and minimal energy requirement process are to be developed to make the process economically viable.

Enzymatic hydrolysis that converts lignocellulosic biomass to fermentable sugars may be the most complex step in this process due to substrate-related and enzyme-related effects and their interactions. Enzymatic hydrolysis shows the potential for higher yields, higher selectivity, lower energy costs, and milder operating conditions than chemical processes, and the mechanism of enzymatic hydrolysis and the relationship between the substrate structure and function of various glycosyl hydrolase components are not well understood. Hence, limited success has been achieved in maximizing sugar yields at very low cost. For improving the hydrolysis rate, various parameters like the role of hemicelluloses and lignin removal, deacetylation, decrystallization, accessible surface area, and the nature of different cellulose components affecting the cellulase accessibility need to be tested to find the actual mechanism taking place during enzymatic saccharification, and also fine-tuning needs to be done to find out the critical factors and also to understand the physical and chemical factors that contribute to the slow rate of biomass deconstruction to fermentable sugars. Better understanding of these mechanisms helps in the minimization of costs of biomass conversion to meet industrial needs (Yang et al. 2011). Emerging modern tools in biotechnology will help in further discovery and characterization of new enzymes with desirable properties and improvement of enzyme characteristics, and production of homologous and heterologous systems can ultimately lead to low-cost conversion of lignocellulosic biomass into fuels and chemicals. Cellulase engineering through directed evolution, rational design, posttranslational modifications, and their combination may greatly increase cellulase performance and dramatically decrease enzyme use.

Although a number of recombinant bacteria and fungi are currently available to ferment xylose, they are all not capable of acclimatizing to fermentation conditions and produce only very low ethanol yields and tolerance.

Several by-products produced during hydrolysis include furans, carboxylic acids, and phenolic compounds (Palmqvist 1998). Levulinic and formic acids are produced by the degradation of hydroxymethylfurfural (HMF). Formic acid is also produced from methoxy groups in the hemicellulose. Acetic acid is produced by the hydrolysis of acetyl groups in the hemicelluloses due to deacetylation of pentosan (Lawford and Rousseau 1993). Several aromatic and polyaromatic compounds are produced during hydrolysis, and these include phenol, vanillin, vanillic acid, vanillyl alcohol, syringic acid, syringaldehyde, catechol, veratrole, homovanillic acid, cinnamaldehyde, dihydroconiferyl alcohol, 4-hydroxybenzoic acid, and 4-hydroxybenzaldehyde. Degradation products are the most important inhibitors of the hydrolyzates. Inhibition of fermentation takes place by disruption of cellular replication, disruption of sugar metabolism, or disruption of membrane integrity. Hence, intense R&D efforts should be carried out to develop strategies with minimum formation of inhibitors of the hydrolyzate. Minimizing inhibitors will lead to a higher sugar yield obtained by decreasing the formation of by-products with a better fermentability of the hydrolyzate.

10.8 Conclusions

Ethanol, produced from lignocellulosic biomass, is a promising alternative source of renewable energy. The conversion of lignocellulosic biomass to ethanol is limited by structural as well as chemical composition of the biomass that limits the usage of this material as a feedstock for cellulosic ethanol production. Though several pretreatment strategies exist, none has been commercialized for cellulosic ethanol production due to economic feasibility. Few pretreatment strategies like dilute acid pretreatment and steam explosion have been used in some pilot plants. Fine-tuning of pretreatment technologies of different biomass types and developing an economically viable process are still needed. Intense research is going on in this direction to develop pretreatment strategies with maximum fermentable sugar recovery, minimal or no inhibitor generation, and low energy input. Another important aspect is the development of efficient biomass-hydrolyzing enzymes. Due to the heterogeneous nature of the lignocellulosic biomass, the efficiency of hydrolysis of the biomass varies. Also, a suitable cocktail of enzymes consisting of cellulase, xylanase, and other accessory enzymes is needed for complete hydrolysis. Another factor affecting enzymatic hydrolysis is the effectiveness of pretreatment. The chemical and structural modifications occurring in the lignocellulosic biomass during pretreatment have a significant effect on sugar release patterns. Even though the individual impacts of these factors on determining the efficiency of enzymatic hydrolysis have not been fully resolved, many of these factors are found to be interrelated during the saccharification process. Recycling of enzymes also helps in reducing the total bioethanol production cost. To make the biofuel economical, the cost of enzymes should be reduced to 10- to 100-fold. The reduction in the cost of enzymes can be achieved by several ways of R&D like manipulation of organisms or enzymes through modern techniques such as genetic engineering, metabolic engineering, strain improvement, and other related approaches. Another strategy is value addition of by-products of enzyme fermentation. Protein engineering is also an effective method where the amino acid residues are modified so that the enzyme shows improved properties such as high specific activity and improved stability. Directed evolution is another approach that is gaining more attention in recent years. This method combines random mutagenesis of the target gene for a particular enzyme with screening and selection of the desired properties. R&D on the genetic improvement of ethanol-fermenting organisms is also necessary. Native ethanol-fermenting microorganisms like common baker's yeast cannot utilize pentose sugars. So development of genetically improved microorganisms that can utilize both hexose and pentose sugars for ethanol production would definitely improve the process. Hence, there is a tremendous need for R&D activities for identifying pathways to lower the cost of pretreatment and enzymatic saccharification systems as well as for developing engineered co-fermenting, inhibitors, and ethanol-tolerant microbes. An integrated biorefinery concept needs to be used for the production of value-added products from each stream, so as to the make the process economically viable.

References

Alizadeh, H., Teymouri, F., Gilbert, T.I., and B.E. Dale. 2005. Pretreatment of switch grass by ammonia fiber explosion (AFEX). *Appl Biochem Biotechnol* 124:1133–1141.

Alterthum, F. and L.O. Ingram. 1989. Efficient ethanol production from glucose, lactose and xylose by recombinant *Escherichia coli*. *Appl Environ Microbiol* 55:1943–1948.

Amiri, H., Karimi, K., and H. Zilouei. 2014. Organosolvent pretreatment of rice straw for efficient acetone, butanol and ethanol production. *Bioresour Technol* 152:450–456.

Balat, M. 2011. Production of bioethanol from lignocellulosic materials via the biochemical pathway: A review. *Energ Convers Manage* 52:858–875.

Binod, P., Janu, K.U., Sindhu, R., and A. Pandey. 2011. Hydrolysis of lignocellulosic biomass for bioethanol production. In *Biofuels: Alternative Feedstock's and Conversion Processes*, eds. A. Pandey, C. Larroche, and S.C. Ricke. Elsevier, Inc., San Diego, CA.

Carpita, N. and D.M. Gibeaut. 1993. Structural models of primary cell walls in flowering plants: Consistency of molecular structure with the physical properties of the walls during growth. *Plant J* 3:1–30.

Chang, V.S. and M.T. Holtzapple. 2000. Fundamental factors affecting biomass enzymatic reactivity. *Appl Biochem Biotechnol* 84:5–37.

Chen, W., Yu, H., Liu, Y., Chen, P., Zhang, M., and Y. Hai. 2011. Individualization of cellulose nanofibers from wood using high-intensity ultrasonication combined with chemical pretreatments. *Carbohydr Polym* 83:1804–1811.

Cheng, K.K., Cai, B.Y., Zhang, J.A. et al. 2008. Sugarcane bagasses hemicellulose hydrolysate for ethanol production by acid recovery process. *Biochem Eng J* 38:105–109.

Demirbas, A. 2003. Energy and environmental issues relating to greenhouse gas emissions in Turkey. *Energ Convers Manage* 44:201–213.

Dias, A., Toullec, J.L., Blandino, A., de Ory, I., and I. Caro. 2013. Pretreatment of rice hulls with alkaline peroxide to enhance enzyme hydrolysis for ethanol production. *Chem Eng Trans* 32:949–954.

Erickson, T., Borjesson, J., and F. Tjerneld. 2002. Mechanism of surfactant effect in enzymatic hydrolysis of lignocelluloses. *Enzyme Microb Technol* 31:353–364.

Fraiture, C., Giordano, M., and Y. Liao. 2008. Biofuels and implications for agricultural water use: Blue impacts of green energy. *Water Pol* 10:67–81.

Guragain, Y.N., De Coninck, J., Husson, F., Durand, A., and S.K. Rakshit. 2011. Comparison of some new pretreatment methods for second generation bioethanol production from wheat straw and water hyacinth. *Bioresour Technol* 102:4416–4424.

Himmel, M.E., Adney, W.S., Baker, J.O., Nieves, R.A., and S.R. Thomas. 1996. Cellulases structure, function and applications. In *Handbook on Bioethanol: Production and Utilization*, ed. C.E. Wyman, pp. 143–161. Taylor & Francis, Boca Raton, FL.

Holtzapple, M.T., Ripley, E.P., and M. Nikolaou. 1994. Saccharification, fermentation, and protein recovery from low-temperature AFEX-treated coastal Bermuda grass. *Biotechnol Bioeng* 44:1122–1131.

Hu, G., Heitmann, J.A., and O.J. Rojas. 2008. Feedstock pretreatments strategies for producing ethanol from wood, bark, and forest residues. *BioResources* 3:270–294.

Hu, J., Arantes, V., Pribowo, A., and J.N. Saddler. 2013. The synergistic action of accessory enzymes enhances the hydrolytic potential of a "cellulase mixture" but is highly substrate specific. *Biotechnol Biofuels* 6:1–12.

Ingram, L.O., Conway, T., Clark, D.P., Sewell, G.W., and J.F. Preston. 1987. Genetic engineering of ethanol production in *Escherichia coli*. *Appl Env Microbiol* 53:2420–2425.

Klinke, H.B., Thomsen, A.B., and B.K. Ahring. 2004. Inhibition of ethanol-producing yeast and bacteria by degradation products produced during pre-treatment of biomass. *Appl Microbiol Biotechnol* 66:10–26.

Lawford, H.G. and J.D. Rousseau. 1993. Effect of pH and acetic acid on glucose and xylose metabolism by a genetically engineered *Escherichia coli*. *Appl Biochem Biotechnol* 39/40:301–322.

Lee, D., Owens, V.M., Boe, A., and P. Jeranyama. 2007. Composition of herbaceous biomass feed stocks. Publication SGINC1-07. South Dakota State University, Brookings, SD.

Limayem, A. and S.C. Ricke. 2012. Lignocellulosic biomass for bioethanol production: Current perspectives, potential issues and future prospects. *Prog Energ Combust Sci* 38:449–467.

Lloyd, T.A. and C.E. Wyman. 2005. Combined sugar yields for dilute sulfuric acid pretreatment of corn stover followed by enzymatic hydrolysis of the remaining solids. *Bioresour Technol* 96:1967–1977.

López-Abelairas, M., Lu-Chau, T.A., and J.M. Lema. 2013. Fermentation of biologically pre-treated wheat straw for ethanol production: Comparison of fermentative microorganisms and process configurations. *Appl Biochem Biotechnol* 170:1838–1852.

Lynd, L.R., van Zyl, W.H., McBride, J.E., and M. Laser. 2005. Consolidated bioprocessing of cellulosic biomass: An update. *Curr Opin Biotechnol* 16:577–583.

Ma, Y., Dong, H., Zou, S., Hong, J., and M. Zhang. 2012. Comparison of glucose/xylose co-fermentation by recombinant *Zymomonas mobilis* under different genetic and environmental conditions. *Biotechnol Lett* 34:1297–1304.

Mosier, N., Wyman, C., Dale, B. et al. 2005. Features of promising technologies for pretreatment of lignocellulosic biomass. *Bioresour Technol* 96:673–686.

Palmqvist, E. 1998. Fermentation of lignocellulosic hydrolysates: Inhibition and detoxification. PhD dissertation, Lund University, Lund, Sweden.

Parawira, W. and M. Tekere. 2011. Biotechnological strategies to overcome inhibitors in lig-nocellulose hydrolysates for ethanol production: Review. *Crit Rev Biotechnol* 31:20–31.

Park, J.M., Oh, B., Seo, J. et al. 2013. Efficient production of ethanol from empty palm fruit bunch fibers by fed-batch simultaneous saccharification and fermentation using *Saccharomyces cerevisiae*. *Appl Biochem Biotechnol* 170:1807–1814.

Pettersen, R.C. 1984. The chemical composition of wood. In *The Chemistry of Solid Wood*, Advances in Chemistry Series, ed. R.M. Rowell, pp. 115–116. Washington, DC: American Chemical Society.

Ravindranath, N.H., Somashekar, H.I., Nagaraja, M.S. et al. 2005. Assessment of sustainable non-plantation biomass resources potential for energy in India. *Biomass Bioenerg* 29:178–190.

Sankh, S.N., Deshpande, P.S., and A.U. Arvindekar. 2011. Improvement of ethanol production using *Saccharomyces cerevisiae* by enhancement of biomass and nutrient supplementation. *Appl Biochem Biotechnol* 164:1237–1245.

Shang, Y., Su, R., Huang, R. et al. 2014. Recycling cellulases by pH-triggered adsorption-desorption during the enzymatic hydrolysis of lignocellulosic biomass. *Appl Microbiol Biotechnol* 98:5765–5774.

Sindhu, R., Kuttiraja, M., Binod, P., Sukumaran, R.K., and A. Pandey. 2014a. Bioethanol production from dilute acid pretreated Indian bamboo variety (*Dendrocalamus* sp.) by separate hydrolysis and fermentation. *Indus Crop Prod* 52:169–176.

Sindhu, R., Kuttiraja, M., Binod, P., Sukumaran, R.K., and A. Pandey. 2014b. Physicochemical characterization of alkali pretreated sugarcane tops and optimization of enzymatic saccharification using response surface methodology. *Renew Energ* 62:362–368.

Singh, A. and N.R. Bishnoi. 2013. Ethanol production from pretreated wheat straw hydro-lyzate by *Saccharomyces cerevisiae* via sequential statistical optimization. *Indus Crop Prod* 41:221–226.

Sukumaran, R.K., Surender, V.J., Sindhu, R. et al. 2010. Lignocellulosic ethanol in India: Prospects, challenges and feedstock availability, *Bioresour Technol* 101:4826–4933.

Sun, Y. and J. Cheng. 2002. Hydrolysis of lignocellulosic materials for ethanol production: A review. *Bioresour Technol* 83:1–11.

Taherzadeh, M.J. and K. Karimi. 2007. Enzyme based hydrolysis processes for ethanol from lignocellulosic materials: A review. *BioResources* 2:707–738.

Tu, M., Chandra, R.P., and J.N. Saddler. 2007. Recycling cellulase during the hydrolysis of steam exploded and ethanol pretreated lodge pole pine. *Biotechnol Prog* 65:1130–1137.

Wan, C. and Y. Li. 2011. Effectiveness of microbial pretreatment by *Ceriporiopis subvermis-pora* on different biomass feedstocks. *Bioresour Technol* 102:7507–7512.

Yang, B., Dai, Z., Ding, S., and C.E. Wyman. 2011. Enzymatic hydrolysis of cellulosic bio-mass. *Biofuel* 2:421–450.

Yang, B., Willies, D.M., and C.E. Wyman. 2006. Changes in the enzymatic hydrolysis rate of avicel cellulose with conversion. *Biotechnol Bioeng* 94:1122–1128.

Zhang, Y.H.P. 2008. Reviving the carbohydrate economy via multi-product biorefineries. *J Indus Microbiol Biotechnol* 35:367–375.

Zhu, J.Y. and H.J. Pan. 2010. Woody biomass treatment for cellulosic ethanol production: Technology and energy consumption evaluation. *Bioresour Technol* 101:4992–5002.

11

Process Design, Flowsheeting, and Simulation of Bioethanol Production from Lignocelluloses

Mohsen Ali Mandegari, Somayeh Farzad, and Johann F. Görgens

Contents

Abstract

Bioethanol is by far the most widely used biofuel for transportation worldwide. In the second generation of bioethanol process, lignocellulosic materials, which are abundant and renewable agricultural wastes, are used to produce bioethanol. Establishment of a few commercial facilities for lignocellulose conversion to bioethanol has proven that the main obstacles of this pathway have been overcome, but more efforts are required to develop this technology for a variety of feedstocks and conditions. Ongoing improvement and enhancement of the conversion processes are expected in the future, due to the development of pertinent technologies and sciences. In this chapter, process design, flowsheeting, and simulations of bioethanol production from lignocellulosic material through fermentation of sugars from lignocellulose hydrolysis will be discussed, taking into account the fundamentals of chemical engineering, bioprocessing, and similar simulations in the petrochemical industry.

Feedstock and plant capacity are the most important components of the design basis, which are related to the type and availability of feedstock. Flowsheeting and simulation are the key elements of the bioethanol process design from lignocellulosic material. The flowsheet design and development procedure will be described using block flow diagram and process flow diagram for bioethanol production (pretreatment, hydrolysis, fermentation, and seed train). The supplementary units of a comprehensive facility for bioethanol production, such as ethanol purification, evaporation, wastewater treatment, and power/steam generation, will be addressed. Simulation development is a crucial step in the process development but requires a realistic approach to the bioprocess simulation. Ultimately techniques such as the process, energy, and water integration will contribute significantly to minimize the capital and operating cost of the plant. The approach and guidelines represented in this chapter are not only limited to the application for bioethanol production but can also be used for the simulation of other biobased processes.

11.1 Introduction

Biofuels—liquid and gaseous fuels derived from organic material—can play an important role in reducing the CO_2 emissions by the transport sector, while enhancing energy security. Biofuels have attracted growing global interest, with some governments announcing commitments to biofuel programs as a way to reduce both greenhouse gas (GHG) emissions and dependence on petroleum-based fuels. The United

States, Brazil, and several EU member states have the largest programs promoting biofuels in the world (Chum et al. 2014; Eggert and Greaker 2014; Tao et al. 2014).

The transportation sector consumes a substantial portion of fossil fuel resources and contributes greatly to atmospheric pollution by releasing GHGs, which are also responsible for the biggest environmental concern today, namely, global warming. With rapid depletion of fossil fuel resources and rising levels of GHGs, it has become vital that these are replaced by alternative fuels that are less hazardous to the environment and also have the same efficiency as conventional fuels (Balat 2007). However, to compete with conventional fuels, the alternative biofuels need to be improved in terms of their properties, production efficiency, and end-use suitability (Sangeeta et al. 2014). Alcohols have been used as fuels since the inception of the automobile. Fuel ethanol blends are successfully used in all types of vehicles and engines that require gasoline, while also having potential for use in compression ignition engines, as blends either with diesel or in ethanol engines. Ethanol can be produced from a variety of materials such as grain, molasses, fruit, cobs, and shell, and its production (excluding that of beverages) has been increasing since the 1930s because of the low cost. With the oil crises of the 1970s, ethanol became established as an alternative fuel, and it has been considered as one possible alternative fuel in some countries since the 1980s (Balat et al. 2008).

Conversion technologies for producing bioethanol from cellulosic biomass resources such as forest materials, agricultural residues, and urban wastes are under development. Although a few commercial plants for cellulosic bioethanol production have been established recently, significant research and development (R&D) is ongoing, to extend large-scale bioethanol production plants (Chum et al. 2014). This chapter will provide insights into flowsheeting and simulation of bioprocessing for the process of bioethanol production from lignocellulose through fermentation of sugars from lignocellulose hydrolysis. Different aspects of a bioethanol production process and their incorporation into a comprehensive and realistic simulation of the conversion process will be discussed.

11.2 Design Basis of Bioethanol Process

Design basis includes specifying and defining the fundamental parameters of the plant. The feedstock characteristics in terms of type of lignocellulose (hard wood, softwood, agricultural, etc.), chemical composition (amount of carbohydrate, lignin, and ash) and available rate of supply for process are the first process parameters to be defined (Sánchez and Cardona 2008; Cherubini 2010). The capacity of the plant as a function of feedstock availability also should be defined (Hess et al. 2007; Seabra et al. 2010). Geographical location of the plant affects some of the design basis parameters such as ambient condition (minimum, maximum, and average temperatures; pressure; and humidity), which will be considered for cooling medium system, equipment design, and other relevant items (Ludwig 1997).

11.2.1 Feedstock

An important stage in bioethanol process is the provision of a renewable, consistent, and regular supply of feedstock. Mechanical processing of feedstock may be required to increase its energy density in order to reduce costs of transport, handling, and storage. The feedstock type and composition have significant impacts on the overall process design and economics. Feedstock type also influences the design of key components in the conversion process, for example, the pretreatment reactor and amount of desired component production. Renewable carbon-based raw materials for bioethanol production are provided from three different sectors (Cherubini 2010):

1. Agriculture (dedicated crops and residues)
2. Forestry
3. Industries (process residues and leftovers)

Lignocellulosic or fibrous plant biomass is composed of carbohydrate polymers (cellulose and hemicellulose), lignin, and a smaller part (extractives, acids, salts, and minerals). The cellulose and hemicellulose, which typically comprise two-thirds of the lignocellulose dry mass, are polysaccharides that can be hydrolyzed to sugars and eventually be fermented to ethanol. Lignin cannot be used to supply sugar for ethanol fermentation (Hamelinck et al. 2005). However, it can be gasified to syngas, which can be used in the synthesis of biological process to produce ethanol (Ragauskas et al. 2014).

There is considerable diversity in the types of biomass feedstock that can be used for ethanol production, depending on geographical location. Ethanol production from some feedstocks such as corn is highly developed, while others such as industrial wastes are in an earlier stage of development. In addition, there are different technologies available for the conversion and separation of these materials into biofuels, including hydrolysis–fermentation and (thermos) chemical and biocatalytic treatment. Consequently, a wide variety of products can be produced with differing values and markets (Alvarado-Morales et al. 2009).

11.2.2 Plant Capacity

Capacity is one of the most important parameters of the plant design, which directly affects its economic viability. However, availability of feedstock is the main constraint to define the plant size. Feedstock availability depends on biomass type, cultivation, and harvesting cycles and differs from one region to another depending on the climate. For instance, the sugarcane cultivation season in South Africa starts in April, being 8–9 months long, and finishes in November/December, thus providing for sugarcane milling of approximately 200 days per year, with most mills averaging approximately 300 t/h or 1.5 Mt/year of cane per annum (Smith et al. 2014; Smithers 2014). The average capacity of a Brazilian sugar mill is 4 Mt/year. Increased transport distances for a larger supply for biomass will also increase the cost of the feedstock, indicating an optimum point between cost of feedstock supply and economics of scale in feedstock conversion to biofuel (Seabra et al. 2010). According to sensitivity analysis carried out by the National Renewable Energy Laboratory, only modest cost reductions due to economies of scale could be achieved beyond 2000 t/day feedstock (Aiden et al. 2002). However, the typical feedstock capacity of the limited number of commercial cellulosic ethanol production plants is about 750–1000 t/day (Banerjee et al. 2010; Chum et al. 2014).

11.3 Bioethanol Flowsheet Design and Development

Process flowsheet can simply be defined as the blueprint of a production plant, or part thereof. It identifies the main feed streams, unit operations, process streams that interconnect the unit operations, and, finally, the product streams. Operating conditions and some technical specifications are included, depending on the detail level of the flowsheet. The complexity of the flowsheet can vary from a rough sketch to a very detailed design specification of a complex plant. In Figure 11.1, a general flowchart for process design and development is shown, which can be used for development of the bioethanol process.

11.3.1 Block Flow Diagram

A block flow diagram (BFD) is a simple flowchart that represents an overview of the process or system and is used to simplify a complex process, without giving a

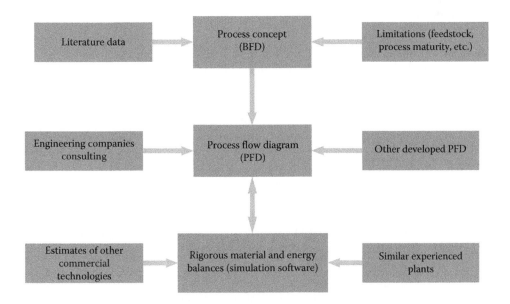

Figure 11.1   General approach for the development of process flow diagram.

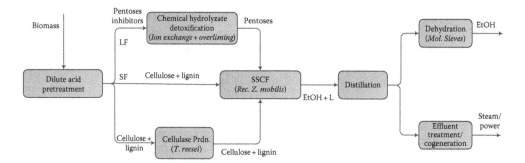

Figure 11.2 A schematic BFD of the bioethanol production. (Adapted from Wooley, R. et al., *Biotechnol. Prog.*, 15, 794, 1999.)

detailed understanding of the process. Each step or activity in a BFD is represented by a rectangle that is descriptive of the process it represents. The rectangles are connected with arrows, which represent the flow of material/energy between the process steps. Alternatively the arrows represent the sequence of the steps or relationships between steps.

Based on the complexity, bioethanol production can be limited to pretreatment, hydrolysis, fermentation, and distillation/purification, or it can be completed by the addition of other units such as feedstock handling, enzyme production, WWT, boiler and steam/power generation, feed/product/chemical storage, and other required utilities (cooling water, chilled water, etc.). In Figure 11.2, a schematic BFD of bioethanol production based on dilute acid pretreatment is illustrated.

11.3.2 Process Flow Diagram

A process flow diagram (PFD) is a diagram commonly used in chemical and process engineering to indicate the general flow of plant processes and equipment. PFD displays the relationship between the major equipment of a plant and does not show minor details such as piping details and designations. Another commonly used term for a PFD is *flowsheet*. Typically, PFDs of a single-unit process will include the following:

1. Process piping
2. Major equipment items
3. Control valves and other major valves
4. Connections with other systems
5. Major bypass and recirculation streams
6. Operational data (temperature [T], pressure [P], mass flow rate, density, etc.), often by stream references to a mass balance
7. Process stream names

PFD is completed by the operating detail of streams such as temperature, pressure, component mass/mole flows, and the main operating parameters of equipment.

As can be seen in Figure 11.1, there is a bilateral relation between PFD and simulation. The data of equipment and input streams will be used in the simulation in order to calculate the unknown condition (temperature, pressure, flow, composition, work, heat flows, etc.) in the plant.

11.3.2.1 Bioethanol Production

The minimum essential process steps for bioethanol production are pretreatment, conditioning (in some cases), hydrolysis, and fermentation. Prior to the enzymatic hydrolysis process, it is necessary to pretreat the material to make the cellulose accessible for the enzymes. In an efficient pretreatment method, both the material loss and coproduct formation are minimal, and a high reaction rate with high sugar yields is achieved. Consequently, pretreatment is the most important and challenging process in the production of cellulosic ethanol. The pretreatment techniques can be divided into four different categories as mechanical, mechanical–chemical, chemical, and biological (Alvira et al. 2010; Mood et al. 2013).

As the pretreatment is finished, the cellulose is prepared for hydrolysis, which means the cleaving of a cellulose molecule by adding a water molecule. This reaction (cellulose to glucose) is catalyzed by dilute acid, concentrated acid, or enzymes. Enzymatic hydrolysis is the critical step in bioethanol production where complex carbohydrates are converted to simple monomers. It requires less energy and milder process conditions than acid hydrolysis (Rabelo et al. 2011; Gupta and Verma 2015). Since enzymatic hydrolysis of native lignocellulose usually results in solubilization of the glucan, some form of pretreatment to increase amenability to enzymatic hydrolysis is included in most process concepts for biological conversion of lignocelluloses (Balat et al. 2008). Where enzymatic hydrolysis is applied, different levels of process integration are possible, mentioned hereafter.

The first application of enzymes for wood hydrolysis in an ethanol process is to simply replace the cellulose acid hydrolysis step with a cellulase enzymatic hydrolysis step, the so-called separate (or sequential) hydrolysis and fermentation (SHF). In the SHF configuration, the liquid flow from hydrolysis reactor first enters into the glucose fermentation reactor. The mixture is then distilled to remove the ethanol, leaving the unconverted xylose behind. In the second reactor, xylose is fermented to ethanol, and ethanol is again distilled. The cellulose hydrolysis and glucose fermentation may also be located parallel to the xylose fermentation (Hamelinck et al. 2005).

Simultaneous saccharification and fermentation (SSF) consolidates hydrolyses of cellulose with the direct fermentation of the produced glucose. This reduces the number of reactors involved, by eliminating the separate hydrolysis reactor. It also avoids the problem of product inhibition associated with enzymes. In SSF, there is a trade-off between the cost of cellulase production and the cost of hydrolysis/fermentation (Hamelinck et al. 2005; Olofsson et al. 2008). Major advantages of SSF include (1) increase of hydrolysis rate by conversion of sugars that inhibit the cellulase activity, (2) lower enzyme requirement, (3) higher product yields, (4) lower requirements for sterile conditions since glucose is removed immediately and bioethanol is produced, (5) shorter process time, and (6) less reactor volume (Sun and Cheng 2002; Balat et al. 2008).

More recently, the SSF technology has proved advantageous for the simultaneous fermentation of hexose and pentose, the so-called simultaneous saccharification and cofermentation (SSCF). In SSCF, the enzymatic hydrolysis continuously releases hexose sugars, which increases the rate of glycolysis such that the pentose sugars are fermented faster and with higher yield. SSCF is preferred since both unit operations (glucose and xylose fermentations) can be performed in the same tank, resulting in lower costs (Mosier et al. 2005; Balat et al. 2008).

In consolidated bioprocessing (CBP), ethanol and all required enzymes are produced by a single-microorganism community, in a single reactor. CBP as a promising approach will circumvent the cost and restrictions of conventional workflow for biofuel production from lignocellulosic biomass and has been investigated increasingly in recent years (Salehi Jouzani and Taherzadeh 2015). CBP technologies using a single organism or consortium of microorganisms combine the enzyme production, hydrolysis, and fermentation stages into a single step. This may enhance processing efficiencies, eliminating the need for added exogenous hydrolytic enzymes and reducing the sugar inhibition of cellulases. CBP systems significantly reduce the number of unit operations (i.e., fewer reactor vessels) and, therefore, reduce maintenance and capital costs (Lynd et al. 2005; Xu et al. 2009). Nonetheless, CBP is still in its early stage of establishment, and hence, more effort is required for it to be practically developed.

A variety of available feedstocks, different pretreatment, and integration of hydrolysis and fermentation result in numerous PFD configuration for bioethanol production. In the following sections, other supplementary units are described comprehensively.

11.3.2.2 Bioethanol Purification and Recovery

The purification and recovery section separates water, anhydrous ethanol, and combustible solids from the fermentation broth. Ethanol distillation is a widely investigated topic in literature, due to its high impact on total energy consumption of the plant. Research studies focus on energy consumption, dehydration techniques, and alternative control strategies for the separation of the ethanol–water binary mixture as well as on the influence of minor components (Dias et al. 2011; Batista et al. 2012; Errico et al. 2013a).

Although many attempts are carried out to develop an efficient bioethanol purification process, in all studies, the main equipment configuration is almost the same. In Figure 11.3, a typical process of bioethanol purification is shown, and the process is described accordingly. There is no solid–liquid separation between fermentation and distillation. Fermentation broth from a SSF process will be stored in the beer tank, and also the outlet water of scrubber column is fed to the beer tank (beer tank is not shown in Figure 11.3). The beer tank is designed for a residence time of 4 h to provide surge capacity between fermentation and distillation (Humbird et al. 2011). Typically beer (culture broth) from the fermentation area (~5 wt% ethanol) is fed to the beer column after passing through an economizing heat exchanger. The beer column contains 32 stages operating at 48% efficiency, with the feed entering the fourth tray from the top. The required molar reflux ratio is 3:1.

The overhead of beer column is vented to the scrubber and contains 85% CO_2, 11% ethanol, and close to 4% water. Most of the ethanol in the vent stream (99%) will be

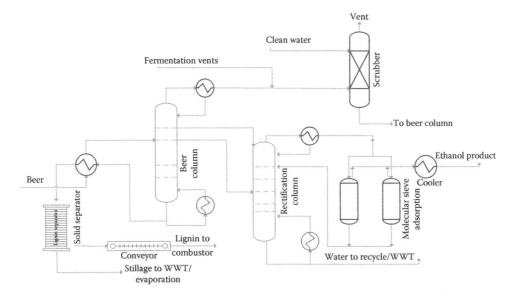

Figure 11.3 A typical process of bioethanol purification. (From Humbird, D. et al., Process design and economics for biochemical conversion of lignocellulosic biomass to ethanol, National Renewable Energy Laboratory [NREL], Golden, CO, 2011.)

recovered in the vent scrubber and recycled to the beer tank. In addition, about 0.8% of the ethanol fed to the beer column is lost in the bottom stream and is considered as permanent loss. More than 99% of ethanol in the feed is removed as a vapor side draw from the column at tray 8 with 40 wt%. The vapor side draw from beer column is fed directly to the rectification column on tray 33 (counting from the top). This column uses 45 stages with an efficiency of 76%.

The recycled material from the molecular sieve, which is enriched in ethanol (72 wt% versus 40 wt%), is fed on tray 14 of rectification column. The required molar reflux ratio is 3.5:1 to obtain a vapor overhead mixture of 92.5% w/w ethanol and a bottom composition of 0.05% w/w ethanol. The rectification column bottom stream is recycled to the pretreatment reactor as dilution water; ethanol in this stream is not considered as loss. The ethanol-containing vent from the beer column overhead, along with the vents from the fermenters (fermentation and seed training), is sent to a water scrubber. The scrubber column recovers 99% of the vented ethanol, while its effluent contains 1.8 wt% ethanol and is returned to the beer well (Humbird et al. 2011). It should be considered that these mentioned data of purification represent a case study. Although the general design of the process is valid, parameters should be adjusted to study other cases.

Due to the presence of binary azeotrope of ethanol–water (95.63 wt% ethanol), ethanol purification should be proceeded via two steps. The first step is typically an ordinary distillation called preconcentration stage (beer and rectification columns in Figure 11.4) that concentrates bioethanol up to the level of 92–94 wt%. The second step is even more complex to dehydrate ethanol up to higher concentrations above the azeotropic composition. Several alternatives are available: pre-evaporation,

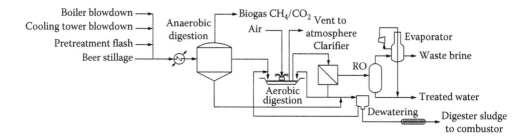

Figure 11.4 A simplified flow diagram of WWT process. (From Humbird, D. et al., Process design and economics for biochemical conversion of lignocellulosic biomass to ethanol, National Renewable Energy Laboratory [NREL], Golden, CO, 2011.)

adsorption, pressure-swing distillation, extractive distillation (ED), azeotropic distillation, and combined hybrid methods (Kiss and Suszwalak 2012), while some efforts were carried out regarding the study of different issues of highly purified ethanol production (Dias et al. 2011; Mulia-Soto and Flores-Tlacuahuac 2011; Errico et al. 2013b). Adsorption with molecular sieves recently became more popular as it requires less energy compared to distillation. However, the desorption step requires high temperature and/or low pressure, thus leading to higher overall equipment costs (Kiss and Suszwalak 2012).

Due to the high maturity of adsorption, the molecular sieve is described hereafter for the ethanol production as illustrated in Figure 11.3. The overhead vapor from rectification column is fed to a molecular sieve adsorption unit, which is a system of columns packed with beds of adsorbent. Water is selectively adsorbed in the beds as the vapor flows through. Saturated vapor from the rectification column is superheated and fed to one of two adsorption columns. The adsorption column removes 95% of the water and a small amount of ethanol. The 99.5% pure ethanol vapor is cooled by heat exchange with the regeneration condensate and finally condensed with cooling water and pumped to storage. While one bed is adsorbing water, the other is being regenerated. The bed is regenerated by passing a small slipstream of pure ethanol vapor back through the water saturated bed while vacuum is applied. The ethanol strips the water off the adsorbent, and the mixture is condensed and returned to the rectification column. The design specifics of this unit general arrangement are mentioned in the literature (Humbird et al. 2011).

The beer column bottom (stillage) stream contains the unconverted insoluble and dissolved solids, which are dewatered by a pressure filter and sent to the combustor. The pressed stillage water is directed to a WWT for cleanup and is used for reutilization in the process, if it contains high levels of organic salts and soluble inorganic compounds. Otherwise, it can be routed to an evaporation unit for more treatment.

11.3.2.3 Evaporation Unit

Evaporation unit is an important unit in which stillage water is purified to be used in the process. Since the amount of stillage water that can be recycled is limited, evaporation unit is more effective than WWT unit. In this context, a multiple effect evaporator is applied to treat the stillage water rather than purifying the stillage water in the WWT unit. The evaporator produces syrup that is burned, in lieu of being sent to the WWT system. The addition of the evaporator does not necessarily reduce the cost. However, the savings in WWT system costs might outweigh the added cost of the evaporator (Wooley et al. 1999a). It is worth noting that when the salt concentrations in the syrup increase, boiler vendors are not comfortable quoting a combustor system to handle it due to the issue of burner fouling (Humbird et al. 2011). Hence, in the case of high-salt production during pretreatment, the stillage evaporators are eliminated, and the stillage is routed to the WWT instead.

The solid–liquid separation of the stillage is carried out with a Pneumapress pressure filter, and the solid residue with a water-insoluble solid concentration of 40% is fed to the dryer. The liquid fraction is concentrated to 50% dry matter (DM) in a multiple-effect evaporation system consisting of three to five effects in a forward-feed arrangement. The key point of the pressure adjustment for intermediate evaporation stages is to have the same size for all evaporators; hence, in the case of a three-stage evaporation, two units are used for each of the first and third effects, whereas one unit is used for the second effect (Aiden et al. 2002). According to steam level definitions, low- or medium-pressure steams are used as heating medium in the first effect, and the produced vapor in the first evaporation is used as heating medium for the next evaporation. The pressure of intermediate effects (considering the number of applied effects) is decreased to reach a pressure of 0.2 bar in the last effect.

Part of the evaporation condensate is recycled to the pretreatment step, and the rest is sent to the WWT facility. Vacuum pumps for the removal of noncondensable gases and a stripper column for the purification of the condensate can be considered (Olsson and Zacchi 2001). In a typical grain-to-ethanol facility, the ratio of the recycled water can be limited to as low as 10% of the centrifuge filtrate stream to the process. Recycling a higher percentage of the treated water to pretreatment may result in levels of ionic strength and osmotic pressures that can be detrimental to the fermenting organism's efficiency. The water that is not recycled is fed to the evaporator, to concentrate the solids in syrup for feeding to the burner, thus minimizing load to the WWT (Wooley et al. 1999b).

11.3.2.4 Wastewater Treatment Unit

The ethanol process generates a number of wastewater streams that must be treated before being recycled to the process or release to the environment. This is accomplished in the WWT unit. The inlet streams to the WWT are condensed pretreatment flash vapor, condensate from the hydrolyzate pneumatic press, rectification column bottom flow, boiler blowdown, cooling tower blowdown, clean-in-place (CIP) waste, and the nonrecycled evaporator condensate. The treated water is assumed clean and fully reusable by the process, which reduces both the fresh makeup water requirement and discharge to the environment. As mentioned in Section 11.3.2.3, when the evaporation unit is not considered for stillage water treatment, WWT should handle the whole produced wastewater plus stillage water; consequently, the size of WWT increases significantly and more facilities (reverse osmosis [RO] and evaporator) are required.

A simplified flow diagram of WWT process is presented in Figure 11.4. The combined wastewater stream is processed by anaerobic digestion and aerobic digestion to digest organic matter in the stream. The flash streams from the pretreatment together with part of the evaporation condensate are cooled to 55°C and treated by anaerobic digestion, in which 50% of the chemical oxygen demand (COD) is assumed to be removed. Methane production is supposed to be 0.35 m³/kg COD consumed, which is used in the steam boiler (Aiden et al. 2002; Humbird et al. 2011). Most of the remaining COD is removed in the subsequent aerobic step (Sassner et al. 2008). Anaerobic digestion produces biogas stream rich in methane, so it is fed to the combustor. According to a developed process, considering the evaporation unit, 3.5 kg methane per ton of DM feedstock is produced, while this value will be 65 without an evaporation unit. Aerobic digestion produces relatively clean water stream that can be reused in the process as well. A sludge that is primarily composed of cell mass is also burned in the combustor. The treated water has a COD of about 0.5 g/L and is routed to an RO membrane system for salt removal (Humbird et al. 2011). Brine is assumed to be wasted, and treated water is used for the water cycle system.

11.3.2.5 Combustor, Boiler, and Power Generation

The purpose of combustor, boiler, and turbo generator subsystem is to burn various organic by-product streams to produce steam and electricity. Combustible by-products include all of the lignin and the unconverted cellulose and hemicellulose from the feedstock, biogas from anaerobic digestion, biomass sludge from WWT, and also concentrated syrup from the evaporator (in the case of using the evaporation unit). However, in some cases, a dryer may be used to evaporate a portion of boiler feed water vapor (He and Zhang 2011; Quintero et al. 2013). Burning these by-product streams to generate steam and electricity allows the plant to be self-sufficient in energy (thermal neutral), reduces solid waste disposal costs, and generates additional revenue through sale of excess electricity.

The fuel streams are fed to a combustor that is capable of handling wet solids, while a fan moves air into the combustion chamber. Treated water enters the heat exchanger circuit in the combustor and is boiled and superheated to high-pressure steam. A multistage back pressure turbine and generator are used to generate electricity. Steam is extracted from the turbine at different conditions that are mostly low- and high-pressure steam to supply process heat demands and steam consumptions. In the last stage, a condensing turbine is used in which the remaining steam is taken down to vacuum and condensed with cooling water for maximum energy conversion. The condensate is returned to the boiler feed water system along with the condensate from the various heat exchangers in the process. The steam turbine uses a generator that produces AC electricity for all users in the plant. The remaining electricity is assumed to be sold back to the grid, providing a coproduct credit.

Boiler efficiency is the most important parameter of the boiler, defined as the percentage of the feed heating value that is converted to steam heat (enthalpy). Boiler efficiency depends on the maximum pressure of produced steam and applied technology for manufacturing of burner, tubes, economizer, and so on. A boiler efficiency of 80% is achievable to produce steam with 60 bar and 450°C. The generator efficiency is another parameter that has a significant effect on the generated power. Considering 85% isentropic efficiency for turbo generators leads to a realistic number. To reach maximum power recovery from produced steam, a condensing turbine is used with an outlet pressure of 0.1–0.2 bar (Humbird et al. 2011).

11.3.2.6 Utility

Utility unit is designed to summarize all the utilities required by the ethanol production facility (except for steam and electricity, which is provided by steam and power generation unit). Utility unit tracks cooling water, chilled water, plant and instrument air, process water, and the CIP system. The process water manifold in this area mixes freshwater with treated wastewater (assumed suitable for all plant users) and provides this water a constant pressure into the facility. Water is provided to the pretreatment, cellulase production, boiler and cooling tower makeup, the CIP system, and the vent scrubber. Freshwater is also mixed with some internally recycled water for dilution before pretreatment and enzymatic hydrolysis.

The plant and instrument air systems provide compressed air for general use (pneumatic tools and cleanup) and instrument operation. Larger users of compressed air, namely, the stillage filter press and cellulase system, have their own compressors specified. The CIP system provides hot cleaning and sterilization chemicals to hydrolysis and fermentation, the enzyme production section, and the distillation system. The utility unit is considered whenever a comprehensive design of biorefinery is required; otherwise, the design of this unit may be ignored and its cost can be considered as a portion of the remaining plant cost.

11.4 Process Simulation

Process simulation is a model-based representation of chemical, physical, biological, and other technical processes as well as unit operations with software, for the design development, analysis, and optimization purposes (Luyben 1989). Process simulation provides an *in silico* model of the operation of an industrial production facility and contains a particular type of processing, such as biological conversion of lignocellulose to bioethanol. Basic prerequisite for calculation of the process parameters in such a facility is a thorough knowledge of chemical and physical properties of pure components and mixtures, reactions, equilibrium data, and mathematical models. Process simulation software describes processes in flowsheets where specific unit operations (steps in the process) are positioned and connected by streams. The software has to solve mass and energy balances around each of these unit operations as well as various combinations of units, to find a stable operating point. The goal of a process simulation is to find optimal conditions for an examined process. In addition, simulated process can be analyzed further in the aspects of energy, economy, and sustainability.

In order to study any process at steady-state condition, a finite set of algebraic equations should be solved. For a simple example with only one reactor with appropriate feed and product streams, the number of equations may be handled using manual calculations or simple computer applications. However, considering distillation columns, heat exchangers, absorbers with many purge, and recycle streams, as is the case with any industrial production plant, the increase in the complexity of a flowsheet, and the number of equations easily approaching many ten thousands, manual solution of algebraic equations is impractical.

Some researchers prefer mathematical representation of the mass and energy balance of a process simulation in a programming language such as MATLAB®, FORTRAN, or C++, although these program scans typically model only a limited number of the main equipment items (especially reactors). However, for process optimization, the effects of all parameters and equipment in terms of mass and energy balances, energy consumption, and economic and environmental effects should be considered.

Hence, applying suitable process simulation software for process modeling is inevitable. In order to solve these large equation sets, a number of computer applications called process flowsheet simulators have been developed, which have highly refined user interfaces and online component databases. These simulators have a number of shortcomings such as lack of detailed interaction with the problem. Although it omits the complexity of the problem, some important concepts of the problem might be missed by the user.

The use of a process flowsheet simulator is beneficial in R&D, design, and operation of an industrial production plant. The use of simulation software in R&D cuts down the process on laboratory experiments and pilot plant runs, while in design stage, it enables a speedier development with simpler comparisons of various alternatives. In the operation stage, it provides a practical means for a risk-free analysis of various what-if scenarios.

11.4.1 Bioethanol Simulation Development

The flowsheet for lignocellulose conversion to bioethanol cannot be developed without using process simulators, due to numerous available pathways, technologies, unit operations, and parameters. Although many simulation package software such as Aspen Plus™, Hysys, PROII, ProSimPlus, CHEMCAD, SimSci, and DWSIM have been developed for chemical process simulation, only a few such as PROII (Alvarado-Morales et al. 2009), UniSim Design (Dias et al. 2009), Hysys (Haelssig et al. 2008), IPSEpro® (Pfeffer et al. 2007), ProSim (Batista et al. 2012), and Aspen Plus (Quintero and Cardona 2011; Zheng et al. 2013; Lassmann et al. 2014) are used for bioprocessing simulation, due to the compatibility of these software with bioprocessing elements. Among these software, Aspen Plus is widely applied due to the support of new solid components, availability of the most relevant component in the data bank, energy analysis, economy analysis, and heat exchanger design capability in Version 8.6. Therefore, many studies of the process for lignocellulosic bioethanol production, especially for techno-economic analysis, were carried out using Aspen Plus (Aiden et al. 2002; Galbe et al. 2005; Cardona and Toro 2006; Cardona and Sánchez 2007; Schausberger et al. 2009; Seabra et al. 2010; Tao et al. 2013).

Since the simulator is a tool to analyze the real behavior of a chemical production process, the developed simulation should be realistic. Realistic simulation must reflect the true industrial behavior of the chemical process in terms of mass and energy balances and equipment operations. In the development of realistic simulation, two important issues can be taken into account: (1) processes should be considered as detailed as possible (considering available data for a selected process through literature) and (2) equipment parameters should be defined according to real operation. This section is a guideline to develop a realistic simulation for bioprocessing, based on the following:

1. Fundamentals of chemical engineering
2. Consideration of practical/industrial aspects of simulation
3. Know-how of similar simulations

Development of a precise and accurate simulation, with a valid mass and energy balance, is essential for reliable energy and economic analysis, which are often used for decision support. Although herein Aspen Plus software is considered as a platform of simulation, the rules and instructions can be used for any other process simulation software as well.

11.4.1.1 Component Specification
Component selection and definition of thermophysical properties for pure components and mixtures is the first vital step in process simulation. Almost all process simulators are supported by built-in component data banks, which contain parameters for numerous components, including organic, inorganic, and salt species (some simulators do not support electrolytes and salts). The species present in the process feed streams and possible products should be defined in the component specification form. Simulation of bioethanol production with most of simulators, is problematic due to lack of data for some components (i.e., solids). In this case, the undefined components should be added to the data bank as user defined by a few representative factors. The main user-defined components (considering Aspen Plus data bank) and relevant parameters are given in Table 11.1.

11.4.1.2 Equation of State Selection
Undoubtedly, the selection of the appropriate equation of state (EOS) for calculation of thermophysical properties of the material streams and equipment significantly affects the accuracy of simulation results and its ability to reflect the real behavior of the process. EOS is a functional relationship between state variables (usually a complete set of such variables), mostly written to express functional relationships between P, T, V, and enthalpy of pure component as well as mixture. EOSs are used to describe the properties of fluids and mixtures of fluids and solids. The selection of EOS depends on the process type, phases (vapor, liquid, and solid), and components.

Bioprocess such as bioethanol production contains a variety of unit operations and requires that suitable EOSs be selected for each section. In general, because of necessary ethanol distillation and inevitable dissolved gases, the standard nonrandom two liquid or Renon (NRTL) route is appropriate. This route includes the NRTL liquid

TABLE 11.1   Parameters of the Some User-Defined Components for Bioethanol Production

Component	Property	Quantity	Units	Remark
Xylan	Formula	$C_5H_8O_4$ (monomer)	—	
	MW	132.117	—	—
	DHSFRM	−182,099.93	cal/mol	
Xylose	DHFORM	−216,752.65	cal/mol	—
Lignin	—	—	—	Used native Aspen component vanillin ($C_8H_8O_3$)
Protein	Formula	$CH_{1.57}O_{0.31}N_{0.29}S_{0.007}$	—	
	MW	22.8396	—	—
	DHSFRM	−17,618	cal/mol	
Ash	—	—	—	Native Aspen component CaO
Enzyme	Formula	$CH_{1.59}O_{0.42}N_{0.24}S_{0.01}$	—	
	MW	24.0156	—	—
	DHSFRM	−17,618	—	
GLUCOLIG	MW	162.1424		Glucose oligomers
	DHSFRM	−192,875.34	cal/mol	
XYLOLIG	MW	162.1424	—	Xylose oligomers
	DHSFRM	−192,875.34	cal/mol	
Extract	—	—	—	Organic extractives, duplicate of glucose
HMF	MW	126.11	—	5-Hydroxymethylfurfural
	TB	532.7	K	
	DHFORM	80,550,000	J/kmol	
ZYMO	Formula	$CH_{1.8}O_{0.5}N_{0.2}$	—	*Z. mobilis* cell mass. Average
	MW	24.6264	—	composition of several
	DHSFRM	−31,169.39	cal/mol	microorganisms

Sources: Adapted from Wooley, R.J. and Putsche, V., Development of an ASPEN PLUS physical property database for biofuels components, National Renewable Energy Laboratory (NREL), Golden, CO, 1996; Humbird, D. et al., Process design and economics for biochemical conversion of lignocellulosic biomass to ethanol, National Renewable Energy Laboratory (NREL), Golden, CO, 2011.

activity coefficient model, Henry's law for the dissolved gases, and Redlich–Kwong–Soave (RKS) EOS for the vapor phase to calculate properties of components in the liquid and vapor phases (Rhodes 1996). Henry's components are defined according to the existence of gases such as oxygen, nitrogen, and carbon dioxide. For flash calculations, the NRTL activity coefficient model is used together with the Hayden-O'Connell EOS to take into account the dimerization of compounds such as acetic acid (Sassner et al. 2008). The ideal gas at 25°C is applied as the standard reference state for normal pressure and temperature calculation of the gas system, while for electrolyte systems, electro-NRTL is the best option whenever properties and equilibrium of electrolytes are considered. It is highly recommended to choose steam/water properties package correlations (water/steam tables) for boiler, steam, and power generation sections.

11.4.1.3 Standard Condition

As an important principle, all streams in the beginning of simulation should be defined in standard condition. The insertion of material, heat, and work streams in the graphic user interface of the software is the starting point of simulation development. Temperature and pressure are the most important intensive parameters of any material stream. An unofficial, but commonly used, standard is standard ambient temperature and pressure (SATP) as a temperature of 298.15 K (25°C, 77°F) and an absolute pressure of 100 kPa (14.504 psi, 0.987 atm), which is used in chemical process development (Gregg and Saddler 1996; Leibbrandt 2010). SATP should also be applied for end process or final streams such as products, vents, drains, and residues. The streams should be in equilibrium with the environment in terms of temperature and pressure, in order to recover maximum energy content of streams considering practical limitations.

11.4.1.4 Pressure and Temperature Alteration

Pressure and temperature alterations are inevitable through the process, but any change of these parameters should be simulated considering the appropriate equipment. Fundamentals of chemical engineering indicate that pressure of gases and liquids can be increased by passing through a compressor and pump, respectively. Some simulation software (particularly Aspen Plus) makes it possible to define the pressure and temperature of each unit operation independent of its upstream and downstream, but it is unrealistic to increase pressure without pump(s) or compressor(s). However, pressure reduction is not a crucial issue and can be achieved using a valve, turboexpander, or even a vessel.

Temperature can also be defined for most of the unit operations such as flash drums, separators, and reactors, but it is highly recommended to use a heater for temperature increase and a cooler for temperature drop. Obviously without considering the necessary equipment for temperature and pressure changes, the capital cost (cost of pump, compressor, heat exchanger, etc.) and also energy consumption (operating cost) of the overall process will not be calculated correctly, which will affect the accuracy of the energy and economic analysis results.

11.4.1.5 Heat and Work Streams

Heat and work streams are imaginary streams used in simulator software to represent the amount as well as direction of energy flows. Energy streams behave differently from material streams. For instance, a splitter may be used to split one material stream into two or more, to be transferred and used in different sections of the process, but the same concept cannot be applied for an energy stream. In another words, heat cannot be split into two or more streams for different usages, since it is not practical. Also it is not possible to use two or more heat streams to supply heat demand of single equipment, due to practical limitations of heat exchanger operation. However, the electrical power (work) stream can be split into supply power demand of more than one equipment.

11.4.1.6 Separators

Some equipment such as two-phase separator, three-phase separator, decanter, and flash drum operate similarly. These types of separators are designed to separate components and phases according to equilibrium; hence, the user-defined parameters should reflect this concept. Since the majority of separators work under adiabatic conditions, the heat duty of the equipment, as a user-defined parameter, should be set to zero. Only where a heating/cooling jacket or interior cooling/heating coils exist, should the temperature or specific heat duty be adjusted, while also taking into account the additional economic cost of these types of equipment. The second parameter of a separator is pressure, which should not be higher than the upstream value. Pressure is equal to upstream (with an assumption of no pressure drop) or less than the upstream (considering realistic pressure drop of the separators).

As mentioned in Section 11.4.1.2, using appropriate EOS is one of the significant considerations especially in three-phase separators or decanters, in that the selected EOS should be capable of identifying system behavior. Three-phase separation requires VLL (vapor–liquid–liquid) EOS such as the NRTL activity coefficient model along with the Hayden-O'Connell for LL (liquid–liquid) system. Subsequently, the NRTL EOS can be insightful for equilibrium data calculation.

11.4.1.7 Heat Exchangers

Heat exchangers are vital element of the process for adjustment of material stream temperatures. Heaters, coolers, and double-fluid heat exchangers are the most popular types, with each having different subcategories. For simulation purposes, it is highly recommended to fix the "outlet temperature" and the "pressure drop" of a heat exchanger, rather than the "heat duty" and "pressure." If there is the probability to reach dew point or bubble point for fluid flow through the heat exchanger, for example, total condensation or kettle type evaporator, it is better to adjust vapor fraction to a value of 1 and 0, respectively. However, for double-fluid heat exchanger, the minimum temperature or logarithmic mean temperature difference (LMTD = 10°C) is a more effective parameter than outlet temperature. Simulation of double-fluid heat exchanger would be easier by considering a cooler and heater while heat flows from cooler to heater. In this case, the second law of thermodynamic shall be checked by the user to avoid temperature cross in heat transfer. Therefore, it is suggested to apply countercurrent exchanger.

11.4.1.8 Component Separator

Component separator can be used to separate partial/total components from the main stream according to desired value. When the desired separation value is defined by the user, the software will apply algebraic methods to calculate the mass flow of each component in the downstreams of separation. It is worth noting that extensive use of component separators does not lead to a realistic simulation, because the effect of removed component through the simulation is ignored, and as a result, the size and operation of downstream equipment will be inaccurate. In the existing plants, component separation through a simple separator (available in simulator software) is not possible, and for any component to be separated, a specific process step and unit operation must be designed (e.g., pressure-swing adsorption for H_2 separation). Therefore, although almost all simulation software has the component separator as a process unit, it is preferred to design the appropriate process for the separation of the desired component rather than using the built-in component separator.

11.4.1.9 Distillation Column

Distillation columns are one of the most important equipment in the process of bio-ethanol production, for product purification. In order to select an appropriate distillation block type, the available data and expected result should be considered. The type called "Radfrac" in Aspen Plus is widely used for distillation simulation within a flowsheeting problem. Simulation of Radfrac requires detailed data for the results to be precise. Accurate simulation of distillation columns is tedious and selection of design parameters is a very important issue. For a distillation column with reboiler and condenser, two process parameters other than pressure must be specified; usually one intensive parameter on the top and one extensive on the bottom of the column, or vice versa, is the best selection for robust simulation. Considering Aspen Plus, the intensive and extensive parameters that may be selected are as follows:

1. *Extensive parameters*: Distillate rate (kg/h or kmol/h), bottom rate (kg/h or kmol/h), reboiler duty (kW), condenser duty (kW), reflux rate, and boiled up rate (kg/h or kmol/h)
2. *Intensive parameters*: Reflux ratio, bottom to feed ratio, temperature (°C or °F), and boiled up ratio

Since in the beer column a variety of components (liquid and solid) exist, it is highly recommended to select nonideal liquid for convergence option of the column.

11.4.1.10 Utilities

The utility terms of a plant generally refer to all supplementary requirements of the process, but hereafter they are limited to steam, cooling water, chilled water, and electricity demands of the plant. In some software, a variety of utilities can be defined in simulation, and the utility requirements of equipment are set according to the pre-defined utilities. This utility definition approach results in easy tracking of the plant energy demand in terms of steam/water tonnage as well as energy flow, which is used for utility design sections.

11.4.2 Automated Simulation Development

Automated simulation development in chemical processes means simulation based on intensive parameters or minimum dependency of simulation to extensive parameters. Simulation requires a lot of basic information that should be specified in the general system model. It would be a waste of resources to start the manual modeling all over again for a small change in the process; hence, it is strongly recommended to pursue a high degree of automation in the information interchange from general system model to domain-specific models (Rudtsch et al. 2013). In an automated simulation, if any process parameter (e.g., feedstock composition and flow) changes, all of the relevant parameters will be updated accordingly. Automated simulation is achievable using different types of automation tools such as calculator blocks, design specifications, and transferring data blocks within simulation, which are known as manipulators.

11.5 Process Integration

Process integration is a holistic approach to process design that emphasizes the unity of the process and considers the interactions between different unit operations from the outset, rather than optimizing them separately. This can also be called integrated process design or process synthesis. Process integration is required between the pretreatment, enzymatic hydrolysis, fermentation, and downstream processing sections of a biobased process. Currently most technologies are developed independently, and many different disciplines are necessary to effectively study the integration of different processes. Particular R&D attention is necessary further downstream on the integration of biotechnological conversion steps with chemical conversion steps, in order to produce target products or monomers of sufficient purity. R&D programs carried out thus far have often focused on either biotechnological conversion or chemical conversion without evaluating the interfaces. These interfaces are crucial for a large number of theoretical pathways, but it is a gap that should be addressed.

Efficiency enhancement (costs reduction) and integration of different processes should be considered for ethanol production from lignocellulose, to improve viability for industrial implementation. The integrated processes can be categorized as simultaneous (when they are accomplished in the same equipment) or conjugated (when they are carried out in different equipment). Additionally the integration of processes are classified as homogenous (when two or more unit operations or chemical reactions are combined) and heterogeneous (when one unit operation and one or more chemical reactions are combined simultaneously). In this context, ED is an example of a simultaneous homogenous process, whereas the coupling of the fermentation with the evaporation is a conjugated heterogeneous process (Cardona and Toro 2006; Cardona and Sánchez 2007). The main integration techniques for bioethanol production are discussed hereafter.

11.5.1 Coproduct Production

The production of coproducts in a so-called biorefinery implies that a plant will integrate various biomass conversion processes and equipment to produce fuels, power, and value-added chemicals (Demirbas 2009; Fatih Demirbas 2009; Cherubini 2010). During the process design, alternative flowsheets should be evaluated, for example, conversions achieved through alternative pathways and also conversion of feedstock to valuable coproducts along with the main product.

Following this approach, economical and even environmental criteria may be met. Various streams, mostly of organic nature, are obtained during the processing of such materials such as sugarcane, corn, or lignocellulosic biomass. These organic materials have an important value either as a fuel or as source for other value-added products. To offset the inherent high cost of processing biological materials, the possibilities for producing coproducts should be taken into account when designing ethanol production processes (Cardona and Sánchez 2007). Therefore, some research studies were conducted regarding production of coproducts such as lactic acid (Thongchul et al. 2010), furfural (Cai et al. 2014), and xylitol (Cheng et al. 2010), but more efforts are still required in this area.

11.5.2 Reaction–Reaction Integration

Integration of two or more reactors in one reactor is considered as reaction–reaction integration. Reaction–reaction integrations have potential to improve the economic viability of the process. The latter is achieved by minimizing enzyme use, lowering the capital costs, improving processing times, and enabling more efficient feedstock use. Simultaneous processes allow a mutual intensification as well as the development of more compact technological schemes. All these designs allow the integration of some of the various steps in an overall process. For instance, a combination of the enzymatic hydrolysis and the microbial transformation corresponds to the reaction–reaction integration (Cardona and Sánchez 2007). Reaction–reaction integration for bioethanol production can consider the following methods:

1. Cofermentation of lignocellulosic hydrolyzates
2. SSF
3. SSCF
4. CBP

11.5.3 Reaction–Separation Integration

Since the major cost in industrial plants is the separation process cost, integration of the separation process with the reaction (reaction–separation integration) may have a more significant impact on the overall process, compared to reaction–reaction integration.

Integration of fermentation step with separation units is a practical integration in bioethanol production process. For this purpose, fermentation and separation should take place in one equipment. Ethanol removal from culture broth is the most possible reaction–separation integration, which is an attractive alternative for the intensification of alcoholic fermentation processes. Ethanol removal from culture is carried out by vacuum, gas stripping, and membrane and liquid extraction (Cardona and Sánchez 2007).

11.5.4 Energy Integration

Similar to the integration of material flows (either for material transformation or for the component separation), energy integration can be performed at different steps of ethanol production process. The key concept of energy integration is to select conditions in individual unit operations to maximize energy efficiency of the overall process, rather than selecting conditions giving the desired performance in the individual unit. Energy integration, in particular heat integration, aims for the best utilization of energy flows (heat, mechanical, and electrical energy) generated or consumed inside the process, to minimize the consumption of external sources of energy such as electricity and fossil fuels (oil and natural gas), mainly used for steam generation and cooling water (Cardona and Sánchez 2007).

Pinch technology is one of the most widely applied approaches for heat integration in process industry, especially in the petrochemical industry (Martinez-Hernandez et al. 2013; Petersen et al. 2015). This technology provides the necessary tools for design of heat exchanger networks, including plant utilities. During the preliminary design of the heat exchanger network, pinch technology allows selection of the best values of many process parameters, such as the types of utilities and their specifications, minimum number of heat exchanger units and their transfer areas, and the estimation of capital and operation costs of these units. Heat integration has now become one of the most important techniques for process integration during the synthesis of diverse technological flowsheets. Developed process simulation can be analyzed by pinch analysis software, to optimize the energy consumption.

11.5.5 Water Integration

The recent systematic methods for water consumption minimization have focused on the optimal synthesis of process water networks. Energy optimization has a major impact on decreasing the required cooling and the water loss (through evaporation and drift in the cooling tower). Therefore, water consumption reduction of the process water networks is achieved through energy optimization. In other words, removing added water to the process typically requires significant energy input (for water separation through purification of the product); therefore, a decrease in process water consumption leads to energy demand reduction. The design of water network can be performed using two different approaches, namely, conceptual engineering approach or systematic methods based on mathematical programming.

Although the low cost of freshwater hinders proper water optimization, few researches have investigated the water integration and minimization of bioethanol plant (Čuček et al. 2011; Martín et al. 2011; Martín and Grossmann 2012). Optimization of the water consumption involves three steps:

1. Optimization of energy consumption in order to reduce the required cooling of the processes and consequently decrease the water losses (evaporation and drift in the cooling tower)

2. Assessment of available technologies to substitute (partially/totally) the use of water as cooling agent
3. Designing the optimal water network by determining water consumption, water recycle, and the required treatment

Further reduction can be obtained if the water released from the crop can be properly recovered and treated.

11.6 Conclusions

The aim of this chapter was to provide insightful knowledge about process design, flowsheeting, and simulation of bioethanol production, which is by far the most widely used biofuel for transportation. Lignocellulosic biomass is the most promising feedstock considering its availability and low cost. Recently a few large-scale plants for commercial production of bioethanol from lignocellulosic materials have been set up worldwide, and more studies are ongoing to extend bioethanol production from a variety of feedstocks. The current chapter can be used as a guideline for bioethanol production simulation as well as other biobased chemical production and also for R&D and practical purposes.

Feedstock type (hard wood, softwood, wastes, etc.) significantly affects the design of the process. Plant size is the most important extensive parameter of the bioethanol plant in terms of operating and fixed capital costs and, consequently, economic viability of the plant. Flowsheeting of the bioethanol plant can be commenced by BFD, which illustrates the outline of the process. PFD is utilized when considering more details containing all major streams, equipment, operating parameters, and energy requirement of the plant. In addition to bioethanol production section (pretreatment, hydrolysis, fermentation, and enzyme production), other supplementary units such as ethanol purification, evaporation, boiler, power and steam generation, and WWT were discussed in detail for PFD development of the bioethanol production. Due to bilateral relation between PFD and simulation, process simulators are used to simulate the process and achieve the mass and energy balances. Also simulation can be used to analyze the effect of different equipment arrangements and operating parameters to promote the flowsheet. Due to the important role of simulation in the process development, some guidelines and technical points to approach realistic simulation were addressed. Also, applicable integration techniques that can be applied for improvement of the developed process were reviewed.

References

Aiden, A., Ruth, M., Ibsen, K. et al. 2002. Lignocellulosic biomass to ethanol process design and economics utilizing co-current dilute acid prehydrolysis and enzymatic hydrolysis for corn stover. NREL, Golden, CO.

Alvarado-Morales, M., Terra, J., Gernaey, K.V., Woodley, J.M., and R. Gani. 2009. Biorefining: Computer aided tools for sustainable design and analysis of bioethanol production. *Chem Eng Res Des* 87:1171–1183.

Alvira, P., Tomás-Pejó, E., Ballesteros, M., and M.J. Negro. 2010. Pretreatment technologies for an efficient bioethanol production process based on enzymatic hydrolysis: A review. *Bioresour Technol* 101:4851–4861.

Balat, M. 2007. Global biofuel processing and production trends. *Energ Explor Exploit* 25:195–218.

Balat, M., Balat, H., and C. Öz. 2008. Progress in bioethanol processing. *Prog Energ Combust Sci* 34:551–573.

Banerjee, S., Mudliar, S., Sen, R. et al. 2010. Commercializing lignocellulosic bioethanol: Technology bottlenecks and possible remedies. *Biofuel Bioprod Bioref* 4:77–93.

Batista, F.R.M., Follegatti-Romero, L.A., Bessa, L.C.B.A., and A.J.A. Meirelles. 2012. Computational simulation applied to the investigation of industrial plants for bioethanol distillation. *Comput Chem Eng* 46:1–16.

Cai, C.M., Zhang, T., Kumar, R., and C.E. Wyman. 2014. Integrated furfural production as a renewable fuel and chemical platform from lignocellulosic biomass. *J Chem Technol Biotechnol* 89:2–10.

Cardona, C.A. and Ó.J. Sánchez. 2007. Fuel ethanol production: Process design trends and integration opportunities. *Bioresour Technol* 98:2415–2457.

Cardona, C.A. and O.J. Toro. 2006. Energy consumption analysis of integrated flowsheets for production of fuel ethanol from lignocellulosic biomass. *Energy* 31:2447–2459.

Cheng, K.K., Zhang, J.A., Chavez, E., and J.P. Li. 2010. Integrated production of xylitol and ethanol using corncob. *Appl Microbiol Biotechnol* 87:411–417.

Cherubini, F. 2010. The biorefinery concept: Using biomass instead of oil for producing energy and chemicals. *Energ Convers Manage* 51:1412–1421.

Chum, H.L., Warner, E., Seabra, J.E.A., and I.C. Macedo. 2014. A comparison of commercial ethanol production systems from Brazilian sugarcane and US corn. *Biofuel Bioprod Bioref* 8:205–223.

Čuček, L., Martín, M., Grossmann, I.E., and Z. Kravanja. 2011. Energy, water and process technologies integration for the simultaneous production of ethanol and food from the entire corn plant. *Comput Chem Eng* 35:1547–1557.

Demirbas, A. 2009. Biorefineries: Current activities and future developments. *Energ Convers Manage* 50:2782–2801.

Dias, M.O.S., Ensinas, A.V., Nebra, S.A., Filho, R.M., Rossell, C.E.V., and M.R.W. Maciel. 2009. Production of bioethanol and other biobased materials from sugarcane bagasse: Integration to conventional bioethanol production process. *Chem Eng Res Des* 87:1206–1216.

Dias, M.O.S., Modesto, M., Ensinas, A.V., Nebra, S.A., Filho, R.M., and C.E.V. Rossell. 2011. Improving bioethanol production from sugarcane: Evaluation of distillation, thermal integration and cogeneration systems. *Energy* 36:3691–3703.

Eggert, H. and M. Greaker. 2014. Promoting second generation biofuels: Does the first generation pave the road? *Energies* 7:4430–4445.

Errico, M., Rong, B.-G., Tola, G., and M. Spano. 2013a. Optimal synthesis of distillation systems for bioethanol separation. Part 2. Extractive distillation with complex columns. *Ind Eng Chem Res* 52:1620–1626.

Errico, M., Rong, B.-G., Tola, G., and M. Spano. 2013b. Optimal synthesis of distillation systems for bioethanol separation. Part 1: Extractive distillation with simple columns. *Ind Eng Chem Res* 52:1612–1619.

Fatih Demirbas, M. 2009. Biorefineries for biofuel upgrading: A critical review. *Appl Energ* 86:S151–S161.

Galbe, M., Lidén, G., and G. Zacchi. 2005. Production of ethanol from biomass-research in Sweden. *J Sci Ind Res* 64:905.

Gregg, D. and J.N. Saddler. 1996. A techno-economic assessment of the pretreatment and fractionation steps of a biomass-to-ethanol process. *Applied Biochemistry and Biotechnology* 57/58:711–727.

Gupta, A. and J.P. Verma. 2015. Sustainable bioethanol production from agro-residues: A review. *Renew Sustain Energ Rev* 41:550–567.

Haelssig, J.B., Tremblay, A.Y., and J. Thibault. 2008. Technical and economic considerations for various recovery schemes in ethanol production by fermentation. *Ind Eng Chem Res* 47:6185–6191.

Hamelinck, C.N., van Hooijdonk, G., and A.P.C. Faaij. 2005. Ethanol from lignocellulosic biomass: Techno-economic performance in short-, middle- and long-term. *Biomass Bioenerg* 28:384–410.

He, J. and W. Zhang. 2011. Techno-economic evaluation of thermo-chemical biomass-to-ethanol. *Appl Energ* 88:1224–1232.

Hess, J.R., Wright, C.T., and K.L. Kenney. 2007. Cellulosic biomass feedstocks and logistics for ethanol production. *Biofuel Bioprod Bioref* 1:181–190.

Humbird, D., Davis, R., Tao, L., Kinchin, C., Hsu, D., and A. Aden. 2011. Process design and economics for biochemical conversion of lignocellulosic biomass to ethanol. National Renewable Energy Laboratory (NREL), Golden, CO.

Kiss, A.A. and D.J.P.C. Suszwalak. 2012. Enhanced bioethanol dehydration by extractive and azeotropic distillation in dividing-wall columns. *Sep Purif Technol* 86:70–78.

Lassmann, T., Kravanja, P., and A. Friedl. 2014. Simulation of the downstream processing in the ethanol production from lignocellulosic biomass with ASPEN Plus® and IPSEpro. *Energ Sustain Soc* 4:1–7.

Leibbrandt, N.H. 2010. Techno echno-economic study for sugarcane bagasse to liquid biofuels in South Africa. PhD dissertation, Stellenbosch University, Stellenbosch, South Africa.

Ludwig, E.E. 1997. *Applied Process Design for Chemical and Petrochemical Plants*. Gulf Professional Publishing. Houston, TX.

Luyben, W.L. 1989. *Process Modeling, Simulation and Control for Chemical Engineers*. McGraw-Hill Higher Education. New York City.

Lynd, L.R., van Zyl, W.H., McBride, J.E., and M. Laser. 2005. Consolidated bioprocessing of cellulosic biomass: An update. *Curr Opin Biotechnol* 16:577–583.

Martín, M., Ahmetović, E., and I.E. Grossmann. 2011. Optimization of water consumption in second generation bioethanol plants. *Ind Eng Chem Res* 50:3705–3721.

Martín, M. and I.E. Grossmann. 2012. Energy optimization of bioethanol production via hydrolysis of switchgrass. *AIChE J* 58:1538–1549.

Martinez-Hernandez, E., Sadhukhan, J., and G.M. Campbell. 2013. Integration of bioethanol as an in-process material in biorefineries using mass pinch analysis. *Appl Energ* 104:517–526.

Mood, S.H., Golfeshan, A.H., Tabatabaei, M. et al. 2013. Lignocellulosic biomass to bioethanol, a comprehensive review with a focus on pretreatment. *Renew Sust Energ Rev* 27:77–93.

Mosier, N., Wyman, C., Dale, B. et al. 2005. Features of promising technologies for pretreatment of lignocellulosic biomass. *Bioresour Technol* 96:673–686.

Mulia-Soto, J.F. and A. Flores-Tlacuahuac. 2011. Modeling, simulation and control of an internally heat integrated pressure-swing distillation process for bioethanol separation. *Comput Chem Eng* 35:1532–1546.

Olofsson, K., Bertilsson, M., and G. Lidén. 2008. A short review on SSF—An interesting process option for ethanol production from lignocellulosic feedstocks. *Biotechnol Biofuels* 1:1–14.

Olsson, J. and G. Zacchi. 2001. Simulation of the condensate treatment process in a kraft pulp mill. *Chem Eng Technol* 24:195–203.

Petersen, A.M., Melamu, R., Knoetze, J.H., and J.F. Görgens. 2015. Comparison of second-generation processes for the conversion of sugarcane bagasse to liquid biofuels in terms of energy efficiency, pinch point analysis and life cycle analysis. *Energ Convers Manage* 91:292–301.

Pfeffer, M., Wukovits, W., Beckmann, G., and A. Friedl. 2007. Analysis and decrease of the energy demand of bioethanol-production by process integration. *Appl Therm Eng* 27:2657–2664.

Quintero, J.A. and C.A. Cardona. 2011. Process simulation of fuel ethanol production from lignocellulosics using aspen plus. *Ind Eng Chem Res* 50:6205–6212.

Quintero, J.A., Moncada, J., and C.A. Cardona. 2013. Techno-economic analysis of bioethanol production from lignocellulosic residues in colombia: A process simulation approach. *Bioresour Technol* 139:300–307.

Rabelo, S.C., Fonseca, N.A.A., Andrade, R.R., Filho, R.M., and A.C. Costa. 2011. Ethanol production from enzymatic hydrolysis of sugarcane bagasse pretreated with lime and alkaline hydrogen peroxide. *Biomass Bioenerg* 35:2600–2607.

Ragauskas, A.J., Beckham, G.T., Biddy, M.J. et al. 2014. Lignin valorization: Improving lignin processing in the biorefinery. *Science* 344:1246843.

Rhodes, C.L. 1996. The process simulation revolution: Thermophysical property needs and concerns. *J Chem Eng Data* 41:947–950.

Rudtsch, V., Bauer, F., and J. Gausemeier. 2013. Approach for the conceptual design validation of production systems using automated simulation-model generation. *Procedia Comput Sci* 16:69–78.

Salehi Jouzani, Gh. and M.J. Taherzadeh. 2015. Advances in consolidated bioprocessing systems for bioethanol and butanol production from biomass: A comprehensive review. *Biofuel Res J* 5:152–195.

Sánchez, Ó.J. and C.A. Cardona. 2008. Trends in biotechnological production of fuel ethanol from different feedstocks. *Bioresour Technol* 99:5270–5295.

Sangeeta, Moka, S., Pande, M. et al. 2014. Alternative fuels: An overview of current trends and scope for future. *Renew Sustain Energ Rev* 32:697–712.

Sassner, P., Galbe, M., and G. Zacchi. 2008. Techno-economic evaluation of bioethanol production from three different lignocellulosic materials. *Biomass Bioenerg* 32:422–430.

Schausberger, P., Bösch, P., and A. Friedl. 2009. Modeling and simulation of coupled ethanol and biogas production. *Clean Technol Environ* 12:163–170.

Seabra, J.E.A., Tao, L., Chum, H.L., and I.C. Macedo. 2010. A techno-economic evaluation of the effects of centralized cellulosic ethanol and co-products refinery options with sugarcane mill clustering. *Biomass Bioenerg* 34:1065–1078.

Smith, G.T., Davis, S.B., Madho, S., and M. Achary. 2014. Eighty-eighth annual review of the milling season in Southern Africa (2012–2013). In *86th Annual Congress of the South African Sugar Technologists' Association (SASTA 2013)*, Durban, South Africa, August 6–8, 2013, pp. 24–54.

Smithers, J. 2014. Review of sugarcane trash recovery systems for energy cogeneration in South Africa. *Renew Sustain Energ Rev* 32:915–925.

Sun, Y. and J. Cheng. 2002. Hydrolysis of lignocellulosic materials for ethanol production: A review. *Bioresour Technol* 83:1–11.

Tao, L., Schell, D., Davis, R., Tan, E., Elander, R., and A. Bratis. 2014. NREL 2012 achievement of ethanol cost targets: Biochemical ethanol fermentation via dilute-acid pretreatment and enzymatic hydrolysis of corn stover. NREL/TP-5100-61563. National Renewable Energy Laboratory (NREL), Golden, CO.

Tao, L., Templeton, D.W., Humbird, D., and A. Aden. 2013. Effect of corn stover compositional variability on minimum ethanol selling price (MESP). *Bioresour Technol* 140:426–430.

Thongchul, N., Navankasattusas, S., and S.-T. Yang. 2010. Production of lactic acid and ethanol by *Rhizopus oryzae* integrated with cassava pulp hydrolysis. *Bioproc Biosyst Eng* 33:407–416.

Wooley, R., Ruth, M., Glassner, D., and J. Sheehan. 1999a. Process design and costing of bioethanol technology: A tool for determining the status and direction of research and development. *Biotechnol Prog* 15:794–803.

Wooley, R., Ruth, M., Sheehan, J., Ibsen, K., Majdeski, H., and A. Galvez. 1999b. Lignocellulosic biomass to ethanol process design and economics utilizing co-current dilute acid prehydrolysis and enzymatic hydrolysis current and futuristic scenarios. National Renewable Energy Laboratory (NREL), Golden, CO.

Wooley, R.J. and V. Putsche. 1996. Development of an ASPEN PLUS physical property database for biofuels components. National Renewable Energy Laboratory (NREL), Golden, CO.

Xu, Q., Singh, A., and M.E. Himmel. 2009. Perspectives and new directions for the production of bioethanol using consolidated bioprocessing of lignocellulose. *Curr Opin Biotechnol* 20:364–371.

Zheng, H., Kaliyan, N., and R.V. Morey. 2013. Aspen plus simulation of biomass integrated gasification combined cycle systems at corn ethanol plants. *Biomass Bioenerg* 56:197–210.

12

Biobutanol Production from Lignocellulosics

Lalitha Devi Gottumukkala and Johann F. Görgens

Contents

Abstract

Next-generation biofuels from renewable sources have gained interest among research investigators, industrialists, and governments due to major concerns on the volatility of oil prices, climate change, and depletion of oil reserves. Biobutanol has drawn significant attention as an alternative transportation fuel due to its superior fuel properties over ethanol. The advantages of butanol are its high energy content, better blending with gasoline, less hydroscopic nature, lower volatility, direct use in convention engines, low corrosiveness, etc. Butanol production through (acetone, butanol, and ethanol) ABE fermentation is a well-established process, but it has several drawbacks like feedstock cost, strain degeneration, product toxicity, and low product concentrations. Lignocellulosic biomass is considered as the most abundant, renewable, low-cost feedstock for biofuels. Production of butanol from lignocellulosic biomass is more complicated due to the recalcitrance of feedstock and inhibitors generated during the pretreatment and hydrolysis process. Advanced fermentation and product recovery techniques are being researched to make biobutanol industrially viable.

12.1 Introduction

Sustainable production of second-generation fuels from lignocellulosic biomass has gained focus due to the abundance of feedstock and its renewability. According to IEA, biomass is the oldest source of energy, and it currently accounts for 10% of the energy consumption. The conventional practice of burning biomass for energy results in severe carbon emissions and hence is not considered environment friendly. Moreover, the current transportation sector that produces about 25% of global-energy-related CO_2 emissions is primarily dependent on liquid transportation fuels generated from crude oil and accounts for approximately 50% of global oil consumption (Eisentraut 2010). This signifies the necessity of renewable, environment-friendly transportation fuels to combat the global greenhouse gas emissions. Though there is huge increase in bioethanol and biodiesel production over the last decade, they are primarily first-generation biofuels. Second-generation biofuel production is still in its demonstration stage, due to the complexity of feedstock and conversion technologies. Bioethanol, biobutanol, biohydrogen, biodiesel, biomethane, and synthetic biofuels are commonly enlisted second-generation fuels from lignocellulosic biomass. While biodiesel and bioethanol are the most discussed alternative transportation fuels, butanol is more efficient when compared to ethanol due to its superior fuel properties and distribution advantages.

Biobutanol has several advantages over ethanol as a fuel, but at the stage of technology and process development, it has several challenges that still need to be addressed. The rigidity of lignocellulosic biomass and the advanced integrated fermentation and product recovery conditions required for *Clostridium* to efficiently convert sugars to butanol hinder the speed of commercial technology development for second-generation butanol production. However, the current support policies that are encouraging second-generation biofuel production globally might bring in the change in biobutanol scenario. This chapter highlights the upstream and downstream process technologies developed for biobutanol production and their feasibility.

12.2 Biobutanol: A Superior Second-Generation Fuel

Butanol, a four-carbon alcohol, is a larger molecule that is converted to more energy than two-carbon molecule, ethanol. Butanol has low vapor pressure than ethanol and hence is safer to use, as it is combustible, but not flammable, while other fuels like methanol, ethanol, and gasoline are flammable and potentially explosive. The energy content, air/fuel ratio, and motor octane ratings of butanol are closer to those of gasoline than to those of ethanol. The energy content of butanol is 11,000 BTU/gallon and for ethanol it is 84,000 BTU/gallon (Szulczyk 2010).

Ethanol can be used in standard car engines with 15% blend in gasoline and blends of up to 85% can be used with modified engines of flexible fuel vehicles (FFVs). Butanol–gasoline blends have no such restrictions, as it resembles gasoline more closely than ethanol. Oxygen content of the fuel is another important fuel parameter, as the presence of oxygen in the fuel allows complete combustion and reduces carbon monoxide emissions (Szulczyk 2010). Methyl tertiary butyl ether (MTBE) is an oxygenate that is mostly used in fuel production but is considered as severe contaminant for underground water resources (EPA 2015). Butanol and ethanol contain 22% and 36% oxygen, respectively, and can be alternatives to MTBE (Szulczyk 2010). Ethanol has advantages over butanol in terms of high oxygen content and octane number, but it has major distribution disadvantages like moisture–fuel contamination and corrosiveness. The less hygroscopic nature of butanol makes it possible to be easily distributed in existing pipelines of gasoline. The major advantage of butanol is that it has thermochemical properties closer to gasoline, hence making it a drop-in fuel to vehicles with unmodified engines. The properties of butanol, ethanol, and gasoline are discussed by Lee et al. (2008) and Szulczyk (2010) and are presented in Table 12.1.

Currently, the European and U.S. regulations allow butanol blend of 15% and 16%, respectively. Butanol blend of 16% by volume with gasoline provides twice as much energy as that of ethanol–gasoline blend 10% (E10) and reduces GHG emissions further (BP Biofuels 2015). With the current focus on E15 (fuel with 15% ethanol) to satisfy renewable fuel standard, the National Marine Manufacturers Association of North America has reported that ethanol blends of more than 10% would cause

TABLE 12.1   Fuel Properties of Butanol and Ethanol Compared with Gasoline

	Units	Butanol	Ethanol	Gasoline
Energy density	MJ/L	29.2	19.6	32
Air/fuel ratio	—	11.2	9	14.6
Heat of vaporization	MJ/kg	0.43	0.92	0.36
Research octane number	100%	96	129	92–99
Motor octane number	100%	78	102	81–89
Reid vapor pressure	Bar	0.023	0.159	0.48–1.03
High heating value	MJ/L	—	23.6	34.8
Low heating value	MJ/L	27.8	21.1–21.3	31.2–32.4
Oxygen content	100%	22	36	0

significant damage to marine engines and identified butanol as the safe alternative fuel for marine engines due to its less hygroscopicity (Green Car Congress 2015).

The fuel properties of biobutanol that are similar to those of gasoline and the process similarity between biobutanol and bioethanol in feedstock requirement and the technology used make biobutanol an efficient drop-in fuel with slight modifications in existing ethanol production facilities. Butanol was traditionally produced by ABE fermentation using *Clostridium acetobutylicum* and is now currently being produced chemically from petroleum sources by oxo or aldol process.

12.2.1 History of ABE Fermentation

The microbial production of butanol goes through a long process, which is referred to as ABE fermentation, due to the mixture of acetone, butanol, and ethanol produced in the ratio 3:6:1. The first report on butanol production was in 1861 by Louis Pasteur, and it was only in 1905 that Schardinger reported acetone production during fermentation. In 1911, Fernbach isolated a culture that was able to ferment potatoes, but not maize starch, to produce butanol. In 1914, Weizmann isolated a culture capable of producing high quantities of butanol from various starchy substrates and this culture was later identified as *C. acetobutylicum* (Jones and Woods 1986).

The microbial production of acetone from potatoes and maize as the substrate was largely commercialized for rubber manufacture. During the World War I, acetone production expanded further for the manufacture of munitions. After the end of the World War I, ABE fermentation was completely ceased due to the inefficient supply of substrates. Later, molasses came into picture as an efficient substrate and ABE fermentation was given top priority once again during the World War II. In the 1960s, ABE fermentation was virtually ceased in the United States and many other countries, as there was acute competition between fermentation and chemical routes due to growth in petrochemical industry and increased cost of the molasses, due to its demand for various processes (Jones and Woods 1986). The ABE fermentation process in South Africa and Russia was continued till the 1980s–1990s (Lee et al. 2008).

12.3 Lignocellulosic Biomass as Feedstock: Availability

Lignocellulosic biomass is considered a popular substrate for second-generation fuels. This is due to the surplus availability and renewability potential of the feedstock. Lignocellulosic biomass can be divided into two categories: (1) dedicated energy crops and (2) agriculture and forest residues. Dedicated energy crops are further divided into woody crops and herbaceous crops. Cultivation and harvesting of herbaceous crops is considered less intensive than the annual woody crops. Agriculture and forest residues can be divided into three categories: primary, secondary, and tertiary residues. Primary residues are leftovers of timber or crop harvesting (e.g., straw, stover, treetops, branches), secondary residues are generated during biomass processing of foods or other biobased products (e.g., nutshells, bagasse, press cake, fruit bunches, saw dusk, bark, and scrap wood), and tertiary residues are waste generated after consumer utilization of the product (e.g., municipal solid waste) (Eisentraut 2010).

The potential of lignocellulosic biomass availability differs from region to region due to change in climatic conditions, land availability for the cultivation of energy crops, food demand, agriculture and forest residues, surplus forest growth, etc. Smeets and Faaij (2007) calculated the possible energy available from the estimates of land area and agriculture/forest residues by 2050 for various regions (Table 12.2). It was reported that biomass potentials are not evenly distributed globally and high amounts are concentrated at regions with favorable climatic conditions and huge land reserves. Global food demand by 2050 was taken into account for the estimations and it was found that Europe has the highest potential for biofuel from dedicated energy crops, while South Asia has a limited potential for the expansion of cropland. It was reported that India does not have surplus agriculture land for the cultivation of dedicated energy crops and also forestry residues are used for fuel and firewood. Hence, agricultural residues have the major potential to serve as lignocellulosic biomass in India.

TABLE 12.2 Comparison of Bioenergy Potential in 2050 in Various Regions, Based on the Availability of Land and Residues

Regions	Agriculture and Forest Residues (EJ)	Surplus Forest Growth (EJ)	Dedicated Energy Crops (EJ)
Europe	11–13	35	53–255
North America	6–17	5	20–174
Latin America	10–12	22	47–221
Africa	16–21	2	31–317
South Asia	10	—	15–25
East Asia	10	—	23–184
Oceania	2–6	—	40–110

Sources: Eisentraut, A., Sustainable production of second-generation biofuels—Potential and perspectives in major economies and developing countries, OECD/IEA, 2010; Smeets, E. and Faaij, A., *Clim. Change*, 81, 353, 2007.

12.4 Recalcitrance of Lignocellulosic Biomass

Lignocellulose is abundant in nature and is mainly composed of cellulose, hemicellulose, and lignin. Cellulose is highly crystalline in nature and is a major component of cell wall with repeated units of glucose. Hemicellulose is amorphous and is a polymer of mainly pentoses and few hexoses. Lignin contains aromatic alcohols and forms a protective layer coating cellulose and hemicellulose. Due to its complex network of lignin, cellulose, and hemicellulose, lignocellulosic biomass is not readily accessible to hydrolysis enzymes, which makes it different from first-generation feedstocks. The reasons for recalcitrance of lignocellulosic biomass and its effects on enzymatic hydrolysis and fermentation include epidermal tissue of plant body, cuticle and epicuticular waxes, relative amount of sclerenchymatous tissue, degree of lignification, cell wall proteins, insolubility of the substrate, and fermentation inhibitors that exist naturally in cell walls to act against bacteria and fungi as defense mechanism (Zhao et al. 2012). The percentage of three main components, cellulose, hemicellulose, and lignin, mainly depends on the source of biomass, whether it is from hardwood, softwood, or grasses. Generally, woody biomass is an abundant source of cellulose and lignin, while grasses have more hemicellulose and ash. The effect of these three components on the enzymatic digestibility of lignocellulosic biomass is discussed in the following.

12.4.1 Cellulose: Crystallinity

Cellulose is a homopolymer of glucose and is made up of crystalline and amorphous regions. Crystalline regions exist in the form of microfibrils with (1,4) β-D glucan chains linked to one another parallel by hydrogen bonding. Because of the hydrogen bonds, crystalline cellulose becomes more resistant to enzymes. Saccharification rate depends on the type of hydrogen bonding (Zhao et al. 2012). It was reported that change in hydrogen bonding from intrachain to interchain by ammonia pretreatment increased the rate of saccharification by fivefold (Chundawat et al. 2011). Due to the difference in hydrogen bonding, amorphous cellulose is more accessible to enzymes and has 3–30 times faster hydrolysis rate (Zhang and Lynd 2004). When the amorphous regions of hemicellulose and lignin are removed by pretreatment, the crystallinity of biomass is increased. But this cannot correspond to the increase in crystallinity of cellulose with the pretreatment. Crystallinity index is not an independent factor. The commonly used crystallinity index determination methods like x-ray diffraction and infrared spectrum can affect the structures of lignin and hemicellulose. Pretreatment also increases the cellulose accessible surface area by reducing the particle size and by breaking open the biomass structure, which is a crucial factor for enzymatic hydrolysis. Hence, the relation between pretreatment, crystallinity, and enzymatic hydrolysis should be carefully analyzed (Zhao et al. 2012).

12.4.2 Hemicellulose: Acetyl Groups

Hemicellulose is made up of glucuronoxylan, glucomannan, and trace amounts of other polysaccharides. Grasses and straws contain mostly xylan and small amount of arabinan, while woody biomass like softwoods and hardwoods contains mannan as the main component. Hemicellulose binds to cellulose through noncovalent interactions and acts as a physical barrier that surrounds the cellulose. Xylan blocks the enzymes access to cellulose, and hence, the percentage of xylan in the biomass influences the pretreatment process and enzyme requirements. Zhao et al. discussed that the pretreatment of biomass with dilute sulfuric acid can remove more than 80% of hemicellulose and the addition of xylanase enzymes during hydrolysis improved the enzymatic digestion of pretreated biomass (Zhao et al. 2012). Hemicellulose is extensively acetylated and the acetyl groups amount to approximately 1%–6% of biomass (Peng et al. 2011). Biomass with more acetyl groups is considered to be more resistant to enzymatic hydrolysis due to the inhibitory activity of acetyl groups on productive binding of cellulase to cellulose. However, the effect is negligible when compared to the effect of lignin and cellulose crystallinity.

12.4.3 Lignin: Rigidity

Lignin is a third crucial biopolymer in plants that provides structural rigidity and mechanical strength to cell wall by occupying the spaces in between carbohydrate polymers by its covalent linkage to hemicellulose. It is a phenolic polymer derived mainly from hydroxycinnamyl alcohols and is an important factor that reduces the accessibility of cellulose to enzymes (Weng and Chapple 2010). Lignin composition and type of its cross-linking to the cellulose and hemicellulose influences the digestibility of cellulose in lignocellulosic biomass. The type of phenolics, composition of lignin, and its linkage varies with the type of biomass. Lignin prevents the cellulose hydrolysis by binding irreversibly to the enzymes, and the removal of lignin by pretreatment increases the rate of hydrolysis and sugar yields. This indicates that lignin in the biomass demands more cellulase dosages for hydrolysis. However, delignification is not a compulsory treatment required for biomass hydrolysis, because it was found that acid hydrolysis that removes high amounts of hemicellulose and very little fraction of lignin still increases the hydrolysis efficiency of biomass (Zhao et al. 2012).

12.4.4 Other Factors: Physical and Chemical

The cell wall proteins have positive and negative effect on hydrolysis enzymes, but in general this effect is not observed during hydrolysis because of inactivation of these proteins during pretreatment. Fiber size, accessible surface area, pore size, and pore area are few physical parameters that have influence on biomass hydrolysis. Removal of lignin and hemicellulose increases the porosity and average pore size of the bio-mass. The reduction in particle size and increase in porosity increases the accessible surface area of the biomass to enzymes and thus improves the hydrolysis rate. Degree of polymerization (DP) is another important factor that influences the cellulose hydrolysis, though it is not an independent factor. Lower DP cellulose is more acces-sible to enzymes due to the more binding sites of cellulose to enzymes. Altering DP also changes crystallinity and porosity of the substrate. The increase in the surface area of small fibers and fines contributes to the higher hydrolysis rate of substrates with smaller fibers (Chandra et al. 2007; Gupta and Lee 2009; Zhao et al. 2012).

12.5 Lignocellulosic Biomass Deconstruction

To increase the enzymatic digestibility of lignocellulosic biomass, several pretreat-ment methods are followed. The type of pretreatment to be used varies with the bio-mass and the end-product requirement. For butanol production from lignocellulosic biomass, various pretreatments like acid, alkali, steam explosion, and liquid hot water (LHW) were tried. Pretreatment is followed by enzymatic hydrolysis to release the monomeric sugars and is one of the most complex steps of the lignocellulosic biomass to butanol production process. Figure 12.1 explains in brief the process of lignocellulosic biomass deconstruction and action of enzymes. The same is discussed as follows in detail.

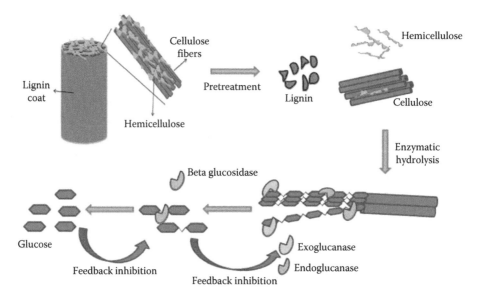

Figure 12.1 Deconstruction of lignocellulosic biomass by pretreatment and enzymatic hydrolysis; mode of action of cellulolytic enzymes.

12.5.1 Mechanical Pretreatment

The lignocellulosic biomass normally obtained in larger particles from the supplier is often reduced to smaller particle size because of ease of handling and increased surface/volume ratio. This can be done by chipping, milling, or grinding. Mechanical pretreatment causes the biomass to shear and reduces the degree of polymerization, which in turn increases the available accessible area for the enzymes. In most cases, milling increased the hydrolysis rate by 5%–25%, but particle size reduction to less than 40 mesh has little effect on the hydrolysis yield and rates (Chang et al. 1997; Chang and Holtzapple 2000).

12.5.2 Thermal Pretreatment

In thermal pretreatment, at temperatures above 150°C–180°C, hemicellulose and lignin present in the lignocellulosic biomass start to solubilize (Bobleter 1994). At temperatures above 180°C, exothermal reaction takes place by breaking down hemicellulose and forming acids that are assumed to catalyze further hydrolysis of hemicellulose. However, temperatures above 160°C lignin and hemicellulose solubilize, which results in the release of toxic compounds like furfurals, hydroxymethyl furfurals, and phenolic compounds (Quintero et al. 2011).

12.5.2.1 Steam Pretreatment and Steam Explosion

Steam pretreatment, steam explosion, and LHW pretreatment are the common thermal pretreatment methods. During steam pretreatment and steam explosion, the moisture content of the biomass influences the pretreatment time. Steam explosion differs from steam pretreatment due to its sudden reduction in pressure and separation of fibers caused by explosive decompression. Steam explosion has both mechanical and chemical effects on the lignocellulosic biomass. The mechanical effects are caused by pressure and the chemical effects by the autohydrolysis of hemicellulose by acetyl groups (Maurya et al. 2015; Mok and Antal 1992). Steam explosion of corn stover and enzymatic hydrolysis resulted in generation of inhibitors in the hydrolysate. Hydrolysate was detoxified with activated charcoal. The detoxified hydrolysate fermented with *C. acetobutylicum* ATCC 824 resulted in 12.38 g/L of ABE, of which butanol was above 8 g/L (Wang and Chen 2011).

12.5.2.2 Liquid Hot Water Pretreatment

In LHW pretreatment to avoid the formation of inhibitors, pH should be maintained between 4 and 7 during the pretreatment. This minimizes the formation of monosaccharides and degradation products (Moiser et al. 2005). The higher water input in LHW pretreatment compared to steam pretreatment results in large amount of solubilized products, like the yield of solubilized xylan. However, it was reported that the increase in solid loading also increases the hydrolytic reactions and formation of xylose and furfural (Quintero et al. 2011). Hot water pretreatment and enzymatic hydrolysis of corn fiber generated less inhibitors, and hence, no detoxification was done. Nondetoxified hydrolysate fermented in batch with *Clostridium beijerinckii* BA101 yielded 8.6 g/L of ABE and 6.5 g/L of butanol (Qureshi et al. 2008b).

12.5.3 Thermochemical/Chemical Pretreatment

12.5.3.1 Acid Pretreatment

Acid pretreatment with dilute acid is preferred over concentrated acids, due to the formation of less inhibitors and its feasibility at industrial scale. Concentrated acids are toxic and are corrosive in nature and are less attractive due to the inhibitors formed and requirement of equipment resistant to corrosion. H_2SO_4 is the common acid used for pretreatment, and dilute acid pretreatment is done either at lower temperatures for longer reaction times (120°C for 30–90 min) or at higher temperatures for shorter reaction times (above 150°C for 2–20 min). Dilute acid pretreatment solubilizes hemicelluloses and lignin, hence resulting in generation of inhibitors due to the degradation of solubilized xylan and lignin (Sun and Cheng 2002). Dilute acid pretreatment is the common pretreatment done for butanol production from lignocellulosic biomass. Corn fiber, wheat straw, barley straw, switchgrass, corn stover, and rice straw were tested for butanol fermentation with dilute acid pretreatment and enzyme hydrolysis. In all the cases, it was observed that detoxification of hydrolysate gave better results indicating the generation of inhibitors during pretreatment (Gottumukkala et al. 2013; Qureshi et al. 2007, 2008b, 2010a,b; Ranjan et al. 2013).

12.5.3.2 Alkali Pretreatment

In alkali pretreatment, solvation and saponification take place, which results in the swelling of biomass. Common catalysts used in alkali pretreatment are hydroxides of sodium, potassium, calcium, and ammonium. This pretreatment can be done at both normal and elevated temperatures and pressures. Alkaline catalysts solubilize lignin and minor portions of hemicellulose and cellulose (Quintero et al. 2011). The use of diluted NaOH is found to be more effective in lignin removal, but it has little or no effect on softwoods that have lignin content higher than 26% (Millet et al. 1976). Generation of toxic compounds with alkali pretreatment is low when compared to acid pretreatment, but the formation of phenolics and lignin condensation and precipitation on the biomass are the drawbacks. Alkali pretreatment of switchgrass and enzymatic hydrolysis gave better ABE and butanol production (22.7 and 13 g/L, respectively), when compared to dilute acid pretreatment and enzymatic hydrolysis (1.48 and 0.97 g/L, respectively). Though the organisms used for both the hydrolysates are different, it still indicates the necessity of choosing right pretreatment for better butanol production (Gao et al. 2014; Qureshi et al. 2010b). Lime [$Ca(OH)_2$] is considered more efficient than NaOH. Lime pretreatment has added advantages of low reagent costs, less safety requirements, and easy recovery, although the downstream processing costs are high (Mosier et al. 2005).

Ammonia-based pretreatments like AFEX are considered more advantageous than other alkali pretreatments due to the less to no inhibitor generation, and hence, no water wash of the pretreated biomass is required. In AFEX, the lignocellulosic biomass is treated with liquid ammonia at mild temperatures (90°C–100°C) for 30–60 min, followed by rapid pressure release that causes swelling and physical disruption of biomass. It reduces cellulose crystallinity and lignin fraction of the biomass (Kim et al. 2011). In AFEX, ammonia must be recycled from the pretreatment reactor for economic and environmental benefits. Ammonia recycled percolation by using packed bed reactor and soaking in aqueous ammonia are two other ammonia-based pretreatment methods.

Pretreatment methods described earlier are the most commonly used methods for pretreating the biomass for butanol fermentation, but there are several other types of pretreatments like oxidative pretreatment, carbon dioxide pretreatment, microwave pretreatment, ozonolysis, ionic liquid pretreatment, and other combination pretreatments. They are clearly described in the review by Maurya et al. (2015).

12.5.4 Enzymatic Hydrolysis: Release of Monomeric Sugars

In enzymatic hydrolysis of lignocellulosic biomass, depolymerization of cellulose to monomeric sugars is acquired by the synergistic action of cellulase components. Endocellulase breaks down the cellulose fibers and exposes reducing and nonreducing ends for the action of exocellulases and cellobiohydrolases. Cellobiose molecules formed by the action of cellobiohydrolyse are broken down to glucose monomers by beta glucosidase. The products being formed have feedback inhibition on the enzymes. Cellobiose has strong inhibition on endo- and exocellulase, but its accumulation beyond a concentration is avoided by the action of beta glucosidase. However, beta glucosidase is feedback inhibited by glucose and hence the cellobiose accumulates and inhibits exocellulase and endocellulase (Figure 12.1). Glucose-tolerant beta glucosidase is necessary for the complete conversion of cellulose to sugars, while addition of xylanase exposes cellulose to cellulases by digesting xylan to its monomers.

Due to substrate-related and enzyme-related effects and their interaction, enzymatic hydrolysis is one of the complex steps of lignocellulosic biomass to butanol or any biofuel process. But enzymatic hydrolysis needs milder operating conditions than chemical hydrolysis and results in low energy costs and less inhibitor formation. Enzymes also have an important advantage of having high selectivity toward the substrate and high yield of sugars.

12.6 Microbial Inhibitors for ABE Fermentation

12.6.1 Source of Inhibitors

Pretreatment and hydrolysis of biomass yield a range of microbial inhibitors like weak acids, furan derivatives, and phenolic compounds. Furfural, hydroxymethyl furfurals, acetic acid, and phenolics are the common inhibitors generated during dilute acid pretreatment. Nonchemical pretreatment like steam explosion results in the generation of formic acid, acetic acid, phenolics, and furfurals. However, the concentration of inhibitors formed by steam explosion is less in comparison to that of the dilute acid pretreatment at high temperature and pressure. Alkali pretreatment was reported to generate few inhibitors, but the salts that are tough to remove after the pretreatment are toxic to clostridia. Hence, alkaline pretreatment is considered less desirable for butanol production by ABE fermentation. There were also reports on the inhibition of enzymatic hydrolysis by the salts formed during the alkaline pretreatment of biomass (Baral and Shah 2014).

12.6.2 Action of Fermentative Inhibitors

12.6.2.1 Clostridium Metabolism and ABE Fermentation Pathway

Clostridia are obligatory anaerobic, spore forming, and hetero fermentative bacteria. C. acetobutylicum and C. beijerinckii are two well-known ABE fermenters and are reported to have a similar metabolism. ABE fermentation takes place in two stages: The first stage is the acidogenic phase, where cells divide in vegetative stage and sugars are converted to acids. The second one is the solventogenic phase, in which cell division stops and solvent formation from sugars or by assimilation of acids occurs. Carboxylic acids, mostly acetic and butyric acid, formed during acidogenic phase lower the pH and induce solventogenic enzymes. Solventogenic phase is an adaptive phase for the low pH conditions and responds by ceasing the acid formation, while cell growth also ceases. In this phase, pH increases slowly as the acids secreted earlier reenter the cells and act as cosubstrates for solvent production. Acetone, butanol, and ethanol are the common solvents formed during solventogenesis in a 3:6:1 ratio, but this can vary with substrate and fermentation conditions.

The biochemical pathways for acidogenesis and solventogenesis are complex and involve several enzymes and cofactors. Gheshlaghi et al. (2009) explained the metabolic pathways of butanol producing Clostridium in detail. Glucose is converted to pyruvate via Embden–Meyerhof–Parnas pathway, where 1 mol of glucose yields 2 mol of pyruvate with the net formation of ATP and NADPH (Gottschalk 1986). Clostridia can utilize pentoses by pentose phosphate pathway and the phosphate intermediates are converted to fructose-6-phosphate and glyceraldehyde-3-phosphate, which enters the glycolytic pathway and forms pyruvate by yielding ATP and NADPH (Rogers 1986). C. butyricum utilizes glycerol by converting it to dihydroxyacetone phosphate, which enters EMP pathway (Saint-Amans et al. 2001). Pyruvate is the key intermediate of ABE pathway and cleaves to form acetyl CoA and CO_2 by the action of pyruvate ferredoxin reductase and concurrent reduction of ferredoxin. Acetyl-CoA is further converted to acetone; acetate and CO_2 (oxidized products); and butyrate, butanol, and ethanol (reduced products) (Gheshlaghi et al. 2009).

12.6.2.2 Effect of Inhibitors on Metabolism

Butanol and ethanol production from glucose is highly dependent on NADPH. In the presence of furfurals and hydroxymethyl furfurals, NADPH is utilized to convert them to less toxic compounds like furfuryl alcohols, and hence, there will be a decrease in ethanol and butanol production (Ujor et al. 2014). Furfurals and HMF also inhibit the cell replications at higher concentrations. Weak acids inhibit fermentation by the accumulation of intracellular anions or uncoupling of cell metabolism (Baral and Shah 2014). Acetates, furfurals, and HMF, at concentrations lower than 1.98 g/L, did not affect the growth of *C. beijerinckii* BA101 and butanol formation, but the concentrations of furfurals and HMF at 0.5 g/L enhanced ABE fermentation and productivity (Ezeji et al. 2007). It was reported by Wang et al. that formic acid at concentrations as low as 0.04 g/L caused acid crash and reduced solvent formation (Wang and Cheng 2011).

Phenolic compounds formed by lignin degradation inhibit acidogenic phase by interfering with the pathway from acetyl CoA to butyryl CoA, which in turn affect solvent formation. The application of vanillin, syringaldehyde, and ferulate decreases the formation of ATP, and these compounds are found to be more toxic than furfurals and HMF, even at lower concentrations. Phenolic compounds have high hydrophobicity potential (log p), an indication of toxicity when compared to weak acids and furfurals. Phenolic compounds, less than 1 g/L, inhibited cell growth by 64%–74% and completely inhibited butanol production (Baral and Shah 2014). Syringaldehyde is not toxic to cell growth but is inhibitory to butanol formation. Ferulic acid is found to be more toxic than coumaric acid for cell growth. ABE fermentation ceased at the concentrations of 0.3 and 0.5 g/L of ferulic acid and coumaric acid, respectively (Ezeji et al. 2007).

Neutralization of alkali- or acid-pretreated substrate for butanol fermentation results in the formation of salts like sodium sulfate and sodium chloride. It has been reported that these salts are toxic to *Clostridium* and the presence of these salts in the fermentation medium inhibited cell growth. In alkali pretreatment, acetate present in the biomass reacts with sodium hydroxide and forms sodium acetate, which is inhibitory for ABE fermentation (Baral and Shah 2014). To avoid the formation of salts, washing of biomass is advised over neutralizing by acids or alkali solutions.

12.7 Detoxification of Hydrolysate for ABE Fermentation

Milder pretreatment conditions and less inhibitor generation are encouraged for the conversion of lignocellulosic biomass to biofuel, but this might result in poor sugar yield and overall low product yield. Detoxification is an additional process step in lignocellulosic biomass conversion to solvents aiming at higher yields of butanol. There are various physical, chemical, and biological methods developed for lignocellulose hydrolysate detoxification. They include alkali treatment, adsorption, extraction, evaporation, and microbial and enzyme catalysts. Detoxification methods followed for different classes of toxic compounds are discussed here.

12.7.1 Alkaline Detoxification

The process of overliming dilute sulfuric acid–pretreated slurry of lignocellulosic biomass leads to the formation of calcium sulfate precipitation. The detoxification is due to the precipitation of toxic compounds and salts. Calcium hydroxide treatment causes extensive loss of sugars by stabilizing enolate ions that degrade to furfurals and hydroxymethyl furfurals (Alriksson et al. 2005; Martinez et al. 2001). Hydrolysate treatment with other alkaline solutions like ammonium hydroxide and sodium hydroxide was found to give better ethanol fermentability (Martinez et al. 2001), but such detoxification method is not yet tried for ABE fermentation. However, alkaline detoxification is known for sugar loss, and Nilvebrant et al. reported the easy degradation of xylose when compared to other monomeric sugars (Nilvebrant et al. 2003). Overliming of dilute acid–pretreated enzymatic hydrolysate of corn stover, switchgrass, and barley straw exhibited improved butanol production when compared with nondetoxified hydrolysates (Qureshi et al. 2010a,b).

12.7.2 Adsorption

Fermentation inhibitors can be removed from the hydrolysate by treating with activated charcoal, ion-exchange resins, and polymeric adsorbents like XAD resins. The effectiveness of adsorption depends on the type of adsorbent, ratio of the adsorbents to the inhibitors, concentration, pH, temperature, and contact time. Activated charcoal treatment is efficient in removing phenolic compounds, acetic acid, furfurals, and hydroxymethyl furfurals without intense effect on sugars in the hydrolysate. Ion-exchange resins and polymeric adsorbents remove most of the inhibitors present in the hydrolysate but are not considered a cost-effective technology. Chandel et al. (2007) reported the removal of 63.4% furans and 75.8% phenolics by anionic resins from the sugarcane bagasse acid hydrolysates. Activated carbon treatment of hydrothermolysed and enzyme-treated switchgrass hydrolysate performed better in butanol production when compared to calcium carbonate pretreatment (Liu et al. 2015). Gottumukkala et al. (2013) reported improved butanol production from rice straw hydrolysate treated with XAD 7, XAD16, and anion-exchange resin when compared to nondetoxified hydrolysate. Among all the resins tested, anion-exchange resin gave maximum butanol production.

12.7.3 Extraction

Ethyl acetate is a well-known liquid extractant of inhibitor compounds and was reported to extract 56% acetic acid, total furans, vanillin, and hydroxybenzoic acid (Wilson et al. 1989). By ethyl acetate extraction, 86% phenolics are removed from eucalyptus wood hemicellulose hydrolysate (Cruz et al. 1999). Though the removal of inhibitors by liquid–liquid extraction is significant, it is not a widely encouraged technique at industrial level due to technical restrictions of the method.

12.7.4 Enzyme-Catalyzed Detoxification

Certain enzymes have the ability to detoxify the hydrolysate by altering the chemical nature of inhibitors in the hydrolysate. Cho et al. (2009) tested peroxidase-catalyzed detoxification of model phenolic compounds for butanol production. Caumoric acid, ferulic acid, vanillic acid, and vanillin were inhibitory to cell growth (64%–74%) at 1 g/L concentration and were 100% removed by treating the model phenolic solution with 0.01 mM of peroxidase enzyme. The detoxified solution exhibited improved cell growth and butanol production. The biological methods of detoxification are more feasible with less energy requirements and are environment friendly. But the slow reaction time of microbial detoxification and the loss of sugars make it unattractive.

Though the discussed detoxification methods can be effectively optimized at industrial scale, even the cheaper technology leads to an additional process step that needs capital cost and running costs. Hence, inhibitor-resistant microbes are being developed by the researchers through genetic engineering and directed evolution methods.

12.8 Butanol Fermentation and Process Improvements

Butanol production from lignocellulosic biomass is being well investigated, and several types of fermentations are tested for improved yield and concentration of butanol. Examples of the various fermentation strategies tried with various substrates for butanol production are given in Table 12.3.

12.8.1 Conventional Fermentation Techniques

Conventional ABE fermentation is a batch process and takes 2–6 days to complete the process depending on the bacterial strain and substrate type. Batch fermentations have disadvantages such as low cell densities, substrate and product toxicities, and low product concentrations. Process productivity in batch operations is limited to 0.5 g/L/h and final butanol concentrations in batch fermentation vary from 12 to 20 g/L depending on the microbial strain and fermentation conditions. Butanol concentration above 15 g/L is achieved with engineered or mutant strains of *Clostridium*, because for the native strains butanol concentration at approximately 16 g/L is toxic and completely ceases the cell growth. *C. beijerinckii* BA101 and *C. acetobutylicum* JB200 are two mutant hyper butanol producers capable of producing 19 and 21 g/L, respectively (Qureshi and Blaschek 2001; Xue et al. 2012).

Standard fed-batch and continuous fermentations are not feasible for butanol production due to the biphasic fermentation, end-product toxicity, and strain degeneration properties of solventogenic *Clostridium*. Hence, advanced fermentation strategies with integrated product removal are being tested to make ABE fermentation industrially feasible. Examples of such strategies are described as follows and given in Figure 12.2.

12.8.2 Advanced Fermentation with Integrated Product Recovery

12.8.2.1 Fed-Batch Fermentation
In fed-batch fermentation, substrate inhibition can be avoided by starting the fermentation with relatively low substrate concentrations. Substrate is fed to the reactor at a

TABLE 12.3   Butanol Production by Various Fermentation Techniques with Simple Sugars and Complex Substrates

Substrate	Organism	Fermentation	Product Recovery	Butanol/ABE	Reference
Maltodextrin/glucose	*C. beijerinckii* BA101	Batch	No	19 g/L	Formanek et al. (1997)
Glucose	*C. acetobutylicum* ATCC824	Batch	No	13.86 g/L	Monot et al. (1982)
Glucose and methyl viologen (reducing compound)	*C. acetobutylicum* B3	Batch (immobilized cells)	No	15.6 g/L	Liu et al. (2013)
Glucose and xylose	*C. acetobutylicum*	Batch (cells pregrown in CaCO₃)	No	11.5 g/L	Kanouni et al. (1998)
Glucose	*C. acetobutylicum*	Two-stage continuous culture under phosphate limitation	No	175 mM	Bahl et al. (1983)
Glucose	*C. acetobutylicum* and *C. beijerinckii*	Batch with CaCO₃ supplementation	No	14.78 and 13.89 g/L	Richmond et al. (2011)
Glucose	*C. beijerinckii NRRL* B592	Two-stage continuous culture in semisynthetic medium	No	9.1 g/L	Mutschlechner et al. (2000)
Glucose	*C. acetobutylicum*	Controlled pH	No	12.3 g/L	Yang et al. (2013)
Glucose	*C. acetobutylicum*	Immobilization in cryogel beads	No	14.47 g/L	Tripathi et al. (2010)
Glucose	*C. acetobutylicum*	Batch	Pervaporation	32.8 g/L (ABE)	Evans and Wang (1988)
Glucose	*C. acetobutylicum*	Fed batch	Pervaporation	165 g/L (ABE)	Evans and Wang (1988)
Glucose	*C. beijerinckii* BA101	Fed batch	Gas stripping	233 g/L (ABE)	Ezeji et al. (2004a)
Glucose	*C. beijerinckii* BA101	Continuous	Gas stripping	460 g/L (ABE)	Ezeji et al. (2004b)
Lignocellulose and starch					
Corn fiber	*C. beijerinckii*	Batch	No	6.5 g/L	Qureshi et al. (2008a)
Rice straw	*C. sporogenes* BE01	Batch	No	5.5 g/L	Gottumukkala et al. (2013)
Wheat straw	*C. beijerinckii* P260	Batch	No	12 g/L	Qureshi et al. (2007)
Wheat straw	*C. beijerinckii* P260	SSF	Gas stripping	7.4 g/L	Qureshi et al. (2008a)
Rice straw	*C. acetobutylicum* MTCC 481	Batch	No	12 g/L	Ranjan et al. (2013)

(Continued)

TABLE 12.3 (*Continued*)   Butanol Production by Various Fermentation Techniques with Simple Sugars and Complex Substrates

Substrate	Organism	Fermentation	Product Recovery	Butanol/ABE	Reference
Barley straw	C. beijerinckii P260	Batch	Gas stripping	47.2 g/L (ABE)	Qureshi et al. (2010a)
Corn stover	C. beijerinckii P260	Batch	Gas stripping	50.14 g/L (ABE)	Qureshi et al. (2010b)
Corn fiber	C. beijerinckii BA101	Batch	None	9.29	Qureshi et al. (2008b)
Cassava bagasse	C. acetobutylicum JB200	Fed batch + cell immobilization	Gas stripping	7.6 g/L	Lu et al. (2012)
Aspen wood	C. acetobutylicum ATCC824	Fed batch	Liquid–liquid extraction	13.5 g/L	Shah ad Lee (1992)
Maple wood spent liquor	C. acetobutylicum ATCC824	Fed batch	None	7.6 g/L	Sun and Liu (2012)
Spruce spent liquor	C. acetobutylicum DSM792	Continuous cell immobilization	None	7.2 g/L	Survase et al. (2011)
Cassava chips	C. saccharoperbutylacetonicum N1-4	Batch	No	15.5 g/L	Thang et al. (2010)
Corn starch	C. saccharoperbutylacetonicum N1-4	Batch	No	16.2 g/L	Thang et al. (2010)
Sago starch	C. saccharoperbutylacetonicum N1-4	Batch	No	15.5 g/L	Thang et al. (2010)
Cassava starch	C. saccharoperbutylacetonicum N1-4	Batch	No	16.9 g/L	Thang et al. (2010)

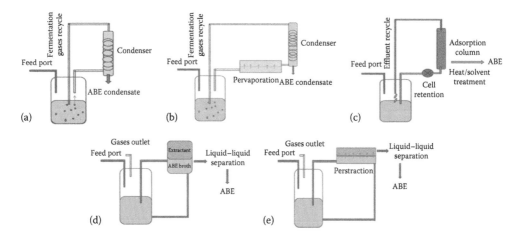

Figure 12.2 Advanced fermentation-integrated product recovery methods for butanol production. (a) Fermentation with *in situ* gas stripping by recycling fermentation gases, (b) *in situ* ABE removal from fermentation broth by pervaporation using fermentation gases, (c) ABE removal by adsorption method with cell retention and effluent recycling, (d) fermentation with *in situ* recovery of ABE by liquid–liquid extraction, and (e) perstraction method for *in situ* ABE recovery from fermentation broth.

slow and controlled rate, based on the rate of consumption. For butanol production, fed-batch fermentation should be integrated with product recovery, as the accumulation of butanol is toxic and fermentation ceases. Ezeji et al. applied integrated fed-batch fermentation-gas stripping product recovery system with pure glucose using *C. beijerinckii* BA101. In this integrated fed-batch system, solvent productivities improved by 400% of the control batch productivities (Ezeji et al. 2004a). Fed-batch butanol fermentation was tried with concentrated cassava bagasse hydrolysate in a fibrous bed reactor with continuous gas stripping using *C. acetobutylicum* JB200. It utilized highly concentrated cassava bagasse hydrolysate and produced 76.6 g/L of butanol, with yield and productivity of 0.23 g/g and 0.29 g/L/h, respectively (Lu et al. 2012). Qureshi et al. used pervaporation method for butanol recovery during fed-batch fermentation with glucose and achieved 154.97 g/L of total solvents, of which 105.35 g/L was butanol. During fermentation, 444.72 g of glucose was utilized and the productivity and yield of total solvents were 0.18 g/L/h and 0.31–0.35 g/g, respectively (Qureshi et al. 2001). Fed-batch fermentation with integrated recovery is a promising technology for higher yields and productivity of butanol. However, in the case of lignocellulosic biomass, high concentrated substrate can be obtained due to the feedback inhibition of enzymatic hydrolysis reaction by the sugars released, and concentrating the hydrolysate would result in the accumulation of inhibitors that negatively affects fermentation.

12.8.2.2 Simultaneous Saccharification and Fermentation

In simultaneous saccharification and fermentation (SSF) process, the substrate is hydrolyzed by enzymes, while the sugar released is simultaneously fermented to solvents. SSF has advantages such as having no feedback inhibition to enzymes and slow release of sugars to the microbes, but it also has major disadvantages like the different optimum conditions for hydrolysis and fermentation. Generally, compromise with hydrolysis is done to run the fermentation at optimum temperature. Hence, thermo-tolerant butanol producers are beneficial for SSF process.

Batch fermentation of wheat straw by SSF with and without product recovery was compared to conventional separate hydrolysis and fermentation process. Gas strip-ping of solvents during SSF with agitation produced the highest concentration of ABE (21.42 g/L). Separate hydrolysis and fermentation of slurry without removing sediments produced 13.12 g/L of ABE, while SSF without agitation and gas stripping produced only 11.93 g/L of ABE (Qureshi et al. 2008a). It shows that during SSF, efficient hydrolysis of the substrate and product recovery is required to achieve maxi-mum yields and productivity of butanol.

12.8.2.3 Two-Stage Continuous Fermentation
In chemostat fermentation, reactor is fed continuously at a particular dilution rate and broth with the product is continuously removed, keeping the reactor volume constant. However, it has drawbacks like cell washout, product recovery from dilute solutions, and disposal of large volumes of effluent. A two-stage continu-ous culture mimics the two stages of ABE fermentation (acidogenic phase and sol-ventogenic phase). In the first-stage turbidostat, cells are grown acidogenically, at faster growth rate, and are transferred to the second stage for solvent production and complete utilization of sugars. Mutschlechner et al. (2000) demonstrated the two-stage continuous culture in a steady-state operation for 1600 h at very low dilution rates to reduce the cell washout. Overall solvent concentration of 15 g/L and productivity of 0.27 g/L/h were achieved. In a study of two-stage fermenta-tion by Bankar et al. (2012), reactor effluent was recycled back with liquid–liquid extraction of the product between two stages and sugarcane bagasse was used as cell holding material. This was to achieve complete sugar utilization for higher sol-vent yields and productivity. In the two-stage fermentation, glucose utilization was maximum (83.21%) at dilution rates of 0.051 h^{-1}, when compared to 54.38% sugar utilization in single-stage chemostat. The highest solvent production of 25.32 g/L (16.9 g/L of butanol) with a productivity of 2.5 g/L/h was achieved, when compared to 15.98 g/L of solvent production in a single stage. This demonstrated that two-stage fermentation is beneficial than single-stage fermentation in terms of sugar utilization and solvent production.

12.8.2.4 Multistage Continuous Fermentation
A four-stage continuous fermentation was tested with cane molasses and corn stover hydrolysate. For cane molasses, gradient modes of dilution rates (0.15, 0.15, 0.125, 0.1 h^{-1}) were used to increase the retention time in reactor 3 and reactor 4, and the average ABE titers of 7.15, 12.24, and 13.75 g/L were obtained in R1, R3, and R4, respectively. ABE productivity decreased in R3 and R4 (0.574 and 0.439 g/L/h) when compared to R1 (1.072). This must be due to the longer retention times achieved with gradient dilution rates. For corn stover hydrolysate, two constant dilution rates (0.1 and 0.15 h^{-1}) were tested for four-stage continuous fermentation. The total ABE

solvent production in R4 was 13.44 and 11.43 g/L with cell densities of 1.37 and 3.79 OD_{600} for the dilution rates of 0.1 and 0.15 h^{-1}, respectively (Ni et al. 2013). It should be noted that higher dilution rates were beneficial to cell growth and lower dilution rates were positive for solvent production.

12.8.2.5 High-Cell-Density Fermentation

ABE fermentation has drawbacks of lower productivity due to low cell concentrations. To overcome this, high-cell-density fermentation with immobilized cells was investigated. *C. beijerinckii* BA101 was immobilized onto clay brick for high-cell-density continuous fermentation. Dilution rates of 0.3–2.5 h^{-1} were tested, and the highest solvent productivity of 15.8 g/L/h was achieved at a dilution rate of 2.5 h^{-1} with the solvent yield of 0.38 g/g. The immobilized cell continuous reactor was stable for 550 h (Qureshi et al. 2000). Huang et al. (2004) demonstrated the high-cell-density fibrous bed reactor by feeding glucose and butyrate as cosubstrates. The optimal butanol productivity of 4.6 g/L/h and a yield of 0.42 g/g were obtained at a dilution rate of 0.9 h^{-1} and pH 4.3. A continuous immobilized cell plug flow (biofilm) reactor was tried with *C. beijerinckii* BA101 immobilized onto clay brick to achieve high reactor productivity. In this fermentation, to improve the sugar utilization, reactor effluent was recycled after removing the butanol by pervaporation. A dilution rate of 2.0 h^{-1} resulted in a reactor productivity of 16.2 g/L/h and sugar utilization of 101.4% (Lienhardt et al. 2002).

Cell recycling is another technique used for high-cell-density fermentations. Cell recycling bioreactors are advantageous due to homogeneity of the broth that facilitates diffusion in the bioreactor. In a high-cell-density continuous ABE fermentation achieved by cell recycling of *C. saccharoper butylacetonicum* N1-4, the maximum ABE productivity of 11.0 g/L/h was obtained at a dilution rate of 0.85 h^{-1}. With continuous recycling of cells, cell concentration greater than 100 g/L resulted in heavy bubbling and broth outflow. Hence, cell bleeding was performed with cell recycling above 100 g/L of cell concentration. Continuous culture with cell recycling and bleeding was operated for more than 200 h without strain degeneration, but the ABE productivity decreased to 7.55 g/L/h (Tashiro et al. 2005).

Though several types of fermentations were tested, most of the studies were done in glucose as carbon source and very few studies were on lignocellulosic biomass hydrolysates. Efforts are being put to improve the butanol production from lignocellulosic biomass hydrolysates by using advanced fermentation technologies.

12.9 Principles of Integrated Product Recovery Techniques

In advanced butanol fermentations, butanol is recovered from the broth during fermentation by *in situ* removal techniques. Conventional distillation is not feasible for butanol removal because of the low product concentrations and its toxicity to cells. The common techniques used for *in situ* product recovery are gas stripping, pervaporation, liquid–liquid extraction, perstraction, and adsorption.

12.9.1 Gas Stripping

Gas stripping is an efficient simple to operate technique for continuous removal of butanol from fermentation broth. After 12–24 h of fermentation, nitrogen gas or fermentation gases CO_2 and H_2 are sparged through the fermentation vessel continually for the whole fermentation period. The solvents stripped off by the gas are condensed at temperatures below 10°C, and the gas with any uncondensed solvents is recycled back to the fermenter. Gas stripping and condensation temperature, gas recycle rate, and bubble size are few important parameters that influence butanol recovery. The solvent recovery rate is increased by applying continuous vacuum or cyclic vacuum during gas stripping (Mariano et al. 2012). The energy required for gas stripping decreases with the increase in solvent concentration in the fermentation broth (Xue et al. 2012). Though gas stripping is cost-effective and easy, it is not a selective recovery technique. The butanol is recovered with other solvents and large amounts of water, which should be further separated by distillation. It is the same with flash fermentation, where solvents are stripped out by boiling under low pressure with less selectivity.

12.9.2 Pervaporation

Pervaporation is a membrane-based technique, in which solvents are selectively removed from the fermentation broth. Fermentation broth is passed through one side of the selective membrane module, while on the other side of the membrane, vacuum or sweep gas is applied. Solvent vapors selectively diffuse through the membrane and are condensed at lower temperatures in the condenser by circulating the coolant through it. Membrane fouling is the common drawback of pervaporation; hence, the type of the membrane and its thickness and surface area should be selected appropriately for the efficient recovery of solvents and to minimize fouling problems. Polypropylene membrane has high flux but low selectivity for butanol (Gapes et al. 1996). PDMS and silicalite and silicone membranes are found to be better than the other membranes (Qureshi and Blaschek 2001; Van et al. 2012). Pervaporation is comparatively cost intensive than the gas stripping technique, due to the membrane installation to the reactor and its maintenance. However, its selectivity toward solvents reduces further butanol separation and purification costs.

12.9.3 Liquid–Liquid Extraction

In liquid–liquid extraction, fermentation broth is layered or mixed with water-insoluble organic extractant. Butanol solubilizes in the organic extractant with high partition coefficient and is removed from the fermentation broth. The concentrated butanol in the organic extractant can be further separated and purified by distillation. Though several extractants were tried, oleyl alcohol is the common extractant used for liquid–liquid extraction due to its nontoxic nature. However, methylated crude palm oil and biodiesel were demonstrated to be efficient and better extractants than oleyl alcohol for butanol recovery (Ishizaki et al. 1999; Li et al. 2010). Liquid–liquid extraction is also one among cost-effective and easy operation *in situ* recovery techniques. Selection of extractant and optimization of extractant to fermentation broth ratio are important to achieve high percentage recovery of butanol. The use of liquid–liquid extraction has disadvantages such as toxicity of extractant to cells, loss of extractant to fermentation broth, and fouling of fermentation broth by forming emulsions.

12.9.4 Perstraction

Perstraction is a membrane-based technique like pervaporation. It is similar to liquid–liquid extraction, but the fermentation broth is separated from the extractant by a membrane. The membrane allows the diffusion of ABE to the extractant and also provides selectivity in recovery of solvents. Loss of extractant to the fermentation broth can be reduced or completely avoided by perstraction method. Silicone rubber and other hydrophobic porous membranes like porous Teflon or polypropylene are used for perstraction (Groot et al. 1990). It has several process variables to be optimized such as the type of the membrane, surface area and porosity of the membrane, type of extractant, and fermentation broth to extractant ratio. Perstraction has disadvantages such as membrane fouling and difficulty in achieving high fluxes due to mass transfer resistance in extractant phase. Though combination of extractant and membrane increases the selectivity of solvent recovery, it also makes it an expensive technique for large-scale operations.

12.9.5 Adsorption

In adsorption, cell-free broth is passed through the packed bed of resins or adsorption membrane for the recovery of solvents. Several adsorbents like activated carbon, XAD resins, bonopore, polyvinylpyridine, and silicalite were tested for butanol recovery (Qureshi et al. 2005; Xue et al. 2014; Yang et al. 1994; Yang and Tsao 1995). Silicalite is found to be more effective with maximum recovery percentage of butanol. Butanol adsorbed onto adsorbents can be recovered by heat treatment or by the application of low-boiling-point solvents like methanol (Qureshi et al. 2005). Adsorption with unclarified or partially clarified broths has a major disadvantage of fouling. Adsorption is comparatively very selective and less energy intensive than other recovery techniques.

Each *in situ* recovery technique has its own advantages and disadvantages in terms of selectivity and energy required (Mariano et al. 2012; Qureshi et al. 2005; Xue et al. 2012, 2014). The selectivity and energy requirements of common butanol recovery methods are given in Table 12.4. From the table, it is clear that the increase in butanol concentration reduces the energy required for butanol recovery, which means reduction in cost. The selection of butanol recovery method with respect to concentration in the fermentation broth is important. Above all the recovery methods, gas stripping

TABLE 12.4 Selectivity and Energy Requirements of Various *In Situ* Recovery Methods

Butanol Recovery Method	Butanol Concentration (g/L)	Selectivity	Energy Requirement (MJ/kg)
Gas stripping and distillation	5	4–22	14–31
	8	4–22	8
Vacuum	26.8	1.55–33.8	32.4
Cyclic vacuum	26.8	1.55–33.8	22
Flash fermentation and distillation	5.6	—	22.3
	10.3	—	13.4
	37.1	—	9.3
Liquid–liquid extraction	—	1.2–4100	7.7–26
Perstraction	—	1.2–4100	7.7
Pervaporation	—	2–209	2–145
Adsorption and desorption	—	130–630	8

and adsorption are less energy intensive, but adsorption with resins leads to higher selectivity of butanol than that by gas stripping. In short, there are several factors to be considered for the selection of an *in situ* recovery method for butanol fermentation.

12.10 Conclusions

Lignocellulosic biomass is a potential feedstock for economical ABE fermentation; however, a proper method of pretreatment, hydrolysis, and detoxification should be identified for all the biomass types available. Understanding the physiology of butanol producing *Clostridium* and developing a hyperbutanol producer will be beneficial to reduce the downstream processing cost of biobutanol, as it is clear that the increase in butanol concentration in the fermentation broth reduces the energy requirement for its recovery. An optimal bioprocess can be developed with improved strain, advanced bioprocess technologies, and efficient downstream processing. Optimum technology include cost-effective breakdown of lignocellulosic biomass to sugars without generating inhibitors, butanol production with concentrations above 20 g/L, and an efficient *in situ* or *ex situ* integrated recovery method. The laboratory-developed process should be studied at pilot scale to understand the challenges presented earlier. Technoeconomic assessment of the developed process technologies is important to assess the process feasibility at industrial scale.

References

Alriksson, B., Horváth, I.S., Sjöde, A., Nilvebrant, N., and L.J. Jönsson. 2005. Ammonium hydroxide detoxification of spruce acid hydrolysates. *Appl Biochem Biotechnol* 121–124:911–922.

Bahl, H., Andersch, W., and G. Gotschalk. 1983. Continuous production of acetone and butanol by *Clostridium acetobutylicum* in a two-stage phosphate limited chemostat. *Eur J Appl Microbiol Biotechnol* 15:201–205.

Bankar, S.B., Survase, S.A., Singhal, R.S., and T. Granström. 2012. Continuous two stage acetone-butanol-ethanol fermentation with integrated solvent removal using *Clostridium
acetobutylicum* B 5313. *Bioresour Technol* 106:110–116.

Baral, N. and A. Shah. 2014. Microbial inhibitors: Formation and effects on acetone-butanol-
ethanol fermentation of lignocellulosic biomass. *Appl Microbiol Biotechnol* 98:9151–9172.

Bobleter, O. 1994. Hydrothermal degradation of polymers derived from plants. *Prog Polym
Sci* 19:797–841.

BP Biofuels. Biobutanol fact sheet. http://www.bp.com/content/dam/bp-alternate-energy/en/
documents/bp_biobutanol_factsheet.pdf. Accessed on December 7, 2015.

Chandel, A.K., Kapoor, R.K., Singh, A., and R.C. Kuhad. 2007. Detoxification of sugarcane
bagasse hydrolysate improves ethanol production by *Candida shehatae* NCIM 3501.
Bioresour Technol 98:1947–1950.

Chandra, R.P., Bura, R., Mabee, W.E., Berlin, A., Pan, X., and J.N. Saddler. 2007. Substrate
pretreatment: The key to effective enzymatic hydrolysis of lignocellulosics? *Adv Biochem
Eng/Biotechnol* 108:67–93.

Chang, V.S., Burr, B., and M.T. Holtzapple. 1997. Lime pretreatment of switchgrass. *Appl
Biochem Biotechnol* 63–65:3–19.

Chang, V.S. and M.T. Holtzapple. 2000. Fundamental factors affecting biomass enzymatic
reactivity. *Appl Biochem Biotechnol* 84–86:5–37.

Cho, D.H., Lee, Y.J., Um, Y., Sang, B.I., and Y.H. Kim. 2009. Detoxification of model phenolic compounds in lignocellulosic hydrolysates with peroxidase for butanol production
from *Clostridium beijerinckii*. *Appl Biochem Biotechnol* 83:1035–1043.

Chundawat, S.P., Bellesia, G., Uppugundla, N. et al. 2011. Restructuring the crystalline cellulose hydrogen bond network enhances its depolymerization rate. *J Am Chem Soc*
133:11163–11174.

Cruz, J.M., Jose, D.M., Herminia, D., and J.C. Parajo. 1999. Solvent extraction of hemicellulose wood hydrolysates: A procedure useful for obtaining both detoxified fermentation
media and polyphenols with antioxidant activity. *Food Chem* 67:147–153.

Eisentraut, A. 2010. Sustainable production of second-generation biofuels—Potential and
perspectives in major economies and developing countries. OECD/IEA. https://www.
iea.org/publications/freepublications/publication/second_generation_biofuels.pdf.
Accessed on December 7, 2015.

EPA (United States Environmental Protection Agency). http://www.epa.gov/mtbe/water.
htm. Accessed on December 7, 2015.

Evans, P.J. and H.Y. Wang. 1988. Enhancement of butanol formation by *Clostridium acetobutylicum* in the presence of decanol-oleyl alcohol mixed extractants. *Appl Environ
Microbiol* 54:1662–1667.

Ezeji, T.C., Qureshi, N., and H.P. Blaschek. 2004a. Acetone butanol ethanol (ABE) production from concentrated substrate: Reduction in substrate inhibition by fed-batch technique and product inhibition by gas stripping. *Appl Microbiol Biotechnol* 63:653–658.

Ezeji, T.C., Qureshi, N., and H.P. Blaschek. 2004b. Butanol fermentation research: Upstream
and downstream manipulations. *Chem Rec* 4:305–314.

Ezeji, T.C., Qureshi, N., and H.P. Blaschek. 2007. Butanol production from agricultural residues: Impact of degradation products on *Clostridium beijerinckii* growth and butanol
fermentation. *Biotechnol Bioeng* 97:14609.

Formanek, J., Mackie, R., and H.P. Blaschek. 1997. Enhanced butanol production by
Clostridium beijerinckii BA101 grown in semidefined p2 medium containing 6 percent
maltodextrin or glucose. *Appl Environ Microbiol* 63:2306–2310.

Gao, K., Boiano, S., Marzocchella, A., and L. Rehmann. 2014. Cellulosic butanol production from alkali-pretreated switchgrass (*Panicum virgatum*) and phragmites (*Phragmites australis*). *Bioresour Technol* 174:176–181.

Gapes, J.R., Nimcevic, D., and A. Friedl. 1996. Long-term continuous cultivation of *Clostridium beijerinckii* in a two-stage chemostat with on-line solvent removal. *Appl Environ Microbiol* 62:3210–3219.

Gheshlaghi, R., Scharer, J.M., Moo-Young, M., and C.P. Chou. 2009. Metabolic pathways of *Clostridia* for producing butanol. *Biotechnol Adv* 27:764–781.

Gottschalk, G. 1986. *Bacterial Metabolism*, 2nd edn. New York: Springer Verlag.

Gottumukkala, L.D., Parameswaran, B., Kuttavan Valappil, S., Mathiyazhakan, K., Pandey, A., and R.K. Sukumaran. 2013. Biobutanol production from rice straw by a non acetone producing *Clostridium sporogenes* BE01. *Bioresour Technol* 145:182–187.

Green Car Congress. National Marine Manufacturers Association endorses use of isobutanol in marine fuel market. June 2015. http://www.greencarcongress.com/2015/06/20150618-nmma.html#more.

Groot, W.J., Soedjak, H.S., Donck, P.B., van der Lans, R.G.J.M., Luyben, K. Ch.A.M., and J.M.K. Timmer. 1990. Butanol recovery from fermentations by liquid–liquid extraction and membrane solvent extraction. *Bioprocess Eng* 5:203–216.

Gupta, R. and Y.Y. Lee. 2009. Mechanism of cellulase reaction on pure cellulosic substrates. *Biotechnol Bioeng* 102:1570–1581.

Huang, W., Ramey, D.R., and S.-T. Yang. 2004. Continuous production of butanol by *Clostridium acetobutylicum* immobilized in a fibrous bed bioreactor. *Appl Biochem Biotechnol* 113–116:887–898.

Ishizaki, A., Michiwaki, S., Crabbe, E., Kobayashi, G., Sonomoto, K., and S. Yoshino. 1999. Extractive acetone-butanol-ethanol fermentation using methylated crude palm oil as extractant in batch culture of *Clostridium saccharoperbutylacetonicum* N1-4 (ATCC 13564). *J Biosci Bioeng* 87:352–356.

Jones, D.T. and D.R. Woods. 1986. Acetone-butanol fermentation revisited. *Microbiol Rev* 50:484–524.

Kanouni, E.A., Zerdani, I., Zaafa, S., Znassni, M., Loutfi, M., and M. Boudouma. 1998. The improvement of glucose/xylose fermentation by *Clostridium acetobutylicum* using calcium carbonate. *World J Microbiol Biotechnol* 14:431–435.

Kim, J.W., Kim, K.S., Lee, J.S. et al. 2011. Two-stage pretreatment of rice straw using aqueous ammonia and dilute acid. *Bioresour Technol* 102:8992–8999.

Lee, S.Y., Park, J.H., Jang, S.H., Nielsen, L.K., Kim, J., and Jung, K.S. 2008. Fermentative butanol production by *Clostridia*. *Biotechnol Bioeng* 101:209–228.

Li, Q., Cai, H., Hao, B. et al. 2010. Enhancing clostridial acetone-butanol-ethanol (ABE) production and improving fuel properties of ABE-enriched biodiesel by extractive fermentation with biodiesel. *Appl Biochem Biotechnol* 162:2381–2386.

Lienhardt, J., Schripsema, J., Qureshi, N., and H.P Blaschek. 2002. Butanol production by *Clostridium beijerinckii* BA101 in an immobilized cell biofilm reactor: Increase in sugar utilization. *Appl Biochem Biotechnol* 98–100:591–598.

Liu, D., Chen, Y., Li, A. et al. 2013. Enhanced butanol production by modulation of electron flow in *Clostridium acetobutylicum* B3 immobilized by surface adsorption. *Bioresour Technol* 129:321–328.

Liu, K., Atiyeh, H.K., Pardo-Planas, O. et al. 2015. Butanol production from hydrothermolysis-pretreated switchgrass: Quantification of inhibitors and detoxification of hydrolyzate. *Bioresour Technol* 189:292–301.

Lu, C., Zhao, J., Yang, S.T., and D. Wei. 2012. Fed-batch fermentation for N-butanol production from cassava bagasse hydrolysate in a fibrous bed bioreactor with continuous gas stripping. *Bioresour Technol* 104:380–387.

Mariano, A.P., Filho, R.M., and T.C. Ezeji. 2012. Energy requirements during butanol production and *in situ* recovery by cyclic vacuum. *Renew Energ* 47:183–187.

Martinez, A., Rodriguez, M.E., Wells, M.L., York, S.W., Preston, J.F., and L.O. Ingram. 2001. Detoxification of dilute acid hydrolysates of lignocellulose with lime. *Biotechnol Prog* 17:287–293.

Maurya, D.P., Singla, A., and S. Negi. 2015. An overview of key pretreatment processes for biological conversion of lignocellulosic biomass to bioethanol. *3 Biotech* 5:597–609.

Millet, M.A., Baker, A.J., and L.D. Scatter. 1976. Physical and chemical pretreatment for enhancing cellulose saccharification. *Biotechnol Bioeng Symp* 6:125–153.

Mok, W.S.L. and M.J. Antal Jr. 1992. Uncatalyzed solvolysis of whole biomass hemicellulose by hot compressed liquid water. *Indus Eng Chem Res* 31:1157–1161.

Monot, F., Martin, J.R., Petitdemange, H., and R. Gay. 1982. Acetone and butanol production by *Clostridium acetobutylicum* in a synthetic medium. *Appl Environ Microbiol* 44:1318–1324.

Mosier, N., Wyman, C., Dale, B. et al. 2005. Features of promising technologies for pretreatment of lignocellulosic biomass. *Bioresour Technol* 96:673–686.

Mutschlechner, O., Swoboda, H., and J.R. Gapes. 2000. Continuous two-stage ABE-fermentation using *Clostridium beijerinckii* NRRL B592 operating with a growth rate in the first stage vessel close to its maximal value. *J Mol Microbiol Biotechnol* 2:101–105.

Ni, Y., Xia, Z., Wang, Y., and Z. Sun. 2013. Continuous butanol fermentation from inexpensive sugar-based feedstocks by *Clostridium saccharobutylicum* DSM 13864. *Bioresour Technol* 129:680–685.

Nilvebrant, N.-O., Persson, P., Reimann, A., de Sousa, F., Gorton, L., and L. Jönsson. 2003. Limits for alkaline detoxification of dilute-acid lignocellulose hydrolysates. *Appl Biochem Biotechnol* 105–108:615–628.

Peng, F., Ren, J.L., Xu, F., and R.C. Sun. 2011. Chemicals from hemicelluloses: A review. In *Sustainable Production of Fuels, Chemicals, and Fibers from Forest Biomass*, eds. J.Y. Zhu, X. Zhang, and X. Pan, pp. 219–259. Washington, DC: ACS Symposium Series.

Quintero, J.A., Rincon, L.E., and C.A Cardona. 2011. Production of bioethanol from agro-industrial residues as feedstocks. In *Biofuels: Alternative Feedstocks and Conversion Processes*, eds. A. Pandey, C. Larroche, S.C. Ricke, C.G. Dussap, and E. Gnansounou, pp. 251–285. Academic Press (Elsevier).

Qureshi, N. and H.P. Blaschek. 2001. Recent advances in ABE fermentation: Hyper-butanol producing *Clostridium beijerinckii* BA101. *J Ind Microbiol Biotechnol* 27:287–291.

Qureshi, N., Ezeji, T.C., Ebener, J., Dien, B.S., Cotta, M.A., and H.P. Blaschek. 2008a. Butanol production by *Clostridium beijerinckii*. Part I: Use of acid and enzyme hydrolyzed corn fiber. *Bioresour Technol* 99:5915–5922.

Qureshi, N., Hughes, S., Maddox, I.S., and M.A. Cotta. 2005. Energy-efficient recovery of butanol from model solutions and fermentation broth by adsorption. *Bioprocess Biosyst Eng* 27:215–222.

Qureshi, N., Meagher, M.M., Huang, J., and R.W. Hutkins. 2001. Acetone butanol ethanol (ABE) recovery by pervaporation using silicalite-silicone composite membrane from fed-batch reactor of *Clostridium* acetobutylicum. *J Membr Sci* 187:93–102.

Qureshi, N., Saha, B.C., and M.A. Cotta. 2007. Butanol production from wheat straw hydrolysate using *Clostridium beijerinckii*. *Bioprocess Biosyst Eng* 30:419–427.

Qureshi, N., Saha, B.C., and M.A. Cotta. 2008b. Butanol production from wheat straw by simultaneous saccharification and fermentation using *Clostridium beijerinckii*: Part II-fed-batch fermentation. *Biomass Bioenerg* 32:176–183.

Qureshi, N., Saha, B.C., Dien, B., Hector, R.E., and M.A. Cotta. 2010a. Production of butanol (a biofuel) from agricultural residues: Part I—Use of barley straw hydrolysate. *Biomass Bioenerg* 34:559–565.

Qureshi, N., Saha, B.C., Hector, R.E. et al. 2010b. Production of butanol (a biofuel) from agricultural residues: Part II—Use of corn stover and switchgrass hydrolysates. *Biomass Bioenerg* 34:566–571.

Qureshi, N., Schripsema, J., Lienhardt, J., and H.P. Blaschek. 2000. Continuous solvent production by *Clostridium beijerinckii* BA101 immobilized by adsorption onto brick. *World J Microbiol Biotechnol* 16:377–382.

Ranjan, A., Khanna, S., and V.S. Moholkar. 2013. Feasibility of rice straw as alternate substrate for biobutanol production. *Appl Energ* 103:32–38.

Richmond, C., Ezeji, T.C., and B. Han. 2011. Stimulatory effects of calcium carbonate on butanol production by solventogenic *Clostridium* species. *Cont J Microbiol* 5:18–28.

Rogers, P. 1986. Genetics and biochemistry of *Clostridium* relevant to development of fermentation processes. *Adv Appl Microbiol* 31:1–60.

Saint-Amans, S., Girbal, L., Andrade, J., Ahrens, K., and P. Soucaille. 2001. Regulation of carbon and electron flow in *Clostridium butyricum* VPI 3266 grown on glucose-glycerol mixtures. *J Bacteriol* 183:1748–1754.

Shah, M.M. and Y.Y. Lee. 1992. Simultaneous saccharification and extractive fermentation for acetone/butanol production from pretreated hardwood. *Appl Biochem Biotechnol* 34–35:557–568.

Smeets, E. and A. Faaij. 2007. Bioenergy potentials from forestry in 2050. *Clim Change* 81:353–390.

Sun, Y. and J. Cheng. 2002. Hydrolysis of lignocellulosic materials for ethanol production: A review. *Bioresour Technol* 83:1–11.

Sun, Z. and S. Liu. 2012. Production of N-butanol from concentrated sugar maple hemicellulosic hydrolysate by *Clostridia acetobutylicum* ATCC824. *Biomass Bioenerg* 39:39–47.

Survase, S.A., Sklavounos, E., Jurgens, G., van Heiningen, A., and T. Granström. 2011. Continuous acetone-butanol-ethanol fermentation using so 2-ethanol-water spent liquor from spruce. *Bioresour Technol* 102:10996–11002.

Szulczyk, K. 2010. Which is a better transportation fuel—Butanol or ethanol? *Int J Energ Environ* 1:501–512.

Tashiro, Y., Takeda, K., Kobayashi, G., and K. Sonomoto. 2005. High production of acetone-butanol-ethanol with high cell density culture by cell-recycling and bleeding. *J Biotechnol* 120:197–206.

Thang, V.H., Kanda, K., and G. Kobayashi. 2010. Production of acetone-butanol-ethanol (ABE) in direct fermentation of cassava by *Clostridium saccharoperbutylacetonicum* N1-4. *Appl Biochem Biotechnol* 161:157–170.

Tripathi, A., Sami, H., Jain, S.R. et al. 2010. Improved biocatalytic conversion by novel immobilization process using cryogel beads to increase solvent production. *Enzyme Microb Technol* 47:44–51.

Ujor, V., Agu, C.V., Gopalan, V., and T.C. Ezeji. 2014. Glycerol supplementation of the growth medium enhances *in situ* detoxification of furfural by *Clostridium beijerinckii* during butanol fermentation. *Appl Microbiol Biotechnol* 98:6511–6521.

Van Hecke, W., Vandezande, P., Claes, S. et al. 2012. Integrated bioprocess for long-term continuous cultivation of *Clostridium acetobutylicum* coupled to pervaporation with PDMS composite membranes. *Bioresour Technol* 111:368–377.

Wang, L. and H. Chen. 2011. Increased fermentability of enzymatically hydrolyzed steam-exploded corn stover for butanol production by removal of fermentation inhibitors. *Process Biochem* 46:604–607.

Weng, J.-K. and C. Chapple. 2010. The origin and evolution of lignin biosynthesis. *New Phytol* 187:273–285.

Wilson, J.J., Deschatelets, L., and N.K. Nishikawa. 1989. Comparative fermentability of enzymatic and acid hydrolysates of steam pretreated aspen wood hemicellulose by *Pichia stipitis* CBS 5776. *Appl Microbiol Biotechnol* 31:592–596.

Xue, C., Zhao, J., Lu, C., Yang, S.T., Bai, F., and I.C. Tang. 2012. High-titer N-butanol production by *Clostridium acetobutylicum* JB200 in fed-batch fermentation with intermittent gas stripping. *Biotechnol Bioeng* 109:2746–2756.

Xue, C., Zhao, J.B., Chen, L.J., Bai, F.W., Yang, S.T., and J.X. Sun. 2014. Integrated butanol recovery for an advanced biofuel: Current state and prospects. *Appl Microbiol Biotechnol* 98:3463–3474.

Yang, X., Tsai, G.J., and G.T. Tsao. 1994. Enhancement of *in situ* adsorption on the acetone-butanol fermentation by *Clostridium acetobutylicum*. *Sep Technol* 4:81–92.

Yang, X. and G.T. Tsao. 1995. Enhanced acetone-butanol fermentation using repeated fed-batch operation coupled with cell recycle by membrane and simultaneous removal of inhibitory products by adsorption. *Biotechnol Bioeng* 47:444–450.

Yang, X., Tu, M., Xie, R., Adhikari, S., and Z. Tong. 2013. A comparison of three pH control methods for revealing effects of undissociated butyric acid on specific butanol production rate in batch fermentation of *Clostridium acetobutylicum*. *AMB Express* 3:3.

Zhang, Y.H.P and L.R. Lynd. 2004. Toward an aggregated understanding of enzymatic hydrolysis of cellulose: Noncomplexed cellulase systems. *Biotechnol Bioeng* 30:797–824.

Zhao, X., Zhang, L., and D. Liu. 2012. Biomass recalcitrance. Part I: The chemical compositions and physical structures affecting the enzymatic hydrolysis of lignocellulose. *Biofuel Bioprod Bioref* 6:465–482.

Section III

Biofuels from Algae

13

Algal Biofuels
Impacts, Significance, and Implications

Brenda Parker, David Benson, and Gill Malin

Contents

Abstract

Recent years have witnessed significant innovation in algal biofuel production technologies and products on a global scale. While these innovations have potentially positive benefits for the sustainable development of the biofuels sector, there are evident caveats. This chapter therefore initially outlines the main emergent algal biofuel production techniques, before examining the significance of their environmental, social, and economic impacts. As our analysis shows, algal biofuels potentially provide alternatives to first-generation biofuels in terms of overall sustainability, but questions still arise over their economic viability and environmental effects, most notably on biodiversity, land use, and water security. The implications for future sectoral expansion and its governance are then discussed.

13.1 Introduction

In the past few years, there has been significant innovation in algal biofuel production technologies and different forms of biofuels globally (Benson et al. 2014). Algae have significant potential for biofuel production for several reasons. This highly diverse, polyphyletic group ranges from single-celled microalgae to multicellular macroalgae or seaweeds that employ light to convert carbon dioxide (CO_2) and nutrients into organic molecules as they grow and reproduce (Driver et al. 2014). The total number of algal species is uncertain, but a recent conservative estimate based on records from the taxonomic database AlgaeBase (Guiry et al. 2014) suggests 72,500, with about half of these yet to be classified (Guiry 2012). By controlling the cultivation strategies of certain algae, biomass can be produced from their photosynthesis that, depending on the specific technique, algal strain, or growth stage, can contain lipids, carbohydrates, proteins, and micronutrients. Lipids, carbohydrates, and proteins can be converted into biofuel products such as biodiesel, bioethanol, hydrocarbons, biogas, and hydrogen (Foley et al. 2011). Algal biofuels consequently have significant potential to contribute to sustainable development through providing alternatives to fossil fuel–derived energy sources, but there are also growing concerns about their overall sustainability (Benson et al. 2014).

Extensive research has already been conducted on the economic, social, and environmental impacts from established forms of biofuels. There is an ongoing worldwide expansion of the so-called *first generation* biofuels, where the source of carbon is sugar, lipids, or starch. For instance, by examining the biodiesel derived from edible vegetable oils produced by agricultural crops such as oilseed rape, researchers have identified multiple concerns for sustainable development. These include environmental threats to water quality and security, land use, increased greenhouse gas (GHG) emissions, and biodiversity loss (Fraiture et al. 2008; Searchinger et al. 2008; Dominguez-Faus et al. 2009; Hoekman 2009; Delucchi 2010; Cai et al. 2011; Demirbas and Demirbas 2011). In addition, such biofuel production, often driven by higher-level subsidies and other policy drivers in developed countries, has supported higher agricultural prices for farmers but disadvantaged the food security of poorer countries that rely on imports of grain (Ewing and Msangi 2009). To an extent, some of these problems are countered through the use of *second-generation* biofuels (Naik et al. 2010). These fuels are produced from the biological or thermochemical processing of lignocellulosic biomass, typically nonedible crops, food crop residues, or forestry by-products (Naik et al. 2010; Bhateria and Dhaka 2015). While these feedstocks avoid the use of food crops, questions still pertain over their economic viability. Academic and political debates have therefore occurred over the promotion of first- and second-generation biofuels, leading some researchers to focus their efforts on developing *third-generation* biofuels from algae as an autotrophic feedstock.

Algal biofuels could have multiple positive impacts in terms of sustainable development, for example, through their potential for reducing GHG emissions, increasing carbon sequestration, and enhancing energy security. While algal biofuel production techniques have existed for several decades (Sheehan et al. 1998), recent years have witnessed significant growth in production processes, types of algal feedstocks, and biofuel products. Genetic modification (GM) of algal strains could be used to extend the types of fuel molecules made in microalgae, overcome biological limitations related to productivity, or ease harvesting and processing, all of which could enhance the competitivity of algal biofuels. However, in the current rush to expand algal biofuel production, there are potential negative externalities such as risks to biodiversity from algae use, which could be further exacerbated by GM deployment. Multiple studies have quantified specific environmental impacts through life cycle assessment (LCA) of production processes, often using process modeling approaches (Lardon et al. 2009; Clarens et al. 2010, 2011; Jorquera et al. 2010; Sander and Murthy 2010;

Stephenson et al. 2010; Campbell et al. 2011; Khoo et al. 2011; Singh and Olsen 2011; Yang et al. 2011; Handler et al. 2012; Liu et al. 2012; Soratana et al. 2012; Ter Veld 2012; Yanfen et al. 2012), along with recent attempts to establish environmental sustainability indicators for production (Efroymson and Dale 2015). Yet, discussion is only just emerging within the scientific literature on the broader sustainability of this developing industrial sector. In addition, there has been rather little discussion to date on just what this expansion in algal biofuels research and production means for its future multilevel governance (Benson et al. 2014). Here, technological development appears to be running ahead of regulatory frameworks. Some impacts may be fully addressed by existing institutional or policy structures, while other, more novel or cross-scale, effects could necessitate entirely new governance mechanisms. As we go on to discuss, GM technologies present particularly unique risks that, to date, have not evidently entered policy discourses let alone constituted a point of debate for researchers involved in their advancement.

In this chapter, we therefore seek to explore potential sustainability impacts—both positive and negative from algal biofuels, their significance, and also the implications for future governance. Three main questions guide our analysis. First, what types of biofuel production technologies and products are evident globally? Initially, we classify the main emerging production techniques, with a focus on open-pond systems (OPS), photobioreactors (PBR) and closed systems, and marine cultivation of macroalgae (Section 13.2). Several different algal biofuel products are then outlined, with a focus on biodiesel, ethanol, and methane (Section 13.3). Second, what are the potential sustainability impacts of algal biofuels? Here, we initially discuss the notions of impact magnitude and significance to then compare algal biofuels to first-generation approaches in terms of their potential contribution to sustainable development (Section 13.4). Finally, what are the implications for future sectoral development and its governance? Depending on the significance of impacts, dedicated policy structures may be required at international and national levels to ensure the socioenvironmental sustainability of production as well as its economic viability (Section 13.5).

13.2 Algal Biofuel Production Processes

There is significant variance in emerging algal biofuel production processes (Brennan and Owende 2010). Nonetheless, three broad categories will be considered in this chapter: OPS, PBR or closed systems, and marine cultivation of macroalgae. We can examine the key characteristics of these forms of cultivation in turn.

The most common production processes worldwide are based on OPS. Here, algal material is mechanically mixed with nutrients and water in open-air artificial ponds or circular tanks driven with a paddle wheel ("raceways") to promote the production of biomass. A review of the literature on algal biofuel production reveals a variety of pond and production pathway designs (Darzins et al. 2010; Handler et al. 2012). Typically, cultivation in open systems takes the form of raceway ponds (Oswald 1988). OPSs have been employed to scale up biofuel production for several reasons, most notably their relative simplicity of construction compared to other approaches, cost efficiency, and flexibility (Wigmosta et al. 2011). While these designs are popular due to their versatility for producing biomass, they do have some limitations including restrictions placed on pond depth by the requirement for natural light that is essential to promote algal growth. Other constraints include variance in external temperatures and pH fluctuations, which can be problematic to control in open systems, thereby influencing the optimization of biomass production. Another obvious problem with OPS is the demand for environmental resources such as water and large areas of land for production facilities. Some operations have significant footprints. For example, Handler et al. (2012) describe single raceway ponds that cover 1 ha and algal pond systems with a total area of several hundred hectares. Research conducted in the United Kingdom shows that using such production processes to meet national biodiesel demand would require 0.5 million hectares of land, even by assuming that 40 tons of biodiesel can be produced per hectare annually (Stephenson et al. 2010). Algal feedstocks in OPS also remain vulnerable to contamination from the external environment, limiting algae strains that can be utilized (Smith et al. 2010; Menetrez 2012). Outdoor cultivation has been most successful for algal species such as *Arthrospira* (*Spirulina*) *platensis* and *Dunaliella salina* because their culture conditions require levels of pH or salinity that are unfavorable to invasive microorganisms.

Many of the factors that can limit OPSs can be controlled in PBRs. While production designs differ significantly (Sevigné Itoiz et al. 2012), PBRs have several common features. Rather than exposing algal material to the open air, PBRs promote its growth within transparent reactors, typically comprising tubes or plates, located indoors or outdoors. In this way, inputs such as water, light, and nutrients can be better controlled to optimize biomass production. Releases of production wastes to the environment can also be more readily contained than in open systems. Such reactors can also reduce land requirements, with some PBR facilities constructed in a vertical fashion. Multiple PBR designs are discussed in the literature, with notable variance in the types of algae grown, nutrient inputs, and fuel product outputs. Closed systems can be used for either phototrophic growth or an organic carbon source such as glucose or glycerol may be supplied for mixotrophic or heterotrophic cultivation of algal strains capable of these trophic modes. Darzins et al. (2010) describe how heterotrophic microalgae can be grown within a closed system to produce fuel from sugars and oxygen. In comparison, PBRs are currently more economically marginal than OPSs due to high capital and running costs, but the latter could have unanticipated environmental effects (Resurreccion et al. 2012).

Current research on algal biofuel production is examining how both PBR and OPS can be used for bioremediation and recycling of environmental wastes. Nutrients are required by algal feedstocks to promote growth, which are primarily supplied from manufactured (e.g., nitrogen derived from the Haber–Bosch process) sources or commercial mining of natural fertilizers. By linking PBR and OPS to wastewater sources of nutrients, such as sewage, anaerobic digestion (AD) systems, or industrial waste, artificial nutrient inputs can be reduced with the additional benefits of bioremediating contaminated water and recycling it for further use (Wang et al. 2008). Carbon dioxide can also be temporarily sequestered by algal biofuel production, thereby providing a means for reducing emissions from polluters such as power stations

(Chung et al. 2010; Jiang et al. 2011). In such approaches, emissions can be directly introduced into production processes, although the efficiency of this technology is still being assessed for commercial purposes (Rosenberg et al. 2011).

Marine macroalgae (seaweed) present another option for biofuel production, although to date most commercial activity has centered on other products. The cultivation of seaweed for food, agricultural fertilizer, and chemicals has a long tradition in European countries such as Ireland and France (NetAlgae 2012; Walsh 2012). These industries still constitute an important economic sector, with over 185 full-time equivalent people employed and around €18 million per year generated by seaweed harvesting in Ireland in 2011 (Walsh 2012). However, most current seaweed production occurs in East Asia, particularly China (Chung et al. 2010). A range of macroalgae products are available, including feed supplements, food additives, human food, fertilizers, chemicals and chemical precursors (dyes, hydrocarbons, alcohols, sugars), pharmaceuticals, and cosmetics (McHugh 2003). While the production of biofuels from macroalgae is comparatively limited in relation to microalgae production processes, several countries are undertaking research into its potential (Parker and Schlarb-Ridley 2013). Several seaweed species contain sufficiently high lipids or carbohydrates to allow processing into bioethanol, biogas, or biobutanol. According to Bhateria and Dhaka (2015), species such as *Macrocystis pyrifera*, *Laminaria digitata*, and *Sargassum* sp. are potentially suitable for this purpose. Typical production involves growing macroalgae on longlines arranged in grids in open offshore waters, with or without the addition of artificial inputs. Using the integrated multitrophic aquaculture, macroalgae are supplied by waste nutrients excreted from fish farming operations (Troell et al. 2009). Once growth maturity is reached, seaweed is then harvested and transported to onshore facilities for dewatering (drying) and processing. As the costs of macroalgal biofuels production are comparatively high, it is typically supplemented by producing higher-value products including pharmaceuticals and food additives within integrated facilities (Chung et al. 2010) or alongside fish and shellfish production.

13.3 Algal Biofuel Products

As discussed earlier, algae are seen as attractive feedstocks for bioenergy as they do not compete for arable land and can be used to upgrade waste streams. Algal biomass or components such as lipids, starch, or protein can be converted into biofuel products. Multiple fuel molecules can be produced from algae, although only the main three biofuel types will be considered in this chapter, namely, biodiesel, ethanol, and methane.*

Biodiesel can be produced from microalgae, by utilizing the ability of certain strains to accumulate relatively high percentages of intracellular lipid. Cells shift

---

* Other biofuels can be produced from microalgae via synthetic biology approaches, including alkanes (Wang et al. 2013), butanol (Lan and Liao 2012; Shen and Liao 2012), ethylene (Guerrero et al. 2012; Ungerer et al. 2012), and hydrogen (Ghirardi et al. 2000).

their metabolism to divert CO_2 into lipid under stress, such as nitrogen or phosphorous limitation (Hu et al. 2008). In the case of microalgae, which have been studied for biofuel production, such as *Chlorella* or *Nannochloropsis*, the cells produce triacylglycerides (Griffiths et al. 2011). These neutral lipids form droplets inside the cell, protected by a phospholipid membrane. Several processing steps are then required to make biodiesel from algal lipids, primarily harvesting, extraction, and transesterification. Algae are first harvested from ponds or PBRs, with the choice of harvesting technique dependent on the characteristics of the microalgae species and the product (Kim et al. 2013). There are several microalgal harvesting or recovery methods, namely, centrifugation, flocculation, filtration, flotation, magnetic separation, electrolysis, ultrasound, and immobilization (Kim et al. 2013). After harvesting, lipids can then be extracted from the biomass with the techniques used depending on whether algal material is wet or requires dewatering (Molina Grima et al. 2003). Cell disruption can be undertaken through several methods, categorized by Kim et al. (2013) into mechanical, chemical, and biological techniques. The efficacy of the disruption process is a critical step in production pipelines for algal products, and methods range from bead milling, through to high-pressure and/or high-speed homogenization, ultrasonication, pulsed electric field, and microwave treatments, along with chemical and biological lysis (Günerken et al. 2015). The recovery of lipid is accomplished by either solvent extraction, supercritical CO_2 extraction, or ultrasonication (Mercer and Armenta 2011). Due to their high viscosity and hence unsuitability for commercial fuel use, extracted lipids require conversion into biodiesel via a chemical process known as transesterification (Meher et al. 2006). This process can vary according to the algal material. For wet biomass, following subsequent wash steps, solvent is recovered and the residual lipid is transesterified into the constituent fatty acid methyl esters (or biodiesel).

Ethanol can be made from micro- and macroalgae via different routes. In microalgae, ethanol can be made directly via genetic engineering, for example, by the introduction of transgenes and diversion of metabolic pathways in *Synechocystis* sp. PCC6803 (Gao et al. 2012). Alternatively, biomass can be fermented directly to ethanol using yeast that can convert starch or glycogen in algal cells, as demonstrated with an *Arthrospira* species (Aikawa et al. 2013). Fermentation of the sugars in macroalgal biomass, for example, alginate, mannitol, and glucan, has also been investigated for ethanol production. There are two potential process routes for this via genetic engineering of microbial strains capable of alginate conversion to ethanol (Wargacki et al. 2012) or via metabolic engineering of *Saccharomyces* strains for improved fermentation performance to enable fractionated sugars to be converted into ethanol (Enquist-Newman et al. 2014).

Biomethane is made from algae using anaerobic digestion (AD). Algal biomass (either macro- or microalgal) is incubated in the presence of an inoculum consortium of methanogenic microorganisms. The resulting biogas is a mixture of methane, hydrogen sulfide, carbon dioxide, and other trace impurities. In most AD operations, the biogas is valorized via a combined heat and power installation for the simultaneous generation of heat and electricity, rather than supplied into the grid due to the impurities present in the biogas mixture. While biomass for AD typically requires little preprocessing, both micro- and macroalgae may contain high concentrations of salt or certain inhibitory compounds such as phenolics or heavy metals that can increase retention times or reduce yield. Efficient hydrolysis of macroalgal polysaccharides, particularly alginates, is seen as a rate-limiting step for efficient AD to proceed (Williams et al. 2013).

13.4 Impact Magnitude, Significance, and the Sustainability of Algal Biofuels

But what are the potential impacts for sustainable development of algal biofuels, particularly when compared to first-generation production? Determining the extent to which industrial processes contribute to sustainable development is typically undertaken by LCA. These assessments seek to quantify environmental impacts across production life cycles, including production inputs, production processes, and end-of-life disposal. Problematically when applied to algal production processes, LCA has focused on quantifiable environmental impacts such as energy use, GHG emissions, and water use, with other critical sustainability issues ignored (Benson et al. 2014). One response is to use other environmental sustainability indicators that include potential impacts such as water quality effects, land use, and biodiversity loss (Efroymson and Dale 2015). Although examining multiple indicators can therefore aid broader assessments of sectoral sustainability, the concepts of impact magnitude and significance require further definition.

Impact magnitude equates to a quantitative measurement of an impact, such as GHG emissions or water use, from a production process, while significance relates to the actual effect on the environment (Benson et al. 2014). The magnitude of an impact can therefore be calculated in terms of its physical, temporal, and spatial scales, probability of occurrence, and reversibility. Once quantified, these impacts can be assessed against the sensitivity of the receiving environment, which in "triple bottom line" sustainability terms should include the prevailing biophysical, social, and economic context to determine their significance. While any industrial processes could have multiple impacts, only some will be *significant* due to their effects on the receptor environment. For example, large-scale water use for OPS biofuel production may not be significant in some contexts but can have serious implications in water-stressed areas. We can therefore use this approach to calculate the potential magnitude and significance of algal biodiesel, ethanol, and methane impacts, compared to first-generation biofuels, using several sustainability indicators.

A review of the literature shows that algal biofuels have potentially positive and negative impacts, when compared with first-generation production, with several categories of effects being potentially significant: water usage, fertilizer inputs, air emissions, land use, global warming potential (GWP), and biodiversity. Such a review is necessarily limited by the propensity of the LCA literature on biofuel production to use varying criteria and approaches to calculating impacts. Nonetheless, several broad trends can be detected for comparison and further discussion.

First, water use could be a significant impact. For first-generation biofuel crops, the global average water footprint per ton of crops varies. Estimates for various crops have been suggested (Mekonnen and Hoekstra 2011) ranging from sugar crops at 200 m^3/ton to cereals at 1600 m³/ton and to oil crops at 2400 m³/ton. Titers of algae in growth media typically range between 1 and 5 g/L. Extrapolating this to a comparable scale suggests that to make 1 ton of microalgal biomass, 200–1000 m³/ton of water is needed for cultivation at current productivities. However, algal growth water can at least be recycled to system postharvest. Furthermore, classification of water use into blue, green, and greywater by Gerbens-Leenes et al. (2014) shows that algal biofuel demand can be lower than cereal biofuel crops. Additional water requirements are needed for OPSs to counter evaporation losses and for PBRs that use water for spray cooling. If seawater is used for the cultivation of marine microalgae, there is still a requirement for freshwater to counter evaporation losses, especially if cultivation is taking place in hot climates. For example, Seambiotic in Israel adds 1670 L per kg dry algae to maintain salinity at correct levels (Passell et al. 2013). Beyond cultivation, and considering fuel production, a National Research Council study in the United States estimated that between 3.15 and 3650 L of freshwater is needed to produce the algal biofuel equivalent to 1 L of gasoline using current technologies. According to their statistics, between 5 and 2140 L of water is needed to produce a liter of corn ethanol and between 1.9 and 6.6 L is needed to produce a liter of petroleum-based gasoline (NRC 2012).

Second, both first-generation and algal biofuel production require fertilizer to maximize biomass production. Agricultural crops require nutrients generally supplied by artificially (Haber–Bosch process) derived fertilizers. These inputs are strongly linked to GWP (see the following text) due to the fossil fuels required for their production. Land-based biofuels vary in their requirement for fertilizers, with oilseed rape, maize, and wheat requiring >100 kg N per hectare (De Vries et al. 2010). Algal biofuels also require significant inputs: for every kg of dry microalgal biomass, 367 g of carbon and 61 g of nitrogen are required (Ras et al. 2011). However, many algal systems can reduce artificial fertilizer inputs by sourcing nutrients from wastewater, such as agricultural wastes or anaerobic digestates. Indeed, they can be used to bioremediate contaminated water and produce recycled water, thereby reducing other environmental impacts (Benson et al. 2014). Using wastewater to cultivate algae gains credits against eutrophication potential in an LCA because of nutrient removal (Mu et al. 2014). Macroalgal production using longlines in proximity to sources of nutrients, for example, fish farms, can be used to enhance growth while also removing nutrients from the marine environment. Calculations from a report commissioned by the UK Crown Estate indicate that a 20 km² kelp farm in the Clyde Sea would extract approximately 480 tons of nitrogen from the marine environment per year if operated at the target yield of 20 tons dry kelp per hectare (Aldridge et al. 2012). They contrast this with a salmon farm that releases 750 tons of nitrogen annually. However, some intensively farmed macroalgae are fertilized to achieve high yields (Aldridge et al. 2012).

Third, air emissions from both types of biofuels are another factor when considering sustainability. The GHG savings by growing bioenergy crops depend on the rate of N fertilization. Currently, energy crops in the United Kingdom are not generally fertilized with N, but it already appears to be standard practice in other countries (Walter et al. 2014). Again, quantification of emissions is difficult as application of fertilizer is dependent on soil type and nutrient levels. For example, annual N_2O emissions from winter oilseed rape were 22% higher than those from winter cereals

fertilized at the same rate (Walter et al. 2014). The same study assumed a fertilization rate of 200 kg N per hectare per year, the mean fraction of fertilizer N that was
lost as N_2O was 1.27% for oilseed rape compared to 1.04% for cereals. The risk of
high yield-scaled N_2O emissions increases as fertilizer loadings increase, after a critical N surplus of about 80 kg N per hectare per year. In contrast to first-generation
production, microalgae are known to produce N_2O at various levels. Quantification
of this has proved difficult. Estimates from axenic cultures of *Chlorella vulgaris*
0.38–10.1 kg N_2O–N per hectare per year in a 0.25 m deep raceway pond operated
under Mediterranean climatic conditions (Guieysse et al. 2013). In reality, however,
maintaining axenic cultures is extremely difficult to impossible in open ponds or outdoor PBRs. In a separate study on a marine microalgal strain, it was suggested that
N_2O production was induced by anoxic conditions when nitrate was present, suggesting that N_2O was produced by denitrifying bacteria within the culture (Fagerstone
et al. 2011). The air–water fluxes of N_2O and CH_4 expressed as CO_2 equivalents of
GWP were 2 orders of magnitude smaller than the overall CO_2 uptake by the microalgae for a diatom culture in an open pond. The average emissions are 19.9 ± 5.6 μmol
CH_4/m^2/day. Upon NO_3^- depletion, the pond shifted from being a source to being a
sink of N_2O, with an overall net uptake during the experimental period of 3.4 ± 3.5
μmol N_2O/m^2/day (Ferrón et al. 2012).

In addition, algae are well known for the emission of trace gases that affect atmospheric chemistry and climate. Few studies have investigated the environmental impacts that may arise through an expansion in algal production. The flux of
biogenic halocarbons, dimethyl sulfide and hydrocarbons requires consideration.
Measurement of halocarbon emissions from tropical seaweeds and estimated
regional emissions concluded that the contribution of aquaculture is small at present (Leedham et al. 2013). Other research has shown that concentrations of halocarbons above intertidal seaweed beds peak at low tide (Carpenter et al. 2000). The
first step in processing seaweed biomass in the tropics is to dry the fronds outdoors,
so Leedham et al. (2015) looked at halocarbon emissions from seaweeds over an 8 h
period of desiccation to mimic the initial stages of this open-air drying (Leedham
Elvidge et al. 2015). A pulse of bromocarbons was seen in 10 min following exposure,
and then emissions reached a plateau within 1–3 h for temperate *Fucus vesiculosus*
but were more sustained for *Ulva intestinalis*. Overall, the data suggested that the
drying step does not add substantially to halocarbon emissions from seaweed farming, but further studies are needed, especially for tropical species. We are not aware of
any research on the trace gas outputs of microalgal cultivation or processing systems.

Fourth, impacts on land use are particularly significant for first-generation biofuel production, with soil quality being a major concern. In particular, soil erosion
during periods of time when crop cover is insufficient. For example, maize is typically planted 0.8 m apart, meaning that it takes up to 3 months to develop adequate
crop cover to protect the soil from washing away (Boardman 2013). Oilseed rape
is planted more densely and covers ground more quickly, thus reducing this issue.
Sediments that wash off from fields cause siltation of rivers, and runoff can carry
pesticides and fertilizers into watercourses (Boardman 2013). Unless a breach in
an algal system occurs, water leaving a biofuel operation should have undergone

effluent treatment to remove nutrients and biomass, though we note that 100% removal of algal cells would be difficult to attain with typical harvesting protocols and to prove. Another significant impact with crop-based biofuels is their requirement for agricultural land that could otherwise be employed for food production; algal biofuels also present problems in this respect. Ponds or PBRs will require space, meaning their impact on land is different in nature. As noted earlier, some pond systems extend over several hectares (Handler et al. 2012), thereby imposing a significant environmental footprint. That said, ponds and PBRs can be situated on non-arable or even degraded land unsuitable for cultivation. Macroalgal sites are generally offshore, in the form of either lines or nets, with only processing or hatchery operations on land. However, material shed from the macroalgal blades falling to the seafloor contributes to the detrital food web in the benthic environment and can have a significant impact on the ecosystem. Longlines may have positive biodiversity impacts as they become habitats, but the effects on the local environment require more characterization (Hughes et al. 2012).

Fifth, some consideration must be given to the GWP of both forms of biofuel production. Significant concerns exist over the impacts of crop-based production on global warming. Muñoz et al. (2013) measure GWP for ethanol production across several crops, with a variance shown from 1.60 kg CO_2-eq for sugarcane to 2.07 kg CO_2-eq for wheat. Research into algal biofuels shows similar divergence in GWP, depending on cultivation method and assumptions made about the process. A meta-analysis of algal biofuel LCA papers found a net energy ratio and normalized GHG emissions of 1.4 MJ produced/MJ consumed and 0.19 kg CO_2-eq/km traveled, respectively (Liu et al. 2012). Cultivation in raceways compared to PBRs can alter the GWP from 2 to 20 kg CO_2-eq/kg algae (Stephenson et al. 2010).

Finally, concerns have been raised over the implications for the biodiversity of both first-generation and algal biofuel production (Campbell and Doswald 2009). The European Union (EU) was forced to amend its Renewable Energy Directive (2009/28/EC) in response to criticisms that its policies were promoting environmental destruction and biodiversity loss, leading to the inclusion of sustainability criteria in its obligations. Under the criteria, biofuels should result in minimum GHG emission savings and cannot be grown on land that has a "high carbon stock," including wetlands and forests, or produced from a material derived from areas of high biodiversity. Algal biofuels would appear to resolve some of these issues but raise other concerns, most notably the implications for the biodiversity of introducing nonnative or even GMO algal species into the environment through accidental release (Benson et al. 2014). Such issues are potentially significant in both freshwater and marine environments, with the potential for transboundary invasive species particularly acute with the latter (Benson et al. 2014).

13.5 Implications for Industrial Development and Governance

Our review of the literature suggests that algal biofuel production may have some implications for governance, both globally and nationally, in terms of promoting the sustainable development of this sector. While the production of biofuels from algae, both micro and macro, at first glance appears to resolve some of the environmental concerns that have emerged over first-generation crop-based biofuels, there are issues that require greater consideration by policy makers. To date, algal biofuel production has been encouraged by policy and research funding in many contexts worldwide, including the EU, but scientific developments appear to be overtaking regulatory frameworks (Benson et al. 2014).

Several impacts require greater consideration in governance frameworks promoting algal biofuel production. In particular, our review shows that significant impacts may become evident over water use. Pond systems in particular require significant freshwater inputs making their future use questionable in water-stressed areas. To an extent, these concerns are offset by PBR systems that employ wastewater bioremediation and recycling, suggesting that integrated production facilities, which combine algal biofuel manufacture with other industrial processes, should receive greater consideration in policy. Another potentially significant issue is land use. Much controversy has surrounded the promotion of first-generation biofuels by governments in the United States, the EU, and developing countries, with its attendant problems of diverting land use away from agricultural production on a global scale. Large-scale open and even PBR systems could cause similar controversies over land take for the development of the algal biofuel sector. While this factor may be relatively unproblematic in some countries, in other more densely populated and intensely farmed contexts, they may prove difficult to establish. Similarly, macroalgal production requires greater consideration within integrated marine planning approaches, particularly in the EU (Benson et al. 2014). A recent UK House of Lords EU Committee Report highlights the need for regional, cross-border cooperation and marine planning for the North Sea Basin given the intense competition for marine space but concluded that neither body nor mechanism exists to implement the critical step-change improvement in management required (House of Lords—European Union Committee 2015). The report makes no mention of algae. Another concern for policy makers is the potential implications for biodiversity. EU policy makers included criteria in the Renewable Energy Directive to counter biodiversity threats, but there appears little consideration of the risks from micro- and macroalgal production. Although GMO use is heavily circumscribed in the EU context, genetically modified forms of algae are being developed worldwide with minimal, if any, political debate over the consequences (Benson et al. 2014). Other factors not actively being considered in policy worldwide are the potential for algal biofuels to reduce or enhance GWP, emissions to air, and also their potential to contribute to wider sustainability within emergent green economies.

13.6 Conclusions

In conclusion, we can return to our research questions. We initially asked: what types of algal biofuel production technologies and products are evident globally?

Research in this sector has developed rapidly in recent years, resulting in multiple forms of production technologies now emerging in many countries. They generally fall into three broad categories, namely, OPSs, PBRs or closed systems, and marine cultivation of macroalgae. While OPSs and macroalgal facilities are relatively straightforward to construct and operate, production in PBRs is easier to control. While these techniques can be utilized to produce a range of fuels, current production is focused on biodiesel, ethanol, and methane. We then asked what are the potential sustainability impacts of algal biofuels? A review of the literature allowed some comparison with first-generation biofuels, although direct comparison is difficult due to the varying criteria employed by LCA studies. Potentially significant impacts from algal biofuels identified include water use, land use, and biodiversity effects. Other concerns are evident as regards air emissions, GWP, and fertilizer use. Of these impacts, water and land use may be particularly significant depending on production techniques employed and the specific production context. Risks to biodiversity in marine and freshwater environments are also potentially significant in terms of invasive algal species and GMOs. Finally, what then are the implications for future sector development and its governance? This question remains difficult to answer as, generally, worldwide there is little actual "policy" regarding the development of the algal biofuels sector. Research into different production techniques has expanded rapidly, driven by economic imperatives. Now these techniques are being scaled up to full production; it is apparent that governance frameworks have not kept pace with sectoral development. Particular concerns can be expressed over the implications for water security, land resources, and biodiversity—concerns that have already clouded the development of first-generation crop-based biofuels in controversy. In terms of future national and international level policy, it could be argued that, on the basis of our analysis, greater consideration be given to these impacts if the industry is to develop in an environmentally, socially, and economically sustainable way.

Acknowledgment

The authors wish to acknowledge funding from the EU INTERREG IVB North West Strategic Initiative "Energetic Algae."

References

Aikawa, S., Joseph, A., Yamada, R. et al. 2013. Direct conversion of spirulina to ethanol without pretreatment or enzymatic hydrolysis processes. *Energ Environ Sci* 6:1844–1849.

Aldridge, J., van de Molen, J. and Forster, R. 2012. 'Wider ecological implications of Macroalgae culitvation' The Crown Estate, 95 pages. ISBN: 978-1-906410-38-4.

Benson, D., Kerry, K., and G. Malin. 2014. Algal biofuels: Impact significance and implications for EU multi-level governance. *J Clean Prod* 72:4–13.

Bhateria, R. and R. Dhaka. 2015. Algae as biofuel. *Biofuel* 5:1–25.

Boardman, J. 2013. Soil erosion in Britain: Updating the record. *Agriculture* 3:418–442.

Brennan, L. and P. Owende. 2010. Biofuels from microalgae—A review of technologies for production, processing, and extractions of biofuels and co-products. *Renew Sust Energ Rev* 14:557–577.

Cai, X., Zhang, X., and D. Wang. 2011. Land availability for biofuel production. *Environ Sci Technol* 45:334–339.

Campbell, A. and N. Doswald. 2009. The impacts of biofuel production on biodiversity: A review of the current literature. UNEP-WCMC, Cambridge, U.K.

Campbell, P.K., Beer, T., and D. Batten. 2011. Life cycle assessment of biodiesel production from microalgae in ponds. *Bioresour Technol* 102:50–56.

Carpenter, L.J., Malin, G., Liss, P.S., and F.C. Küpper. 2000. Novel biogenic iodine-containing trihalomethanes and other short-lived halocarbons in the coastal East Atlantic. *Global Biogeochem Cycle* 14:1191–1204.

Chung, I.K., Beardall, J., Mehta, S., Sahoo, D., and S. Stojkovic. 2010. Using marine macroalgae for carbon sequestration: A critical appraisal. *J Appl Phycol* 23:877–886.

Clarens, A.F., Nassau, H., Resurreccion, E.P., White, M.A., and L.M. Colosi. 2011. Environmental impacts of algae-derived biodiesel and bioelectricity for transportation. *Environ Sci Technol* 45:7554–7560.

Clarens, A.F., Resurreccion, E.P., White, M.A., and L.M. Colosi. 2010. Environmental life cycle comparison of algae to other bioenergy feedstocks. *Environ Sci Technol* 44:1813–1819.

Darzins, A., Pienkos, P., and L. Edye. 2010. Current status and potential for algal biofuels production. A Report to IEA Bioenergy Task 39. Report T39-T2. Accessed on June 21, 2016. http://www.fao.org/uploads/media/1008_IEA_Bioenergy_-_Current_status_ and_potential_for_algal_biofuels_production.pdf.

de Fraiture, C., Giordano, M., and Y. Liao. 2008. Biofuels and implications for agricultural water use: Blue impacts of green energy. *Water Pol* 10:67–81.

De Vries, S.C., van de Ven, G.W.J., van Ittersum, M.K., and K.E. Giller. 2010. Resource use efficiency and environmental performance of nine major biofuel crops, processed by first-generation conversion techniques. *Biomass Bioenerg* 34:588–601.

Delucchi, M.A. 2010. Impacts of biofuels on climate change, water use, and land use. *Ann N Y Acad Sci* 1195:28–45.

Demirbas, A. and M.F. Demirbas. 2011. Importance of algae oil as a source of biodiesel. *Energ Convers Manage* 52:163–170.

Dominguez-Faus, R., Powers, S.E., Burken, J.G., and P.J. Alvarez. 2009. The water footprint of biofuels: A drink or drive issue? *Environ Sci Technol* 43:3005–3010.

Driver, T., Bajhaiya, A., and J.K. Pittman. 2014. Potential of bioenergy production from microalgae. *Curr Sust/Renew Energ Rep* 1:94–103.

Efroymson, R.A. and V.H. Dale. 2015. Environmental indicators for sustainable production of algal biofuels. *Ecol Indic* 49:1–13.

Enquist-Newman, M., Faust, A.M.E., Bravo, D.D. et al. 2014. Efficient ethanol production from brown macroalgae sugars by a synthetic yeast platform. *Nature* 505:239–243.

Ewing, M. and S. Msangi. 2009. Biofuels production in developing countries: Assessing trade-offs in welfare and food security. *Environ Sci Pol* 12:520–528.

Fagerstone, K.D., Quinn, J.C., Bradley, T.H., De Long, S.K., and A.J. Marchese. 2011. Quantitative measurement of direct nitrous oxide emissions from microalgae cultivation. *Environ Sci Technol* 45:9449–9456.

Ferrón, S., Ho, D.T., Johnson, Z.I., and M.E. Huntley. 2012. Air-water fluxes of N_2O and CH_4 during microalgae (*Staurosira* sp.) cultivation in an open raceway pond. *Environ Sci Technol* 46:10842–10848.

Foley, P.M., Beach, E.S., and J.B. Zimmerman. 2011. Algae as a source of renewable chemicals: Opportunities and challenges. *Green Chem* 13:1399–1405.

Gao, Z., Zhao, H., Li, Z., Tan, X., and X. Lu. 2012. Photosynthetic production of ethanol from carbon dioxide in genetically engineered cyanobacteria. *Energ Environ Sci* 5:9857–9865.

Gerbens-Leenes, P.W., Xu, L., de Vries, G.J., and A.Y. Hoekstra. 2014. The blue water footprint and land use of biofuels from algae. *Water Resour Res* 50:8549–8563.

Ghirardi, M.L., L. Zhang, J.W. Lee et al. 2000. Microalgae: A green source of renewable H(2). *Trends in Biotechnology* 18(12):506–511.

Griffiths, M.J., van Hille, R.P., and S.T.L. Harrison. 2011. Lipid productivity, settling potential and fatty acid profile of 11 microalgal species grown under nitrogen replete and limited conditions. *J Appl Phycol* 24:989–1001.

Guerrero, F., V. Carbonell, M. Cossu, D. Correddu, and P.R. Jones. 2012. Ethylene synthesis and regulated expression of recombinant protein in *Synechocystis* sp. PCC 6803. *PloS One* 7(11):e50470.

Guieysse, B., Plouviez, M., Coilhac, M., and L. Cazali. 2013. Nitrous oxide (N_2O) production in axenic *Chlorella vulgaris* microalgae cultures: Evidence, putative pathways, and potential environmental impacts. *Biogeosciences* 10:6737–6746.

Guiry, M.D. 2012. How many species of algae are there? *J Phycol* 48:1057–1063.

Guiry, M.D., Guiry, G.M., Morrison, L. et al. 2014. AlgaeBase: An on-line resource for algae. *Cryptogamie, Algologie* 35:105–115.

Günerken, E., D'Hondt, E., Eppink, M.H.M. et al. 2015. Cell disruption for microalgae biorefineries. *Biotechnol Adv* 33:243–260.

Handler, R.M., Canter, C.E., Kalnes, T.N. et al. 2012. Evaluation of environmental impacts from microalgae cultivation in open-air raceway ponds: Analysis of the prior literature and investigation of wide variance in predicted impacts. *Algal Res* 1:83–92.

Hoekman, S.K. 2009. Biofuels in the U.S.—Challenges and opportunities. *Renew Energ* 34:14–22.

House of Lords—European Union Committee. 2015. The north sea under pressure: Is regional marine co-operation the answer HL 137, tenth report of session 2014–15. http://www.publications.parliament.uk/pa/ld201415/ldselect/ldeucom/137/13702.htm. Accessed on June 21, 2016.

Hu, Q., Sommerfeld, M., Jarvis, E. et al. 2008. Microalgal triacylglycerols as feedstocks for biofuel production: Perspectives and advances. *Plant J Cell Mol Biol* 54:621–639.

Hughes, A.D., Kelly, M.S., Black, K.D., and M.S. Stanley. 2012. Biogas from macroalgae: Is it time to revisit the idea? *Biotechnol Biofuel* 5:86.

Jiang, L., Luo, S., Fan, X., Yang, Z., and R. Guo. 2011. Biomass and lipid production of marine microalgae using municipal wastewater and high concentration of CO_2. *Appl Energ* 88:3336–3341.

Jorquera, O., Kiperstok, A., Sales, E.A., Embiruçu, M., and M.L. Ghirardi. 2010. Comparative energy life-cycle analyses of microalgal biomass production in open ponds and photobioreactors. *Bioresour Technol* 101:1406–1413.

Khoo, H.H., Sharratt, P.N., Das, P. et al. 2011. Life cycle energy and CO_2 analysis of microalgae-to-biodiesel: Preliminary results and comparisons. *Bioresour Technol* 102:5800–5807.

Kim, J., Yoo, G., Lee, H. et al. 2013. Methods of downstream processing for the production of biodiesel from microalgae. *Biotechnol Adv* 31:862–876.

Lan, E.I. and J.C. Liao. 2012. ATP drives direct photosynthetic production of 1-butanol in cyanobacteria. *Proceedings of the National Academy of Sciences of the United States of America* 109(16):6018–6023.

Lardon, L., Hélias, A., Sialve, B., Steyer, J.-P., and O. Bernard. 2009. Life-cycle assessment of biodiesel production from microalgae. *Environ Sci Technol* 43:6475–6481.

Leedham, E.C., Hughes, C., Keng, F.S.L. et al. 2013. Emission of atmospherically significant halocarbons by naturally occurring and farmed tropical macroalgae. *Biogeosciences* 10:3615–3633.

Leedham, E.C., Phang, S.-M., Sturges, W.T., and G. Malin. 2015. The effect of desiccation on the emission of volatile bromocarbons from two common temperate macroalgae. *Biogeosciences* 12:387–398.

Liu, X., Clarens, A.F., and L.M. Colosi. 2012. Algae biodiesel has potential despite inconclusive results to date. *Bioresour Technol* 104:803–806.

McHugh, D. 2003. A guide to the seaweed industry. FAO fisheries technical paper 441. Rome, Italy. http://www.fao.org/docrep/006/y4765e/y4765e00.HTM. Accessed on June 21, 2016.

Meher, L., Vidyasagar, D., and S. Naik. 2006. Technical aspects of biodiesel production by transesterification—A review. *Renew Sust Energ Rev* 10:248–268.

Mekonnen, M.M. and A.Y. Hoekstra. 2011. The green, blue and grey water footprint of crops and derived crop products. *Hydrol Earth Syst Sci* 15:1577–1600.

Menetrez, M.Y. 2012. An overview of algae biofuel production and potential environmental impact. *Environ Sci Technol* 46:7073–7085.

Mercer, P. and R.E. Armenta. 2011. Developments in oil extraction from microalgae. *Eur J Lipid Sci Technol* 113:539–547.

Molina Grima, E., Belarbi, E.-H., Acién Fernández, F., Robles Medina, A., and Y. Chisti. 2003. Recovery of microalgal biomass and metabolites: Process options and economics. *Biotechnol Adv* 20:491–515.

Mu, D., Min, M., Krohn, B. et al. 2014. Life cycle environmental impacts of wastewater-based algal biofuels. *Environ Sci Technol* 48:11696–11704.

Muñoz, I., Flury, K., Jungbluth, N. et al. 2013. Life cycle assessment of bio-based ethanol produced from different agricultural feedstocks. *Int J Life Cycle Assess* 19:109–119.

Naik, S.N., Goud, V.V., Rout, P.K., and A.K. Dalai. 2010. Production of first and second generation biofuels: A comprehensive review. *Renew Sust Energ Rev* 14:578–597.

National Research Council. Sustainable Development of Algal Biofuels in the United States. Washington, DC: The National Academies Press, 2012. doi:10.17226/13437.

NetAlgae. 2012. Overview of the seaweed industry by country. http://www.netalgae.eu/uploadedfiles/posters_final_UK_1.pdf. Accessed on June 21, 2016.

Oswald, W.J. 1988. Large-scale algal culture systems: Engineering aspects. In *Microalgal Biotechnology*, M.A. Borowitzka and L.J. Borowitzka (eds.). Cambridge University Press, Cambridge, U.K.

Parker, B. and B. Schlarb-Ridley. 2013. A UK roadmap for algal technologies. https://connect.innovateuk.org/documents/3312976/3726818/AB_SIG+Roadmap.pdf. Accessed on June 21, 2016.

Passell, H., Dhaliwal, H., Reno, M. et al. 2013. Algae biodiesel life cycle assessment using current commercial data. *J Environ Manage* 129:103–111.

Ras, M., Lardon, L., Bruno, S., Bernet, N., and J.-P. Steyer. 2011. Experimental study on a coupled process of production and anaerobic digestion of *Chlorella vulgaris*. *Bioresour Technol* 102:200–206.

Resurreccion, E.P., Colosi, L.M., White, M.A., and A.F. Clarens. 2012. Comparison of algae cultivation methods for bioenergy production using a combined life cycle assessment and life cycle costing approach. *Bioresour Technol* 126:298–306.

Rosenberg, J.N., Mathias, A., Korth, K., Betenbaugh, M.J., and G.A. Oyler. 2011. Microalgal biomass production and carbon dioxide sequestration from an integrated ethanol biorefinery in Iowa: A technical appraisal and economic feasibility evaluation. *Biomass Bioenerg* 35:3865–3876.

Sander, K. and G.S. Murthy. 2010. Life cycle analysis of algae biodiesel. *Int J Life Cycle Assess* 15:704–714.

Searchinger, T., Heimlich, R., Houghton, R.A. et al. 2008. Use of U.S. croplands for biofuels increases greenhouse gases through emissions from land-use change. *Science* 319:1238–1240.

Sevigné Itoiz, E., Fuentes-Grünewald, C., Gasol, C.M. et al. 2012. Energy balance and environmental impact analysis of marine microalgal biomass production for biodiesel generation in a photobioreactor pilot plant. *Biomass Bioenerg* 39:324–335.

Sheehan, J., Dunahay, T., Benemann, J., and P. Roessler. 1998. A look back at the U.S. department of energy's aquatic species program: Biodiesel from algae. http://www.nrel.gov/docs/legosti/fy98/24190.pdf. Accessed on June 21, 2016.

Shen, C.R. and J.C. Liao. 2012. Photosynthetic production of 2-methyl-1-butanol from CO_2 in cyanobacterium *Synechococcus elongatus* PCC7942 and characterization of the native acetohydroxyacid synthase. *Energy and Environmental Science* 5(11). The Royal Society of Chemistry: 9574.

Singh, A. and S.I. Olsen. 2011. A critical review of biochemical conversion, sustainability and life cycle assessment of algal biofuels. *Appl Energ* 88:3548–3555.

Smith, V.H., Sturm, B.S.M., Denoyelles, F.J., and S.A. Billings. 2010. The ecology of algal biodiesel production. *Trend Ecol Evol* 25:301–309.

Soratana, K., Harper Jr., W.F., and A.E. Landis. 2012. Microalgal biodiesel and the renewable fuel standard's greenhouse gas requirement. *Energ Pol* 46:498–510.

Stephenson, A.L., Kazamia, E., Dennis, J.S. et al. 2010. Life-cycle assessment of potential algal biodiesel production in the United Kingdom: A comparison of raceways and air-lift tubular bioreactors. *Energ Fuel* 24:4062–4077.

Ter Veld, F. 2012. Beyond the fossil fuel era: On the feasibility of sustainable electricity generation using biogas from microalgae. *Energ Fuel* 26:3882–3890.

Troell, M., Joyce, A., Chopin, T. et al. 2009. Ecological engineering in aquaculture—Potential for Integrated Multi-Trophic Aquaculture (IMTA) in marine offshore systems. *Aquaculture* 297:1–9.

Ungerer, J., L. Tao, M. Davis et al. 2012. Sustained photosynthetic conversion of CO_2 to ethylene in recombinant *Cyanobacterium Synechocystis* 6803. *Energy and Environmental Science* 5(10). The Royal Society of Chemistry: 8998.

Walsh, M. 2012. Seaweed harvesting in Ireland. Presentation to the NetAlgae project. http://www.netalgae.eu/uploadedfiles/WALSH_M_(EN).pdf. Accessed on June 21, 2016.

Walter, K., Don, A., Fub, R. et al. 2014. Direct nitrous oxide emissions from oilseed rape cropping: A meta-analysis. *Glob Change Biol Bioenerg* 7:1260–1271.

Wang, B., Li, Y., Wu, N., and C.Q. Lan. 2008. CO_2 bio-mitigation using microalgae. *Appl Microbiol Biotechnol* 79:707–718.

Wang, W., X. Liu, and X. Lu. 2013. Engineering cyanobacteria to improve photosynthetic production of alka(e)nes. *Biotechnology for Biofuels* 6(1):69.

Wargacki, A.J., Leonard, E., Win, M.N. et al. 2012. An engineered microbial platform for direct biofuel production from brown macroalgae. *Science* 335:308–313.

Wigmosta, M.S., Coleman, A.M., Skaggs, R.J., Huesemann, M.H., and L.J. Lane. 2011. National microalgae biofuel production potential and resource demand. *Water Resour Res* 47:1–13.

Williams, A.G., Withers, S., and A.D. Sutherland. 2013. The potential of bacteria isolated from ruminal contents of seaweed-eating North Ronaldsay sheep to hydrolyse seaweed components and produce methane by anaerobic digestion *in vitro*. *Microb Biotechnol* 6:45–52.

Yanfen, L., Zehao, H., and M. Xiaoqian. 2012. Energy analysis and environmental impacts of microalgal biodiesel in China. *Energ Pol* 45:142–151.

Yang, J., Xu, M., Zhang, X. et al. 2011. Life-cycle analysis on biodiesel production from microalgae: Water footprint and nutrients balance. *Bioresour Technol* 102:159–165.

14

Microalgal Biofuels
Flexible Bioenergies for Sustainable Development

Jasvirinder Singh Khattar, Yadvinder Singh,
Shahnaz Parveen, and Davinder Pal Singh

Contents

Abstract

Major sources of energy that drive the world are fossil fuels. Fossil fuel reserves are declining at a fast rate, and burning of these fuels also contributes to increased production of greenhouse gases leading to global warming. Thus, fossil fuels are considered to be unsustainable. In order to meet the environmental and economic sustainability, renewable, carbon-neutral fuels are necessary. Biofuels are suggested to be a good alternative for sustainable development. Competition with food crops leading to increase in food prices and lack of suitable and economical technology has hampered the use of first-generation and second-generation biofuels by developing countries. Nowadays, microalgal biofuels are considered to be flexible energy source for sustainable development. Algae-based biofuels are technically and economically viable and cost competitive, require no additional land, require minimal water use, and mitigate atmospheric CO_2. Microalgae can be mass produced in raceway ponds or photobioreactors (PBRs). To reduce the cost of biomass production, microalgae can also be grown in wastewater that has the advantage of treatment of wastewater as well. Microalgal biomass may be harvested by flocculation, centrifugation, filtration, and flotation. Harvested biomass is then dried by sun or by freeze, drum drying, oven drying, or spray drying. Dried biomass is subjected to biofuel production such as biodiesel, biomethane, bioethanol, and biohydrogen. The viability of microalgal biodiesel production can be achieved by designing advanced PBRs and developing low-cost technologies for biomass harvesting, drying, and oil extraction. Commercial microalgal biodiesel production can also be accomplished by employing genetic engineering strategies to improve microalgal strains capable of withstanding environmental stress conditions and by engineering metabolic pathways for high lipid production. Biodiesel is a sustainable fuel, as it is available throughout the year and can be used to run any engine. Bioethanol from microalgae is a good alternative to bioethanol from food grains. Microalgal residue after the extraction of lipids or starch can be used for the production of biomethane. Microalgal biofuels can satisfy the energy needs of future generations.

14.1 Introduction

14.1.1 Energy Crisis

Energy is the basic need for the development of any nation. The major source of energy to drive the world is fossil fuels, but the world is experiencing decline in fossil fuel reserves at a time when energy demand is increasing day by day. Fossil fuels are formed from the geologic transformation of buried organic materials (i.e., dead plants and animals) over millions of years, but continuously growing human population and increase in industry and transportation lead to increasing energy demands all over the world. The consumption of fossil fuel is 105 times faster than what nature can generate (Satyanarayana et al. 2011). Adversely, fossil fuel reserves are being exhausted day by day (Nomanbhay et al. 2012). Energy demand of every nation can't be met only from fossil fuels, with the result almost 1.4 billion people of our planet face daily shortage of energy (Komerath and Komerath 2011). The basic sources of energy are coal, petroleum oil, and natural gas (Surendhiran and Vijay 2012; Innocent et al. 2013) and fulfill 90% of the world's energy needs (Dale 2008; Huang et al. 2012). Among them, petroleum oil fulfills 40% of the energy needs (Sivakumar et al. 2010). Energy requirement has increased 2.4-fold from 5,000 million tons of oil equivalent (Mtoe) in 1971 to 11,700 Mtoe in 2010; 70% of this increase was recorded in Asia (Van Lienden et al. 2010). This trend of increasing energy demand will continue and is expected to rise nearly two-fold in the near future (Matsuo et al. 2013; Sani et al. 2013). Comparing the forecasted energy demand and available resources of crude oil, it is undeniable that future energy demand cannot solely be met by fossil fuels (Saito 2010). In 2014, the quantity of accessible crude oil resources was estimated to be about 1700.1 billion barrels. Based on the current consumption of about 93 million barrels of crude oil per day all over the world (Figure 14.1), it is expected that the entire available resources will suffice only for a fairly short time period (BP 2015). According to Ong et al. (2011), the global proven reserves for crude oil and natural gas are estimated to last for 41.8 and 60.3 years, respectively. It is therefore questionable whether there will be enough fossil fuels for every requirement in the near future.

14.1.2 Global Warming and Climate Change

Conventional energy sources based on oil, coal, and natural gas have proven to be highly effective drivers of economic progress, but at the same time, these have caused immense damage to the environment by increasing the CO_2 level in the atmosphere resulting in global warming (Sawayama et al. 1995; Hafizan and Zainura 2013). The level of CO_2 emissions have significantly increased in the last 35–40 years, and the total amount of CO_2 emissions due to burning of fossil fuels has reached about 26 billion tons (Saito 2010). Statistical data show that at present CO_2 concentration in the atmosphere is about 380 ppm, compared to 280 ppm before the Industrial Revolution. The total global CO_2 emissions in 2030 will be 1.6 times higher than that of 2004 (Saito 2010). The use of conventional fossil fuels can cause fast-rising CO_2 emissions, and with the ever-increasing pace of modernization, this trend will continue if a feasible alternative energy source is not found in time (Roman-Leshkov et al. 2007; Krumdieck et al. 2008). According to the U.S. Energy Information Administration (2011), more than 90% of the greenhouse gas emissions are from the burning of fossil fuels that contribute 42% of the total CO_2 emission besides producing harmful gases

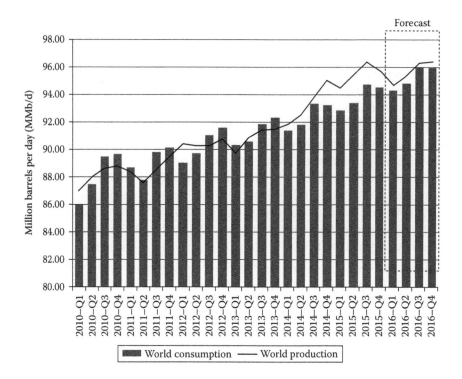

Figure 14.1 World's liquid fuel production and consumption. (From U.S. Energy Information Administration, Short-term energy outlook, 2015, https://www.eia.gov/forecasts/steo/report/global_oil.cfm, accessed on September 15, 2015.)

like nitrogen oxides, sulfur dioxide, volatile organic compounds, and heavy metals. This has resulted in environmental impacts such as global warming, acid rain, and air quality deterioration and desertification in many parts of the world (Bayram and Ozturk 2014).

14.1.3 Sustainable Development

Recently, more and more concerns have been expressed regarding sustainable development. The simplest definition of sustainable development was given by the World Commission on Environment and Development (WCED 1987): "development that meets the needs of the present without compromising the ability of future generations to meet their own needs." The concept of sustainable development can be divided into four parts: environmental sustainability, economic sustainability, social sustainability, and cultural sustainability (Figure 14.2). Presently, the environmental issues that strongly bind economic, social, and cultural impacts are dominating the international agenda, and much importance has been attached in particular to industrial sustainability. Identifying the core environmental, economic, social, and cultural impacts is the first step in supporting the development of a sustainable

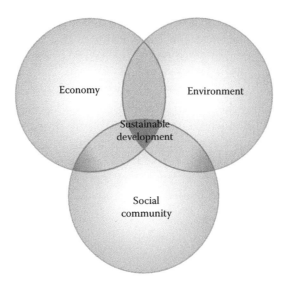

Figure 14.2 Sustainability concepts and their interrelationships.

industry. Unsustainable aspects can be identified using the techniques of risk assessments (Gupta et al. 2002) and environmental impact assessments (Salvador et al. 2000). Potential risks can thus be forecasted and then either mitigated or eliminated to some degree. Achieving sustainable economic development along with sustainability of energy is a long-term goal. Many significant problems lie in energy production and consumption, such as shortage of resources, low energy efficiency, high emissions, damage to environment, and lack of effective management systems (Zhang et al. 2011). As fossil fuels' supply dwindles and their cost rises, nations will be forced to utilize alternative energy sources (Ren et al. 2012; Xie et al. 2013). In order to achieve a secure and stable energy supply that does not cause environmental damage, renewable energy sources must be explored and promising technologies should be developed (Wargacki et al. 2012). Thus, economic and policy mechanisms are required to support the widespread dissemination of and sustainable markets for renewable energy systems. Renewable energy sources such as biomass, wind, solar, hydropower, and geothermal can provide sustainable energy services, based on the use of routinely available indigenous resources. The potential of renewable energy sources is enormous as these can meet many times the world's energy demand. Furthermore, these are cyclical in nature due to the effects of oligopoly in production and distribution. This will decrease the dependence of many countries on fuel imports as well as reduce their foreign exchange bills (Moller et al. 2014; Prakasham et al. 2014). It is becoming clear that future growth in the energy sector is primarily in the new regime of renewable, and to some extent natural gas–based systems, and not in conventional oil and coal sources. Renewable energy sources currently supply somewhere between 15% and 20% of the world's total energy demand that is dominated by traditional biomass, mostly fuel wood used for cooking and heating, especially in developing countries. A major contribution is also obtained from the use of hydropower, with nearly 20% of the global electricity supply being obtained from this source. A number of studies have indicated that in the second half of the twenty-first century, contribution of renewable energy to total energy might range from 20% to more than 50% with the right policies in place. Biomass fuels or biofuels may offer a promising alternative to fossil fuels as these are witnessing increasing attention as evidenced by the growing research and development efforts all over the world (Solomon 2010). Biofuels can play a vital role in mitigating energy crisis and environmental pollution (Huang et al. 2010; Kumar 2013). Moreover, biofuels release almost the same amount of CO_2 that was fixed during photosynthesis; hence, they hold great potential to fulfill/supplement future energy requirements with minimum hazards to the environment provided their production is carefully regulated and planned (Gul et al. 2013).

14.2 Biofuels

Biofuels are a renewable solution to replace petroleum-derived liquid fuels. The term *biofuel* is referred to as solid (biocoal), liquid (bioethanol and biodiesel), or gaseous (biogas, biosyngas, and biohydrogen) fuel that is produced predominantly from biomass (Demirbas 2010). The advantages of using biofuels can be on three aspects—environment, energy security, and economy (Demirbas 2009)—while a biofuel investment is driven mainly by two fundamental factors: market developments and policy levers (Huang et al. 2012). Biofuels are environmentally friendly; generate much less air pollution, carbon dioxide, and greenhouse gas emissions compared to petroleum oil (Huang et al. 2012). Global biofuel production grew from 16 billion liters in 2000 to more than 120 billion liters in 2014 and with annual average increase of 2.7% may rise to 135 billion liters in 2019 (International Energy Agency 2014). The most common biofuels are ethanol, produced from sugar and starch crops, and biodiesel, produced from vegetable oils and animal fats (Bajpai and Tyagi 2006). Liquid biofuels, bioethanol and biodiesel, may offer a promising alternative because of characteristics similar to those of petroleum fuel (Antolin et al. 2002). Corn is the main source of ethanol in the United States. However, corn ethanol creates a number of serious issues, including competing with the food industry and increasing food prices (Farrell et al. 2006; Searchinger et al. 2008; BP 2015). Often, the crops used for renewable feedstocks for biofuels occur abundantly in the country of production. For example, in the United States ethanol is produced primarily from corn, whereas in Brazil ethanol is produced primarily from sugarcane. Empirical analysis of the link between biofuels and food production has so far been focused mainly on developed countries, such as the United States and those in the European Union, and the use of maize and wheat for ethanol production and the use of vegetable oil for biodiesel impact global food prices (de Gorter and Just 2010; Helming et al. 2010; DEFRA 2012). A number of developing countries have also started investing in feedstocks for the production of biofuels. Tanzania and several other sub-Saharan nations have started producing biofuels in order to replace imports of petroleum and save on foreign exchange reserves (Jumbe et al. 2009). Some other countries like

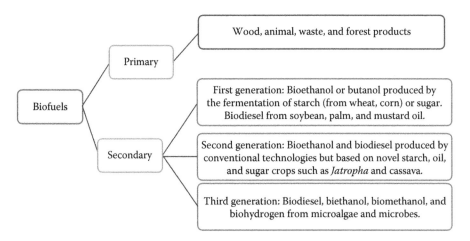

Figure 14.3   Classification of biofuels based on different sources.

Malaysia, Indonesia, and Argentina are producing biofuels for export to the attractive European and American markets (Gerasimchuk et al. 2012).

Biofuels are generally divided into two categories: primary and secondary. Primary biofuels, such as fuel wood, are used in an unprocessed form primarily for heating, cooking, or electricity production, while secondary biofuels such as bioethanol and biodiesel are produced by processing biomass and are suitable for use in vehicles and various industrial processes. The secondary biofuels are further categorized into three generations based on processing technology, type of feedstock, or their level of development (Figure 14.3).

14.2.1 First-Generation Biofuels

First-generation biofuels are produced from food crops like corn, soybean, rapeseed, palm oil, and mustard. First-generation bioethanol, produced from corn in the United States and from sugarcane in Brazil, is now widely produced and used, but the first-generation biofuel production systems have considerable economic and environmental limitations (Milledge et al. 2014). The main concern related to the first-generation biofuels is that as their production capacity increases, their competition with agriculture crops for arable land also increases, which results in the loss of biodiversity, excess utilization of water, and increased greenhouse gas emissions (Schenk et al. 2008). The burning of rapeseed oil results in nitrous oxide emissions leading to contamination of air (Sawyer et al. 2003). Also, if these fuels become lucrative for the farmers, they would start growing crops for fuel production instead of food production, which would lead to shortage of food, which in turn can lead to price hikes (Al-Mulali 2015). The World Bank has estimated that the food prices increased 83% between 2005 and 2008 due to the use of food crops as biofuel crops. As a result, 100 million people have fallen into poverty. The amount of biomass required to replace a significant proportion of the fossil fuel used in the transportation sector runs into millions of tons. In order to grow crops for fuel, intensive farming techniques should be employed. These farming methods may have a wide range of negative impacts on the environment like soil erosion and excess use of pesticides and fertilizers. The increased use of pesticides and fertilizers leads to environmental problems like eutrophication and water shortage. Another downside of this approach is that monocropping can lead to loss of soil fertility (Lang et al. 2001).

14.2.2 Second-Generation Biofuels

Second-generation biofuels are produced from lignocellulosic biomass, that is, the woody part of plants that do not compete with food production, and are divided into three types: energy crops, agricultural residue, and forest residue such as leaves, straw, or wood chips as well as the nonedible components of corn or sugarcane (Brennan and Owende 2010; Carriquiry et al. 2011). The second-generation biofuels used presently are bioethanol, isobutanol, and wood diesel. Polysaccharides in the lignocellulosic material are either broken down into sugars and processed to biofuel or gasified into syngas (Timilsina and Shrestha 2011). There is still a great deal of work to be done to make this fuel economically viable, as cellulosic ethanol currently costs two to three times more on an energy equivalency basis with fossil fuel (Carriquiry et al. 2011). The issue with second-generation fuel processes is extraction of useful sugars locked within lignin and cellulose and the required enzymes for freeing these sugar molecules from cellulose. Enzymatic hydrolysis required to convert lignocellulose to ethanol is an expensive process and also poses a technical challenge.

14.2.3 Third-Generation Biofuels

Third-generation biofuels include algae-derived fuels such as biodiesel, bioethanol, and biohydrogen (Dragone et al. 2010). Third-generation biofuels are considered to be a viable alternative energy resource devoid of the major drawbacks associated with first- and second-generation biofuels (Hyka et al. 2013). Large-scale cultivation of microalgae may be 10–20 times more productive on per-hectare basis than other biofuel crops. These organisms are able to use a wide variety of water sources and may be cultivated on waste and barren land without any competition for food production (Chisti 2007). Triacylglycerols (TAGs) generally serve as energy storage in microalgae that, once extracted, can be easily converted into biodiesel through transesterification reactions (Christenson and Sims 2011). Several species of microalgae, such as *Botryococcus braunii*, *Nannochloropsis* sp., *Dunaliella primolecta*, *Chlorella* sp., and *Crypthecodinium cohnii*, produce large quantities of hydrocarbons and lipids. *B. braunii*, the colonial green microalga, has the capability to produce a large amount of hydrocarbons as compared to its biomass (Cohen 1999; Li and Qin 2005; Qin 2005; Chisti 2006; Qin and Li 2006; Rao et al. 2006). The oil content in *B. braunii* may reach up to 80% and the levels from 20% to 50% are quite common (Chisti 2006; Powell and Hill 2009; Mata et al. 2010). Various *Chlorella* strains have proved to be suitable for biodiesel production (Liu et al. 2010; Feng et al. 2011).

14.3 Microalgae as Source of Biofuel

The algae for fuel concept has gained renewed interest with wide fluctuations in energy prices (Hu et al. 2008). First-generation biofuels have created a lot of disputes due to their negative impacts on food security, global food markets, water scarcity, and deforestation (FAO 2009). In addition, the second-generation biofuels derived from nonedible oils (*Jatropha curcas, Pongamia pinnata, Simarouba glauca*, etc.), lignocellulose biomass, and forest residues require huge areas of land that otherwise could be used for food production. Currently, the second-generation biofuel production also lacks efficient technologies for commercial exploitation of wastes as source for biofuel generation (Tabatabaei et al. 2011). Based on the drawbacks associated with the first- and second-generation biofuels, microalgal biofuels seem to be a viable alternative source of energy to replace or supplement the fossil fuels. Compared to the terrestrial crops investigated until now, one unit of growing area of microalgae can produce much more biomass and oil (Table 14.1).

Microalgal biofuel production is commercially viable because it is cost competitive with fossil-based fuels, does not require extra land, improves the air quality by absorbing atmospheric CO_2, and utilizes minimal water (Wang and Lan 2011). A number of algae such as *Chlamydomonas reinhardtii, Dunaliella salina, Chlorella* spp., *B. braunii, Phaeodactylum tricornutum*, and *Thalassiosira pseudonana* have been reported as promising biofuel candidates (Scott et al. 2010). A variety of high-value biofuels like biohydrogen, bioethanol, biomethane, and biodiesel may be produced from algae on a commercial scale (Harun et al. 2010b; John et al. 2011).

TABLE 14.1 Comparison of Oil Content, Oil Yield, and Biodiesel Productivity of Microalgae with the First- and Second-Generation Biodiesel Feedstock Sources

Feedstock Source	Oil Content (% Oil by wt. in Biomass)	Oil Yield (Oil in L/ha/ Year)	Biodiesel Productivity (kg Biodiesel/ha/Year)
Oil palm	36	5,366	4,747
Maize	44	172	152
Physic nut	41–59	741	656
Castor	48	1,307	1,156
Microalgae with low oil content	30	58,700	51,927
Microalgae with medium oil content	50	97,800	86,515
Microalgae with high oil content	70	136,900	121,104

Source: Medipally et al., *BioMed. Res. Int.*, 2015, Article ID 519513, 2015.

14.3.1 Biohydrogen

Biohydrogen can be produced by the process of biophotolysis or photofermentation (Fedorov et al. 2005; Kapdan and Kargi 2006; Ferreira et al. 2012; Shaishav et al. 2013). Biohydrogen production from microalgae has been known for more than 65 years and was first observed in the green alga *Scenedesmus obliquus* (Boichenko and Hoffmann 1994). The green alga *C. reinhardtii* is most commonly used for biohydrogen production (Melis 2000). *Chlorella sorokiniana* strain ce generate biogas containing hydrogen that has been successfully used as fuel for powering small polymer electrolyte membrane fuel cell system (Chader et al. 2011). Park et al. (2011) found *Gelidium amansii*, a red alga, as the potential source of biomass for the production of biohydrogen through anaerobic fermentation. Researchers are now focusing on enzymes that can be genetically modified to enhance the biohydrogen production (Gavrilescu and Chisti 2005; Hankamer et al. 2007; Wecker et al. 2011; Yacoby et al. 2011; Rajkumar et al. 2014).

14.3.2 Bioethanol

Attempts have been made for bioethanol production through fermentation using algae as the feedstock to make it as an alternative to conventional crops (Singh et al. 2011; Chaudhary et al. 2014). Different microalgae and macroalgae, such as *Chlorococcum* sp., *Prymnesium parvum*, *Gracilaria amansii*, *Gracilaria* sp., *Laminaria* sp., *Sargassum* sp., and *Spirogyra* sp., have been used for the bioethanol production (Eshaq et al. 2011; Miranda et al. 2012; Rajkumar et al. 2014). A number of reports are available on starch production by cultivating microalgae and its fermentation by yeasts and bacteria to yield ethanol (Ergas et al. 2010; Harun et al. 2010a). Ethanol production by dark fermentation by using marine green alga *Chlorococcum littorale* was also investigated (Ueno et al. 1998). Singh and Trivedi (2013) used *Spirogyra* biomass for the production of bioethanol using *Saccharomyces cerevisiae* and *Zymomonas mobilis*. Although brown, green, and red algae have been fermented to ethanol, brown algae are considered as the principal feedstock for bioethanol production as they have high carbohydrate content and can be readily mass cultivated (Jung et al. 2013). Carbohydrate components, such as sulfated polysaccharides, mannitol, alginate, agar, and carrageenan of seaweeds, can be fermented to bioethanol (Yoon et al. 2010; Yanagisawa et al. 2011, 2013).

14.3.3 Biomethane

Methane is another biofuel from algae produced by anaerobic digestion of biomass (Lee et al. 2002; Spolaore et al. 2006; Li et al. 2008). Some species of *Laminaria* (Chynoweth 2005) and *Ulva* (Harun et al. 2014) are promising source for biomethane production. Various species of algae such as *Chlamydomonas, Scenedesmus,*

Spirulina, *Euglena*, and *Ulva* are known for the production of biogas through fermentation (Yen and Brune 2007; Mussgnug et al. 2010; Ras et al. 2011; Zhong et al. 2012; Saqib et al. 2013).

14.3.4 Biodiesel

The use of microalgae for biodiesel production has long been recognized, and its potential has been widely reported (Gouveia and Oliveira 2009; Rodolfi et al. 2009; Afify et al. 2010; Damiani et al. 2010; Liu et al. 2010; Abou-Shanab et al. 2011; Ahmad et al. 2011). Microalgae are capable of producing oil all year long and productivity of microalgae is greater compared to conventional crops (Baliga and Powers 2010). TAGs are the best substrates for biodiesel production (Hu et al. 2008). TAG content varies among different species of algae and can be in the range of 20%–50% that is greater than other competitors (Leite and Hallenbeck 2012). Biodiesel produced from algal lipids is nontoxic and highly biodegradable.

Algae as feedstock for biodiesel have several advantages over conventional crops (Figure 14.4). Because of their simple cellular structure, algae have higher rates of biomass and oil production than conventional crops (Becker 1994). The Aquatic Species Program sponsored by the U.S. Department of Energy estimated that algal oil yield of 5,000–10,000 gallons per acre per year is possible compared to 50–100 gallons per acre per year for traditional oil crops (Li et al. 2010). Microalgae produce 15–300 times more oil for biodiesel production than traditional crops on an area basis (Karmakar et al. 2010; Andrade et al. 2011; Atabani et al. 2012; Kirrolia et al. 2013; Saharan et al. 2013). The other advantages of using algae are that they can be grown in freshwater, seawater, wastewater, or nonarable

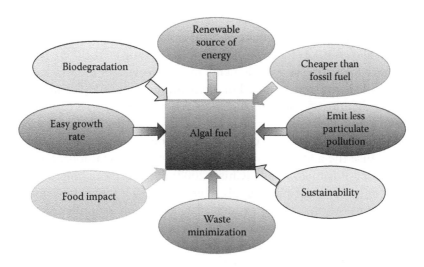

Figure 14.4 Advantages of algal fuel.

lands (Abdel-Raouf et al. 2012; Lewicki et al. 2013), use less water than the traditional oil seed crops, and have no competition with food crops (Tsukahara and Sawayama 2005; Sharma et al. 2012). Algal aquatic farms can be built in deserts or on land that is otherwise of little value (Chisti 2007; Dismukes et al. 2008; Li et al. 2008; Pienkos and Darzin 2009; Amaro et al. 2011). Thus, microalgae are an alternative fuel feedstock that could avoid fuel versus food conflict (Sawaengsak et al. 2014). Microalgal lipids are mostly neutral lipids with lower degree of unsaturation, which makes them a potential replacement for fossil fuel. Biomass productivity, total lipid content, and lipid productivity are among the main parameters that determine the economic feasibility of using microalgae as a source of biofuels (Wijffels and Barbosa 2010).

14.4 Microalgal Cultivation Technology

14.4.1 Microalgal Biology

Microalgae are able to fix CO_2 efficiently from different sources, including the atmosphere, industrial exhaust gases, and soluble carbonates. Fixation of CO_2 from the atmosphere is probably the most basic method to sink carbon and relies on the mass transfer from air to the microalgae in their aquatic growth environments. However, because of the relatively small percentage of CO_2 in the atmosphere (approximately 0.036%), it becomes a limiting factor for microalgal biomass production (Brennan and Owende 2010). On the other hand, industrial exhaust gases such as flue gas contain up to 15% CO_2, providing a CO_2-rich source for microalgal cultivation and a potentially more efficient route for CO_2 biofixation. Many microalgal species have been shown to utilize carbonates such as Na_2CO_3 and $NaHCO_3$ for growth. Some of these species typically have high activity of extracellular carboanhydrase, which is responsible for the conversion of carbonates to free CO_2 to facilitate CO_2 assimilation. In addition, the direct uptake of bicarbonate by an active transport system has also been shown in several microalgal species (Wang et al. 2008). Nitrogen (N) and phosphorus (P) are essential elements for the growth of microalgae. Nitrogen is mostly supplied as nitrate (NO_3^-), but often ammonium (NH_4^+) and urea are also used. Urea is the most favored nitrogen source because, for an equivalent nitrogen concentration, it gives higher yields and causes smaller pH fluctuations in the medium during algal growth (Shi et al. 2000). On the other hand, P must be supplied in excess because the added phosphate complexes with metal ions, and therefore, not all the added P is bioavailable (Chisti 2007). Furthermore, microalgal growth depends not only on an adequate supply of essential macronutrient elements (C, N, P) and major ions (Mg^{2+}, Ca^{2+}, Cl^-, and SO_4^{2-}) but also on a number of micronutrients such as iron, manganese, zinc, cobalt, copper, and molybdenum (Sunda et al. 2005). Minimal algal nutritional requirements can be estimated using the approximate molecular formula of the microalgal biomass, which is $CO_{0.48}H_{1.83}N_{0.11}P_{0.01}$ (Li et al. 2008).

14.4.2 Culture Conditions

Specific environmental conditions, which vary among microalgal species, are required in order to successfully cultivate microalgae. Factors that influence microalgal growth include abiotic factors such as light intensity, temperature, O_2, CO_2, pH, salinity, nutrients, and toxins; biotic factors such as bacteria, fungi, viruses, and competition with other microalgae for nutrients; and operational factors such as mixing and stirring degree, width and depth of cultivation tank, dilution rate, harvest frequency, and addition of bicarbonate (Mata et al. 2010). The most important factors affecting microalgal growth are described in the following sections.

14.4.2.1 Light

Light is the basic energy source for photoautotrophic organisms. Light intensity and photoperiod are one of the most important factors influencing the growth rate and biomass composition and, hence, production of high-value microalgal products in a wide range of algal species (Fabregas et al. 2002; Khoeyi et al. 2012; Wahidin et al. 2013). Total biomass of *B. braunii* increased under a 24:0 light cycle, compared to control 12L:12D condition (Ruangsomboon 2012). Alterations in the photoperiod induced changes in the total protein, pigment, and fatty acid content in *C. vulgaris* (Gardner et al. 2011) and *Porphyridium cruentum* (Gutierrez 2009) and cell density, cell growth rate, and total lipid content in *Nannochloropsis* sp. (Wahidin et al. 2013), *C. vulgaris* (Khoeyi et al. 2012), *Dunaliella* sp. (Chisti 2007), *S. obliquus* (Prescott 1982), and *C. kessleri* (Li et al. 2012). The growth rate of the microalgal cultures increased with increased light intensities, until saturating light (generally around 200–400 µE) was reached (Radakovits et al. 2010). Oversaturating light can lead to the formation of reactive oxygen species, which is harmful for microalgae (photoinhibition), and thereby decreases the biomass productivity (Stanier and Cohenbazire 1977; Richmond 2000). Cheirsilp and Torpee (2012) reported that the growth of *Nannochloropsis* sp. continuously increased up to the maximum level with light intensity up to 10,000 lux. A study on *Chlorella* sp. showed an increase in cell density when the population was exposed to flashing light (Lananan et al. 2013). Several studies have shown that the typical standing biomass of *C. vulgaris* under both phototrophic and heterotrophic growth condition after 12–15 days of incubation was between 250 and 1700 mg L^{-1} (Liang et al. 2009; Bhola et al. 2011). In a recent study, biomass production of *C. vulgaris* reached to 2730 mg L^{-1} under blue LED at 200 µmol m^{-2} s^{-1} and photoperiod of 12 h L:12 h D for 312 h (Atta et al. 2013).

14.4.2.2 Temperature

Temperature is another key limiting factor, especially for outdoor algal cultivation systems. Generally, microalgal growth increases exponentially as temperature increases to an optimal level, after which the growth rate declines. Keeping cultures at temperatures above the optimum will result in total culture loss (Alabi et al. 2009). Although microalgae are able to survive at a variety of temperatures, optimal temperature for growth is limited to a narrow range (20°C–30°C) (Chisti 2007; Singh et al. 2012). Depending upon the geographical location, overheating issues might occur in outdoor systems, thus water-cooling systems should be considered to ensure that temperature does not exceed the optimal range. Generally, in an optimal temperature range, rise in temperature leads to improved microalgal biomass production. Temperatures above the optimal range cause decline in growth, and in several conditions can even kill microalgal cells. However, low temperatures seem to reduce the biomass loss caused by respiration during dark periods (Raven and Geider 2006; Chisti 2007). Therefore, high biomass production can be achieved by increasing the temperature to optimal in the morning (to enhance productivity during the day) and decreasing the temperature at night (to avoid biomass loss) (Hu 2004). Temperature has been shown to have a major effect on the fatty acid composition of microalgae (Guschina and Harwood 2006; Morgan-Kiss et al. 2006).

14.4.2.3 Nutrients

In addition to macronutrients nitrogen and phosphorus, trace metals such as Fe, Mg, Mn, B, Mo, K, Co, and Zn, are also required for the proper growth of microalgae. The nutrients can be supplied in the form of simple, easily available agricultural fertilizers. Several studies have reported that N or P deficiency or limitation during microalgal cultivation can improve the lipid accumulation (Khozin-Goldberg and Cohen 2006; Hu et al. 2008; Devi and Mohan 2012). In practice, microalgae are cultured in media with enough nutrients in the early stages, while during later stages nutrient deficiency or limitation needs to be designed to improve the lipid content. Ito et al. (2013) observed that nitrogen deficiency could cause a decrease in amino acids, while the quantities of neutral lipids increased greatly. Recently, most kinds of wastewater have been tested for microalgal cultivation. This results in the treatment of wastewater and production of algal biomass. The N and P removal from wastewaters mainly results from the uptake by microalgal cells during growth (Su et al. 2011). Moreover, bacteria in the wastewater can also contribute to the nutrient degradability. Ammonium, nitrate, and nitrite can be degraded by bacteria via nitrification and denitrification (Zhu et al. 2011). Phosphate can also be degraded to some degree through microbial activities (Kim et al. 2005; Oehmen et al. 2007). In addition, if the pH of microalgae culture increases, it will also contribute to the P removal via P precipitation (Ruiz-Marin et al. 2010). Heavy metals like cadmium, iron, copper, and zinc have also been found to increase the lipid content in some microalgae (Einicker et al. 2002). The effect of iron on growth and lipid accumulation in *Chlorella vulgaris* has been investigated (Liu et al. 2008). The cultures in the late exponential growth phase when supplemented with Fe^{3+} at different concentrations increased the lipid content up to 56.6% of dry biomass (Liu et al. 2008).

14.4.2.4 CO_2

Algal growth limitation might occur if algal culture is supplied only with air, which contains nearly 0.036% CO_2. Extra CO_2 can be blended with air and injected into algae cultures via gas addition facilities (Mata et al. 2010). Different amounts and sources of carbon have been shown to affect both the content and the composition of lipids in microalgae. A significant effect of CO_2 on the composition of the polyunsaturated fatty acids and alkenones in *Emiliania huxleyi* was reported. Specifically, low CO_2 concentrations led to an increase in 22:6 fatty acids, whereas 14:0 fatty acids were found to be predominant at higher concentrations of CO_2 (Riebesell et al. 2000). Increased CO_2 was observed to not only increase the amount of fatty acid accumulation but changed the fatty acid composition as well in *D. salina* (Muradyan et al. 2004). High concentration of carbon dioxide induced the accumulation of saturated fatty acids, whereas low concentration of carbon dioxide facilitated the production of unsaturated fatty acids in microalgae (Tsuzuki et al. 1990; Riebesell et al. 2000; Hu and Gao 2003). Cheng et al. (2006) reported that the high biomass of *C. vulgaris* was obtained in cultures aerated with 2% CO_2. Standing biomass of 2 g L^{-1} was obtained in *Chlorella* sp. when aerated with 5% CO_2 (Ryu et al. 2009). CO_2 at 2% and 10% resulted in biomass synthesis of 1.5–1.67 g L^{-1} after 6 days of incubation (Chiu et al. 2011). Exposure of the algae to white LED light and supplementation of cultures with 8.5% CO_2 resulted in the highest standing biomass of 1.6 g L^{-1} in *Scenedesmus dimorphus* (Lunka and Bayless 2013).

14.4.2.5 pH

pH is one of the most important factors in algal cultivation since it determines the solubility and availability of CO_2 and essential nutrients (Juneja et al. 2013). Alkaline pH increases the flexibility of the cell wall of mother cells, which prevents its rupture and inhibits autospore release, thus increasing the time for cell cycle completion (Guckert and Cooksey 1990). Alkaline pH indirectly results in an increase in triglyceride accumulation but decrease in membrane-associated polar lipids because of cell cycle inhibition. Under alkaline conditions whereby the extracellular pH is higher than intracellular pH, the cell must rely on active transport of HCO_3^- and not on passive flux of CO_2 for inorganic carbon accumulation. Higher pH thus limits the availability of CO_2, which, in turn, suppresses algal growth (Azov 1982). Most species of algae grow maximally in slightly alkaline pH (7.0–7.6) as observed in the case of *Ceratium lineatum*, *Heterocapsa triquetra*, *Prorocentrum minimum* (Hansen 2002), and *Chlamydomonas applanata* (Visviki and Santikul 2000).

14.4.2.6 Salinity

Salinity is another important factor that alters the biochemical composition of algal cells. Salinity is considered as one of the most significant ecological factor affecting the growth and metabolic activities of plants and microorganisms (Carrieri et al. 2010). Exposing algae to lower or higher salinity levels than their natural (or adapted) levels can change growth rate and alter their composition. High salinity has been shown to increase the algal lipid content in *B. braunii* (Zhila et al. 2011). In *Dunaliella*, a marine alga, saturated and monounsaturated fatty acids increased with an increase in NaCl concentration from 0.4 to 4 M (Xu and Beardall 1997). Increasing the NaCl level in cultures of *B. braunii*, a freshwater alga, showed an increase in growth rate, carbohydrate content, and lipid content. The coupled effect of nitrogen starvation and salinity stress increased the lipid content in *Chlamydomonas* sp. JSC4 (Ho et al. 2014). In addition, high salinity tends to induce the saturation of fatty acids, thus increasing the productivity of biodiesel (Chen et al. 2008).

14.4.2.7 Mixing

When the culture density is high, light cannot penetrate to the bottom of culture vessels, resulting in reduced biomass productivity. Therefore, mixing is necessary to ensure that all algal cells are suspended with equal access to light. Mixing is also useful to mix nutrients which helps the cells to take up these nutrients. Additionally, mixing can also make gas exchange more efficient.

14.4.3 Mass Cultivation of Microalgae

The growth rate and biomass production of microalgae in culture systems are affected by abiotic (light, temperature, pH, salinity, O_2, CO_2, nutrient stress, and toxic chemicals), biotic (pathogens and competition by other algae), and operational (shear produced by mixing, dilution rate, depth, harvest frequency, and addition of bicarbonate) factors. Usually, microalgae can be mass cultivated using four methods: phototrophic, heterotrophic, mixotrophic, and photoheterotrophic (Wang et al. 2014). Among these, only phototrophic cultivation is commercially feasible for large-scale microalgal biomass production (Borowitzka 2005). Phototrophic microalgae capture atmospheric carbon dioxide during cultivation and act as a potential carbon sink.

14.4.3.1 Phototrophic Cultivation

In phototrophic method, microalgae are cultivated in open ponds and enclosed photobioreactors.

14.4.3.1.1 Open Pond Production These are the traditional and simplest systems used for large-scale microalgal production. Currently, about 98% of commercial algae are produced in open pond systems. There are diverse types of open pond systems that are mainly differentiated based on their size, shape, and material used for construction, type of agitation, and inclination (Borowitzka 2005). Some common ones include raceways stirred by a paddle wheel, wide shallow unmixed ponds, circular ponds mixed with a rotating arm, and sloping thin-layer cascade systems. Among these systems, raceways are the most commonly used artificial system (Jimenez et al. 2003). Open pond system is the cheapest method for mass cultivation of microalgae compared to close PBRs. Open pond systems do not compete with agricultural crops for land, since these can be established in minimal crop production areas (Chisti 2008). The construction, regular maintenance, and cleaning of these systems are easy and they also consume relatively less energy (Rodolfi et al. 2009). Open pond systems are less technical in design and are more scalable; however, they are limited by abiotic growth factors like temperature, pH, light intensity, and dissolved oxygen content and are easily subjected to contamination. Contamination from the air and ground is often a serious limiting factor for the cultivation of algae in open pond systems, and therefore, most of the species cultured in these systems are grown under selective environments such as high alkalinity and high salinity (Cysewski and Lorenz 2004; Moheimani and Borowitzka 2006; Borowitzka 2010; Medipally et al. 2015).

14.4.3.1.2 Enclosed Photobioreactors These systems are available in the form of tubes, bags, or plates, which are made up of glass, plastic, or other transparent materials. Algae are cultivated in these systems with sufficient supply of light, nutrients, and carbon dioxide (Pulz 2001; Carvalho et al. 2006). Although many PBR designs are available, only a few are practically used for the mass production of algae (Ugwu and Aoyagi 2012). Some common PBR designs include annular, tubular, and flat-panel reactors, with large surface area (Carvalho et al. 2006; Chisti 2006).

Annular PBRs: These PBRs are more commonly used as bubble columns or airlift reactors. In column PBRs, the columns are arranged vertically and aeration is provided from below and light illumination is supplied through transparent walls. Column PBRs have the advantages of best controlled growth conditions, efficient mixing, and highest volumetric gas transfer rates (Eriksen et al. 2007).

Tubular PBRs: In these reactor systems, algal cultures are pumped through long and transparent tubes. The mechanical pumps or airlifts create the pumping force, and the airlift also allows the exchange of CO_2 and O_2 between the liquid medium and the aeration gas (Converti et al. 2006; Medipally et al. 2015).

Flat-panel PBRs: Flat-panel PBRs support higher growth densities and promote higher photosynthetic efficiency (Eriksen 2008). In flat-panel systems, a thin layer of more dense culture is mixed or sailed across a flat clear panel, and the incoming light is absorbed within the first few millimeters at the top of the culture.

As compared to open pond systems, PBRs have many advantages such as controllable growth, system efficiency, and algal purity. However, there are some disadvantages such as high costs of construction, operation, and maintenance. Though these drawbacks can be partially compensated by higher productivities, cost-effective production of microalgae biomass on required scale for biodiesel production is still a limiting factor.

14.4.3.1.3 Hybrid Production Systems In hybrid systems, both open ponds and close PBRs are used together in combination to get better results. In these systems, the required amount of contamination-free inocula obtained from PBRs is transferred to open ponds or raceways to get maximum biomass yield (Medipally et al. 2015).

14.4.3.2 Genetic Engineering for Enhanced Biomass Production

Microalgal growth is usually influenced by various environmental stress conditions such as temperature, light, salt concentration, and pH. The growth of microalgae can be enhanced by genetic engineering and manipulations of growth characteristics, but these manipulations increase the cultivation costs of microalgae. Thus, it will be advantageous if genetic engineering strategies are employed to control these environmental stress conditions. The average light intensity that provides the maximum photosynthesis in most microalgae is around 200–400 μM photons m^{-2} s^{-1}. Light intensity above this level reduces the microalgal growth. During midday, the maximum light intensity reaches up to 2000 μM photons m^{-2} s^{-1} (Melis 2000). Because of this, microalgal growth efficiency during daytime is less. Therefore, several studies were carried out to improve the microalgae photosynthetic efficiency and also to reduce the effect of photoinhibition. These studies were aimed at reducing the number of light-harvesting complexes (LHC) or to lower the chlorophyll antenna size to decrease light-absorbing capacity of individual chloroplasts (Mussgnug et al. 2007). LHC expression in transgenic *C. reinhardtii* was downregulated to increase the resistance to photooxidative damage and to enhance the efficiency of photosynthesis by 50% (Mussgnug et al. 2007; Beckmann et al. 2009). This alteration allowed *C. reinhardtii* to tolerate photoinhibition. Genes able to withstand stress conditions such as temperature, pH, salt concentration, and other stimuli have also been identified. The application of modern metabolic engineering tools in photosynthetic microalgae has the potential to create important sources of renewable fuels that will not compete with food production or require freshwater and arable land (Radakovits et al. 2010).

14.4.4 Biomass Harvesting and Drying Technology

Harvesting and dewatering are one of the challenging areas of current biofuel technology as microalgae have small size and low density. The most challenging job is the release of lipids from intracellular location in the most energy efficient and economical way possible, avoiding the use of large amounts of solvent and utilizing as much of the carbon in the biomass as liquid biofuel as possible, potentially with the recovery of minor high-value products (Scott et al. 2010). The major techniques presently applied in microalgae harvesting and recovery include flocculation, centrifugation, filtration, and flotation.

14.4.4.1 Flocculation
Flocculation, a process of forming aggregates known as algal flocs, is often performed as a pretreatment to destabilize algal cells from water and to increase the cell density by natural, chemical, or physical means. Chemicals called flocculants are usually added to induce flocculation, and commonly used flocculants are inorganic flocculants such as alum or organic flocculants such as chitosan or starch (Morales et al. 1985; Knuckey et al. 2006; Vandamme et al. 2010). The surface charge of microalgal cells in general is negative due to the ionization of functional groups on the cell walls that can be neutralized by applying positively charged electrodes and cationic polymers. This harvesting method is pretty expensive because of the cost of flocculants; hence, flocculants need to be inexpensive, easily produced, and nontoxic.

14.4.4.2 Centrifugation
Centrifugation is an extensively used method of microalgal separation on the basis of particle size and density. Separation efficiency depends on the size of desired algal species. Numerous centrifugal techniques in various types and size have been employed, depending on the use, such as tubular centrifuge, multichamber centrifuges, imperforate basket centrifuge, decanter, solid retaining disc centrifuge, nozzle-type centrifuge, solid ejecting–type disc centrifuge, and hydrocyclone (Shelef et al. 1984).

Despite being an energy-intensive method, it is rapid and a preferred method for microalgal cell recovery. Despite being very effective, centrifugation is considered unfeasible in large-scale algal culture systems due to the high capital and operational costs.

14.4.4.3 Filtration

Filtration harvests microalgal biomass through filters on which the algae accumulate forming thick algal paste and allow the liquid medium to pass through. Filtration systems can be classified as macrofiltration (pore size of >10 μm), microfiltration (pore size of 0.1–10 μm), ultrafiltration (pore size of 0.02–2 μm), and reverse osmosis (pore size of <0.001 μm) (Gultom and Hu 2013). Nevertheless, filtration entails extensive running costs and is also time-consuming.

14.4.4.4 Flotation

Microalgal cells are trapped on microair bubbles and float at the surface of water (Sharma et al. 2013). Usually, the flotation efficiency depends on the size of the created bubble: nanobubbles (<1 μm), microbubbles (1–999 μm), and fine bubbles (1–2 mm) (Kim et al. 2013). Dissolved air flotation is an extensively used technique in which microalgal cells are usually flocculated first and then air is bubbled through the liquid, causing the flocs to float to the surface for easier harvesting. Hydrophobic interaction and surface charge of microalgae play a crucial role in attachment of microalgae to the bubbles.

14.4.4.5 Drying of Harvested Biomass

Harvesting is followed by drying of the wet biomass. Drying of harvested biomass is essential to increase the viability of biomass for lipid extraction. Drying methods include natural sun drying or using advanced techniques like freeze-drying, drum drying, oven drying, spray drying, and fluidized bed drying. Despite being among the slower methods, sun drying is cost- and energy-effective as compared to other techniques. Freeze-drying is widely used for dewatering of microalgal biomass. Freeze-drying is a gentle process in which all the cell constituents are preserved without rupturing the cell wall (Chen et al. 2009; Halim et al. 2012).

14.5 Downstream Processing

14.5.1 Microalgal Lipid Extraction

Lipids can be extracted from the biomass by both mechanical and chemical methods. Mechanical disruption is a method that is initially employed to disrupt the cell wall by grinding, pressing, beating, or crushing prior to the application of the extraction solvents (U.S. Department of Energy 2010). Solvent extraction has proved to be successful in the extraction of lipids from microalgae. In this approach, organic solvents, such as benzene, cyclohexane, hexane, acetone, and chloroform, are added to algal paste. Solvents destroy algal cell wall and extract oil from aqueous medium because of their higher solubility in organic solvents. Solvent extract is then subjected to distillation process to separate oil from solvent. Hexane is reported to be the most efficient solvent in the extraction of lipids based on its highest extraction capability and low cost. Supercritical extraction such as subcritical water extraction and super-critical methanol extraction is also employed (U.S. Department of Energy 2010). In addition, extraction methods such as ultrasonic and microwaves have also been tried for oil extraction (Mata et al. 2010).

14.5.2 Conversion of Microalgal Oil to Biodiesel

After the extraction processes, the resulting microalgal oil is converted into bio-diesel through a process called transesterification. The transesterification reaction consists of transforming triglycerides into fatty acid alkyl esters, in the presence of an alcohol, such as methanol or ethanol, and a catalyst, such as an alkali or acid, with glycerol as a by-product (Vasudevan and Briggs 2008). For user acceptance, microalgal biodiesel needs to comply with the existing standards, such as ASTM Biodiesel Standard D 6751 (United States) or Standard EN 14214 (European Union). Some microalgal oils contain high degree of polyunsaturated fatty acids (with four or more double bonds) when compared to vegetable oils, which make these susceptible to oxidation during storage and therefore reduces their acceptability as biodiesel. However, the extent of unsaturation and content of fatty acids with more than four double bonds can be reduced easily by partial catalytic hydrogena-tion of the oil, the same technology that is commonly used in making margarine from vegetable oils (Chisti 2007). Nevertheless, microalgal biodiesel has physical and chemical properties similar to petroleum diesel and first-generation biodiesel and compares favorably with the international standard EN 14214 (Brennan and Owende 2010).

14.5.3 Technologies for Conversion of Microalgal Biomass to Biofuels

Technologies for microalgal biomass conversion to biofuels are classified into bio-chemical conversion, thermochemical conversion, chemical reaction, and direct combustion (Pena 2008) (Figure 14.5). Biochemical conversion can be applied to produce methanol (anaerobic digestion) and ethanol (fermentation) from microalgal biomass (Spolaore et al. 2006). Thermochemical conversion processes can be catego-rized into pyrolysis (bio-oil, charcoal), gasification (fuel gas), and liquefaction (bio-oil) (Hirano et al. 1998; Minowa and Sawayama 1999; Chiaramonti et al. 2007). The energy stored in microalgal cells can be converted into electricity by using direct combustion process. In chemical conversion technologies, transesterification process can be employed for the conversion of extracted lipids into biodiesel (Chisti 2007). Catalytic processes are more appropriate in converting biomass to biodiesel, espe-cially nanocatalysts that are good at improving product quality and attaining best operating conditions (Akia et al. 2014).

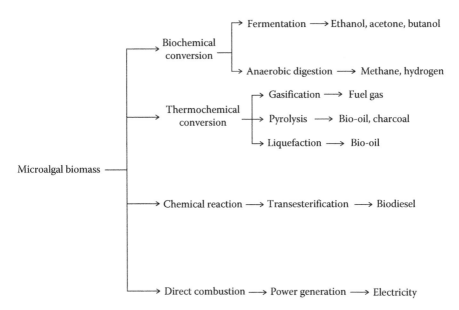

Figure 14.5 Principal microalga biomass transformation processes for biofuel production.

14.6 Conclusions

Algae-based biofuels are technically and economically viable and cost competitive, require no additional land, require minimal water use, and mitigate atmospheric CO_2. However, commercial production of microalgae biodiesel is still not feasible due to the low biomass production and costly downstream processes. Microalgal biofuel production can be achieved by designing advanced PBRs and developing low-cost technologies for biomass harvesting, drying, and oil extraction. There is a need to identify fast-growing microalgal strains containing high amounts of lipids/starch. Strain improvement through genetic engineering strategies or engineering of metabolic pathways for high lipid production can bring down the cost of microalgal biofuel production. Thus, current researches should focus on the development of novel upstream and downstream technologies to make commercial production of biofuels from microalgae viable.

References

Abdel-Raouf, N., Al-Homaidan, A.A., and I.B.M. Ibraheem. 2012. Microalgae and wastewater treatment. *Saudi J Biol Sci* 19:257–275.

Abou-Shanab, R.A.I., Hwang, J.H., Cho, Y., Min, B., and B.H. Jeon. 2011. Characterization of microalgal species isolated from fresh water bodies as a potential source for biodiesel production. *Appl Energ* 88:3300–3306.

Afify, A.E.M.R., Shalaby, E.A., and S.M.M. Shanab. 2010. Enhancement of biodiesel production from different species of algae. *Grasas Y Aceites* 61:416–422.

Ahmad, A.L., Yasin, N.H.M., Derek, C.J.C., and J.K. Lim. 2011. Microalgae as a sustainable energy source for biodiesel production: A review. *Renew Sust Energ Rev* 15:584–593.

Akia, M., Yazdani, F., Motaee, E., Han, D., and H. Arandiyan. 2014. A review on conversion of biomass to biofuel by nanocatalysts. *Biofuel Res J* 1:16–25.

Alabi, A.O., Tampier, M., and E. Bibeau. 2009. Microalgae technologies and processes for biofuels/bioenergy production in British Columbia. Current technology, suitability and barriers to implementation. Final report submitted to The British Columbia Innovation Council, Seed Science Press, Nanaimo, BC.

Al-Mulali, U. 2015. The impact of biofuel energy consumption on GDP growth, CO_2 emission, agricultural crop prices, and agricultural production. *Int J Green Energ* 12:1100–1106.

Amaro, H.M., Guedes, A.C., and F.X. Malcata. 2011. Advances and perspectives in using microalgae to produce biodiesel. *Appl Energ* 88:3402–3410.

Andrade, J.E., Perez, A., Sebastian, P.J., and D. Eapen. 2011. A review of biodiesel production processes. *Biomass Bioenerg* 35:1008–1020.

Antolin, G., Tinaut, F.V., Briceno, Y., Castano, V., Perez, C., and A.L. Ramirez. 2002. Optimization of biodiesel production by sunflower oil transesterification. *Bioresour Technol* 83:111–114.

Atabani, A.E., Silitonga, A.S., Irfan, A.B., Mahlia, T.M.I., Masjuki, H.H., and S. Mekhilef. 2012. A comprehensive review on biodiesel as an alternative energy resource and its characteristics. *Renew Sust Energ Rev* 16:2070–2093.

Atta, M., Idris, A., Bukhari, A., and S. Wahidin. 2013. Intensity of blue LED light: A potential stimulus for biomass and lipid content in fresh water microalgae *Chlorella vulgaris*. *Bioresour Technol* 148:373–378.

Azov, Y. 1982. Effect of pH on inorganic carbon uptake in algal cultures. *Appl Environ Microbiol* 43:1300–1306.

Bajpai, D. and V.K. Tyagi. 2006. Biodiesel: Source, production, composition, properties and its benefits. *J Oleo Sci* 55:487–502.

Baliga, R. and S.E. Powers. 2010. Sustainable algae biodiesel production in cold climates. *Int J Chem Eng* 2010:1–13.

Bayram, H. and A.B. Ozturk. 2014. Global climate change, desertification, and its consequences in Turkey and the Middle East. In *Global Climate Change and Public Health*, eds. Pinkerton, K.E. and W.N. Rom, Vol. 7, pp. 293–305. Humana Press, New York.

Becker, E.W. 1994. Measurement of algal growth. In *Microalgae: Biotechnology and Microbiology*, eds. Chen, X., He, G., Deng, Z., Wang, N., and W. Jiang, pp. 52–62. Cambridge University Press, Cambridge, U.K.

Beckmann, J., Lehr, F., Finazzi, G. et al. 2009. Improvement of light to biomass conversion by de-regulation of light-harvesting protein translation in *Chlamydomonas reinhardtii*. *J Biotechnol* 142:70–77.

Bhola, V., Desikan, R., Santosh, S.K., Subburamu, K., Sanniyasi, E., and F. Bux. 2011. Effects of parameters affecting biomass yield and thermal behaviour of *Chlorella vulgaris*. *J Biosci Bioeng* 111:377–382.

Boichenko, V.A. and P. Hoffmann. 1994. Photosynthetic hydrogen-production in prokaryotes and eukaryotes: Occurrence, mechanisms and functions. *Photosynthetica* 30:527–552.

Borowitzka, M.A. 2005. Culturing microalgae in outdoor ponds. In *Algal Culturing Techniques*, ed. Anderson, R.A., pp. 205–218. Elsevier, London, U.K.

Borowitzka, M.A. 2010. Single cell oils. Microbial and algal oils. In *Algae Oils for Biofuels: Chemistry, Physiology, and Production*, eds. Cohen, Z. and C. Ratledge, pp. 271–289. AOCS Urbana, IL.

BP. 2015. Statistical review of world energy, June 2015. http://www.bp.com/content/dam/bp/ pdf/energy-economics/statistical-review-2015/bp-statistical-review-of-world-energy- 2015-full-report.pdf. Accessed on September 15, 2015.

Brennan, L. and P. Owende. 2010. Biofuels from microalgae-A review of technologies for production, processing, and extractions of biofuels and co-products. *Renew Sust Energ Rev* 14:557–577.

Carrieri, D., Momot, D., Brasg, I.A., Ananyev, G., Lenz, O.B., and D.A. Bryant. 2010. Boosting autofermentation rates and product yields with sodium stress cycling: Application to production of renewable fuels by cyanobacteria. *Appl Environ Microbiol* 76:6455–6462.

Carriquiry, M.A., Du, X., and G.R. Timilsina. 2011. Second generation biofuels: Economics and policies. *Energ Pol* 39:4222–4234.

Carvalho, A.P., Meireles, L.A., and F.X. Malcata. 2006. Microalgal reactors: A review of enclosed system designs and performances. *Biotechnol Progr* 22:1490–1506.

Chader, S., Mahmah, B., Chetehouna, K., Amrouche, F., and K. Abdeladim. 2011. Biohydrogen production using green microalgae as an approach to operate a small l proton exchange membrane fuel cell. *Int J Hydrogen Energ* 36:4089–4093.

Chaudhary, L., Pradhan, P., Soni, N., Singh, P., and A. Tiwari. 2014. Algae as a feedstock for bioethanol production: New entrance in biofuel world. *Int J Chem Technol Res* 6:1381–1389.

Cheirsilp, B. and S. Torpee. 2012. Enhanced growth and lipid production of microalgae under mixotrophic culture condition: Effect of light intensity, glucose concentration and fed batch cultivation. *Bioresour Technol* 110:510–516.

Chen, W., Zhang, C., Song, L., Sommerfeld, M., and Q. Hu. 2009. A high through put Nile red method for quantitative measurement of neutral lipids in microalgae. *J Microbiol Methods* 77:41–47.

Chen, Y., Cheng, J.J., and K.S. Creamer. 2008. Inhibition of anaerobic digestion process: A review. *Bioresour Technol* 99:4044–4064.

Cheng, L.H., Zhang, L., Chen, H.L., and C.J. Gao. 2006. Carbon dioxide removal from air by microalgae cultured in a membrane photobioreactor. *Sep Purif Technol* 50:324–329.

Chiaramonti, D.A., Oasmaa, A., and Y. Solantausta. 2007. Power generation using fast pyrolysis liquids from biomass. *Renew Sust Energ Rev* 11:1056–1086.

Chisti, Y. 2006. Microalgae as sustainable cell factories. *Environ Eng Manage J* 5:261–274.

Chisti, Y. 2007. Biodiesel from microalgae. *Biotechnol Adv* 25:294–306.

Chisti, Y. 2008. Biodiesel from microalgae beats bioethanol. *Trends Biotechnol* 26:126–131.

Chiu, S.Y., Kao, C.Y., Huang, T.T. et al. 2011. Microalgal biomass production and on-site bioremediation of carbon dioxide, nitrogen oxide and sulfur dioxide from flue gas using *Chlorella* sp. cultures. *Bioresour Technol* 102:9135–9142.

Christenson, L. and R. Sims. 2011. Production and harvesting of microalgae for wastewater treatment, biofuels, and bioproducts. *Biotechnol Adv* 29:686–702.

Chynoweth, D.P. 2005. Renewable biomethane from land and ocean energy crops and organic wastes. *Hort Sci* 40:283–286.

Cohen, Z. 1999. *Chemicals from Microalgae*, 1st edn. Ben-Gurion University of the Negev, Beersheba, Israel.

Converti, A., Lodi, A., Del Borghi, A., and C. Solisio. 2006. Cultivation of *Spirulina platensis* in a combined airlift-tubular reactor system. *Biochem Eng J* 32:13–18.

Cysewski, G.R. and R.T. Lorenz. 2004. Industrial production of microalgal cell-mass and secondary products-species of high potential: *Haematococcus*. In *Microalgal Culture: Biotechnology and Applied Phycology*, ed. Richmond, A., pp. 281–288. Blackwell Science, Oxford, U.K.

Dale, B. 2008. Biofuels: Thinking clearly about the issues. *J Agric Food Chem* 56:3885–3891.

Damiani, M.C., Popovich, C.A., Constenla, D., and P.I. Leonardi. 2010. Lipid analysis in *Haematococcus pluvialis* to assess its potential use as biodiesel feedstock. *Bioresour Technol* 101:3801–3807.

De Gorter, H. and D.R. Just. 2010. The social costs and benefits of biofuels: The intersection of environmental, energy and agricultural policy. *Appl Econ Persp Policy* 32:4–32.

DEFRA (Department for Environment, Food and Rural Affairs). 2012. Can biofuels policy work for food security? Draft discussion paper. DEFRA, London, U.K.

Demirbas, A. 2009. Political, economic and environmental impacts of biofuels: A review. *Appl Energ Suppl* 86:108–117.

Demirbas, A. 2010. Biodiesel for future transportation energy needs. *Energ Source* 32:1490–1500.

Devi, M.P. and S.V. Mohan. 2012. CO_2 supplementation to domestic wastewater enhances microalgae lipid accumulation under mixotrophic microenvironment: Effect of sparging period and interval. *Bioresour Technol* 112:116–123.

Dismukes, G.C., Carrieri, D., Bennette, N., Ananyev, G.H., and M.C. Poseit. 2008. Aquatic phototrophs: Efficient alternatives to land-based crops for biofuels. *Curr Opin Biotechnol* 19:235–240.

Dragone, G., Fernandes, B., Vicente, A.A., and J.A. Teixeira. 2010. Third generation biofuels from microalgae. *Appl Microbiol* 2:1355–1366.

Einicker-Lamas, M., Mezian, G.A., Fernandes, T.B. et al. 2002. *Euglena gracilis* as a model for the study of Cu^{2+} and Zn^{2+} toxicity and accumulation in eukaryotic cells. *Environ Pollut* 120:779–786.

Ergas, S., Yuan, X., Sahu, A. et al. 2010. Enhanced CO_2 fixation and biofuel production via microalgae: Recent developments and future directions. *Trends Biotechnol* 28:371–780.

Eriksen, N.T. 2008. The technology of microalgal culturing. *Biotechnol Lett* 30:1525–1536.

Eriksen, N.T., Riisgard, F.K., Gunther, W.S., and J.J. Lønsmann Iversen. 2007. On-line estimation of O_2 production, CO_2 uptake, and growth kinetics of microalgal cultures in a gas-tight photobioreactor. *J Appl Phycol* 19:161–174.

Eshaq, F.S., Ali, M.N., and M.K. Mohd. 2011. Production of bioethanol from next generation feed-stock alga *Spirogyra* species. *Int J Eng Sci Technol* 3:1749–1755.

Fabregas, J., Maseda, A., Dominguez, A., Ferreira, M., and A. Otero. 2002. Changes in the cell composition of the marine microalga *Nannochloropsis gaditana*, during a light: Dark cycle. *Biotechnol Lett* 24:1699–1703.

FAO (Food and Agriculture Organization). 2009. Algae-based biofuels: A review of challenges and opportunities for developing countries. FAO, Rome, Italy. www.fao.org/bioenergy/aquaticbiofuels. Accessed on July 10, 2015.

Farrell, A.E., Plevin, R.J., Turner, B.T., Jones, A.D., O'Hare, M., and D.M. Kammen. 2006. Ethanol can contribute to energy and environmental goals. *Science* 311:506–508.

Fedorov, A.S., Kosourov, S., Ghirardi, M.L., and M. Seibert. 2005. Continuous hydrogen photoproduction by *Chlamydomonas reinhardtii*: Using a novel two-stage, sulfate limited chemostat system. *Appl Biochem Biotechnol A Enzyme Eng Biotechnol* 121:403–412.

Feng, Y., Li, C., and D. Zhang. 2011. Lipid production of *Chlorella vulgaris* cultured in artificial wastewater medium. *Bioresour Technol* 102:101–105.

Ferreira, A.F., Marques, A.C., Batista, A.P., Marques, P., Gouveia, L., and C. Silva. 2012. Biological hydrogen production by *Anabaena* sp. yield, energy and CO_2 analysis including fermentative biomass recovery. *Int J Hydrogen Energ* 37:179–190.

Gardner, R., Peters, P., Peyton, B., and K. Cooksey. 2011. Medium pH and nitrate concentration effects on accumulation of triacylglycerol in two members of the Chlorophyta. *J Appl Phycol* 26:1005–1016.

Gavrilescu, M. and Y. Chisti. 2005. Biotechnology—A sustainable alternative for chemical industry. *Biotechnol Adv* 23:471–499.

Gerasimchuk, I., Bridle, R., Charles, C., and T. Moerenhout. 2012. *Cultivating Governance: Cautionary Tales for Biofuel Policy Reformers*. IISD-GSI, Geneva, Switzerland.

Gouveia, L. and A.C. Oliveira. 2009. Microalgae as a raw material for biofuel production. *J Ind Microbiol Biotechnol* 36:269–274.

Guckert, J.B. and K.E. Cooksey. 1990. Triglyceride accumulation and fatty acid profile changes in *Chlorella* (Chlorophyta) during high pH induced cell cycle inhibition. *J Phycol* 26:72–79.

Gul, B., Abideen, Z., Ansari, R., and M.A. Khan. 2013. Halophytic biofuels revisited. *Biofuels* 4:575–577.

Gultom, S.O. and B. Hu. 2013. Review of microalgae harvesting via co-pelletization with filamentous fungus. *Energies* 6:5921–5939.

Gupta, A.K., Suresh, I.V., Misra, J., and M. Yunus. 2002. Environmental risk mapping approach: Risk minimization tool for development of industrial growth centres in developing countries. *J Clean Prod* 10:271–281.

Guschina, I.A. and J.L. Harwood. 2006. Lipids and lipid metabolism in eukaryotic algae. *Prog Lipid Res* 45:160–168.

Gutierrez, C.C. 2009. Algal biofuels: The effect of salinity and pH on growth and lipid content of algae. MA dissertation, University of Texas, Austin, TX.

Hafizan, C. and N.Z. Zainura. 2013. Biofuel: Advantages and disadvantages based on life cycle assessment (LCA) perspective. *J Environ Res Dev* 7:1444–1449.

Halim, R., Danquah, M.K., and P.A. Webley. 2012. Extraction of oil from microalgae for biodiesel production: A review. *Biotechnol Adv* 30:709–732.

Hankamer, B., Lehr, F., Rupperecht, J., Mussgung, C.P., and O. Kruse. 2007. Photosynthetic biomass and H$_2$ production by green algae: From bioengineering to bioreactor scale-up. *Physiol Planta* 131:10–21.

Hansen, P.J. 2002. Effect of high pH on the growth and survival of marine phytoplankton: Implications for species succession. *Aquat Microb Ecol* 28:279–288.

Harun, R., Danquah, M.K., and G.M. Forde. 2010a. Microalgal biomass as a fermentation feedstock for bioethanol production. *J Chem Technol Biotechnol* 85:199–203.

Harun, R., Singh, M., Forde, G.M., and K.M. Danquah. 2010b. Bioprocess engineering of microalgae to produce a variety of consumer products. *Renew Sust Energ Rev* 14:1037–1047.

Harun, R., Yip, J.W.S., Thiruvenkadam, S., Ghani, W.A.W.A.K., Cherrington, T., and M.K. Danquah. 2014. Algal biomass conversion to bioethanol a step-by-step assessment. *Biotechnology* 9:73–86.

Helming, J.F., Pronk, M., Pronk, A., and G. Woltjer, 2010. Stabilisation of the grain market by flexible use of grain for bioethanol, LEI Report 2010-039. Wageningen University, Wageningen, the Netherlands.

Hirano, A., Hon-Nami, K., Kunito, S., Hada, M., and Y. Ogushi. 1998. Temperature effect on continuous gasification of microalgal biomass: Theoretical yield of methanol production and its energy balance. *Catal Today* 45:399–404.

Ho, S.H., Nakanishi, A., Ye, X. et al. 2014. Optimizing biodiesel production in marine *Chlamydomonas* sp. JSC4 through metabolic profiling and an innovative salinity-gradient strategy. *Biotechnol Biofuels* 7:97.

Hu, H. and K. Gao. 2003. Optimization of growth and fatty acid composition of a unicellular marine picoplankton, *Nannochloropsis* sp., with enriched carbon sources. *Biotechnol Lett* 25:421–425.

Hu, Q. 2004. Environmental effects on cell composition. In *Handbook of Microalgal Culture*, ed. Richmond, A., pp. 83–93. Blackwell Publishing Ltd., Oxford, U.K.

Hu, Q., Sommerfeld, M., Jarvis, E. et al. 2008. Microalgal triacylglycerols as feedstocks for biofuel production: Perspectives and advances. *Plant J* 54:621–639.

Huang, G.H., Chen, F., Wei, D., Zhang, X.W., and G. Chen. 2010. Biodiesel production by microalgal biotechnology. *Appl Energ* 87:38–46.

Huang, J., Yang, J., Msangi, S., Rozelle, S., and A. Weersink. 2012. Biofuels and the poor: Global impact pathways of biofuels on agricultural markets. *Food Pol* 37:439–451.

Hyka, P., Lickova, S., Ribyl, P., Melzoch, K., and K. Kovar. 2013. Flow cytometry for the development of biotechnological processes with microalgae. *Biotechnol Adv* 31:2–16.

IEA (International Energy Agency). 2014. World energy outlook, 2014. http://www.worldenergyoutlook.org/publications/weo-2014/. Accessed on June 9, 2015.

Innocent, U., Behnaz, H.Z., Trina, H. et al. 2013. Harvesting microalgae grown on wastewater. *Bioresour Technol* 139:101–110.

Ito, T., Tanaka, M., Shinkawa, H. et al. 2013. Metabolic and morphological changes of an oil accumulating trebouxiophycean alga in nitrogen-deficient conditions. *Metabolomics* 9:S178–S187.

Jimenez, C., Cossio, B.R., Labella, D., and F.X. Niell. 2003. The feasibility of industrial production of *Spirulina* (*Arthrospira*) in Southern Spain. *Aquaculture* 217:179–190.

John, R.P., Anisha, G., Nampoothiri, K.M., and A. Pandey. 2011. Micro and macroalgal biomass: A renewable source for bioethanol. *Bioresour Technol* 102:186–193.

Jumbe, C., Msisska, F., and M. Madjera. 2009. Biofuel development in sub-Saharan Africa: Are the policies conducive? *Energ Pol* 37:4980–4986.

Juneja, A., Ceballos, R.M., and G.S. Murthy. 2013. Effects of environmental factors and nutrient availability on the biochemical composition of algae for biofuels production: A review. *Energies* 6:4607–4638.

Jung, K.A., Lim, S.R., Kim, Y., and J.M. Park. 2013. Potentials of macroalgae as feedstocks for biorefinery. *Bioresour Technol* 135:182–190.

Kapdan, I.K. and F. Kargi. 2006. Biohydrogen production from waste materials. *Enzyme Microb Technol* 38:569–582.

Karmakar, B., Karmakar, S., and S. Mukherjee. 2010. Properties of various plants and animals feedstocks for biodiesel production. *Bioresour Technol* 101:7201–7210.

Khoeyi, Z.A., Seyfabadi, J., and Z. Ramezanpour. 2012. Effect of light intensity and photoperiod on biomass and fatty acid composition of the microalgae, *Chlorella vulgaris*. *Aquacult Int* 20:41–49.

Khozin-Goldberg, I. and Z. Cohen. 2006. The effect of phosphate starvation on the lipid and fatty acid composition of the fresh water eustigmatophyte *Monodus subterraneus*. *Phytochemistry* 67:696–701.

Kim, H.J., Turner, T.L., and Y.S. Jin. 2013. Combinatorial genetic perturbation to refine metabolic circuits for producing biofuels and biochemical. *Biotechnol Adv* 31:976–985.

Kim, Y., Bae, B., and Y. Choung. 2005. Optimization of biological phosphorus removal from contaminated sediments with phosphate-solubilizing microorganisms. *J Biosci Bioeng* 99:23–29.

Kirrolia, A., Narsi, R.B., and R. Singh. 2013. Microalgae as a boon for sustainable energy production and its future research and development aspects. *Renew Sust Energ Rev* 20:642–656.

Knuckey, R.M., Brown, M.R., Robert, R., and D.M.F. Frampton. 2006. Production of microalgal concentrates by flocculation and their assessment as aquaculture feeds. *Aquacult Eng* 35:300–313.

Komerath, N.M. and P.P. Komerath. 2011. Terrestrial micro renewable energy applications of space technology. *Phys Procedia* 20:255–269.

Krumdieck, S., Wallace, J., and O. Curnow. 2008. Compact, low energy CO_2 management using amine solution in a packed bubble column. *Chem Eng J* 135:3–9.

Kumar, S.K. 2013. Performance and emission analysis of diesel engine using fish oil and bio-diesel blends with isobutanol as an additive. *Am J Eng Res* 2:322–329.

Lananan, F., Jusoh, A., Ali, N., Lam, S.S., and A. Endut. 2013. Effect of Conway medium and f/2 medium on the growth of six genera of South China Sea marine microalgae. *Bioresour Technol* 141:75–82.

Lang, X., Dalai, A.K., Bakhshi, N.N., Reaney, M.J., and P.B. Hertz. 2001. Preparation and characterization of biodiesels from various biooils. *Bioresour Technol* 80:53–62.

Lee, J.S., Kim, D.K., Lee, J.P. et al. 2002. Effects of SO_2 and NO on growth of *Chlorella* species KR-1. *Bioresour Technol* 82:1–4.

Leite, G. and P.C. Hallenbeck. 2012. Algae oil. In *Microbial Technologies in Advanced Biofuels Production*, ed. Hallenbeck, P.C., pp. 231–259. Springer, Boston, MA.

Lewicki, A., Dach, J., Janczak, D., and W. Czekala. 2013. The experimental photoreactor for microalgae production. *Procedia Technol* 8:622–627.

Li, X., Hu, H.Y., Gan, K., and Y.X. Sun. 2010. Effects of different nitrogen and phosphorus concentrations on the growth, nutrient uptake, and lipid accumulation of a freshwater microalga *Scenedesmus* sp. *Bioresour Technol* 101:5494–5500.

Li, Y., Horsman, M., Wu, N., Lan, C.Q., and N. Dubois-Calero. 2008. Biofuels from microalgae. *Biotechnol Progr* 24:815–820.

Li, Y. and J.G. Qin. 2005. Comparison of growth and lipid content in three *Botryococcus braunii* strains. *J Appl Phycol* 17:551–556.

Li, Y., Zhou, W., Hu, B., Min, M., Chen, P., and R. Ruan. 2012. Effect of light intensity on algal biomass accumulation and biodiesel production for mixotrophic strains *Chlorella kessleri* and *Chlorella protothecoide* cultivated in highly concentrated municipal waste water. *Biotechnol Bioeng* 109:2222–2229.

Liang, Y., Sarkany, N., and Y. Cui. 2009. Biomass and lipid productivities of *Chlorella vulgaris* under autotrophic, heterotrophic and mixotrophic growth conditions. *Biotechnol Lett* 31:1043–1049.

Liu, J., Huang, J.C., Fan, K.W. et al. 2010. Production potential of *Chlorella zofingiensis* as a feedstock for biodiesel. *Bioresour Technol* 101:8658–8663.

Liu, Z.Y., Wang, G.C., and B.C. Zhou. 2008. Effect of iron on growth and liquid accumulation in *Chlorella vulgaris*. *Bioresour Technol* 99:4717–4722.

Lunka, A.A. and D.J. Bayless. 2013. Effects of flashing light-emitting diodes on algal biomass productivity. *J Appl Phycol* 25:1679–1685.

Mata, T.M., Martins, A.A., and N.S. Caetano. 2010. Microalgae for biodiesel production and other applications: A review. *Renew Sust Energ Rev* 14:217–232.

Matsuo, Y., Yanagisawa, A., and Y. Yamashita. 2013. A global energy outlook to 2035 with strategic considerations for Asia and Middle East energy supply and demand interdependencies. *Energ Strategy Rev* 2:79–91.

Medipally, S.R., Yusoff, F.M., Banerjee, S., and M. Shariff. 2015. Microalgae as sustainable renewable energy feedstock for biofuel. *BioMed Res Int* 2015:Article ID 519513.

Melis, A. 2000. Sustained photobiological hydrogen gas production upon reversible inactivation of oxygen evolution in the green alga *Chlamydomonas reinhardtii*. *Plant Physiol* 122:127–136.

Milledge, J.J., Smith, B., Dyer, P.W., and P. Harvey. 2014. Macroalgae-derived biofuel: A review of methods of energy extraction from seaweed biomass. *Energies* 7:7194–7222.

Minowa, T. and S. Sawayama. 1999. A novel microalgal system for energy production with nitrogen cycling. *Fuel* 78:1213–1215.

Miranda, J.R., Passarinho, P.C., and L. Gouveia. 2012. Bioethanol production from *Scenedesmus obliquus* sugars: The influence of photobioreactors and culture conditions on biomass production. *Appl Microbiol Biotechnol* 96:555–564.

Moheimani, N.R. and M.A. Borowitzka. 2006. The long-term culture of the coccolithophore *Pleurochrysis carterae* (Haptophyta) in outdoor raceway ponds. *J Appl Phycol* 18:703–712.

Moller, F., Slento, E., and P. Frederiksen. 2014. Integrated well to wheel assessment of biofuels combining energy and emission LCA and welfare economic cost benefit analysis. *Biomass Bioenerg* 60:41–49.

Morales, J., de la Noue, J., and G. Picard. 1985. Harvesting marine microalgae species by chitosan flocculation. *Aquacult Eng* 4:257–270.

Morgan-Kiss, R.M., Priscu, J.C., Pocock, T., Gudynaite-Savitch, L., and N.P.A. Huner. 2006. Adaptation and acclimation of photosynthetic microorganisms to permanently cold environments. *Mol Biol Rev* 70:222–252.

Muradyan, E., Klyachko-Gurvich, G., Tsoglin, L., Sergeyenko, T., and N. Pronina. 2004. Changes in lipid metabolism during adaptation of the *Dunaliella salina* photosynthetic apparatus to high CO_2 concentration. *Russ J Plant Physiol* 51:53–62.

Mussgnug, J.H., Klassen, V., Schluter, A., and O. Kruse. 2010. Microalgae as substrates for fermentative biogas production in a combined biorefinery concept. *J Biotechnol* 150:51–56.

Mussgnug, J.H., Thomas-Hall, S., Rupprecht, J. et al. 2007. Engineering photosynthetic light capture: Impacts on improved solar energy to biomass conversion. *Plant Biotechnol J* 5:802–814.

Nomanbhay, S., Hussain, R., Rahman, M., and K. Palanisamy. 2012. Integration of biodiesel and bioethanol processes: Conversion of low cost waste glycerol to bioethanol. *Adv Nat Appl Sci* 6:802–818.

Oehmen, A., Lemos, P.C., Carvalho, G. et al. 2007. Advances in enhanced biological phosphorus removal: From micro to macro scale. *Water Res* 41:2271–2300.

Ong, H.C., Mahlia, T.M.I., and H.H. Masjuki. 2011. A review on energy scenario and sustainable energy in Malaysia. *Renew Sust Energ Rev* 15:639–647.

Park, J.H., Yoon, J.J., Park, H.D., Kim, Y.J., Lim, D.J., and S.H. Kim. 2011. Feasibility of biohydrogen production from *Gelidium amansii*. *Int J Hydrogen Energ* 36:13997–14003.

Pena, N. 2008. *Biofuels for Transportation: A Climate Perspective*. Pew Centre on Global Climate Change, Arlington, VA.

Pienkos, P.T. and A.L. Darzin. 2009. The promise and challenges of microalgal derived biofuels. *Biofuels Bioprod Bioref* 3:431–441.

Powell, E.E. and G.A. Hill. 2009. Economic assessment of an integrated bioethanol-biodiesel-microbial fuel cell facility utilizing yeast and photosynthetic algae. *Chem Eng Res Design* 87:1340–1348.

Prakasham, R.S., Nagaiah, D.K.S., Vinutha, A. et al. 2014. *Sorghum* biomass: A novel renewable carbon source for industrial bioproducts. *Biofuels* 5:159–174.

Prescott, G.W. 1982. *Algae of the Western Great Lakes Areas*. W.C. Brown Co., Dubuque, IA.

Pulz, O. 2001. Photobioreactors: Production systems for phototrophic microorganisms. *Appl Microbiol Biotechnol* 57:287–293.

Qin, J. 2005. Biohydrocarbons from algae-Impacts of temperature, light and salinity on algae growth. A report for Rural Industries Research and Development, Barton, Australian Capital Territory, Australia.

Qin, J.G. and Y. Li. 2006. Optimization of the growth environment of *Botryococcus braunii* Strain CHN 357. *J Freshwater Ecol* 21:169–176.

Radakovits, R., Jinkerson, R.E., Darzins, A., and M.C. Posewitz. 2010. Genetic engineering of algae for enhanced biofuel production. *Eukaryotic Cell* 9:486–501.

Rajkumar, R., Yaakob, Z., and M.S. Takriff. 2014. Potential of the micro and macro algae for biofuel production: A brief review. *Bioresour Technol* 9:1606–1633.

Rao, A.R., Sarada, R., Baskaran, V., and G.A. Ravishankar. 2006. Antioxidant activity of *Botryococcus braunii* extract elucidated in vitro models. *J Agric Food Chem* 54:4593–4599.

Ras, M., Lardon, L., Bruno, S., Bernet, N., and J.P. Steyer. 2011. Experimental study on a coupled process of production and anaerobic digestion of *Chlorella vulgaris*. *Bioresour Technol* 102:200–206.

Raven, J.A. and R.J. Geider. 2006. Temperature and algal growth. *New Phytol* 110:441–461.

Ren, H.Y., Liu, B.F., Ding, J. et al. 2012. Enhanced photo-hydrogen production in *Rhodopseudomonas faecalis* RLD-53 by EDTA addition. *Int J Hydrogen Energ* 37:8277–8281.

Richmond, A. 2000. Microalgal biotechnology at the turn of the millennium: A personal view. *J Appl Phycol* 12:441–451.

Riebesell, U., Revill, A.T., Holdsworth, D.G., and J.K. Volkman. 2000. The effects of varying CO_2 concentration on lipid composition and carbon isotope fractionation in *Emiliania huxleyi*. *Geochim Cosmochim Acta* 64:4179–4192.

Rodolfi, L., Zittelli, G.C., Bassi, N., Padovani, G., Biondi, N., and G. Bonini. 2009. Microalgae for oil: Strain selection, induction of lipid synthesis and outdoor mass cultivation in a low-cost photobioreactor. *Biotechnol Bioeng* 102:100–112.

Román-Leshkov, Y., Barrett, C.J., Liu, Z.Y., and J.A. Dumesic. 2007. Production of dimethyl-furan for liquid fuels from biomass-derived carbohydrates. *Nature* 447:982–985.

Ruangsomboon, S. 2012. Effect of light, nutrient, cultivation time and salinity on lipid production of newly isolated strain of the green microalga, *Botryococcus braunii* KMITL 2. *Bioresour Technol* 109:261–265.

Ruiz-Marin, A., Mendoza-Espinosa, L.G., and T. Stephenson. 2010. Growth and nutrient removal in free and immobilized green algae in batch and semicontinuous cultures treating real wastewater. *Bioresour Technol* 101:58–64.

Ryu, H.J., Oh, K.K., and Y.S. Kim. 2009. Optimization of the influential factors for the improvement of CO_2 utilization efficiency and CO_2 mass transfer rate. *J Ind Eng Chem* 15:471–475.

Saharan, B.S., Sharma, D., Sahu, R., Sahin, O., and A. Warren. 2013. Towards algal biofuels production: A concept of green bioenergy development. *Innov Roman Food Biotechnol* 12:1–21.

Saito, S. 2010. Role of nuclear energy to a future society of shortage of energy resources and global warming. *J Nucl Mater* 398:1–9.

Salvador, N.N.B., Glasson, J., and J.M. Piper. 2000. Cleaner production and environmental impact assessment: A UK perspective. *J Clean Prod* 8:127–132.

Sani, Y.M., Wan Daud, W.M.A., and A.R. Aziz. 2013. Solid acid-catalyzed biodiesel production from microalgal oil-The dual advantage. *J Environ Chem Eng* 1:113–121.

Saqib, A., Tabbssum, M.R., Rashid, U., Ibrahim, M., Gill, S.S., and M.A. Mehmood. 2013. Marine macroalgae *Ulva*: A potential feed-stock for bioethanol and biogas production. *Asian J Agric Biol* 1:155–163.

Satyanarayana, K.G., Mariano, A.B., and J.V.C. Vargas. 2011. A review on microalgae, a versatile source for sustainable energy and materials. *Int J Energ Res* 35:291–311.

Sawaengsak, S.H., Silalertruksa, T., Bangviwat, A., and S.H. Gheewala. 2014. Life cycle cost of biodiesel production from microalgae in Thailand. *Energ Sust Develop* 18:67–74.

Sawayama, S., Inoue, S., Dote, Y., and S.Y. Yokoyama. 1995. CO_2 fixation and oil production through microalga. *Energ Convers Manage* 36:729–731.

Sawyer, C.N., McCarty, P.L., and G.F. Parkin. 2003. *Chemistry for Environmental Engineering and Science*, 5th edn. McGraw-Hill Companies, Inc., Boston, MA.

Schenk, P.M., Thomas-Hall, S.R., Stephens, E., Marx, U.C., Mussgnug, J.H., and C. Posten. 2008. Second generation biofuels: High efficiency microalgae for biodiesel production. *Bioenerg Res* 1:20–43.

Scott, S.A., Davey, M.P., Dennis, J.S. et al. 2010. Biodiesel from algae: Challenges and prospects. *Curr Opin Biotechnol* 21:277–286.

Searchinger, T., Heimlich, R., Houghton, R.A. et al. 2008. Use of US croplands for biofuels increases greenhouse gases through emissions from land-use change. *Science* 319:1238–1240.

Shaishav, S., Singh, R.N., and T. Satyendra. 2013. Biohydrogen from algae: Fuel of the future. *Int Res J Environ Sci* 2:44–47.

Sharma, K.K., Garg, S., Li, Y., Malekizadeh, A., and P.M. Schenk. 2013. Critical analysis of current microalgae dewatering techniques. *Biofuels* 4:397–407.

Sharma, K.K., Schuhmann, H., and P.M. Schenk. 2012. High lipid induction in microalgae for biodiesel production. *Energies* 5:1532–1553.

Shelef, G., Sukenik, A., and M. Green. 1984. Microalgae harvesting and processing: A literature review. Technion Research and Development Foundation, Haifa, Israel.

Shi, X.M., Zhang, X.W., and F. Chen. 2000. Heterotrophic production of biomass and lutein by *Chlorella protothecoides* on various nitrogen sources. *Enzyme Microb Technol* 27:312–318.

Singh, A., Nigam, P.S., and J.D. Murphy. 2011. Mechanism and challenges in commercialization of algal biofuels. *Bioresour Technol* 102:26–34.

Singh, A., Pant, D., Olsen, S.I., and P.S. Nigam. 2012. Key issues to considering microalgae based biodiesel production. *Energ Educ Sci Technol A Energ Sci Res* 29:687–700.

Singh, D.P. and R.K. Trivedi. 2013. Production of biofuel from algae: An economic and eco-friendly resource. *Int J Sci Res* 2:352–357.

Sivakumar, G., Vail, D.R., Xu, J. et al. 2010. Bioethanol and biodiesel: Alternative liquid fuels for future generations. *Eng Life Sci* 10:8–18.

Solomon, B.D. 2010. Biofuels and sustainability. *Ann N Y Acad Sci* 1185:119–134.

Spolaore, P., Joannis-Cassan, C., Duran, E., and A. Isambert. 2006. Commercial applications of microalgae. *J Biosci Bioeng* 101:87–96.

Stanier, R.Y. and G. Cohen-Bazire. 1977. Phototrophic prokaryotes: The cyanobacteria. In *Annual Review of Microbiology*, eds. Starr, M.P., Ingraham, J.L., and A. Balows, pp. 225–274. Annual Reviews Inc., Palo Alto, CA.

Su, H., Zhang, Y., Zhang, C., Zhou, X., and J. Li. 2011. Cultivation of *Chlorella pyrenoidosa* in soybean processing wastewater. *Bioresour Technol* 102:9884–9890.

Sunda, W.G., Price, N.M., and F.M.M. Morel. 2005. Trace metal ion buffers and their use in culture studies. In *Algal Culturing Techniques*, ed. Andersen, R.A., pp. 35–63. Elsevier, Amsterdam, the Netherlands.

Surendhiran, D. and M. Vijay. 2012. Microalgal biodiesel—A comprehensive review on the potential and alternative biofuel. *Res J Chem Sci* 11:71–81.

Tabatabaei, M., Tohidfar, M., Jouzani, G.S., Safarnejad, M., and M. Pazouki. 2011. Biodiesel production from genetically engineered microalgae: Future of bioenergy in Iran. *Renew Sust Energ Rev* 15:1918–1927.

Timilsina, G. and A. Shreshtha. 2011. How much hope should we have for biofuels. *Energies* 36:2055–2069.

Tsukahara, K. and S. Sawayama. 2005. Liquid fuel production using microalgae. *J Jpn Petrol Inst* 48:251–259.

Tsuzuki, M., Ohnuma, E., Sato, N., Takaku, T., and A. Kawaguchi. 1990. Effects of CO_2 concentration during growth on fatty acid composition in microalgae. *Plant Physiol* 93:851–856.

Ueno, Y., Kurano, N., and S. Miyachi. 1998. Ethanol production by dark fermentation in the marine green alga, *Chlorococcum littorale*. *J Ferment Bioeng* 86:38–43.

Ugwu, C.U. and H. Aoyagi. 2012. Microalgal culture systems: An insight into their designs, operation and applications. *Biotechnology* 11:127–132.

US Department of Energy. 2010. National algal biofuels technology roadmap. Energy Efficiency and renewable Energy. Biomass Program. www1.eere.energy.gov/biomass/pdfs/algal biofuelsroadmap.pdf. Accessed on June 10, 2015.

US Energy Information Administration. 2011. Carbon dioxide emissions from energy consumption by source. http://www.eia.gov/totalenergy/data/annual/index.cfm#environment. Accessed on June 10, 2015.

US Energy Information Administration. 2015. Short-term energy outlook. https://www.eia.gov/forecasts/steo/report/global_oil.cfm. Accessed on September 18, 2015.

Van Lienden, A.R., Gerbens-Leenes, P.W., Hoekstra, A.Y., and T.H. Meer. 2010. Biofuel scenarios in a water perspective: The global blue and green water footprint of road transport in 2030. *Global Environ Change* 22:764–775.

Vandamme, D., Foubert, I., Meesschaert, B., and K. Muylaert. 2010. Flocculation of microalgae using cationic starch. *J Appl Phycol* 22:525–530.

Vasudevan, P.T. and M. Briggs. 2008. Biodiesel production-current state of the art and challenges. *J Ind Microbiol Biotechnol* 35:421–430.

Visviki, I. and Santikul, D. 2000. The pH tolerance of *Chlamydomonas applanata* (Volvocales, Chlorophyta). *Arch Environ Contam Toxicol* 38:147–151.

Wahidin, S., Idris, A., and S.R. Shaleh. 2013. The influence of light intensity and photoperiod on the growth and lipid content of microalga *Nannochloropsis* sp. *Bioresour Technol* 129:7–11.

Wang, B. and C.Q. Lan. 2011. Biomass production and nitrogen and phosphorus removal by the green algae *Neochloris oleoabundans* in simulated wastewater and secondary municipal waste water effects. *Bioresour Technol* 102:5639–5644.

Wang, B., Li, Y., Wu, N., and Q.L. Christopher. 2008. CO_2 biomitigation using microalgae. *Appl Microbiol Biotechnol* 79:707–718.

Wang, J., Yang, H., and F. Wang. 2014. Mixotrophic cultivation of microalgae for biodiesel production: Status and prospects. *Appl Biochem Biotechnol* 172:3307–3329.

Wargacki, A.J., Leonard, E., Win, M.N. et al. 2012. An engineered microbial platform for direct biofuel production from brown macroalgae. *Science* 335:308–313.

Wecker, M.S.A., Meuser, J.E., Posewitz, M.C., and M.L. Ghirardi. 2011. Design of a new biosensor for algal H_2 production based on the H_2-sensing system of *Rhodobacter capsulatus*. *Int J Hydrogen Energ* 36:11229–11237.

Wijffels, R.H. and M.J. Barbosa. 2010. An outlook on microalgal biofuels. *Science* 329:796–799.

World Commission on Environment and Development. 1987. *Our Common Future*. Oxford University Press, Oxford, U.K.

Xie, G.J., Liu, B.F., Xing, D.F., Nan, J., Ding, J., and N.Q. Ren. 2013. Photo-fermentative bacteria aggregation triggered by L-cysteine during hydrogen production. *Biotechnol Biofuels* 6:64–70.

Xu, X.Q. and J. Beardall. 1997. Effect of salinity on fatty acid composition of a green micro-alga from an antarctic hypersaline lake. *Phytochemistry* 45:655–658.

Yacoby, I., Pochekailov, S., Topotrik, H., Ghirardi, M.L., King, P.W., and S. Zhang. 2011. Photosynthetic electron partitioning between hydrogenase and ferredoxin: NADP-oxidoreductase (FNR) enzymes *in vitro*. *Proc Natl Acad Sci USA* 108:9396–9401.

Yanagisawa, M., Kawai, S., and K. Murata. 2013. Strategies for the production of high concentrations of bioethanol from seaweeds: Production of high concentrations of bioethanol from seaweeds. *Bioengineered* 4:224–235.

Yanagisawa, M., Nakamura, K., Ariga, O., and K. Nakasaki. 2011. Production of high concentrations of bioethanol from seaweeds that contain easily hydrolyzable polysaccharides. *Process Biochem* 46:2111–2116.

Yen, H.W. and D.E. Brune. 2007. Anaerobic co-digestion of algal sludge and waste paper to produce methane. *Bioresour Technol* 98:130–134.

Yoon, J.J., Kim, Y.J., Kim, S.H. et al. 2010. Production of polysaccharides and corresponding sugars from red seaweed. *Adv Mater Res* 93–94:463–466.

Zhang, N., Lior, N., and H. Jin. 2011. The energy situation and its sustainable development strategy in China. *Energy* 36:3639–3649.

Zhila, N.O., Kalacheva, G.S., and T.G. Volova. 2011. Effect of salinity on the biochemical composition of the alga *Botryococcus braunii* Kutz IPPAS H-252. *J Appl Phycol* 23:47–52.

Zhong, W., Zhang, Z., Luo, Y., Oiao, W., Xiao, M., and M. Zhang. 2012. Biogas productivity by co-digesting Taihu blue algae with corn straw as an external carbon source. *Bioresour Technol* 114:281–286.

Zhu, L., Li, Z., and T. Ketola. 2011. Biomass accumulations and nutrient uptake of plants cultivated on artificial floating beds in China's rural area. *Ecol Eng* 37:1460–1466.

15

Microalgal Cultivation for Biofuels
Cost, Energy Balance, Environmental Impacts, and Future Perspectives

M. Muthukumaran and V. Sivasubramaian

Contents

Abstract

This study forms the basis for the work to reduce our dependency on fossil fuels and to find out alternative renewable fuel sources. The most alarming facts today are the very high emission of CO_2 and the reduction in forest covers and cultivable lands due to urbanization. Our use of fossil fuels has become inevitable in certain scenarios. With a search for alternative energy source, microalgae were found to be promising. Biofuel from algae does not increase CO_2 levels and is found to be more efficient than any terrestrial plants. The global consciousness on ecological balance, pollution abatement, and environmental friendliness increases the need for technology that caters to modern trends and necessities on rise. This chapter deals with the microalgal cultivation for biofuels and so the significant topics associated with it: cost, energy balance, environmental impacts, future perspectives, and investigation on biomass production and biofuel potentials of microalgae, for example, using open raceway pond chrome sludge that was treated with the microalga *Desmococcus olivaceus* (Persoon ex Acharius) J.R. Laundon. There was a considerable amount of sludge reduction and biomass production in open raceway pond amended with chrome sludge. A remarkable reduction was found in total dissolved solid, sodium, potassium, and phosphate. With the aim to identify a potential species of microalgae for biodiesel production on a nitrate–bicarbonate dye effluent, an investigation was done and *Chlorococcum humicola* was found to be a promising species with 13% of oil in effluents. Through the fatty acid methyl esters (FAME) analysis, it was found that it contains a maximum of 75.8% of heneicosanoic acid methyl ester (C_{21}:0). The total amount of saturated fatty acid content comes to a maximum of 94% in the algae grown in combined dyeing effluent, which is greater than the content in algae grown under other condition.

15.1 Introduction

Microalgae have long been cultivated commercially for human nutritional products around the world in several small- to medium-scale production systems, producing a few tons to several hundred tons of biomass annually. It is estimated that the total world commercial microalgal biomass production is at about 10,000 tons per year. The predominant microalgal species currently cultivated photosynthetically for the production of nutritional products are *Spirulina* sp., *Chlorella* sp., *Dunaliella* sp., and

Haematococcus sp. Microalgae are a rich source of protein and other nutritional substances such as carbohydrates and fatty acids (FAs), as can be found in higher plants (Hiremath and Mathad 2010). They play an important role as primary producers of various consumers such as rotifer, copepods, and shrimps, which are, in turn, fed to late larval and juvenile fish and crustaceans (Richmond 1996; Seyfabadi et al. 2011). *Chlorella vulgaris* is one such green microalga found cosmopolitan in occurrence. Microalgae are a diverse group of prokaryotic or eukaryotic photosynthetic microorganisms with natural growing requirements such as light, sugars, CO_2, N, P, and K that can produce pigments and a variety of biochemicals in large amounts over short periods of time; they are an untapped resource with more than 25,000 species.

Microalgae have also long been recognized as a potentially good source for biofuel production due to their high oil content and rapid biomass production. Algal biomass is composed of three main components, namely, carbohydrates, proteins, and lipids/natural oils. The bulk of the natural oil made by microalgae is in the form of triacylglycerols (TAGs), which is the right kind of oil for producing biodiesel from microalgae; they are the exclusive focus of the algae-to-biofuel industry. There are two methods used for large-scale production of microalgae, namely, raceway pond and tubular photobioreactor (PBR)—of the two raceway pond is the best system for mass cultivation of microalgae. Moreover, they are the most profitable too (Ugwu et al. 2008).

Microalgae have been studied in laboratories and in mass outdoor cultures for more than a century. The breakthrough concept of photosynthesis that can be traced back in the laboratories of Otto Warburg, wherein *Chlorella* was initially used as the model organism, became unrivaled (Nickelsen 2007). As Grobbelaar (2004) pointed out, applied phycology and the mass production of microalgae became a reality in the 1940s. Since then, microalgae have been grown for a variety of potential applications, such as the production of lipids for energy using flue gases, antimicrobial substances, and cheap proteins for human nutrition and the production of various biochemicals. At present, the focus is on bioenergy (Tan and Amthor 1988); however, their only real success has been in wastewater treatment. A major frustration for microalgal biotechnologists has been the realization of much lower yields than what is potentially possible from laboratory measurements. The inability to operate PBRs including raceway ponds, at maximal photosynthetic efficiencies, impacts directly the economies of scale; due to this, many large-scale projects were not able to deliver what was predicted and many investors lost their investments. Richmond's (1996) observation that "microalga culture, however, is yet very far from supplying any basic human needs" is as true today as then and he concluded that "the major reason for this stems from the failure to develop production systems which utilize solar energy efficiently." The consequence of low yields and high production costs renders this technology suitable only for exclusive high-priced products. The methods such as flocculation, centrifugation, and filtration are used unaided to dewater algal biomass.

Rosana et al. (2012) reported that microalgae are promising organisms capable of CO_2 mitigation by CO_2 fixation, biofuel production, and wastewater treatment. CO_2 fixation by photoautotrophic algal cultures has the potential to diminish the release of CO_2 into the atmosphere, thereby helping to alleviate the trend toward global warming (Figure 15.1).

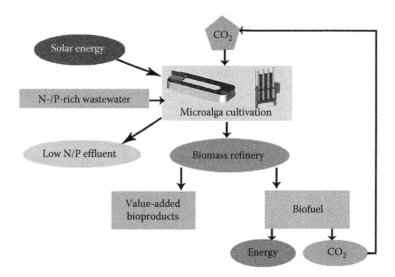

Figure 15.1 A combined biofuel production with reference to CO_2 biomitigation and N/P removal from wastewater using the microalgal system.

Microalgae, when fed with CO_2 and sunlight, produced large amounts of lipids and hence increase the output of algal oil. The enzyme acetyl CoA carboxylase from microalgae helps to catalyze and transform CO_2 in the synthesis of oils in algae. Biofuels are derived from microbes that can live on land unfit for crops and generate nearly engine-ready chemicals, which are considered to be third-generation biofuels (New Scientist 2011). A conceptual microalgal system for combined biofuel production, combining CO_2 biomitigation and N/P removal from wastewater (Wang et al. 2008).

Tsukahara and Sawayama (2005) reported that the technological developments including advances in PBR design, microalgal biomass harvesting, drying, and other downstream processing technologies are important areas that may lead to enhanced cost-effectiveness and, therefore, effective commercial implementation of the biofuel using a microalgal strategy (Figure 15.2).

15.1.1 Algal Cultivation and Harvesting Techniques

Many of the methods and basic culture medium concepts that are used today were developed in the late 1800s and the early 1900s.

15.1.1.1 Culture Media

15.1.1.1.1 Nutrified Agar Pour Plates
Some algae will not grow on the surface of agar, but they will grow if embedded in the agar. To embed the cells, one usually prepares 1%–2% concentration agar, and then as the agar reaches its gelling temperature, a liquid suspension of cells is mixed aseptically with the sterile agar. The mixture is swirled to distribute the algae, and then the agar is poured into petri plates.

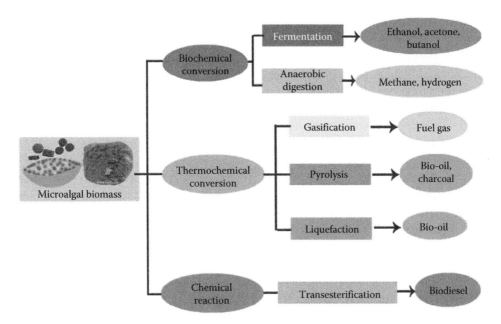

Figure 15.2   Secondary products from microalgal biomass.

15.1.1.1.2 Freshwater Culture Media   Various culture media have been developed and used for isolation and cultivation of freshwater algae, derived from the analysis of the water in the native habitat, some are formulated after detailed study on the nutrient requirement of the organism, and some are established after consideration of ecological parameters.

15.1.1.1.3 Marine Culture Media   Preoccupation with making a complete medium autoclavable without precipitation led to the following extensive modifications in recips: Addition of synthetic metal chelators such as EDTA or nitrilotriacetic acid to decrease metal precipitates. Addition of a pH buffer such as Tris or glycylglycine (7.0–8.5 range), because the amount of precipitate increased as the pH rose during autoclaving. Reduction in salinity, thereby reducing the amount of salts available for precipitation. Replacement of Mg^{2+} and Ca^{2+} with more soluble univalent salts. Replacement of inorganic phosphorus with an organic source (e.g., sodium glycerophosphate) to avoid the precipitation of $Ca_3(PO_4)_2$ (Droop 1969). Introduction of weak solubilizers that are acids (e.g., citric acid) having highly soluble salts with calcium.

15.1.1.1.4 Artificial Seawater   Artificial or synthetic seawater consists of two parts: the basal (main) salts that form the "basal seawater" and the enrichment solution (often the same as the enrichment solution that is added to NSW).

15.1.1.2 Sterilization
Sterilization is a process for establishing an aseptic condition, that is, the removal or killing of all microorganisms. Sterilization is very important in phycological research, especially when maintaining living organisms as isolated strains in culture. There are

various sterilization methods that can be roughly classified into four categories: heat sterilization, electromagnetic wave sterilization, sterilization using filtration, and chemical sterilization.

Heat sterilization is the most common of the general categories and usually requires high temperatures ($\geq 100°C$), implying that the materials to be sterilized can resist high temperatures (e.g., glassware, metallic instruments, and aluminum foil). Liquids are filter sterilized when the liquid contains fragile components that are destroyed by high temperature.

Electromagnetic waves (e.g., ultraviolet rays, gamma rays, x-rays, and microwaves) are used as an alternative for materials that cannot be exposed to high temperature (e.g., many plastic products or liquids with a labile component).

Finally, many different types of chemicals have been used for the purpose of sterilization; however, chemical traces may remain after the sterilization treatment, and those chemicals may be detrimental to living algae and to the investigator.

15.1.1.3 Microalgal Biomass Production

Microalgal cultivation can be done in open culture systems such as lakes or ponds and in highly controlled, closed culture systems called PBRs (Mata et al. 2010). The photosynthetic growth of microalgal biomass requires light, CO_2, water, organic salts, and temperature of 20°C–30°C. The microalgal biomass can be achieved by different cultivating methods. Chisti (2007) has described producing microalgal biomass is much more expensive than growing crops. Photosynthetic growth requires light, carbon dioxide, water, and inorganic salts. Temperature must remain generally within 20°C–30°C. To minimize expense, biodiesel production must rely on freely available sunlight, despite daily and seasonal variations in light levels.

Growth medium must provide the inorganic elements that constitute the algal cell. Essential elements include nitrogen, phosphorus, iron, and, in some cases, silicon. Minimal nutritional requirements can be estimated using the approximate molecular formula of the microalgal biomass that is CO0.48H1.83N0.11P0.01. This formula is based on data presented by Grobbelaar (2004). Nutrients such as phosphorus must be supplied in significant excess because the phosphates added complex with metal ions; therefore, not all the added phosphorus is bioavailable to enhance biomass production of *Scenedesmus* spp. (green microalgae) using a fermented swine wastewater (Kim et al. 2007).

Microalgal biomass offers a number of advantages over conventional biomass such as higher productivities, use of nonproductive land, reuse and recovery of waste nutrients, use of saline or brackish waters, and reuse of CO_2 from power plant flue gas. The most commonly used marine algal cultures are *Botryococcus braunii, Ch. vulgaris, Chaetoceros muelleri, Dunaliella salina, Nannochloropsis oculata, Arthrospira maxima,* and *Scenedesmus quadricauda.* The current review provides details of the microalgal biomass with emphasis on strain selection, cultivation, strain improvement, and biotechnological potentials (Raja et al. 2014).

15.1.1.4 Harvesting Techniques of Microalgal Biomass

Harvesting biomass represents one of the significant cost factors in the production of biomass. Efficient biomass harvesting from cultivation broth is essential for mass production of biodiesel from microalgae. The selection of harvesting techniques is dependent on the properties of microalgae such as density, size, and value of the desired products. The major techniques presently applied to the harvesting of microalgae include centrifugation, flocculation, filtration, screening, gravity sedimentation, immobilization, and electrophoresis.

15.1.1.4.1 Centrifugation   Most microalgae can be recovered from the liquid broth using centrifugation, because it is rapid, efficient, and universal. A report by Chen et al. (2011) stated that the laboratory centrifugation tests on pond effluent at 500–1000 g recovered about 80%–90% microalgae within 2–5 min.

15.1.1.4.2 Flocculation   Flocculation is the process in which dispersal particles are aggregated together to form large particles for setting. There are two types of flocculation: autoflocculation, which occurs as a result of precipitation of carbonate salts with algal cells, and chemical flocculation in which adding chemicals to microalgal culture induces flocculation. Autoflocculation is also termed as *bioflocculation* (Pittman et al. 2011). Flocculation may be achieved by the use of cationic polymers or the addition of alkali substances (Rawat et al. 2010).

15.1.1.4.3 Bioflocculation   A novel harvesting method is presented as a cost- and energy-efficient alternative: the bioflocculation by using one flocculating microalga to concentrate the nonflocculating microalga of interest. The advantages of this method are that no addition of chemical flocculants is required and that similar cultivation conditions can be used for the flocculating microalgae as for the microalgae of interest that accumulate lipids. This method is as easy and effective as chemical flocculation, which is applied at industrial scale; however, in contrast, it is sustainable and cost-effective as no costs are involved for the pretreatment of the biomass for oil extraction and for the pretreatment of the medium before it can be reused (Salim et al. 2011).

15.1.1.4.4 Gravity Sedimentation   Gravity sedimentation is a common method of harvesting biomass. The process is rudimentary but works for various types of algae and is highly energy efficient. Gravity sedimentation can be effective for separating larger and smaller organisms (Mutanda et al. 2011). Enhanced microalgal harvesting by sedimentation can be achieved through lamella separators and sedimentation tanks (Chen et al. 2011).

15.1.1.4.5 Filtration   Filtration is a method commonly used for solid–liquid separation (Rawat et al. 2010). Vacuum filtration is effective in the recovery of larger algae (greater 70 μm), when used with the aid of filters. For small cells, membrane microfiltration and ultrafiltration are alternative methods (Pittman et al. 2011). In addition, microstrainers have several advantages such as simplicity in function and construction, easy operation, low investment, energy intensive, and having high filtration ratios (Chen et al. 2011).

15.1.1.4.6 Floatation   Floatation is a gravity separation process in which air or gas bubbles attach to solid particles and then carry them to the liquid surface. Floatation

TABLE 15.1   Common Algal Harvesting Techniques

Microalgal Species	Algal Harvesting Method	Relative Cost Effect
Scenedesmus sp., *Chlorella* sp.	Foam fractionation	Very high
Scenedesmus sp., *Chlorella* sp.	Ozone flocculation	Very high
Scenedesmus sp., *Chlorella* sp.	Centrifugation	Very high
Scenedesmus sp., *Chlorella* sp.	Electrofloatation	High
Scenedesmus sp., *Chlorella* sp.	Inorganic chemical flocculation	High
Dunaliella sp.	Polyelectrolyte flocculation	High
Spirulina sp., *Coelastrum* sp.	Filtration	High
Spirulina sp.	Microstrainers	Unknown
Micractinium sp.	Tube settling	Low
Coelastrum sp.	Discrete sedimentation	Low
Euglena sp., *Dunaliella* sp.	Phototactic autoconcentration	Very low
Micractinium sp.	Autoflocculation	NA
Micractinium sp.	Bioflocculation	NA
Scenedesmus sp., *Chlorella* sp.	Tilapia-enhanced sedimentation	NA

is more beneficial and effective than sedimentation with regard to removing micro-algae. Floatation can capture particles with a diameter of less than 500 μm by collision between a bubble and a particle and the subsequent adhesion of the bubble and particle (Chen et al. 2011). The mechanism of action is interaction with the negatively charged hydrophilic surfaces of algal cells (Rawat et al. 2010). A disadvantage of this method is that it is an expensive process.

15.1.1.4.7 Magnetic Separation   A simple and rapid harvesting method by *in situ* magnetic separation with naked Fe_3O_4 nanoparticles has been developed for the microalgal recovery of *B. braunii* and *Chlorella ellipsoidea*. The developed *in situ* magnetic separation technology provides a great potential for saving time and energy associated with improving microalgal harvesting (Xu et al. 2011; Cerff et al. 2012; Hu et al. 2013). The other common microalgal harvesting techniques showed comparative cost effect (Table 15.1).

15.1.1.5 Chemical Composition of Microalgae

Microalgae are able to enhance the nutritional content of conventional food preparations and, hence, positively affect the health of humans and animals. The high protein content of various microalgal species is one of the main reasons to consider them as an unconventional source of protein (Cornet 1998; Soletto et al. 2005). In addition, the amino acid pattern of almost all algae compares favorably with that of other food proteins. As the cells are capable of synthesizing all amino acids, they can provide the essential ones to humans and animals (Guil-Guerrero et al. 2004). However, to completely characterize the protein and determine the amino acid content of microalgae, information on the nutritive value of the protein and the degree of availability of amino acids should be studied comprehensively (Becker 1988). Carbohydrates in microalgae can be found in the form of starch, glucose, sugars, and other polysaccharides. Their overall digestibility is high, which is why there is no limitation to use dried whole microalgae in foods or feeds (Becker 2004). The average lipid content of algal cells varies between 1% and 70% but can reach 90% of dry weight under certain conditions. Metting (1996) found that the algal lipids are composed of glycerol, sugars, or bases esterified to saturated or unsaturated FAs. Among all the FAs in microalgae, some FAs of the ω3 and ω6 families are of particular interest. The total amount and relative proportion of FAs can be affected by nutritional and environmental factors and nitrogen limitation (Borowitzka 1988). Microalgae also represent a valuable source of nearly all essential vitamins (e.g., A, B_1, B_2, B_6, B_{12}, C, E, nicotinate, biotin, folic acid, and pantothenic acid) (Becker 2004). Vitamins improve the nutritional value of algal cells, but their quantity fluctuates with environmental factors, the harvesting treatment, and the method of drying the cells (Brown et al. 1999). Microalgae are also rich in pigments like chlorophyll (0.5%–1% of dry weight), carotenoids (0.1%–0.2% of dry weight on average and up to 14% of dry weight for β-carotene of *Dunaliella* sp.), and phycobiliproteins. These molecules have a wide range of commercial applications. Thus, their composition gives microalgae interesting qualities, which can be applied in human and animal nutrition. However, prior to commercialization, algal material must be analyzed for the presence of toxic compounds to prove their harmlessness. In this content, recommendations published by different international organizations and additional national regulations often exist. They are concerned on the level of nucleic acids, toxins, and heavy metal toxicity. The safe level is about 20 g of algae per day or 0.3 g of algae per kg of body weight (Becker 2004). Finally, many metabolic studies have confirmed the capacities of microalgae as a novel source of protein: the average quality of most of the algae examined is equal or even superior to that of other conventional high-quality plant proteins (Becker 2004).

15.1.1.5.1 Bioactive Molecule Several environmental factors influence the proportion of different constituents of algal biomass, and very extreme variations in composition were reported in response to parameters such as temperature, illumination, pH value of the medium, mineral nutrients, and CO_2 supply.

To obtain an algal biomass with a desired composition, the proportion of the different constituents of several algae can be modified very specifically by varying the culture conditions, for instance, nitrogen or phosphorus depletion on the medium or changes of physical factors such as osmotic pressure, radiation intensity, population density, and light or dark growth. Spoehr and Milner (1949) were probably among the first who published detailed information on the effects of environmental conditions on algal composition and described the effect of varying nitrogen supply on the lipid chlorophyll content of *Chlorella* and some diatoms. Spoehr and Milner (1949) were reported of changes in the composition pattern and amino acid profile of basic proteins during the growth cycle of algae were described using the techniques of synchronous culture.

As with any higher plant, the chemical composition of algae is not an intrinsic constant factor but varies over a wide range. Environmental factors, such as temperature, illumination, pH value, mineral contents, CO_2 supply, population density, growth phase, and algal physiology, can greatly modify chemical composition. It presents indicative values of a gross chemical composition of different algae and is compared with the composition of selected conventional foodstuffs. Microalgae can biosynthesize, metabolize, accumulate, and secrete a great diversity of primary and secondary metabolites, many of which are valuable substances with potential applications in the food, pharmaceutical, and cosmetics industries (Yamaguchi 1997).

15.1.1.5.2 Pigments   One of the most obvious and arresting characteristics of the algae is their color. In general, each phylum has its own particular combination of pigments and an individual color. Aside from chlorophylls, the primary photosynthetic pigment, microalgae also form various accessory or secondary pigments, such as phycobilin proteins and a wide range of carotenoids. These natural pigments are able to improve the efficiency of light energy utilization of the algae and protect them against solar radiation and related effects. Their function as antioxidants in the plant shows interesting parallels with their potential role as antioxidants in foods and humans (Van den Berg et al. 2000). Therefore, microalgae are recognized as an excellent source of natural colorants and nutraceuticals. It is expected that they will surpass synthetics as well as other natural sources due to their sustainability of production and renewable nature (Dufossé et al. 2005).

15.1.1.5.3 Carotenoids   The second important groups of pigments found in algae are the carotenoids. Carotenoids are yellow, orange, or red lipophilic pigments of aliphatic or acyclic structure composed of eight, five-carbon (isoprenoid) units, which are linked so that the methyl groups nearest the center of molecule are in the 1,5-positions, whereas all the other lateral methyl groups are in 1,6-position. Lycopene, synthesized by stepwise desaturation of the first 40-carbon polyene phytoene, is the precursor of all carotenoids found in algae. The carotenoids can be divided into two main groups, that is, pigments composed of oxygen-free hydrocarbons, the carotenes, and their oxygenated derivatives, the xanthophylls, which contain epoxy, hydro, carboxylic, glycosidic, allenic, or acetylene groups. All algae contain carotenoids, each species usually between 5 and 10 major forms, the variety of which in algae is greater than in higher plants.

15.1.1.5.4 Phycobiliprotein   Phycobiliproteins are the major photosynthetic accessory pigments in cyanobacteria (blue-green algae, prokaryotic), rhodophytes (red algae, eukaryotic), cryptomonads (biflagellate unicellular eukaryotic algae), and cyanelles (endosymbiotic plastid-like organelles). Phycobiliproteins are brilliant-colored and water-soluble antenna-protein pigments organized in supramolecular complexes, called phycobilisomes, which are assembled on the outer surface of the thylakoid membranes. The colors of phycobiliproteins originate mainly from covalently bound prosthetic groups that are open-chain tetrapyrrole chromophores bearing A, B, C, and D rings named "phycobilins." There are four main classes of phycobiliproteins—allophycocyanin (bluish green), phycocyanin (PC; deep blue), phycoerythrin (deep red), and phycocyanobilin (orange). Blue-green algae with high levels of specific phycobiliproteins are of commercial interest. The primary potential of these molecules is as natural dyes in food industry. A number of investigations have shown their health-promoting properties and pharmaceutical applications. Among different phycobiliproteins, PC is of greater importance because of its various biological and pharmacological properties. Recent studies have demonstrated its antioxidant (Miranda et al. 1998), antimutagenic (Chamorro et al. 1996), antiviral (Ayehunie et al. 1998), anticancer (Schwartz et al. 1988), and immune-enhancing (Qureshi et al. 1996) properties.

15.1.1.5.5 Biochemicals

15.1.1.5.5.1 Proteins The high protein content of several microalgal species was one of the main reasons for considering these organisms as unconventional sources of proteins. Algae that are generally considered as the most valuable sources of protein are of good nutritional quality, provided the algal material is processed by proper treatments and is fully digestible. The protein quality is high compared with other plant protein and is about 80% of casein. Proteins are estimated using the procedures, based on color reactions. Other protein estimation procedures, based on colour reactions with defined protein constituents and which do not react with other nitrogen containing compounds, are the method after Lowry (Lowry et al. 1951).

15.1.1.5.5.2 Carbohydrates Apart from protein, other components of the algal biomass such as carbohydrate and fibers will affect the overall digestibility (Venkataraman and Becker 1985). Carbohydrates of algae exist as starch, cellulose, sugars, and other polysaccharides. On the *in vitro* digestibility of drum-dried *Scenedesmus obliquus* and sun-dried *Spirulina* spp., the tests are based on enzymatic amylolysis by α-amylase and subsequent colorimetric estimation of the amount of maltose released by test analyzed process.

15.1.1.5.5.3 Lipids Extraction of microalgal lipid is an important process in the production of biodiesel. Lipid extraction is performed by chemical methods in the form of solvent extractions, physical methods, or a combination of the two. Extraction methods used should be fast, effective, and nondamaging to lipids extracted and easily scaled up (Rawat et al. 2010). The harvested biomass must be dried because intracellular elements such as oils are difficult to extract from wet biomass (Melis 2005). Methods that have been used include sun drying, low pressure, shelf drying, spray drying, drum drying, fluidized bed drying, freeze-drying, and Refractance Window™ drying technology (Olguin 2003). Freeze-dried cells are preferable for biodiesel production.

Various methods are available for the extraction of algal oil, such as expeller/press, enzymatic extraction, chemical extraction through different organic solvents, ultrasonic extraction, and supercritical extraction using carbon dioxide. A simple process is to use a press to extract an average percentage (70%–75%) of the oils out of algae. In enzymatic extraction, cell wall–degrading enzymes are used to release the intracellular protein and oil. The lipids produced by algae are often accumulated intracellularly, which would require extraction of the lipids from crude pastes. Different cell disruption techniques such as autoclaving, osmotic stress, sonication, microwaves and bead beating, high-pressure homogenization, addition of hydrochloric acid, sodium hydroxide, and alkaline lysis have been evaluated in order to increase lipid extraction efficiency (Rawat et al. 2010).

In chemical extraction, many methods for algal lipid extraction have been recommended; the most popular is the slightly modified method of Bligh and Dyer (1959), Soxhlet method, and Folch method. The solvent extraction was still the main extraction procedure used by many researchers due to its simplicity and because it is relatively inexpensive requiring almost no investment for equipment. The choice of solvent for lipid extraction, as with harvesting, will depend on the type of the microalgae grown. Other preferred characteristics of the solvents are that they should be inexpensive, volatile, nontoxic and nonpolar, and poor extractors of other cellular components. The most popular inexpensive chemical used for solvent extraction is hexane.

Lipids produced by microalgae generally include neutral lipids, polar lipids, wax esters, sterols, and hydrocarbons, as well as phenyl derivatives such as tocopherols, carotenoids, terpenes, quinines, and pyrrole derivatives such as the chlorophylls. Lipids produced by microalgae can be grouped into two categories, storage lipids (nonpolar lipids) and structural lipids (polar lipids). Storage lipids are mainly in the form of TAG made of predominately saturated FAs and some unsaturated FAs, which can be transesterified to produce biodiesel. Structural lipids typically have a high content of polyunsaturated fatty acids (PUFAs), which are also essential nutrients for aquatic animals and humans. Polar lipids (phospholipids) and sterols are important structural components of cell membranes, which act as a selective permeable barrier for cells and organelles. These lipids maintain specific membrane functions, providing the matrix for a wide variety of metabolic processes and participate directly in membrane fusion events. In addition to a structural function, some polar lipids may act as key intermediates (or precursors of intermediates) in cell signaling pathways (e.g., inositol lipids, sphingolipids, oxidative products) and play a role in responding to changes in the environment. Of the nonpolar lipids, TAGs are abundant storage products, which can be easily catabolized to provide metabolic energy. In general, TAGs are mostly synthesized in the light, stored in cytosolic lipid bodies, and then reutilized for polar lipid synthesis in the dark. Microalgal TAGs are generally characterized by both saturated and monounsaturated FAs. However, some oil-rich species have demonstrated a capacity to accumulate high levels of long-chain polyunsaturated fatty acids (LC-PUFAs) as TAG. A detailed study on both accumulation of TAG in the green microalga *Parietochloris incisa* and storage into chloroplastic lipids (following recovery from nitrogen starvation) led to the conclusion that TAGs may play an additional role beyond being an energy storage product in this alga (Burlew 1953; Cohen 2002). Hence, PUFA-rich TAGs are metabolically active and are suggested to act as a reservoir for specific FAs. In response to a sudden change in the environmental condition, when the *de novo* synthesis of PUFA may be slower, PUFA-rich TAG may donate specific acyl groups to monogalactosyldiacylglycerol and other polar lipids to enable rapid adaptive membrane reorganization.

15.1.1.5.5.4 Fatty Acids   Some microalgae synthesize FAs with particular interest, namely, γ-linolenic acid (18:3w6; *Arthrospira*), arachidonic acid (20:4w6; *Porphyridium*), eicosapentaenoic acid (EPA, 20:5w3; *Nannochloropsis*, *Phaeodactylum*, *Nitzschia*, *Isochrysis*, *Diacronema*), and docosahexaenoic acid (DHA, 22:6w3; *Crypthecodinium*, *Schizochytrium*) (Sánchez Mirón et al. 2003; Spolaore et al. 2006). These LC-PUFAs (more than 18 carbons) cannot be synthesized by higher plants and

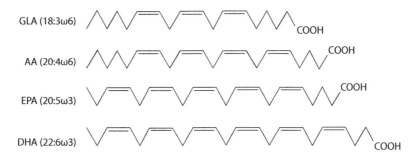

GLA (18:3ω6)

AA (20:4ω6)

EPA (20:5ω3)

DHA (22:6ω3)

Figure 15.3 Chemical structure of polyunsaturated fatty acids of pharmaceutical and nutritional interest.

animals, only by microalgae, which supply whole food chains. It is estimated that only healthy human adults are able to elongate 18:3w3 to EPA to an extent lower than 5% and convert EPA to DHA in a rate inferior to 0.05%, being inhibited in childhood and elderly life (Wang and Conover 1986). This statement confirms the importance of the inclusion of these long-chain FAs in daily diet.

Fish and fish oils are the main sources of LC-PUFAs (Figure 15.3), still global fish stocks are declining due to general fishing methods and overfishing, and the derived oils are sometimes contaminated with a range of pollutants, heavy metals, toxins and typical fishy smell, unpleasant taste, and poor oxidative stability (Luiten et al. 2003). The production of LC-PUFA from microalgal biotechnology is an alternative approach, and currently microalgal DHA from *Crypthecodinium* and *Ulkenia* is commercially available by the Martek (USA) and Nutrinova (Germany) companies (respectively), for application in infant formulas, nutritional supplements, and functional foods (Spolaore et al. 2006). PUFA's ω-3, especially docosahexaenoic acid (DHA), are essential in infant nutrition, being important building blocks in brain development, retinal development and ongoing visual, cognitive, as well as important fatty acids in human breast milk (Ghys et al. 2002; Wroble et al. 2002; Arteburn et al. 2007). Long-chain *n*-3 FA consumption has been associated with the regulation of eicosanoid production (prostaglandins, prostacyclins, thromboxanes, and leukotrienes), which are biologically active substances that influence various functions in cells and tissues (e.g., inflammatory processes) being important in the prophylaxis and therapy of chronic and degenerative diseases including reduction of blood cholesterol and protection against cardiovascular and coronary heart diseases, atherosclerosis, diabetes, hypertension, rheumatoid arthritis, rheumatism, skin diseases, digestive and metabolic diseases, and cancer (Simopoulos 2002; Sidhu 2003; Thies et al. 2003). Other important role is attributed to gene expression regulation as well as cholesterol and fasting TAG decreases (Calder 2004). The evidence of a dietary deficiency in long-chain omega-3 FAs is firmly linked to increased morbidity and mortality from coronary heart disease.

Alkali-catalyzed transesterification is carried out at approximately 60°C under atmospheric pressure, as methanol boils off at 65°C at atmospheric pressure. Under these conditions, reaction takes about 90 min to complete. A higher temperature can be used in combination with higher pressure, but this is expensive. Methanol and oil do not mix; hence, the reaction mixture contains two liquid phases. Other alcohols can be used, but methanol is the least expensive.

Shay reported that algae were one of the best sources of biodiesel. In fact, algae are the highest yielding feedstock for biodiesel. It can produce up to 250 times the amount of oil per acre as soybeans. In fact, producing biodiesel from algae may be the only way to produce enough automotive fuel to replace current gasoline usage. Algae produce 7–31 times greater oil than palm oil. It is very simple to extract oil from algae. The best algae for biodiesel would be microalgae.

Compare oil yields of microalgae with other oil feedstocks. It shows that there are significant variations in biomass productivity, oil yield, and biodiesel productivity. Microalgae are more advantageous due to their higher biomass productivity, oil yield, and biodiesel productivity. The table shows that low, medium, and high oil content microalgae have high oil yield/ha/year and hence higher biodiesel productivities (L/ha/year), which is much more than the productivities of oilseed crops. This is one of the most important reasons that microalgae have attracted the attention of researchers in India to scientifically grow, harvest, extract oil, and convert it to biodiesel (Table 15.2) (Peterson and Hustrulid 1998; Kulay and Silva 2005; Vollmann et al. 2007; Kheira and Atta 2008; Nielsen 2008; Reijnders and Huijbregts 2008; Mubee et al. 2010; Cenciani et al. 2011; Rajvanshi and Sharma 2012; Medipally et al. 2015).

Microalgae have been investigated for the production of numerous biofuels including biodiesel, which is obtained by the extraction and transformation of the lipid material; bioethanol, which is produced from the sugars, starch, and carbohydrate residues in general; biogas; and biohydrogen, among others (Demirbas 2011). Algae thrive in nutrient-rich waters like municipal waste waters (sewage), animal wastes and some industrial effluents, at the same time purifying these wastes while producing a

TABLE 15.2   Relative Comparison of Microalgal Biomass with Other Biodiesel Feedstocks

Organism Name (Oil Feedstocks)	Land Use (m²/Year/L Biodiesel)	Oil Yield (L Oil/ ha/Year)	Oil Content (% Dry Weight Biomass)	Biodiesel Productivity (L Biodiesel/ ha/Year)
Sunflower (*Helianthus annuus* L.)	9	1,070	40	1,113
Castor (*Ricinus communis*)	8	1,307	48	1,360
Palm oil (*Elaeis guineensis*)	2	5,366	36	5,585
Physic nut (*Jatropha curcas* L.)	13	741	41–59	656
Camelina (*Camelina sativa* L.)	10	915	42	952
Canola/rapeseed (*Brassica napus* L.)	10	974	41	1,014
Corn/maize (*Zea mays* L.)	56	172	44	179
Hemp (*Cannabis sativa* L.)	26	363	33	378
Soybean (*Glycine max* L.)	15	636	18	661
Microalgae (low oil content)	0.2	58,700	30	61,091
Microalgae (medium oil content)	0.1	97,800	50	101,782
Microalgae (high oil content)	0.1	136,900	70	142,475

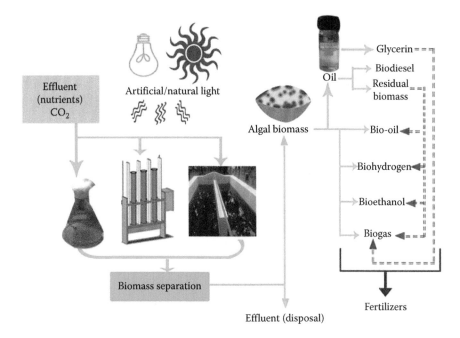

Figure 15.4 Effluent treatment using algae, biomass harvesting, and utilization of various biofuels with reference to CO_2.

biomass suitable for biofuels production. The flow chart showed that after phycoremediation the grew algal biomass harvested and utilized various biofuels with reference to CO_2 (Figure 15.4).

15.2 Mass Cultivation of Microalgae

Algal cultivation is an environmentally friendly process for the production of organic material by photosynthesis from carbon dioxide, light energy, and water. The water used by algae can be of low quality, including industrial process water, effluent of biological water treatment, or other wastewater streams. The open systems, in order to increase their efficiency, are generally designed as a continuous culture in which a fixed supply of culture medium or influent ensures constant dilution of the system. The organisms adapt their growth rate to this dilution regime, with the organism best adapted to the environment prevailing in the system winning the competition with the other organisms. A drawback of the common open algal culture systems is the major risk of contamination by undesirable photosynthetic microorganisms, which can be introduced via air or rain.

Algal cultivation can be done in a variety of environments. Algal cultivation in various environments is discussed in the following pages:

1. Cultivation in open pond
2. Cultivation in closed ponds

    3. Cultivation in PBRs
    4. Desert-based algal cultivation
    5. Cultivation in wastewater
    6. Marine algal cultivation
    7. Cultivation next to power plants (http://www.oilgae.com)

15.2.1 Viable Microalgal Cultivation Methods

15.2.1.1 Open Pond System

Advantages for utilizing the open pond system include low initial and operational costs (Figure 15.5). Disadvantages of open pond system include enormous land or area, which is not affordable in many regions, and a huge requirement of water.

15.2.1.2 Closed Photobioreactor System

However, the "best bioreactor," which would be able to achieve maximum productivity and maximum energy efficiency under a given set of operational costs, does not exist. Refer to Figures 15.6 and 15.7.

Figure 15.5 Algal production in open pond.

Figure 15.6 Closed photobioreactor system.

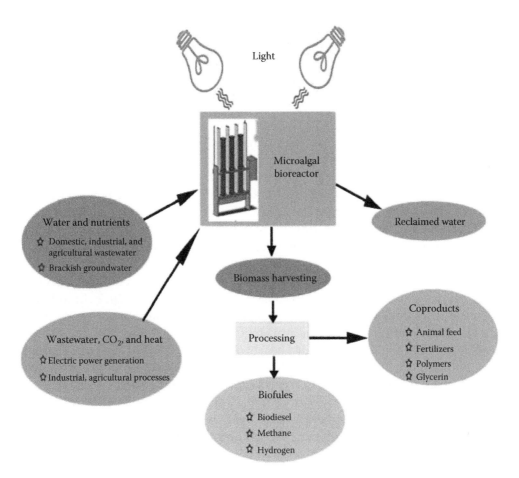

Figure 15.7 Integration of microalgal bioreactors into existing wastewater and power generation infrastructures.

Following are the advantages:

1. The photobioreactor system has a higher potential productivity due to better environmental control and harvesting efficiency.
2. Even though the open pond systems seem to be favored for commercial cultivation of microalgae at present due to their low capital costs, closed systems offer better control over contamination, mass transfer, and other cultivation conditions.
3. Closed photobioreactors require less freshwater than open ponds.

15.2.2 Successfully Employed Mass Cultivation Methods

15.2.2.1 High-Rate Algal Ponds
The concept of the high-rate algal pond (HRAP) was developed by Oswald and coworkers in the mid-1950s and is in place in various countries around the world (Oswald and Gotaas 1957). The system typically consists of a primary settlement lagoon with a shallow (0.2–0.6 m depth) meandering open channel in which the effluent is propelled by a paddle wheel to prevent settling and compensate for solid removal device (Oswald 1988). Most ponds are operated at average velocities from 10 to 30 cm/s to avoid deposition of algal cells (Dodd 1986). HRAPs are very appropriate for the sanitation of small rural communities because of their simplicity of operation in comparison to conventional technologies such as activated sludge treatment facilities.

15.2.2.2 Raceway Ponds
The ponds in which the alga is cultivated are called the raceway ponds. In these ponds, the algae, water, and nutrients circulate around a racetrack. A raceway pond is made of a closed-loop recirculation channel that is typically about 0.3 m deep with a paddle wheel (Chisti 2007) and illustrates the working principle of raceway pond. Raceways are perceived to be less expensive than PBRs, because they cost less to build and operate. Economically, it is 10 times costly in comparison with PBRs. The waste of chrome sludge from the electroplating industry supported algal growth when using open raceway pond chrome sludge was treated with microalga, *Desmococcus olivaceus*, there was a considerable amount of sludge reduction and biomass production in open raceway pond amended with chrome sludge (Muthukumaran et al. 2012). In addition, it consumes minimal power (Ananadhi and Stanley 2012; Bobade and Khyade 2012). The advantage of this method is the availability of domestic municipal wastewater as medium for cultivation with an added benefit of bioremediation. If a system is located near a power plant, cheaply available flue gas can be used to speed up the photosynthetic rates in the pond or pure CO_2 can be bubbled into the pond (Tsukahara and Sawayama 2005). But it has some drawbacks. The environment in and around the pond is not completely understood. Uneven light intensity and difficulty in maintaining the temperature and atmospheric evaporation are also drawbacks.

15.2.2.3 Photobioreactor
A bioreactor that is used for cultivating algae and fixing CO_2 producing biomass is called an algal bioreactor or algal PBR. The main advantage of the PBR is that it

can produce a large amount of biomass. The biomass recovery from PBR-cultured broth costs only a fraction of the recovery cost for broth produced in raceways. This is because the typical biomass concentration that is produced in PBRs is nearly 30 times the biomass concentration that is generally obtained in raceways (Kumar et al. 2014).

15.2.3 Nutrient Challenge

Algae require nutrients, light, water, and a carbon source, most often CO_2, for efficient growth. The major nutrients required by most algae include phosphorus, nitrogen, iron, and sulfur.

In the open oceans, iron is a major limiting nutrient for algal growth, as demonstrated by the induction of algal blooms by the addition of exogenous iron to open oceans (Coale et al. 2004). Interestingly, the addition of iron to induce an algal bloom has been considered and tested as a strategy to sequester CO_2 (Coale et al. 1996, 2004; Boyd et al. 2004). Biologically, iron is required for electron transport in all known photosynthetic organisms, including *Chlamydomonas reinhardtii*, and is typically found in iron–sulfur clusters in a variety of photosynthetic proteins (Godman and Balk 2008).

These strategies appear to be viable at some scale; however, alternative possibilities must also be developed. Ultimately, a combination of methods may be required, and perhaps a recycling of micro- and macronutrients will have to be developed for alga-based biofuels to reach a capacity that impacts present fossil fuel use. One of the most promising techniques for recycling nutrients in algal ponds is to use anaerobic digestion (Sialve et al. 2009). This bacterial process produces methane gas, while keeping the majority of the nutrients in a bacterial slurry that can be killed and the mix used for algal fertilizer. Methane gas is not currently a high-value commodity but can help provide energy to operate algal farms and cheap anaerobic digestion. Therefore, a balance should be reached between efficient anaerobic digestion and high-value coproducts, as shown in Figure 15.8.

This chapter deals with the microalgal cultivation for biofuels: cost, energy balance, environmental impacts, future perspectives, and investigation on biomass production and biofuel potentials of microalgae, using open raceway pond chrome sludge, which was treated with the microalga *D. olivaceus* (Persoon ex Acharius) J.R. Laundon. There was a considerable amount of sludge reduction and biomass production in open raceway pond amended with chrome sludge. A remarkable reduction was found in total dissolved solid (TDS), sodium, potassium, and phosphate. With an aim to identify a potential species of microalgae for biodiesel production on a nitrate–bicarbonate–dye effluent, investigation was done and *Chlorococcum humicola* was found to be a promising species with 13% of oil in effluents. Through the FAME analysis, it was found that it contains maximum of 75.8% of heneicosanoic acid methyl ester (C_{21}:0). Total amount of saturated FA content comes to maximum of 94% in the algae grown in combined dyeing effluent, which is greater than the content in algae grown in other condition.

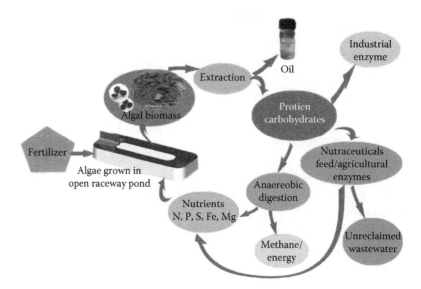

Figure 15.8   Nutrient recycling to maximize algal biofuel sustainability.

15.3 Microalgal Strains for Biomass and Biofuel Production

One likely source of biomass for alternative fuel production is microalgae that have the ability to grow rapidly and synthesize and accumulate large amounts (20%–50% of dry mass) of neutral lipid (mainly in the form of TAG) stored in cytosolic lipid bodies (Day et al. 1999; Hu et al. 2008; Duffy et al. 2009; Pienkos and Darzins 2009). Some species of diatoms (e.g., *Cha. muelleri*) and green microalgae (e.g., *Ch. vulgaris, Chlorococcum littorale, B. braunii, Nannochloris*) have been considered to be candidate strains for the production of neutral lipids for conversion to various types of biofuels (e.g., biodiesel, kerosene, gasoline) (Mcginnis et al. 1997; Illman et al. 2000; Hu et al. 2008; Berberoglu et al. 2009). The biodiesel is having less CO_2 and NO_x emissions. The transesterification of micro- and macroalgae and their yield and the property of the biodiesel produced are compared between the *Oedogonium* sp. and the *Nannochloropsis* sp., which are both microalgae and are grown in CO_2-rich environment; it helps us to reuse the CO_2 in the air (Manikandan et al. 2014). Additionally, microalgae could grow under harsher conditions and have reduced needs for nutrients so that they could be grown in areas unsuitable for agricultural purposes independently of the seasonal weather changes, without competing for arable land use. Compared with other biomass-derived biofuels, alga-based biodiesel is receiving increasing attentions worldwide in the recent years.

The starting point for this process is the identification of suitable algal strains that possess high composition of total lipids, in general, and neutral lipids, in particular, and/ or are capable of rapid accumulation of large quantities of neutral lipids under various culture conditions. In the past decade, many investigations with the aim of screening oil-producing microalgae were implemented in North America, Europe, the Middle East, and Australia, as well as in many other parts of the world (Sheehan et al. 1998;

Blackburn et al. 2009; Carioca et al. 2009; Rodolfi et al. 2009; Vijayaraghavan and Hemanathan 2009). A good example is Aquatic Species Program funded by the U.S. Department of Energy (U.S. DOE) from 1978 to 1996 representing the most comprehensive research efforts to date on fuels from microalgae (Sheehan et al. 1998).

Liu et al. (2011) studied accumulations of biomass for 43 algal strains. The results ranged from 0.53 to 6.07 g/L during the experiments, with the highest biomass of 6.07 g/L for the green alga *Scenedesmus bijuga*. The lipid content for the tested algal strains varied from 20% to 51% of the dry biomass at the end of cultivation experiments. The green alga *Chlorella pyrenoidosa* was one of the best oil producers based on our investigations, with the total lipid content of 51% of dry biomass. Taking the growth rates and the accumulations of intracellular lipids into consideration, 10 strains were considered to have significant potential for biofuel applications (Liu et al. 2011). Comparative data of some algal species and their lipid content are shown in Table 15.3 (Darzins et al. 2010b; Mata et al. 2010; Emad 2011; Liu et al. 2011; Rajvanshi and Sharma 2012; Manikandan et al. 2014).

15.4 Phycoremediation: An Emerging Technology

Phycoremediation is defined as the use of algae to remove pollutants from the environment or to render them harmless (Dresback et al. 2001). Olguin (2003) defines phycoremediation in a much broader sense as the use of macroalgae or microalgae for the removal or biotransformation of pollutants, including nutrients and xenobiotics from wastewater and CO_2 from waste air.

Phycoremediation comprises several applications: (1) oxygenation of the atmosphere, (2) nutrient removal from municipal wastewaters and effluents rich in organic matter, (3) nutrient and xenobiotic compounds removal by biosorption using algae, (4) treatment of acidic and metal wastewaters, (5) CO_2 sequestration, (6) transformation and degradation of xenobiotics, and (7) biosensing of toxic compounds by algae.

15.4.1 Microalgal Biomass Production Achieved through Phycoremediation

Adaptability of algae to the effluent was studied under laboratory conditions by employing selected microalgae, such as blue-green algae (Cyanophyceae) including *Chroococcus turgidus* and *Dactylococcopsis raphioides;* green algae (Chlorophyceae) including *Chlamydomonas pertusa, Chlorella vulgaris, Chlorococcum humicola, Chlorococcum vitiosum, Desmococcus olivaceus, Scenedesmus dimorphus, Scenedesmus accuminatus,* and *Scenedesmus incrassatulus;* and diatom (Bacillariophyceae) including *Amphora leaves, Amphora turgida, Amphiprora paludosa, Navicula pennata, Synedra tabulata,* and *Thalassiosira weissflogii,* which had vast potential to treat various industrial effluents (phycoremediation), such as textile dyeing, soft drink, chemical, detergent, oil drilling, leather processing, alginate, and petrochemical industrial effluents. The various effluents from industries was screened using micro algae at the laboratory level. After selection, the microalgal strain is implemented into the field

TABLE 15.3   Comparative Data of Some Algal Species and Their Lipid Content

Microalgal Species (Marine and Freshwater)	Lipid Content (% Dry Weight Biomass)
Ankistrodesmus sp.	28–40
Botryococcus braunii	29–75
Chaetoceros muelleri	33.6
Chaetoceros calcitrans	14.6–16.4
Chlorella emersonii	25.0–63.0
Chlorella protothecoides	14.6–57.8
Chlorella sorokiniana	19.0–22.0
Chlorella vulgaris	5.0–58.0
Chlorella pyrenoidosa	18.67–52.08
Chlorella ellipsoidea	27.5–45.35
Chlorella saccharophila	45.56
Chlorella sorokiniana	28.91
Chlorococcum sp.	19.3–24.81
Crypthecodinium cohnii	20.0–51.1
Crypthecodinium cohnii	20
Cyclotella sp.	9–59
Dunaliella salina	6.0–25.0
Dunaliella primolecta	23.1–0.09
Dunaliella tertiolecta	16.7–71.0
Ellipsoidion sp.	27.4
Euglena gracilis	14.0–20.0
Haematococcus pluvialis	25.0
Hantzschia sp.	66
Isochrysis galbana	7.0–40
Isochrysis sp.	7.1–33
Monodus subterraneus	16.0
Monallanthus salina	20.0–22.0
Nannochloris sp.	20.0–56.0
Nannochloropsis oculata	22.7–29.7
Neochloris oleoabundans	29.0–65.0
Nitzschia sp.	16.0–47.0
Oocystis pusilla	10.5
Pavlova salina	30.9
Pavlova lutheri	35.5
Phaeodactylum tricornutum	18.0–57.0
Porphyridium cruentum	9.0–18.8
Scenedesmus obliquus	11.0–55.0
Scenedesmus bijuga	34.10–35.24
Scenedesmus dimorphus	26.35–48.35
Scenedesmus quadricauda	15.16–27.61
Scenedesmus sp.	19.6–21.1

(*Continued*)

**TABLE 15.3 (*Continued*)   Comparative Data of Some Algal Species and
Their Lipid Content**

Microalgal Species (Marine and Freshwater)	Lipid Content (% Dry Weight Biomass)
Skeletonema sp.	13.3–31.8
Skeletonema costatum	13.5–51.3
Spirulina platensis	4.0–16.6
Spirulina maxima	4.0–9
Stichococcus sp.	33
Schizochytrium sp.	50–77
Thalassiosira pseudonana	20.6–31
Tetraselmis suecica	15–32

level, both pilot and scaled up, to treat a specific industrial effluent. The microalgae
C. humicola, *D. olivaceus*, *Ch. vulgaris*, *Chr. turgidus*, and *A. turgida* were involved
in significant correction of pH, BOD, chemical oxygen demand (COD), removal of
color, TDS, nitrate, phosphate, etc. (around 60%–80%). The immobilized cells of *S.
accuminatus* could remove nitrate, phosphate (44%–76%), and TDS (28%–54%) significantly in chemical industrial effluents. The utilization of algal biomass after it was
harvested from the treated effluents for commercial applications has been reported
(Muthukumaran et al. 2005, 2012; Sivasubramanian 2006; Muthukumaran 2009).

15.4.1.1 Raceway Pond Study

In the preliminary lab trials, *D. olivaceus* showed better survival in 0.2% sludge concentration than other algae. So it was selected for further field experiments in open
raceway pond, and the phycoremediation of sludge was carried out. Initially, 1 kL
of modified CFTRI medium was taken in the tank (along with EDTA to chelate the
metal ions (Kedziorek and Bourg 2000) in the ratio of 3 mg/g of sludge) in which
D. olivaceus was inoculated. TDS and sludge showed gradual reduction and cell count
increased (Muthukumaran et al. 2012). The effluent was analyzed and the reduction
percentage was calculated.

When the raw and alga-treated electroplating industrial chrome sludge was analyzed, the phycoremediated effluent showed considerable reduction in all parameters
analyzed. Remarkable reduction was found in TDS (74.24%), sodium (50%), potassium (60%), and phosphate (86.16%). BOD and COD showed a reduction percentage
of 22.70% and 23.80%, respectively. Heavy metals such as copper, zinc, and chromium showed a reduction percentage of 8.38%, 16.49%, and 33.33%, respectively. The
results are given in Table 15.4 and Figures 15.9 through 15.11.

15.4.1.1.1 Growth and Sludge Reduction by Desmococcus olivaceus   This experiment
was performed basically to analyze the growth of the microalga *D. olivaceus* in the
chrome sludge and simultaneous reduction of sludge in pilot tank.

In the experiment conducted for 12 days, 0.2% chrome sludge was mixed with
modified CFTRI medium along with metal-chelating agent EDTA (60 g/kL) in an
open raceway pond. On the first day of the experiment, 2 L of *D. olivaceus* culture

TABLE 15.4   Growth and Sludge Reduction by
Desmococcus olivaceus **Grown in Open**
Raceway Pond

Day	Cell Count (×10⁴ Cells/mL)	% Sludge Reduction
1st	3	0
2nd	8	—
3rd	12	04.98
4th	40	—
5th	82	09.41
6th	128	—
7th	143	13.87
8th	161	—
9th	260	—
10th	308	20.95
11th	298	—
12th	280	11.21

Figure 15.9   Algal biomass production by *Desmococcus olivaceus* grown in chrome sludge of electroplating industry in open raceway pond. (From Muthukumaran, M. et al., *Int. J. Curr. Sci.*, 52, 2012.)

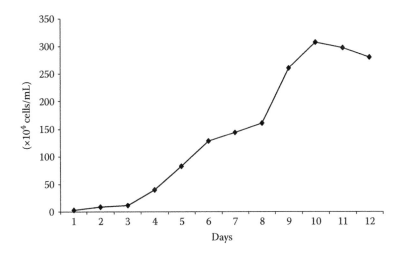

Figure 15.10 Growth of *Desmococcus olivaceus* in electroplating chrome sludge: pilot tank.

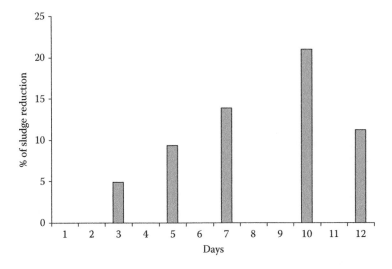

Figure 15.11 Sludge removal efficiency of *Desmococcus olivaceus* grown in electroplating chrome sludge: pilot tank.

was added. The parameters such as pH, TDS, growth rate, and sludge reduction were monitored. There was a considerable sludge reduction (20.95%), when the cell count of *D. olivaceus* was also found to be maximum (308×10^4 cells/mL) on 10th day. The results are given in Table 15.4 and Figures 15.9 through 15.11.

15.4.1.1.2 Algal Biomass Production by Desmococcus olivaceus *Grown in Open Raceway Pond* The microalga *D. olivaceus* was treated in the chrome sludge and simultaneously the algal biomass harvested and daily estimated by dry weight method in open raceway pond. There was a considerable biomass production (110.5412×10^4 dry wt. g/kL), when the cell count of *D. olivaceus* was also found to be maximum

TABLE 15.5 Algal Biomass Production by *Desmococcus olivaceus* Grown in Open Raceway Pond

Day	Algal Biomass in Dry Weight ($\times 10^4$ g/kL)
1st	001.0766
2nd	002.8712
3rd	004.3068
4th	014.3560
5th	029.4298
6th	045.9392
7th	051.3227
8th	057.7829
9th	093.3140
10th	110.5412
11th	106.9522
12th	100.4920

(308×10^4 cells/mL) on 10th day (Muthukumaran et al. 2012). The results are given in Table 15.5 and Figures 15.9 through 15.11. The harvested algal biomass was analyzed by various useful biochemical parameters.

15.4.2 Biodiesel Potentials of Microalgae

Biodiesel derived from oil crops is a potential renewable and carbon-neutral alternative to petroleum fuels. Unfortunately, biodiesel from oil crops, waste cooking oil, and animal fat cannot realistically satisfy even a small fraction of the existing demand for transport fuels, whereas microalgae appear to be the only source of renewable biodiesel that is capable of meeting the global demand for transport fuels (Sheehan et al. 1998). Like plants, microalgae use sunlight to produce oils, but they do so more efficiently than crop plants. Oil productivity of many microalgae greatly exceeds the oil productivity of the best producing oil crops.

Microalgae are found to be the most suitable for oil production as they are unicellular and they can yield 30 times more oil than any terrestrial plant. They have high photosynthetic efficiency as they trap more sunlight owing to their very high surface area. Microalgae commonly double their biomass within 24 h. Biomass doubling times during exponential growth are commonly as short as 3.5 h. Most algae are rich in protein and lipids providing high-energy diet even when consumed in low quantity. Microalgae appear to be the only source of biodiesel that has the potential to completely displace petrodiesel. Unlike other oil crops, microalgae grow rapidly of which many are exceedingly rich in oil. Large-scale biodiesel production from microalga *Chlorella protothecoides* was done through heterotrophic cultivation in bioreactors (Li et al. 2008).

Oil content in microalgae can exceed 80% by weight of dry biomass. Oil levels of 20%–50% are quite common. Oil productivity, that is, the mass of oil produced per unit volume of the microalgal broth per day, depends on the algal growth rate and the oil content of the biomass. Microalgae with high oil productivities are desired for producing biodiesel (Banerjee et al. 2002; Gavrilescu and Chisti 2005).

The development of biodiesel production from microalgae presents an important move to address the limitations posed by current first-generation biodiesel crops. Microalgae, once developed for commercial biodiesel production, may offer many economical and environmental advantages. Current biodiesel production from microalgae is in the research phase but is being developed to commercial scale in many countries. Finding promising microalgae for commercial cultivation is multifaceted and challenging because particular microalgal strains have different requirements in terms of nutrient intake, environmental and culturing conditions, and lipid extraction technology. However, diversity of lipid-producing microalgal species is one of the major advantages of this group of organisms that is likely to lead to selection of suitable algal crops to achieve algal biodiesel production in different regions. A combination of conventional and modern techniques is likely the most efficient route from isolation to large-scale cultivation. Careful initial analyses and farsighted selection of microalgae with a view toward downstream processing and large-scale production with potential value-added products are important prerequisites to domesticate and develop algal crops for biodiesel production (Duong et al. 2012). Industrial and municipal wastewaters can be potentially utilized for the cultivation of microalgal oil that can be used for the production of biodiesel to completely displace petrodiesel. The microalgal biomass has been reported to yield high oil contents and have the diesel production (Rajvanshi and Sharma 2012).

Biodiesel production can potentially use some of the carbon dioxide that is released in power plants by burning fossil fuels (Sawayama et al. 1995; Yun et al. 1997). This carbon dioxide is often available at little or no cost.

Dimitrov (2007) in his works and investigation has given the idea of generation of algal biomass using the flue gas from a fossil fuel–based power plant and solar energy as inputs. The idea is extremely attractive as flue gas is CO_2-containing pollutants that need to be dealt with and sunlight is free and abundant. The resulting biomass can be monetized via the following mechanisms (Figure 15.12):

1. Converting it to biodiesel via transesterification
2. Converting it to bioethanol via fermentation
3. Converting it to liquid/gas fuels via pyrolysis
4. Generating heat/electricity by burning it (with or without gasification)
5. Selling it as feed protein
6. Disposing it in a landfill and receiving credits for avoided emissions

Production of biofuels will also have carbon mitigation potential by the virtue of avoided fossil fuel use and thus will benefit from future carbon credits. The view that algal biofuels are superior to terrestrial biofuels in the displacement of fossil fuels has wider support.

Dimitrov (2007) argues that producing bioethanol is a very lucrative option too, considering its high selling prices. Production of ethanol from corn or sucrose by fermentation is an established process; however, it is significantly more exothermic

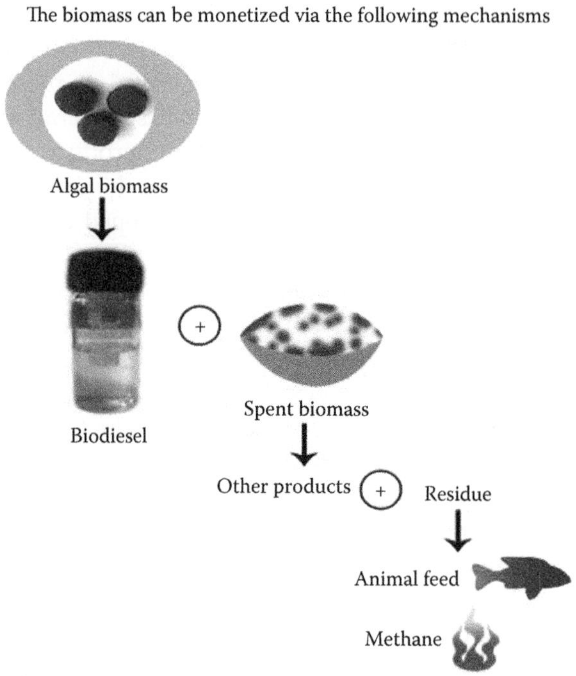

The biomass can be monetized via the following mechanisms

Figure 15.12   Algal biomass can be monetized via the following mechanisms.

(losses energy) than biodiesel production from lipids, and in addition, it has not been proven commercially with algal feedstock.

As demonstrated here, microalgal biodiesel is technically feasible. It is the only renewable biodiesel that can potentially completely displace liquid fuels derived from petroleum. Economics of producing microalgal biodiesel need to improve substantially to make it competitive with petrodiesel, but the level of improvement necessary appears to be attainable. Producing low-cost microalgal biodiesel requires primarily improvements to algal biology through genetic and metabolic engineering. Chisti (2000) emphasized the use of the biorefinery concept and advances in phycoremediation will further lower the cost of production. The oil contained a very high percentage of saturated lipids. The oil has a high boiling point and would prove good for biodiesel preparations.

The effect of nitrogen stress is a major study on green alga; many researchers have studied the effect of nitrogen on various algae and have found variations in oil content and also in composition. The effect of nitrate concentration on *Chlorococcum humicola*. The results have been illustrated in Tables 15.1 and 15.3. The optimum concentration of nitrate required for better growth of this alga is found to be 0.25 g/L. The growth of algae under optimum condition was monitored periodically for the biomass generation, nutrient utilization, etc. The cell count went as high as 140×10^4 cells/mL. The corresponding percentage of oil content was 12.4%.

The study has established the role of nitrate in the modification of lipid content in the alga *C. humicola*. But the efforts to bring about a higher yield of oil in the algae under lab conditions based on nitrate concentration proved futile. Under lab conditions, 1 L of culture gives 1.5 g of algal biomass. The analysis of total oil content of the algae shows the presence of 11.8% of oil. For producing 1 kg of the oil, we need about 12.7 kg of algae, which can be obtained from approximately 8500 L of lab culture of *C. humicola* for every 10 days (Muthukumaran 2009).

The growth under added bicarbonate concentrations did not show any profitable yields; also the oil content in the biomass was drastically reduced to about 8.3%; the culture density was low compared to the culture in standard bold basal medium. The search for an alternative means for the mass production of algae ended at the phycoremediation plants. Here algae grow in a highly stressed environment. The alga *C. humicola* does survive well in these effluents, tolerating stressed conditions like very high TDS and adverse pH conditions based on the dye used and also giving higher oil content. The oil content also proved its worth on analysis for the composition.

The analyses of total oil content of the algae show the presence of 12% and 13% of oil in algae grown in dyebath and wash effluent, respectively. Considering the maximum yield, we get 13 mg of oil from 1 g of biomass. In case of algae growing in effluents for 1 kg of oil, we need approximately 7.7 kg of biomass, which is an economically feasible process in case of phycoremediation plants that operate continuously (Muthukumaran 2009). Results are shown in Table 15.5.

15.4.2.1 Biodiesel Potentials of Chlorococcum humicola

The search for an alternative means for the mass production of algae ended at the phycoremediation plants. Here algae grow in a highly hostile environment. The alga *C. humicola* did survive well in these effluents, tolerating stressed conditions like very high TDS and adverse pH conditions producing high lipids.

The analysis shows the existence of a single FA in major composition indicating that it is highly suitable for biodiesel production and a very little unsaturation is a good sign of hope in the process of biodiesel production from algae, and the *C. humicola* has a great potential of being a feedstock for biodiesel production in spite of its low lipid content (Muthukumaran 2009).

15.4.2.2 Variation in Oil Content of Chlorococcum humicola

The lipid and oil content was the same as other plants. The algal biomass also showed the presence of oil. It showed noticeable change in the composition. The oil proved its worth on analysis. The oil contained a very high percentage of saturated lipids. The oil has a high boiling point and proved good for biodiesel preparations.

The analysis shows the existence of a single FA in major composition and little unsaturation is a good sign indicating that it is highly suitable for biodiesel production. The alga *C. humicola* has a great potential of being a feedstock for biodiesel production in spite of low lipid content (Muthukumaran 2009). The results are given in Tables 15.6 and 15.7 and Figures 15.13 through 15.16.

15.4.3 Calorific Value Analysis of Algal Biomass Pellet

Algae growing on industrial effluents are harvested and dried. These dried algae, with the help of a pelletizer (fluid power hydraulics—S. No. 20617), were converted

TABLE 15.6   Analysis of Oil Content and Biomass Yield

	0.25 g NaNO$_3$/L	0.50 g NaHCO$_3$/L	Dyebath Effluent	Wash Effluent
Cell count (cells/mL)	142×10^4	127×10^4	72×10^4	120×10^4
Oil content	12.4	08.3	12	13
Biomass (g/L)	1.8	0.7	1	1.4

TABLE 15.7   Gas Chromatographic Analysis of Algal Oil Extracted from *Chlorococcum humicola*

Oil from Alga	Alga Grown in Lab (%)	Alga Grown in Combined Dyeing Effluent (Field, %)	Alga Grown in Dyebath Effluent (Field, %)
Heneicosanoic acid methyl ester (C$_{21}$:0)	67	73.3	75.8
Arachidic acid methyl ester (C$_{20}$:0)	14.7	13	10.7
cis-5,8,11,14,17- Eicosapentaenoic acid methyl ester (C$_{20}$:5n3)	3.15	5.14	6.3
Lignoceric acid methyl ester (C$_{24}$:0)	10.35	8.5	6.7
Total amount of saturated fatty acid content comes to around	92	94	91

Figure 15.13 *Chlorococcum humicola*, grown in laboratory with reference to nitrate and bicarbonate stress condition.

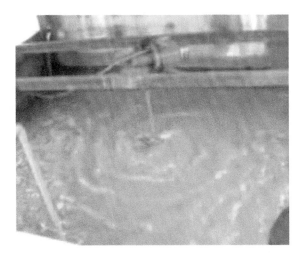

Figure 15.14 *Chlorococcum humicola*, grown in dyeing effluent—a pilot study.

Figure 15.15 Pelletized dry algal biomass—*Chlorococcum humicola*.

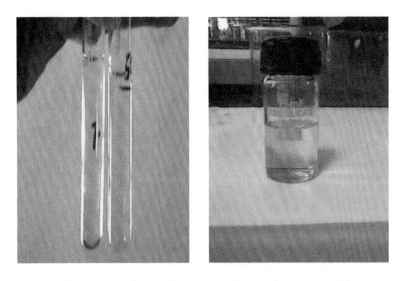

Figure 15.16 Lipid extraction from *Chlorococcum humicola*, grown in laboratory and field.

into pellets (diameter size 8 mm). Calorific value analysis was performed using these algal pellets (Figure 15.15) (Muthukumaran 2009).

15.4.3.1 Calorific Value of Algal Biomass

The calorific value of five strains of *Chlorella* grown in low-nitrogen medium was determined (Illman et al. 2000). The algae were grown in small (2 L) stirred tank bioreactors and the best growth was obtained with *Ch. vulgaris* with a growth rate of 0.99 d^{-1} and the highest calorific value of 29 kJ/g was obtained with *Chlorella emersonii*. The cellular components were assayed at the end of the growth period and the calorific value appears to be linked to the lipid content rather than any other component (Illman et al. 2000).

15.4.3.1.1 Thermal Gravimetric and Differential Thermogravimetry (DTG) (%) Analysis Used Static Air Flow (Nitrogen Gas)

To measure the calorific value, 21.17 mg of the dried alga pellet was taken. It was determined using NETZSCH STA 409 C/CD using static air flow (nitrogen gas), by the determination of thermal gravimetric (TG) analysis (%) and DTG (%/min). The TG showed a decrease in mass by −12% at an inflation temperature of 82°C. The maximum reduction in mass of 49% was observed at inflation temperature of 600°C. A further reduction of −15% was observed at inflation temperature of 938.3°C. The DGA percentage reduction per minutes showed an overall decrease of −77% (Figure 15.17).

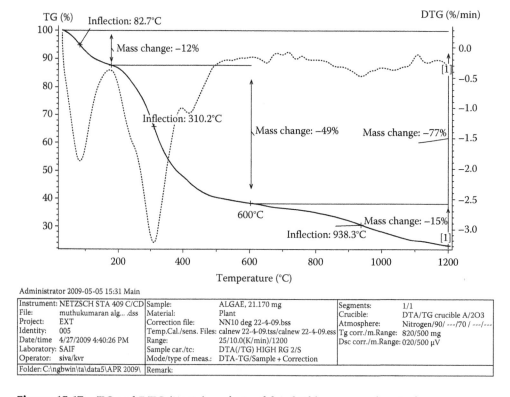

Figure 15.17 TG and DTG (%min) analysis of dried *Chlorococcum humicola*.

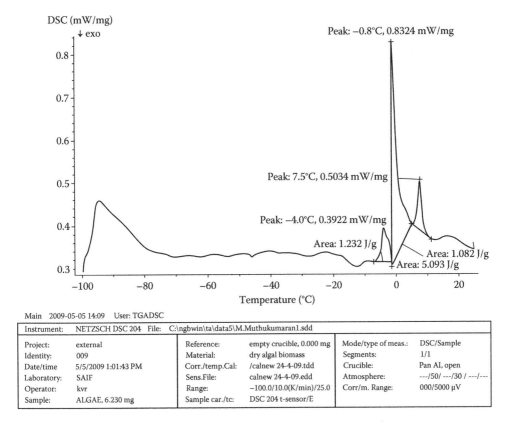

DSC (mW/mg)

↓ exo

Peak: −0.8°C, 0.8324 mW/mg

Peak: 7.5°C, 0.5034 mW/mg

Peak: −4.0°C, 0.3922 mW/mg

Area: 1.232 J/g

Area: 1.082 J/g

Area: 5.093 J/g

Temperature (°C)

Main   2009-05-05 14:09   User: TGADSC

Instrument:	NETZSCH DSC 204 File:	C:\ngbwin\ta\data5\M.Muthukumaran1.sdd			
Project:	external	Reference:	empty crucible, 0.000 mg	Mode/type of meas.:	DSC/Sample
Identity:	009	Material:	dry algal biomass	Segments:	1/1
Date/time	5/5/2009 1:01:43 PM	Corr./temp.Cal:	/calnew 24-4-09.tdd	Crucible:	Pan Al, open
Laboratory:	SAIF	Sens.File:	calnew 24-4-09.edd	Atmosphere:	---/50/ ---/30 / ---/---
Operator:	kvr	Range:	−100.0/10.0(K/min)/25.0	Corr/m. Range:	000/5000 µV
Sample:	ALGAE, 6.230 mg	Sample car./tc:	DSC 204 t-sensor/E		

Figure 15.18 DSC (mW/mg) analysis of dried alga, *Chlorococcum humicola* grown on industrial effluents.

15.4.3.1.2 DSC/(mW/mg) Analysis of Algal Sample   Dried algal pellet of 6.230 mg was taken and was measured in NITZSCH DSC 204, range −100/10 (K/min)/25. There were three peaks found: the first peak at −4°C, 0.3922 mW/mg (area 5.093 J/g); the second peak at −0.8°C, 0.8324 mW/mg (area 5.093 J/g); and the third peak at 7.5°C, 0.5034 mW/mg (area 1.082 J/g) (Figure 15.18).

The results indicate that the algal biomass pellet was high in calorific value and seems to be equal to the heat energy from coals. The results are shown in Figures 15.17 and 15.18. The alga *C. humicola* used in the present study showed a high calorific value showing that after harvesting, they can be used for fuel (Muthukumaran 2009).

15.4.4 Importance and Challenges of Algal Biofuels

The global economy is in need of fossil hydrocarbons to function, from producing plastics and fertilizers to providing the energy required for lighting, heating, and transportation. With the increasing population and expanding economy, there will be an increased fossil fuel use. As countries improve their gross domestic product per capita, data suggest that their fossil fuel use will increase significantly, and competition for these limited resources will increase. In addition, there is an increasing concern over atmospheric CO_2 concentration and the potential for significant greenhouse gas (GHG)–mediated climate change (Parry 2007), which now seems likely to affect the biodiversity of the planet. Finally, petroleum, which is partially derived from ancient algal deposits, is a limited resource that will eventually run out or become too expensive to recover (Dyni 2006; Schindler and Zittel 2008; Energy Information Administration 2009).

15.4.5 Challenges for Algal Fuel Commercialization

The high growth rates, reasonable growth densities, and high oil contents have all been cited as reasons to invest significant capital to turn algae into biofuels. However, for algae to mature as an economically viable platform to offset petroleum and, consequently, mitigate CO_2 release, there are a number of hurdles to overcome right from the place and method to grow these algae to improving oil extraction and fuel processing. The algal biofuel production chain shows that the major challenges include strain isolation, nutrient sourcing and utilization, production management, harvesting, coproduct development, fuel extraction, refining, and residual biomass utilization.

15.4.6 Advantages of Using Microalgae for Biodiesel Production

Many research reports and articles have pointed out many advantages of using microalgae for biodiesel production in comparison with other available feedstocks. From a practical point of view, they are easy to cultivate and easy to obtain nutrients and can grow with less attention using wastewater. Microalgae self-reproduce using photosynthesis to convert sun energy into chemical energy, completing an entire growth cycle every few days. Moreover, they can grow almost anywhere, requiring sunlight and some simple nutrients, and the growth rates can be accelerated by the addition of specific nutrients and sufficient aeration.

Different microalgal species can adapt to live in a variety of environmental conditions. Thus, it is possible to find species best suited to local environments or specific growth characteristics, which is impossible to do with other biodiesel feedstocks (e.g., soybean, rapeseed, sunflower, and palm oil) as shown in Table 15.2.

They have much higher growth rates and productivity when compared to conventional forestry, agricultural crops, and other aquatic plants, requiring much less land area when compared to other biodiesel feedstocks of agricultural origin, with up to 49 or 132 times less when compared to rapeseed or soybean crops, for a 30% (w/w) of oil content in algal biomass. Therefore, the competition for arable soil with other crops, in particular for human consumption, is mitigated.

Microalgae can provide feedstock for several different types of renewable fuels such as biodiesel, methane, hydrogen, and ethanol. Algal biodiesel contains no sulfur and performs as well as petrodiesel while reducing emissions of particulate matter, CO, hydrocarbons, and SO_x. However, emissions of NO_x may be higher in some engine types.

Microalgae for biofuel production can be utilized for other purposes. Some possibilities currently being considered are listed in the following:

1. Removal of CO_2 from industrial flue gases by algal biofixation, reducing the GHG emissions of a company or process while producing biodiesel.
2. Wastewater treatment by the removal of NH_4^+, NO_3.
3. PO_4 making algae to grow using these water contaminants as nutrients.
4. After oil extraction, the resulting algal biomass can be processed into ethanol, methane, and livestock feed, used as organic fertilizer due to its high N:P ratio, or simply burned for energy cogeneration (electricity and heat).
5. Combined with their ability to grow under harsher conditions and their reduced needs for nutrients, they can be grown in areas unsuitable for agricultural purposes independently of the seasonal weather changes, thus not competing for arable land use, and can use wastewaters as the culture medium, not requiring the use of freshwater.
6. Depending on the microalgal species, other compounds may also be extracted, with valuable applications in different industrial sectors, including a large range of fine chemicals and bulk products, such as fats, PUFAs, oil, natural dyes, sugars, pigments, antioxidants, high-value bioactive compounds, and other fine chemicals and biomass.
7. Because of this variety of high-value biological derivatives, with many possible commercial applications, microalgae can potentially revolutionize a large number of biotechnology areas including biofuels, cosmetics, pharmaceuticals, nutrition and food additives, aquaculture, and pollution prevention.

15.5 Cost Performance

Cost analysis is a powerful tool that can be used to both estimate the ultimate costs of algal biofuels and identify the process elements, which contribute most to the production cost—thereby helping with the focus on future research and design. The limitations of algal production cost assessments are similar to those faced by life cycle assessments (LCAs) and include data constraints and reliance on parameters extrapolated from lab-scale analyses. The current state of the art for microalgal culture may also be uncaptured. For instance, one of the most frequently cited sources of cost modeling parameters is a paper published in 1996 (Benemann and Oswald 1996), which in turn contains assumptions going back to the mid-1970s. Estimates for algal productivity, CO_2 capture efficiency, and system availability may also reflect future aspirations rather than current achievable results. As with LCA studies, the production of coproducts, or provision of coservices, greatly affects the economic viability.

Algal biofuels may provide a viable alternative to fossil fuels; however, this technology must overcome a number of challenges before it can compete in the fuel market and be broadly deployed. These challenges include strain identification and improvement, both in terms of crop protection and oil productivity, nutrient and resource allocation and use, and the production of coproducts to improve the economics of the entire system. Although there is much excitement about the potential of algal biofuels, much work is still required to be accomplished in this field. In this article, we attempt to elucidate the major challenges in the economics of algal biofuels at various scales and improve the focus of the scientific community to address these challenges and move algal biofuels from promise to reality (Hannon et al. 2010).

Alga-based biofuel production has a number of potential advantages:

1. Biofuels and by-products can be synthesized from a large variety of algae.
2. Algae have a rapid growth rate.
3. Algae can be cultivated in brackish coastal water and seawater.
4. Some land areas that are unsuitable for agriculture can be used to cultivate algae.
5. Algal nutrient uptake uses high nitrogen, silicon, phosphate, and sulfate nutrients from human or animal waste.
6. Algae can sequester carbon dioxide (CO_2) from industrial sources.

15.6 Energy and Carbon Balance of Microalgal Production

If microalgae are to be a viable spring for biofuel production, the overall energy (and carbon balance) must be favorable. There have been efforts to evaluate this for large-scale microalgal biofuel production using LCA methods to describe and quantify inputs and emissions from the production process. Attempts have been hindered by the fact that no industrial-scale process is designed specifically for biofuel production. Consequently, the data that reinforce microalgal LCA must be extrapolated from laboratory-scale systems or from commercial schemes that have been designed to produce high-value products such as pigments and health food supplements. In spite of this limitation, it is anticipated that LCA can still serve as a tool to assist with system design.

The microalgal biomass is a potential energy source but has adversaries like the imperative decrease of energy, fertilizer, and other inputs, like flocculants. The starting point for the economic and environmental viability of energy production from microalgal biomass is the achievement of a favorable energy balance. The use of residual CO_2 and residual nutrients from various processes and energy sources significantly improves the energy balance and reduces GHG emissions, increasing the chances for the commercial application of this technology in a competitive way. Some topics may be in depth through further research, such as evaluation of other environmental categories, local productivity rates and land use impacts at different regions, comparison with other renewable sources, and assessment of other routes for biomass energy conversion like biodiesel, biomethane, and bioethanol (Medeiros et al. 2013).

Nevertheless, many research and detailed analysis are required to measure in which situations they can contribute to it. Microalgal biomass combusts to produce heat, and comparisons with the use of different sources of electricity with respect to GHG emissions and net energy ratio (NER) are being assessed. With some fossil sources as reference, methodologies adhered to ISO 14040/44 standards and most of the data were obtained from scientific publications. The results showed that NER from microalgal combustion is still a disadvantage compared to fossil options. Microalgal GHG emissions were higher than fossil using U.S. electrical grid but lower using the

Brazilian one. Although the fossil options show slightly better yields related to microalgae in the two categories analyzed, the fossil energy technology is more mature and has less scope for improvements while microalgae are in its infancy and have many technological innovations being developed (Medeiros et al. 2013).

The U.S. DOE (2010) states that in favorable climates, microalgae represent a more advantageous technological alternative over conventional tertiary treatment technologies for the removal of nutrient. According to Park et al. (2011), the use of wastewater is shown as the most economical way to produce microalgal biomass in a sustainable way with the minimal environmental impact. On the other hand, Christenson and Sims (2011) claim that only a few preliminary studies were conducted on the microalgal biofuels and bioproduct production grown in wastewater. Compare different routes of energy production from microalgal biomass such as biogas electricity and biodiesel, direct biomass combustion for electricity and biodiesel, and biomass combustion for electricity only. The most energy outcome, or best NER, comes from utilizing biomass for combustion. Most importantly, fuel is a low value compound, yet there is a tendency, particularly in the algae biofuel space, to prioritize high yields without sufficient regard to the ultimate cost (Clarens 2011).

15.7 Environmental Impacts and Constraints

Environmental impact assessment (EIA) is the evaluation of the effects likely to arise from a major project (or other action) significantly affecting the environment. It is a systematic process for considering possible impacts prior to a decision being taken on whether or not a proposal should be given approval to proceed. EIA requires, inter alia, the publication of an EIA report describing the probable significant environmental impacts in detail. Consultation and public participation are integral to this evaluation. EIA is thus an anticipatory, participatory environmental management tool.

Large-scale microalgal production may have a wide variety of environmental impacts beyond the intake of energy in the making process. These impacts could constrain system design and operation. The impacts presented here are the ones most prominent in the existing literature and identified as important in discussion with stakeholders. The conceptual and often incomplete nature of algal production systems investigated within the existing literature, together with limited sources of primary data for process and scale-up assumptions, highlights future uncertainties around microalgal biofuel production. Environmental impacts from water management, carbon dioxide handling, and nutrient supply could constrain system design and implementation options. Cost estimates need to be improved, and this will require pragmatic data on the performance of systems designed specifically to produce biofuels. Significant (>50%) cost reductions may be achieved if CO_2, nutrients, and water can be obtained at low cost. This is a very demanding requirement, however, and it could dramatically restrict the number of production locations available (Raphael and Bauen 2013).

EIA was taken up assessed because any effluent treatment methodology should be evaluated for environmental safety. Hence, the environmental safety credentials of

C. humicola were analyzed. The microalga was successfully used for seed germination and also as a feed for fishes. Thus, the algae not only are safe to the environment but can also be used as food, feed, and biofertilizers (Muthukumaran 2009).

15.7.1 Water Resources

A reliable, low-cost water supply is critical to the success of biofuel production from microalgae. Freshwater needs to be added to raceway pond systems to compensate evaporation; water may also be used to cool some PBR designs. One suggestion is that algal cultivation could use water with few competing uses, such as seawater and brackish water from aquifers. Brackish water, however, may require pretreatment to remove growth-inhibiting components and this could raise the energy demand of the process (Darzins et al. 2010).

15.7.2 Land Use and Location

One of the suggested benefits of algal production is that it could use marginal land, thereby minimizing competition with food production. Topographic and soil constraints limit the land availability for raceway pond systems as the installation of large shallow ponds requires relatively flat terrain. Soil porosity/permeability will also affect the need for pond lining and sealing (Lundquist et al. 2010).

15.7.3 Nutrient and Fertilizer Use

Algal cultivation requires the addition of nutrients, primarily nitrogen, phosphorus and potassium (some species, e.g. diatoms, also require silicon). Fertilization cannot be avoided as the dry algal mass fraction consists of ~7% nitrogen and ~1% phosphorus. Substituting fossil fuels with algal biomass would require a lot of fertilizer. As an illustration, if the EU substituted all existing transport fuels with algae biofuels this would require ~25 million tonnes of nitrogen and 4 million tonnes of phosphorus per annum (Wijffels and Barbosa 2010).

15.7.4 Carbon Fertilization

Algal cultivation requires a source of carbon dioxide. Assuming algae have a carbon mass fraction of 50%, it follows that producing 1 kg dry algal biomass requires at least 1.83 kg CO_2. In reality, however, CO_2 usage will be several times of this. For raceway ponds, the rate of outgassing is a function of the pond depth, friction coefficient of the lining, mixing velocity, pH, and alkalinity. Depending on operational conditions, the theoretical efficiency of CO_2 use can range from 20% to 90% (Van Egmond et al. 2002).

15.7.5 Fossil Fuel Inputs

The majority of the fossil fuel inputs to algal cultivation come from electricity consumption during cultivation and, where included, from natural gas used to dry the algae. Algae are temperature sensitive and maintaining high productivity (particularly in PBRs) may require temperature control. Both heating and cooling demand could increase fossil fuel use. The environmental performance could, however, be improved by integration options such as using waste heat from power generation to dry the algal biomass. System optimization to minimize energy demand will be essential (Acien et al. 2012).

15.7.6 Eutrophication

Nutrient pollution (eutrophication) can lead to undesirable changes in ecosystem structure and function. The impact of algal aquaculture could be positive or negative. Negative impacts could occur if residual nutrients in spent culture medium are allowed to leach into local aquatic systems. On the other hand, positive impacts could occur if algal production were to be integrated into the treatment of water bodies already suffering from excess nutrient supply. For example, Agricultural Research Service scientists found that 60%–90% of nitrogen runoff and 70%–100% of phosphorus runoff can be captured from manure effluents using an algal turf scrubber (AquaFUELs 2011). Remediation of polluted water bodies suffering from algal blooms may also provide locally significant amounts of free waste biomass, and this could be used for biofuel production on a small scale.

15.7.7 Genetically Modified Algae

In the search for algae that can deliver high biomass productivity and lipid content simultaneously, genetic modification is one possible option (Lundquist et al. 2010). Applications of molecular genetics range from speeding up the screening and selection of desirable strains to cultivating modified algae on a large scale. Traits that might be desirable include herbicide resistance to prevent the contamination of cultures by wild-type organisms and increased tolerance to high light levels. Containment of genetically modified algae poses a major challenge. In open pond systems, culture leakage and transfer (e.g., by waterfowl) is unavoidable. Closed bioreactors appear more secure, but Lundquist et al. (2010) commented that as far as containment is concerned, PBRs are only cosmetically different from open ponds and some culture leakage is inevitable.

15.7.8 Algal Toxicity

At certain stages of their life cycle, many algal species can produce toxins ranging from simple ammonia to physiologically active polypeptides and polysaccharides.

Toxic effects can range from the acute (e.g., the algae responsible for paralytic shell-fish poison may cause death) to the chronic (e.g., carrageenan toxins produced in red tides can induce carcinogenic and ulcerative tissue changes over long periods of time). Toxin production is species and strain specific and may also depend on environmental conditions. The presence or absence of toxins is thus difficult to predict (Collins 1978; Rellan et al. 2009).

15.7.9 Insights on Environmental Impacts

Microalgal culture can have a diverse range of environmental impacts, many of which are location specific. Depending on how the system is configured, the balance of impacts may be positive or negative. Impacts such as the use of genetic engineering are uncertain but may affect what systems are viable in particular legislatures. Possibly the most important environmental aspect of microalgal culture that needs to be considered is water management: both the water consumed by the process and the emissions to water courses from the process. In any algal cultivation scheme, it should be anticipated that environmental monitoring will play an important role and will be an ongoing requirement.

15.8 Future Perspective of Algae

Algal biofuel production process involves the growth, concentration, separation, and conversion of microalgal biomass, some of which can be genetically altered. After the extraction of the desired biofuel product, a significant portion of by-product remains (e.g., ethanol). The remaining by-products should have a useful and safe purpose for the economic feasibility and environmental sustainability. The waste streams may include biological toxins, allergens, and carcinogens produced by microorganisms, antibiotics, enzymes, chemicals (e.g., wastewater high in nutrients, BOD), and acids and bases. Exposure to GMOs carries possible human health and environmental risks. An evaluation methodology should be employed for better understanding of the algal production and waste streams and the associated risks to humans and the environment.

As a sustainable source of energy, algae and its feedstocks have great potential to meet the demand of petroleum-based fuels. The versatility of algae to produce lipids, carbohydrates, and protein can be used to create multiple products in various markets to successfully satisfy economic demand. Currently, biotechnology firms and the algal industry are focused on producing relatively low volumes of high-value products such as pharmaceuticals or nutritional supplements. These same industries must refocus on high volumes of biofuel production at low, competitive prices, as well as using by-products such as the protein for distiller's grains and carbohydrates for ethanol. Postextraction by-products must be used efficiently.

Alga-derived biofuel will directly impact the current generation of transportation fuels, and as the major part of the future of renewable fuel, it will also impact many environmental and economic resources. Examples of these impacts are the treatment of wastewater; capture of carbon dioxide from power plants; production of human and animal food, cosmetics, and organic fertilizers; aquaculture; and soil nutrient recovery. Ultimately, the need to decrease fossil fuel dependence makes it important that algae and alga-derived products must be safe to humans and the environment. The rapid commercial expansion of the algal biofuel industry is an excellent example of sustainable product development with great future potential in contributing to the fuel supplies, yet many questions regarding algal production remain unanswered. The state of knowledge regarding the potential environmental impact of the production of algae and alga-derived biofuels continues to be incomplete, fragmented, and largely obscured by proprietary concerns. This knowledge is rapidly changing, which is facilitated by research and industry driven by economics. Commercialization of the production of alga-derived biofuels as part of the overall biofuel industry will have a profound future impact on society. Waste products that are currently discharged into the environment as contaminants will be utilized to produce much energy in a renewable way. Now is the time to initiate the development of an algal industry evaluation methodology that allows for the advancement of knowledge and evaluation tools for authorities for clear understanding of the potential implications.

We have discussed strategies to make alga-based fuels costs competitive with petroleum. Bioprospecting is of much importance to identify algal species growing on low-cost media that have desired traits (e.g., high lipid content, growth rates, growth densities, and/or the presence of valuable coproducts). In spite of the potential of this strategy, the most likely scenario is that bioprospecting will fail to identify species that are cost competitive with petroleum, and subsequent genetic engineering and breeding will be required to bring these strains to economic viability. The potential for engineering algae is just beginning to be realized, from improving lipid biogenesis and crop protection to producing valuable enzyme or protein coproducts. All sustainable technology has challenges, but blind promotion of those technologies without honest consideration of the long-term implications may lead to the acceptance of strategies whose long-term consequences outweigh their short-term benefits. We have presented what we view as the most significant current and upcoming challenges of algal biofuels, but, as with any new industry, the more we learn, the more we will realize the challenges that exist without being identified yet. Despite these uncertainties, we believe that fuel production from algae can be cost-effective, scalable, and deployable in the near future. Only if we continue to expand our understanding of these organisms, we will expand our ability to engineer them for the development of a new energy industry.

15.9 Conclusions

The word "sustainability" is given such importance in the current world that it has become the central theme around which various nations build their policies. The ability of a person to consume or create things that can be replenished at a rate at which it has been consumed is generally what sustainability means. Energy is the most important aspect of sustainability and is the field in which the whole world looks to improve upon. The three most persistent global problems at this point in human history are climate change, poverty, and water scarcity. In various ways, energy is the cause and its sustainability will be the solution for these problems. Global population has increased exponentially from the days of industrial revolution around 1750, and the growth has accelerated from the 1950s when the oil economy started to show its dominance in the world economic structure. Energy and population are directly proportional to each other, and historically, they have both increased and decreased along with each other. During the period of "great acceleration" (1950s onward), the oil production and consumption increased at an unprecedented rate and the emissions and results of consumption were ignored by people in pursuit of comfort and ease of work. Shocking revelations of average global temperature rise, ozone layer depletion, and diseases and environmental hazards due to the release of GHGs by the consumption of fossil fuels have provided a topic on which the whole world now unites and tries solving. The Copenhagen Summit, the Kyoto Protocol, and the Montreal Protocol are some prominent examples where the whole world united and made decisions on cutting down carbon footprints. Energy is perhaps the most significant topic and is the reason for the chaotic situation that we face today.

Renewable energy has been a topic that has found the interest of most students and researches at the present. Solar, wind, hydroelectric, and nuclear power plants are the topic that runs hot in many countries. The Indian government has announced an ambitious target of 100 GW of solar energy and considerable rise in electric cars. The United States and European Union are setting their own targets on increasing the green energy sources. Though these sources of energy are appreciated and supported, the fact that they need storage system that can convert the produced power to chemical energy, which can be used when needed, has raised the fact that there is considerable loss in the power that reaches for consumption. Batteries are the most prominent storage system and are costly and have hazardous chemicals in them. This creates doubts over its eco-friendliness. Fuel cells that are currently present are too costly for commercialization. On completing the LCA, it is clear that even the products, which are developed to reduce carbon emissions, create those emissions during its production and manufacturing stage, which is meaningless at this point since the energy used for such processes is from fossil fuels. The greatest disadvantage of these technologies comes from the transportation sector of energy, which makes up to more than 30% of current global energy consumption. The promotion of battery cars and vehicles that are built with storage system is something that must not be promoted since they all have a lifetime after which the disposal or recycle is a risk. The current set of petroleum-based vehicles must be dumped when these cars replace the former. The scenario is now demanding sources that are similar to the fossil fuel and that replace them without costing the expense on infrastructure. The most apt answer for the question that has been raised is biofuels. These fuels are made from plants, which release carbon compounds that equal the amount that they removed from air by consumption, thus neutralizing the carbon emissions. On the other hand, fossil fuels are those plants or organism that consumed carbon compounds but failed to release it at a time where the effect would have negated any impacts, but consuming them millions of years afterward is the reason for all the problems that were mentioned earlier. Biofuels are stable and can fit in the current infrastructure. They can even promote the cultivation of fuel crops, and with greater production rates, the cost of the fuel would come down significantly. The future seems very bright for biofuels because they answer the question of storage that other sources have answered less effectively or provide answers that cannot be commercialized. Thus, biofuels are the subject that needs to be studied, researched, and debated upon for the benefit of those who live on this planet. This chapter covers various aspects and the scope of biofuels. A deep understanding of the cost, production methods, energy balance, and the impacts of the algal biofuels is provided in this chapter. The demand for this revolutionary technology is on an unprecedented rise at this moment and is only projected to increase through the coming years. The fact that this technology can treat waste products through phycoremediation from various aspects of human activities to produce biofuels with greater oil content at a very low cost makes algal biofuels a technology that will shape the future of human lifestyle and energy demand in the future. No one energy source can be the solution for the global problem. The innovation and application must be from all the sides. The global population that is expected to grow to around 16 billion from the current 7.2 billion puts a greater stress for such sustainable technologies. Making these technologies cheaper, efficient, and more accessible to everyone will aid in improving the life of the bottom half billion of the global population. Let us try to improve and innovate the current technology and make life of the citizens of this globe a sustainable one.

References

Acien, F.G., Fernandez, J.M., Magan, J.J., and E. Molina. 2012. Production cost of a real microalgae production plant and strategies to reduce it. *Biotechnol Adv* 30:1344–1353.

Ananadhi, P.M.R. and S.A. Stanley. 2012. Microalgae as an oil producer for biofuel applications. *Res J Recent Sci* 1:57–62.

AquaFUELs. 2011. Report on biology and biotechnology of algae with indication of criteria for strain selection. http://www.aquafuels.eu/deliverables.html.

Arteburn, L.A., Oken, H.A., Hoffman, J.P. et al. 2007. Bioequivalence of docosahexaenoic acid from different algal oils in capsules and in DHA-fortified food. *Lipids* 42:1011–1024.

Ayehunie, S., Belay, A., Baba, T.W., and R.M. Ruprecht. 1998. Inhibition of HIV-1 replication by an aqueous extract of *Spirulina platensis* (*Arthrospira platensis*). *J Acquir Immune Defic Syndr Hum Retroviral* 18:7–12.

Banerjee, A., Sharma, R., Chisti, Y., and U.C. Banerjee. 2002. *Botryococcus braunii*: A renewable source of hydrocarbons and other chemicals. *Crit Rev Biotechnol* 22:45–79.

Becker, E.W. 1988. Micro-algae for human and animal consumption. In *Micro-algal Biotechnology*, eds. M.A. Borowitzka and L.J. Borowitzka, pp. 222–256. Cambridge, U.K.: Cambridge University Press.

Becker, W. 2004. Microalgae in human and animal nutrition. In *Handbook of Microalgal Culture*, ed. A. Richmond, pp. 312–351. Oxford, U.K.: Blackwell.

Benemann, J.R. and W.J. Oswald. 1996. Systems and economic analysis of microalgae ponds for conversion of CO_2 to biomass. Report number DOE/PC/93204/T5. Berkley, CA: Department of Civil Engineering, University of California, p. 215.

Berberoglu, H., Gomez, P.S., and L. Pilon. 2009. Radiation characteristics of *Botryococcus braunii*, *Chlorococcum littorale*, and *Chlorella* sp. used for CO_2 fixation and biofuel production. *J Quant Spectrosc Radiat Transf* 110:1879–1893.

Blackburn, S.I., Dunstan, G.A., and D.M.F. Frampton. 2009. Australian strain selection and enhancement for biodiesel from algae. *Phycology* 48:8–9.

Bligh, E.G. and W.J. Dyer. 1959. A rapid method for total lipid extraction and purification. *Can J Biochem Physiol* 37:911–917.

Bobade, S.N. and V.B. Khyade. 2012. Detail study on the properties of *Pongamia pinnata* (Karanja) for the production of biofuel. *Res J Chem Sci* 2:16–20.

Borowitzka, M.A. 1988. Vitamins and fine chemicals from microalgae. In *Micro-algal Biotechnology*, eds. M.A. Borowitzka and L.J. Borowitzka. Cambridge, U.K.: Cambridge University Press.

Boyd, P.W., Law, C.S., and C.S. Wong. 2004. The decline and fate of an iron-induced subarctic phytoplankton bloom. *Nature* 428:549–553.

Brown, M.R., Mular, M., Miller, I., Farmer, C., and C. Trenerry. 1999. The vitamin content of microalgae used in aquaculture. *J Appl Phycol* 11:247–255.

Burlew, J.S. 1953. *Algae Culture—From Laboratory to Pilot Plant*. Washington, DC: Carnegie Institution of Washington.

Calder, P.C. 2004. Review—n-3 fatty acids and cardiovascular disease: Evidence explained and mechanisms explored. *Clin Sci* 107:1–11.

Carioca, J., Hiluy, J.J., and M. Leal. 2009. The hard choice for alternative biofuels to diesel in Brazil. *Biotechnol Adv* 27:1043–1050.

Cenciani, K., Oliveira, M.C.B., Feigl, B.J., and C.C. Cerri. 2011. Sustainable production of biodiesel by microalgae and its application in agriculture. *Afr J Microbiol Res* 5:4638–4645.

Cerff, M., Morweiser, M., Dillschneider, R., Michel, A., and K. Menzel. 2012. Harvesting fresh water and marine algae by magnetic separation: Screening of separation parameters and high gradient magnetic filtration. *Bioresour Technol* 118:289–295.

Chamorro, G., Salazar, M., Favila, L., and H. Bourges. 1996. Pharmacology and toxicology of *Spirulina* alga. *Rev Invest Clin* 48:389–399.

Chen, C.Y., Yeh, K.L., Aisyah, R., Lee, D.J., and J.S. Chang. 2011. Cultivation, photobioreactor design and harvesting of microalgae for biodiesel production: A critical review. *Bioresour Technol* 102:71–81.

Chisti, Y. 2000. Animal-cell damage in sparged bioreactors. *Trends Biotechnol* 18:420–432.

Chisti, Y. 2007. Biodiesel from microalgae: A review. *Biotechnol Adv* 25:294–306.

Christenson, L. and R. Sims. 2011. Production and harvesting of microalgae for wastewater treatment, biofuels, and bioproducts. *Biotechnol Adv* 29:686–702.

Clarens, A.F. 2011. Environmental impacts of algae-derived biodiesel and bioelectricity for transportation. *Environ Sci Technol* 45:7554–7560.

Coale, K.H., Johnson, K.S., and F.P. Chavez. 2004. Southern ocean iron enrichment experiment: Carbon cycling in high- and low-Si waters. *Science* 304:408–414.

Coale, K.H., Johnson, K.S., and S.E. Fitzwater. 1996. A massive phytoplankton bloom induced by an ecosystem-scale iron fertilization experiment in the equatorial pacific ocean. *Nature* 383:495–501.

Cohen, Z. 2002. Lipid and fatty acid composition of the green oleaginous alga *Parietochloris incisa*, the richest plant source of arachidonic acid. *Phytochemicals* 60:497–503.

Collins, M. 1978. Algal toxins. *Microbiol Rev* 42:726e46.

Cornet, J.F. 1998. Le technoscope: les photobioréacteurs. *Biofutur* 176:1–10.

Darzins, A., Pienkos, P., and L. Edye. 2010a. Current status and potential for algal biofuels production: International energy agency (IEA) bioenergy task 39: Commercializing—1st and 2nd—Generation of liquid biofuels from biomass. Golden, CO: National Renewable Energy Laboratory, pp. 1–146.

Darzins, A., Pienkos, P., and L. Edye. 2010b. Current status and potential for algae biofuels production: A report to IEA Bioenergy Task 39. Paris, France: International Energy Agency (IEA). www.task39.org (accessed August 2010).

Day, J.G., Benson, E.E., and R.A. Fleck. 1999. *In vitro* culture and conservation of microalgae: Applications for aquaculture, biotechnology and environmental research. *Vitro Cell Dev Biol Plant* 35:127–136.

Demirbas, M.F. 2011. Biofuels from algae for sustainable development. *Appl Energy* 88:3473–3480.

Dimitrov, K. 2007. Green fuel technologies: A case study for industrial photosynthetic energy capture. http://www.nanostring/Algae/CaseStudy.pdf.

Dodd, J.C. 1986. Elements of pond design and construction. In *Handbook of Microalgal Mass Culture*, ed. A. Richmond, pp. 265–283. Boca Raton, FL: CRC Press.

Dresback, K., Ghoshal, D., and A. Goyal. 2001. Phycoremediation of trichloroethylene (TCE). *Physiol Mol Biol Plants* 7:117–123.

Droop, M.R. 1969. Algae. In *Methods in Microbiology*, eds. J.R. Norris and D.W. Ribbon, pp. 1–324. New York: Academic Press.

Duffy, J.E., Canuel, E.A., and W. Adey. 2009. Biofuels: Algae. *Science* 326:1345.

Dufossé, L., Galaup, P., Yaron, A. et al. 2005. Microorganisms and microalgae as sources of pigments for food use: A scientific oddity or an industrial reality? *Trend Food Sci Technol* 16:389–406.

Duong, V.T., Li, Y., Nowak, E., and P.M. Schenk. 2012. Microalgae isolation and selection for prospective biodiesel production. *Energies* 5:1835–1849.

Dyni, J.R. 2006. Geology and resources of some world oil-shale deposits. Scientific Investigations Report 2005-5294. Reston, VA: U.S. Geological Survey.

Emad, A.S. 2011. Algal biomass and biodiesel production. In *Biodiesel—Feedstocks and Processing Technologies*, pp. 1–52. InTech Publishing: Croatia.

Energy Information Administration. 2009. International energy outlook, Vol. 284. Washington, DC: EIA.

Gavrilescu, M. and Y. Chisti. 2005. Biotechnology—A sustainable alternative for chemical industry. *Biotechnol Adv* 23:71–99.

Ghys, A., Bakkere, E., Hornstra, G., and M. Van der Hout. 2002. Red blood cell and plasma phospholipid arachidonic and docosahexaenoic acid levels at birth and cognitive development at 4 years of age. *Early Hum Dev* 69:83–90.

Godman, J. and J. Balk. 2008. Genome analysis of *Chlamydomonas reinhardtii* reveals the existence of multiple, compartmentalized iron-sulfur protein assembly machineries of different evolutionary origins. *Genetics* 179:59–68.

Grobbelaar J.U. 2004. Algal nutrition: Mineral nutrition. In: *Handbook of Microalgal Culture: Biotechnology and Applied Phycology* (ed. A. Richmond), pp. 97–115. Blackwell Publishing: Oxford, UK.

Guil-Guerrero, J.L., Navarro-Juárez, R., López-Martínez, J.C., Campra-Madrid, P., and M.M. Rebolloso-Fuentes. 2004. Functional properties of the biomass of three microalgal species. *J Food Eng* 65:511–517.

Hannon, M., Gimpel, J., Tran, M., Rasala, B., and S. Mayfield. 2010. Biofuels from algae: Challenges and potential. *Biofuels* 1:763–784.

Hiremath, S. and P. Mathad. 2010. Impact of salinity on the physiological and biochemical traits of *Chlorella vulgaris* beijerinck. *J Algal Biomass Utln* 1:51–59.

Hu, Q., Sommerfeld, M., and E. Jarvis. 2008. Microalgal triacylglycerols as feedstocks for biofuel production: Perspectives and advances. *Plant J* 54:621–639.

Hu, Y.R., Wang, F., Wang, S.K., Liu, C.Z., and C. Guo. 2013. Efficient harvesting of marine microalgae *Nannochloropsis* maritime using magnetic nanoparticles. *Bioresour Technol* 138:387–390.

Illman, A.M., Scragg, A.H., and Shales, S.W. 2000. Increase in *Chlorella* strains calorific values when grown in low nitrogen medium. *Enzyme Microb Technol* 27:631–635.

Kedziorek, M.A.M. and A.C.M. Bourg. 2000. Solubilization of lead and chromium during the percolation of EDTA through a soil polluted by smelting activities. *J Contamin Hydrol* 40:381–392.

Kheira, A.A.A. and N.M.M. Atta. 2008. Response of *Jatropha curcas* L. to water deficit: Yield, water use efficiency and oilseed characteristics. *Biomass Bioenerg* 33:1343–1350.

Kim, M.K., Park, J.W., Park, C.S. et al. 2007. Enhanced production of *Scenedesmus* spp. (green microalgae) using a new medium containing fermented swine wastewater. *Bioresour Technol* 98:2220–2228.

Kulay, L.A. and G.A. Silva. 2005. Comparative screening LCA of agricultural stages of soy and castor beans. Paper presented at *Second International Conference on Life Cycle Management*, Barcelona, Spain, pp. 5–7.

Kumar, A., Ergas, S., Yuan, X. et al. 2014. Enhanced CO_2 fixation and biofuel production via microalgae: Recent developments and future directions. *Trends Biotechnol* 28:371–380.

Li, Y., Horsman, M., Wu, N., Lan, C.Q., and N. Dubois-Calero. 2008. Biofuels from microalgae. *Biotechnol Prog* 24:815–820.

Liu, A., Chen, W., Zheng, L., and Song L. 2011. Identification of high-lipid producers for biodiesel production from forty-three green algal isolates in China. *Prog Nat Sci Mater Int* 21:269–276.

Lowry, O., Rosebrough, H.N.J., Farr, A.L., and R.J. Randall. 1951. Protein measurement with the Folin phenol reagent. *J Biol Chem* 193:265–275.

Luiten, E.E.M., Akkerman, I., Koulman, A. et al. 2003. Realizing the promises of marine biotechnology. *Biomol Eng* 20:429–439.

Lundquist, T.J., Woertz, I.C., Quinn, N.W.T., and J.R. Benemann. 2010. A realistic technology and engineering assessment of algae biofuel production. Berkeley, CA: Energy Biosciences Institute, University of California. http://works.bepress.com/tlundqui/5.

Manikandan, G., Kumar, P.S., and R. Prakalathan. 2014. Comparison of biodiesel production from macro and micro algae. *Int J Chem Tech Res* 6:4143–4147.

Mata, T.M., Martins, A.A., and N.S. Caetano. 2010. Microalgae for biodiesel production and other applications. *Renew Sust Energ Rev* 14:217–232.

Mcginnis, K.M., Dempster, T.A., and M.R. Sommerfeld. 1997. Characterization of the growth and lipid content of the diatom *Chaetoceros muelleri*. *J Appl Phycol* 9:19–24.

Medeiros, D.L., Sales, E.A., and A. Kiperstok. 2013. Energy production from microalgae biomass: The carbon footprint and energy balance. Paper presented at *Fourth International Workshop Integrating Advances Cleaner Production into Sustainability Strategies*, São Paulo, Brazil.

Medipally, S.R., Yusoff, F.M.D., Banerjee, S., and M. Shariff. 2015. Microalgae as sustainable renewable energy feedstock for biofuel production. *BioMed Res Int.* Volume 2015 (2015), Article ID 519513, 13 pages. http://dx.doi.org/10.1155/2015/519513.

Melis, T. 2005. Integrated biological hydrogen production. Paper presented at *Proceedings International Hydrogen Energy Congress and Exhibition IHEC*, Istanbul, Turkey.

Metting, F.B. 1996. Biodiversity and application of microalgae. *J Ind Microbiol* 17:477–489.

Miranda, M.S., Cintra, R.G., Barros, S.B., and F.J. Mancini. 1998. Antioxidant activity of the micro alga *Spirulina maxima. Braz J Med Biol Res* 31:1075–1079.

Mubee, U., Zia-ul-Islam, M., Hussain, W., and K.A. Malik. 2010. Future of your fuel tank. Lahore, Pakistan: Department of Biological Sciences, p. 131.

Mutanda, T., Ramesh, D., Karthikeyan, S., Kumari, S., Anandraj, A., and F. Bux. 2011. Bioprospecting for hyper-lipid producing microalgal strains for sustainable biofuel production. *Bioresour Technol* 102:57–70.

Muthukumaran, M. 2009. Studies on the phycoremediation of industrial effluents and utilization of algal biomass. PhD, thesis. University of Madras, Chennai, India.

Muthukumaran, M., Raghavan, B.G., Subramanian, V.V., and V. Sivasubramanian. 2005. Bioremediation of industrial effluent using microalgae. *Indian Hydrobiol* 7(S):105–122.

Muthukumaran, M., Thirupathi, P., Chinnu, K., and V. Sivasubramanian. 2012. Phycoremediation efficiency and biomass production by micro alga *Desmococcus olivaceus* (Persoon et Acharius) J.R. Laundon treated on chrome-sludge from an electroplating industry—A open raceway pond study. *Int J Curr Sci* (Special Issue): 52–62. http://www.currentsciencejournal.info/issuespdf/Muthu.pdf.

New Scientist. 2011. The rush towards renewable oil. *New Sci* 203:61–62.

Nickelsen, K. 2007. Otto Warburg's first approach to photosynthesis. *Photosynth Res* 92:109–120.

Nielsen, D.C. 2008. Oilseed productivity under varying water availability. Paper presented at *Proceedings of 20th Annual Central Plains Irrigation Conference and Exposition*, Akron, OH, Vol. 2, pp. 30–33.

Olguin, E.J. 2003. Phycoremediation: Key issues for cost-effective nutrient removal processes. *Biotechnol Adv* 22:81–91.

Oswald, W.J. 1988. Micro-algae and waste-water treatment. In *Micro-Algal Biotechnology*, eds. M.A. Borowitzka and L.J. Borowitzka, pp. 305–328. Cambridge, U.K.: Cambridge University Press.

Oswald, W.J. and H.B. Gotaas. 1957. Photosynthesis in sewage treatment. *Trans Am Soc Civil Eng* 122:73–105.

Park, J.B.K., Craggs, R.J., and A.N. Shilton. 2011. Wastewater treatment high rate algal ponds for biofuel production. *Bioresour Technol* 102:35–42.

Parry, M.L. 2007. Intergovernmental panel on climate change, working group II, world meteorological organization, United Nations Environment Programme. Summary for policymakers. In *Climate Change 2007: Impacts, Adaptation and Vulnerability Contribution of Working Group II to the Fourth Assessment Report of the Intergovernmental Panel on Climate Change*, eds. M.L. Parry, O.F. Canziani, J.P. Palutikof, P.J. Van der Linden, and C.E. Hanson. Cambridge, U.K.: Cambridge University Press.

Peterson, C.L. and T. Hustrulid. 1998. Carbon cycle for rapeseed oil biodiesel fuels. *Biomass Bioenerg* 14:91–101.

Pienkos, P.T. and A. Darzins. 2009. The promise and challenges of microalgal-derived biofuels. *Biofuels Bioprod Biorefining* 3:431–440.

Pittman, J.K., Dean, A.P., and O. Osundeko. 2011. The potential of sustainable algal biofuel production, using wastewater resources. *Bioresour Technol* 102:17–25.

Qureshi, M.A., Garlich, J.D., and M.T. Kidd. 1996. Dietary *Spirulina platensis* enhances humoral and cell-mediated immune functions in chickens. *Immunopharmacol Immunotoxicol* 18:465–476.

Raja, R., Shanmugam, H., Ganesan V., and I.S. Carvalho. 2014. Biomass from microalgae: An overview. *Oceanography* 2:1–7.

Rajvanshi, S. and M.P. Sharma. 2012. Microalgae: A potential source of biodiesel. *J Sustain Bioenerg Syst* 2:49–59.

Raphael, S. and A. Bauen. 2013. Micro-algae cultivation for biofuels: Cost, energy balance, environmental impacts and future prospects. *Biomass Bioenerg* 53:29–38.

Rawat, I., Kumar, R., Mutanda, T., and F. Bux. 2010. Dual role of microalgae: Phycoremediation of domestic wastewater and biomass production for sustainable biofuels production. *Appl Energ* 88:3411–3424.

Reijnders, L. and M.A.J. Huijbregts. 2008. Biogenic green-house gas emissions linked to the life cycles of biodiesel derived from European rapeseed and Brazilian soybeans. *J Clean Prod* 16:1943–1948.

Rellan, S., Osswald, J., Saker, M., Gago-Martinez, A., and V. Vasconcelos. 2009. First detection of anatoxin-a in human and animal dietary supplements containing cyanobacteria. *Food Chem Toxicol* 47:2189e95.

Richmond, A. 1996. Efficient utilization of high irradiance for production of photoautotrophic cell mass: A survey. *J Appl Phycol* 8:381–387.

Rodolfi, L., Zittelli, G.C., and N. Bassi. 2009. Microalgae for oil: Strain selection, induction of lipid synthesis and outdoor mass cultivation in a low-cost photobioreactor. *Biotechnol Bioeng* 102:100–112.

Rosana, C.S.S., Thiago, R.B., Pablo, D.G., Maiara, P.S., Valeriano, A.C., and A.L. Eduardo. 2012. Potential production of biofuel from microalgae biomass produced in wastewater. http://www.intechopen.com/books/biodiesel-feedstocks-production-and-applications/potential-biofuel-from-the-microalgae-biomass-produced-in-wastewater.

Salim, S., Bosma, R., Vermuë, M.H., and R.H. Wijffels. 2011. Harvesting of microalgae by bioflocculation. *J Appl Phycol* 23:849–855.

Sánchez Mirón, A., Cerón García, M.-C., Contreras Gómez, A., García Camacho, F., Molina Grima, E., and Y. Chisti. 2003. Shear stress tolerance and biochemical characterization of *Phaeodactylum tricornutum* in quasi steady-state continuous culture in outdoor photobioreactors. *Biochem Eng J* 16:287–297.

Sawayama, S., Inoue, S., Dote, Y., and S.-Y. Yokoyama. 1995. CO_2 fixation and oil production through microalga. *Energ Convers Manage* 36:729–731.

Schindler, J. and W. Zittel. 2008. Crude oil—The supply outlook, Vol. 102. Ottobrunn, Germany: Energy Watch Group.

Schwartz, J., Shklar, G., Reid, S., and D. Trickler. 1988. Prevention of experimental oral cancer by extracts of *Spirulina-Dunaliella* algae. *Nutr Cancer* 11:127–134.

Seyfabadi, J., Ramezanpour, Z., and Z.A. Khoeyi. 2011. Protein, fatty acid, and pigment content of *Chlorella vulgaris* under different light regimes. *J Appl Phycol* 23:721–726.

Sheehan, J., Dunahay, T., Benemann, J., and P. Roessler. 1998. A look back at the U.S. Department of Energy's Aquatic Species Program—Biodiesel from algae. Golden, CO: National Renewable Energy Laboratory.

Sialve, B., Bernet, N., and O. Bernard. 2009. Anaerobic digestion of microalgae as a necessary step to make microalgal biodiesel sustainable. *Biotechnol Adv* 27:409–416.

Sidhu, K.S. 2003. Health benefits and potential risks related to consumption of fish or fish oil. *Regul Toxicol Pharmacol* 38:336–344.

Simopoulos, A.P. 2002. The importance of the ratio of omega-6/omega-3 essential fatty acids. *Biomed Pharmacother* 56:365–379.

Sivasubramanian, V. 2006. Phycoremediation—Issues and challenges. *Indian Hydrobiol* 9:13–22.

Soletto, D., Binaghi, L., Lodi, A., Carvalho, J.C.M., and A. Converti. 2005. Batch and fed-batch cultivations of *Spirulina platensis* using ammonium sulphate and urea as nitrogen sources. *Aquaculture* 243:217–224.

Spoehr, H.A. and H.W. Milner. 1949. The chemical composition of *Chlorella*. Effect of environmental conditions. *Plant Physiol* 24:120–149.

Spolaore, P., Joannis-Cassan, C., Duran, E., and A. Isambert. 2006. Commercial applications of microalgae. *J Biosci Bioeng* 101:87–96.

Sreevatsan, S. Oilgae: Comprehensive report on attractive algae product opportunities. http://www.oilgae.com.

Tan, D.K.Y. and J.S. Amthor. 1988. *Bioenergy*. Cambridge, U.K.: Cambridge University Press.

Thies, F., Garry, J.M., Yagoob, P. et al. 2003. Association of n-3 polyunsaturated fatty acids with stability of atherosclerotic plaques: A randomized controlled trial. *Lancet* 361:477–485.

Tsukahara, K. and S. Sawayama. 2005. Liquid fuel production using microalgae. *J Jpn Petrol Inst* 48:251–259.

U.S. DOE. 2010. National algal biofuels technology roadmap. U.S. Department of Energy, Office of Energy Efficiency and Renewable Energy, Biomass Program. D. Fishman, R. Majumdar, J. Morello, R. Pate, and J. Yang (Eds.), December 9–10, 2008, College Park, MD. http://biomass.energy.gov.

Ugwu, C.U., Aoyagi, H., and H. Uchiyama. 2008. Photobioreactors for mass cultivation of algae. *Bioresour Technol* 99:4021–4028.

Van den Berg, H., Faulks, R., Granado, H.F. et al. 2000. The potential for the improvement of carotenoid levels in foods and the likely systemic effects. *J Sci Food Agric* 80:880–912.

Van Egmond, K., Bresser, T., and L. Bouwman. 2002. The European nitrogen case. *AMBIO J Hum Environ* 31:72e8.

Venkataraman, L.V. and E.W. Becker. 1985. Biotechnology and utilization of algae—The Indian experience. New Delhi, India: Department of Science and Technology.

Vijayaraghavan, K. and K. Hemanathan. 2009. Biodiesel production from freshwater algae. *Energy Fuel* 23:5448–5453.

Vollmann, J., Moritz, T., Karg, C., Baumgartner, S., and H. Wagentrist. 2007. Agronomic evaluation of camelina genotypes selected for seed quality characteristics. *Ind Crop Prod* 26:270–277.

Wang, B., Li, Y., Wu, N., and C.Q. Lan. 2008. CO_2 bio-mitigation using microalgae. *Appl Microbiol Biotechnol* 79:707–718.

Wang, R. and R.J. Conover. 1986. Dynamics of gut pigment in the copepod *Temora longicornis* and the determination of *in situ* grazing rates. *Limnol Oceanogr* 31:867–877.

Wijffels, R.H. and M.J. Barbosa. 2010. An outlook on microalgal biofuels. *Science* 329:796e9.

Wroble, M., Mash, C., Williams, L., and R.B. McCall. 2002. Should long chain polyunsaturated fatty acids be added to infant formula to promote development? *Appl Dev Phychol* 23:99–112.

Xu, L., Guo, C., Wang, F., Zheng, S., and C.Z. Liu. 2011. A simple and rapid harvesting method for microalgae by *in situ* magnetic separation. *Bioresour Technol* 102:10047–10051.

Yamaguchi, K. 1997. Recent advances in microalgal bioscience in Japan, with special reference to utilization of biomass and metabolites: A review. *J Appl Phycol* 8:487–502.

Yun, Y.S., Lee, S.B., Park, J.M., Lee, C.I., and Yang, J.W. 1997. Carbon dioxide fixation by algal cultivation using wastewater nutrients. *J Chem Technol Biotechnol* 69:451–455.

16

Photobioreactors for Microalgal Cultivation and Influence of Operational Parameters

M.V. Rohit, P. Chiranjeevi, C. Nagendranatha Reddy, and S. Venkata Mohan

Contents

Abstract

Photobiotechnology research is gaining interest in the recent years for the production of sustainable and eco-friendly fuels and chemicals. Phototrophic microalgae produce dense biomass and high-value products for pharmaceutical, health, cosmetic, and biofuel purposes. The microalgal cultivation requires efficient photobioreactors (PBRs), which can harvest maximum sunlight in the form of photosynthetically active radiation. Carbon recycling and positive net energy ratios are often required by large-scale PBR operations in order to offset the production costs and increase energy efficiency and economic profitability. An overview of various types and designs of PBRs and factors affecting the scale of operation is presented. Designs based on hydrodynamic flow parameters, lighting, and surface-to-volume ratio are also discussed.

16.1 Introduction

Microalgal cultivations have enormous futuristic potential to meet carbon neutral energy supply as well as food supplementation. Photobiotechnology that employs phototrophic microalgae to produce dense biomass and high-value products for pharmaceutical, health, cosmetic, and biofuel purposes is gaining rapid attention (Socher et al. 2015). Microalgae can be cultivated in photoautotrophic, mixotrophic, and heterotrophic modes of cultivation (Devi et al. 2013; Venkata Mohan et al. 2015) based on the availability of light (photoautotrophic) and carbon source. Heterotrophic growth is based on the cellular utilization of organic carbon instead of light, and mixotrophic growth uses the combination of these energy sources (Rohit and Venkata Mohan 2016). Mixotrophic mode of cultivation is receiving increased attention in recent years (Wang et al. 2013; Chandra et al. 2014; Venkata Mohan et al. 2015). Production of high-value products, namely, pharmaceuticals, nutraceuticals, and cosmetics, is much more feasible in well-controlled photobioreactors (PBRs) with closed system operations.

The microalgal production requires efficient PBRs in order to meet the constraints of energy efficiency and economic profitability (Morweiser et al. 2010). Current cultivation systems are designed for high-value products rather than for mass production (Borowitzka 2013). High mass transfer coefficient KLa is one of the important criteria for PBR design, especially for CO_2 sequestration (Ugwu et al. 2008). Closed photobioreactor systems facilitate better control of culture environment, such as carbon supply, water supply, optimal temperature, efficient exposure to light, pH levels, gas supply rate, and mixing regime, and can achieve high growth rates (Sierra et al. 2008; Mata et al. 2010). Closed systems avoid contamination and direct exchange of gases between the cultivation systems and the external environment. PBRs facilitate the minimization of contamination over open pond systems. The advantage of an enclosed system under outdoor conditions is the lower energy costs (Sevigné et al. 2012). This chapter summarizes the types of PBRs and factors affecting their performance.

16.2 Types of Photobioreactors

PBRs have several advantages (Table 16.1) over open systems, namely, they reduce contamination, have better control over operational parameters, avoid water evaporation and CO_2 loss, allow higher biomass productivity and are flexible to operate

TABLE 16.1 Advantages and Limitations of Different Photobioreactors

Production Systems	Advantages	Limitations
Tubular	Large illumination surface area	Biofilm wall growth on walls
	Good biomass productivity	Biofouling issues
	High gas transfer	Gradients of pH, DO, and CO_2 along tubes
		High shear stress
Airlift	Good and uniform mixing	Hydrodynamic stress
	Efficient light distribution	Uneven mass transfer
Bubble column	High mass transfer	Small illumination area
	Low energy consumption	High shear stress
	Good mixing with low shear stress	Sophisticated construction
	Reduced photoinhibition and photooxidation	
Helical	High degree of control	Low light-harvesting efficiency
	High gas transfer efficiency	Self-shading
Flat panel	High biomass productivity	Difficulty in scale-up
	Low O_2 buildup	Difficult to control temperature
	Good light path and large illumination surface area	Hydrodynamic stress
		High surface to volume ratio (SVR)
Stirred tank	Good mixing with low shear stress	Low surface-to-volume ratio (SVR)
	High mass transfer	Low lighting efficiency

with optimum process parameters. Low contamination from predators and pathogens along with higher biomass productivity is an advantage of closed systems. Different closed systems/PBRs, namely, vertical (tubular) columns, airlift, bubble column, helical/flat panel, and stirred tank, are used for microalgal cultivation (Singh and Sharma 2012).

16.2.1 Tubular Photobioreactors

A closed tubular PBR (Figure 16.1) is made up of transparent plastic or glass tubes that consist of an array of straight, coiled, or looped tubing to allow the penetration of light for microalgal cultivation and are suitable for outdoor mass culture (Molina et al. 2001; Richmond 2004; Perner and Posten 2007; Ribeiro et al. 2009). Algae are circulated through the tubes by airlift mechanism using a gas-sparging pump installed at the bottom of the reactor which converts the inlet gas into tiny bubbles and provides the driving force for mixing, mass transfer of CO_2 and removing O_2 produced during photosynthesis (Molina et al. 2000). The temperature of the tubular PBR is better controlled by floating or submerging the tubes in a pool of water (Pulz and Scheibenbogen 1998). The design of vertical column PBRs does not employ a mechanical agitation system. Depending upon the liquid flow patterns inside the reactor, the vertical tubular PBRs are categorized into airlift and bubble column reactors.

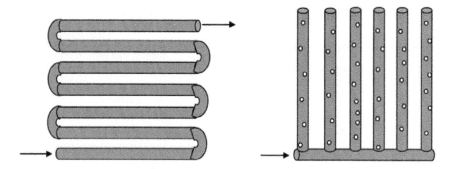

Figure 16.1   Tubular photobioreactors.

16.2.2 Airlift Photobioreactors

Airlift PBRs, which are common in conventional bioreactor designs, are made of a vessel comprising two interconnecting zones called the riser, where the gas mixture flows upward to the surface from the sparger, and the downcomer, which does not receive the gas, but the medium flows down toward the bottom and circulates within the riser (Figure 16.2). Based on the circulation mode, the design of an airlift reactor can be further classified into two forms: internal loop where regions are separated by either a draft tube or a split cylinder and external loop where the riser and downcomer are separated physically by two different tubes with no physical agitation (Chisti 1989; Miron et al. 2000). The riser is similar to a bubble column, where the gas moves upward haphazardly. The fluid dynamics of the airlift reactor are significantly influenced by gas held up in the downcomer. The difference between a riser and a downcomer, by increasing the gas holdup, is an important criterion in designing the airlift systems (Chisti 1989; Kaewpintong et al. 2007).

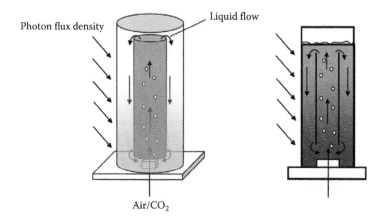

Figure 16.2   Airlift photobioreactors.

An airlift reactor has the characteristic advantage of creating circular mixing patterns where liquid culture passes continuously through dark and light phases, giving a flashing light effect to the microalgal cells (Barbosa et al. 2005). Gas–liquid mass transfer, heat transfer, mixing, and turbulence are influenced by the gas retention in various zones.

16.2.3 Bubble Column Photobioreactors

Bubble column PBRs are characterized by low capital cost, high surface-area-to-volume ratio (SVR), suitable heat and mass transfer, relatively homogeneous culture environment, and efficient release of O_2 and residual gas mixture (Tredici et al. 2010). Usually, they are cylindrical with height greater than twice their diameter, and gas bubbling upward from the sparger provides the required mixing and CO_2 mass transfer with an external light supply. Designing of gas sparger is important for efficient performance of a bubble column (Doran 1995; Nigar et al. 2005). Photosynthetic efficiency mainly depends on the gas flow rate, which further depends on the light and dark cycle as the liquid is circulated regularly from central dark zone to external photozone at higher gas flow rate (Janssen et al. 2003). Increase in gas flow rate will increase the photosynthetic efficiency leading to shorter light and dark cycles. Increase of biomass productivity was observed in bubble column PBR with batch mode operation using *Chlorella vulgaris* (Degen et al. 2001).

16.2.4 Helical-Type Photobioreactors

The helical type of PBR is a coiled transparent and flexible tube of small diameter with separate or attached degassing unit. For better photosynthetic efficiency, a centrifugal pump is used to drive the culture through a long tube to the degassing unit where CO_2 gas mixture and feed can be circulated from either direction (Morita et al. 2011). Degasser removes the oxygen produced during photosynthesis and residual gas of the injected gas stream. This reactor provides efficient mass transfer of CO_2 from gas to liquid phase (Watanabe et al. 1995).

16.2.5 Flat Panel Photobioreactors

The flat panel photobioreactors (FPPBRs) (Figure 16.3) are one of the efficient systems fabricated with transparent materials, namely, Perspex and glass, and characterized with higher penetration of solar irradiance, high SVR, minimal light path, and open gas diffusion systems (Xu and Xiong 2009). Flat panel design inspired by nature considers laminar morphology of plant leaves having high SVR and is evolved for capturing maximum solar energy (Barbosa et al. 2005). SVR, agitation, aeration, illumination, less space utilization, uniform gas diffusion throughout the length of

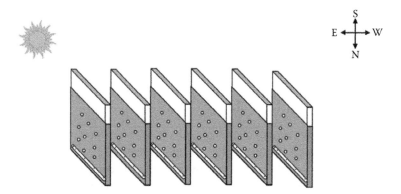

Figure 16.3 Flat panel photobioreactors.

the bioreactor, etc., are some of the parameters considered for design. Mixing/agitation is provided either by bubbling air through a perforated tube from either bottom or side of the reactor or by rotating it mechanically through a motor. The light distribution is mainly from the sides by direct sunlight with an advantage of positioning the reactor vertically. Barbosa et al. (2005) have illuminated with fluorescent tubes on one side of a reactor accounting for 1000 μmol photons/m²/s total light intensity. Baffles were also included for proper mixing but with limitations of low mixing rate and retaining escape corners and dead zones thereby increasing shear stress and cell adhesion to reactor walls inhibiting sunlight penetration. Zhang et al. (2002) suggested increment in liquid height and widening of light path for scale-up of the process rather than increasing the length of the bioreactor. Circulation of cells and media in terms of airlift mode is gaining more prominence because of smaller downcomer zone and large riser zone provided by the injection of compressed air into the bioreactor (Degen et al. 2001).

16.2.6 Stirred-Tank Photobioreactors

Microalgae were initially cultivated photoautotrophically in stirred-tank photobioreactors (STPBRs) by using either sunlight or artificial light (Figure 16.4). STPBRs are categorized under conventional reactor setup wherein impellers and baffles are used for agitation and light is provided by illumination externally. Air (enriched with CO_2) is sparged from the bottom of the reactor (Demessie and Bekele 2003). In STPBR systems, the biomass growth was enhanced by 3.7 times when compared to static cultures (Camacho et al. 2011). A hydraulically integrated serial turbidostat algal reactor with two sealed turbidostats and a series of open, hydraulically connected continuous flow stirred-tank reactors (CFSTRs) were developed to grow *Selenastrum capricornutum* sp. (Benson et al. 2007). CFSTR was used as a biomass amplifier of initial inoculums. This system was used to establish a prediction-based model for determining microalgal productivity for large-scale applications. The drawback of this system is

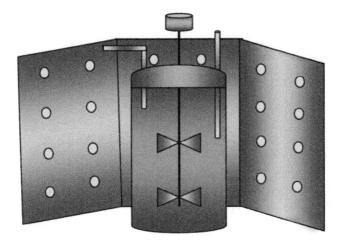

Figure 16.4   Stirred-tank reactor.

low SVR, which leads to low lighting efficiency. Also, mechanical agitation results in high shear stress, which restricts their use in CO_2 sequestration.

16.3 Factors Influencing Photobioreactor Operations

To obtain maximum productivity when cultivated in PBR, abiotic parameters should be controlled in the optimum range. The abiotic parameters, namely, pH, salinity, temperature, nutrients, CO_2 concentration, light intensity, agitation, air sparging, and PBR design, are controlled to allow for efficient biomass growth and augment the synthesis of high-quality products (Norsker et al. 2011).

16.3.1 pH

The availability of nutrients and CO_2 can affect pH of the system. High CO_2 concentrations will lower the pH and will cease the biological activity of microalgae by potentially provoking physiological effects. The optimum pH for most microalgae is 7 (Olivieri et al. 2012), and it varies with species to species (Dasgupta et al. 2010). Few strains like *Stichococcus bacillaris* are able to grow in extreme pH like 3 and have achieved lipid content of 28% (Durmaz 2007). The microalgae grown at pH 7 showed higher motility and appeared uniform, whereas in the case of lower pH (pH 3) operation, the microalgae are aggregated, nonmotile, and not uniform in cell shape and size. With elevated CO_2 concentrations, pH drops down to pH 5, and with higher SO_x concentrations, pH drops even down to pH 2.6 (Maeda et al. 1995; Westerhoff et al. 2010). When highly concentrated CO_2 gases like flue gases are sparged to PBRs, there is a possibility of microalgal growth inhibition as the medium pH reaches very low values based on the concentration. But this provides an inexpensive way of CO_2 utilization for value addition and its removal from the atmosphere (Concas et al. 2013).

16.3.2 CO_2 Concentration

CO_2 has a significant role in the photosynthetic process. Dissolved CO_2 exists in the form of H_2CO_3, HCO_3^-, and CO_3^{2-} ions in aqueous environment. The concentration of CO_2 dissolved is influenced by the pH and temperature of the medium. The optimum CO_2 concentration for microalgal growth is 5%–10% by volume, and it varies with PBR and algae type cultivated (Yang and Gao 2003). Microalgae have the ability to uptake certain levels of CO_2 concentrations after which it inflicts deleterious effects on the growth and biomass enhancement. CO_2 concentration not only affects the biomass growth and productivity of microalgae but also shows impact on lipids and its fatty acid profile. The productivity and lipid content increase with an increase in CO_2 concentration up to a certain extent and beyond, which shows drastic drop in the performance (Kumar et al. 2011). The activity of the biocatalyst will be reduced at higher CO_2 concentrations as the bicarbonate formed in the medium will inhibit its performance. CO_2 sequestration experiments conducted by Chiu et al. (2008) at a flow rate of 0.25 vvm reported that 2% (v/v) of CO_2 is optimum for the growth of *Chlorella*, while at 10% (v/v), specific growth rate becomes insignificant. Due to fast interconvertible reactions among them, consumption of any inorganic carbon does not affect the equilibrium. Microalgal cells preferentially uptake HCO_3 over CO_2 despite the fact that the former is a poor source of carbon than the latter (Carvalho et al. 2006). The sequestration of CO_2 from flue gas emitted by coal-fired thermal power plant confirms that *Chlorella* sp. T-1 can tolerate up to 100% CO_2 concentration, but the maximum growth rate was obtained when using 10% CO_2 with no significant decrease in growth rate up to 50% CO_2 concentration (Maeda et al. 1995). They also concluded that preadaptation of cells with lower percentage of CO_2 concentration leads the tolerability of cells in higher percentage of CO_2. According to Tang et al. (2011), the utilization of flue gas for algal growth will yield higher biomass and lipid productivities. Supplying higher CO_2 concentration to the PBR will increase the mass transfer from gas phase to the culture medium enhancing the polyunsaturated fatty acid accumulation. CO_2 sparging time and intervals have significant influence on biodiesel production (Devi and Venkata Mohan 2012).

16.3.3 Temperature

The optimum temperature for algal growth may vary with culture composition and PBR types and is usually 20°C–24°C for most of the cultures (Varshney et al. 2015). The temperature regulation in open ponds is difficult and cannot be modified according to the requirement, but in the case of PBRs, there is a possibility of maintaining the temperature close to the optimum temperature that further yields higher biomass productivity. There are several species of microalgae that can tolerate higher (60°C) and lower (16°C) temperatures, but in the case of extreme temperatures, the microalgae will slow down its growth to survive in those hostile environments (Varshney et al. 2015). Higher cell suspension, cell density, and biomass productivity were achieved by regulating the temperature when compared to uncontrolled temperature fluctuations in open ponds (Garcia Gonzalez et al. 2005). The utilization of flue gas in PBRs may alter the internal temperature leading to culture death. So this requires heat exchange installation for lowering the temperature of flue gas or usage of thermophilic species for experiments. Temperature control during the light and dark phases (day and night) has positive effect on biomass growth to prevent biomass loss due to dark respiration (Torzillo et al. 2003). Temperature in the PBRs can be controlled and regulated by spraying water on the tube surfaces (evaporative cooling), regulating feed temperature, regulating recirculation stream, placing light-harvesting unit inside a pool, and using proper construction material that can endure high temperature (Siddiqui et al. 2015). Higher lipids were produced during lipid induction phase when operated at 30°C (Subhash et al. 2014).

16.3.4 Irradiance

The microalgal growth rate is directly proportional to incident light (Buehner et al. 2009). Light serves as an important factor in attaining high-density algal culture. Low-cost, durable, efficient, and reliable light source is required to exploit the commercial potential of microalgae. Light is required to generate ATP and also requires dark phase for biochemical reactions simultaneously. The lower light intensity becomes limiting factor for microalgal growth and higher light causes photoinhibition leading to the inactivation of other systems along with photosystems (Dao and Beardall 2016). The amount of incident light is dependent on many factors such as light wavelength, cell concentration, PBR type, and depth of reactor or pond. The light that is incident will get absorbed or get scattered when the amount of algal cell density is higher in a system. Light penetration is higher in circular type of PBR because of its uniform distribution rather than in plane PBR highlighting the importance of engineering and design of PBR (Tredici et al. 2010). Light utilization efficiency and overall photosynthetic efficiency are determined by saturation light intensity (Is). Light utilization efficiency (Es) and overall photosynthetic efficiency (Er) are greatly dependent on the ratio of incident light intensity (Io) and saturation light intensity (Is). The geographical areas that have optimum temperatures throughout the year and high irradiance are optimum for microalgal cultivation.

Oversaturating light conditions will overload the pigments present in the photosystem and cease the light-harvesting process thereby resulting in photoinhibition by producing reactive oxygen species. Photosynthetic cells have a regulatory process called nonphotochemical quenching through which the excess light incident is dissipated as heat. The microalgal productivity in outdoor conditions can be enhanced by making practical modifications in the PBR. To overcome the limitations like incident light, uniform distribution of light, and high irradiance present in the design of open ponds, modifications were done to the PBR engineering in terms of limitations mentioned earlier thereby enhancing the biomass productivity. Placing the PBR vertical will enhance the incident light to spread uniformly on the overall surface area, which will result in enhancing the photosynthetic efficiency (Cuaresma et al. 2011; Rosch and Posten 2012). Light distribution and light delivery are key factors in enhancing photosynthetic mechanism. The factors that should be taken care of for effective light delivery are minimization of photon loss; high electrical efficiency; low heat dissipation; long lifetime; reasonable compactness; spectral absorption range; low cost; filtration of wavelengths that are harmful; and elimination of heat generation. The light energy that is not utilized in photosynthesis can be converted to heat energy, and hence to overcome the photon loss, the usage of internal lighting is suggested (Lee and Palsson 1994). In *Dunaliella* cells, higher irradiance enhances the accumulation of carotene in the PBR from morning to evening and decreases at night.

16.3.5 Mixing Systems

Uniform distribution of nutrients is important for ideal growth and lipid productivities of microalgae cultivated in closed systems. Mixing is the most important factor that affects the growth rates of algae when compared to other operational parameters (Suh and Lee 2003). Ideal mixing pattern helps in both homogeneous distribution of nutrients and enhanced distribution of light over cells. Mixing helps in activation of flashing light effect, which uniformly distributes algal cells in light and dark zones and reduces the intensity of incident light resulting in 40% increase in biomass productivity (Ugwu et al. 2002). The lighting/irradiation efficiency has direct influence on mixing regime and does not work at low light intensities. In tubular reactors, setting up of static mixers assists in increasing the light consumption and biomass yields in scaled-up reactors with tube diameter. Mixing helps not only in consistent light flow throughout the reactor but also in decreasing the diffusion barriers around the cells resulting in enhanced mass transfer (Richmond 2004). The coefficient of mass transfer (KLa) inside a reactor depends on the agitation rate, type of sparger, surfactants or antifoam agents, and temperature. The size of the bubbles and the gas bubble velocity are dependent on the liquid flow rate. The use of fine spargers could result in the formation of large bubbles, which leads to poor mass transfer because of the reduced contact area between liquid and gas. A closed reactor with a dual orifice and perforated membrane sparger system, which separates the CO_2 supply from the air supply, was used for mixing resulting in five times higher magnitude of CO_2 transfer from gas phase to liquid phase relative to conventional sparging (Eriksen et al. 1998). In airlift bioreactors, modifications of sparging design to the annulus provide improved mixing when compared to bubble column reactors (Janssen et al. 2003). Optimization of mixing levels is a prerequisite as high levels of mixing can lead to cell death from shear and bubble formation, which is the major cause for cell death in gas-sparged reactors. By increasing the number of nozzles or increasing the nozzle diameter, gas velocity at the sparger can be minimized, which can drastically reduce shear-related cell death (Barbosa et al. 2003).

16.3.6 Strain Selection

Microalgal species are a result of persistent evolution over a billion years, which is reflected in the vast diversity of algal species being explored for fuel production, which includes green algae, diatoms, and cyanobacteria (Tirichine and Bowler 2011). Considering this enormous genetic pool that is larger than animals and land plants, finding the right strain for production becomes a herculean task. Strain selection is crucial in determining the success of a microalgal industrial operation. Most of the industrial processes target specific metabolites of high value such as lipids, hydrocarbons, vitamins, omega-3 fatty acids, pigments, antioxidants, and sterols. The main species used commercially are *Chlorella* and *Spirulina* for health food supplements, *Dunaliella salina* for β-carotene production, *Haematococcus pluvialis* for astaxanthin production, and several other species for aquaculture feed (Borowitzka 1999).

One of the major obstacles in commercial-scale production of microalgal biofuels is to find a single novel species with the following features, namely, high photoconversion efficiency, rapid and stable biomass growth, high lipid content and valuable coproducts, high CO_2-absorbing capacity and limited nutrient requirements, robustness toward shear stresses in PBRs, tolerance to temperature variations resulting from the diurnal cycle and seasons, capability for live extraction ("milking") of valuable secondary metabolites, and self-flocculation ability (Tirichine and Bowler 2011). The secondary metabolite production like astaxanthin depends on the various stages of cell cycle (Kumar et al. 2011). *Spirulina* can tolerate higher stress levels, whereas *Dunaliella* is extremely fragile due to the lack of cell walls. The high biomass productivity of *Scenedesmus* sp. has higher CO_2 mitigation potential due to its efficient and high CO_2 mitigation potential and higher carbon-fixing abilities. The isolated species of *Scenedesmus* and *Chlorella* sp. from limestone mineral hot spring can tolerate up to 40% CO_2 levels without any major changes in growth rates (Westerhoff et al. 2010).

In addition to improved lipid production, other strain-specific improvements are being considered, including growth at high pH, improved nutrient utilization, and traits that lead to more efficient harvesting, such as flocculation (Georgianna and Stephen 2012). The economic impact from these improvements will allow for decrease in operational costs. Flocculation will allow improved harvesting by making it easier to concentrate algae, decreasing the cost of water extraction.

16.3.7 Biomass Density

In large-scale operations, a standardized and stable environment is highly essential for a high biomass yields in microalga-based biofuel production (Tredici and Zittelli 1998). After strain selection, the important parameter for determining the newly identified species is their growth kinetics. The growth rates and maximum biomass densities monitoring should be carried out in real-world growth conditions than laboratory conditions. Replicated conditions can be used to some extent in laboratory setup using suitable media, mixing/agitation and light regimes, programmable temperature and pH control relative to external environments, etc. Numerous engineering strategies have been studied to increase both light and nutrient utilization (Greenwell et al. 2009). For the assessment of true economic viability of microalgal cultivation systems, total physiological aspects of microalgae must be studied under real-world conditions rather than idealized set of lab-scale studies (Slade and Bauen, 2013).

16.3.8 O_2 Accumulation

Higher concentrations of oxygen along with sunlight can lead to photooxidative stress and damage to algal cells (Chisti 2007). A high presence of oxygen around algal cells is undesirable. The combination of intense sunlight and high oxygen concentration results in photooxidative damage to algal cells. Open ponds operate better as they do not build up significant amount of oxygen when compared to closed systems. The oxygen bound in the liquid media is evolved by the water-splitting reactions at the site of PSII and causes photobleaching and reduces the photosynthetic efficiency of algal cells. A degassing system can be employed to flush out the entrapped O_2 (Miron et al. 1999).

16.3.9 Optimal Design

PBR design also plays a significant role in effective photosynthesis resulting in enhanced biomass growth and productivity. In this context, many researchers have developed various types of PBRs (tubular, flat panel, helical, stirred tank, bubble column, etc.) by optimizing parameters, which were limitations in older systems. Each PBR has its own advantages and disadvantages, but according to the requirement, the parameters considered are given priority (Buehner et al. 2009). Operating parameters, hydrodynamic parameters, and geometric parameters are considered for design optimization (Siddiqui et al. 2015). Hydrodynamics play an important role in achieving the goal and can be accomplished by using modeling and simulation technique (Rosch and Posten 2012). This can be considered as a new approach to innovative design that overcomes the problem of scalability. To cultivate microalgae, the wavelength of sunlight is also an important parameter to design the PBR, and the microalgal growth will vary depending on the type of light and color (Suh and Lee 2003). Flat plate photobioreactor was constructed by Zijffers et al. (2008) in which the sunlight is focused on top of the PBR by utilizing dual axis lens so that the light is uniformly distributed in the whole compartment, thus enhancing the overall activity of PBR. The oxygen consumption rate will be enhanced by fivefold when proper precautions with regard to light and other factors are taken into consideration (Lee and Palsson 1994).

16.4 Scale-Up Criteria

When biomass with higher lipid content is produced, it results in enhanced oil extraction efficiency and downstream processing (Rodolfi et al. 2009). The importance of integrating reactor design with downstream processing is one of the key parameters for scale-up studies along with efficient light provision, minimal carbon dioxide losses, and ensuring efficient mixing and removal of photosynthetically generated oxygen (Molina et al. 1999; Richmond 2004). Tubular and flat panel designs have been readily applied in industries for large-scale cultivation process, while bubble column is more widely used in aquaculture (Barbosa et al. 2003). Computational fluid dynamics can be used to optimize the structural configuration of PBRs for scale-up. The inner structure parameter optimization of an airlift FPPBR has resulted in increased reactor volume from 15 to 300 L capacity (Yu et al. 2009). Modularization is an additional option for scale-up, which uses a reasonably small-scale reactor where light limitations, carbon dioxide losses, and inefficient mixing can be reduced or even eliminated (Loubiere et al. 2009). A 120 L multimodule PBR with 18 elementary modules of 6.1 L working volume was commercialized. A modular system has components that can be assembled/disassembled and provides consistency in results. Even if one module has a problem or suffers mechanical hitch, the entire process is not compromised (Loubiere et al. 2009).

16.5 Advances in Photobioreactor Design

The most common types of PBRs can be categorized into tubular and flat panel designs (Rosch and Posten 2012). The reactor material includes glass or Perspex/polypropylene/polyvinylchloride with transparent surface, which will ensure good inflow of the light (Morweiser et al. 2010). However, there are various possibilities to expand the scope of design and process engineering of PBRs by the integration of novel approaches. The areas that are highly appropriate for design innovation are geometric and hydrodynamic parameters, measurable performance criteria, and mode/stability of operation (Rosch and Posten 2012). By considering the drawbacks of light penetration, the concept of internally illuminated photobioreactors (IIPBRs) was developed. The concept of IIPBR design was prevailing since the 1990s and has recently started attracting limelight due to increased research activities on high-value products and recombinant proteins from microalgae (Borowitzka 2013; Rasala and Mayfield 2015). Though lab-scale results showed promising results, lack of scalable PBRs has inspired researchers to build new systems like IIPBRs, which enable high light penetration and improved SVR leading to photoautotrophic or mixotrophic production of these biofuel compounds (Heining and Buchholz 2015). In the context of wastewater treatment, rapid utilization of nutrients from wastewater can be accomplished by microalgae with the help of membrane PBRs (Marbelia et al. 2014). The setback of biofilm formation of suspended algae on the walls of PBRs was transformed into a novel design and scale-up opportunity. Biofilm PBRs are attracting interests as a cultivation platform as they can be harvested with high dry solid content, a decreased energy requirement, and short hydraulic retention times without washout of the microalgae (Zitelli et al. 2013).

Kim et al. (2015) developed semipermeable membrane photobioreactors (SPM-PBRs) to produce biodiesel with low-energy inputs in the ocean. The SPM-PBRs are capable of transferring nutrients dissolved in seawater into the algal broth while containing the cells inside. The parameters like biomass dynamics, variations in nutrient concentration, pH, and salinity were monitored in reservoir model prior to operating in ocean (Kim et al. 2015). Attached-growth PBR systems are gaining more interests. A novel attached-growth PBR was designed to facilitate the harvesting of cyanobacterial biomass and to maximize biomass and lipid production compared to suspended-growth cultivation systems (Economou et al. 2015). Higher biomass was achieved in attached-growth mode of PBR with *Limnothrix* sp. for biodiesel production along with simultaneous treatment of a model wastewater.

16.6 Conclusions

Efficient PBRs are essential production platform for the generation of high-value products and biofuels from microalgae. Future bioreactors will imply innovative solutions in terms of efficient mixing and lighting systems, enhanced mass and gas transfer, efficient degassing systems, temperature control, etc. The energy requirements can be fulfilled with sustainable sources of power like solar/wind and photovoltaic cells to enable zero energy operation. Strain selection and optimization based on desired product, geographical area, and sufficient sunlight availability are quintessential for a successful operation of a commercial-scale PBR. Synthesis of value-added products and biofuels can be integrated into closed-loop biorefinery for carbon and nutrient recovery.

Acknowledgments

We thank the director of CSIR-IICT, Hyderabad, for encouragement and support. We acknowledge funding from Council of Scientific and Industrial Research (CSIR) in the form of 12th five-year plan project BioEn (CSC-0116). MVR acknowledges University Grants Commission (UGC) for providing research fellowship. PC and CNR acknowledge CSIR for providing research fellowship.

References

Barbosa, M.J., Hoogakker, J., and R.H. Wijffels. 2005. Optimization of cultivation parameters in photobioreactors for microalgae cultivation using the A-stat technique. *Biomol Eng* 20:115–123.

Barbosa, M.J., Janssen, M., Ham, N., Tramper, J., and R.H. Wijffels. 2003. Microalgae cultivation in air-lift reactors: Modeling biomass yield and growth rate as a function of mixing frequency. *Biotechnol Bioeng* 82:170–179.

Benson, B.C., Gutierrez-Wing, M.T., and K.A. Rusch. 2007. The development of a mechanistic model to investigate the impacts of the light dynamics on algal productivity in a hydraulically integrated serial turbidostat algal reactor (HISTAR). *J Aquac Eng* 36:198–211.

Borowitzka, M. 1999. A commercial production of microalgae: Ponds, tanks, tubes and fermenters. *J Biotechnol* 70:313–321.

Borowitzka, M. 2013. High-value products from microalgae-their development and commercialisation. *J Appl Phycol* 25:743–756.

Buehner, M.R., Young, P.M., Willson, B. et al. 2009. Microalgae growth modeling and control for a vertical flat panel photobioreactor. Paper presented at *American Control Conference Hyatt Regency Riverfront*, St. Louis, MO.

Camacho, F.G., Gallardo, R.J.J., Sanchez, M.A., Belarbi, E.H., Chisti, Y., and G.E. Molina. 2011. Photobioreactor scale-up for a shear-sensitive dinoflagellate microalga. *Proc Biochem* 46:936–944.

Carvalho, A.P., Meireles, L.A., and F.X. Malcata. 2006. Microalgal reactors: A review of enclosed system designs and performances. *Biotechnol Prog* 22:1490–1506.

Chandra, R., Rohit, M.V., Swamy, Y.V., and S. Venkata Mohan. 2014. Regulatory function of organic carbon supplementation on biodiesel production during growth and nutrient stress phases of mixotrophic microalgae cultivation. *Bioresour Technol* 165:279–287.

Chisti, Y. 1989. *Airlift Bioreactors*. London, U.K.: Elsevier.

Chisti, Y. 2007. Biodiesel from microalgae. *Biotechnol Adv* 25:294–306.

Chiu, S.Y., Kao, C.Y., Chen, C.H., Kuan, T.C., Ong, S.C., and C.S. Lin. 2008. Reduction of CO_2 by a high-density culture of *Chlorella* sp. in a semicontinuous photobioreactor. *Bioresour Technol* 99:3389–3396.

Concas, A., Pisu, M., and G. Cao. 2013. Mathematical modelling of *Chlorella vulgaris* growth in semi-batch photobioreactors fed with pure CO_2. *AIDIC Conference Series*, Milan, Italy, Vol. 11, pp. 121–130.

Cuaresma, M., Janssen, M., Vílchez, C., and R.H. Wijffels. 2011. Horizontal or vertical photobioreactors? How to improve microalgae photosynthetic efficiency. *Bioresour Technol* 102:5129–5137.

Dao, L.H.T. and Beardall, J. 2016. Effects of lead on two green microalgae Chlorella and Scenedesmus: photosystem II activity and heterogeneity. *Algal Res* 16:150–159.

Dasgupta, C.N., Gilbert, J.J., Lindblad, P. et al. 2010. Recent trends on the development of photobiological processes and photobioreactors for the improvement of hydrogen production. *Int J Hydro Energ* 35:10218–10238.

Degen, J., Uebele, A., Retze, A., Schmid-Staiger, U., and W. Trosch. 2001. A novel airlift photobioreactor with baffles for improved light utilization through the flashing light effect. *J Biotechnol* 92:89–94.

Demessie, E.S. and S.U.R. Bekele. 2003. Pillai residence time distribution of fluids in stirred annular photoreactor. *Catal Today* 88:61–72.

Devi, M.P., Swamy, Y.V., and S. Venkata Mohan. 2013. Nutritional mode influences lipid accumulation in microalgae with the function of carbon sequestration and nutrient supplementation. *Bioresour Technol* 142:278–286.

Devi, M.P. and Venkata Mohan, S. 2012. CO_2 supplementation to domestic wastewater enhances microalgae lipid accumulation under mixotrophic microenvironment: effect of sparging period and interval. *Bioresour Technol* 112:116–123.

Doran, P.M. 1995. *Bioprocess Engineering Principles*. London, U.K.: Academic Press Limited.

Durmaz, Y. 2007. Vitamin E (α-tocopherol) production by the marine microalgae *Nannochloropsis oculata* (Eustigmatophyceae) in nitrogen limitation. *Aquaculture* 272:717–722.

Economou, C.N., Moustaka-gouni, M., Kehayias, G., Aggelis, G., and D.V. Vayenas. 2015. Lipid production by the filamentous cyanobacterium. *Ann Microbiol.* 65:1941–1948.

Eriksen, N.T., Poulsen, B.R., and Iversen, J.J.L. 1998. Dual sparging laboratory-scale photobioreactor for continuous production of microalgae. *J Appl Phycol* 10:377–382.

Garcıa-Gonzalez, M., Moreno, J., Manzano, J.C., Florencioa, F.J., and M.G. Guerrero. 2005. Production of *Dunaliella salina* biomass rich in 9-*cis*-carotene and lutein in a closed tubular photobioreactor. *J Biotechnol* 115:81–90.

Georgianna, D.R. and Stephen, P. 2012. Exploiting diversity and synthetic biology for the production of algal biofuels, *Nature* 488:329–335.

Greenwell, H.C., Laurens, L.M., Shields, R.J., Lovitt, R.W., and K.J. Flynn. 2009. Placing microalgae on the biofuels priority list: A review of the technological challenges. *J R Soc Interface* 7:703–726.

Heining, M. and R. Buchholz. 2015. Photobioreactors with internal illumination—A survey and comparison. *Biotechnol J* 10:1–7.

Janssen, M., Tramper, J., Mur, L.R., and R.H. Wijffels. 2003. Enclosed outdoor photobioreactors: Light regime, photosynthetic efficiency, scale-up, and future prospects. *Biotechnol Bioeng* 81:193–210.

Kaewpintong, K., Shotipruk, A., Powtongsook, S., and P. Pavasant. 2007. Photoautotrophic high-density cultivation of vegetative cells of *Haematococcus pluvialis* in airlift bioreactor. *Bioresour Technol* 9:288–295.

Kim, Z., Park, H., and Y. Ryu. 2015. Algal biomass and biodiesel production by utilizing the nutrients dissolved in seawater using semi-permeable membrane photobioreactors. *J Appl Phycol.* 27:1763–1773.

Kumar, K., Dasgupta, C.N., Nayak, B., Lindblad, P., and D. Das. 2011. Development of suitable photobioreactors for CO_2 sequestration addressing global warming using green algae and cyanobacteria. *Bioresour Technol* 102:4945–4953.

Lee, C.G. and B. Palsson. 1994. High-density algal photobioreactors using light-emitting diodes. *Biotechnol Bioeng* 44:1161–1167.

Loubiere, K., Olivo, E., Bougaran, G., Pruvost, J., Robert, R., and Legrand, J. 2009. A new photobioreactor for continuous micro-algal production in hatcheries based on external-loop airlift and swirling flow, *Biotechnol Bioeng* 102:132–147.

Maeda, K., Owada, M., Kimura, N., Omata, K., and I. Karube. 1995. CO_2 fixation from flue gas on coal fired thermal power plant by microalgae. *Energ Convers Manage* 36:717–720.

Marbelia, L., Bilad, M.R., Passaris, I. et al. 2014. Membrane photobioreactors for integrated microalgae cultivation and nutrient remediation of membrane bioreactors effluent. *Bioresour Technol* 163:228–235.

Mata, T.M., Martins, A.A., and N.S. Caetano. 2010. Microalgae for biodiesel production and other applications: A review. *Renew Sust Energ Rev* 14:217–232.

Miron, S., Camacho, G.F., Gomez, C.A., Grima, M.E., and Y. Chisti. 2000. Bubble column and airlift photobioreactors for algal culture. *AIChe J* 46:1872–1887.

Miron, S., Gomez, A.C., Camacho, F.G., Grima, E.M., and Y. Chisti. 1999. Comparative evaluation of compact photobioreactors for large-scale monoculture of microalgae. *J Biotechnol* 70:249–270.

Molina, E., Fernandez, J., Acien, F.G., and Y. Chisti. 2001. Tubular photobioreactor design for algal cultures. *J Biotechnol* 92:113–131.

Molina, E.M., Fernandez, F.G., Camacho, F.G., Rubio, F.C., and Y. Chisti. 2000. Scale-up of tubular photobioreactors. *J Appl Phycol* 12:355–368.

Molina-Grima, E., Garcia Camacho, F., and F.G. Acien Fernandez. 1999. Production of EPA from *Phaeodactylum tricornutum*. In *Chemicals from Microalgae*, ed. Z. Cohen, pp. 57–92. London, U.K.: Taylor & Francis.

Morita, M., Watanabe, Y., Okawa, T., and H. Saiki. 2011. Photosynthetic productivity of conical helical tubular photobioreactors incorporating *Chlorella* sp. under various culture medium flow conditions. *Biotechnol Bioeng* 74:136–144.

Morweiser, M., Kruse, O., Hankamer, B., and C. Posten. 2010. Developments and perspectives of photobioreactors for biofuel production. *Appl Microbiol Biotechnol* 87:1291–1301.

Nigar, K., Fahir, B., and O.U. Kutlu. 2005. Bubble column reactors. *Proc Biochem* 40: 2263–2283.

Norsker, N.H., Barbosa, M.J., Vermuë, M.H., and R.H. Wijffels. 2011. Microalgal production—A close look at the economics. *Biotechnol Adv* 29:24–27.

Olivieri, G., Gargano, I., Andreozzi, R. et al. 2012. Effects of CO_2 and pH on *Stichococcus bacillaris* in laboratory scale photobioreactors. *Chem Eng Trans* 27:127–132.

Perner-Nochta, I. and C. Posten. 2007. Simulations of light intensity variation in photobioreactors. *J Biotechnol* 131:276–285.

Pulz, O. and K. Scheibenbogen. 1998. Photobioreactors: Design and performance with respect to light energy input. *Adv Biochem Eng Biotechnol* 59:124–154.

Rasala, B.A. and S.P. Mayfield. 2015. Photosynthetic biomanufacturing in green algae; production of recombinant proteins for industrial, nutritional, and medical uses. *Photosynth Res* 123:227–239.

Ribeiro, R.L.L., Mariano, A.B., Dilay, E., Souza, J.A., Ordonez, J.C., and J.V.C. Vargas. 2009. The temperature response of compact tubular microalgae photobioreactors. *Therm Eng* 8:50–55.

Richmond, A. 2004. Biological principles of mass cultivation. In *Handbook of Microalgal Culture: Biotechnology and Applied Phycology*, ed. A. Richmond, pp. 125–177. Oxford, U.K.: Wiley-Blackwell.

Rodolfi, L., Zittelli, G.C., Bassi, N. et al. 2009. Microalgae for oil: Strain selection, induction of lipid synthesis and outdoor mass cultivation in a low-cost photobioreactor. *Biotechnol Bioeng* 102:100–112.

Rohit, M.V. and Venkata Mohan, S. 2016. Tropho-metabolic transition during *Chlorella* sp. cultivation on synthesis of biodiesel. *Renew Energy* doi:10.1016/j.renene.2016.03.041.

Rosch and Posten, C. 2012. Technikfolgenabschätzung—Theorie und Praxis 21. Jg., Heft 1, July 2012.

Sevigné Itoiz, E., Fuentes-Grünewald, C., Gasol, C.M. et al. 2012. Energetic balance and environmental impact analysis of marine microalgal biomass production for biodiesel generation in a photobioreactor pilot plant. *Biomass Bioenerg* 39:324.

Siddiqui, S., Rameshaiah, G.N., and G. Kavya. 2015. Development of photobioreactors for improvement of algal biomass production. *Int J Sci Res* 4:220–226.

Sierra, E., Acien, F.G., Fernandez, J.M., Garcia, J.L., Gonzalez, C., and E. Molina. 2008. Characterization of a flat plate photobioreactor for the production of microalgae. *Chem Eng J* 138:136–147.

Singh, R.N. and S. Sharma. 2012. Development of suitable photobioreactor for algae production—A review. *Renew Sust Energ Rev* 16:2347–2353.

Slade, R. and Bauen, A. 2013. Microalgae cultivation for biofuels: Cost, energy balance, environmental impacts and future prospects. *Biomass and Bioenerg* 53:29–38.

Socher, M.L., Loser, C., Schott, C., Bley, T., and Steingroewer, J. (2016). The challenge of scaling up photobioreactors: Modeling and approaches in small scale. *Eng Life Sci*, doi: 10.1002/elsc.201500134.

Subhash, G.V., Rohit, M.V, Devi, M.P., Swamy, Y.V, and Mohan, S.V. (2014). Bioresource Technology Temperature induced stress influence on biodiesel productivity during mixotrophic microalgae cultivation with wastewater. *Bioresour Technol* 169:789–793.

Suh, S. and C.-G. Lee. 2003. Photobioreactor engineering: Design and performance. *Biotechnol Bioprocess Eng* 8:313–321.

Tang, D., Han, W., Li, P., Miao, X., and J. Zhong. 2011. CO_2 biofixation and fatty acid composition of *Scenedesmus obliquus* and *Chlorella pyrenoidosa* in response to different CO_2 levels. *Bioresour Technol* 102:3071–3076.

Tirichine, L. and C. Bowler. 2011. Decoding algal genomes: Tracing back the history of photosynthetic life on earth. *Plant J* 66:45–57.

Torzillo, G., Pushparaj, B., Masojidek, J., and A. Vonshak. 2003. Biological constraints in algal biotechnology. *Biotechnol Bioproc Eng* 8:338–348.

Tredici, M.R. and G.C. Zittelli. 1998. Efficiency of sunlight utilization: Tubular versus flat photobioreactors. *Biotechnol Bioeng* 57:187–197.

Tredici, M.R., Chini Zittelli, G., and Rodolfi, L. 2010. Photo-bioreactors. In: *Encyclopedia of Industrial Biotechnology: Bioprocess, Bioseparation, and Cell Technology*. Vol 6 (eds M.C. Flickinger & S. Anderson), pp. 3821–3838.

Ugwu, C.U., Aoyagi, H., and H. Uchiyama. 2008. Photobioreactors for mass cultivation of algae. *Bioresour Technol* 99:4021–4028.

Ugwu, C.U., Ogbonna, J.C., and H. Tanaka. 2002. Improvement of mass transfer characteristics and productivities of inclined tubular photobioreactors by installation of internal static mixers. *Appl Microbiol Biotechnol* 58:600–607.

Venkata Mohan, S., Rohit, M.V., Chiranjeevi, P., Chandra, R., and B. Navaneeth. 2015. Heterotrophic microalgae cultivation to synergize biodiesel production with waste remediation: Progress and perspectives. *Bioresour Technol* 184:169–178.

Wang, Y., Rischer, H., Eriksen, N.T., and M.G. Wiebe. 2013. Mixotrophic continuous flow cultivation of *Chlorella protothecoides* for lipids. *Bioresour Technol* 144:608–614.

Watanabe, Y., de la Noue, J., and D.O. Hall. 1995. Photosynthetic performance of a helical tubular photobioreactor incorporating the cyanobacterium *Spirulina platensis*. *Biotechnol Bioeng* 47:261–269.

Westerhoff, P., Hu, Q., Esparza-Soto, M., and W. Vermaas. 2010. Growth parameters of microalgae tolerant to high levels of carbon dioxide in batch and continuous-flow photobioreactors. *Environ Technol* 31:523–580.

Xu, L. and X. Xiong. 2009. Microalgal bioreactors: Challenges and opportunities. *Eng Life Sci* 9:178–189.

Yang, Y., and Gao, K. 2003. Effects of CO_2 concentrations on the freshwater microalgae, Chlamydomonas reinhardtii, Chlorella pyrenoidosa and Scenedesmus obliquus (Chlorophyta). *J Appl Phycol* 15:379–389.

Yu, H., Jia, S., and Y. Dai. 2009. Growth characteristics of the cyanobacterium *Nostoc flagelliforme* in photoautotrophic, mixotrophic and heterotrophic cultivation. *J Appl Phycol* 21:127–133.

Zhang, K., Kurano, N., and S. Miyachi. 2002. Optimized aeration by carbon dioxide gas for microalgal production and mass transfer characterization in a vertical flat-plate photobioreactor. *Bioproc Biosyst Eng* 25:97–101.

Zijffers, J.W.F., Janssen, M., Tramper, J., and R.H. Wijffels. 2008. Design process of an area-efficient photobioreactor. *Mar Biotechnol* 10:404–415.

Zittelli, G.C., Biondi, N., Rodolfi, L., and Tredici, M.R. 2013. Photobioreactors for Mass Production of Microalgae. In *Handbook of Microalgal Culture: Applied Phycology and Biotechnology*, Second Edition. ed. A. Richmond and Q. Hu. doi:10.1002/9781118567166.ch13.

17

Sustainable Biorefinery Design
for Algal Biofuel Production

*Didem Özçimen, Benan İnan, Anıl Tevfik Koçer,
and Zubaidai Reyimu*

Contents

Abstract

Algae are the most promising biomass resources for biofuel production when compared with first and second generations of biomass resources. In recent years, algae have been widely studied for biofuel production such as biodiesel, biogas, bioethanol, biohydrogen, and biochar because of their promising advantages. Although the first algal studies were carried out on biodiesel production from algal oil, energy demand results in the production of different algal fuels such as bioethanol and biogas. At the same time, technological developments are continued for algal biodiesel production with high productivity by using novel cultivation and harvesting methods and in situ biodiesel production. In addition to these traditional products, recent trends on utilizing biomass for biohydrogen production are also implemented for algae, which can be used as substrate or microorganism that performs photolysis process for biohydrogen production, even though biohydrogen production from various biomass sources is still a need for more innovations; algal biohydrogen production comes into prominence due to their advantages and physiological properties. In addition to direct usage of algae for biofuel production, the remains of algae, which are obtained after various biofuel production processes, can be utilized for different purposes such as fertilizer and animal feed. It is clear that the implementation of algal biomass for different fuel production is promising, but how to fully use this biomass in industrial level and how to reduce waste and costs need more investigation. In this chapter, the detailed process of liquid, gas, and solid fuels from algae will be mentioned with their pros and cons, and most importantly, algal biorefinery concept through production process will be introduced for the purpose of effective usage of algal biomass as a feedstock for biofuel production.

17.1 Introduction

Today, our energy demand is greatly met by fossil sources; however, it is a fact that their environmental effects, global warming, climate change, and the limited reserved of these fossil sources, make countries to take serial precautions. Instead of using petroleum-based fuels, various actions such as informing society about using biofuels and incentives made for their production have been performed. However, it is seen that evaluating traditional feedstock (first and second generations) for biofuel production resulted in plenty debates since first-generation feedstocks are also used for food and second-generation feedstock requires many processes before biofuel production process. Algae as third-generation feedstock have been used for the last two decades for producing biofuels in order to overcome the cons of first- and second-generation biofuel feedstocks. Interest on algal biotechnology has been starting to grow with realizing the potential of algal biomass for biofuel production. Since algae are environmentally sustainable sources, they lead promising researches and applications. However, considering the operational and capital cost of harvesting and dewatering steps of algal bioprocesses, currently biofuel production from algae alone is not economically feasible. Due to their physiology and morphology, algae are excellent organisms for obtaining various products. In order to utilize this potential, biorefinery concept for algal biomass should be taken into account. A biorefinery is a facility that integrates biomass conversion processes and equipment to produce fuel, power, and value-added chemicals from biomass, a concept analogous to today's petroleum refineries, which produce multiple fuels and products from crude petroleum (Subhadra 2010). Since it is very important to consider biorefinery concept and put it into practice in algal productions, biorefinery processes present new and different products in broad perspectives and also these processes reduce the energy requirement. For instance, biogas production from algal wastes, which are remained after oil extraction using for biodiesel, is one of the most preferred methods since the produced biogas can be used in the same facility in order to provide the necessary energy. With another perspective, these algal wastes can be utilized for bioethanol production and released CO_2 gas can be sent to the part of algal cultivation units. Creating a cycle like this will decrease the cost of main supplement for algal growth. Figure 17.1 shows the production of algal biofuels and other valuable products in biorefinery concept.

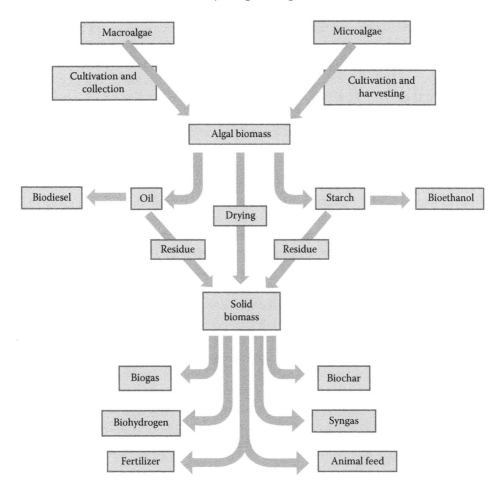

Figure 17.1 Production of algal biofuels and other valuable products with biorefinery concept.

Today, industrial biorefineries like algal biorefineries have been identified as the most promising route for sustainable bioprocesses, and because of the demand of biobased products, they will play a major role in the future by replacing petroleum-based refinery. In this chapter, algal fuels and their production process have been described, and a biorefinery design has been presented. Considering this design, various production options and utilization ways have been offered. With the idea of zero waste, processes have been connected to each other in different ways in order to use algal biomass completely.

17.2 Algal Biorefinery

In this section of study, algal production stages that cover cultivation in photobioreactor, harvesting, oil extraction, and biodiesel production and then according to desired product, bioethanol, biooil, biogas, syngas, biohydrogen, and biochar production processes from microalgal remains are presented by using ChemCad design program. All stages are given in process flow diagram (PFD) in detail (Figure 17.2). As it is seen in a PFD, process starts with microalgal cultivation in photobioreactor (R-101), and then with stream 1–4, cultivated microalgae are filtered and sent for the oil extraction in (6) extractor. In this step, hexane is used as a solvent and microalgal oil obtained from this process is sent for fuel conversion to heat-jacketed transesterification reactor (R-201) (stream 11–13). Transesterification reaction is carried out by reactants with potassium hydroxide as a catalyst and methanol as an alcohol. Once transesterification product is obtained, it is sent for separation and purification steps. While in stream 17–18, impurities are removed from biodiesel, in 17–22, glycerin as by-product and excess methanol are separated and methanol is collected in a tank (TK-201), and from here, it is sent to the mixer (M-201) where methanol and KOH are gathered. The remaining microalgal biomass is pretreated in order to remove moisture inside, and then it is sent to hydrolysis reactor (R-301) for bioethanol production (stream 28–31). After hydrolysis, biomass is used for bioethanol production via *Saccharomyces cerevisiae* yeast in a fermenter (R-302). Product is sent for distillation to obtain 99% pure bioethanol (stream 39–43). Carbon dioxide as by-product is also collected in a tank (TK 301), and it is sent to use for microalgal cultivation in photobioreactor. Unconverted glucose in algal remains is collected in a tank for sugar industry. Water released from this process is sent to water–gas shift reactor (R-502). For other options, dried algal biomass is sent for liquefaction, gasification, pyrolysis, and biogas production. For liquefaction process, oxygen, which is heated with a heat exchanger to the temperature of 345°C for process (stream 92–95), Na_2CO_3 as a catalyst, and biomass are gathered in heat-jacketed reactor (R-701) for biooil production in stream 64–65–69.

In stream 32, it can be seen that the process is divided into two branch streams, which are 57 and 58. Stream 58 is leading to produce biogas by adding TK-bacterial culture (stream 59) and fertilizer (stream 61), respectively, with anaerobic reactor (R-401). Another parallel stream, 57, is just for gathering water, which is produced through different processes.

Stream 33 is a main process, which will lead to gasification and pyrolysis after drying the biomass (stream 62–65). There are four different branches for gasification and pyrolysis, which are from stream 66 to 69 for biomass to produce different products. In stream 66, the biomass is mainly used to produce animal feed. While in stream 67, pyrolysis is implemented to produce biochar from stream 73 to 75 with pyrolysis reactor (R-501). In this reactor, syngas is also produced as by-product (stream 76–77). Syngas production process can also be used to produce other gases like hydrogen (stream 79–82) and carbon dioxide (stream 83–84), respectively, with membrane (SP-501) and (SP-502). Stream 68, on the other hand, leads biomass to gasification reactor (R-601) to implement gasification. This reactor is also contributed for producing syngas (stream 86–88).

The algal biomass either can be led to transesterification process or can be used for producing hydrogen gas through stream 100–104 via biophotolysis reactor (R-801). Oxygen is produced as by-product in stream 105 in this reactor.

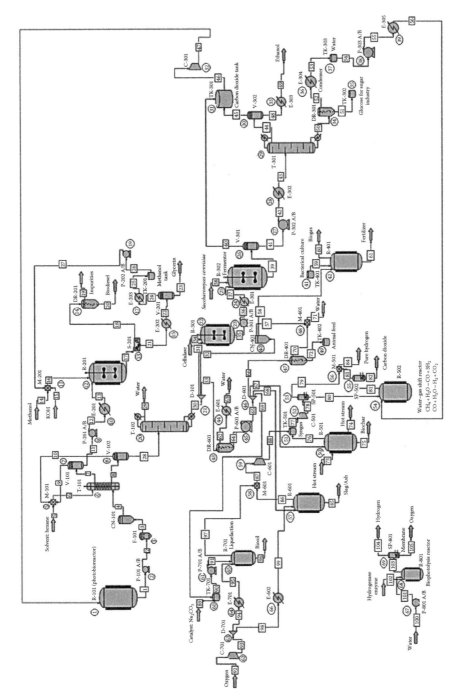

Figure 17.2 Design of the algal biorefinery system with various algal biofuel productions.

17.2.1 Algal Liquid Fuels

Algal liquid fuels are the most investigated and used alternative sources in comparison with other algal fuels. Especially, biodiesel and bioethanol, which are the main renewable fuels used all over the world, form the basis of algal biotechnological researches.

17.2.1.1 Algal Biodiesel

Biodiesel production from algae is the most widely used method for algal feedstock to obtain algal biofuels. Similar to other first- and second-generation biodiesel feedstocks, they can be converted into biodiesel via transesterification process after an extraction process applied to biomass to obtain algal oils. There are a lot of methods for extraction of lipid from algal biomass, but the most common techniques are oil presses, liquid–liquid extraction (solvent extraction), supercritical fluid extraction, and ultrasonic techniques (Özçimen et al. 2012). Oil presses are usually used for first-generation feedstocks such as nuts, corns, and any other vegetable seeds for extracting their oils. Similarly, the same technique can be applied to algal biomass; however, algal biomass must be dried in order to achieve high yields. Pressure breaks the cells and removes algal oil (Popoola and Yangomodou 2006). Although it is simple and can extract 75% of oil, this method is less effective in longer extraction times (Borowitzka 1999). Solvent extraction is a more successful method for extracting algal oils. In this method, organic solvents such as hexane, acetone, and chloroform are added in the algal paste. Since the solubility of oil is higher in organic solvents than water, solvent breaks the cell wall and easily extracts algal oil. Yet, solvent extraction continues with distillation process for separating oil from the solvent, which increases the extraction cost (Serrato 1981). As an alternative, recently, some new methods have been used for extraction such as supercritical extraction and ultrasound methods. Supercritical extraction uses high pressure and temperature for breaking cells. It is widely used and provides efficient extraction time. Another novel technique, the ultrasound method, uses high-intensity ultrasonic waves that create bubbles around the cell. Shock waves are emitted by collapsing bubbles. It breaks cell wall and desired components are released to the solution. This method also improves the extraction rate like supercritical extraction in the same way. It is widely used in laboratory scale, but in commercial scale, there is not enough information about cost and applicability (Borowitzka 1999; Wiltshire et al. 2000).

As it is mentioned in the beginning, algal biodiesel can be produced with transesterification process by providing the most promising solution to the high-viscosity problem in biodiesel product (Balat and Balat 2010). Transesterification reactions require 3 moles of alcohol and 1 mole of glyceride to obtain 3 moles of fatty acid ester and 1 mol of glycerol, stoichiometrically. According to the alcohol used in the process, obtained product is defined as fatty acid methyl ester (FAME) or fatty acid ethyl ester (FAEE). In this process, triglycerides are converted to diglycerides, then the diglycerides are converted to monoglycerides, and the monoglycerides are converted to esters (biodiesel) and glycerol (by-products), respectively (Ahmad et al. 2011). Although there are various types of catalyst used in transesterification, the three common kinds of catalysts used in this process are lipase catalysts, acid catalysts, and alkali catalysts.

Each of these catalysts has advantages and disadvantages (Lin et al. 2011). While acid-catalyzed reactions need higher molar ratio, similar results can be achieved by using low alcohol–oil ratio in alkali-catalyzed reactions. Xu et al. studied acid-catalyzed transesterification of microalgal (heterotrophically cultivated *Chlorella protothecoides*) oil. They used methanol as alcohol and it is reported that 80% of FAME yield have been achieved (Xu et al. 2006). Johnson (2009) also made a study on *Schizochytrium limacinum* microalgal oil with acid-catalyzed transesterification and 82.6% of biodiesel yield was obtained. Velasquez-Orta et al. (2012) studied on biodiesel production from *Chlorella vulgaris* by using alkali-catalyzed transesterification for conversion and they achieved 71% of FAME yield. Ferrentino et al. (2006) conducted alkali-catalyzed transesterification of *Chlorella* sp. oil and have obtained high yield values from their experiments. In another study, Carvalho et al. used alkali-catalyzed transesterification for biodiesel production from algal oil. In their study, *Chlorella emersonii* oil was used and 93% conversion yield was obtained (Carvalho et al. 2011). Unlike acid or alkali-catalyzed transesterification, lipase-catalyzed transesterification is considered to be a green method for the production of biodiesel (Balat and Balat 2010). On the other hand, the high cost of enzyme production is the main obstacle to the commercialization of enzyme-catalyzed processes. Tran et al. used *C. vulgaris ESP-31* oil for producing biodiesel via lipase-catalyzed transesterification. As a result, they reported that they achieved 94.78% of FAME yield (Tran et al. 2012). In addition to traditional transesterification methods, novel methods like supercritical process, microwave-assisted method, and ultrasonic-assisted process have also been started to be used.

Biodiesel production can be easily achieved without using any catalysts by supercritical process. Supercritical fluid is a substance whose temperature and pressure are above the critical point, environmentally friendly, and economic. Although water, carbon dioxide, and alcohol are common supercritical fluids, in biodiesel production, generally supercritical methanol and supercritical ethanol are used. Advantages of this process include easier purification, shortened reaction time, and more effective reaction (Özçimen and Yücel 2011). Patil et al. have used wet algae to perform supercritical biodiesel production with the alcohol–oil ratio of 9:1. The temperature of the reaction occurred at 255°C and 1200 psi and it resulted in 90% of FAME yield (Patil et al. 2009).

Microwave-assisted transesterification process is based on the microwaves that activate differences in small degrees of polar molecules and ions to start chemical reactions. Molecules do not have enough time to relax and heat generation occurs in a short period of time because energy interacts with molecules faster. Due to the fact that reaction is conducted in a short period of time and results in an efficient manner, high-yield product is obtained in a short-term separation. Besides, cost of production and formation of by-product are reduced (Özçimen and Yücel 2011). Today, microwave-assisted transesterification is one of the most used techniques in laboratory studies. Patil et al. studied biodiesel production from dry microalgae by using microwave-assisted process. KOH was used as a catalyst, and microwave condition was set to 800 W. The performance of the study was around 80% (Patil et al. 2011). Another study showed that in methanol–macroalga ratio of 1:15, 2 wt% sodium hydroxide concentration and a reaction time of 3 min were the best condition (Cancela et al. 2012).

Another frequently used technique, the ultrasonic-assisted process, carries out the reaction with effective mixing method. Powerful mixing creates smaller droplets than the conventional mixing and increases the contact areas between the oil phases. In addition to that, it provides the activation energy, which is needed for initiating transesterification reactions (Özçimen and Yücel 2011). Ultrasonic cleaning bath and ultrasonic probe, which are usually operated at a fixed frequency, are mainly used as ultrasonic apparatus in the studies. Frequency is dependent on particular type of transducer, which is 20 kHz for probes and 40 kHz for bath. Ultrasonic-assisted transesterification of oil presents some advantages compared to conventional stirring methods such as reducing reaction time, increasing the chemical reaction speed, and decreasing molar ratio of methanol, increasing yield and conversion (Özçimen and Yücel 2011). Although there are lots of studies about microwave- and ultrasonic-assisted biodiesel production from various vegetables or waste raw materials, there are just a few studies on microalgal biodiesel production via these processes directly. In order to achieve high-yield product and reduce energy input and cost, in situ transesterification process is performed mostly. In the study of Ehimen et al., it is focused on the in situ transesterification of microalgae by ultrasonic technique. The reaction takes 1 h with the use of methanol–oil ratio of 315:1. The result was 0.295 ± 0.003 g biodiesel/g dry *Chlorella*, which shows that the result is higher than mechanically stirred in situ technique (Ehimen et al. 2012).

Koberg et al. investigated *Nannochloropsis* for algal biodiesel production with microwave-assisted method. The higher biodiesel yield was observed as 37.1% with microwave technique. The same conditions for sonication technique resulted in lower yield (Koberg et al. 2011). In another study that *Chlorella* sp. was used as a feedstock for in situ transesterification, 96.2% FAME conversion has been achieved with ultrasonic technique. Tamilarasan and Sahadevan performed an in situ transesterification with *Caulerpa peltata* under the methanol–oil ratio condition of 12:1. It was seen that 98.11% conversion has occurred (Tamilarasan and Sahadevan 2014).

Biodiesel, which is synthesized by algal feedstock, involves no sulfur and performs like petroleum diesel. Microalgal biodiesel reduces emissions of CO, SO_x, and hydrocarbons (Mata et al. 2010). However, emissions of NO_x can be higher according to engine types.

There are several advantages listed in the following for other purposes when microalgae are used for biodiesel production:

1. CO_2 is removed from industrial flue gases for algal photosynthesis. Greenhouse gas emissions of a company are reduced while biodiesel is being produced.
2. Water contaminants such as NH_4^+, NO_3^-, and PO_4^{3-} can be used as nutrients for microalgae.
3. Microalgae can grow under harsh conditions and need less nutrients. Therefore, the growth of microalgae is possible for many areas.
4. There is no need to use freshwater for microalgal cultivation because wastewaters can be used for photosynthesis.
5. Different compounds can be extracted from microalgae depending on the species. These compounds can be used in different industrial processes.

17.2.1.2 Algal Bioethanol

Although early studies of algal biofuels are focused on biodiesel production, there is a potential for carbohydrates in the structure of algae that can be utilized for ethanol production after various hydrolysis processes. Terrestrial plants are composed of structural biopolymers such as hemicelluloses and lignin. On the contrary, algal cells do not have these polymers, which makes it difficult for the bioethanol production (John et al. 2011). Marine algae are capable of producing high amount of carbohydrate every year (Kraan 2012). Besides, energy is not required for distribution and transportation of molecules like starch in microalgae. Not only microalgae but also macroalgae can be used for bioethanol production due to the absence of lignin or having less lignin in their structures, which simplifies the hydrolysis stages (John et al. 2011; Özçimen et al. 2012). Depending on the macroalgal species, they have various amounts of heteropolysaccharides in their structures. Whereas red algae contain carrageenan and agar, brown algae have laminaran and mannitol in their structure (Daroch et al. 2013).

Bioethanol production from algal biomass consists of three stages, which are known as pretreatments, enzymatic hydrolysis, and fermentation. Although there are various pretreatment methods for various biomass sources, chemical pretreatments are the most used techniques for the pretreatment of algal biomass. Chemical pretreatment processes are easy to perform and also good conversion yields can be achieved in a short period of time (Hill et al. 2006). Chemical pretreatments used for algal bioethanol production can be divided into acid pretreatment and alkaline pretreatments. Acid pretreatments can be performed with concentrated or dilute acid. Acid, which is used as a catalyst, makes cellulose more accessible to the enzymes. Using concentrated acid is less preferable than dilute acid because it forms high amount of inhibiting components, causing corrosion in the equipment (Martin et al. 2007). When it comes to the acid pretreatment, sulfuric acid, hydrochloric acid, nitric acid, and phosphoric acid are the most used chemicals. Since dilute acid is applied at moderate temperatures to convert lignocellulosic structures to soluble sugars, pretreatment with dilute sulfuric acid is more preferred to hydrolyze hemicelluloses and facilitate enzymatic hydrolysis (Lee et al. 1999; Balat et al. 2008). Despite the effectiveness of dilute sulfuric acid, it hydrolyzes biomass to hemicelluloses and then hydrolyzes to xylose and other sugars and breaks xylose down to furfural, which is a toxic compound, and it affects ethanol production yield. Miranda et al. have studied the effects of acid pretreatments with the concentrations between 0.05 and 10 N and have achieved the highest sugar yield under the condition of 2 N acid pretreatment. In their experiments, the effects of 2 and 10 N acid pretreatments are observed, and a decrease has been observed in sugar yields as a result (Miranda et al. 2012). Larsson et al. also mentioned that a decrease in ethanol yields has been observed with an increasing acid concentration due to the formation of formic acid, which inhibits the fermentation (Larsson et al. 1999). Unlike acid pretreatments, lignin can be removed without major effects on the other components via alkaline pretreatments. On the other hand, there are limitations on the formation of unrecoverable salts due to the transformation of some alkaline. Hemicelluloses and celluloses are less soluble in this pretreatment compared to acid pretreatment (Carvalheiro et al. 2008). The most used alkaline products in this method are sodium hydroxide, potassium hydroxide, calcium hydroxide, and ammonia (Wan et al. 2011). The effects of alkaline pretreatments can be changed according to biomass. Like dilute acid pretreatments, dilute alkaline pretreatments can also form inhibitory by-products such as furfural, hydroxymethylfurfural, and formic acid (Vincent et al. 2014). Although it was reported that under the conditions of increasing alkaline concentrations, glucose yields increased (Wang et al. 2012), in a study of coastal Bermuda grass, reducing sugar yields decreases with an increasing alkaline concentration (Wang et al. 2010).

Fermentation is the last and the main stage of bioethanol production process. It consists of conversion of glucose to alcohol and carbon dioxide. In this process, 0.51 kg bioethanol and 0.49 kg carbon dioxide are obtained from per kg of glucose in theoretical maximum yield. Yet, because microorganisms use glucose not only for bioethanol production but also for their growth, the actual yield is achieved at less than 100% (Balat and Balat 2009). One of the most effective microorganisms that produces bioethanol is *S. cerevisiae*. In addition to having high bioethanol production yields, *S. cerevisiae* has a resistance to high bioethanol concentration and inhibiting components, which can be formed after acid hydrolyzation of lignocellulosic biomass. Fermentation is an anaerobic process; therefore, oxygen molecules must be removed with nitrogen gas as a swept gas. Bioethanol yields of microorganisms depend on temperature, pH level, alcohol tolerance, osmotic tolerance, resistance for inhibitors, growth rate, and genetic stability (Balat and Balat 2009). Yeast and fungi used for bioethanol production have toleration for 3.5–5.0 pH ranges (Balat et al. 2008). *S. cerevisiae* has high osmotic resistance and can tolerate low pH levels like 4.0. *Zymomonas* stands out with rapid bioethanol production and high productivity compared to other traditional yeasts. But it cannot tolerate the toxic effects of acetic acid and various phenolic compounds in the lignocellulosic hydrolysate (Doran-Peterson et al. 2008). In spite of developing various designed production processes, fermentation processes are carried out with two basic processes as *simultaneous saccharification and fermentation* (SSF) and *separate hydrolysis and fermentation* (SHF) (Hill et al. 2006). In SHF, enzymatic hydrolysis is carried out separately from fermentation. The liquid comes from hydrolysis reactor is first converted to ethanol in a reactor that glucose was fermented in, and then ethanol is distilled and remained unconverted; xylose is converted into ethanol in a second reactor. Performing reactions in optimum conditions is the main advantage of this process. Conversely, cost increases with the usage of different reactors. Also glucose and cellulose units that are obtained after hydrolysis may inhibit enzyme activity and decrease hydrolysis rate (Sarkar et al. 2012). In contrast with SHF, pretreatment and enzymatic hydrolysis steps are carried out with fermentation step in the same reactor in SSF. High bioethanol yields can be achieved with SSF process. In addition to that, due to fermenting glucose and cellulose units in the same media by yeast, inhibition of enzyme activity is very low. Consequently, this process needs low amount of enzyme. Another advantage of this process is reduction of process cost because the reactions are carried out in one reactor. However, as a disadvantage, temperature differences between saccharification and fermentation may cause various effects in the growth of microorganisms (Balat et al. 2008; Harun et al. 2011; Sarkar et al. 2012). There are various studies about bioethanol production from algal biomass (micro and macro) with different pretreatment methods and fermentation processes. However, there is still a need for increasing bioethanol yields with current methods.

17.2.1.3 Algal Biooil

Algal biooil production via liquefaction process is based on heating of biomass at high temperatures between 200°C and 500°C with the presence of catalysts. Liquefaction pressure should be greater than 20 bars and optimum temperature and residence time for liquefaction process should be 340°C and 30 min, respectively. Biooil yield can be changed between 9% to 72% and gas mixture changes between 6% and 20%. Microalgal biooil, which is obtained by liquefaction, has an energy content of 30–39 MJ/kg (Suali and Sarbatly 2012). When comparing liquefaction to other thermochemical processes, it can be seen that liquefaction has an advantage, which is being tolerant to moisture content up to 65%, and it is also the only thermochemical process that does not require drying of biomass.

Microalgal biooil production by liquefaction involves treating algal biomass with a catalyst to obtain biomass slurry. Biomass slurry is pumped into a liquefaction reactor where oxygen enters and partial oxidation takes place. Products of liquefaction process are biooil and synthesis gas (Suali and Sarbatly 2012). Liquefaction products are affected by biomass composition, temperature, pressure, residence time, and catalyst type. Catalyst type has an important effect on the synthesis gas liquefaction. On the other hand, biooil is less sensitive to the types of catalyst. If catalyst is not used for liquefaction, biooil becomes highly viscous, dark brown liquid with a foul odor.

17.2.2 Algal Gas Fuels

17.2.2.1 Biogas Production from Algae

Biogas is a mixture of different types of gas and can be produced through the breakdown of organic matter in the absence of oxygen, which is referred to as anaerobic digestion (AD). It contains around 65%–70% of methane and 30%–35% of CO_2 (Passos et al. 2014). Methane is a gas fuel that produces fewer atmospheric pollutants and CO_2 emission than other fossil fuels. It can be used in various places such as vehicles, industrial applications, and power generation. Methane also has a higher heating value than biodiesel, bioethanol, and biomethanol. Researchers reported the theoretical yields of methane from lipids, proteins, and carbohydrates. Results showed that lipids have the highest value of 1.014 L CH_4/g VS, which is the most effective when compared with proteins and carbohydrates (Dębowski et al. 2013). Algal biomass, especially the microalgae, presents high lipid content, so it is clear that microalgae can be regarded as a valid feedstock for AD applications. As for macroalgae, although its lipid contents are not as high as microalgae, it contains higher amount of carbohydrates that can also be an alternative for AD process.

Algal biomass can be found in natural, eutrophicated, and degraded water bodies. Eutrophication, on the other way, intensive blooming of blue-green algae, which contains an excessive amount of N, P, CO_2, and insufficient amount of dissolved O_2, becomes a threatening problem in coastal seawater environment (Dębowski et al. 2013). One of the algal cultivation techniques, seaweed cultivation, can be used to remove nutrients such as N, P, and C, and at the same time, it can be used as a feedstock for bioenergy conversion process.

Since that algal biomass cannot resist low temperature and climate change, special cultivation techniques should be investigated. These special cultivation methods include open and closed installation of different designs (Montingelli et al. 2015). The open systems include raceway ponds, traditional ground, concrete ponds, and circular ponds with mechanical stirring. As for closed systems, it can be one of the different types of photobioreactors. The advantage of closed system is that it allows constant control of temperature, pH, illumination time, etc. As for microalgae, raceway ponds and photobioreactors are mostly the preferred cultivation systems. Although the raceway ponds are less expensive to build and operate, the productivity of photobioreactors is much higher.

Pretreatment process is a necessity for cell disruption of microalgae as well as for biogas production. The effectiveness of pretreatment highly depends on the algal species and the types of cell wall and the molecular composition of the cell.

As for algal biomass, the pretreatment process can be cheaper than that of the first- and second-generation biomass sources due to containing less or no lignin in their structure. Pretreatment can be divided into four categories: (1) thermal, (2) mechanical, (3) chemical, and (4) biological methods (Passos et al. 2014).

When to use which type of pretreatment method mostly depends on the species. Thermal pretreatments are widely studied and regarded as a feasible method. It was indicated by Ward et al. (2014) that the most efficient thermal pretreatment for microalgal biomass required heating to 100°C for 8 h without an increase in pH using the addition of sodium hydroxide. Mechanical pretreatments, although not related with the types of algal species, require high energy input when comparing with other pretreatment methods. A mechanical pretreatment method indicated that ultrasound treatment shows similar results with the heat treatment, which are conducted at the temperature of 150°C. Besides that, chemical pretreatments are indeed successful, but it may lead to contamination at downstream process. As enzymatic pretreatment seems to have great effect on hydrolysis, it is regarded as promising method with its low energy consumption.

Biogas is produced through digestion process under anaerobic condition. AD process is implemented under the collaboration of different types of microorganisms and with providing continuous energy into the system. Generally, codigestion of algae with waste activated sludge is used to improve the productivity when the lipid composition in the algal cell is not over 40%. According to Montingelli et al. (2015), AD consists of four main steps: hydrolysis, acidogenesis, acetogenesis, and methanogenesis. Organic substrates such as fats/lipids, carbohydrates, and proteins can be degraded into simple sugars, amino acids, and fatty acids through pretreatment process. After pretreatment, with the help of acidogenic bacteria, substrates are turned into volatile organic acids and alcohol. At acetogenesis step, hydrogengenic bacteria that are produced from acetogenic bacteria play an important role in the transformation of acetics, H_2, and CO_2. At methanogenesis process, methanogenic bacteria convert acetics, H_2, and CO_2 into methane. Although the whole process is taken under specific condition, the microbial community is sensitive to variations in systems. Thus, the AD process must be managed and controlled properly or the process would become unstable and the production yield would be reduced. There are several factors that affect AD process in biogas production, including C/N ratio, organic loading rate (OLR), hydraulic retention time (HRT), and pH.

17.2.2.1.1 Important Factors That Affect the Anaerobic Digestion Process

17.2.2.1.1.1 C/N Ratio
According to several articles, an optimal C/N ratio for AD process is 20–30. If the ratio is too low, it may lead to the production of ammonia and volatile fatty acids (VFA) that produce toxicity. Generally, the C/N ratio in algal biomass is about 10:1, which is apparently too low for the implementation of AD and will lead to toxic effect. To deal with the problem, additional organic carbon–rich materials should be added to the substrate to enhance digestion performance.

17.2.2.1.1.2 Organic Loading Rate
The OLR is defined as the amount of volatile solids (VS) or chemical oxygen demand components fed per day per unit digester volume. Montingelli et al. (2015) show that the higher OLR can help to reduce the digester's size and its capital cost. Methane yield is highest when the process is operating at low OLR and high HRT (Montingelli et al. 2015). So the suitable OLR and HRT must be chosen depending on the nature and composition of algal substrate.

17.2.2.1.1.3 Hydraulic Retention Time
HRT is defined as the time required to complete the degradation of organic matter (Mao et al. 2015). The retention time for AD process must be optimum as the short HRT may lead to VFA accumulation, whereas longer HRT will result in insufficient utilization of digester components.

17.2.2.1.1.4 Ammonia
Ammonia inhibition occurs during AD because of the effect of some parameters such as ammonia concentrations, temperature, pH, and the presence of other ions and acclimation. It is indicated that ammonia toxicity affects the methane production in two ways: (1) the ammonium ion may inhibit the methane-synthesizing enzyme directly and (2) the hydrophobic ammonia nitrogen molecule may diffuse passively into the cell, causing proton imbalance and/or potassium deficiency (Ward et al. 2014). Ammonia inhibition of macroalgae during digestion may not occur due to the high dilution factor used in the digester. But when nitrogen content is between 3.5% and 8.7%, it may lead to methanogenesis inhibition (Ward et al. 2014). As for microalgae, ammonia toxicity may occur because microalgal cells contain high protein (50%–60%).

17.2.2.1.1.5 Volatile Fatty Acids
VFA produced in the process could be used to predict the digester's performance. Researches show that low concentrations of VFA will not lead to inhibitory phenomena or toxicity. Some studies also refer that VFA accumulation could also be affected by the OLR, which can increase the VFA concentrations with the rising of OLR, and as a result, it reduces the production yield. Due to the articles, this problem could be solved with longer HRT use.

The green microalga *Chlamydomonas reinhardtii* was allowed to produce H_2 under sulfate-deprived conditions as established by Melis et al. (2000). After the microalga had stopped producing H_2, the microalgal cells were harvested and used as a substrate in a subsequent anaerobic fermentation process. However, the targeted fermentative product was methane, which is an alternative biogas to H_2. Through this integrated system, the methane yield increased to 123% compared to the control value (Lam and Lee 2013).

17.2.2.2 Biohydrogen Production from Algae

Biohydrogen is one of the most promising alternative energy resources, which can be produced by electrolysis, pyrolysis, gasification, and steam-reforming methods (Lam and Lee 2013). But these processes are energy consuming and not economical. So finding a renewable and nonpolluting production method is urgent. Current developments indicate that biological production of hydrogen can be the most promising and sustainable way to fulfill the requirement. There are lots of microorganisms that can use solar energy for photosynthesis and producing H_2. As for green algae, according to its advantages and benefits listed in the previous part, it is the optimal biomass for hydrogen production regarding the studies. In this chapter, the whole hydrogen production from green alga will be introduced.

Algal harvesting is the first step for hydrogen production, and it can be implemented by the two different methods called the open system and closed system, which was introduced in the earlier parts. To establish hydrogen production easily and effectively, pretreatment of algae is recommended as for the different cell wall structures of algae than other hydrogen productive microorganisms. Algae, especially the green microalgae, can use carbon dioxide and sunlight as a carbon source and energy source to produce hydrogen. There are two methods of H_2 production via microalgae. One is direct biophotolysis and the other is indirect biophotolysis. The key enzymes for producing hydrogen are hydrogenase and nitrogenase that each enzyme works in different conditions and stages. Hydrogen production can be affected by different factors such as oxygen, temperature, and pH. To gain maximum yield of H_2, these factors must be controlled in a proper way and the energy consumption must be reduced to minimum.

Direct photolysis is the simplest way to produce H_2 by splitting water into proton (H^+) and oxygen (O_2) in the presence of light. According to Melis et al. (2000), photosynthetic microalgae can absorb the light and enhance the oxidation of H_2O molecules by photosystem II (PSII, or water–plastoquinone oxidoreductase (Show and Lee 2013)), and at the same time, electrons are derived from water. Derived electrons then transferred through thylakoid membrane electron transport chain and then via photosystem I (ferredoxin oxidoreductase) and ferredoxin (Fd) to the hydrocarbon cluster of [Fe]-hydrogenase (Show and Lee 2013). [Fe]-hydrogenase is an essential enzyme that can reduce the proton (H^+) to H_2 molecules. The pathway of direct photolysis can be presented as

$$H_2O \rightarrow H^+ + O_2$$

$$2H^+ + 2Fd^- \rightarrow H_2 + 2Fd$$

Ghirardi et al. (2009) regarded that O_2 is a powerful inhibitor of the [Fe]-hydrogenase enzymatic reaction and is also a suppressor of [Fe]-hydrogenase gene expression. As a result, direct photolysis might be suppressed by produced oxygen and proton (H^+) that produced mixes back to water again. There are several methods to solve this problem: (1) The reaction must be implemented in anaerobic condition so that hydrogenase can be activated during the reaction. The anaerobic condition can be achieved by sending inert gas into the microalgal cultivation system to maintain the activity of [Fe]-hydrogenase to produce H_2 for a long period of time. But the extra cost of the method is high, and it requires extra photobioreactor volume to reach the ideal inert gas quantity that is infeasible to commercialize. (2) Another method is called sulfur deprivation, which will be introduced in *sulfur deprivation process*.

The advantage of direct photolysis is that the energy driver and substrates of the direct photolysis are readily available from the sunlight and water. But the challenge point is how to eliminate the oxygen inhibition to [Fe]-hydrogenase through the reaction, and additionally, the direct photolysis requires significant algal cultivation area to collect sufficient light (Show and Lee 2013).

In indirect biophotolysis, the problem of oxygen suppression can be eliminated by separating O_2 and H_2 by sulfur deprivation. Indirect photolysis consists of two stages. In the first stage, microalgal cells accumulate organic compounds such as carbohydrates through photosynthesis and by-product oxygen is produced. In the second stage, the collected organic compounds are degraded under anaerobic condition and produce hydrogen. The anaerobic condition can be achieved via sulfur deprivation process. The second stage of indirect photolysis has two processes: photofermentation and dark fermentation. In photofermentation, after organic compounds are achieved, microalgae use these compounds as substrates under light condition. As for dark fermentation, the process is undertaken without light and anaerobic condition. Here are the specific pathways of the two fermentation methods.

17.2.2.2.1 Photofermentation The feature of photofermentation is that organic compounds are degraded into small molecules in the presence of light. In this process, organic substrates are converted into hydrogen and carbon dioxide through ferredoxin and nitrogenase (Rashid et al. 2013). The reaction of the process is as follows:

Organic compounds (CH_2O) + light + water $\rightarrow$ Ferredoxin $\rightarrow$ Nitrogenase $\rightarrow H_2$

So, the process can be designed to produce hydrogen in the presence of light by degrading organic wastes and pollutions. Uyar et al. (2007) regarded in their present study that the photofermentation can be influenced by these several factors such as light intensity, wavelength, and illumination pattern on growth and hydrogen production of microorganisms and for the purpose of improving the photobioreactor activity during photofermentation.

17.2.2.2.2 Dark Fermentation Green algae can also produce hydrogen via dark or heterotropic fermentation without providing any light source. But the process efficiency is very low due to the fact that in dark fermentation, methane can be produced with hydrogen because of anaerobic condition. According to Show and Lee (2013), hydrogen is produced in the first stage of methane production called acidogenesis, which is also used as an electron donor by methanogens at the second-stage methanogenesis. So harvesting hydrogen gas at the first stage of dark fermentation is important for hydrogen production through inhibiting the other process of the fermentation. But for biorefinery concept, it is suitable to harvest both methane and hydrogen in different processes of the fermentation, as it will increase the energy efficiency and reduce the cost through eliminating the process of methane inhibition.

17.2.2.2.3 Key Enzymes in Algal Hydrogen Production There are two important enzymes that affect hydrogen production significantly, hydrogenase and nitrogenase. Each enzyme is activated in different conditions. For example, nitrogenase activates in the presence of light and the absence of nitrogen, while hydrogenase activates at high light intensity and pH.

17.2.2.2.3.1 Hydrogenase There are two different types of hydrogenase, which are uptake hydrogenase and reversible hydrogenase. (1) Uptake hydrogenase catalyzes the unidirectional uptake of hydrogen. That is to say, uptake hydrogenase consumes hydrogen. So it is considered not suitable for hydrogen production. (2) Reversible hydrogenase, on the other hand, is a nuclear-encoded, monomeric enzyme, which is located in the chloroplast of microalgae and cytoplasm of cyanobacteria. It is called reversible because it catalyzes both uptake and production of hydrogen. To increase the hydrogen yield, anaerobic adaptation can be used to inhibit the reversion of produced hydrogen into the water. According to Winkler et al. (2002), the reversible hydrogenases of *Ch. reinhardtii* and *Scenedesmus obliquus* are strongly and rapidly induced. *Ch. reinhardtii* shows the maximum rate. It showed that reversible hydrogenase activity can be increased rapidly during anaerobic condition (Winkler et al. 2002). So it is desirable for hydrogen production through anaerobic operations.

17.2.2.2.3.2 Nitrogenase Nitrogenase is found in prokaryotes (cyanobacteria), and it converts N_2 to ammonia and hydrogen, with the requirements of ATP, in the process called nitrogen fixation. The conversion of hydrogenase in the presence of light by nitrogenase is low, because prokaryotic microorganisms such as cyanobacteria, which contain nitrogenase, can also uptake hydrogenase that reduces the hydrogen yield by consuming it as substrate.

17.2.2.2.4 Immobilization To improve the stability of enzymes and increase the light utilization efficiency, immobilization of cells is recommended in some studies. Rashid et al. (2013) have used two-staged hydrogen production for immobilization. In the first stage, immobilized cells are suspended in growth medium under the light for photosynthesis. In the second stage, the cells are subjected to anaerobic condition for hydrogen production. Investigations showed that immobilized cells can produce hydrogen for three cycles (Rashid et al. 2013).

17.2.2.2.5 Sulfur Deprivation   Sulfur deprivation is a pathway that transfers the cells from sulfur-rich medium to sulfur-deprived medium. According to Show and Lee (2013), the photochemical activity of PSII declines, and the absolute activity of photosynthesis becomes less than that of respiration under sulfur deprivation. As a result, the rates of photosynthetic oxygen evolution drop below those of oxygen consumption by respiration. And it causes net consumption of oxygen by the cells, which can help to achieve anaerobic condition in medium (Show and Lee 2013). But the sulfur-deprived conditions must be altered into nutrient-rich condition due to the fact that substrates are utilized for catabolism, and as a result, the protein and starch produced by algae will be consumed as a substrate after 30 h of sulfur deprivation (Lam and Lee 2013). So it is important to recultivate the biomass in a nutrient-rich medium to replace the needed metabolites before implementing sulfur deprivation process again.

17.2.2.2.6 Integrated System for Effective Production of H_2   The bottlenecks for biohydrogen production process include low hydrogen yields and high energy cost. To solve the first problem, photofermentation can be used; however, the energy efficiency is low at this method. To reduce the high energy cost, dark fermentation can be implemented, but the production yield is low. So to better solve the problem, researchers proposed a two-stage process by the integration of photofermentation and dark fermentation (Rashid et al. 2013).

17.2.2.3 Syngas Production from Algae

As biogas production of algae through anaerobic process and biohydrogen production through photosynthesis are introduced, there are also other ways of producing biofuel through thermochemical conversion processes like direct combustion, pyrolysis, gasification, and liquefaction. It can be either directly burnt to produce energy or can be used as a feedstock in the production of chemicals such as methanol (Cherad et al. 2014). Therefore, it is urgent to do more research on the thermochemical process of algal biomass for different gas fuel productions.

Among the gases produced from thermochemical processes, syngas appeals to be the most outstanding and perspective one. Syngas consists of gases like methane, hydrogen, carbon dioxide, and nitrogen (Trivedi et al. 2015). Another study indicates that the most valuable content of syngas is CO and hydrogen (Ferrera-Lorenzo et al. 2014).

According to the relative studies, there are several methods for syngas production through algal biomass: pyrolysis and gasification. In the following section, these two important syngas production methods will be introduced.

17.2.2.3.1 Pyrolysis   Pyrolysis is one of the methods used for thermochemical conversion, and the products obtained through conversion are solid, liquid, and gaseous products. It is regarded as an efficient method for waste treatment, as it is able to process a wide variety of residues such as municipal solid waste, plastic waste, agricultural residues, and sludges (Ferrera-Lorenzo et al. 2014). There are two types of pyrolysis according to Ferrera-Lorenzo et al. (2014): conventional and microwave pyrolysis. The main difference of these two methods is heating pattern. According to Ferrera-Lorenzo et al. (2014), the study group has worked on two different pyrolyses of macroalgal biomass *Gelidium*. They compared the gas production yield of conventional pyrolysis and microwave pyrolysis. The results showed that the microwave pyrolysis produces a gas with lower CO_2, CH_4, C_2H_6, and C_2H_4 contents than conventional pyrolysis as well as a higher content in H_2 and CO. This finding shows that microwave pyrolysis contributes to an increase in the production of syngas (H_2 + CO) (Ferrera-Lorenzo et al. 2014).

Another study has focused on the microwave and conventional pyrolysis of microalgae. Quartz reactor with a single-mode microwave oven was used in the experiments. Because of the poor microwave absorption of biomass, it is recommended to mix with suitable microwave absorber to achieve high temperatures required for pyrolysis. In this study, the biochar that is obtained at the temperature of 800°C through pyrolysis is used as an absorber. As a conclusion, it showed that the potentials of microalgae and their residues on syngas production via microwave-induced pyrolysis are huge, even at low temperature (Beneroso et al. 2013). There is also another type of pyrolysis (Hu et al. 2014), which is catalytic pyrolysis that produces syngas from algae.

Researchers are working deeply on different pyrolysis methods and found that catalytic pyrolysis is also efficient for syngas production. The study tries to find optimal catalyst for pyrolysis by using *C. vulgaris*. There are several catalysts such as activated carbon, $ZnCl_2$, SiC, and MgO. After harvesting and pretreatment process, different catalysts with biomass are put into quartz tube reactor and heated from room temperature to 800°C. The results showed that both catalysts are efficient for syngas production, but when it comes to the consideration of industrial production, activated carbon is suitable due to being cheap and easy to obtain (Hu et al. 2014). Besides that, the study also focused on the effect of different activated carbons on syngas production. The results showed that due to the strong adsorption capacity of activated carbon, more than 5% of activated carbon adsorbs the H_2 and CO that is produced. Contrarily, the effects of less than 5% activated carbon on H_2 and CO are stronger and syngas emission is higher than adsorption. In conclusion, the optimal content of activated carbon is found as 3% (Hu et al. 2014).

17.2.2.3.2 Gasification The other method for syngas production is gasification. Because of the high moisture and ash content of macroalgae and gasification also needs relatively dry feedstock, it is recommended to implement gasification with water, which is called hydrothermal process. The study (Cherad et al. 2014) introduced three different types of hydrothermal process, which are hydrothermal carbonization (HTC), hydrothermal liquefaction (HTL), and hydrothermal gasification (HTG).

HTC is implemented under the temperature less than 200°C and the obtained product mainly is solid char. Similarly, HTL is carried out between 200°C and 375°C, and the obtained product is biooil. As for HTG, the process is performed with the temperature higher than 375°C and produces syngas (Cherad et al. 2014).

HTG of macroalgae has a number of advantages. The process is not only tolerant to the ash content but also added alkali salts can be used as a catalyst, which can increase the production yield of syngas. So the definition of supercritical water gasification (SCWG) is the process that utilizes supercritical water as gasifying medium (water above 374°C and 22 MPa) and provides high hydrogen and carbon dioxide yields and low char and tar formation in almost complete gasification of the macroalgal feedstock (Cherad et al. 2014). But the drawback of the process is that it is still in its early stage and to aggrandize the process needs more time to study deeply. Besides that, from algal biorefinery aspects, the produced by-product CO_2 can be collected and used as a feedstock for microalgal cultivation as well as the process water from SCWG, which is rich in nutrients. According to Cherad et al. (2014), the study investigated the factors that could affect the SCWG process of *Laminaria hyperborean*, such as effects of different catalysts, catalyst loading, feed concentration, hold time, and temperature, and results showed that the hydrogen yield is optimal at 30 min of reaction time and 6.66% of feed concentration with 20% of Ru/Al_2O_3 as catalyst at temperature of 550°C.

17.2.3 Algal Solid Fuels and Products

Algal biomass remains after extracting oil and starch or dried algal biomass can be used as animal feed and organic fertilizer because of its high N/P ratio, or it can be simply burned for energy cogeneration (pellet or briquette). Further, it can be used in the production of biochar and activated carbon (Wang et al. 2008; Gouveia 2011). In many studies, activated carbon, produced from the algal biomass, has been reported to be very advantageous due to the porous structure (Kirtania et al. 2014).

17.2.3.1 Algal Biochar
Biochar is a solid product that has high carbon content and is obtained via thermal decomposition of organic or inorganic material subjected to low temperature in the absence of oxygen (Lehmann and Joseph 2009). Various organic and inorganic materials such as woody material, algae, grasses, corn stover, straw, peanut shells, sorghum, olive pits, bark, and sewage wastes can be used for biochar production. Scientific researches on biochar are especially related with wood materials due to its consistency and relatively low ash content. The sector of wood products such as paper and furnitures is the primary source of biochar's raw materials (Winsley 2007).

Biochar can be produced by pyrolysis or gasification systems. Pyrolysis systems produce biochar by degrading biomass thermally in the absence of oxygen. Generally, there are two types of pyrolysis systems in use today: fast pyrolysis and slow pyrolysis. Fast pyrolysis tends to produce more biooils, while more biochar is obtained with slow pyrolysis. Gasification systems produce smaller quantities of biochar in a directly heated reaction vessel. The main goal in gasification system is to produce biogas; biochar is obtained as a by-product in this process (Özçimen 2013).

Biochar is generally composed of aromatic structures and it is similar to the structure of graphite, but it shows difference with uneven settlement of the aromatic ring

(Lehmann and Joseph 2009). Carbon in the feedstock is converted to stable or unstable biochar structures as the final product. Because of having the aromatic ring structure, the stable biochar can remain in the soil for hundreds of years, whereas unstable compounds remain in the soil for weeks or years, depending on climate changes (Jirka and Tomlinson 2014). In addition to using biochar as an energy source, charcoal has been used for centuries for improving soils. Charcoal can be found in many places of the world due to forest fires and historical soil improvement practices. Biochar usage to improve soil as traditional farming practices is applied in the Amazon, Japan, China, Africa, North America, and Europe (Jirka and Tomlinson 2014).

Factors affecting the efficiency and quality of biochar can be analyzed under two titles as characteristic of biomass and process parameters. Process parameters consist of temperature, heating rate, and properties of gas atmosphere (inert gas, reactive gas, the pressure); biomass characteristics consist of types of biomass, mineral content, particle size, and moisture content (Özçimen 2007; Demirbaş 2011).

In literature, there are many studies about the relationship between biochar and process temperatures. These studies showed that increase in temperature affects biochar yields and reactivity of biochar negatively. In these studies, the maximum yields of microalgal or macroalgal biochar were obtained at the temperatures of 350°C–550°C (Peng et al. 2001; Ross et al. 2008; Grierson et al. 2009; Bird et al. 2011; Maddi et al. 2011; Rizzo et al. 2013; Watanabe et al. 2014). Kirtania et al. (2014) indicated that the reactivity of algal biochar that is produced at a temperature of 800°C–950°C is similar to the reactivity of woody biochar that is produced under similar conditions, and Yanik et al. (2013) calculated biochar yields that are produced at a temperature of 500°C as 29%–36%. However, these researchers found that the reactivity of biochar decreases with an increase in temperature. Rizzo et al. (2013) produced biochar using microalgae (*Nannochloropsis* and *Chlorella* sp.) by pyrolysis process. They calculated biochar yields as 29% at a temperature of 450°C. Garciano et al. (2012) made similar experiments using *Botryococcus braunii*, one of the microalgal species, and reached similar conclusions. Different species of microalgae such as *Ch. reinhardtii*, *Synechocystis* sp., and *Dunaliella salina* were used in similar studies by different researchers (Heilmann et al. 2010, 2011; Agrawal and Chakraborty 2013).

The heating rate is important for the structure of biochar. For slow heating rates, volatile components are released from the solid structure and no important change happens in the particle structure. However, for fast heating rates, the originality of cellular structure has disappeared (Demirbaş 2011; Kirtania et al. 2014). Kirtania et al. (2014) and Ross et al. (2008) found similar results in their studies. Kirtania et al. accentuated that the reactivity of biochar decreases with increasing heating rate. Ross et al. pointed out that interaction between biochar yield and heating rate is similar to the interaction between the reactivity of biochar and heating rate.

One of the most important parameters affecting the yields and quality of biochar is particle size of the feedstock. As the particle size of biomass samples used in carbonization process is increased, solid yield increases, because decaying the large diameter of the particles completely is difficult and takes time as thermal degradation leads from the surface to the center of particles. High yield of solid product obtained if there is particle size reduction. The reason for this, is the reduction of mass transfer resistance and impacts as a result of heating of small particles is more uniform. Thus, mass loss of small particles is greater than big particles because of the heat effect in the process of carbonization carried out in the same conditions and the solid product yield is lower. In contrast, temperature profile that is formed in the coarse particles by heat transfer resistance raises the efficiency of the solid product while decreasing the yield of volatile substances (Knight 1976; White and Plaskett 1981; Goldstein 1983). For these reasons, to increase the yield of the solid product in carbonization process, big particles may be preferred instead of smaller ones (Özçimen 2007). Based on this information, increasing of the solid yield value can be observed when macroalgal biomass samples are preferred instead of microalgae.

Mineral content, moisture content, and alkali metal content are also important parameters for biochar production from algal biomass. Mineral content of algae can change according to their species, cultivation sites, physical effects and chemical effects like wave exposure and process type, and mineralization method (Ruperez 2002; Ross et al. 2008). Due to high moisture content, biochar production from algal biomass is usually not considered. However, for the biochar production, the availability of alkali metals in algal biomass improves the biochar yield. Therefore, algae should not be disregarded as biochar feedstock and should be considered for utilization (Ross et al. 2009; Haykiri-Acma et al. 2013).

17.2.3.2 Algal Biofertilizer

Fertilizers are vital materials for the growth of plants because of the valuable ingredients. Due to being environmentally friendly and cost-effective, biofertilizers are preferred rather than chemical fertilizers recently. Biofertilizers contain microorganisms that can fix nitrogen, solubilize phosphate, and promote plant growth. Algae can be helpful in agriculture with these functions and have high N/P ratio in structure. For this reason, the leftover algal biomass from the production of liquid biofuels can be utilized for the growth of plants as fertilizer. Macro- and microalgae especially blue-green algae are used as nutritional supplements and as biofertilizers to improve the growth of plant and production yield (Guedes et al. 2014). Algal biomass includes regulatory macro- and micronutrients like cytokinins, auxins, gibberellins, and betaines, which can increase plant growth by inducing (Valente et al. 2006). Although there are some fertilizers that are produced from marine algae, which are used in agriculture, researches for utilizing algae as biofertilizer continue to develop this application. Thorsen et al. (2010) have investigated the effects of utilizing *Laminaria digitata*, one of the brown seaweed species, on plant growth, and it is reported that *Laminaria digitata* increases seed germination and improves rooting in terrestrial plants. Macroalgae and microalgae are found beneficial in cultivating plants and improving the productivity with a number of substances like vitamins, amino acids, and antibacterial and antifungal matter, which exist in composition of algal biomass (de Mule et al. 1999). Schwartz and Krienitz (2005) also implied that different indirect growth promotion effects may have various influences such as enhancing the water-holding capacity of soils or substrates and producing antifungal and antibacterial compounds.

17.2.3.3 *Algae as Animal Feed*

The content of algae comprises important nutritional elements, which can meet the requirement for animal feed. Therefore, the evaluation of algal biomass for animal feed is an important and viable option (Zubia et al. 2008). Protein content of plants is low for aquatic animals and demand of protein source for aquatic feed cannot be met with these materials. To address this issue, algal biomass has been used as protein source traditionally. Besides protein content, algae contain other important nutrients, which can participate in food chain later, such as vitamins, essential polyunsaturated fatty acids, and pigments such as carotenoid, and so fishes gain resistance to bacterial contamination (Guedes and Malcata 2012). Microalgal fatty acids, which have longer than 10 carbon atoms, can stimulate the lysis of bacterial protoplasts and such bacterial infections can be avoided (Guedes et al. 2011). A list of valuable nutrients from microalgal biomass is given in Figure 17.3 (Khan and Rashmi 2010).

Studies that have been performed to evaluate nutritional and toxicological values of algal biomass showed that it is a convenient feed source for animals (Gendy and El-Temtary 2013). Nowadays, most species of algae are used as livestock feed. The most used algal species are *Spirulina*, *Chlorella*, and *Scenedesmus* species. Positive results on health have been obtained from the usage of algal biomass to feed some animals such as cows, horses, pigs, poultry, cats, and dogs (Spolaore et al. 2006). Besides that, it has beneficial effects in livestock raising, for example; in a research, *Laminaria digitata* is used as animal feed for pigs, and it resulted with an increase of 10% in the weight of pigs on a daily basis (Harun et al. 2010). However, in some cases like poultry feed, the utilization of algal biomass in higher concentrations can cause reduction in growth rate and color, and flavor changes can be seen in chicken eggs (Hudek et al. 2014).

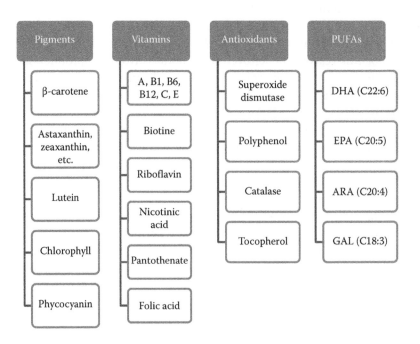

Figure 17.3 Valuable nutrients from microalgal biomass.

17.3 Conclusions

Nowadays, it can be said that algae are the best biomass sources for a future carbon-neutral biofuel feedstock because of their several advantages. Nevertheless, in spite of having been studied for more than 50 years now, there are still only just a few corporations that are cultivating algae for biofuel production on a large or commercial scale. The economics of producing algae for biofuel production are not viable and do not currently justify the intensity of the numerous processing stages and current practicalities. In order to carry out sustainable and economical productions, algal biorefinery studies have increased rapidly in the last decade due to reduced production cost and obtained various products with maximum efficiency.

In this chapter, a design of an algal biorefinery system with various choices according to desired algal product by using ChemCad program, which shows a simple PFD to produce biofuel from microalgae, is presented. The design of this system starts with biodiesel production, and depending on the microalgal species and desired product, processes involve bioethanol, biohydrogen, biogas, biooil, biochar, and syngas productions basically. These products are mentioned in detail with their different production methods. These methods are also explained with hitting the high spots. Advantages and disadvantages of these methods are mainly discussed. Although biorefinery system can be altered according to algal species and desired product, this design, which is mostly based on algal fuel production, will help to carry out future practices. In addition to this, researches should also be carried on with genetic improvements and other alternatives such as microbial fuel cell concept and integrations with utilizing wastewater or flue gas should be considered.

References

Agrawal, A. and S. Chakraborty. 2013. A kinetic study of pyrolysis and combustion of microalgae *Chlorella vulgaris* using thermo-gravimetric analysis. *Bioresour Technol* 128:72–80.

Ahmad, A.L., Yasin, N.H., and J.K. Derek. 2011. Microalgae as a sustainable energy source for biodiesel production: A review. *Renew Sust Energ Rev* 15:584–593.

Balat, M. and H. Balat. 2009. Recent trends in global production and utilization of bioethanol fuel. *Appl Energ* 86:2273–2282.

Balat, M. and H. Balat. 2010. Progress in biodiesel processing. *Appl Energ* 87:1815–1835.

Balat, M., Balat, H., and C. Öz. 2008. Progress in bioethanol processing. *Prog Energ Combust Sci* 3:551–573.

Beneroso, D., Bermúdez, J.M., Arenillas, A., and J.A. Menéndez. 2013. Microwave pyrolysis of microalgae for high syngas production. *Bioresour Technol* 144:240–246.

Bird, M.I., Wurster, C.M., de Paula Silva, P.H. et al. 2011. Algal biochar—Production and properties. *Bioresour Technol* 102:1886–1891.

Borowitzka, M.A. 1999. Commercial production of microalgae: Ponds, tanks, and fermenters. *Prog Ind Microbiol* 35:313–321.

Cancela, A., Maceiras, R., Urrejola, S., and A. Sanchez. 2012. Microwave-assisted transesterification of macroalgae. *Energies* 5:862–871.

Carvalheiro, F., Duarte, L.C., and F.M. Gírio. 2008. Hemicellulose biorefineries: A review on biomass pretreatments. *J Sci Ind Res* 67:849–864.

Carvalho, J., Ribiero, A., Castro, J., Vilarinho, C., and F. Castro. 2011. Biodiesel production by microalgae and macroalgae from north littoral Portuguese coast. Paper presented at *First International Conference of Wastes: Solutions, Treatments and Opportunities*. Guimarães, Portugal.

Cherad, R., Onwudili, J.A., Williams, P.T., and A.B. Ross. 2014. A parametric study on supercritical water gasification of *Laminaria hyperborea*: A carbohydrate-rich macroalga. *Bioresour Technol* 169:573–580.

Daroch, M., Geng, S., and G. Wang. 2013. Recent advances in liquid biofuel production from algal feedstocks. *Appl Energ* 102:1371–1381.

de Mule, M.C.Z., de Caire, G.Z., de Cano, M.S. et al. 1999. Effect of cyanobacterial inoculation and fertilizers on rice seedlings and postharvest soil structure. *Commun Soil Sci Plant Anal* 30:97–107.

Debowski, M., Zielinski, M., Grala, A., and M. Dudek. 2013. Algae biomass as an alternative substrate in biogas production technologies—Review. *Renew Sust Energ Rev* 27:596–604.

Demirbas, M.F. 2011. Biofuels from algae for sustainable development. *Appl Energ* 88:3473–3480.

Doran-Peterson, J., Cook, D.M., and S.K. Brandon. 2008. Microbial conversion of sugars from plant biomass to lactic acid or ethanol. *Plant J* 54:582–592.

Ehimen, E.A., Sun, Z.F., and C.G. Carrington. 2012. Use of ultrasound and co-solvents to improve the in-situ transesterification of microalgae biomass. *Procedia Environ Sci* 15:47–55.

Ferrentino, J.M., Farag, I.H., and L.S. Jahnke. 2006. Microalgal oil extraction and *in-situ* transesterification. Chemical Engineering, University of New Hampshire (UNH), Durham, NH. http://www.nt.ntnu.no/users/skoge/prost/proceedings/aiche-2006/data/papers/P69332.pdf (accessed February 8, 2012).

Ferrera-Lorenzo, N., Fuente, E., Bermúdez, J.M., Suárez-Ruiz, I., and B. Ruiz. 2014. Conventional and microwave pyrolysis of a macroalgae waste from the agar-agar industry. Prospects for biofuel production. *Bioresour Technol* 151:199–206.

Garciano, L.O., Tran, N.H., Kannangara, G.S.K. et al. 2012. Pyrolysis of a naturally dried *Botryococcus braunii* residue. *Energ Fuel* 26:3874–3881.

Gendy, T.S. and S.A. El-Temtamy. 2013. Commercialization potential aspects of microalgae for biofuel production: An overview. *Egypt J Pet* 22:43–51.

Ghirardi, M.L., Dubini, A., Yu, J., and P.C. Maness. 2009. Photobiological hydrogen-producing systems. *Chem Soc Rev* 38:52–61.

Goldstein, I.S. 1983. *Organic Chemicals from Biomass*. Boca Raton, FL: CRC Press.

Gouveia, L. 2011. *Microalgae as a Feedstock for Biofuels*. Verlag Berlin Heidelberg: Springer.

Grierson, S., Strezov, V., Ellem, G. et al. 2009. Thermal characterisation of microalgae under slow pyrolysis conditions. *J Anal Appl Pyrol* 85:118–123.

Guedes, A.C., Amaro, H.M., and F.X. Malcata. 2011. Microalgae as sources of high added-value compounds—A brief review of recent work. *Biotechnol Prog* 27:597–613.

Guedes, A.C., Amaro, H.M., Sousa-Pint, I. et al. 2014. Applications of spent biomass. In *Biofuels from Algae*, eds. A. Pandey, D.J. Lee, Y. Chisti, and C. Rsoccol, pp. 205–233. Elsevier B.V.

Guedes, A.C. and F.X. Malcata. 2012. Nutritional value and uses of microalgae in aquaculture. In *Aquaculture*, ed. Z.A. Muchlisin, pp. 59–79. Rijeka, Croatia: InTech.

Harun, R., Liu, B., and M.K. Danquah. 2011. Analysis of process configurations for bioethanol production from microalgal biomass. In *Progress in Biomass and Bioenergy Production*, ed. S. Shaukat, pp. 395–409. Rijeka, Croatia: Intech Science, Technology & Medicine.

Harun, R., Singh, M., Forde, G.M. et al. 2010. Bioprocess engineering of microalgae to produce a variety of consumer products. *Renew Sust Energ Rev* 14:1037–1047.

Haykiri-Acma, H., Yaman, S., and S. Kucukbayrak. 2013. Production of biobriquettes from carbonized brown seaweed. *Fuel Process Technol* 106:33–40.

Heilmann, S.M., Davis, H.T., Jader, L.R. et al. 2010. Hydrothermal carbonization of microalgae. *Biomass Bioenerg* 34:875–882.

Heilmann, S.M., Jader, L.R., and L.A. Harned. 2011. Hydrothermal carbonization of microalgae II. Fatty acid, char, and algal nutrient products. *Appl Energ* 88:3286–3290.

Hill, J., Nelson, E., Tilman, D., Polasky, S., and D. Tiffany. 2006. Environmental, economic, and energetic costs and benefits of biodiesel and ethanol biofuels. *Proc Natl Acad Sci USA* 103:11206–11210.

Hu, Z., Ma, X., Li, L., and J. Wu. 2014. The catalytic pyrolysis of microalgae to produce syngas. *Energ Convers Manage* 85:545–550.

Hudek, K., Davis, L.C., Ibbini, J. et al. 2014. Commercial products from algae. In *Algal Biorefineries: Cultivation of Cells and Products*, eds. B. Rakesh, P. Aleš, and Z. Mark, pp. 275–295. Netherlands, Springer.

Jirka, S. and T. Tomlinson. 2014. *2013 State of the biochar industry, a survey of commercial activity in the biochar field*. International Biochar Initiative (IBI).

John, R.P., Anisha, G.S., Nampoothiri, K.M., and A. Pandey. 2011. Micro and macroalgal biomass: A renewable source for bioethanol. *Bioresour Technol* 102:186–193.

Johnson, M.B. 2009. Microalgal biodiesel production through a novel attached culture system and conversion parameters. MS thesis, Virginia Polytechnic Institute and State University, Blacksburg, VA.

Khan, S.A. and S.A. Rashmi. 2010. Algal biorefinery: A road towards energy independence and sustainable future. *Int Rev Chem Eng* 2:63–68.

Kirtania, K., Joshua, J., Kassim, M.A. et al. 2014. Comparison of CO_2 and steam gasification reactivity of algal and woody biomass chars. *Fuel Process Technol* 117:44–52.

Knight, J.A. 1976. Pyrolysis of fine sawdust. Paper presented at *172nd American Chemical Society National Meeting*, San Francisco, CA.

Koberg, M., Cohen, M., Ben-Amotz, A., and A. Gedanken. 2011. Biodiesel production directly from the microalgae biomass of nannochloropsis by microwave and ultrasound radiation. *Bioresour Technol* 5:4265–4269.

Kraan, S. 2012. Algal polysaccharides, novel applications and outlook. In *Carbohydrates-Comprehensive Studies on Glycobiology and Glycotechnology*. ed. C.F. Chang, Rijeka, Croatia: InTech.

Lam, M.K. and K.T. Lee. 2013. Biohydrogen Production from Algae, In *Biohydrogen*, eds. P. Ashok, C. Jo-Shu, H. Patrick and L. Christian, pp. 161–184. Poland: Elsevier B.V.

Larsson, S., Palmqvist, E., Hahn-Hägerdal, B. et al. 1999. The generation of fermentation inhibitors during dilute acid hydrolysis of softwood. *Enzyme Microb Technol* 24:151–159.

Lee, Y.Y., Iyer, P., and R.W. Torget. 1999. Dilute-acid hydrolysis of lignocellulosic biomass. *Adv Biochem Eng/Biotechnol* 65:93–115.

Lehmann, J. and S. Joseph. 2009. *Biochar for Environmental Management*. London, U.K.: Earthscan.

Lin, L., Cunshan, Z., Vittayapadung, S., Xiangqian, S., and D. Mingdong. 2011. Opportunities and challenges for biodiesel fuel. *Appl Energ* 88:1020–1031.

Maddi, B., Viamajala, S., and S. Varanasi. 2011. Comparative study of pyrolysis of algal biomass from natural lake blooms with lignocellulosic biomass. *Bioresour Technol* 102:11018–11026.

Mao, C., Feng, Y., Wang, X., and G. Ren. 2015. Review on research achievements of biogas from anaerobic digestion. *Renew Sust Energ Rev* 45:540–555.

Martin, C., Klinke, H.B., and A.B. Thomsen. 2007. Wet oxidation as a pretreatment method for enhancing the enzymatic convertibility of sugarcane bagasse. *Enzyme Microb Technol* 40:426–432.

Mata, M.T., Martins, A.A., and S.N. Caetano. 2010. Microalgae for biodiesel production and other applications: A review. *Renew Sust Energ Rev* 14:217–232.

Melis, A., Zhang, L., Forestier, M., Ghirardi, M.L., and M. Seibert. 2000. Sustained photobiological hydrogen gas production upon reversible inactivation of oxygen evolution in the green alga *Chlamydomonas reinhardtii*. *Plant Physiol* 122:127–135.

Miranda, J.R., Passarinho, P.C., and L. Gouveia. 2012. Pre-treatment optimization of *Scenedesmus obliquus* microalga for bioethanol production. *Bioresour Technol* 104:342–348.

Montingelli, M.E., Tedesco, S., and G. Olabi. 2015. Biogas production from algal biomass: A review. *Renew Sust Energ Rev* 43:961–972.

Özçimen, D. 2007. Evaluation of various vegetable residues by carbonization. PhD thesis, Istanbul Technical University, Istanbul, Turkey.

Özçimen, D. 2013. An approach to the characterization of biochar and biooil. In *Renew Renewable Energy for Sustainable Future*. ed. S.P. Lohani, pp. 41–58. Hong Kong: iConcept Press.

Özçimen, D., Gülyurt, M.Ö., and B. İnan. 2012. Algal biorefinery for biodiesel production. In *Biodiesel—Feedstocks, Production and Applications*, ed. Z. Fang. Rijeka, Croatia: InTech.

Özçimen, D. and S. Yücel. 2011. Novel methods in biodiesel production. In *Biofuel's Engineering Process Technology*, ed. M.A. dos Santos Bernardes, pp. 353–384. Rijeka, Croatia: InTech.

Passos, F., Uggetti, E., Carrère, H. et al. 2014. Pretreatment of microalgae to improve biogas production: A review. *Bioresour Technol* 172:403–412.

Patil, P., Deng, S., Isaac, R.J., and P.J. Lammers. 2009. Conversion of waste cooking oil to biodiesel using ferric sulfate and supercritical methanol processes. *Fuel* 89:360–364.

Patil, P.D., Gude, V.G., Mannarswamy, A. et al. 2011. Optimization of microwave-assisted transesterification of dry algal biomass using response surface methodology. *Bioresour Technol* 102:1399–1405.

Peng, W., Wu, Q., Tu, P. et al. 2001. Pyrolytic characteristics of microalgae as renewable energy source determined by thermogravimetric analysis. *Bioresour Technol* 80:1–7.

Popoola, T.O.S. and O.D. Yangomodou. 2006. Extraction, properties and utilization potentials of cassava seed oil. *Biotechnology* 5:38–41.

Rashid, N., Rehman, M.S.U., Memon, S. et al. 2013. Current status, barriers and developments in biohydrogen production by microalgae. *Renew Sust Energ Rev* 22:571–579.

Rizzo, A.M., Prussi, M., Bettucci, L. et al. 2013. Characterization of microalga *Chlorella* as a fuel and its thermogravimetric behavior. *Appl Energ* 102:24–31.

Ross, A.B., Anastasakis, K., Kubacki, M.L. et al. 2009. Investigation of the pyrolysis behaviour of brown algae before and after pre-treatment using PY-GC/MS and TGA. *J Anal Appl Pyrol* 85:3–10.

Ross, A.B., Jones, J.M., Kubacki, M.L. et al. 2008. Classification of macroalgae as fuel and its thermochemical behavior. *Bioresour Technol* 99:6494–6504.

Ruperez, P. 2002. Mineral content of edible marine seaweeds. *Food Chem* 79:23–26.

Sarkar, N., Ghosh, S.K., Bannerjee, S., and K. Aikat. 2012. Bioethanol production from agricultural wastes: An overview. *Renew Energ* 37:19–27.

Schwartz, D. and L. Krienitz. 2005. Do algae cause growth-promoting effects on vegetables grown hydroponically? In *Fertigation: Optimizing the Utilization of Water and Nutrients*, ed. M.R. Price, pp. 161–170. Beijing, China: International Potash Institute.

Serrato, A.G. 1981. Extraction of oil from soybeans. *J Am Oil Chem* 58(3):157–159.

Show, K.Y. and D.J. Lee. 2013. Production of biohydrogen from microalgae. In *Biofuels from Algae*, ed. A. Pandey, D.J. Lee, Y. Chisti and C.R. Socco, pp. 189–204. Great Britain: Elsevier B.V.

Spolaore, P., Joannis-Cassan, C., Duran, E. et al. 2006. Commercial applications of microalgae. *J Biosci Bioeng* 101:87–96.

Suali, E. and R. Sarbatly. 2012. Conversion of microalgae to biofuel. *Renew Sust Energ Rev* 16:4316–4342.

Subhadra, B.G. 2010. Sustainability of algal biofuel production using integrated renewable energy park (IREP) and algal biorefinery approach. *Energ Pol* 38:5892–5901.

Tamilarasan, S. and R. Sahadevan. 2014. Ultrasonic assisted acid base transesterification of algal oil from marine macroalgae *Caulerpa peltata*: Optimization and characterization studies. *Fuel* 128:347–355.

Thorsen, M., Woodward, S., and B.M. McKenzie 2010. Kelp (*Laminaria digitata*) increases germination and affects rooting and plant vigour in crops and native plants from an arable grassland in the Outer Hebrides. *Scot J Coast Conserv* 14:239–247.

Tran, D., Yeh, K., Chen, C., and J. Chang. 2012. Enzymatic transesterification of microalgal oil from *Chlorella vulgaris* ESP-31 for biodiesel synthesis using immobilized *Burkholderia* lipase. *Bioresour Technol* 108:119–127.

Trivedi, J., Aila, M., Bangwal, D.P., Kaul, S., and M.O. Garg. 2015. Algae based biorefinery—How to make sense? *Renew Sust Energ Rev* 47:295–307.

Uyar, B., Eroglu, I., Yücel, M., Gündüz, U., and L. Türker. 2007. Effect of light intensity, wavelength and illumination protocol on hydrogen production in photobioreactors. *Int J Hydro Energ* 32:4670–4677.

Valente, L.M.P., Gouveia, A., Remaa, P., Matosa, J., Gomesa, E.F., and I.S. Pintoa. 2006. Evaluation of three seaweeds *Gracilaria bursa-pastoris*, *Ulva rigida* and *Gracilaria cornea* as dietary ingredients in European sea bass (*Dicentrarchus labrax*) juveniles. *Aquaculture* 252:85–91.

Velasquez-Orta, S.B., Lee, J.G.M., and A. Harvey. 2012. Alkaline in situ transesterification of *Chlorella vulgaris*. *Fuel* 94:544–550.

Vincent, M., Pometto, A.L., and J. van Leeuwen. 2014. Ethanol production via simultaneous saccharification and fermentation of sodium hydroxide treated corn stover using *Phanerochaete chrysosporium* and *Gloeophyllum trabeum*. *Bioresour Technol* 158:1–6.

Wan, C., Zhou, Y., and Y. Li. 2011. Liquid hot water and alkaline pretreatment of soybean straw for improving cellulose digestibility. *Bioresour Technol* 102:6254–6259.

Wang, B., Wu, N., and C.Q. Lan. 2008. CO_2 biomitigation using microalgae. *Appl Microb Biotechnol* 79:707–718.

Wang, Z., Keshwani, D.R., Redding, A.P., and J.J. Cheng. 2010. Sodium hydroxide pretreatment and enzymatic hydrolysis of coastal Bermuda grass. *Bioresour Technol* 101:3583–3585.

Wang, Z., Li, R., Xu, J. et al. 2012. Sodium hydroxide pretreatment of genetically modified switchgrass for improved enzymatic release of sugars. *Bioresour Technol* 110:364–370.

Ward, J., Lewis, D.M., and F.B. Green. 2014. Anaerobic digestion of algae biomass: A review. *Algal Res* 5:204–214.

Watanabe, H., Li, D., Nakagawa, Y. et al. 2014. Characterization of oil-extracted residue biomass of *Botryococcus braunii* as a biofuel feedstock and its pyrolytic behavior. *Appl Energy* 132:475–484.

White, L.P. and L.G. Plaskett. 1981. *Biomass as Fuel*. London, U.K.: Academic Press.

Wiltshire, K.H., Boersma, M., Moller, A., and H. Buhtz. 2000. Extraction of pigments and fatty acids from the green alga *Scenedesmus obliquus* (Chlorophyceae). *Aquat Ecol* 34:119–126.

Winkler, M., Hemschemeier, A., Gotor, C., Melis, A., and T. Happe. 2002. [Fe]-hydrogenases in green algae: Photo-fermentation and hydrogen evolution under sulfur deprivation. *Int J Hydro Energ* 27:1431–1439.

Winsley, P. 2007. Biochar and bioenergy production for climate change mitigation. *NZ Sci Rev* 64:1.

Xu, H., Miao, X., and Q. Wu. 2006. High quality biodiesel production from a micro-alga *Chlorella protothecoides* by heterotrophic growth in fermenters. *J Biotechnol* 126:499–507.

Yanik, J., Stahl, R., Troeger, N. et al. 2013. Pyrolysis of algal biomass. *J Anal Appl Pyrol* 103:134–141.

Zubia, M., Payri, C., and E. Deslandes. 2008. Alginate, mannitol, phenolic compounds and biological activities of two range-extending brown algae, *Sargassum mangarevense* and *Turbinaria ornata* (Phaeophyta: Fucales), from Tahiti (French Polynesia). *J Appl Phycolloid* 20:1033–1043.

Section IV

Future Perspectives of Biofuels

18

Novel Enzymes in Biofuel Production

Ranjeeta Bhari and Ram Sarup Singh

Contents

Abstract

The growing concern for environmental sustainability and energy security had made the search of alternative energy sources the utmost need of the hour to synchronize environmental friendliness with biodegradability and renewability, thereby reducing reliance on petroleum products. Numerous catalytic approaches have been investigated for the production of ethanol, high-chain alcohols, alkanes, fatty acids or their esters, hydrogen, methane, and hydrocarbons. Liquid biofuels work well with the current infrastructure and internal combustion engines, representing a solution to the transportation sector. Innovations in biocatalysis and advanced biochemical processes using enzymes have contributed to the launch of biorefineries using cheaper substrates for the production of biofuels. The usefulness of amylases and now cellulases, xylanases, and monooxygenases has been much investigated for the production of alcohol-based biofuels. Researchers after exploring enzymatic transesterification of edible oils are now focusing on the transesterification of waste oils and animal tallow to biodiesel using lipases. Though much research is carried out for the production of liquid biofuels, these represent only short-term solution to the transportation sector. The use of hydrogen as biofuel, however, represents a long-term solution, and enzymatic production of hydrogen using hydrogenases has been looked forward. This chapter focuses on the developments in biocatalysis for the production of alcohol fuels, biodiesel, and biohydrogen.

18.1 Introduction

According to the factsheet released by the International Energy Agency (2006), fossil fuels account for about 80.3% of the primary energy consumed worldwide, with 57.7% being used in transportation sector alone. The global energy system continues to face a major crisis worldwide, and the primary energy demand is assumed to increase 37% by 2040 (International Energy Agency, 2014). Sustainable development of the humankind needs the production of renewable energy at affordable costs. Biofuels represent sustainable and renewable sources of energy, which when burnt emit reduced levels of particulates, carbon oxides, and sulfur oxides and, therefore, hold promise as the ideal fuels of the future to completely replace the petroleum fuels. Biofuels have been produced from starch and vegetable oils. Innovations in biocatalysis and advanced biochemical processes have contributed to the launch of biorefineries using biomass as a source of sugars for the production of second-generation or so-called advanced biofuels. Numerous biocatalytic approaches have focused on the use of biomass sugars for the production of biofuels such as ethanol (Taherzadeh and Karimi, 2007; Singhania et al., 2013), butanol (Qureshi et al., 2008a,b), and hydrogen (Zhang et al., 2007). Biocatalysts allow the use of unrefined feedstock, including waste oil, without the need to separate free fatty acids that may be present in large amounts in the feedstock (Nielsen et al., 2008). This chapter attempts to provide a comprehensive review on enzymes with potential applications in biofuel industry, particularly for the generation of alcohol fuels, biohydrogen, and biodiesel. Various advances in biocatalysis for biofuel production have been highlighted.

18.2 Alcohol Fuels

Bioethanol is the most widely produced biofuel today with an annual production of 84 billion liters projected to reach 125 billion liters by 2017 (Walker, 2011). Butanol is an intensively investigated emerging transportation fuel having higher energy density than methanol and ethanol and is less readily contaminated with water (Ezeji et al., 2007). Butanol is a cleaner and superior fuel extender or oxygenate compared to ethanol with octane numbers 113 and 94 in comparison to 111 and 94 for ethanol (Ladisch, 1991). With these superior fuel properties and recent advances in biotechnology and bioprocessing for the development of superior strains and advanced process technology, commercial interest has returned to butanol fermentation. Earlier alcohol fuels were mainly produced from starch or sugars, which are potential food source, and thus, the need for exploring nonfood biomass like lignocellulosics was realized in the light of process economy. The complex polymers present in these substrates can be converted to fermentable sugars using enzymes that can be acted upon by suitable microbial strains to form respective alcohols, but the heterogeneity of plant cell wall, inaccessibility, and recalcitrance of its individual components present limitation in biorefinery approaches (Klein-Marcuschamer et al., 2012). The important enzymes in substrate pretreatment for bioethanol and biobutanol production will be discussed in the subsequent sections.

18.2.1 First-Generation Alcohol Fuels and Amylases

Starch, a conventional substrate for ethanol and butanol production, is a long-chain polymer of glucose that cannot be directly fermented to ethanol. The macromolecular structure first has to be broken down to glucose. Starch feedstocks are ground and mixed with water to form mash containing 15%–20% starch. The mash is then cooked at or above its boiling point and subsequently treated with amylases; α-amylases hydrolyze α-1,4 linkages of starch molecules to short chains of dextrins, whose chain length depends on the treatment time and hydrolysis temperature. The dextrins are further hydrolyzed by glucoamylases to glucose, maltose, and isomaltose that can be readily fermented to ethanol or butanol with suitable microbial strain (Lee et al., 2007; Visioli et al., 2014). Chheda et al. (2007) proposed a novel hybrid route that depolymerizes starch to glucose, which can be isomerized to fructose using isomerases. The fructose so obtained can be converted chemically to 2,5-dimethylfuran that has higher energy density and octane number supporting its superiority as a motor fuel (Wackett, 2008). Since these substrates are potential food sources, their use in biofuel industry is not very promising.

18.2.2 Proteases

Proteases are typically used in biofuel production to supply nitrogen source to fermenting yeast. Proteases may benefit the fermentation by changing the chemistry of the grain by dismantling starch–gluten complexes, making starch more accessible to the action of amylases (Alvarez et al., 2010). Novozymes described another benefit of proteases by hydrolyzing oleosins present in corn kernels, releasing more oil from kernels (Huang, 1996). Oil extraction can be simultaneously carried out in ethanol plants, creating a high-value coproduct that can be used for biodiesel production (Wisner, 2013). The use of appropriate protease under optimized conditions can improve ethanol yield (Harris et al., 2014).

18.2.3 Lignocellulosics and Production of Advanced Alcohol Fuels

A considerable amount of lignocellulosic materials are generated from agro-based industries and have received increasing research attention as raw material for high-value products like biofuels, value-added fine chemicals, and cheap energy sources (Asgher et al., 2013). Recently, Mielenz et al. (2015) stimulated the development of agave as dedicated feedstock for biofuel production. Lignocellulosic biomass mainly comprises interlinked network of cellulose (40%–50%), hemicellulose (20%–40%), and lignin (20%–30%); the relative abundance of each varies depending on the source (Pauly and Keegstra, 2008). Cellulose is a linear polymer of β-1,4-linked glucose units and chains aggregate into crystalline and insoluble microfibrils via hydrogen bonds and van der Waals interactions. Consecutive sugars along the chains are rotated by 180° so that cellobiose is the repeating unit. Complete depolymerization of cellulose yields glucose (Horn et al., 2012). Common hemicelluloses are xylan, mannan, and glucan, with the former two being more common. Xylans are mainly contained in hardwoods, while softwoods have glucomannans. These are heteropolymers with varying degree of branching. Xylans have β-1,4-linked xylose units with high degree of acetylesterifications. Glucomannans contain mixed β-1,4-linked mannose/glucose backbone substituted with α-1,6-linked galactose and some acetylated mannose units. Depolymerization of hemicelluloses yields a mixture of hexoses and pentoses. Lignin, another component of lignocellulosics, is a relatively hydrophobic and aromatic heteropolymer consisting of monolignols commonly p-hydroxyphenyl, guaiacyl and syringyl monolignols, coniferyl alcohol, sinapyl alcohol, and p-coumaryl alcohol that are methoxylated to varying degree and form a protective seal around cellulose and hemicellulose. The relative abundance of these constituents varies with species and forms a complex highly branched less reactive network (Ralph et al., 2004).

The cellulose–hemicellulose–lignin matrix is highly recalcitrant and the conversion of this matrix to biofuels requires a multistep processing intended to break down the lignin barrier so as to release cellulose and hemicellulose that can be subjected to enzymatic hydrolysis or modify the porosity of the material to allow enzyme penetration for sufficient hydrolysis (Galbe and Zacchi, 2002; Xiao et al., 2012). An effective pretreatment (physical, chemical, or biological) protocol should preserve hemicellulose fractions, yield maximum fermentable sugars, limit loss of carbohydrate, avoid the formation of inhibitory by-products, involve minimum energy input, and be economically efficient (Asgher et al., 2013). Physical methods include the use of ball mills, hammer mills, high-pressure steaming, pyrolysis, or irradiation. Chemical pretreatment with dilute or concentrated hydrochloric or sulfuric acids faces serious limitation due to the formation of secondary products that can lower sugar yield. Bases such as sodium and ammonium hydroxide can also be used for pretreatment. The biological methods of delignification employ bacteria or fungi to modify the structure and composition of lignocellulosic biomass (Anwar et al., 2014). The enzymatic hydrolysis and factors affecting the hydrolysis of lignocelluloses are discussed in the following sections.

18.2.3.1 Degradation of Lignocelluloses Using Enzymes

Cellulose degradation is accomplished by synergestic action of the three classes of hydrolases together referred to as cellulase or cellulolytic enzymes: endo-β-1,4-glucanase that randomly cleaves the internal bonds in cellulose chain, exo-β-1,4-glucanase that acts on the reducing or non-reducing end of the cellulose polymer, and β-glucosidase that converts cellobiose to glucose (Chandra et al., 2007). Cellulases are currently the third largest industrial enzymes because of their extensive use and would become the largest volume industrial enzymes if ethanol or butanol becomes the major transportation fuels (Wilson, 2009).

Hemicellulolytic enzymes are more complex and involve endo-β-1,4-xylanase, exo-β-1,4-xylanase, endo-β-1,4-mannanase, β-mannosidase, acetyl xylan esterase, α-glucuronidase, α-arabinofuranosidase, and α-galactosidase (Jorgensen et al., 2003). Several species of bacteria such as *Acetivibrio, Bacillus, Bacteroides, Clostridium, Cellulomonas, Erwinia, Thermomonospora, Ruminococcus,* and *Streptomyces* and fungi such as *Fusarium, Humicola, Penicillium, Phanerochaete, Trichoderma,* and *Schizophyllum* are known to produce cellulases and hemicellulases (Sun and Cheng, 2002). Fischer et al. (2014) identified crude cellulase preparations from *Aspergillus niger* and *Mucor racemosus* as potential candidates for the hydrolysis of sugarcane bagasse for subsequent ethanol production.

Lignin is a major component of plant cell wall that contributes to recalcitrance of biomass and consequently increases the cost associated with conversion of ligno-cellulose (Chen and Dixon, 2007). Current methods of delignification include treatment with organic solvents, ionic liquids, dilute acid, ammonia fiber expansion, or hydrothermolysis, which are expensive and increase the overall cost of the process. Thus, increasing efficiency and decreasing cost of pretreatment are high priority. White-rot fungi deploy peroxidases and laccases to generate reactive molecules that subsequently degrade lignin (Martinez et al., 2005). Genome analysis of *Phanerochaete chrysosporium* reveals hundreds of lignin-degrading enzymes encoded in clusters in the genome (Martinez et al., 2004; Vandan et al., 2006). Termites are capable of completely digesting hemicellulose to acetate and molecular hydrogen. Understanding lignocellulosic digestion by termites could help to get an insight into the enzymes responsible for biomass delignification (Weng et al., 2008).

Hydrolysis of cellulose yields hexoses, while hemicelluloses form pentoses that can be fermented using simultaneous saccharification and co-fermentation by microbes capable of utilizing six- and five-carbon sugars, respectively (Teixeira et al., 2000). The hydrolysis and fermentation can also be carried out simultaneously using genetically modified organisms that can ferment both hexoses and pentoses in the same media in which enzymes for the hydrolysis of cellulose and hemicellulose are present, making the process economically favorable (Olsson and Hahn-Hagerdal, 1996). A novel approach to use homoacetogenic bacteria that can use both hexoses and pentoses simultaneously to produce acetic acid has been described. Acetic acid in the broth can be esterified to produce ethyl acetate that is relatively insoluble. This ethyl acetate can be easily recovered from broth and hydrogenated to produce 2 mol of ethanol; thus, 1 mol of hexose yields 3 mol of ethanol resulting in improved yield (Eggeman and Verser, 2006). Qureshi et al. (2008) demonstrated butanol production from cellulase- and cellobiase-hydrolyzed corn fiber. Butanol production in a fed-batch fermentation using wheat straw hydrolyzed by cellulase, β-glucosidase, and xylanase has also been reported (Qureshi et al., 2008). Wheat straw is composed of 35%–45% cellulose, 20%–30% hemicellulose, and relatively low lignin content of about 20%

that makes its bioconversion to biofuels particularly attractive. Other workers have also reported the use of wheat straw as a potential substrate for various bioconversion processes (Saha et al., 2005). The addition of surfactants during hydrolysis has been described to modify the surface properties of cellulose and lower enzyme loading. Non-ionic surfactants like Tween 20, Tween 80, and polyethylene glycol have been found to be particularly effective in improving the hydrolysis process by preventing non-productive adsorption of cellulase to lignin (Alkasrawi et al., 2003; Börjesson et al., 2007). The choice of surfactant is a critical parameter that might have a negative effect on the fermentation of hydrolysate (Taherzadeh and Karimi, 2007).

Commercially available cellulase cocktails are usually derived from the fungus *Hypocrea jecorina* (*Trichoderma reesei*), and the enzyme has pH optima between 4.5 and 5.0 and temperature optima between 40°C and 50°C (Horn et al., 2012). Though these preparations show fairly good degradation potential, the major disadvantage of *Trichoderma* cellulases is suboptimal levels of β-glucosidases. However, *Aspergillus* species are good β-glucosidase producers. The hydrolytic efficiency of *Trichoderma* cellulases can be improved by supplementing β-glucosidases (Ortega et al., 2001; Itoh et al., 2003; Sukumaran et al., 2009). Wen et al. (2005) suggested the co-cultivation of *T. reesei* and *Aspergillus phoenicis*. Nakazawa et al. (2012) constructed a recombinant *T. reesei* strain expressing *Aspergillus aculeatus* β-glucosidase. *Acremonium cellulolyticus* strains have been reported to produce not only cellulase and β-glucosidase but also carboxymethyl cellulose–degrading enzymes and small amount of xylanase, β-1,3-glucanase, and amylase. Gusakov (2011) has reviewed the potential sources of cellulase as alternatives to *T. reesei*. Cellulase expression in microbes is inducible and repressible. Lv et al. (2015) developed a copper-responsive expression system based on *T. reesei* copper transporter gene *tcu1* and achieved a constitutive expression of cellulase on non-inducible media containing glucose. Recently, improved cellulolytic enzyme production has been reported in *Penicillium oxalicum* by the manipulation of genes aimed to amplify gene expression and relieve repression (Yao et al., 2015).

The existing enzymatic hydrolysis is carried out at temperatures below 50°C, but slow hydrolysis rates, low sugar yield, high enzyme dose required, and susceptibility to microbial contamination are few limitations that can be overcome by using thermophilic biocatalysts (Bhalla et al., 2013). A number of thermophiles belonging to the genera *Bacillus*, *Geobacillus*, *Acidothermus*, *Caldocellum*, *Clostridium*, *Thermotoga*, *Anaerocellum*, *Rhodothermus*, *Caldicellulosiruptor*, *Sporotrichum*, *Scytalidium*, and *Thermomonospora* are reported to produce thermostable cellulases (Kumar et al., 2008; Bhalla et al., 2013). Thermostable endoglucanases from *Bacillus subtilis* (Yang et al., 2010), *Fervidobacterium nodosum* (Wang et al., 2010), and *Thermoanaerobacter tengcongensis* (Liang et al., 2011) have been cloned and expressed in *Escherichia coli*.

Reese et al. (1950) suggested the requirement of a nonhydrolytic component for the hydrolysis of cellulose that could disrupt polymer packing in the substrate, increasing its accessibility for hydrolytic enzymes. Moser et al. (2008) showed that carbohydrate-binding module (CBM33) from *Thermobifida fusca* potentiates cellulase-catalyzed hydrolysis of cellulose. In 2011, CelS2, a CBM33 from *Streptomyces coelicolor*, was reported to cleave cellulose producing aldonic acids, and the activity of CelS2 was found to be inhibited by EDTA and restored by divalent ions. Later on, these proteins were shown to be copper-dependent monooxygenases (Forsberg et al., 2011; Horn et al., 2012). Many fungi produce glycosyl hydrolases (GH61) that are structurally similar to CBM33 and act synergistic to cellulases (Harris et al., 2010). These have been shown to catalyze the oxidative cleavage of cellulose in the presence of external electron donors like gallic acid, ascorbic acid, and reduced glutathione (Quinlan et al., 2011) and have also been described as copper-dependent lytic polysaccharide monooxygenases (Horn et al., 2012). These monooxygenases have been proposed to bear flat substrate binding sites, which fit well to crystalline cellulose surfaces, where they might disrupt packing and increase substrate accessibility to hydrolases (Horn et al., 2012). These novel monooxygenases have been shown to speed up the enzymatic conversion, thus reducing processing times and enzyme loading rate. Cellic CTec2, a commercial cellulase preparation from Novozymes, contains extra GH61 that may be responsible for improved performance of enzyme preparation (Cannella et al., 2012). An expansin-like protein identified in *B. subtilis* has also been reported to stimulate the cellulase-mediated hydrolysis of corn stover (Kim et al., 2009). Swollenin, another expansin-like protein found in a number of cellulolytic fungi, also disrupts hydrogen bonds in native cellulose, facilitating enzyme binding (Chen et al., 2010).

Cellulose degradation by a combination of cellulases and oxygenases yields oxidized sugars that can evade the problem of product inhibition. Cellobionic acid is known to exhibit less inhibitory effects on cellulase as compared to cellobiose. However, cellobionic acid is less readily hydrolyzed by β-glucosidases and the resulting gluconic acid shows stronger product inhibition than glucose (Cannella et al., 2012). However, gluconic acid can be directly fermented to ethanol using suitable strains (Fan et al., 2012).

18.2.3.2 Factors Affecting Enzymatic Hydrolysis of Lignocellulose

Major factors affecting the hydrolysis of lignocellulose can be enzyme related or substrate related and mainly include substrate concentration and quality, pretreatment method employed, enzyme characteristics, and hydrolysis conditions. Optimum pH and temperature are functions of raw material, enzyme source, and hydrolysis time. Substrate concentration is an important factor governing the yield and initial rate of hydrolysis. High substrate concentration affects mass transfer, results in enzyme inhibition, and subsequently lowers yields. The ratio of substrate to enzyme concentrations is another important factor governing the rate and extent of hydrolysis (Sun and Cheng, 2002). Combining the enzymatic hydrolysis of lignocellulosics and subsequent fermentation in a single step is one of the most successful methods of ethanol production that allows simultaneous consumption of sugars being hydrolyzed, thereby keeping the sugar concentration low at all times that minimizes product inhibition effects (Sun and Cheng, 2002). However, for simultaneous hydrolysis and fermentation, optimum conditions need to be as close as possible, and since hydrolysis is usually accomplished at elevated temperatures, employing thermotolerant organisms for fermentation offers an advantage (Taherzadeh and Karimi, 2007). Alternatively, nonisothermal simultaneous saccharification and fermentation can be performed in separate reactors each operating at different temperatures (Wu and Lee, 1998). Varga et al. (2004) have suggested a multistep approach for simultaneous saccharification and fermentation, in which small amount of cellulase is added at 50°C and prehydrolysis step is carried out in fed-batch mode to maximize solid concentration and obtain better mixing conditions. In the second step, more cellulase is combined with fermenting organism at 30°C.

Genetic engineering techniques have enabled a consolidated bioprocessing that combines cellulose degradation and fermentation in a single step by engineering cellulose-degrading yeast strains. Yeast strains with surface expression of endoglucanase and exoglucanase from *T. reesei* and β-glucosidase from *As. aculeatus* have been engineered (Murai et al., 1998; Fujita et al., 2004; Wen et al., 2010). Engineered yeast cells expressing cellulases on the cell surface have been reported to exhibit better hydrolysis potential. Tsai et al. (2009, 2010) have purified cellulases of *Clostridium thermocellum* CelA and BglA and *Clostridium cellulolyticum* CelE and CelG from *E. coli*. Baek et al. (2012) demonstrated ethanol production from phosphoric acid–swollen cellulose by combining three types of recombinant yeast cells displaying *T. aurantiacus* endoglucanase EGI, *T. reesei* exoglucanase CBHII, and *As. aculeatus* β-glucosidase BGLI, respectively, instead of co-expressing these enzymes in a single cell. This system allows optimization of ethanol production by adjusting the relative ratio of different cell types. The workers have achieved 1.3-fold higher ethanol production with EGI/CBHII/BGLI ratio of 6:2:1 as compared to equimolar concentration of each cell type.

18.2.3.3 Xylanases and Xylan Degradation

Complete enzymatic hydrolysis of hemicellulose requires synergistic action of several enzymes. Endo-β-1,4-xylanase cleaves β-1,4-linked xylose residues in the backbone of xylans. β-1,4-Xylosidase releases xylose monomers and α-glucuronidase cleaves o-methyl-α-glucuronic acid from non-reducing ends of xylooligosaccharides. α-L-Arabinofuranosidase removes arabinose side chains from xylose backbone of arabinoglucuronoxylan. These arabinoglucuronoxylans are typically acetylated at o-2 and o-3 positions of xylose chain and frequently have ferulic acid or coumaric acid groups esterified to 5'OH of arabinofuranosyl groups, which can be hydrolyzed by esterases. Acetyl xylan esterase and feruloyl esterase produce short substituted xylooligosaccharides with concomitant release of ferulic acid and acetic acid. The substituted xylooligosaccharides are acted upon by arabinofuranosidase and glucuronidase that liberate arabinose and glucuronic acid and form linear non-branched xylooligosaccharides that can be acted upon by xylosidase. The released xylose residues can be readily fermented to ethanol (Saha, 2003; Dodd and Cann, 2009).

Xylan-degrading enzymes are produced by fungi, bacteria, yeast, marine algae, protozoa, crustaceans, snails, and insects. Thermostable xylanases are produced by a number of fungi, namely, *Laetiporus sulphureus*, *Talaromyces thermophilus*, *Thermomyces lanuginosus*, *Nomura flexuosa*, *Thermoascus aurantiacus*, and *Rhizomucor miehei*, and bacteria, namely, *Bacillus*, *Geobacillus*, *Thermotoga*, *Acidothermus*, *Cellulomonas*, *Paenibacillus*, *Thermoanaerobacterium*, *Actinomadura*, *Alicyclobacillus*, *Anoxybacillus*, *Nesterenkonia*, and *Enterobacter* (Bhalla et al., 2013). Bacterial xylanases have higher temperature optima and thermostability and therefore preferred over fungal xylanases, but their low-level expression limits their use that can be enhanced by cloning and overexpression. Filamentous fungi produce multiple xylanases in their cellulase system and some of these have modular structures consisting of a catalytic domain and CBM1. Inoue et al. (2015) have engineered *Talaromyces cellulolyticus* endo-β-1,4-xylanase and reported that combination of CBM containing cellulases and xylanases could contribute to the reduction of enzyme loading in hydrolysis of pretreated lignocelluloses. A xylanase gene from *Geobacillus* sp. (Wu et al., 2006) and *Actinomadura* sp. (Sriyapai et al., 2011) has been cloned and expressed in *E. coli* and *Pichia pastoris*, respectively.

18.2.3.4 Pectins and Pectinolytic Enzymes

Pectins are the structural components of plant cell walls, mainly abundant in sugar beet pulp, apple, and citrus fruits, where they constitute around 50% of the polymeric content of the cell wall (Brummell, 2006). The architectural properties of cell walls suggest that pectins might mask cellulose–hemicellulose, blocking their exposure to hydrolytic enzymes. Pectin has the backbone of homogalacturonic acid with neutral side chains of L-rhamnose, arabinose, galactose, and xylose. Pectin-degrading enzymes include polymethylgalacturonase, endopolygalacturonase, pectinase, exo-polygalacturonase, and exopolygalacturanosidase that hydrolyze pectin polymer; α-L-rhamnosidase that hydrolyzes rhamnogalacturonan; α-L-arabinofuranosidase that hydrolyzes L-arabinose side chains; endoarabinase that acts on arabinan side chains in pectin; and pectin lyase, pectate lyase, and pectate disaccharide lyase that cleave galacturonic acid polymer by β-elimination (Jayani et al., 2005). Pectin-rich materials have low lignin content that make them attractive substrates for bioethanol production (Xiao and Anderson, 2013). Many pectin-rich materials like sugar beet pulp, citrus waste, apple pomace, and potato pulp have been analyzed as energy feedstocks (Lesiecki et al., 2012; Xiao and Anderson, 2013).

18.2.4 Methanol Production Using Dehydrogenases

Due to the abundance of CO_2 as the major greenhouse gas and the most oxidized form of carbon, the efficient utilization of CO_2 has gained much research interest. Jiang et al. (2003) have explored a novel approach to reduce carbon dioxide to methanol by consecutive action of three dehydrogenases: reduction of CO_2 to formate catalyzed by formate dehydrogenase, reduction of formate to formaldehyde by formaldehyde dehydrogenase, and reduction of formaldehyde to methanol by alcohol dehydrogenase. Reduced nicotinamide adenine dinucleotide (NADH) acts as a terminal electron donor for each dehydrogenase-catalyzed reduction. However, this method has not gained much industrial significance.

18.3 Biodiesel Production

Biodiesel is a mixture of monoalkyl esters obtained from vegetable oils, animal fats, waste cooking oil, grease, and algae (Pearl, 2002; Deba et al., 2015). Biodiesel produces approximately 80% less CO_2 emissions and 100% less SO_2 (Krawczyk, 1996). Its energy content and physicochemical properties are similar to conventional diesel fuel, which makes it compatible with existing fuel distribution infrastructure. Better lubricant properties enhance engine yield and extend shelf life (Vasudevan and Briggs, 2008). Biodiesel does not produce explosive vapors and has high flash point, which makes transportation, storage, and handling safer compared to conventional diesel. Other advantages include biodegradability and miscibility in all ratios with petrodiesel (Moser, 2009).

Although direct usage of vegetable oils as biodiesel is possible by blending it in suitable ratio with conventional diesel fuels, high viscosity, acid contamination, free fatty acid formation, gum formation by oxidation and polymerization, and carbon deposition make long-term storage of these triglyceride esters impractical. To acquire viscosity and volatility similar to fossil fuels, vegetable oils need to be processed to permit their direct usage. Available processing techniques involve pyrolysis, microemulsification, and transesterification (Al-Zuhair, 2007). Pyrolysis, also called cracking, involves heating that forms alkanes, alkenes, and carboxylic acids, thereby reducing viscosity. Microemulsification includes the use of diesel, vegetable oil, alcohols, and alkyl nitrates in suitable proportions (Ma and Hanna, 1999). These two processes are unsatisfactory for commercial biodiesel production. Transesterification is the method of choice that involves reaction between triglycerides and an acyl acceptor (carboxylic acid, alcohol, or another ester) in the presence of a catalyst, resulting in a mixture of low-viscosity monoalkyl esters and glycerol as shown in Figure 18.1 (Vasudevan and Briggs, 2008; Robles-Medina et al., 2009). High-viscosity component (glycerol) is removed, and the resulting biodiesel has low viscosity. Alcohols are most frequently used acyl acceptors, especially methanol and ethanol; the former is cheaper and more reactive. Moreover, fatty acid methyl esters are more volatile than fatty acid ethyl esters justifying the superiority of the former. The use of other higher alcohols is discouraged due to their high cost (Bozbas, 2008).

Though transesterification process can be carried out in the absence of a catalyst at temperatures exceeding 350°C, thermal degradation of esters at elevated temperature limits the process (Demirbas, 2006). Transesterification can be catalyzed by an alkaline catalyst (sodium hydroxide, potassium hydroxide, sodium methoxide), acid catalyst (sulfuric acid, hydrochloric acid, sulfonic acid, phosphoric acid), enzyme catalyst (lipase), or inorganic heterogeneous catalyst (solid-phase catalyst). Demirbas (2006) has demonstrated better yield using methanol in supercritical state, with calcium oxide as catalyst. The alkali process has only found industrial application due

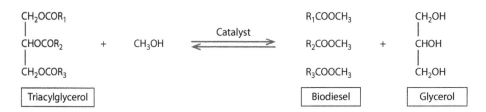

Figure 18.1   Lipase-catalyzed transesterification reaction for biodiesel production.

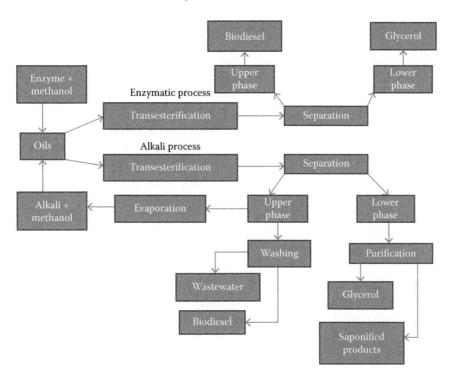

Figure 18.2 Comparison of alkali and enzymatic process for biodiesel production.

to its cost-effectiveness and high efficiency. However, the resulting soap formation makes the separation of esters and glycerol difficult, and extensive downstream operations are required for the removal of catalyst and unreacted methanol (Figure 18.2). On the other hand, simpler downstream operations involved in enzyme-mediated transesterification have attracted the interest of most researchers toward enzymatic approach to biodiesel production.

18.3.1 Lipase-Mediated Transesterification Process

Lipases (triacylglycerol lipases, EC 3.1.1.3) are carboxylic ester hydrolases that display catalytic activity in non-aqueous solvents as well (Rousseau and Marangoni, 2002). These catalysts display remarkable chemoselectivity, regioselectivity, and enantioselectivity and catalyze a forward ester hydrolysis reaction and a reverse condensation reaction (Kirk et al., 2002). The equilibrium can be controlled by water and alcohol concentrations; high alcohol/water ratio shifts the equilibrium towards condensation of free fatty acids and alcohols to biodiesel. Mild reaction conditions required are the key advantage of an enzymatic production process. Lipases are found in almost all living organisms, including plants (papaya latex, oat seed lipase, castor seed lipase), animals (pancreatic lipase), and microorganisms. Microbial lipases are of prime industrial importance due to their selectivity, stability, broad substrate specificity, and availability (Hwang et al., 2014). These lipases may be extracellular or intracellular. Major microbial producers include *A. niger, Bacillus thermoleovorans, Candida antarctica, Candida cylindracea, Chromobacterium violaceum, Geotrichum candidum, Fusarium heterosporum, Fusarium oxysporum, Humicola lanuginosa, Mucor miehei, Penicillium cyclopium, Penicillium roqueforti, Pseudomonas cepacia, Pseudomonas fluorescens, Rhizopus oryzae, Rhizopus thermosus, Rhodotorula rubra, Staphylococcus hyicus,* and *Th. lanuginosus* (Akoh et al., 2007; Bisen et al., 2010). *C. antarctica* lipase B, the most studied enzyme for the transesterification of various oils, has broad substrate specificity and exhibits high stability in organic solvents. Table 18.1 lists the lipases having the potential application for biodiesel production.

Vegetable oils have been most studied for enzymatic biodiesel production (Shah et al., 2003), though few reports document the use of waste frying oil as the substrate for biodiesel production, thus converting a pollutant waste to sustainable energy source (Maceiras et al., 2009). However, insufficient supply of waste oils and grease limits their industrial applicability. Animal fats having high proportion of saturated fatty acids can also be used, though these are of minor importance (Demirbas, 2009).

Non-edible oils, unfit for human consumption due to the presence of anti-nutritional factors or toxic components, provide good alternative substrates for bio-diesel production. Biodiesel from jatropha oil using lipases from *Chromobacterium viscosum* (Shah et al., 2004) and *Enterobacter aerogenes* (Annapurna et al., 2009) has been reported.

The enzyme-catalyzed lipolytic reactions are complicated by the immiscibility of lipid substrate and the use of organic solvents to dissolve lipids. Water is required to activate and stabilize lipase activity making the reaction media biphasic where the reaction takes place at liquid–liquid interface, whose interfacial properties influence enzyme activity (Akoh et al., 2007). The active site of lipases is covered by amphiphilic peptide loop that acts like a lid, disabling the substrate to bind to enzyme active site in aqueous media. Upon contact with lipid–water interface, the lid undergoes a conformational change allowing the binding of substrate to the active site (Schmid and Verger, 1998). Other parameters known to influence yield and rate of lipase-mediated transesterification reaction are solvent used, reaction temperature and time, type and concentration of alcohol, units of enzyme used in the reaction, and water content (Mittelbach and Remschmidt, 2004). Lipases show good stability properties in hydro-phobic solvents with log P values between 2 and 4 (Laane et al., 1987). The highest lipase activity has been reported in a media with water activity between 0.25 and 0.45, corresponding to water content of 0.5%–1.0% (Villeneuve, 2007).

A serious limitation of enzymatic transesterification is enzyme deactivation due to the insolubility of methanol. Methanol/oil ratio above 1.5:1 has been reported to cause serious inhibition. *Ps. cepacia* and *Ps. fluorescens* lipases have been reported to exhibit greatest methanol resistance among the most known lipases, making them attractive for use in methanolysis reactions (Kaieda et al., 2001; Soumanou and Bornscheuer, 2003). To address the problem of enzyme deactivation, a number of workers have suggested a stepwise methanol addition (Shimada et al., 1999; Watanabe et al., 2000). Crude lipase from *Cryptococcus* sp. has been reported to catalyze the methanolysis of vegetable oil in a water continuing system in a single step, without the need of a step-wise methanol addition (Kamini and Iefuji, 2001). Several workers have focused their

TABLE 18.1 Lipases Used for Biodiesel Production

Lipases	Oil	Acyl Acceptor	Solvent	Yield (%)	Reference
Bacillus subtilis	Waste cooking oil	Methanol	Solvent-free	>90	Ying and Chen (2007)
Burkholderia cepacia	Soybean oil	Methanol	Solvent-free	>80	Kaieda et al. (2001)
Candida antarctica	Cottonseed oil	Methanol	t-Butanol	97	Royon et al. (2007)
	Sunflower oil	Methyl acetate	Solvent-free	>95	Ognjanović et al. (2009)
	Cottonseed oil	Methanol, propanol, butanol, amyl alcohol	Solvent-free	91.5	Kose et al. (2002)
	Rapeseed oil	Methanol	Solvent-free	91.1	Watanabe et al. (2007)
	Jatropha seed oil, karanj oil, sunflower oil	Solvent-free	Solvent-free	>90	Modi et al. (2007)
	Soybean oil	Methanol	Solvent-free	93.8	Watanabe et al. (2002)
	Fish oil	Ethanol	Solvent-free	100	Breivik et al. (1997)
	Tallow, soybean, rapeseed oil	Secondary alcohol	Hexane	61.2–83.8	Nelson et al. (1996)
Candida antarctica + *Pseudomonas cepacia*	Recycled restaurant grease	Ethanol	Solvent-free	85.4	Wu et al. (1999)
Candida antarctica B	Waste cooking palm oil	Methanol	t-Butanol	79.1	Halim et al. (2009)
Candida antarctica B	Soybean oil	Methyl acetate	Solvent-free	92	Du et al. (2004)
Candida rugosa	*Jatropha* seed oil	Ethanol	Solvent-free	98	Shah and Gupta (2007)
Candida rugosa	Rapeseed oil	2-Ethyl-1-hexanol	Solvent-free	97	Linko et al. (1998)
Candida sp. 99-125	Waste cooking oil	Methanol	n-Hexane	91.08	Chen et al. (2009)
Cryptococcus SP-2	Rice bran oil	Methanol	Solvent-free	80.20	Kamini and Iefuji (2001)
Enterococcus aerogenes	*Jatropha* oil	Methanol	t-Butanol	94	Kumari et al. (2009)
Penicillium expansum	Waste oil	Methanol	t-Amyl alcohol	92.8	Li et al. (2009)
Pseudomonas cepacia	Palm kernel oil	Ethanol	Solvent-free	72	Abigor et al. (2000)
		t-Butanol		62	
		1-Butanol		42	
		n-Propanol		42	
		Isopropanol		24	
		Methanol		15	

(Continued)

TABLE 18.1 (Continued) Lipases Used for Biodiesel Production

Lipases	Oil	Acyl Acceptor	Solvent	Yield (%)	Reference
	Coconut oil	1-Butanol		40	Kumari et al. (2007)
		Isobutanol		40	
		1-Propanol		16	
		Ethanol		35	
		Methanol		Traces	
	Mahua oil	Ethanol		96	Noureddini et al. (2005)
	Soybean oil	Methanol		67	Deng et al. (2005)
	Sunflower oil	1-Butanol		88.4	Mittelbach (1990)
Pseudomonas fluorescens	Sunflower oil	Methanol	Solvent-free	3	
		Methanol	Petroleum ether	79	
		Ethanol	Solvent-free	82	
Rhizomucor miehei *Penicillium cyclopium*	Soybean oil	Methanol	Solvent-free	68–95	Guan et al. (2010)
Rhizomucor miehei	Sunflower oil	Methanol	n-Hexane	>80	Soumanou and Bornscheuer (2003)
Rhizopus miehei	Soybean oil	Methanol	n-Hexane	92.2	Shieh et al. (2003)
	Tallow, soybean, rapeseed oil	Primary alcohol	Hexane	94.8–98.5	Nelson et al. (1996)
		Methanol	Solvent-free	19.4	
		Ethanol	Solvent-free	65.5	
Rhizopus oryzae	Soybean oil	Methanol	Solvent-free	80–90	Kaieda et al. (1999)
	Palm oil	Methanol	Waste-activated bleaching earth	55	Pizzaro and Park (2003)
Thermomyces lanuginose	Soybean oil	Ethanol	n-Hexane/solvent-free	70–100	Rodrigues et al. (2010)
	Soybean oil Waste cooking oil	Methanol	Solvent-free	90–97	Dizge et al. (2009)
	Sunflower oil	Methanol	Solvent-free	>90	Soumanou and Bornscheuer (2003)

attention on branched and long-chain alcohols as better acyl acceptors for trans-esterification reactions (Watanabe et al., 2007). Kose et al. (2002) analyzed the effect of different alcohols on the esterification of cottonseed oil and observed the highest yield with isoamyl alcohol.

The deactivating effect of methanol can also be overcome by using solvents for both methanol and oil (Iso et al., 2001). Depending on the type of lipase, a variety of solvents like petroleum ether, hexane, isooctane, 1,4-dioxane, t-butanol, ionic liquids, or supercritical carbon dioxide have been suggested (Mittelbach and Remschmidt, 2004). Many workers have achieved methanolysis of respective oils using t-butanol as organic solvent (Li et al., 2006; Royon et al., 2007; Jeong and Park, 2008). Mittelbach (1990) has reported substantial yield using hexane as solvent for the transesterifica-tion of tallow oil. Park et al. (2008) and Kojima et al. (2004) reported the use of con-ventional diesel as solvent for *Candida rugosa* lipase. Bako et al. (2002) pointed out that the inhibitory effects on lipase-mediated transesterification reaction may also be contributed by glycerol accumulation. The workers have recommended *in situ* glycerol removal by dialysis or extraction using isopropanol (Xu et al., 2004). Wang et al. (2006) suggested that transesterification in t-butanol as solvent dissolves glyc-erol and enhances alcohol solubility. Few others have proposed the use of methyl and ethyl acetates for interesterification of vegetable oils and observed no inhibitory effects on lipase (Du et al., 2004; Modi et al., 2007). Methyl acetate enhances lipase stability as triacetin is produced instead of glycerol (Figure 18.3), which has a less negative effect and higher commercial value as compared to glycerol. It can be used as a fuel additive, or antiknock agent, and can be used to improve cold and viscos-ity properties of biodiesel (Tan et al., 2010). However, a higher amount of lipase and methyl acetate is required to shift the esterification reaction in forward direction. The use of dimethyl carbonate for the esterification of oil (Figure 18.3) has also been suggested. It offers advantage in being neutral, cheap, and nontoxic (Xu et al., 2005; Modi et al., 2007; Su et al., 2007). Despite the promising results, solvent removal from the final product poses additional difficulty, which increases the overall production cost. Alternatively, pretreatment of lipases can affect the progress of methanolysis.

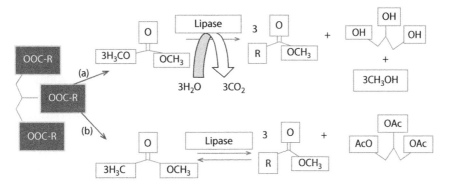

Figure 18.3 Transesterification of triglyceride for the synthesis of biodiesel using (a) dimethyl carbonate and (b) methyl acetate as acyl acceptor.

Pretreatment of immobilized Novozym 435 with methyl oleate for 30 min and then in soybean oil for 12 h has been reported to display more than 97% esterification yield in 3.5 h (Samukawa et al., 2000).

18.3.1.1 Process Considerations

The cost of lipase is the major obstacle for the commercialization of enzymatic transesterification process. Solid-state fermentation utilizes agro-industrial residues or by-products for enzyme production that can partly reduce the final product cost (Rodriguez et al., 2006). Castilho et al. (2000) compared the production economics of *Penicillium restrictum* lipase in submerged and solid-state fermentation and found the latter to be economically more attractive. Salum et al. (2010) produced a lipase from *Burkholderia cepacia* using solid-state fermentation on sugarcane bagasse and sunflower meal oil. Beef tallow as substrate for the production of *Pseudomonas gessardii* lipase has been used by Ramani et al. (2010). Recently, Kumar et al. (2015) have reported the transesterification of animal tallow using immobilized lipase from Novozymes in a solvent-free system.

A number of factors are known to influence enzymatic transesterification reaction. Optimal parameters vary depending on enzyme origin, source of oil, and reactor type. The use of waste oil can economize biodiesel production process. These have a high free fatty acid content that is favorable for enzymatic catalysis supported by higher yield and longer half-life of enzymes in free fatty acids medium as compared to triglyceride-rich substrates (Watanabe et al., 2001). Currently, the most widely used reactor type for biodiesel production is the batch stirred tank reactor (Balcao et al., 1996). Packed bed, fluidized bed, and continuous stirred tank reactors are used for immobilized lipases (Luković et al., 2011). Few workers have demonstrated fed-batch fermentation to support better lipase productivity. Show and coworkers (2012) have developed extractive fermentation method wherein a two-phase system has been used for simultaneous cell growth and recovery of extracellular lipase from *Bu. cepacia*.

Lipases have shorter half-life and poor catalytic activity in non-aqueous media. The relevant properties of lipases need to be improved for successful industrial applications. Protein engineering and site-directed mutagenesis allow tailoring of lipase for improved substrate specificity, stereoselectivity, thermostability, alcohol tolerance, and catalytic efficiency. *Pseudomonas aeruginosa* lipase has been modified with higher half-life in DMSO, cyclohexane, n-octane, and n-decane (Kawata and Ogino, 2009). Many workers have attempted to improve the specificity of lipases toward long-chain substrates (Santarossa et al., 2005; Hidalgo et al., 2008). Volpato et al. (2008) reported the use of glycerin as an alternative feedstock for the production of organic solvent–tolerant lipase by submerged culture of *Staphylococcus caseolyticus* EX17. Mangas-Sanchez and Adlercreutz (2015) have reported the conversion of triacylglycerol oils to ethyl esters using *Th. lanuginosus* lipase as catalyst in aqueous/organic two-phase system. The workers have reported high mass transfer using silica particles that decrease the size of emulsion droplets.

Tools of rDNA technology can be used to overexpress lipases for biofuel production, thus reducing the overall process cost at industrial scale. Heterologous expression systems have been most promising to lower the cost of lipases (Valero, 2012). The most common expression systems include *E. coli, Saccharomyces cerevisiae, P. pastoris,* and *Aspergillus* sp. that have been extensively reviewed by Hwang et al. (2014). Although lipases from natural sources can be used for biodiesel production, they have maximum catalytic efficiency at 30–50°C and low reaction rate at these temperatures, which make the process economically less viable (Hwang et al., 2014). Although lipases from a few thermophiles like *B. subtilis, Thermomyces lanuginosa, R. oryzae,* and *Pseudomonas* sp. have been reported to withstand elevated temperatures (Haki and Rakshit, 2003; Bouzas et al., 2006), long-term thermostability required for industrial application is still missing. Santarossa et al. (2005) identified key residues that can be modified to enhance the thermostability of *Pseudomonas fragi* lipase. Using error-prone PCR and DNA shuffling approaches, Yu et al. (2012) improved the thermostability of *Rhizopus chinensis* lipase. Reetz and Carballeira (2007) were able to shift the thermal stability of *B. subtilis* lipase from 48°C to 93°C by mutagenesis. Lipase productivity can also be substantiated by metabolic engineering techniques. The pioneering work in this respect has been initiated by Son et al. (2012) who attempted secretion of a thermostable lipase from *Ps. fluorescens.*

18.3.2 Enzymatic Reactions in Compressed and Supercritical Fluids

Potential advantages of biocatalysis in supercritical fluid include high diffusivity and low surface tension leading to low mass transfer limitations. Enzymes show low solubility in supercritical fluids, and therefore, heterogeneous systems are normally used (Adlercreutz, 1996). Moisture content, particle size, and solvent flow rate are important parameters that need to be optimized (Dunford, 2004). The most widely used medium is supercritical carbon dioxide that allows easy removal by lowering the pressure and exhibits lower toxicity compared to organic solvents. Lipase catalysis in pressurized fluids has been demonstrated as a potential technology for biodiesel production (Hildebrand et al., 2009). Lipase stability and activity may depend on the characteristics of enzyme and compressed fluid, water content of enzyme/carrier/ reaction mixture, and process variables. Lanza et al. (2004) have demonstrated the stability of *C. antarctica* lipase in supercritical carbon dioxide. The study of phase behavior of high-pressure multicomponent systems found in transesterification reactions is, however, essential for the design of supercritical fluid reaction system (Feltes et al., 2011).

18.4 Biohydrogen Production and Enzyme Hydrogenases

Hydrogen (H_2) is an alternative fuel that can be produced from domestic resources and represents long-term solution to the transportation sector. It is considered as less polluting and CO_2 neutral. It can be burnt to produce heat energy and used directly in internal combustion engine, completely replacing gasoline, or used in fuel cells to generate electricity. The energy in 1 kg of hydrogen gas is about the same as in 1 gallon of gasoline. It can also be used to turn the turbines on for generating electricity. Although in its market infancy as a transportation fuel, industries and researchers are working toward clean and economical hydrogen production and its distribution for use as transportation fuel (Mudhoo et al., 2011). Because hydrogen has a low volumetric energy density, additional steps are needed to store enough hydrogen onboard fuel cell vehicles to achieve the driving range of conventional vehicles. Most current applications use high-pressure tanks capable of storing hydrogen at either 5,000 or 10,000 psi. Other storage technologies are under development, including bonding hydrogen chemically with a material such as metal hydride or low-temperature sorbent materials. The National Aeronautics and Space Administration, United States, has used hydrogen for space flight since the 1950s (www.nasa.gov/topics/technology/hydrogen/hydrogen_fuel_of_choice.html).

Hydrogen is locked up in enormous quantities in water (H_2O), hydrocarbons, and other organic matter. One of the challenges of using hydrogen as a fuel comes from being able to efficiently extract hydrogen from these compounds. Currently, steam reforming or combining high-temperature steam with natural gas accounts for the majority of hydrogen produced in the United States. But these methods of hydrogen generation from methane or petroleum emit carbon monoxide or carbon dioxide. Therefore, low energy costs and environment friendliness prompt hydrogen gas production from biomass. There are two different methods of producing hydrogen from renewable resources: dark fermentation of cheap substrates or biophotolysis (Benemann, 1996). In dark fermentation, cheap substrates like glucose have been reported to yield about 0.37–3.3 mol H_2 per mole of glucose (Mertens and Liese, 2004). The biophotolysis process first demonstrated in green algae *Scenedesmus obliquus* and *Chlamydomonas reinhardtii* involves photosynthetic splitting of water to hydrogen and oxygen (Homann, 2003). The reaction is highlighted in Figure 18.4. The process comprises light reaction and dark reaction; the former involves capture of energy by chlorophyll molecules that split water. The electrons generated ferredoxin and the reducing power is used to fix CO_2 to carbohydrates. Ferredoxin is also an important mediator that helps to oxidize H^+ generated from water to H_2 by enzyme hydrogenases.

Hydrogenases (EC 1.12) were first described by Stephenson and Stickland (1931) and are widespread in prokaryotes and lower eukaryotes. According to their metal content, three types of hydrogenases have been identified: (NiFe) hydrogenases, (Fe)

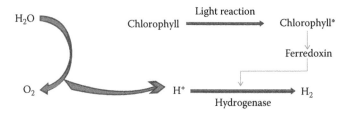

Figure 18.4 Biophotolysis of water for hydrogen production. * signifies excited chlorophyll molecule.

hydrogenases, and metal-free hydrogenases, more appropriately referred to as (FeS) cluster-free hydrogenase (Vignais et al., 2001; Lyon et al., 2004). Semin et al. (2013) reported a correlation in the accumulation of ferrous ions and induction of hydrogenase activity in *Ch. reinhardtii* under anaerobic conditions. Hydrogenase enzyme is inhibited in the presence of oxygen (Lee and Greenbaum, 2003). Schuetz et al. (2004) pointed out that mutants deficient in hydrogen uptake are more efficient hydrogen producers; otherwise, the produced hydrogen is reoxidized by the nitrogenase activity of nitrogen-fixing blue-green algae.

In place of whole cells, isolated hydrogenase can also be used for H_2 generation, but the enzyme requires cofactor $NADP^+$ that has to be regenerated during the production process. The hydrogenase from anaerobic thermophile *Pyrococcus furiosus* has been coupled *in vitro* with glucose dehydrogenase to produce hydrogen from glucose (Woodward et al., 1996; Inoue et al., 1999). Both the enzymes use $NADP^+$ as cofactor that can be recycled during oxidation of glucose and one mol H_2 is produced per mol of glucose. This system can also be coupled to invertase and cellulose to generate H_2 from sucrose and cellulose, respectively (Woodward and Orr, 1998; Woodward et al., 2000). Woodward et al. (2000) successfully coupled hydrogenase from *Py. furiosus* with pentose phosphate pathway enzymes and yield of 11.6 mol H_2 per mole glucose-6-phosphate could be achieved. Woodward et al. (2002) produced hydrogen by enzymes in cell-free extracts of *Thermotoga maritima* and *Py. furiosus*. The workers have successfully cloned and expressed two enzymes of pentose phosphate pathway in *E. coli*. Studies have been carried out on the cloning and overexpression of hydrogenases (Ohmiya et al., 2003).

18.5 Conclusions

With continuous developments in advanced biochemical processes, enzymes have proved themselves to be a vital component in biofuel production. There are, however, several obstacles in the use of enzymes as catalysts for industrial biofuel production; the most important are related to the instability and functionality of enzymes making the whole process cost extensive. Research attention is now focusing on genetic and metabolic engineering approaches for the production and overexpression of enzymes. Genes of various enzymes have successfully been cloned, and still some more are being explored for cloning in the future. Recombinant DNA technology holds promise to produce large quantities of recombinant enzymes at lower costs. Protein engineering can help to create novel biocatalysts with favorable properties. The introduction of a new generation of cheap enzymes, with enhanced activities and resilience, should change the economic balance in favor of enzyme use. The rational design of proteins requires a prior knowledge to study the structure–function relationship. Brady et al. (1990) were the first to study the x-ray crystallographic structure of *Rhizopus miehei* lipase. Later, the structure of other lipases, including *Bacillus thermocatenulatus*, *C. antarctica*, *Ps. cepacia*, and *B. subtilis*, was also solved. Statistical tools can also help to optimize the culture conditions and reaction conditions to substantiate yield at lower process cost for possible industrial adaptation.

Almost all of the hydrogen produced in the United States each year is used for refining petroleum, treating metals, producing fertilizer, and processing foods. Although the production of hydrogen may produce emissions affecting air quality depending on the source, fuel cell vehicles are zero-emission vehicles that emit water vapor and warm air as exhaust. Major research and development efforts are aimed at making these vehicles practical for widespread use and have led to the initial rollout of production vehicles entering the consumer market in the international scenario.

References

Abigor, R.D., Uaudia, P.O., Foglia, T.A. et al. 2000. Lipase-catalyzed production of biodiesel fuel from some Nigerian lauric oils. *Biochem Soc Trans* 28:979–981.

Adlercreutz, P. 1996. Modes of using enzyme in organic media. In *Enzymatic Reaction in Organic Media*, eds. A.M.P. Koskinen and A.M. Klibanov, pp. 9–42. Glasgow, Scotland: Blackie Academic and Professional.

Akoh, C.C., Chang, S.-W., Lee, G.-C., and J.-F. Shae. 2007. Enzymatic approach to biodiesel production. *J Agric Food Chem* 55:8995–9005.

Alkasrawi, M., Eriksson, T., Börjesson, J. et al. 2003. The effect of Tween-20 on simultaneous saccharification and fermentation of softwood to ethanol. *Enzyme Microb Technol* 33:71–78.

Alvarez, M.M., Carrillio, P.E., and S.S. Saldivar. 2010. Effect of decortications and protease treatment on the kinetics of liquefaction, saccharification and ethanol production from sorghum. *J Chem Technol Biotechnol* 85:1122–1130.

Al-Zuhair, S. 2007. Production of biodiesel: Possibilities and challenges. *Biofuels Bioprod Biorefin* 1:57–66.

Annapurna, K., Mahapatra, P., Garlapati, V.K., and R. Banerjee. 2009. Enzymatic transesterification of jatropha oil. *Biotechnol Biofuels* 2:1.

Anwar, Z., Gulfraz, M., and M. Irshad. 2014. Agro-industrial lignocellulosic biomass a key to unlock the future bioenergy: A brief review. *J Radiat Res Appl Biosci* 7:163–173.

Asgher, M., Ahmad, Z., and H.M.N. Iqbal. 2013. Alkali and enzymatic delignification of sugarcane bagasse to expose cellulose polymers for saccharification and bioethanol production. *Ind Crop Prod* 44:488–495.

Baek, S.-H., Kima, S., Leea, K., Leeb, J.-K., and J.-S. Hahna. 2012. Cellulosic ethanol production by combination of cellulase-displaying yeast cells. *Enzyme Microb Technol* 51:366–372.

Bako, K.B., Kova, F.C.S., Gubicza, L., and J.K. Hansco. 2002. Enzymatic biodiesel production from sunflower oil by *Candida antarctica* lipase in a solvent-free system. *Biocatal Biotransform* 20:437–439.

Balcao, V.M., Paiva, A.L., and F.X. Malcata. 1996. Bioreactors with immobilized lipases: State of art. *Enzyme Microb Technol* 18:392–416.

Benemann, J. 1996. Hydrogen biotechnology: Progress and prospects. *Nat Biotechnol* 14:1101–1103.

Bhalla, A., Bansal, N., Kumar, S.M., Bischoff, K.K., and R. Sani. 2013. Improved lignocellulose conversion to biofuel with thermophilic bacteria and thermostable enzymes. *Bioresour Technol* 128:751–759.

Bisen, P.S., Sanodiya, B.S., Thakur, G.S., Baghel, R.K., and G.B.K.S. Prasad. 2010. Biodiesel production with special emphasis on lipase-catalyzed transesterification. *Biotechnol Lett* 32:1019–1030.

Börjesson, J., Peterson, R., and F. Tjerneld. 2007. Enhanced enzymatic conversion of softwood lignocellulose by poly(ethylene glycol) addition. *Enzyme Microb Technol* 40:754–762.

Bouzas, T.D., Barros-Velazquez, J., and T.G. Villa. 2006. Industrial applications of hyperthermophilic enzymes: A review. *Protein Pept Lett* 13:645–651.

Bozbas, K. 2008. Biodiesel as an alternative motor fuel: Production and policies in the European Union. *Renew Sust Energ Rev* 12:542–552.

Brady, L., Brzozowski, A.M., Derewenda, Z.S. et al. 1990. A serine protease triad forms the catalytic center of a triacylglycerol lipase. *Nature* 343:767–770.

Breivik, H., Haraldsson, G.G., and B. Kristinsson. 1997. Preparation of highly purified concentrates of eicosapentaenoic acid and docosahexaenoic acid. *J Am Oil Chem Soc* 74:1425–1429.

Brummell, D.A. 2006. Cell wall disassembly in ripening fruit. *Funct Plant Biol* 33:103–119.

Cannella, D., Hsieh, C.-W., Felby, C., and H. Jorgensen. 2012. Production and effect of aldonic acids during enzymatic hydrolysis of lignocellulose at high dry matter content. *Biotechnol Biofuels* 5:26.

Castilho, L.R., Polato, C.M.S., Baruque, E.A., Sant'anna, Jr. G.L., and D.M.G. Freri. 2000. Economic analysis of lipase production by *Penicillium restriction* in solid state and submerged fermentation. *Biochem Eng J* 4:239–247.

Chandra, R.P., Bura, R., Mabee, W.E., Berlin, A., Pan, X., and J.N. Saddler. 2007. Substrate pretreatment: The key to effective enzymatic hydrolysis of lignocellulosics. *Biofuels* 108:67–93.

Chen, F. and R.A. Dixon. 2007. Downregulation of six lignin genes produces plants with improved saccharifiability, in some cases even without pretreatment (although pretreatment remains necessary to obtain the highest sugar yields). *Nat Biotechnol* 25:759.

Chen, X.A., Ishida, N., Todaka, N. et al. 2010. Promotion of efficient saccharification of crystalline cellulose by *Aspergillus fumigatus* Swo1. *Appl Environ Microbiol* 76:2556–2561.

Chen, Y., Xiao, B., Chang, J., Fu, Y., Lv, P., and X. Wang. 2009. Synthesis of biodiesel from waste cooking oil using immobilized lipase in fixed bed reactor. *Energ Convers Manage* 50:668–673.

Chheda, J.N., Roman-Leshkov, Y., and J.A. Dumesic. 2007. Production of 5-hydroxymethyl-furfural and furfural by dehydration of biomass-derived mono- and polysaccharides. *Green Chem* 9:342–350.

Deba, A.A., Tijani, H.I., Galadima, A.I., Mienda, B.S., Deba, F.A., and L.M. Zargoun. 2015. Waste cooking oil: A resourceful waste for lipase catalysed biodiesel production. *Int J Sci Res Publ* 5:1–12.

Demirbas, A. 2006. Biodiesel from sunflower in supercritical methanol with calcium oxide. *Energ Convers Manage* 48:937–941.

Demirbas, A. 2009. Progress and recent trends in biodiesel fuels. *Energ Convers Manage* 50:14–34.

Deng, L., Xu, X.B., Haraldsson, G.G., Tan, T.W., and F. Wang. 2005. Enzymatic production of alkyl esters through alcoholysis: A critical evaluation of lipases and alcohols. *J Am Oil Chem Soc* 82:341–347.

Dizge, N., Aydiner, C., Imer, D.Y., Bayramoglu, M., Tanriseven, A., and B. Keskinler. 2009. Biodiesel production from sunflower, soybean and waste cooking oils by transesterification using lipase immobilized onto a novel microporous polymer. *Bioresour Technol* 100:1983–1991.

Dodd, D. and I.K.O. Cann. 2009. Enzymatic deconstruction of xylan for biofuel production. *Glob Change Biol Bioenerg* 1:2–17.

Du, W., Xu, Y., Liu, D., and J. Zeng. 2004. Comparative study on lipase-catalyzed transformation of soybean oil for biodiesel production with different acyl acceptors. *J Mol Catal B Enzym* 30:125–129.

Dunford, N.T. 2004. Utilization of supercritical fluid technology for oil and oilseed processing. In *Nutritional Enhanced Edible Oil and Oilseed Processing*, eds. N.T. Dunford and H.B. Dunford, pp. 100–116. Champaign, IL: AOCS Press.

Eggeman, T. and D. Verser. 2006. The importance of utility systems in today's biorefineries and a vision for tomorrow. *Appl Biochem Biotechnol* 129–132:361–381.

Ezeji, T.C., Quereshi, N., and H.P. Blaschek. 2007. Bioproduction of butanol from biomass from genes to bioreactors. *Curr Opin Biotechnol* 18:220–227.

Fan, Z., Wu, W., Hildebrand, A., Kasuga, T., Zhang, R., and X. Xiong. 2012. A novel biochemical route for fuels and chemicals production from cellulosic biomass. *PLOS ONE* 7:e31693.

Feltes, M.M.C., de Oliveira, D., Ninow, J.L., and J.V. de Oliveira. 2011. An overview of enzyme-catalyzed reactions and alternative feedstock for biodiesel production. In *Alternative Fuel*, ed. M. Manzanera. InTech, Croatia, European Union.

Fischer, J., Lopes, V., Santos, E.F.Q., Coutinho, F.U., and V. Cardoso. 2014. Second generation ethanol production using crude enzyme complex produced by fungi collected in Brazilian cerrado (brazilian savanna). *Chem Eng Trans* 38:487–492.

Forsberg, Z., Vaaje-Kolstad, G., Westereng, B. et al. 2011. Cleavage of cellulose by a CBM33 protein. *Prot Sci* 20:1479–1483.

Fujita, Y., Ito, J., Ueda, M., Fukuda, H., and A. Kondo. 2004. Synergistic saccharification, and direct fermentation to ethanol, of amorphous cellulose by use of an engineered yeast strain codisplaying three types of cellulolytic enzyme. *Appl Environ Microbiol* 70:1207–1212.

Galbe, M. and G. Zacchi. 2002. A review of the production of ethanol from softwood. *Appl Microbiol Biochem* 59:618–628.

Guan, F., Peng, P., Wang, G. et al. 2010. Combination of two lipases more efficiently catalyzes methanolysis of soybean oil for biodiesel production in aqueous medium. *Process Biochem* 45:1667–1682.

Gusakov, A.V. 2011. Alternatives to *Trichoderma reesei* in biofuel production. *Trends Biotechnol* 29:419–425.

Haki, G.D. and S.K. Rakshit. 2003. Developments in industrially important thermostable enzymes: A review. *Bioresour Technol* 89:17–34.

Halim, S.F.A., Kamaruddin, A.H., and W.J.N. Fernando. 2009. Continuous biosynthesis of biodiesel from waste cooking palm oil in a packed bed reactor: Optimization using response surface methodology (RSM) and mass transfer studies. *Bioresour Technol* 100:710–716.

Harris, P.V., Welner, D., McFarland, K.C. et al. 2010. Stimulation of lignocellulosic biomass hydrolysis by proteins of glycoside hydrolase family 61: Structure and function of a large, enigmatic family. *Biochemistry* 49:3305–3316.

Harris, P.V., Xu, F., Kreel, N.E., Kang, C., and S. Fukuyama. 2014. New enzyme insights drive advances in commercial ethanol production. *Curr Opin Chem Biol* 19:162–170.

Hidalgo, A., Schliessmann, A., Molina, R., Hermoso, J., and U.T. Bornscheuer. 2008. A one-pot, simple methodology for cassette randomisation and recombination for focused directed evolution. *Protein Eng Des Sel* 21:567–576.

Hildebrand, C., Dalla Rosa, C., Freira, D.M.G. et al. 2009. Fatty acid methyl esters production using a non-commercial lipase in pressurized propane medium. *Cienc Tecnol Aliment* 29:603–608.

Homann, P.H. 2003. Hydrogen metabolism to green algae: Discovery and early research—A tribute to Hans Gaffron and his coworkers. *Photosynth Res* 76:93–103.

Horn, S.J., Vaaje-Kolstad, G., Westereng, B., and V.G.H. Eijsink. 2012. Novel enzymes for the degradation of cellulose. *Biotechnol Biofuels* 5:45–57.

Huang, A.H. 1996. Oleosins and oil bodies in seeds and other organs. *Plant Physiol* 110:1055–1061.

Hwang, H.T., Qi, F., Yuan, C. et al. 2014. Lipase-catalyzed process for biodiesel production: Protein engineering and lipase production. *Biotechnol Bioeng* 111:639–653.

Inoue, H., Kishishita, S., Kumagai, A., Katoaka, M., Fujii, T., and K. Ishikawa. 2015. Contribution of family 1 carbohydrate binding module in thermostable glycoside hydrolase to xylanase from *Talaromyces cellulolyticus* toward synergistic enzymatic hydrolysis of lignocellulose. *Biotechnol Biofuels* 8:77.

Inoue, T., Kumar, S.N., Kamachi, T., and I. Okura. 1999. Hydrogen evolution from glucose with the combination of glucose dehydrogenase and hydrogenase from *A. eutrophus* H16. *Chem Lett* 2:147–148.

International Energy Agency. 2006. World energy outlook 2006. https://www.iea.org/publications/freepublications/publication/weo2006.pdf. Accessed on September 24, 2015.

International Energy Agency. 2014. World energy investment outlook. A special report. www.iea.org/publications/freepublications/publication/weio2014.pdf. Accessed on September 24, 2015.

Iso, M., Chen, B., Eguchi, M., Kudo, T., and S. Shrestha. 2001. Production of biodiesel fuel from triglycerides and alcohol using immobilized lipase. *J Mol Catal B Enzym* 16:53–58.

Itoh, H., Wada, M., Honda, Y., Kuwahara, M., and T. Watanabe. 2003. Bioorganosolve pre-treatments for simultaneous saccharification and fermentation of beech wood by ethanolysis and white rot fungi. *J Biotechnol* 103:273–280.

Jayani, R.S., Saxena, S., and R. Gupta. 2005. Microbial pectinolytic enzymes: A review. *Process Biochem* 40:2931–2944.

Jeong, G.-T. and D.-H. Park. 2008. Lipase-catalyzed transesterification of rapeseed oil for biodiesel production with tert-butanol. *Appl Biochem Biotechnol* 148:131–139.

Jiang, Z., Wu, H., Xu, S., and S. Huang. 2003. Enzymatic conversion of carbon dioxide to methanol by dehydrogenases encapsulated in sol–gel matrix. In *Utilization of Greenhouse Gases*, Vol. 852, ACS Symposium Series, eds. C.-J. Lui, R.G. Mallinson, and M. Aresta, pp. 212–218. Washington, DC: American Chemical Society.

Jorgensen, H., Kutter, J.P., and L. Olsson. 2003. Separation and quantification of cellulases and hemicellulases by capillary electrophoresis. *Anal Biochem* 317:85–93.

Kaieda, M., Samukawa, T., Kondo, A., and H. Fukuda. 2001. Effect of methanol and water contents on production of biodiesel fuel from plant oil catalyzed by various lipases in a solvent-free system. *J Biosci Bioeng* 91:12–15.

Kaieda, M., Samukawa, T., Matsumoto, T. et al. 1999. Biodiesel fuel production from plant oil catalyzed by *Rhizopus oryzae* lipase in a water-containing system without an organic solvent. *J Biosci Bioeng* 88:627–631.

Kamini, N.R. and H. Iefuji. 2001. Lipase catalyzed methanolysis of vegetable oils in aqueous medium by *Cryptococcus* spp. S-2. *Process Biochem* 37:405–410.

Kawata, T. and H. Ogino. 2009. Enhancement of the organic solvent-stability of the LST-03 lipase by directed evolution. *Biotechnol Prog* 25:1605–1611.

Kim, E.S., Lee, H.J., Bang, W.G., Choi, I.G., and K.H. Kim. 2009. Functional characterization of a bacterial expansion from *Bacillus subtilis* for enhanced enzymatic hydrolysis of cellulose. *Biotechnol Bioeng* 102:1342–1353.

Kirk, O., Borchert, T.V., and C.C. Fuglsang. 2002. Industrial enzyme application. *Curr Opin Biotechnol* 13:345–351.

Klein-Marcuschamer, D., Oleskowicz-Popiel, P., Simmons, B.A., and H.W. Blanch. 2012. The challenge of enzyme cost in the production of lignocellulosic biofuels. *Biotechnol Bioeng* 109:1083–1087.

Kojima, S., Du, D., Sato, M., and E.Y. Park. 2004. Efficient production of fatty acid methyl ester from waste activated bleaching earth using diesel oil as organic solvent. *J Biosci Bioeng* 98:420–424.

Kose, O., Tuter, M., and H.A. Aksoy. 2002. Immobilized *Candida antarctica* lipase-catalyzed alcoholysis of cotton seed oil in a solvent-free medium. *Bioresour Technol* 83:125–129.

Krawczyk, T. 1996. Biodiesel—Alternative fuel makes inroads but hurdles remain. *Inform* 7:801–829.

Kumar, R., Singh, S., and O.V. Singh. 2008. Bioconversion of lignocellulosic biomass: Biochemical and molecular perspectives. *J Ind Microbiol Biotechnol* 35:377–391.

Kumar, S., Ghaly, A.E., and M.S. Brooks. 2015. Production of biodiesel from animal tallow via enzymatic transesterification using the enzyme catalyst NS88001 with methanol in a solvent-free system. *J Fundam Renew Energ Appl* 5:156.

Kumari, A., Mahapatra, P., Garlapati, V.K., and R. Banerjee. 2009. Enzymatic transesterification of Jatropha oil. *Biotechnol Biofuels* 2:1–7.

Kumari, V., Shah, S., and M.N. Gupta. 2007. Preparation of biodiesel by lipase-catalyzed transesterification of high free fatty acid containing oil from *Madhuca indica*. *Energ Fuels* 21:368–372.

Laane, C., Boeren, S., Vos, K., and C. Veegk. 1987. Rules for optimization of biocatalysis in organic solvent. *Biotechnol Bioeng* 30:81–87.

Ladisch, M.R. 1991. Fermentation derived butanol and scenarios for its use in energy derived applications. *Enzyme Microb Technol* 13:280–283.

Lanza, M., Priamo, W.L., Oliveria, J.V., Dariva, C., and D. Oliveria. 2004. The effect of temperature, pressure, exposure time, and pressurization rate on lipase activity in $SCCO_2$. *Appl Biochem Biotechnol* 113–116:181–187.

Lee, J.W. and E. Greenbaum. 2003. A new oxygen sensitive and its potential application in photosynthesis H_2 production. *Appl Biochem Biotechnol* 105–108:303–313.

Lee, S., Speight, J.G., and S.K. Loyalka. 2007. *Handbook of Alternative Fuel Technologies*. Boca Raton, FL: CRC, Taylor & Francis Group.

Lesiecki, M., Bialas, W., and G. Lewandowicz. 2012. Enzymatic hydrolysis of potato pulp. *Acta Sci Pol Technol Aliment* 11:53–59.

Li, L., Du, W., Liu, D., Wang, L., and Z. Li. 2006. Lipase-catalyzed transesterification of rapeseed oils for biodiesel production with a novel organic solvent as the reaction medium. *J Mol Catal B Enzym* 43:58–62.

Li, N., Zong, M., and H. Wu. 2009. Highly efficient transformation of waste oil to biodiesel by immobilized lipase from *Penicillium expansum*. *Process Biochem* 44:685–688.

Liang, C., Xue, Y., Fioroni, M. et al. 2011. Cloning and characterization of thermostable and halotolerant endoglucanase from *Thermoanaerobacter tengcongensis* MB4. *Appl Microbiol Biotechnol* 89:315–326.

Linko, Y.Y., Lansa, M., Wu, X., Uosukainen, E., Seppala, J., and P. Linko. 1998. Biodegradable products by lipase biocatalysis. *J Biotechnol* 66:41–50.

Luković, N., Knežević-Jugović, Z., and D. Bezbradica. 2011. Biodiesel fuel production by enzymatic transesterification of oils: Recent trends, challenges and future perspectives. In *Alternative Fuel*, ed. M. Manzanera. InTech, Croatia, European Union.

Lv, X., Zheng, F., Li, C., Zhang, W., Chen, G., and W. Liu. 2015. Characterization of a copper responsive promoter and its mediated overexpression of the xylanase regulator 1 results in an induction-independent production of cellulases in *Trichoderma reesei*. *Biotechnol Biofuels* 8:67.

Lyon, E.J., Shima, S., Buurman, G. et al. 2004. uv-A/blue light inactivation of the metal free hydrogenase from methanogenic archaea. *Eur J Biochem* 271:195–204.

Ma, F. and M.A. Hanna.1999. Biodiesel production: A review. *Bioresour Technol* 70:1–15.

Maceiras, R., Vega, M., Costa, C., Ramos, P., and M.C. Márquez. 2009. Effect of methanol content on enzymatic production of biodiesel from waste frying oil. *Fuel* 88:2130–2134.

Mangas-Sanchez, J. and P. Adlercruetz. 2015. Highly efficient enzymatic biodiesel production promoted by particle-induced emulsification. *Biotechnol Biofuels* 8:58.

Martinez, A.T., Speranza, M., Ruiz-Duenas, F.J. et al. 2005. Biodegradation of lignocellulosic: Microbial, chemical, and enzymatic aspects of the fungal attack of lignin. *Int Microbiol* 8:195–204.

Martinez, D., Larrondo, L.F., Putnam, N. et al. 2004. Genome sequence of the lignocelluloses degrading fungus *Phanerochaete chrysosporium* strain RP78. *Nat Biotechnol* 22:695–700.

Mertens, R. and A. Liese. 2004. Biotechnological applications of hydrogenases. *Curr Opin Biotechnol* 15:343–348.

Mielenz, J.A., Rodriguez, M., Thompson, O.A., Yang, X., and H. Yin. 2015. Development of agave as dedicated biomass source: Production of biofuels from whole plants. *Biotechnol Biofuels* 8:79.

Mittelbach, M. 1990. Lipase catalyzed alcoholysis of sunflower oil. *J Am Oil Chem Soc* 67:168–170.

Mittelbach, M. and C. Remschmidt. 2004. *Biodiesel—The Comprehensive Handbook*, pp. 69–80. Vienna, Austria: Boersedruck Ges.m.b.H.

Modi, M.K., Reddy, J.R.C., Rao, B.V.S.K., and R.B.N. Prasad. 2007. Lipase-mediated conversion of vegetable oils into biodiesel using ethyl acetate as acyl acceptor. *Bioresour Technol* 98:1260–1264.

Moser, B.R. 2009. Biodiesel production, properties and feed stocks. *In Vitro Cell Dev Biol Plant* 45:229–266.

Moser, F., Irwin, D., Chen, S.L., and D.B. Wilson. 2008. Regulation and characterization of *Thermobifida fusca* carbohydrate-binding module proteins E7 and E8. *Biotechnol Bioeng* 100:1066–1077.

Mudhoo, A., Forster-Carneiro, T., and A. Sánchez. 2011. Biohydrogen production and bioprocess enhancement: A review. *Crit Rev Biotechnol* 31:250–263.

Murai, T., Ueda, M., Kawaguchi, T., Arai, M., and A. Tanaka. 1998. Assimilation of cellooligosaccharides by a cell surface-engineered yeast expressing beta-glucosidase and carboxymethylcellulase from *Aspergillus aculeatus*. *Appl Environ Microbiol* 64:4857–4861.

Nakazawa, H., Kawai, T., Ida, N. et al. 2012. Construction of a recombinant *Trichoderma reesei* strain expressing *Aspergillus aculeatus* β-glucosidase 1 for efficient biomass conversion. *Biotechnol Bioeng* 109:92–99.

Nelson, L.A., Foglia, T.A., and W.N. Marmer. 1996. Lipase-catalyzed production of biodiesel. *J Am Oil Chem Soc* 73:1191–1195.

Nielsen, P.M., Brask, J., and L. Fjerbaek. 2008. Enzymatic biodiesel production: Technical and economical considerations. *Eur J Lipid Sci Technol* 110:692–700.

Noureddini, H., Gao, X., and R.S. Philkana. 2005. Immobilized *Pseudomonas cepacia* lipase for biodiesel fuel production from soybean oil. *Bioresour Technol* 96:769–777.

Ognjanović, N., Bezbradica, D., and Z. Knežević-Jugović. 2009. Enzymatic conversion of sunflower oil to biodiesel in a solvent-free system: Process optimization and immobilized system stability. *Bioresour Technol* 100:5146–5154.

Ohmiya, K., Sakka, K., Kimura, T., and K. Morimoto. 2003. *Clostridium paraputrificum* hydrogenase gene hyd and recombinant expression for enhanced hydrogen gas production. US Patent 2003, 2001-300527.

Olsson, L. and B. Hahn-Hagerdal. 1996. Fermentation of lignocellulosic hydrolysates for ethanol production. *Enzyme Microb Technol* 18:312–331.

Ortega, N., Busto, M.D., and M. Perez-Mateos. 2001. Kinetics of cellulose saccharification by *Trichoderma reesei* cellulases. *Int Biodeterior Biodegrad* 47:7–14.

Park, Y.M., Lee, D.W., Kim, D.K., Lee, J.S., and K.Y. Lee. 2008. The heterogeneous catalyst system for the continuous conversion of free fatty acids in used vegetable oils for the production of biodiesel. *Catal Today* 131:238–243.

Pauly, M. and K. Keegstra. 2008. Cell-wall carbohydrates and their modification as a resource for biofuels. *Plant J* 54:559–568.

Pearl, G.G. 2002. Animal fat potential for bioenergy use. Paper presented at *Bioenergy 2002, The Biennial Bioenergy Conference*, Boise, ID.

Pizzaro, A.V.L. and E.Y. Park. 2003. Lipase-catalyzed production of biodiesel fuel from vegetable oils contained in waste activated bleaching earth. *Process Biochem* 38:1077–1082.

Quinlan, R.J., Sweeney, M.D., Lo Leggio, L. et al. 2011. Insights into the oxidative degradation of cellulose by a copper metalloenzyme that exploits biomass components. *Proc Natl Acad Sci USA* 108:15079–15084.

Qureshi, N., Ezeji, T.C., Ebener, J., Dien, B.S., Cotta, M.A., and H.P. Blaschek. 2008a. Butanol production by *Clostridium beijerinckii*. Part I: Use of acid and enzyme hydrolyzed corn fiber. *Bioresour Technol* 99:5915–5922.

Qureshi, N., Saha, B.C., and M.A. Cotta. 2008b. Butanol production from wheat straw by simultaneous saccharification and fermentation using *Clostridium beijerinckii*: Part II—Fed-batch fermentation. *Biomass Bioenerg* 32:176–183.

Ralph, J., Lundquist, K., Brunow, G. et al. 2004. Lignins: Natural polymers from oxidative coupling of 4-hydroxyphenyl-propanoids. *Phytochem Rev* 3:29–60.

Ramani, K., Kennedy, L.J., Ramakrishnan, M., and G. Sekarn. 2010. Purification, characterization and application of acidic lipase from *Pseudomonas gessardii* using beef tallow as a substrate for fats and oil hydrolysis. *Process Biochem* 45:1683–1691.

Reese, E.T., Siu, R.G.H., and H.S. Levinson. 1950. The biological degradation of soluble cellulose derivatives and its relationship to the mechanism of cellulose hydrolysis. *J Bacteriol* 59:485–497.

Reetz, M.T. and J.D. Carballeira. 2007. Iterative saturation mutagenesis (ISM) for rapid directed evolution of functional enzymes. *Nat Protoc* 2:891–903.

Robles-Medina, A., Gonzalez-Moreno, P.A., Esteban-Cerdan, L., and E. Molina-Grima. 2009. Biocatalysis: Towards ever greener biodiesel production. *Biotechnol Adv* 27:308–408.

Rodrigues, R.C., Pessela, B.C.C., Volpato, G., Fernandez-Lafuente, R., Guisan, J.M., and M.A.Z. Ayub. 2010. Two step ethanolysis: A simple and efficient way to improve the enzymatic biodiesel synthesis catalyzed by an immobilized-stabilized lipase from *Thermomyces lanuginosus*. *Process Biochem* 45:1268–1273.

Rodriguez, J.A., Mateos, J.C., Nungaray, J. et al. 2006. Improving lipase production by nutrient source modification using *Rhizopus homothallicus* cultured in solid state fermentation. *Process Biochem* 41:2264–2269.

Rousseau, D. and A.G. Marangoni. 2002. The effect of interesterification on the physical properties of fats. In *Physical Properties of Lipids*, eds. A.G. Marangoni and S.S. Narine, pp. 1348–1354. Boca Raton, FL: CRC Press.

Royon, D., Daz, M., Ellenrieder, G., and S. Locatelli. 2007. Enzymatic production of biodiesel from cottonseed oil using t-butanol as a solvent. *Bioresour Technol* 98:648–653.

Saha, B.C. 2003. Hemicellulose bioconversion. *J Ind Microbiol Biotechnol* 30:279–291.

Saha, B.C., Iten, L.B., Cotta, M.A., and Y.V. Wu. 2005. Dilute acid pretreatment, enzymatic saccharification and fermentation of wheat straw to ethanol. *Process Biochem* 40:3693–3700.

Salum, T.F.C., Villeneuve, P., Barea, B. et al. 2010. Synthesis of biodiesel in column fixed bed bioreactors using the fermented solid produced by *Burkholderia cepacia* LTEB11. *Process Biochem* 45:1345–1348.

Samukawa, T., Kaieda, M., Matsumoto, T. et al. 2000. Pretreatment of immobilized *Candida antarctica* lipase for biodiesel fuel production from plant oil. *J Biosci Bioeng* 90:180–183.

Santarossa, G., Lafranconi, P.G., Alquati, C. et al. 2005. Mutations in the "lid" region affect chain length specificity and thermostability of a *Pseudomonas fragi* lipase. *FEBS Lett* 579:2383–2386.

Schmid, R.D. and R. Verger. 1998. Lipases: Interfacial enzymes with attractive applications. *Angew Chem Int Ed* 37:1608–1633.

Schuetz, K., Happe, T., Toshima, O. et al. 2004. Cyanobacteria H_2 production—A comparative analysis. *Planta* 218:350–359.

Semin, B.K., Davlestshin, L.N., Novakova, A.A. et al. 2013. Accumulation of ferrous iron in *Chlamydomonas reinhardtii* influence of CO_2 and anaerobic induction of reversible hydrogenase. *Plant Physiol* 131:1756–1764.

Shah, S. and M.N. Gupta. 2007. Lipase catalyzed preparation of biodiesel from Jatropha oil in a solvent free system. *Process Biochem* 42:409–414.

Shah, S., Sharma, S., and M.N. Gupta. 2003. Enzymatic transesterification for biodiesel production. *Indian J Biochem Biophys* 40:392–399.

Shah, S, Sharma, S., and M.N. Gupta. 2004. Biodiesel preparation by lipase-catalysed transesterification of jatropha oil. *Energ Fuels* 18:154–159.

Shieh, C.J., Liao, H.F., and C.C. Lee. 2003. Optimization of lipase-catalyzed biodiesel by response surface methodology. *Bioresour Technol* 88:103–106.

Shimada, Y., Watanabe, Y., Samukawa, T. et al. 1999. Conversion of vegetable oil biodiesel using immobilized *Candida antarctica* lipase. *J Am Oil Chem Soc* 76:789–793.

Show, P.L., Tan, C.P., Anuar, M.S. et al. 2012. Extractive fermentation for improved production and recovery of lipase derived from *Burkholderia cepacia* using a thermoseparating polymer in aqueous two-phase systems. *Bioresour Technol* 116:226–233.

Singhania, R.R., Patel, A.K., Sukumaran, R.K., Larroche, C., and A. Pandey. 2013. Role and significance of beta-glucosidase in hydrolysis of cellulose for bioethanol production. *Bioresour Technol* 127:500–507.

Son, M., Moon, Y., Oh, M.J. et al. 2012. Lipase and protease double-deletion mutant of *Pseudomonas fluorescens* suitable for extracellular protein production. *Appl Environ Microbiol* 78:8454–8462.

Soumanou, M.M. and U.T. Bornscheuer. 2003. Lipase-catalyzed alcoholysis of vegetable oils. *Eur J Lipid Sci Technol* 105:656–660.

Sriyapai, T., Somyoonsap, P., Matsui, K., Kawai, F., and K. Chansiri. 2011. Cloning of thermostable xylanase from *Actinomadura* sp. S14 and its expression in *Escherichia coli* and *Pichia pastoris*. *J Biosci Bioeng* 111:528–536.

Stephenson, M. and L.H. Stickland. 1931. Hydrogenase: A bacterial enzyme activating molecular hydrogen. I. The properties of hydrogenase. *Biochem J* 25:205–214.

Su, E.-Z., Zhang, M.-J., Zhang, J.-G., Gao, J.-F., and D.-Z. Wei. 2007. Lipase catalysed irreversible transesterification of vegetable oils for fatty acid methyl ester production with dimethyl carbonate as the acyl acceptor. *Biochem Eng J* 36:167–173.

Sukumaran, R.K., Singhania, R.R., Mathew, G.M., and A. Pandey. 2009. Cellulase production using biomass feed stock and its applications in lignocelluloses saccharification for bioethanol production. *Renew Energ* 34:421–424.

Sun, Y. and C. Cheng. 2002. Hydrolysis of lignocellulosic materials for ethanol production: A review. *Bioresour Technol* 83:1–11.

Taherzadeh, M.J. and K. Karimi. 2007. Enzyme-based hydrolysis processes for ethanol production from lignocellulosic materials: A review. *BioResources* 2:707–738.

Tan, K., Lee, K.T., and A.R. Mohamed. 2010. A glycerol free process to produce biodiesel by supercritical methyl acetate technology: An optimization study via response surface methodology. *Bioresour Technol* 101:965–969.

Teixeira, L.C., Linden, C.J., and H.A. Schroeder. 2000. Simultaneous saccharification and cofermentation of peracetic acid-pretreated biomass. *Appl Biochem Biotechnol* 84–86:111–127.

Tsai, S.L., Goyal, G., and W. Chen. 2010. Surface display of a functional minicellulosome by intracellular complementation using a synthetic yeast consortium and its application to cellulose hydrolysis and ethanol production. *Appl Environ Microbiol* 76:7514–7520.

Tsai, S.L., Oh, J., Singh, S., Chen, R., and W. Chen. 2009. Functional assembly of minicellulosomes on the *Saccharomyces cerevisiae* cell surface for cellulose hydrolysis and ethanol production. *Appl Environ Microbiol* 75:6087–6093.

Valero, F. 2012. Heterologous expression systems for lipases: A review. In *Methods in Molecular Biology, Lipases and Phospholipases: Methods and Protocol*, ed. G. Sandowal, pp. 161–178. Totowa, NJ: Humana Press.

Vandan, W.A., Sabat, G., Mozuch, M., Kersten, P.J., Cullen, D., and R.A. Blanchette. 2006. Structure, organization and transcriptional regulation of a family of copper radical oxidase genes in the lignin-degrading basidiomycetes *Phanerochaete chrysosporium*. *Appl Environ Microbiol* 72:4871–4877.

Varga, E., Klinke, H.B., Reczey, K., and A.B. Thomsen. 2004. High solid simultaneous saccharification and fermentation of wet oxidized corn stover to ethanol. *Biotechnol Bioeng* 88:567–574.

Vasudevan, P.T. and M. Briggs. 2008. Biodiesel production-current state of the art and challenges. *J Ind Microbiol Biotechnol* 35:421–430.

Vignais, P., Billoud, B., and J. Meyer. 2001. Classification and phylogeny of hydrogenases. *FEMS Microbiol Rev* 25:455–501.

Villeneuve, P. 2007. Lipase in lipophilization reaction. *Biotechnol Adv* 25:515–536.

Visioli, L.J., Enzweiler, H., Kuhn, R.C., Schwaab, M., and M.A. Mazutti. 2014. Recent advances on biobutanol production. *Sustain Chem Process* 2:15.

Volpato, G., Rodrigues, R.C., Heck, J.X., and M.A.Z. Ayub. 2008. Production of organic solvent tolerant lipase by *Staphylococcus caseolyticus* EX17 using raw glycerol as substrate. *J Chem Technol Biotechnol* 83:821–828.

Wackett, P.L. 2008. Biomass to fuels via microbial transformations. *Curr Opin Chem Biol* 12:187–193.

Walker, G.M. 2011. Fuel alcohol: Current production and future challenges. *J Inst Brew* 117:3–22.

Wang, J., Bai, Y.P., Luo, H. et al. 2010. A new xylanase from thermoalkaline *Anoxybacillus* sp. E2 with high activity and stability over a broad pH range. *World J Microbiol Biotechnol* 26:917–924.

Wang, L., Du, W., Liu, D., Wang, L., Li, L., and N. Dai. 2006. Lipase-catalyzed biodiesel production from soybean oil deodorizer distillate with absorbent present in t-butanol system. *J Mol Catal B Enzym* 43:29–32.

Watanabe, Y., Pinsirodom, P., Nagao, T. et al. 2007. Conversion of acid oil by-produced in vegetable oil refining to biodiesel fuel by immobilized *Candida antarctica* lipase. *J Mol Catal B Enzym* 44:99–105.

Watanabe, Y., Shimada, Y., Sugihara, A., Noda, H., Fukuda, H., and Y. Tominaga. 2000. Continuous production of biodiesel fuel from vegetable oil using immobilized *Candida antarctica* lipase. *J Am Oil Chem Soc* 77:355–360.

Watanabe, Y., Shimada, Y., Sugihara, A., and Y. Tominaga. 2001. Enzymatic conversion of waste edible oil to biodiesel fuel in a fixed bed bioreactors. *J Am Oil Chem Soc* 78:703–707.

Watanabe, Y., Shimada, Y., Sugihara, A., and Y. Tominaga. 2002. Conversion of degummed soybean oil to biodiesel fuel with immobilized *Candida antarctica* lipase. *J Mol Catal B Enzym* 17:151–155.

Wen, F., Sun, J., and H. Zhao. 2010. Yeast surface display of trifunctional minicellulosomes for simultaneous saccharification and fermentation of cellulose to ethanol. *Appl Environ Microbiol* 76:1251–1260.

Wen, Z., Liao, W., and S. Chen. 2005. Production of cellulose/beta-glucosidase by the mixed fungi culture *Trichoderma reesei* and *Aspergillus phoenicis* on dairy manure. *Process Biochem* 40:3087–3094.

Weng, K.-J., Li, X., Bonawitz, D.N., and C. Chapple. 2008. Emerging strategies of lignin engineering and degradation for cellulosic biofuel production. *Curr Opin Biotechnol* 19:166–172.

Wilson, B.D. 2009. Cellulases and biofuels. *Curr Opin Biotechnol* 20:295–299.

Wisner, R. 2013. Feedstocks used for U.S. biodiesel: How important is corn oil? AgMRC Renewable Energy Climate Change. *Newsletter*, April 2013. http://www.agmrc.org/renewable_energy/biodiesel/feedstocks-used-for-us-biodiesel-how-important-is-corn-oil. Accessed on September 24, 2015.

Woodward, J., Cordray, K.A., Edmonston, R.J., Blanco-rivera, M., Mattingly, S.M., and B.R. Evan. 2000. Enzymatic hydrogen production: Conversion of renewable resource for energy production. *Energ Fuels* 14:197–201.

Woodward, J., Heyer, N.I., Getty, J.P., O'Neill, H.M., Pinkhassik, E., and B.R. Evans. 2002. Efficient hydrogen production using enzymes of the pentose phosphate pathway. *Proceedings of the 2002 U.S. DOE Hydrogen Program Review*, NREL/CP-610-32405, Golden, CO.

Woodward, J., Mattingly, S.M., Damson, M., Hough, D., Ward, N., and M. Adams. 1996. *In vitro* hydrogen production by glucose by dehydrogenase and hydrogenase. *Nat Biotechnol* 14:872–874.

Woodward, J. and M. Orr. 1998. Enzymatic conversion of sucrose to hydrogen. *Biotechnol Prog* 14:897–902.

Woodward, J., Orr, M., Cordray, K., and E. Greenbaum. 2000. Enzymatic production of bio-hydrogen. *Nature* 405:1014–1015.

Wu, A. and Y.Y. Lee. 1998. Nonisothermal simultaneous saccharification and fermentation for direct conversion of lignocellulosic biomass to ethanol. *Appl Biochem Biotechnol* 70–72:479–492.

Wu, S., Liu, B., and X. Zhang. 2006. Characterization of a recombinant thermostable xyla-nase from deep-sea thermophilic *Geobacillus* sp. MT-1 in East Pacific. *Appl Microbiol Biotechnol* 72:1210–1216.

Wu, W.H., Foglia, T.A., Marmer, W.N., and J.G. Phillips. 1999. Optimizing production of ethyl esters of grease using 95% ethanol by response surface methodology. *J Am Oil Chem Soc* 76:517–521.

Xiao, C. and C.T. Anderson. 2013. Roles of pectin in biomass yield and processing for biofu-els. *Plant Sci* 4:67.

Xiao, W., Wang, Y., Xia, S., and P. Ma. 2012. The study of factors affecting the enzymatic hydrolysis of cellulose after ionic liquid pretreatment. *Carbohydr Polym* 87:2019–2023.

Xu, Y., Du, W., and D. Liu. 2005. Study on the kinetics of enzymatic interesterification of tri-glycerides for biodiesel production with methyl acetate as the acyl acceptor. *J Mol Catal B Enzym* 32:241–245.

Xu, Y., Du, W., Zeng, J., and D. Liu. 2004. Conversion of soybean oil to biodiesel fuel using Lipozyme TL IM in a solvent free medium. *Biocatal Biotransform* 22:45–48.

Yang, D., Weng, H., Wang, M., Xu, W., Li, Y., and H. Yang. 2010. Cloning and expression of a novel thermostable cellulose from newly isolated *Bacillus subtilis* strain 115. *Mol Biol Rep* 37:1923–1929.

Yao, G., Li, Z., Gao, L. et al. 2015. Redesigning the regulatory pathway to enhance cellulase production in *Penicillium oxalicum*. *Biotechnol Biofuels* 8:71.

Ying, M. and G. Chen. 2007. Study on the production of biodiesel by magnetic cell biocatalyst based on lipase-producing *Bacillus subtilis*. *Appl Biochem Biotechnol* 137–140:793–804.

Yu, X.W., Wang, R., Zhang, M., Xu, Y., and R. Xiao. 2012. Enhanced thermostability of a *Rhizopus chinensis* lipase by *in vivo* recombination in *Pichia pastoris*. *Microb Cell Fact* 11:1–11.

Zhang, Y.-H.P., Evans, B.R., Mielenz, J.R., Hopkins, R.C., and M.W.W. Adams. 2007. High-yield hydrogen production from starch and water by a synthetic enzymatic pathway. *PLOS ONE* 2:e456.

19

New Trends in Enzyme Immobilization and Nanostructured Interfaces for Biofuel Production

Ranjeeta Bhari, Manpreet Kaur, and Ram Sarup Singh

Contents

Abstract

Biofuels have provided a solution to world's energy crisis and tolling due to population invasion and depleting fossil fuels. Catalysts are potential molecules for the rapid conversion of cheaper feedstock to valuable fuels like bioethanol or biodiesel. Immobilized catalysts offer advantages over free enzymes due to their reusability and easy separation from product. Immobilization of enzyme on nanomaterials increases the surface area and dispersability of immobilized biocatalyst, thus signifying the commercial viability of nanotechnology in the biofuel industry. Waste oils and lignocellulosic feedstock are currently being explored for biodiesel and bioethanol production, respectively. Enzyme lipase and cellulase are the key catalyst. Algae are potential source of oils and lipids that can be effectively extracted and used as biodiesel, and this extraction can be accelerated using nanoscale particles. However, there are challenges in the use of nanomaterials, including stability and handling. The need for risk assessment and safety analysis of nanomaterials has been realized in the present scenario. Understanding the biocompatibility of nanomaterials is a prerequisite for the development of safe nanotechnology. The application of nanotechnology for the immobilization of biocatalysts and biofuel production is highlighted in this chapter. The applicability of nanoemulsions as future biofuels is also included.

19.1 Introduction

Rapidly growing population and urbanization, leading to increasing emission of greenhouse gases, has prompted the need for sustainable fuels and a breakthrough in research on biofuels (Srivastava and Prasad 2000). A consistent improvement in the production of biofuels through successive generations has been witnessed in the past two decades (Demirbas 2008). Commercially, biofuels are generated from feedstock such as vegetable oils, animal fats, corn, and sugarcane. However, increased cost of food and feedstuffs in competition to mass production of biofuels has diverted the focus on technology that allows the use of nonfood sources such as cellulosic biomass and biowastes for biofuel production (Farell et al. 2006). Enzymes have been attractive biomolecules for the production of biofuels due to their high specificity, mild reaction conditions, and absence of secondary reactions and reduced energy consumption as compared to chemical processes (Szczesna-Antczak et al. 2009).

Cellulases and lipases are the primary candidates used for the large-scale production of biofuels. Lipases are highly enantioselective enzymes that catalyze transesterification reactions for the production of biodiesel (Goswami et al. 2013; Kralovec et al. 2010, 2012; Wang et al. 2011). Cellulases catalyze saccharification of the lignocellulosic materials for ethanol production (Puri et al. 2012). Cellulases are a mixture of three different enzymes—endoglucanase, exoglucanase, and β-glucosidase—that work synergistically to produce glucose that can be subsequently fermented into ethanol using suitable microbial strains (Chandel et al. 2012; Garvey et al. 2013). However, the use of free enzymes involves limitations such as enzyme inactivation by solvents, high enzyme costs, instability, efficient recovery from substrates, and scale-up barriers that can be partially reduced by using immobilized enzymes (Verma et al. 2015). Immobilized enzymes have thermal and operational stability at extreme conditions of pH, temperature, and organic solvents (Ansari and Husain 2012; Li et al. 2011; Verma et al. 2012). Immobilized forms result in improved product quality and lower processing cost (Verma et al. 2015). However, shear stress to immobilized enzymes in stirred tank bioreactors might disrupt enzyme–carrier interaction, although packed bed reactors offer solution to this problem (Verma et al. 2012). The limitations of free and immobilized enzymes can be overcome using nanomaterials as physical support structures for enzymes. These nanosupports possess several favorable characteristics such as high surface area, effective enzyme loading, and good biocatalytic potential that strengthen their commercial viability (Feng et al. 2011; Kim et al. 2006). Binding of cells to nanosupports has been reported to improve ethanol tolerance and enhanced thermal stability and, therefore, present novel candidates for biocatalysis in aqueous/nonaqueous media (Ivanova et al. 2011; Verma et al. 2015). The present chapter is focused on the viability of nanomaterials for the immobilization of enzymes and the commercial applicability of nanotechnology in biofuel industry. Public concerns and safety issues regarding the use of nanomaterials are highlighted.

19.2 Nanomaterials as Immobilization Support for Enzymes

Materials at nanoscale exhibit properties different than they have in bulk such as strength, conductivity, reflectivity, and chemical reactivity (Verma et al. 2013). Nanoparticles represent current forefront of nanotechnology and have a diameter ranging between 1 and 100 nm (Ngo et al. 2013). Nanostructured supports such as Fe_3O_4, Fe_2O_3, polystyrene, silica, chitosan, gold, carbon, and peptide have been employed and are found more suitable for enzyme immobilization than micron-sized particles due to their high surface area to volume ratio, high enzyme loading, better biocatalytic activity, and stability (Andrade et al. 2010; Dyal et al. 2003; Gupta et al. 2011; Kim et al. 2006; Pavlidis et al. 2012). Low mass transfer resistance of nanomaterial-bound enzymes enhances enzyme activity (Kim et al. 2008). The use of magnetic nanoparticles offers easy separation of magnetic nanomaterial-bound enzymes, thus lowering enzyme instability as well as enzyme costs, which offers advantages over the conventional methods of separation (centrifugation and filtration). Enzyme-immobilized nanofiber membrane bioreactors are more advantageous to use as compared to traditional enzyme-immobilized membranes and fixed bed reactors due to small pressure drop and high flow rate of nanofibers (Nair et al. 2007; Ren et al. 2011; Safarik and Safarikova 2009).

However, there are a few challenges in the use of nanomaterial-immobilized enzymes such as formation, monodispersity, thermodynamic stability, aggregation or precipitation of particles, toxicity, and handling of nanomaterials (Arico et al. 2005; Krumov et al. 2009).

There are no specific regulatory requirements for the use of nanomaterials now, but the use of nanomaterials requires selection of representative for each nanomaterial based on their characteristics as size, shape, surface chemistry, and property. Toxicity testing for the selected nanomaterials is required and finally the implementation of good manufacturing practices ensures the quality, handling, and stability of nanomaterials (Chan 2006). Assessment of environmental impact of nanomaterials is equally important. Therefore, understanding the biological interactions of nanomaterials and biocompatibility is prerequisite to develop safe future nanotechnology (Verma et al. 2013).

19.3 Functionalization of Nanomaterials

Nanomaterials such as nanoparticles, nanofibers, nanotubes, nanopores, nanosheets, and nanocomposites have many applications in biofuel production (Verma et al. 2013). To provide stability, biocompatibility, and functionality, functionalization of nanomaterials is an important step during which functional groups are grafted onto the surface of nanoparticle on which enzyme can bind (Shim et al. 2002; Verma et al. 2013). A variety of organic and inorganic materials such as silica, natural polymers (dextran, starch, gelatin, chitosan), synthetic polymers (polyethylene glycol, polyvinyl alcohol, polylactic acid, polymethylmethacrylate, polyacrylic acid, bio-polymers, dendrimers) and small molecules are used for the functionalization of nanoparticles (Jiang et al. 2013). These functional groups play important roles in enzyme immobilization and control electrostatic interaction, create links with the amino acid groups of the targeted enzyme, and decrease the size of pore entrance for the entrapment of enzymes in the nanochannels (Verma et al. 2015). Various physical and chemical approaches like co-precipitation, electrospinning, dealloying, and thermal annealing are well documented for the synthesis of nanomaterials (Borlido et al. 2013; Wu et al. 2008).

In the co-precipitation method for the preparation of nanoparticles, ferric chloride and ferrous chloride are dissolved in deionized water and stirred in a nitrogen environment, followed by the addition of ammonium or sodium hydroxide solution, resulting in the formation of a black precipitate that is collected with a magnet and washed repeatedly with water to remove nonmagnetic by-products (Kalantari et al. 2012; Wu et al. 2008). Electrospinning process is a simplified technique for nanofiber preparation. The electrospinning system basically comprises three components: a high-voltage power supply, a spinneret, and a grounded collecting plate; it utilizes a high-voltage source to inject charge of a certain polarity into a polymer solution or melt, which is then accelerated toward a collector of opposite polarity (Sill and Recum 2008). Arc discharge, laser ablation, and chemical vapor deposition techniques have been used for the preparation of nanotubes, and electricity, heat, or high light intensity is used for their respective functionalization. For the synthesis of carbon nanotubes, a carbon source is required to produce fragments of groups or single carbon atoms (Saifuddin et al. 2013). Different pore sizes of nanoporous gold have been synthesized by using dealloying and thermal annealing processes. In this method, Ag and Au chemically dealloy in concentrated nitric acid (Qiu et al. 2011). Thermal exfoliation of graphite oxide is used to obtain nanosheets (Kishore et al. 2012).

Magnetic nanoparticles are being increasingly used in biofuel production (Ngo et al. 2013). Difficulty in separation of non-magnetic nanoparticles for reuse offers nanofibers as substitutes. Nanofibers have large surface area and interconnected pores for enzyme attachment (Verma et al. 2015). The surface of native nanoparticles can be deposited with organic or inorganic layer of silica, silane, or oleic acid. The hybrid nanomaterial formed, referred to as nanocomposite, facilitates grafting of more functional groups, thus acting as better supports for enzyme immobilization (Tran et al. 2012). Georgelin et al. (2010) have reported covalent grafting of β-glucosidase enzyme on twice-functionalized γ-Fe_2O_3 @ SiO_2 core–shell magnetic nanocomposite.

19.4 Strategies for Immobilization of Enzymes on Nanomaterials

Nanostructured forms are preferred for enzyme immobilization and stabilization. Though expensive, these robust nanoscaffolds have ideal characteristics to balance key factors governing the efficiency of biocatalysts like high surface area, minimum diffusion limitations, and high volumetric enzyme loading rate (Mohamad et al. 2015). Co-immobilization of multienzymes on nanomaterials can also be achieved.

Enzymes can be immobilized using various techniques such as adsorption, covalent binding, cross-linking, and encapsulation (Datta et al. 2013). Covalent immobilization improves enzyme stability, thus leading to easy recovery and reuse. However, steric hindrance, diffusion, and structural changes during immobilization process are some of its limitations that can be overcome by cross-linking (Zhu et al. 2010).

Some of the commonly used nanomaterials are magnetic particles like silicon, iron oxide, zinc oxide, chitosan, gold, and silver that allow easy separation of immobilized enzyme from reaction mixture (Ansari and Husain 2012). However, aggregation, precipitation, thermodynamic stability, and monodispersity of nanomaterials are some of their limitations (Verma et al. 2013). Studies have demonstrated the use of redox polymer microparticles to maintain activity for longer time. Microparticles are required to have some of the essential factors such as transfer of electrons generated in enzymatic reaction and to maintain longer activity and should be attached to the surface firmly (Lin et al. 2013). Nanomaterial-immobilized enzymes having potential application in biofuel industries are highlighted in Table 19.1.

Enzymes undergo substantial changes in the surface microenvironment and confirmation after immobilization. Hence, the analysis of surface orientation and functionality of biomolecules upon immobilization is the key for the development of biocatalysts for wide biotechnological applications. There are various analytical techniques to characterize the nanosupport, successful binding of enzyme to support. Some of them include Fourier transform infrared, thermogravimetric analysis, Raman spectroscopy, transmission electron microscopy, scanning electron microscopy, surface plasmon resonance, circular dichroism spectroscopy, atomic force microscopy, Forster resonance energy transfer, and x-ray photoelectron spectroscopy (Mohamad et al. 2015).

19.5 Nanotechnology in Biorefineries

19.5.1 Biomass to Biofuels

Search for sustainable alternatives to petroleum fuel has catapulted biofuels, particularly ethanol, on the top of the list. Bioethanol was commercially produced from sugar, starch, and oilseed-based feedstocks, which being food sources have created a surge for nonfood feedstocks. Therefore, biomass or lignocellulosics are explored for biofuel production (Guo et al. 2012). Nanotechnology involved in the conversion of fossil feedstocks into conventional fuels or biomass feedstocks into biofuels is the application of nanostructure solids to catalyze the desired chemical transformations. The use of heterogeneous catalyst is the oldest commercial application of nanotechnology. During the designing of effective catalyst, two fundamental parameters must be controlled, that is, creation of active centers and transport of reactants to and products away from the active center. Both these aspects can be controlled by manipulating catalyst structure only at nanoscale (Weisz et al. 1962; Weisz and Miale 1965). During the initial discovery of crystalline, microporous zeolite solids, crystal size (10^1–10^4 nm) as well as pore size and shape (0.3–50 nm) has been engineered

TABLE 19.1 Nanomaterial-Immobilized Enzymes in Biofuel Production

Enzyme	Nanomaterial	Type of Bonding	Salient Features	Reference
Nanoparticles				
Candida rugosa lipase	γ-Fe$_2$O	Covalent	Long-term stability	Dyal et al. (2003)
Candida rugosa lipase	Fe$_3$O$_4$	Covalent	Increased activity and affinity of enzyme	Huang et al. (2003)
Candida rugosa lipase	Polylactic acid	Adsorption	Enhancement of enzyme activity	Chronopoulou et al. (2011)
Candida rugosa lipase	Magnetic	Covalent	Higher esterification efficiency, good reusability	Dandavate et al. (2011)
Candida rugosa lipase	Polyvinyl alcohol	Covalent	Equivalent activity to that of commercially immobilized lipase	Nakane et al. (2007)
Candida rugosa lipase	Fe$_3$O$_4$	Covalent	High conversion rate of biodiesel production	Wang et al. (2009)
Pseudomonas cepacia lipase	Zirconia	Covalent	High stability	Chen et al. (2009)
Burkholderia cepacia lipase	Magnetic	Adsorption	Good reusability	Rebelo et al. (2010)
Mucor javanicus lipase	Silica	Covalent	High loading and reusability	Kim et al. (2006)
Thermomyces lanuginose lipase	Fe$_3$O$_4$	Covalent	High conversion rate of biodiesel production	Xie and Ma (2009)
Trichoderma viride cellulase	Silica	Adsorption	Ethanol production from microcrystalline Cellulose	Lupoi and Smith (2011)
Aspergillus niger β-glucosidase	Iron oxide	Covalent	High conversion rate of cellobiose to glucose	Verma et al. (2013)
Nanofibers				
Candida antartica lipase	Polysulfone	Adsorption	Enhanced thermal stability	Wang (2006)
Candida rugosa lipase	Polyvinyl alcohol	Covalent	High enzyme loading and catalytic rate	Wang and Hsieh (2008)

(Continued)

TABLE 19.1 (Continued) Nanomaterial-Immobilized Enzymes in Biofuel Production

Enzyme	Nanomaterial	Type of Bonding	Salient Features	Reference
Pseudomonas cepacia lipase	Polyvinyl alcohol	Adsorption	High initial transesterification rate	Sakai et al. (2010)
Aspergillus niger β-glucosidase	Polystyrene	Covalent	Biodiesel from cellulose	Lee et al. (2010)
Nanopores				
Pseudomonas cepacia lipase	Nanoporous gold	Adsorption	Biodiesel production from soybean oil and methanol	Wang et al. (2011)
Porcine pancreas lipase	Mesoporous silica SBA-15	Adsorption	Enhanced lipase activity	Zou et al. (2010)
Phycomyces nitens	Mesoporous silica	Encapsulation	High activity and stability	Itoh et al. (2010)
Nanotubes				
Rhizopus arrhizus	Carbon nanotube	Covalent	Improved resolution efficiency	Ji et al. (2010)
Candida rugosa	Peptide	Adsorption	High thermal stability	Yu et al. (2005)
Nanosheets				
Candida rugosa	Graphene oxide	Adsorption	High activity and stability	Pavlidis et al. (2012)
Nanocomposite				
Burkholderia cepacia lipase	Ferric-silica	Adsorption	High conversion rate of biodiesel production	Tran et al. (2012)
Candida rugosa lipase	Magnetic Fe_3O_4–chitosan	Covalent	High enzyme loading and activity	Wu et al. (2009)

to alter the distribution of reaction products during the catalytic cracking of large petroleum-derived hydrocarbons (Corma et al. 2004).

During the processing of raw biomass, techniques have been developed to transform biomass into fluid media that can be converted in subsequent steps into fuels (Huber et al. 2006). Biomass can be gasified to yield synthesis gas (syngas mixture of H_2/CO), which is important feedstock for conversion into other fuels. But the major problem during biomass gasification is the production of tar that might be addressed through the design of new, efficient nanostructure catalysts (Huber et al. 2006). Nano-sized NiO (nano-NiO) particles have received research attention for their catalytic properties. Li et al. (2008a) investigated the effect of nano-NiO particles and micro-NiO particles as catalysts for biomass gasification and reported that the decomposition of cellulose in the presence of micro-NiO was accomplished at 10°C lower than that of the pure cellulose, while the decomposition of cellulose with nano-NiO started at 294°C, which was 19°C lower than that of the pure cellulose. Using nano-Ni catalyst (NiO supported on gamma alumina) in direct gasification of sawdust improved the quality of the produced gas while significantly eliminating tar production (Li et al. 2008a,b). Now, research has focused on using microbial systems for the biosynthesis of liquid biofuels. According to Rude and Schirmer (2009), there are basically three routes to convert renewable resources into energy-rich, fuel-like molecules, or fuel precursors. These include direct production from CO_2 by photosynthetic organisms, such as plants, algae, and photosynthetic bacteria; fermentative or non-fermentative production from cheap organic compounds (e.g., sugars) by heterotrophic microorganisms, such as bacteria, yeast, or fungi; and chemical conversion of biomass to fuels.

Cellulose comprises about 40% of all biomass dry matter. It has a polymer structure of β-linked glucose molecules that can be hydrolyzed by sequential action of endoglucanase, exoglucanase, and β-glucosidase. The glucose so formed can be readily fermented to ethanol using suitable microbial strains. Immobilization of cellulase enzymes on nanoparticles increases the biocatalytic property of enzyme, enabling stability and reusability (Ansari and Hussain 2012). Immobilization of cellulase on magnetic nanoparticles offers easy recovery of immobilized cellulase after hydrolysis. Alftren and Hobley (2013) immobilized β-glucosidase covalently on nonporous magnetic particles to enable the reuse of the enzyme. Iron oxide nanoparticles have been incorporated into single-walled carbon nanotubes (SWCNTs) to produce magnetic single-walled carbon nanotubes (mSWCNTs). Amyloglucosidase has been immobilized onto mSWCNTs using physical adsorption and covalent immobilization. Workers report that the enzyme retains up to 40% of its catalytic efficiency after repeated use up to 10 cycles. Additionally, immobilized enzyme has been found to be stable at 4°C for at least 1 month (Goh et al. 2012). Using physical adsorption method cellulase has been immobilized to silica nanoparticles. Immobilized cellulase was used to increase ethanol yields in the simultaneous saccharification and fermentation reaction of cellulose as compared to free enzymes (Lupoi and Smith 2011). Thus, by applications of nanomaterial-bound enzymes, biocatalytic efficiency in biofuel production has been improved, which is a crucial step toward future applications in biofuel production.

19.5.1.1 Biodiesel Production

Biodiesel can be produced from vegetable oils, animal fats, microalgal oils, waste products of vegetable oil refinery or animal rendering, and waste frying oils. Biodiesel is the most widely accepted alternative fuel for diesel engines due to its technical, environmental, and strategic advantages (Enweremadu et al. 2011). Biodiesel is produced by the transesterification of triglycerides using enzyme or chemical catalyst. Industrially, chemical catalyst is acceptable for its high conversion and reaction rates. However, downstream processing costs, environmental issues associated with biodiesel production, and by-product recovery have led to the development of enzymatic production process (Bisen et al. 2010).

A number of researchers have studied the preparation of nanosized heterogeneous catalysts to increase the catalytic activity due to high surface area and high catalytic activity of nanocatalysts. Transesterification reaction of sunflower oil with magnetic $Cs/Al/Fe_3O_4$ showed high catalytic activity for biodiesel production and the biodiesel yield reached 94.8% (Feyzi et al. 2013). Obadiah et al. (2012) used calcined Mg–Al hydrotalcite for the transesterification of Pongamia oil with methanol as a solid base catalyst and obtained 90.8% conversion of methyl esters. Biodiesel production from rapeseed oil and sunflower oil using different nanocrystalline MgO catalysts in nanosheets has resulted in 98.03% yield (Verziu et al. 2008). Kaur and Ali (2011) studied lithium ion–impregnated calcium oxide as a nanocatalyst for biodiesel production from karanja and jatropha oils, and 99% yield has been reported by using Li–CaO nanocatalyst after 1 and 2 h, respectively. Nanomagnetic solid base catalyst $KF/CaO–Fe_3O_4$ has been used for the transesterification of stillingia oil extracted from the seeds of Chinese tallow (*Sapium sebiferum*), and 95% biodiesel production has been reported (Hu et al. 2011).

Dyal et al. (2003) studied higher enzyme loading (5.6 wt.%) on γ-Fe_2O_3 nanoparticles (20 nm). The Fe_3O_4 magnetic nanoparticle (12.7 nm)-bound *Candida rugosa* lipase activated by carbodiimide exhibited 1.41-fold enhanced activity, 31-fold improved stability, and better tolerance to pH variation as compared to free enzyme (Huang et al. 2003). Chronopoulou et al. (2011) immobilized *Candida rugosa* lipase on poly-DL-lactic acid (PDLLA) nanoparticles (220 nm) and immobilized enzyme showed improved activity and stability as compared to free enzyme. Dandavate et al. (2011) immobilized *Candida rugosa* lipase on magnetic nanoparticles (15 nm) and showed higher esterification efficiency compared to free lipase for the synthesis of ethyl isovalerate. Nakane et al. (2007) showed *Candida rugosa* lipase immobilized on polyvinyl alcohol and the enzyme exhibited equivalent activity to that of commercially immobilized lipase Novozym-435. High conversion rate for biodiesel production has been observed by immobilizing *Candida rugosa* lipase on Fe_3O_4 (Wang et al. 2009). *Pseudomonas cepacia* lipase (PCL) immobilized on modified zirconia nanoparticles has been used for asymmetric synthesis in organic media (Chen et al. 2009). *Burkholderia cepacia* lipase (BCL) was immobilized on superparamagnetic nanoparticles (10 nm) using adsorption and chemisorption methodologies (Rebelo et al. 2010). Lipase has been applied as a recyclable biocatalyst in the enzymatic kinetic resolution of (RS)-1-(phenyl) ethanols via transesterification reactions. Covalent immobilization of silica nanoparticles supported higher enzyme loading with *Mucor javanicus* lipase (Kim et al. 2006).

A high conversion rate for biodiesel production (90%) was reported from covalently immobilized *Thermomyces lanuginosus* lipase to the amino-functionalized magnetic nanoparticles after 12 h. Xie and Ma (2009) showed high conversion of soybean oil to biodiesel. Using physical adsorption method, cellulase was immobilized to silica nanoparticles and the immobilized cellulase was used to increase ethanol yields in the simultaneous saccharification and fermentation reaction of cellulose as compared to free enzymes (Lupoi and Smith 2011). β-Glucosidase from *Aspergillus niger* immobilized covalently on iron oxide nanoparticles has been used for the conversion of cellobiose to glucose (Verma et al. 2013).

Wang (2006) reported high thermal stability of *Candida rogusa* lipase immobilized covalently on polysulfone nanofibers as compared to free enzymes. *Candida rogusa* lipase immobilized onto polyvinyl alcohol nanofibers showed high enzyme loading (Wang and Hsieh 2008). Lipase entrapped in electrospun polyvinyl alcohol nanofibers enhanced the initial transesterification rate (Sakai et al. 2010). Lee et al. (2010) studied biodiesel production from cellulose by covalently immobilizing β-glucosidase from *Aspergillus niger* onto polystyrene nanofibres.

Tran et al. (2012) synthesized core–shell nanoparticles by coating Fe_3O_4 core with silica gel, treated the nanoparticles with dimethyl octadecyl [3-(trimethoxysilyl) propyl], and used ammonium chloride as immobilization supporters for *Burkholderia* lipase. This has been used to catalyze the transesterification of olive oil. High enzyme loading and activity have been reported for *Candida rugosa* lipase covalently immobilized onto magnetic Fe_3O_4-chitosan nanocomposites (Wu et al. 2009). Wang et al. (2011) reported biofuel production from soybean oil and methanol using *Pseudomonas cepacia* lipase immobilized onto nanoporous gold. Porcine pancreas lipase immobilized onto mesoporous silica by physical adsorption has been reported to exhibit higher lipase activity as compared to free enzymes (Zou et al. 2010).

In the United States, oil extracted from soybeans has been the primary source of biodiesel. However, the use of a crop-based feedstock such as soybeans has inherent limitations, such as a limited cultivation area and potential negative impacts on the food source market. A more appropriate feedstock for biodiesel production would utilize nonfood sources, such as waste animal fat, waste biomass, or microalgae. Algae are known to produce a wide variety of high-value and value-added hydrocarbons and lipids (Metzger and Largeau 2005; Zhila et al. 2005). Additionally, soybeans offer a lower yield of oil per hectare than microalgae. Currently, biodiesel is produced from very low free fatty acid percentage (<1%) crop oils by the conversion of triacylglycerides through a base-catalyzed transesterification. Algae can be used as a potential feedstock source for the production of biofuels, as many species of algae grow in ocean water or wastewater and can produce more energy than other biofuel crops (Chisti 2008). Li and coworkers demonstrated large-scale production of biodiesel by successfully growing algae to a density of 14.2 g/L in an 11,000 L bioreactor with a lipid content of 44.3% dry weight. The workers demonstrated the conversion of algal oil to fatty acid methyl esters through transesterification using immobilized lipase (Li et al. 2007). However, the potential for employing microalgae-derived oil as a source for biodiesel is currently limited by the cost and technical complexity associated with the cultivation of the microalgae, extraction, and refining of the oil and its conversion into biodiesel (Angenent et al. 2004). This necessitates the development of technology to separate algae-produced molecules to optimize the economics of converting algal oil to energy. Nanoscale materials provide increased surface area for enzyme loading and help to increase the diffusion rate of substrates to the enzymes and this property helps to ease oil extraction from algae (Cruz et al. 2007). Furthermore, the extracted oil can be converted into biodiesel by transesterification (Lin et al. 2013). Nanomaterials act as better catalyst during transesterification. Calcium oxide nanoparticles with crystalline size 20 nm and surface area 90 m^2/g are more efficient than conventionally sized particles (43 nm). Similarly, tungstated zirconia nanocatalysts also influence the catalytic behavior (Tran et al. 2012).

The discovery and development of surfactant micelle–templated mesoporous silica nanoparticles (MSN) with high surface area and pore volume have advanced the utilization of these materials for applications in catalysis, sensors, delivery vessels, and adsorbents. Structurally ordered nanoparticles have attracted attention because of their tailored porosity and surface chemistry; high specific surface area; high thermal, chemical, and mechanical stabilities; and large pore volume (Chen et al. 2005). Ordered mesoporous catalytic solid (MCS) nanoparticles can be used as a heterogeneous catalyst to enable cleaner biodiesel production. They aim to selectively isolate fuel-relevant hydrocarbons from live microalgae by using mesoporous material to convert microalgae-based hydrocarbons and waste oils to biodiesel in a single step using a mesoporous mixed metal-oxide catalyst. This is expected to increase the extraction efficiency of fuel-relevant hydrocarbons from feedstock and the establishment of extraction conditions appropriate for a large-scale oil production.

Sharma et al. (2015) used electroplating waste for the extraction of nickel nanoparticles by chemical precipitation and sol–gel method. They used these nanoparticles as heterogeneous catalyst to carry out transesterification from the butchery waste (Semwal et al. 2011). A high conversion of biodiesel production (90%) has been reported from covalently immobilized *Thermomyces lanuginosus* lipase to amino-functionalized magnetic nanoparticles for 12 h. Xie and Ma (2009, 2010) showed the high conversion of soybean oil to biodiesel. Liu et al. (2012) immobilized *Burkholderia* sp. lipase onto hydrophobic magnetic particles (HMP) via an adsorption method for 8 h. About 90% biodiesel production has been achieved by the immobilized *Pseudomonas cepacia* lipase via an adsorption method onto the nanopores (35 nm) of a nanoporous gold support (Wang et al. 2011); however, only 74% conversion was achieved with free lipase in 24 h.

Waste grease has also been employed for the production of biodiesel using novel magnetic nanoparticles (Ngo et al. 2013). *Thermomyces lanuginosus* lipase and *Candida antartica* lipase B were covalently immobilized to magnetic nanoparticles for 4 h, followed by freeze-drying to give high enzyme loading of 61 or 22 mg, respectively. Immobilized *Thermomyces lanuginosus* lipase showed best performance for the transesterification of grease with methanol, giving high yield (99%) in 12 h. The immobilized lipase retained about 88% productivity even after 11 cycles.

19.6 Hydrogenation

Hydrogen energy system is being looked upon as long-term solution to energy crisis (Veziroğlu and Şahin 2008). But the major problem in a future hydrogen economy is the development of a safe, compact, robust, and efficient means of hydrogen storage, in particular for mobile applications. Industrially, thermochemical and biochemical methods are available for biomass-based hydrogen production that causes environment pollution. For the production of useful fuels, nanostructure catalysts have been tested, which can increase the efficiency of reforming and gasification processes (Colmenares et al. 2009). De et al. (2011) successfully synthesized 5-hydromethylfurfural from fructose, glucose, and cellulose using nanocatalysts. This 5-hydromethylfurfural can be converted to 2,5-dimethylfuran or to levulinic acid (Dutta et al. 2012). Biomass-derived levulinic acid serves as precursor for liquid fuels like γ-valerolactone (Manzer 2006). For efficient aerobic oxidation of 5-hydromethylfurfural, highly dispersed metal nanoparticulates on metal oxide supports have been used. 5-Hydromethylfurfural oxidation/esterification process was carried out for the production of 2,5-furandimethylcarboxylate by using Au on a TiO_2 support, and the reactions were run at 130°C and 4 bar O_2 in methanol with MeONa as base (Taarning et al. 2008). Biomass-derived 5-hydromethylfurfural oxidation into furandimethylcarboxylate has been carried out by using Au on CeO_2 but without a base (Casanova et al. 2009). Workers concluded that the catalytic efficiency of Au/CeO_2 depends on the particle size of the catalyst. Testing of aqueous oxidation of benzyl alcohol and hydrogenation of furfural in water by using microwave irradiation using carbon-supported Pd nanoparticles showed that the most active nanocatalysts are trioctylphosphine and triphenylphosphine that stabilize Pd nanoparticles on oxidized carbon support. The oxygen groups present on the surface of the carbon support improve water affinity as well as Pd nanoparticle immobilization affinity and, consequently, the catalytic performances of the system (Garcia-Suarez et al. 2012).

Metal nanoparticles, including Pt, Pd, Rh, and Ru (5% loading), supported on a mesoporous material derived from biomass have been used for the hydrogenation of succinic acid in aqueous ethanol because these nanoparticles would be evenly dispersed due to their smaller size (Clark et al. 2009). Hydrogenation can be used to convert cellobiose to C_6-alcohols by breaking glycosidic bonds using water-soluble transition metal (Ru, Rh, Pd, or Pt) nanocluster catalysts under H_2 pressure in ionic liquid, and results revealed that Ru was the only metal that exhibited high stability under the reaction conditions (120°C and 40 bar of H_2). Ru nanoparticles gave high activity and selectivity for the conversion of cellobiose to C_6-alcohols. This concept created new avenues to transform cellulose into biofuels and other useful value-added chemicals (Yan et al. 2006).

19.7 Nanotech Liquid Additives

The preceding discussion referred to practical opportunities for nanotechnology related to the use of solid nanoparticles. However, the performance of additive biofuels or fuel blends can be enhanced by using liquid nanoparticles or droplets (Guo et al. 2012). Nanoemulsions are nanotech-based liquid additives that can improve detergency and water cosolvency and result in more complete combustion, consequently enhancing fuel efficiency. These nanodroplets are derived from the surfactant action of additive in the fuel formulation. The presence of some water in all commercial fuel systems is usually due to condensation (Trindade 2011).

Nanoemulsions with fuel, water, and surfactant are thermodynamically stable, microscopically isotropic, and nanostructured. The nanostructures with fuel, water, and surfactant are able to reduce soot and NO_x emissions, resulting in higher fuel efficiency (Wulff et al. 2008). Strey (2007) filed a patent on nanoemulsions fuels and interpreted the behavior of stable diesel–water–surfactant nanoemulsions. The surfactant components (oleic acid) and nitrogen-containing compounds (amines) dissolve readily in diesel (and possibly in biodiesel) fuel and bind water to it without stirring. The nanometer water droplets help to stabilize the emulsion. The resulting "liquid sponge" can be stored indefinitely, like ordinary diesel fuel, without the risk of phase separation. This fuel formulation, when burned, results in the near-complete elimination of soot and a reduction of up to 80% in NO_x emissions. The surfactant in the formulation also burns without creating emissions beyond water, carbon dioxide, and nitrogen.

19.8 Public Concerns over Nanotechnology

Despite promising results in the field of sustainable energy applications, nanotechnology is a newer area of technology and faces many challenges like reliability, lifetime, safety, and costs. In an attempt to assess possible consequences of the deployment of nanotechnology to humans and environment, Woodrow Wilson Center carried out a nanotechnology project from 2005 and concluded that "manipulating materials at the atomic level can have astronomic repercussions, both positive and negative. The problem is no one really knows exactly what these effects may be" (http://www.loe.org/shows/segments.htm?program-00050&segmentID=3). The International Risk Governance Council also undertook a nanotechnology project and experts concluded that care should be taken to address the possible impact of nanomaterials on society and environment. Due to untested and unpredictable risks of nanotechnology, UK Soil Association banned the use of nanomaterials in all the certified products (Trindade, 2011). At nanoscale, large surface area is due to smaller size; thus, its electronic structure also changes that not only affects its catalytic activity but can also lead to aggressive chemical reactivity.

Nanoparticles in free form can be released into the environment during the production process and ultimately accumulate in the soil, water, or plant life. In fixed form, where they are part of a manufactured substance or product, they will ultimately have to be recycled or disposed of as waste. But unknowingly maybe new classes of nonbiodegradable pollutants are generated. Moreover, no method has been discovered that can remove pollutants from environment because most traditional filters are not suitable for such tasks. Health and environmental issues combine in the workplace of companies engaged in producing or using nanomaterials and in the laboratories engaged in nanoscience. To properly assess the health hazards of engineered nanoparticles, the whole cycle of these particles needs to be evaluated, including their fabrication, storage and distribution, application and potential abuse, and disposal. The impact on humans or the environment may vary at different stages of the cycle (Guo 2012).

19.9 Conclusions

Despite efforts at improving energy efficiency and diversification of energy systems, the demand for energy services is increasing and biofuels play a key role in this scenario. However, limited supply of biofuels in future has necessitated the use of nonfood feedstocks and new technologies for the production of biofuels. Nanotechnologies are the candidate technology for the production of biofuels from lignocellulosic feedstock, algal feedstock, oily seeds, and biomass without jeopardizing security, public health, or the environment. The development of nanoparticles or nanoparticle-coated catalytic membrane may render efficient sugar recovery without enzymatic hydrolysis and also play a role in augmenting the efficiency of using current and future liquid fuels, especially biofuels, by providing improved combustion of nanodroplets. While every new technology is subject to new risks, advancement in science and technology offers better ways to risk assessment and according actions that it seems possible to advance nanotechnology-based biofuels. But the reach of nanotechnology is vast and goes much beyond biofuels and offers hopes in many areas, including human health. Immobilizing enzymes onto a variety of nanostructured materials provides several benefits when compared to immobilization on larger materials or unimmobilized enzymes. The application of nanomaterials in enzyme immobilization results in higher enzyme loading, multiple recycling, and protection from denaturation of enzymes particularly in a packed bed reactor. Nanoemulsions, though not applied as commercial fuels, can be explored as potential fuels of future. Thus for the sustainable energy production, nanotechnology is one of the fastest-growing research fields in the world and will hopefully lead to the development of a renewable energy economy to produce more valuable chemicals. Nanotechnological developments may provide new inputs to the next generation of biofuels. However, a lot of research inputs are required to understand the technical challenges related to synthesis, toxicity, and safety evaluation of nanomethods. Studies are also required for process development and scale-up of nanomaterials as novel matrices to exploit their potential at industrial scale.

References

Alftren, J. and T.J. Hobley. 2013. Covalent immobilization of β-glucosidase on magnetic particles for lignocellulose hydrolysis. *Res Gate* 1:169.

Andrade, L.H., Rebelo, L.P., Netto, C.G.C.M., and H.E. Toma. 2010. Kinetic resolution of a drug precursor by *Burkholderia cepacia* lipase immobilized methodologies on superparamagnetic nanoparticles. *J Mol Catal B Enzym* 66:55–62.

Angenent, L.T., Karim, K., Al-Dahhan, M.H., and B.A. Wrenn. 2004. Production of bioenergy and biochemicals from industrial and agriculture wastewater. *Trend Biotechnol* 22:477–485.

Ansari, S.A. and Q. Husain. 2012. Potential applications of enzymes immobilized on/in nanomaterials: A review. *Biotechnol Adv* 30:512–523.

Arico, A.S., Bruce, P., and B. Scrosati. 2005. Nanostructured materials for advanced energy conversion and storage devices. *Nat Mater* 4:366–377.

Bisen, P.S., Sanodiya, B.S., Thakur, G.S., Baghel, R.K., and G.B. Prasad. 2010. Biodiesel production with special emphasis on lipase-catalyzed transesterification. *Biotechnol Lett* 32:1019–1030.

Borlido, L., Azevedo, A.M., Roque, A.C.A., and M.R. Aires-Barros. 2013. Magnetic separations in Biotechnology. *Biotechnol Adv* 31:1374–1385.

Casanova, O., Iborra, S., and A. Corma. 2009. Biomass into chemicals: Aerobic oxidation of 5-hydroxymethyl-2-furfural into 2,5-furandicarboxylic acid with gold nanoparticle catalysts. *Chem Sust Energ Mater* 2:1138–1144.

Chan, V.S.W. 2006. Nanomedicine: An unresolved regulatory issue. *Regul Toxicol Pharmacol* 46:218–224.

Chandel, A.K., Chandrasekhar, G., Silva, M.B., and S.S.D. Silva. 2012. The realm of cellulases in biorefinery development. *Crit Rev Biotechnol* 32:187–202.

Chen, M., Kumar, D., Yi, C.W., and D.W. Goodman. 2005. The promotional effect of gold in catalysis in palladium-gold. *Science* 310:291–293.

Chen, Y.Z., Ching, C.B., and R. Xu. 2009. Lipase immobilisation on modified zirconia nanoparticles: Studies on the effects of modifiers. *Process Biochem* 44:1245–1251.

Chisti, Y. 2008. Biodiesel from microalgae beats bioethanol. *Trends Biotechnol* 26:126–131.

Chronopoulou, L., Kamel, G., Sparago, C. et al. 2011. Structure-activity relationships of *Candida rugosa* lipase immobilised on polylactic acid nanoparticles. *Soft Matter* 7:2653–2662.

Clark, J.H., Yoshida, K., and P.L. Gai. 2009. Efficient aqueous hydrogenation of biomass platform molecules using supported metal nanoparticles on Starbons®. *Chem Comm* 35:5305–5307.

Colmenares, J.C., Aramendía, M.A., Marinas, A., Marinas, J.M., and F.J. Urbano. 2009. Nanostructured photocatalysts and their applications in the photocatalytic transformation of lignocellulosic biomass: An overview. *Materials* 2:2228–2258.

Corma, A, Atienzar, P., Garcia, H., and J.Y. Chane-Ching. 2004. Hierarchically mesostructured doped CeO_2 with potential for solar-cell use. *Nat Mater* 3:394–397.

Cruz, A.D., Kulkarni, M.G., Meher, L.C., and A. Dalai. 2007. Synthesis of biodiesel from canola oil using heterogeneous base catalyst. *J Am Oil Chem Soc* 84:937–943.

Dandavate, V., Keharia, H., and D. Madamwar. 2011. Ester synthesis using *Candida rugosa* lipase immobilised on magnetic nanoparticles. *Biocatal Biotransform* 29:37–45.

Datta, S., Christena, L.R., and Y.R.S. Rajaram. 2013. Enzyme immobilization: An overview on techniques and support materials. *3 Biotech* 3:1–9.

De, S., Dutta, S., Patra, A.K., Bhaumik, A., and B. Saha. 2011. Self-assembly of mesoporous TiO_2 nanospheres via aspartic acid templating pathway and its catalytic application for 5-hydroxymeth-yl-furfural synthesis. *J Mater Chem* 21:17505–17510.

Demirbas, A. 2008. Hazardous emission from combustion of biomass. *Energ Source A* 42:1357–1378.

Dutta, S., De, S., and B. Saha. 2012. A Brief summary of the synthesis of polyester building-block chemicals and biofuels from 5-hydroxyme-thylfurfural. *ChemPlusChem* 77:259–272.

Dyal, A., Loos, K., Noto, M. et al. 2003. Activity of *Candida rugosa* lipase immobilised on gamma-Fe_2O_3 magnetic nanoparticles. *J Am Chem Soc* 125:1684–1685.

Enweremadu, C.C., Rutto, H.L., and J.T. Oladeji. 2011. Investigation of the relationship between some basic flow properties of shea butter biodiesel and their blends with diesel fuel. *Int J Phys Sci* 6:758–767.

Farell, A.E., Plevin, R.J., Turner, B.T., Jones, A.D., and M.O. Hare. 2006. Ethanol can contribute to energy and environmental goals. *Science* 311:506–508.

Feng, W., Feng, Y., Wu, Z., Fujii, A., Ozaki, M., and K. Yoshino. 2011. Optical and electrical characterizations of nanocomposite film of titania adsorbed onto oxidized multiwalled carbon nanotubes. *J Phys Condens Matter* 17:4361–4368.

Feyzi, M., Hassankhani, A., and H. Rafiee. 2013. Preparation and characterization of $CsAlFe_3O_4$ nanocatalysts for biodiesel production. *Energ Convers Manage* 71:62–68.

Garcia-Suarez, E.J., Balu, A.M., Tristany, M., Garcia, A.B., Philippot, K., and R. Luque. 2012. Versatile dual hydrogenation-oxidation nanocatalysts for the aqueous transformation of biomass-derived platform molecules. *Green Chem* 14:1434–1439.

Garvey, M., Klose, H., and R. Fischer. 2013. Cellulases for biomass degradation: Comparing recombinant cellulose expression platforms. *Trend Biotechnol* 31:581–590.

Georgelin, T., Maurice, V., Malezieux, B., Siaugue, J.M., and V. Cabuil. 2010. Design of multifunctionalized γ-Fe_2O_3@SiO_2 core–shell nanoparticles for enzymes immobilisation. *J Nanoparticle Res* 12:675–680.

Goh, W.J., Makam, V.S., Hu, J. et al. 2012. Iron oxide filled magnetic carbon nanotube-enzyme conjugates for recycling of amyloglucosidase: Toward useful applications in biofuels production process. *Langmuir* 28:16864–16873.

Goswami, D., Basu, J.K., and S. De. 2013. Lipase applications in oil hydrolysis with a case study on castor oil: A review. *Crit Rev Biotechnol* 33:81–96.

Guo, K.W. 2012. Green nanotechnology of trends in future energy: A review. *Int J Energ Res* 36:1–7.

Guo, Y., Azmat, M.U., Liu, X., Wang, Y., and G. Lu. 2012. Effect of support's basic properties on hydrogen production in aqueous-phase reforming of glycerol and correlation between WGS and APR. *Appl Energ* 92:218–223.

Gupta, M.N., Kaloti, M., Kapoor, M., and K. Solanki. 2011. Nanomaterials as matrices for enzyme immobilisation. *Artif Cell Blood Subst Biotechnol* 39:98–109.

Hu, Sh., Guan, Y., Wang, Y., and H. Han. 2011. Nano-magnetic catalyst KF/CaO-Fe_3O_4 for biodiesel production. *Appl Energ* 88:2685–2690.

Huang, S.H., Liao, M.H., and D.H. Chen. 2003. Direct binding and characterization of lipase onto magnetic nanoparticles. *Biotechnol Prog* 19:1095–1100.

Huber, G.W., Iborra, S., and A. Corma. 2006. Synthesis of Transportation fuels from biomass: Chemistry, catalysts and engineering. *Chem Rev* 106:4044–4098.

Itoh, T., Ishii, R., Matsuura, S.-I. et al. 2010. Enhancement in thermal stability and resistance to denaturants of lipase encapsulated in mesoporous silica with alkyltrimethylammonium (CTAB). *Colloid Surf B Biointerface* 75:478–482.

Ivanova, V., Petrova, P., and J. Hristov. 2011. Application in the ethanol fermentation of immobilized yeast cells in matrix of alginate/magnetic nanoparticles on chitosan-magnetic microparticles and cellulose coated magnetic nanoparticles. *Int Rev Chem Eng* 3:289–299.

Ji, P., Tan, H.S., Xu, X., and W. Feng. 2010. Lipase covalently attached to multiwalled carbon nanotubes as efficient catalyst in organic solvent. *AIChE J* 56:3005–3011.

Jiang, S., Win, K.Y., and S. Liu. 2013. Surface-functionalized nanoparticles for biosensing and imaging-guided therapeutics. *Nanoscale* 5:3127–3148.

Kalantari, M., Kazemeini, M., Tabandeh, F., and A. Arpanaei. 2012. Lipase immobilization on magnetic silica nanocomposite particles: Effects of the silica structure on properties of the immobilized enzyme. *J Mater Chem* 22:8385–8394.

Kaur, M. and A. Ali. 2011. Lithium ion impregnated calcium oxide as Nano catalyst for the biodiesel production from Karanja and Jatropa oils. *Res Gate* 36:2866–2871.

Kim, J., Grate, J.W., and P. Wang. 2008. Nanobiocatalysis and its potential applications. *Trend Biotechnol* 26:639–646.

Kim, M.I., Ham, H.O., Oh, S.D., Park, H.G., Chang, H.N., and S.H. Choi. 2006. Immobilisation of *Mucor javanicus* lipase on effectively functionalized silica nanoparticles. *J Mol Catal B Enzym* 39:62.

Kishore, D., Talat, M., Srivastava, O.N., and A.M. Kayastha. 2012. Immobilization of β-galactosidase onto functionalized graphene nanosheets using response surface methodology and its analytical applications. *PLOS ONE* 7:e40708.

Kralovec, J.A., Wang, W., and C.J. Barrow. 2010. Production of omega-3 triacylglycerol concentrates using a new food grade immobilised *Candida antartica* lipase B. *Aust J Chem* 63:922–928.

Kralovec, J.A., Zhang, S., and C.J. Barrow. 2012. A review of the progress in enzymatic concentration and microencapsulation of omega-3 rich oil from fish and microbial sources. *Food Chem Toxicol* 131:639–644.

Krumov, N., Perner-Nochta, I., and S. Oder. 2009. Production of inorganic nanoparticles by microorganisms. *Chem Eng Technol* 32:1026–1035.

Lee, C.H., Lin, T.S., and C.Y. Mou. 2010. Mesoporous materials for encapsulating enzymes. *Nano Res* 4:165–179.

Li, J., Yan, R., and B. Xiao. 2008a. Development of Nano-NiO/Al$_2$O$_3$ catalyst to be used for tar removal in biomass gasification. *Environ Sci Technol* 42:6224–6229.

Li, J., Yan, R., and B. Xiao. 2008b. Preparation of Nano-NiO particles and evaluation of their catalytic activity in pyrolyzing biomass components. *Energ Fuel* 22:16–23.

Li, S.F., Chen, J.P., and W.T. Wu. 2007. Electrospun polyacrylonitrile nanofibrous membranes for lipase immobilisation. *J Mol Catal B Enzym* 47:117–124.

Li, S.F., Fan, Y.H., Hu, R.F., and W.T. Wu. 2011. *Pseudomonas cepacia* lipase immobilized onto the electrospun PAN nanofibrous membranes for biodiesel production from soybean oil. *J Mol Catal B Enzym* 72:40–45.

Lin, C.W., Loughran, M., Tsai, T.Y., and S.W. Tsai. 2013. Evaluation of convenient extraction of chicken skin collagen using organic acid and pepsin combination. *J Chin Soc Anim Sci* 42:27–38.

Liu, C.H., Huang, C.C., and Y.W. Wang. 2012. Biodiesel production by enzymatic transesterification catalyzed by *Burkholderia* lipase immobilized on hydrophobic magnetic particles. *Appl Energ* 100:41–46.

Lupoi, J.S. and E.A. Smith. 2011. Evaluation of nanoparticle-immobilized cellulase for improved yield in simultaneous saccharification and fermentation reactions. *Biotechnol Bioeng* 108:2835–2843.

Manzer, L.E. 2006. Biomass derivatives: A sustainable source of Chemicals. In *Feedstocks for the Future: Renewables for the Production of Chemicals and Materials*, ACS Symposium Series, Vol. 921, eds. J.J. Bozell and M.K. Patel, pp. 40–51. Washington, DC: American Chemical Society.

Metzger, P. and C. Largeau. 2005. *Botryococcus braunii:* A rich source for hydrocarbons and related ether lipids. *Appl Microbiol Biotechnol* 66:486–496.

Mohamad, N.R., Marzuki, N.H.C., Nor, N.A., Huyop, F., and R.A. Wahab. 2015. An overview of technologies for immobilization of enzymes and surface analysis techniques for immobilized enzymes. *Biotechnol Biotechnol Equip* 29:205–220.

Nair, S., Kim, J., Crawford, B., and S.H. Kim. 2007. Improving biocatalytic activity of enzymeloaded nanofibres by dispersing entangled nanofiber structure. *Biomacromolecules* 8:1266–1270.

Nakane, K., Hotta, T., Ogihara, T., Ogata, N., and S.J. Yamaguchi. 2007. Synthesis of (Z)-3-hexen-1-yl acetate by lipase immobilised in poly-vinyl alcohol nanofibers. *J Appl Polym Sci* 106:863–867.

Ngo, T.P.N., Li, A., Tiew, K.W., and Z. Li. 2013. Efficient transformation of grease to biodiesel using highly active and easily recyclable magnetic nanobiocatalyst aggregates. *Bioresour Technol* 145:233–239.

Obadiah, A., Kannan, R., Ravichandran, P., Ramasubbu, A., and S.V. Kumar. 2012. Nano hydrotalcite as a novel catalyst for biodiesel conversion. *Digest J Nanomater Biostruct* 7:321–327.

Pavlidis, I.V., Vorhaben, T., Gournis, D., Papadopoulos, G.K., Bornscheuer, U.T., and H. Stamatis. 2012. Regulation of catalytic behavior of hydrolases through interactions with functionalized carbon based nanomaterials. *J Nanoparticle Res* 14:842–851.

Puri, M., Abraham, R.E., and C.J. Barrow. 2012. Biofuel production: Prospects, challenges and feedstock in Australia. *Renew Sust Energ Rev* 16:6022–6031.

Qiu, F., Li, Y., Yang, D., Li, X., and P. Sun. 2011. Heterogeneous solid base nanocatalysts: Preparation, characterization and application in biodiesel production. *Bioresour Technol* 102:4150–4156.

Rebelo, L.P., Netto, C.G.C.M., Toma, H.E., and L.H. Andrade. 2010. Enzymatic kinetic resolution of (RS)-1-(Phenyl)ethanols by *Burkholderia cepacia* lipase immobilised on magnetic nanoparticles. *J Braz Chem Soc* 21:1537–1542.

Ren, Y., Rivera, J.G., He, L., Kulkarni, H., Lee, D.K., and P.B. Messersmith. 2011. Facile, high efficiency immobilization of lipase enzyme on magnetic iron oxide nanoparticle via a biomimetic coating. *BMC Biotechnol* 11:63.

Rude, M.A. and A. Schirmer. 2009. New microbial fuels: A biotech perspective. *Curr Opin Microbiol* 12:274.

Safarik, I. and M. Safarikova. 2009. Magnetic nano and microparticles in biotechnology. *Chem Paper* 63:497–505.

Saifuddin, N., Raziah, A.Z., and A.R. Junizah. 2013. Carbon nanotubes: A review on structure and their interaction with proteins. *J Chem Educ* 2013:676815.

Sakai, S., Liu, Y.P., and T. Yamaguchi. 2010. Production of butyl-biodiesel using lipase physically-adsorbed onto electrospun polyacrylonitrile fibers. *Bioresour Technol* 101:7344–7349.

Semwal, S., Arora, A.K., Badoni, R.P., and D.K. Tuli. 2011. Biodiesel production using heterogeneous catalysts. *Bioresour Technol* 102:2151–2161.

Sharma, A.K., Desnavi, S., Dixit, C., Varshney, U., and A. Sharma. 2015. Extraction of nickel nanoparticles from electroplating waste and their applications in production of biodiesel from biowastes. *Int J Chem Eng Appl* 6:156–159.

Shim, M., Kam, N.W.S., and R.J. Chen. 2002. Functionalisation of carbon nanotubes for biocompatibility and biomolecular recognition. *Nano Lett* 2:285–288.

Sill, T.J. and H.A.V. Recum. 2008. Electrospinning: Applications in drug delivery and tissue engineering. *Biomaterials* 29:1989–2006.

Srivastava, A. and R. Prasad. 2000. Triglycerides-based diesel fuels. *Renew Sust Energ Rev* 4:111–133.

Strey, R. 2007. Microemulsions and use thereof as a fuel. U.S. Patent Application 2007/028507.

Szczesna-Antczak, M., Kubiak, A., Antczak, T., and S. Bielecki. 2009. Enzyme Biodiesel synthesis—Key factors affecting efficiency of the process. *Renew Energ* 34:1185–1194.

Taarning, E., Nielsen, I.S., Egeblad, K., Madsen, R., and C.H. Christensen. 2008. Chemicals from renewables: Aerobic oxidation of furfural and hydroxymethylfurfural over gold catalysts. *Chem Sust Energ Mater* 1:75–78.

Tran, D.T., Chen, C.L., and J.S. Chang. 2012. Immobilization of *Burkholderia* sp. lipase on a ferric silica nanocomposite for biodiesel production. *J Biotechnol* 158:112–119.

Trindade, S.C. 2011. Nanotech biofuels and fuel additives. In *Biofuel's Engineering Process Technology*, ed. M.A.S. Bernades. In Tech, Croatia, European Union.

Verma, M.L., Barrow, C.J., Kennedy, J.F., and M. Puri. 2012. Immobilization of β-D-galactosidase from *Kluyveromyces lactis* on functionalized silicon dioxide nanoparticles: Characterization and lactose hydrolysis. *Int J Biol Macromol* 50:432–437.

Verma, M.L., Barrow, C.J., and M. Puri. 2013. Nanobiotechnology as a novel paradigm for enzyme immobilisation and stabilisation with potential applications in biodiesel production. *Appl Microbiol Biotechnol* 97:23–39.

Verma, M.L., Puri, M., and C.J. Barrow. 2015. Recent trends in nanomaterials immobilized enzymes for biofuels production. *Crit Rev Biotechnol* 23:1–12.

Verziu, M., Cojocaru, B., Hu, J. et al. 2008. Sunflower and rapeseed oil transesterification to biodiesel over different nanocrystalline MgO catalysts. *Green Chem* 10:373–378.

Veziroğlu, T.N. and S. Şahin. 2008. 21st Century's energy: Hydrogen energy system. *Energ Convers Manage* 49:1820–1831.

Wang, P. 2006. Nanoscale biocatalyst systems. *Curr Opin Biotechnol* 17:574–579.

Wang, X., Dou, P., Zhao, P., Zhao, C., Ding, Y., and P. Xu. 2009. Immobilisation of lipases onto magnetic Fe_3O_4 nanoparticles for application in biodiesel production. *Chem Sustain Energ Mater* 2:947–50.

Wang, X., Liu, X., and X. Yan. 2011. Enzyme-nanoporous gold biocomposite: Excellent biocatalyst with improved biocatalytic performance and stability. *PLOS ONE* 6:e24207.

Wang, X., Liu, X., and C. Zhao. 2011. Biodiesel production in packed-bed reactors using lipase-nanoparticle biocomposite. *Bioresour Technol* 102:6352–6355.

Wang, Y. and Y.L. Hsieh. 2008. Immobilisation of lipase enzyme in polyvinyl alcohol (PVA) nanofibrous membranes. *J Membr Sci* 309:73–81.

Weisz, P.B., Frilette, V.J., Maatman, R.W., and E.B. Mober. 1962. Novel metal containing catalyst. *J Catal* 1:307–312.

Weisz, P.B. and J.N. Miale. 1965. Superactive crystalline aluminosilicate hydrocarbon catalysts. *J Catal* 4:527–529.

Wu, W., He, Q., and C. Jiang. 2008. Magnetic iron oxide nanoparticles: Synthesis and surface functionalization strategies. *Nanoscale Res Lett* 3:397–415.

Wu, Y., Wang, Y., Luo, G., and Y. Dai. 2009. *In situ* preparation of magnetic Fe_3O_4-chitosan nanoparticles for lipase immobilization by cross-linking and oxidation in aqueous solution. *Bioresour Technol* 100:3459–3464.

Wulff, P., Lada, B., Sandra, E., and S. Reinhard. 2008. Water-biofuel microemulsions. Institute for Physical Chemistry, University of Cologne, Cologne, Germany. Available from: http://strey.unikoeln.de/fileadmin/user_upload/Download/WATER___BIOFUEL_MICROEMULSIONS.pdf. Accessed on December 25, 2015.

Xie, W. and N. Ma. 2009. Immobilised lipase on Fe_3O_4 nanoparticles as biocatalyst for biodiesel production. *Energ Fuel* 23:1347–1353.

Xie, W. and N. Ma. 2010. Enzymatic transesterification of soybean oil by using immobilized lipase on magnetic nano-particles. *Biomass Bioenerg* 34:890–896.

Yan, N., Zhao, C., Luo, C., Dyson, P.J., Liu, H., and Y. Kou. 2006. One-step conversion of cellobiose to C6-alcohols using a ruthenium nanocluster catalyst. *J Am Chem Soc* 128:8714–8715.

Yu, L., Banerjee, I.A., Gao, X.Y., Nuraje, N., and H. Matsui. 2005. Fabrication and application of enzyme-incorporated peptide nanotubes. *Bioconjugate Chem* 16:1484–1487.

Zhila, N.O., Kalacheva, G.S., and T.G. Volova. 2005. Effect of nitrogen limitation on the growth and lipid composition of the green alga *Botryococcus braunii* Kutz IPPAS H-252. *Russ J Plant Physiol* 52:357–365.

Zhu, G., Yang, R., Wang, S., and Z.L. Wang. 2010. Flexible high-output nanogenerator based on lateral Zno nanowire array. *Nano Lett* 10:3151–3155.

Zou, B., Hu, Y., Yu, D. et al. 2010. Immobilization of *porcine pancreatic* lipase onto ionic liquid modified mesoporous silica SBA-15. *Biochem Eng J* 53:150–153.

20

Current Insights into Proteomics of Biofuel Crops and Cyanobacteria

Balwinder Singh Sooch, Manpreet Kaur Mann, and Ram Sarup Singh

Contents

Abstract

Sustainable energy is considered as one of the most important concerns of the twenty-first century. Man has been dependent on fossil fuel reserves in the past centuries, but current scenario presents the diminishing behavior of these fossil fuel reserves. There is an urgent need for alternative sources to cope up with the diminishing fossil fuel reserves that promise a good yield of biomass for biofuel production. The energy requirements can be fulfilled with the production of bioenergy from the existing natural sources. Some crops consisting high sucrose content have the potential to produce biofuel by the process of conversion of sugars into bioethanol. Some chloroplast-containing microorganisms are also able to synthesize biofuels in their cells. There is a great potential to improve the yield and quality of biofuel from these natural resources for commercial success. To achieve this goal, it is important to understand the biological mechanisms like metabolomics, transcriptomics, and proteomics involved in these crops to visualize the scope for the enhancement of their yield and to formulate the strategies for further improvement. Proteomics deals with the identification and characterization of expressed proteins, which reveals their biological functions. A large number of databases are available to access the information deposited regarding proteomics and genomics data of biofuel crops and microorganisms. In this chapter, various studies conducted on proteomics in the past for plant-based sources like maize, sugarcane, sugar beet, sorghum, jatropha, and microorganisms (*Synechocystis* and *Chlamydomonas reinhardtii*) have been discussed.

20.1 Introduction

Sustainable energy ranks as one of the most significant concerns of the twenty-first century. The diminishing fossil fuel reserves, mounting crude oil prices, and intensifying threats of climatic variation have encouraged the present generation for the use of alternate energy sources for a balanced future (Stone 2006, 2011; Hepbasli 2008). One of the promising substitutes for fuels is nuclear energy, but recent disastrous events occurred at Chernobyl (Stone 2006) and Fukushima (Stone 2011) have forced the researchers to explore some other options for fuels. An alternate energy source, which can be relied upon in the present era, is bioenergy, owing to abundance of biological sources that promise to produce a good yield of energy. Biofuels can embrace a reduction of greenhouse gas emissions and also have a positive energy balance. It is established from various studies that extensive reduction in emission of greenhouse gases can be achieved by using ethanol as fuel made from sugarcane (Macedo et al. 2008). Additionally, bioenergy is also advantageous over combustion engines by presenting electrical vehicles.

There was a 109% increase in worldwide biofuel production from 2008 to 2013, and the Organization for Economic Cooperation and Development (OECD) and Food and Agriculture Organization (FAO) indicated a further increase by 60% in biofuel production by 2021. There are numerous biological materials and photosynthesizing organisms like algae, cyanobacteria, grasses, and plants that are already being used for biofuel production.

Presently, few European countries are producing biofuels and many more developing countries are working on biofuel production. It is mandatory by regulations passed by some countries to replace the use of gasoline with bioethanol that involves a necessary contribution from advanced fuels to promise a further reduction of greenhouse gas emissions. However, there is an imperative need to understand the biological mechanisms of bioresources to accelerate the biofuel production process and yield for successful commercialization. Hence, it is very important to explore the postgenomics and systems biology approaches (metabolomics, transcriptomics, and proteomics) to investigate how these organisms respond and adapt to the changing environmental conditions. Proteomics is defined as the systematic representation of the protein family in subcellular and cellular portions or tissue of a plant or a microbe. Proteomics is a means to recognize the biology and new methods to modify plant and other biobased sources for an increased yield of biofuels. Efforts have been made in the past to study the proteomics of some biofuel crops and microorganisms to make them more efficient.

20.2 Sources of Biofuels

Biofuels as a source of sustainable energy are in great demand nowadays, because fossil fuel reserves are diminishing at greater rates and overutilization of these fossil fuels has also posed some of the environmental problems like pollution and global warming. In this regard, recently photosynthetic plants/organisms have attracted significant attention as they are capable of utilizing solar energy and CO_2 to produce renewable biofuels. Prominent plant-based biofuel sources include maize, sorghum, sugarcane, jatropha, and sugar beet, whereas microbe-based sources include algae or cyanobacteria, which have a good capacity to directly convert solar energy into substrates for biofuels (Waditee et al. 2005; Ruffing 2011). *Synechocystis* and *Chlamydomonas reinhardtii* are common cyanobacteria used for biofuel production. The growth rate of these organisms is very high with simple nutritional requirements and requires very less land area as compared to plants. In addition, genetic manipulation to improve the growth and cultivation under controlled conditions is very easy in microorganisms (Golden et al. 1987; Waditee et al. 2005; Ruffing 2011). All microalgal cells have the capacity to synthesize and accumulate high lipids and triglycerides in their cells. These lipids are used further for direct conversion to biodiesel. Cyanobacteria possess an internal membrane system, that is, thylakoid membrane, due to which they are phototrophic in nature. Cyanobacteria can thrive well in high temperature and extreme pH conditions. Some strains of cyanobacteria have adapted to extreme environmental conditions including high salinity. This property of the microorganisms has made them a good candidate for application in saline areas, because nearly one-fifth of the world's irrigated land is saline at present (Boyer 1982; Nelson et al. 1998), and the condition is predicted to get worse in the coming 40–50 years (Hart et al. 2003).

20.3 Proteomics in Biofuel Sources

The revolution of genomics has changed the logic of analysis of biological systems and processes. Proteomics is considered as one of the most important areas of research, which involves advances in bioinformatics, genome sequencing, and analytical tools (Bergh and Archens 2009; Penque 2009). Proteomics involves the identification and characterization of expressed proteins and is a powerful tool to analyze the function and biological role of various proteins (Lin et al. 2003a,b). A large number of DNA sequences have been produced from a wide number of organisms generated through multiple genome projects. It is assumed that biological systems and processes can be defined upon comparison between global quantitative gene expression patterns obtained from different cells or tissues presenting their different states. Posttranscriptional mechanisms, which control the rate of synthesis and half-life of proteins and gene expression, involve the unknown relation between mRNA and protein levels. Hence, direct measurement of protein expression is also important for analyzing biological processes and systems. An overview of the protocol followed in any proteomics study has been shown in Figure 20.1. Standard analytical methods for proteome analysis involve separation of proteins by high-resolution SDS-PAGE, 2D gel electrophoresis (2DE) with mass spectrometric (MS)

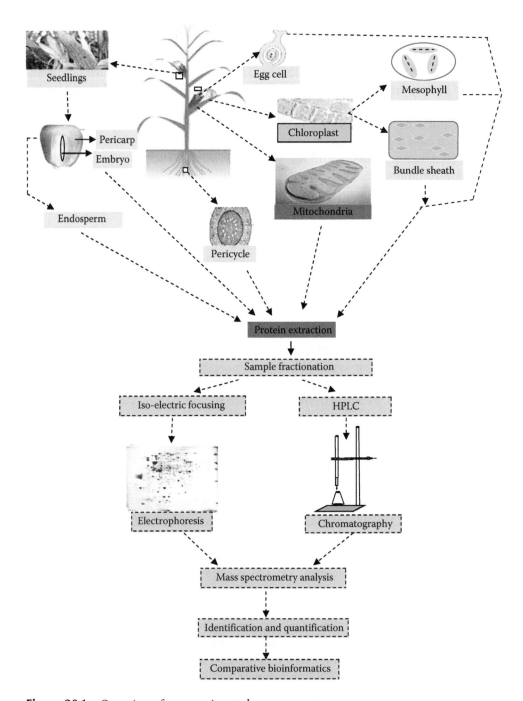

Figure 20.1 Overview of proteomics study.

identification of spots of selected proteins (Carrette et al. 2006; Lopez 2007; Penque 2009). The response of the cells to their changing environmental conditions can be identified through quantitation of expression of proteins in a proteome. The translational proteomics to solve the issues related to plant proteomics has been explored to promise food safety, energy maintenance, and human health (Aggarwal et al. 2012). A large amount of data have been generated from numerous studies regarding proteomics, and these data have been deposited and organized in the form of databases that are available to scientists through web services. The list of important databases is given in Table 20.1.

20.3.1 Proteomics for Plant-Based Sources of Biofuels

Various plant-based crops like maize, sorghum, sugarcane, jatropha, and sugar beet are currently used for biofuel production. Members of the Saccharine subtribe, generally *Saccharum*, *Sorghum*, and *Miscanthus*, best suit to the requirements for biofuel production from plants. These crops are being grown in short growing seasons even under adverse environmental conditions such as low temperature, low nutrient input, and periodic droughts (Waclawovsky et al. 2010). In tropical climates, high yield of *Saccharum* can be obtained, whereas *Sorghum* and *Miscanthus* can grow well under temperate climatic conditions. Various proteomics studies have been conducted on these crops to understand the underlying mechanism for their adaptations under different soil, environmental, and climatic conditions.

20.3.1.1 Proteomics of Maize

Various proteomics studies have been conducted on maize to understand the underlying mechanisms for their adaptations under different environments. Maize is one of the widely cultivated cereal crops, followed by wheat and rice. Maize is a staple food for most of the population of the world. But it has also been utilized as an alternate energy crop and used as a source for bioethanol production in the North Central and Midwestern United States (Chen et al. 2013). At the beginning of the twentieth century, maize became a model crop especially to study the genetics of the monocotyledonous plants and is the most studied genetic organization (Strable and Scanlon 2009). This is because maize is a cross-pollinating species and bears an extraordinary level of genetic variations, and this helps to study various biological events such as recombination, evolution, and heterosis (McClintock 1930).

Research in maize has been accelerated by the availability of the maize genomic sequence of the B73 line in 2009 (Schnable et al. 2009). The genomic and genetic information of maize is now accessible in various public repositories like maize GDB, plant GDB, plant proteome database, TIGR maize database, and maize assembled genomic island (Table 20.1).

These databases have further facilitated the research work on maize to improve its nutritional value, resistance to pests and diseases, enhanced crop yields, and environment adaptability. Proteome profiling is done to identify and classify gene products from various cells, organs, and tissues. Protein extracts from whole cells contain a mixture of the entire cellular and subcellular proteins, which provides us little information

TABLE 20.1   List of Online Proteomics/Genomics Databases of Bioenergy Sources

Name of Database	Website Link
Plant databases	
Maize genome/proteome databases	
Maize GDB	www.maizegdb.org
ZmGDB-Plant GDB	www.plantgdb.org/ZmGDB
Plant proteome database	www.ppdb.tc.cornell.edu
Plant GDB	www.plantgdb.org
Maize Genome Sequencing project	www.maizesequence.org
TIGR Maize Database	www.maize.jcvi.org
Maize Assembled Genomic Island	www.magi.plant genomics.iastate.edu
Maize Genome Sequencing Project	www.maizesequence.org
Sugarcane proteomics/genomics databases	
Sugarcane—GMO Database—GMO compass	www.gmo.compass.org/eng/database/ plants/TE.sugar_cane.html
SymGRASS	www.ncbi.nlm.nih.gov/pubmed/23368899
Sugar beet proteomics databases	
Sugar beet phylome	www.phylomedb.org/phylome_152
Sorghum proteomics/genomics databases	
Sorghum bicolor Genome project	www.phytozome.net/sorghum
Gramene: A comparative resource for plants	www.gramene.org
Sorghum bicolor Genome DB	www.plantgdb.org/SbGDB
Sorghum genomics at Phytozome	www.phytozone.jgi.doe.gov
Jatropha proteomics/genomics databases	
Matrix Science	www.matrixscience.com
KEGG pathway database	www.genome.jp/Kegg
GO/Gene Ontology Consortium	www.gene ontology.org/page/go-database
COG, National Center for Biotechnology Information	www.ncbi.nlm.nih.gov/COG
Other plant-based proteomics databases	
The *Arabidopsis* Information Resource (TAIR)	www.arabidopsis.org
MSU Rice Genome Annotation Project	www.rice.plantbiology.msu.edu
MIPS *Oryza sativa* database (MOsDB)	www.mips.gsf.de/proj/plant/jsf/rice
Rice Genome Automated Annotation System (RiceGAAS)	www.ricegaas.dna.affrc.go.jp
CGF, College of Agricultural and Environmental Sciences	www.cgf.ucdavis.edu
HarvEST, EST database	www.harvest.ucr.edu
UniProtKB/Swiss-Prot Plant Proteome Annotation Program (PPAP)	www.ca.expasy.org/sprot/ppap
Plant proteomics in Europe, COST F0603	www.costfa0603.com
Multinational *Arabidopsis* Steering Committee Proteomics (MASCP)	www.masc-proteomics.org/
Arabidopsis proteome database	www.fgcz-atproteome.unizh.ch
NASC Proteomics database for *Arabidopsis* data	www.proteomics.arabidopsis.info
GabiPD, proteomic data—*Arabidopsis thaliana* and *Brassica napus*	www.gabi.rzpd.de/projects/ Arabidopsis_Proteomics

(Continued)

TABLE 20.1 (*Continued*)   List of Online Proteomics/Genomics Databases of Bioenergy Sources

Name of Database	Website Link
Plant Proteome DataBase (PPDB) for *Arabidopsis thaliana* and maize	www.ppdb.tc.cornell.edu
Soybean Proteome Database (2DE)	www.proteome.dc.affrc.go.jp/soybean
Center for Medicago Genomics Research: Proteomics	www.noble.org/medicago/proteomics.html
Rice Proteome Database (2DE)	www.gene64.dna.affrc.go.jp/RPD
Plant proteome–specialized databases	
Protein Mass spectra Extraction (ProMEX) MSMS spectral database	www.promex.mpimp-golm.mpg.de/ home.shtml
Plant Organelles Database 2 (PODB2)	www.podb.nibb.ac.jp/organellome
SUBA, subcellular location of *Arabidopsis* proteins	www.suba.bcs.uwa.edu.au
AraPerox, a database of putative proteins of *Arabidopsis* peroxisomes	www.araperox.uni-goettingen.de
PCLR, a method for chloroplast localization prediction	www.andrewschein.com/pclr/index.html
ChloroP, a server for chloroplast transit peptides	www.cbs.dtu.dk/services/ChloroP
Arabidopsis mitochondrial proteome project	www.gartenbau.uni-hannover.de/genetik/ AMPP
Arabidopsis Mitochondrial Protein Database	www.plantenergy.uwa.edu.au/ampdb
Plprot, a plastid protein database	www.plprot.ethz.ch
Arabidopsis seed proteome	www.seed-proteome.com
Aramemnon, a plant membrane protein database	www.aramemnon.botanik.uni-koeln.de
Putative Orthologous Groups and Plant RNA-Binding Protein Database	www.plantrbp.uoregon.edu
An *Arabidopsis* coiled-coil protein database (ARABI COIL)	www.www.coiled-coil.org/arabidopsis
Arabidopsis Nucleolar Protein Database (AtNoPDB)	www.bioinf.scri.sari.ac.uk/cgi-bin/atnopdb
An *Arabidopsis* phosphoproteome database (PhosPHat)	www.phosphat.mpimp-golm.mpg.de
Plant Protein Phosphorylation Database (P3DB)	www.digbio.missouri.edu/p3db

regarding less abundant subcellular proteins that often remain undetected due to the masking effect of highly abundant proteins. Protein mapping of several parts of maize plant is possible with the use of advanced tools of proteomics. Various proteomics studies were carried out for the root tip, root hairs, primary root, primary root pericycle, endosperm, leaf, seeds, rachis, starch granules, and also subcellular organelles like mitochondria, cell wall of the primary root, and chloroplast. Proteomics for further subsections of chloroplast like envelope membrane, nucleoids, thylakoids, and stroma was also investigated in the past as shown in Table 20.2. During proteome profiling in maize, 10 tissues, 7 subcellular compartments, and 2 secretomes were profiled and 13 2D proteome reference maps were created. These profiles were composed of the egg cell (Okamoto et al. 2004), leaf (Porubleva et al. 2001; Majeran et al. 2010), root hair (Nestler et al. 2011), primary root proteome associated with starch granules (Koziol et al. 2012), seed flour (Albo et al. 2007), primary root (Hochholdinger et al. 2004, 2005), primary root pericycle (Dembinsky et al. 2007), root tip (Chang et al. 2000), rachis (Pechanova et al. 2010), and endosperm (Mechin et al. 2004).

TABLE 20.2　Summary of Maize Tissues, Organelles, and Secretomes That Were Mapped/Profiled Using Proteomic Approaches

Profiled Plant Material	Age[a]	Separating Technique	Spots/Proteins Identified	Identification Method	Annotated 2D Map	Reference
Tissue						
Leaf	14 DAG	2DE	149/73	MALDI-MS	Yes	Porubleva et al. (2001)
Leaf	9-day-old seedling	1DE	NA/>4300	Nano LC-MS/MS	NA	Majeran et al. (2010)
Endosperm	14 DAP	2DE	496/NA	LC-MS/MS	Yes	Mechin et al. (2004)
Egg cell	NA[b]	1DE	NA/5	Nano LC-MS/MS	NA	Okamoto et al. (2004)
Egg cell	NA	2DE	3/3	Nano LC-MS/MS	Yes	Okamoto et al. (2004)
Primary root	5 DAG	2DE	156/74	MALDI-MS	Yes	Hochholdinger et al. (2005)
Primary root	9 DAG	2DE	NA/47	MALDI-MS	No	Hochholdinge et al. (2004)
Root hair	4 DAG	1DE	NA/2573	Nano LC-MS/MS	NA	Nestler et al. (2011)
Root tip	6 DAG	2DE	46/31	MALDI-MS	Yes	Chang et al. (2000)
Primary root pericycle	2.5 days old	2DE	NA/20	Nano LC-MS/MS	Yes	Dembinsky et al. (2007)
Rachis	21 DAS	2DE	416/517	1D LC-MS/MS	Yes	Pechanova et al. (2010)
Rachis	21 DAS	2D LC	NA/701	2D LC-MS/MS	NA	Pechanova et al. (2010)
Starch granule	NA	1D LC	NA/109	Nano LC MS/MS	NA	Koziol et al. (2012)
Grain flour	Mature seeds	2DE	40/25	MALDI-MS, Nano LC-MS/MS	Yes	Albo et al. (2007)
Organelle						
Chloroplast	0–48 HOG	2DE	54/26	MALDI-MS	No	Lonosky et al. (2004)
Mitochondria	Immature ears	2DE	100/58	MALDI-MS	Yes	Hochholdinger et al. (2004)
Cell walls of the primary root elongation zone						
Cell wall protein fraction	28 HAG	2DE	43/60	MALDI-MS	Yes	Zhu et al. (2006)
Total soluble fraction	28 HAG	2DE	36/34	MALDI-MS	Yes	Zhu et al. (2006)

(Continued)

TABLE 20.2 (Continued) Summary of Maize Tissues, Organelles, and Secretomes That Were Mapped/Profiled Using Proteomic Approaches

Profiled Plant Material	Age[a]	Separating Technique	Spots/Proteins Identified	Identification Method	Annotated 2D Map	Reference
Stroma of chloroplasts (total protein)			400[c]			
Stroma	14-day-old seedling	2DE	NA/221[c]	MALDI-MS, Nano LC-MS/MS	Yes	Majeran et al. (2005)
Stroma	14-day-old seedling	1D LC	NA/100[c]	Nano LC-MS/MS	NA	Majeran et al. (2005)
Stroma	14-day-old seedling	1D LC	NA/305[c]	Nano LC-MS/MS	NA	Majeran et al. (2005)
Membranes of BS chloroplast	12–14-day-old seedling	1D BN	NA/393	Nano LC-MS/MS	No	Majeran et al. (2008)
Membranes of M chloroplasts	12–14-day-old seedling	1D BN	NA/483	Nano LC-MS/MS	No	Majeran et al. (2008)
Chloroplast (BS and M)	12–14-day-old seedling	1DE	NA/1105[c]	Nano LC-MS/MS	No	Friso et al. (2010)
Nucleoids of chloroplast (total protein)			1092[c]		No	Majeran et al. (2012)
Nucleoid leaf base	9–10-day-old seedling	1DE	NA/678[c]	Nano LC-MS/MS	No	Majeran et al. (2012)
Nucleoid leaf tip	9–10-day-old seedling	1DE	NA/710[c]	Nano LC-MS/MS	No	Majeran et al. (2012)
Nucleoid young seedlings	7–8 day-old seedling	1DE	NA/829[c]	Nano LC-MS/MS	No	Majeran et al. (2012)
Proplastids	9–10-day-old seedling	1DE	NA/1717	Nano LC-MS/MS	No	Majeran et al. (2012)
Secretome						
Xylem sap	23 DAS	2DE	154/NA	Nano LC-MS/MS	Yes	Alvarez et al. (2006)
Root mucilage	3-day-old seedling	1DE	NA/2848	Nano LC-MS/MS	No	Ma et al. (2010)

Source: Pechanova, O. et al., *Proteomics*, 13, 637, 2013.

[a]DAG, days after germination; DAP, days after pollination; DAS, days after silking (rachis); DAS, days after sawing (xylem sap); HOG, hours of greening; HAG, hours after germination.

[b]NA, data not available.

[c]The number reflects the combined proteome from chloroplasts suborganelles of both bundle sheath (BS) and mesophyll (M) cells.

Different data sets for proteins are generated from the mitochondria (Hochholdinger et al. 2004), cell wall of primary root elongation zone (Zhu et al. 2006), chloroplast (Lonosky et al. 2004; Friso et al. 2010), and its several subcompartments such as stroma (Majeran et al. 2005), membranes (Majeran et al. 2008), and nucleoids (Majeran et al. 2012). Through 2DE survey, 58 proteins were identified from normal and T-cytoplasm of the maize plant. It was observed that ATPase alpha F1 protein was only encoded by the mitochondrial genome, but the remaining 57 proteins were encoded by the nuclear genome. This data set contains mitochondrial proteome consisting of HSPs, which is voltage-dependent anion channel proteins, prohibitins, and subunits of various ATPases. In addition to it, superoxide dismutase and inorganic pyrophosphatase are some of the differentially accumulated mitochondrial proteins between normal and T-cytoplasm mitochondrial proteomes. This showed that the genotype of the mitochondrion can control the accumulation of the nuclear-encoded fraction of the mitochondrial proteome (Hochholdinger et al. 2004). Analysis was carried out for the comparison of TSP with the water-soluble and ionically bound proteins located in the walls of the maize primary root for the classification of root cells (Zhu et al. 2006). Among them, 60 proteins were examined, which were found associated with type II cells of maize.

Lonosky et al. (2004) were the first ones to characterize maize chloroplast proteome, an important organelle of the photosynthetic system, and 26 proteins were examined by them in the five stages of greening of a leaf. Among C4 plants, the chloroplasts are distinguished between highly specialized bundle sheath (BS) and mesophyll (M) cells. The distribution of the photosynthetic apparatus between these two types of cells was studied extensively and truly; comprehensive proteome profiles for membranes, stroma, and nucleoids of chloroplasts of both BS and M were obtained. Multiple gel-based and gel-free proteomic approaches coupled with high-resolution and high-sensitivity MS instrumentation were used in these studies to attain maximum coverage (Majeran et al. 2005, 2008, 2012). Similarly, chloroplasts from M and BS cells were analyzed, and a data set of 1105 proteins was obtained. Analysis was also carried out for fatty acids, detoxification, redox regulation, sulfur, nitrogen, and amino acid metabolism of the cells. Nucleoids were also analyzed for mature and immature chloroplasts that resulted in the description of 1092 proteins, which helped in understanding the plastid gene expression machinery (Majeran et al. 2012). The identified proteins were deposited into the Plant Proteome Database (Sun et al. 2009). It was found that the enzymes involved in nitrogen import, lipid biosynthesis, isoprenoid biosynthesis, and tetrapyrrole are located in the M chloroplasts. Further, enzymes that are involved in the synthesis of starch and sulfur accumulate in the chloroplasts of BS. Differential accumulation of proteins was observed to be tangled in the expression of plastid-encoded proteins (EF-Tu-, EF-G-, and mRNA-binding proteins) and formation of thylakoid (VIPP1). Enzymes associated with triose phosphate reduction and triose phosphate isomerase are mainly located in the chloroplasts of M, signifying the involvement of M-restricted triose phosphate shuttle in BS-confined Calvin cycle (Majeran et al. 2005). In plastid, there are multiple copies of the plastid chromosome, along with folding of proteins and RNA into nucleoids, but the factors that are tangled in plastid DNA organization, replication and repair are not understood well to date. To understand the function of the nucleoid, certain isolates from the maize tip and leaf base were characterized by proteomics. The comparisons made between proteomes of unfractionated proplastids and chloroplasts facilitated the determination of nucleoid-enriched proteins by MS. Some proteins of unidentified function, including DnaJ, pentatricopeptide repeat, tetratricopeptide repeat, and mitochondrial transcription termination factor domain proteins were also identified with the help of proteomics (Majeran et al. 2012). It is reported by Majeran et al. (2012) that mRNA processing, splicing and editing, and ribosome assembly occur in association with the nucleoid and concluded that these processes occur cotranscriptionally. The analysis of stromal proteome was also carried out through 400 identified proteins responsible for a large range of functions such as primary carbon metabolism, photosynthesis (Calvin cycle, oxidative phosphor pentose pathway, glycolysis, starch metabolism, amino acid metabolism, lipid metabolism, nitrogen sulfur assimilation), and protein synthesis along with folding and degradation (Majeran et al. 2005). In thylakoid and envelope membranes, 393 and 483 proteins were investigated, respectively, by using 1D blue native gel electrophoresis (Majeran et al. 2008). Much higher proteome coverage with 954 stromal and 882 membrane proteins was recorded by Friso et al. (2010) through MS data (Majeran et al. 2010). By comparing the grain crop to tropical maize, it is better to use tropical maize as their stalks contain high amount of soluble sugars and an increased biomass, but there is no report available on the proteomics of stalks of maize, which remains an open material for study on proteomics.

20.3.1.2 Proteomics of Sugarcane

Among all the cash crops, sugarcane is of greater importance and cultivated for its stalks, which has abundant sucrose content and provides 60% of the white sugar worldwide. Some commercial varieties of sugarcane have amazing capability of storing high levels of sucrose (40% of their dry weight) in their stems (Moore 1995). Average dry mass production is found to be 39 t/ha in sugarcane, whereas other bioenergy crops like *Miscanthus* yield around 29.6 t/ha and switchgrass yields around 10.4 t/ha (Heaton et al. 2008).

Molasses are by-product of sugar industry and widely used as substrate for bioethanol production. The biomass left is also used for the production of bioethanol and power generation. The largest producers of sugarcane in the world include Brazil and India, while more than 90 countries all over the world are cultivating sugarcane. Sugarcane is found to be at the top among all the other crops for biomass production (Vettore et al. 2003; Ma et al. 2004). High yield of sugar content has changed sugarcane into a very competitive source of sucrose for food, fiber, and ethanol production (Calsa and Figueira 2007). Sugarcane has wide applications like production of some natural pharmaceutical compounds like dextrans and furfural (Menendez et al. 1994). Very small data are available for proteomics on sugarcane. Various protein extraction methods were used for 2DE investigation in many plants (Agrawal and Rakwal 2006, 2008; Suokko et al. 2008), but for sugarcane, only phenol–SDS protein extraction method was surveyed. Most of the study published so far highlighted protein extraction only from leaf tissues (Ramagopal and Carr 1991; Ramagopal 1994;

Sugiharto et al. 2002). There is no appropriate method for the extraction of proteins in sugarcane, and none of the high-throughput proteomic concept has been used to raise the expression of proteins or 2D gel reference mapping. Current diversities in sugarcane are formed from crosses of the sucrose-gathering relative *Saccharum officinarum* and the wild relative *Saccharum spontaneum*, with contributions from *Saccharum robustum*, *Saccharum sinense*, *Saccharum barberi*, *Erianthus*, and *Miscanthus* (Daniels and Roach 1987; Ming et al. 2006).

Some strategies were developed to integrate crossbreeding between sugarcane and *Miscanthus* to achieve cold- and draught-resistant traits in sugarcane from *Miscanthus* (Burner et al. 2009). To produce biofuels from sugarcane, certain proteomic approaches were suggested to achieve lower biomass resistance, expression, and assembly of microbial cellulolytic enzymes in the leaves of sugarcane to improve the hydrolysis of the biomass (Harrison et al. 2011; Jung et al. 2012). The first proteome analysis of sugarcane stalk was carried out by Sugarcane Breeding Institute, Tamil Nadu, and a 2D gel proteome map was constructed (Amalraj et al. 2010). It was observed that those proteins involved in sugar metabolism like glyceraldehyde 3-phosphate, glyceraldehyde 3-phosphate dehydrogenase (GAPDH), UDP-glucose 6-dehydrogenase, and triose phosphate isomerase contain the most abundant sugar content and can be used for bioethanol production (Wu and Birch 2007). It is being felt that for the realistic study of both molecular and transgenic-assisted methodology for the proteomics of sugarcane, there is a need for a reference genome sequence, because researchers are presently sequencing the whole genome with bacterial artificial chromosomes to produce a reference genome (Mendes et al. 2011). One of the greatest differences between sequencing of *Saccharum* and *Miscanthus* from sorghum genome is that the latter is large in size and has large copy numbers. Molecular markers showing linkage maps identify the collinearity of sorghum and sugarcane genomes, but the same are much complicated in sugarcane due to polyploidy and the absence of inbred lines (Pastina et al. 2012). This problem was overcome by using single-dose markers, which are differentiated in the ratio of 1:1 in the gametes of a heterozygous genotype and are responsible for 70% polymorphic loci in sugarcane (Wu et al. 1992; Pastina et al. 2010). Recently, an algorithm and software (ONE MAP) has been developed, which helps the breeders in creating the linkage maps of outcrossing plant species that is successfully functional to sugarcane (Margarido et al. 2007). Although alterations in the gene expression are known well, sugarcane genomics is a challenge due to the complex large autoploid and aneuploid genome alternating and the absence of reference sequences. Alternatively, sorghum genome was used as a reference genome previously and is studied by different technologies like cDNA microarrays to understand the profile expression and function of genes responsive to biotic and also abiotic stresses in different tissues (Nishiyma et al. 2013). Recently, resources and fundamental databases have been used to study functional genomes of sugarcane (Menossi et al. 2008; Casu 2010; Manners and Casu 2011), and a comprehensive database named SUCEST-FUN has been developed for storage, integration, and retrieval of genome sequencing and transcriptome and gene profiling. Gene catalogues with major contribution to ESTs comes from the SUCEST database project, and the remainder comes from India, the United States,

Australia, etc. (Vettore et al. 2003; Manners and Casu 2011; Nishiyama et al. 2013). The SUSCEST-FUN project database and SAS sequences were also updated with EST of sugarcane from the National Center for Biotechnology Information, and it was compared with SoGI assembly. A total of 282,683 ESTs were currently catalogued in the SUCEST-FUN database. Comparisons of ESTs from sorghum with rice, sugarcane, and maize were done, and it is reported that they are 86%, 97%, and 93% identical, respectively, which determines a close similarity between sugarcane and sorghum. A total of 39,021 proteins of sugarcane were identified from 43,141 clusters (Khan et al. 2011) with the help of ESTScan (Iseli et al. 1999) and *Oryza sativa* matrix. Orthologous relationships among multiple proteomics were analyzed for five plant species, namely, *Sorghum bicolor, Zea mays, Saccharum, Arabidopsis thaliana*, and *Oryzae sativa*, with the multiparanoid software, and 18,611 orthologous clusters were made with main proteins grouped into them. A total of 16,723 proteins were found in sugarcane, while 13,804 orthologs were found in sorghum, 16,913 in rice, 13,998 in *Arabidopsis*, and 22,312 in maize by using BLOSUM80 matrix (Alexeyenko et al. 2006). The work done on transcriptomics of sugarcane has been reviewed by many workers (Menossi et al. 2008; Casu et al. 2010; Manners and Casu 2011; Nishiyama et al. 2013). In addition to this, the use of oligoarrays is included in the studies on the control of antisense genes expression in sugarcane, which indicates the role of transcripts in drought conditions (Lembke et al. 2012). Like sorghum and sugarcane, transcriptomics and genomic studies on *Miscanthus* have been started lately. The sequencing of its genome and transcriptome recognizes the presence of repeats that vigorously produce sRNA (Swaminathan et al. 2010), and the creation of genetic map sugarcane finds simple sequence repeats (SSRs). The creation of Miscanthus genetic map has helped the scientists to identify some useful similar sequence repeats in Miscanthus and this application was found useful to compare Miscanthus maps to the sorghum genome (Kim et al. 2012). These types of findings will enhance the knowledge and understanding of complex genomes of sugarcane, which would be helpful to understand the proteomics of sugarcane (Swaminathan et al. 2012).

Thus, plants belonging to the Saccharine family are quite promising for bioenergy production as they are relatively cheap and reliable. However, efforts are continuously being carried out by researchers to improve their traits like drought tolerance, adaptability to the changing environment, sucrose concentration, and yield. The genomics and proteomics studies can speed up the breeding mechanisms to improve various traits for better and efficient biofuel production.

20.3.1.3 Proteomics of Sugar Beet

Sugar beet (*Beta vulgaris*) is a dicotyledonous plant of Amaranthaceae family and is of great commercial importance for ethanol production, because it is also the great source of sucrose after sugarcane. Catusse et al. (2008) identified more than 750 proteins that have allowed to reconstitute the detailed metabolism of seeds of sugar beet during its germination. Quantitative results of the 2D analysis determine some similarity between proteomics of the root and cotyledons, but with respect to accumulation of certain proteins, both the tissues appear differently. It was concluded that during germination, mature seeds are well prepared to mobilize its reserved compounds. It was further revealed from proteomics studies that 157 proteins were more abundant in the proteome of the cotyledon, whereas 76 proteins were found to be more abundant in the proteome of the root. A total of 759 proteins were identified during proteomics studies of mature sugar beet seeds and their specific tissue expression in the root, perisperm, and cotyledons were also studied. The proteomics of the storage tissue of perisperm exposed certain pathways that contributed to produce vigorous seedlings like glycine betaine accretion in seeds. Sugar beet seeds can recruit translation through the traditional cap-dependent mechanism or by a certain cap-independent process. The analysis of the tissue-specific proteome of the seed established a distinction from metabolic activity between the cotyledons, roots, and perisperm, which signifies the segmentation of metabolic jobs between various tissues of the plant. Perisperm, which is mainly considered as a dead tissue, appears to be biochemically active and plays multiple roles in assigning sugars and several metabolites to embryo's tissues. Characterization of sugar beet seed vigor by proteomics is quite challenging because there are virtually no genomics data available on this plant that can conclude protein separation, but recent research proved the ability of MS to recognize and enumerate thousands of profiles of proteins from various classes of sugar beet (Ma et al. 2003; Delalande et al. 2005). The tissue specificity for the accumulation of the seed proteins is determined with the help of this approach. Further, the proteomics of the storage tissue of the perisperm was determined, and around 750 proteins were identified. The effect of salinity on protein expression in sugar beet was also studied using 2DE, and most of the observed proteins showed no alteration under salt stress. It was concluded through the statistical analysis of detected proteins that the expression of only 6 and 3 proteins from the shoots and roots, respectively, is altered significantly, and these altered expressions could not be attributed to sugar beet adaptation under salt stress (Wakeel et al. 2011). These datasets also provide information regarding the metabolism of sugar beet seeds and proteomics of the metabolic activities that occurred during the development. With the availability of MS, it is now possible to study the profiles of various proteins of sugar beet. Further, the quality and yield of the crop can be enhanced with the establishment of reference maps that can provide an evidence for developing mechanisms helpful for improving the crop yield and seed germination.

20.3.1.4 Proteomics of Sorghum

Sorghum is another important plant that has the potential for biofuel production because of its high sucrose content, which can be exploited for the production of bioethanol (Prasad et al. 2007). Salinity and drought are the major abiotic stresses that affect the growth of this plant due to ion toxicity and nutrient deficiency (Wyn Jones 1981; Zhu 2001; Koller et al. 2002). But at the same time, sorghum is one of the most widely grown grain crops and regarded as more tolerant to salt stress compared to maize (Krishnamurthy et al. 2007). Sorghum has four main varieties of interest including grain crops, sweet stems, high-energy fibers, and other multipurpose grains (Woods 2001). Among these, sweet-stemmed varieties are of great value for bioethanol production (Prasad et al. 2007).

Metabolic pathways and biological processes involved in response to salt stress were examined, and it was reported that salt stress and abiotic stress cause some change in gene expression (Hasegawa et al. 2000), which further leads to changes in the expression of proteins (Kasuga et al. 1999; Shinozaki et al. 2003). Proteomics studies are being used to study the salt stress–responsive expression of proteins in plants like potato (Aghaei et al. 2008) and rice (Parker et al. 2006). Sorghum proteomics was fortified by the sorghum genome sequence that was published in the year 2009, which further helped in the identification of proteins through MS (Bowers et al. 2009). The first sorghum proteome analysis was carried out at the University of the Western Cape in Cape Town, South Africa (Ngara et al. 2008). The vast source of homogeneous plant material was provided by cell suspension cultures, and these studies showed the 2DGE protein patterns of secreted culture filtrate protein and the total soluble proteins (TSP). Mapping and characterization of cell suspension culture of sorghum secretome were also done during these studies (Ngara and Ndimba 2011). In the past, it has been described that there are unique changes in the proteome of certain food crops under salt stress, but limited studies are available on the proteomics of sorghum (Sobhanian et al. 2011). Swami and his coworkers (2011) were the first to report changes in proteome expression in sorghum leaf tissue under salt stress. They further described the proteomics analysis of 21 salt stress–responsive proteins of sorghum in an experiment using 2DE and MS (Swami et al. 2011). Ngara et al. (2012) carried out a similar investigational arrangement and identified 55 separated protein spots by 2DE. These studies used bioinformatics and proteomics tools to categorize sorghum leaf proteins into six broad functional categories, that is, proton transport, protein synthesis, hydrolytic enzymes, carbohydrate metabolism, detoxifying enzymes, and nucleotide metabolism. It has been reported that proteins related to carbohydrate metabolism were the most characterized proteins in sorghum (Ngara et al. 2012; Ndimba and Ngara 2013). RNA-sequencing methods have overcome many problems associated with DNA microarray technology and allowed to reveal sorghum genes and its genomic network (Dugas et al. 2011).

It was concluded that from various genomic and proteomics studies conducted on sorghum by various workers that in response to the salt stress, there are some changes in the genes of these crops that further cause changes in the expression of proteins. However, very less work has been done in this direction due to the limited availability of genomics data on sorghum plants.

20.3.1.5 Proteomics of Jatropha

Jatropha curcas belongs to the Euphorbiaceae family, which is a tropical shrub, used for biodiesel production. It is found to be an attractive crop due to its short incubation period, capability to adapt to various kinds of soil conditions, and drought tolerance. There are some other species of *Jatropha* like *Jatropha gossypiifolia*, *Jatropha glandulifera*, *Jatropha multifida*, and *Jatropha podagrica*, but *J. curcas* is the most widely studied. *J. curcas* has some medicinal properties (including antifungal and antibacterial) and also produces a number of secondary metabolites like tannins, flavonoids, phytosterols, and glycosides. It has been found that biodiesel obtained from *J. curcas* reduces the problem of global warming as compared to diesel obtained from fossil fuels, but it does not reduce environmental problems like eutrophication, acidification, water depletion, and ecotoxicity (Gmunder et al. 2012).

The seeds of the jatropha plant are the major source of oil that can be further converted into biodiesel (Takeda 1982; Francis et al. 2005; Berchmans and Hirata 2008). This commercially important plant bears the property to synthesize fatty acids in their seeds, and the protein present in seeds contributes to the synthesis of fatty acids. Further, these fatty acids contribute to the formation of triacylglycerol, which helps in obtaining biodiesel from plants. Hsieh and Huang (2004) detected that both embryo and endosperm consist of oleosin-enriched seeds with a high ratio of oils and tiny oil bodies. Bewley (1997) reported that in mature seeds, certain enzymes and cellular structures that are required for metabolic processes are presumed to be synthesized during grain filling that survived the desiccation process. In this process, there are certain signal-related and stress-related proteins, and biosynthesis of isoleucine, leucine, and valine is related to branched-chain amino acid transaminases (Diebold et al. 2002). Several genes like *JcERF*, stearoyl-acyl carrier protein desaturase genes, and curcin genes have been isolated and characterized by different research groups (Lin et al. 2003a; Tang et al. 2007). A complete system of *Agrobacterium*-mediated transformation using cotyledon disc was described in detail by Li et al. (2008). Certain techniques like SSR, amplified fragment length polymorphism markers, intersimple sequence repeat, and random amplified polymorphic DNA analyses were performed for the determination of molecular diversity in the past (Vettore et al. 2003; Ma et al. 2004). The 2DE analysis revealed that the endosperm and the embryo from dry seed have a similar spot dispersal of proteins, and the same facts were also verified by LC-ESI-Q/TOF MS/MS and MALDI-TOF-MS. It was reported that there were 533 and 380 protein spots in endosperm and embryo, respectively, while 180 protein spots were matched, whereas unmatched proteins show diverse purpose of embryo and endosperm. In response to high-temperature stress, 18.5 kDa class I heat shock protein was reported and it was postulated that various other heat shock proteins reveal relationship for *stress* adaptations (244). GTP-bound protein Ras found in eukaryotic cells controls the signal processing like the survival of cells and its differentiation (Anne and Channing 1998). Yang et al. (2009) described the proteomics of endosperm and oil mobilization during seed germination and postgermination. It was analyzed that oil mobilization occurs at the time of germination, and consequently, the oil gets consumed during early stages of seed development. Certain pathways are found to be involved in oil mobilization like glycolysis, glyoxylate cycle, beta oxidation, trichloroacetic acid cycle, pentose phosphate pathways, and gluconeogenesis. Popluechai et al. (2011) described the composition of proteomics of various *Jatropha* species and oil bodies of *J. curcas*. The proteins were also isolated with isoelectric precipitation, and an in-depth study for identifying stress-responsive proteins was carried out by Eswaran and his coworkers (2012). Metallothioneins, which are proteins lavishly characterized during the stress response, are cysteine rich and of low molecular weight that help in metal detoxification. Some plant proteins like annexins, aquaporins, and thioredoxins were analyzed, wherein their major share in abiotic stress was indicated. Certain studies showed that the enzymes are associated with diterpenoid biosynthetic pathway. In 2009, Lin and his colleagues characterized an isoprenoid biosynthetic gene, known as 3-hydroxy-3-methylglutaryl coenzyme A reductase. This enzyme catalyzes the first committed step in mevalonic acid synthesis, which further aids in the production of phorbol esters.

20.3.2 Proteomics of Algae/Cyanobacteria

The problem of increasing salinity affects nearly one-fifth of the world's irrigated land, which affects the overall productivity of the world, and the problem is going to get worse in the coming years (Boyer 1982; Nelson et al. 1998). Blue-green algae, generally known as cyanobacteria, bear an internal photosynthetic thylakoid membrane system (Bryant 1986). These groups of organisms have dominated a wide range of ecosystems including air, dry rock aquatic systems, and soil (Joset et al. 1996). Some of the species are able to tolerate extreme environmental conditions like pH, high temperatures, and high salinity (Stal 2007). These bacteria are categorized into three types in association with their salt tolerance, which include extremely halotolerant, moderately halotolerant, and salt sensitive (Reed and Stewart 1988). The example genera for each group include *Aphanothece*, *Synechocystis*, and *Anabaena*, respectively. These cyanobacterial species have several advantages like easy to culture, low culture costs, and rapid growth rate. It acts as a model to analyze the configuration of salt tolerance (Hart et al. 2003; James et al. 2003). These organisms have the capability to form their own solutes and can also uptake from their surroundings. They contain ectoine to protect the cells from stress and UVA-induced rays (Desmarais et al. 1997; Roberts 2005). These are able to tolerate high salt conditions by accumulating solutes within their cells (Hagemann and Zuther 1992; Gabbay-Azaria and Tel-Or 1993; Joset et al. 1996). The advancement of proteomics for algae is according to the tendencies set for higher plants and animals. Various sequenced proteomic/genomic online databases related to cyanobacteria for the production of biofuels are given in Table 20.3.

TABLE 20.3   List of Online Proteomics/Genomics Databases of Cyanobacteria

Cyanobacterial Genomic/Proteomics Databases	
Proteome Sciences	www.proteomics.com
ProteinProspector at UCSF	www.prospector.ucsf.edu/
Swiss-2D PAGE	www.expasy.ch/ch2d/
UniProt	http://www.uniprot.org
Algal Functional Annotation Tool	http://pathways.mcdb.ucla.edu/algal/index.html
BiGG	http://bigg.ucsd.edu/
BioCyc	http://biocyc.org/
Biomart	http://www.biomart.org/index.html
BRENDA	http://www.brenda-enzymes.info/
COBRA	http://opencobra.sourceforge.net/openCOBRA/
ExPASy	http://www.expasy.org/
KBase	http://kbase.us
KEGG	http://www.genome.jp/kegg/
Model SEED	http://www.theseed.org/wiki/Main Page
MetaCyc	http://metacyc.org/
Pathway Tools	http://pathwaytools.org/
Reactome	http://www.reactome.org/PathwayBrowser

Recently, two methods have been utilized for biofuel production from cyanobacteria: the first approach is to isolate fatty acids from lipid-rich biomass and transform it chemically to various other products like biodiesel (Rittmann 2008; Sheng et al. 2011), and the second approach leads to the production of a number of fuel products like free fatty acids (Sheng et al. 2011), ethylene (Liu and Curtiss 2009), ethanol (Takahama et al. 2003; Liu et al. 2011), isobutyraldehyde (Deng and Coleman 1999), isoprene (Dexter and Fu 2009), hydrogen (Lan and Liao 2012), fatty alcohols (Lindberg et al. 2010), and 1-butanol (Lee et al. 2010; McNeely et al. 2010). Recently, a biosynthetic pathway for acyl–acyl carrier protein reductase and aldehyde decarbonylase in cyanobacteria *Synechococcus elongatus* has been expressed heterologously in *Escherichia coli*, which yielded C13–C17 mixtures of alkanes at ~0.3 g/L after 40 h cultivation of *E. coli* (Jang et al. 2012). Some studies were conducted on *Synechocystis* sp. PCC 6803, which is a freshwater and unicellular cyanobacterium having fully annotated and sequenced genome (Schirmer et al. 2010). It has the capability to grow heterotrophically in the presence of sunlight and can be transformed naturally, which boosts up its reliability for proteomics and metabolomic studies (Grigorieva and Shestakov 1982; Kaneko et al. 1996; Burja et al. 2003). Salt studies on this organism have raised several kinds of basic information involving several fields like physiology and biochemical studies (McNeely et al. 2010; Zhang et al. 2011), gene level responses and salt-regulated genes (Vinnemeier et al. 1998; Bohnert et al. 2001), salt intake and cell signaling (Convin and Anderson 1967; Harnisch et al. 1983), microarrays (Hayashi et al. 2003; Jude et al. 2004), postgenomics (Vemuri and Aristidou 2005; Murata et al. 2011; Stancu 2011), and mutational analysis of salt tolerance determinants (Shigi et al. 2008). Joset et al. (1996) reported that available data have enhanced our perception of adaptation of cyanobacteria to high salinity. It is reported that the *Synechocystis* sp. PCC 6803 and *S. elongatus* sp. PCC 7942 have the capability to get transformed naturally and are suitable for metabolic engineering (Doukyu et al. 2012). Isobutanol, with the help of metabolic engineering of 2-keto acid–based pathway, was obtained from *S. elongatus* sp. PCC 7942 at a very high yield (Joset et al. 1996). Lan and Liao (2011) described great production of 1-butanol by altering the CoA-dependent pathway from *S. elongatus* sp. PCC 7942 (Atsumi et al. 2009). Instead of high lipid content, they undergo several limitations relating to the biocatalyst and the bioreactor, which obstruct the possibility of scale-up economically (Parthibane et al. 2012). Through proteomics, protein study was carried out on certain microalgal model organisms like *Chlamydomonas reinhardtii* and *Synechocystis* sp. PCC 6803 to investigate genome under stress responses (Pandhal et al. 2008; Lan and Liao 2011). It was suggested by some workers that genetic engineering in algae is important for fundamental use and should be studied with lipid metabolism (Stauber and Hippler 2004; Katavic et al. 2006). Lipidomics and proteomics details identify the importance of certain molecules and proteins in biofuel production, which will play a major role in encouraging the production of energy from algae in the coming years. May et al. (2008) reported a combined examination of the molecular collection of *C. reinhardtii* with certain reference parameters. Annotation methods of bioinformatics in combination with LC/MS- and GC/MS-based metabolomics and shotgun profiling of proteomics were used to describe 159 metabolites and 1069 proteins in *C. reinhardtii* (May et al. 2008).

Among all the identified proteins, 204 proteins do not have EST sequence support; hence, some portion of the proteomics-identified proteins becomes evidence for *in silico* gene models. With the help of this information obtained from genomic annotations of bioinformatics on the metabolites and proteins, a metabolic network for *Chlamydomonas* was constructed. The data sets were available through web-accessible databases within *Chlamydomonas*-adapted MapMan annotation analysis. Information of known peptides is available directly at JGI *Chlamydomonas* genomic resource database (Merchant et al. 2007). Guarnieri et al. (2011) reported the study of triacylglycerol biosynthetic pathway in the unsequenced oleaginous microalga known as *Chlorella vulgaris,* and a strategy was established to store genomic sequence information with the help of transcriptome as a reference. The results specify an upregulation of both triacylglycerol and fatty acid biosynthesis machinery under oil-accumulating situations and laid down the usefulness of a *de novo* assembled transcriptome for the search of unsequenced microalga proteomic examination (May et al. 2008). Lan and Lio (2011) described the fermentation production of 1-butanol from carbohydrates with the help of *Clostridium* species and many other engineered hosts. A modified CoA-dependent 1-butanol production pathway was transferred into a cyanobacterium *Sy. elongatus* PCC 7942 for the production of butanol from CO_2. The activity of each enzyme related to the pathway of expression of the genes and chromosomal integration was described by Lan and Lio (2011). It was observed that in *Treponema denticola*, trans-enoyl-CoA reductase (Ter) is present, which uses NADH as a reducing source, which further reduces crotonyl-CoA to butyryl-CoA rather than the conversion of butyryl-CoA dehydrogenase to dodge the need of clostridial ferredoxins of *Clostridium acetobutylicum*. The added polyhistidine tag raised the global activity of Ter, and the production of 1-butanol was enhanced. While understanding the proteomics of cyanobacteria, *Synechocystis* species is found to be the most promising candidate regarding salt stress tolerance, but qualitative proteome data are still required to study the salt stress response in cyanobacteria in detail.

20.4 Conclusions and Future Prospects

Sustainable energy is in great demand due to the depleting fossil fuel reserves. There is a need to think it over and to develop alternate natural energy sources, which can replace the fossil fuels and reduce environmental problems such as pollution. So bioenergy is capable of resolving all these problems, and several natural sources like plants, algae, microorganisms, and animals can be exploited for the production of biofuels to meet the present need of sustainable energy. Although photosynthetic organisms are capable of utilizing solar energy and CO_2 to produce bioenergy, but their efficiency needs to be improved for successful commercialization. Presently, many plants like maize, sugarcane, sugar beet, sorghum, and jatropha and microorganisms like algae or cyanobacteria are used to convert solar energy into biofuels. But it is important to understand the biological mechanisms through genomics and proteomics, which improve the traits required for the efficient production of biofuels. Certain approaches involving genomics, transcriptomics, and metabolomics can lead to construction of new pathways to develop crops with high energy. To date, plant proteomics involves both qualitative and quantitative research leading to a new phase named as "second-generation plant proteomics," which involves quantitative and gel-free proteomic techniques. Some approaches regarding gene expression profiling with the help of transcriptomics are used to identify enzymes responsible for stress tolerance, metabolism, and transcription factor proteins, etc., through advanced tools of proteomics. Different databases for the proteomics analysis of various plants are now available online. Various proteomic approaches are being studied to achieve lower biomass resistance, expression, and assembly of microbial cellulolytic enzymes in the leaves of sugarcane to improve the hydrolysis of the biomass for biofuel production. The efforts are being made to use proteomics studies to improve the yield of biofuel from sorghum after mapping extracted proteins. Cyanobacteria are also capable of producing biofuels as they require simple nutrition input and can be genetically manipulated easily. Although various online databases for the sequenced proteomics and genomics of cyanobacteria are available on web services, there are not enough qualitative proteome data available, and there is a requirement to study the salt stress response in cyanobacteria in detail. It is mainly due to the complex understandings and diversification of proteomes in addition to some other technical limitations in sensitivity, speed of data capture, resolution, quantitation, and analysis. After reviewing data on research on proteomics and analyzing the previous publications on proteomics with recent methods, it is inferred by some researchers that it is feasible to recognize the errors obtained from improper experimental design and data analysis. Improper statistical interpretation of the results can be found clearly in the literature, and regarding this, MIAPE documents have been developed by HUPO's Proteomics Standards Initiative. Research based on proteomics for the production of biofuels is growing slowly. The majority of the research involves recognizing the various components of proteins and generating database for plant- and algal-based biofuel sources. The high resolving power of 2DE displays different proteins on a single gel, which facilitates the direct overview of proteomics of several proteins at a time. Nowadays, there are published reference maps of several proteins, which help in comparison of the unknown protein with the known ones. There are some limitations of 2DE like variability in the expression of the same set of protein samples or low solubility of the membrane proteins, which makes it difficult to identify the lowest possible difference among proteins. But with the use of difference gel electrophoresis, it is possible to detect quantitative proteomics by studying the different patterns of protein expression identified with the help of MS. Even though there is a constant advancement and upgrading of influential proteomic techniques, equipments, protocols, and bioinformatics tools, only a negligible portion of the cell proteome and very few organisms are characterized to date. Some other problems are required to be defined precisely like potential number of protein species per gene as a consequence of posttranscriptional and posttranslational modifications and interaction events in protein trafficking.

However, advances in the proteomic techniques that include automated robotic sample preparation, improved resolution in protein separation, and fast, sensitive, and accurate spectrometers will further facilitate the future developments in this field by countering the present drawbacks. But there is a huge scope of more investigations on bioenergy-producing plants and microbes to completely understand the phenomena of producing biofuel practically. Finally, the most important aspect to be kept in view is the proper selection of plant or microbe to obtain a better yield of bioenergy from these biobased sources.

References

Aghaei, K., Ehsanpour, A.A., and S. Komatsu. 2008. Proteome analysis of potato under salt stress. *J Proteome Res* 7:4858–4868.

Agrawal, G.K., Pedreschi, R., Barkla, B.J., Bindschedler, L.V., Cramer, R., and A. Sarkar. 2012. Translational plant proteomics: A perspective. *J Proteom* 75:4588–4601.

Agrawal, G.K. and R. Rakwal. 2006. Rice proteomics: A cornerstone for cereal food crop proteomes. *Mass Spectrom Rev* 25:1–53.

Agrawal, G.K. and R. Rakwal. 2008. *Plant Proteomics: Technologies, Strategies and Applications*. Hoboken, NJ: John Wiley & Sons.

Albo, A.G., Mila, S., Digilio, G., Motto, M., Aime, S., and D. Corpillo. 2007. Proteomic analysis of a genetically modified maize flour carrying Cry1Ab gene and comparison to the corresponding wild-type. *Maydica* 52:443–455.

Alexeyenko, A., Tamas, I., Liu, G., and E.L. Sonnhammer. 2006. Automatic clustering of orthologs and in paralogs shared by multiple proteomes. *Bioinformatics* 22:e9–e15.

Alvarez, S., Goodger, J.Q.D., Marsh, E.L., Chen, S., Asirvatham, V.S., and D.P. Schachtman. 2006. Characterization of the maize xylem sap proteome. *J Proteome Res* 5:963–972.

Amalraj, R.S., Selvaraj, N., Veluswamy, G.K. et al. 2010. Sugarcane proteomics: Establishment of a protein extraction method for 2-DE in stalk tissues and initiation of sugarcane proteome reference map. *Electrophoresis* 31:1959–1974.

Anne, B.V. and J.D. Channing. 1998. Increasing complexity of the Ras signaling pathway. *J Biol Chem* 273:19925–19928.

Atsumi, S., Higashide, W., and J.C. Liao. 2009. Direct photosynthetic recycling of carbon dioxide to isobutyraldehyde. *Nat Biotechnol* 27:1177–1180.

Berchmans, H.J. and S. Hirata. 2008. Biodiesel production from crude *Jatropha curcas L.* seed oil with a high content of free fatty acids. *Bioresour Technol* 99:1716–1721.

Bergh, G. and L. Arckens. 2009. High resolution protein display by two-dimensional electrophoresis. *Curr Anal Chem* 5:106–115.

Bewley, J.D. 1997. Seed germination and dormancy. *Plant Cell* 9:1055–1066.

Bohnert, H.J., Ayoubi, P., Borchert, C. et al. 2001. A genomics approach towards salt stress tolerance. *Plant Physiol Biochem* 39:295–311.

Bowers, J.E., Bruggmann, R., Dubchak, I., Grimwood, J., and H. Gundlach. 2009. The *Sorghum bicolor* genome and the diversification of grasses. *Nature* 457:551–556.

Boyer, J.S. 1982. Plant productivity and environment. *Science* 218:443–448.

Bryant, D.A. 1986. The cyanobacterial photosynthetic apparatus: Comparisons to those of higher plants and photosynthetic bacteria. In *Photosynthetic Picoplankton*, Canadian Bulletin of Fisheries and Aquatic Sciences, eds. T. Platt and W.K.W. Li, pp. 423–500. Ottawa, Ontario, Canadian Government Publishing Centre, Canada.

Burja, A.M., Dhamwichukorn, S., and P.C. Wright. 2003. Cyanobacterial postgenomic research and systems biology. *Trends Biotechnol* 21:504–511.

Burner, D.M., Tew, T.L., Harvey, J.J., and D.P. Belesky. 2009. Dry matter partitioning and quality of *Miscanthus, Panicum,* and *Saccharum* genotypes in Arkansas, USA. *Biomass Bioenerg* 33:610–619.

Calsa, T. and A. Figueira. 2007. Serial analysis of gene expression in sugarcane (*Saccharum* spp.) leaves revealed alternative C4 metabolism and putative antisense transcripts. *Plant Mol Biol* 63:745–762.

Carrette, O., Burkhard, P.R., Sanchez, J.C., and D.F. Hochstrasser. 2006. State-of-the-art two-dimensional gel electrophoresis: A key tool of proteomics research. *Nat Protocols* 1:812–823.

Casu, R. 2010. Role of bioinformatics as a tool for sugarcane research. In *Genetics, Genomics and Breeding of Sugarcane,* eds. R. Henry and C. Kole, pp. 229–248. Enfield, NH: Science Publishers.

Casu, R., Hotta, C.T., and G.M. Souza. 2010. Functional genomics: Transcriptomics of sugarcane-Current status and future prospects. In *Genetics, Genomics and Breeding of Sugarcane,* eds. R. Henry and C. Kole, pp. 167–191. Enfield, NH: Science Publishers.

Catusse, J., Strub, J., Job, C.M., Van Dorsselaer, A., and D. Job. 2008. Proteome-wide characterisation of sugarbeet seed vigor and its tissue specific expression. *Proc Natl Acad Sci USA* 105:10262–10267.

Chang, W.W.P., Huang, L., Shen, M., Webster, C., Burlingame, A.L., and J.K. Roberts. 2000. Patterns of protein synthesis and tolerance of anoxia in root tips of maize seedlings acclimated to a low-oxygen environment and identification of proteins by mass spectrometry. *Plant Physiol* 122:295–317.

Chen, M.H., Kaur, P., Dien, B., Below, F., Vincent, M.L., and V. Singh. 2013. Use of tropical maize for bioethanol production. *World J Microbiol Biotechnol* 29:1509–1515.

Convin, H. and M. Anderson. 1967. The effect of intramolecular hydrophobic bonding on partition coefficients. *J Org Chem* 32:2583–2586.

Daniels, J. and B.T. Roach. 1987. Taxonomy and evolution. In *Sugarcane Improvement through Breeding,* ed. D.J. Heinz, pp. 7–84. Amsterdam, the Netherlands: Elsevier.

Delalande, F., Carapito, C., Brizard, J.P., Brugido, C., and V.A. Dorsselaer. 2005. Multigenic families and proteomics: Extended protein characterization as a tool for paralog gene identification. *Proteomics* 5:450–460.

Dembinsky, D., Woll, K., Saleem, M., and Y. Liu. 2007. Transcriptomic and proteomic analyses of pericycle cells of the maize primary root. *Plant Physiol* 145:575–588.

Deng, M.D. and J.R. Coleman. 1999. Ethanol synthesis by genetic engineering in cyanobacteria. *Appl Environ Microb* 65:523–528.

Desmarais, D., Jablonski, P.E., Fedarko, N.S., and M.F. Roberts. 1997. 2-Sulfotrehalose, a novel osmolyte in haloalkaliphilic archaea. *J Bacteriol* 179:3146–3153.

Dexter, J. and P. Fu. 2009. Metabolic engineering of cyanobacteria for ethanol production. *Energ Environ Sci* 2:857–864.

Diebold, R., Schuster, J., Dachner, K., and S. Binder. 2002. The branched chain amino acid transaminase gene family in *Arabidopsis* encodes plastid and mitochondrial proteins. *Plant Physiol* 129:540–550.

Doukyu, N., Ishikawa, K., Watanabe, R., and H. Ogino. 2012. Improvement in organic solvent tolerance by double disruptions of proV and marR genes in *Escherichia coli. J Appl Microbiol* 112:464–474.

Dugas, D.V., Monaco, M.K., Olsen, A. et al. 2011. Functional annotation of the transcriptome of *Sorghum bicolor* in response to osmotic stress and abscisic acid. *BMC Genomics* 12:514–520.

Eswaran, N., Sriram, P., Balaji, S., Raja, K.K., Bhagyam, A., and T.S. Johnson. 2012. Generation of expressed sequence tag (EST) library from salt stressed roots of *Jatropha curcas* for the identification of abiotic stress responsive genes. *Plant Biol* 14:428–437.

Francis, G., Edingger, R., and K. Becker. 2005. A concept for simultaneous wasteland reclamation fuel production and socio economic development in degraded areas in India need potential & perspectives of Jatropha plantation. *Nat Resour Forum* 29:12–24.

Friso, G., Majeran, W., Huang, M., Sun, Q., and K.J. van Wijk. 2010. Reconstruction of metabolic pathways, protein expression, and homeostasis machineries across maize bundle sheath and mesophyll chloroplasts: Large-scale quantitative proteomics using the first maize genome assembly. *Plant Physiol* S152:1219–1250.

Gabbay-Azaria, R. and E. Tel-Or. 1993. Mechanisms of salt tolerance in cyanobacteria. In *Plant Responses to the Environment*, ed. P.M. Gresshoff, pp. 692–698. Boca Raton, FL: CRC Press.

Gmunder, S., Singh, R., Pfister, S., Adheloya, A., and R. Zah. 2012. Environmental impacts of *Jatropha curcas* biodiesel in India. *J Biomed Biotechnol* 2012:1–10.

Golden, S.S., Brusslan, J., and R. Haselkorn. 1987. Genetic engineering of the cyanobacterial chromosome. *Methods Enzymol* 153:215–231.

Grigorieva, G. and S.V. Shestakov. 1982. Transformation in the cyanobacterium *Synechocystis* sp. PCC6803. *FEMS Microbiol Lett* 13:367–370.

Guarnieri, M.T., Nag, A., Smolinski, S.L., Darzins, A., Seibert, M., and P.T. Pienkos. 2011. Examination of triacylglycerol biosynthetic pathways via de novo transcriptomic and proteomic analyses in an unsequenced microalga. *PLOS ONE* 6:25851.

Hagemann, M. and E. Zuther. 1992. Selection and characterization of mutants of the cyanobacterium *Synechocystis* sp. PCC 6803 unable to tolerate high salt concentrations. *Arch Microbiol* 158:429–434.

Harnisch, M., Mockel, H., and G. Schulze. 1983. Relationship between log PO, shake-flask values and capacity factors derived from reversed phase high-performance liquid chromatography for n alkyl benzenes and some OECD reference substance. *J Chromatogr* 282:315–332.

Harrison, M.D., Geijskes, J., Coleman, H.D. et al. 2011. Accumulation of recombinant cellobiohydrolase and endoglucanase in the leaves of mature transgenic sugarcane. *Plant Biotechnol J* 9:884–896.

Hart, B.T., Lake, P.S., Webb, J.A., and M.R. Grace. 2003. Ecological risk to aquatic systems from salinity increases. *Aust J Bot* 51:689–702.

Hasegawa, P.M., Bressan, R.A., Zhu, J.K., and H.J. Bohnert. 2000. Plant cellular and molecular responses to high salinity. *Annu Rev Plant Physiol Plant Mol Biol* 51:463–499.

Hayashi, S., Aono, R., Hanai, T., Mori, H., Kobayashi, T., and H. Honda. 2003. Analysis of organic solvent tolerance in *Escherichia coli* using gene expression profiles from DNA microarrays. *J Biosci Bioeng* 95:379–383.

Heaton, E.A., Dohleman, F.G., and S.P. Long. 2008. Meeting US biofuel goals with less land: The potential of *Miscanthus*. *Global Change Biol* 14:2000–2014.

Hepbasli, A. 2008. A key review on exergetic analysis and assessment of renewable energy resources for a sustainable future. *Renew Sust Energ Rev* 12:593–661.

Hsieh, K. and A.H.C. Huang. 2004. Endoplasmic recticulum, oleosins and oils in seeds and tapetum cells. *Plant Physiol* 136:3427–3434.

Hochholdinger, F., Guo, L., and P.S. Schnable. 2004. Cytoplasmic regulation of the accumulation of nuclear-encoded proteins in the mitochondrial proteome of maize. *Plant J* 37:199–208.

Hochholdinger, F., Woll, K., Guo, L., and P.S. Schnable. 2005. The accumulation of abundant soluble proteins changes early in the development of the primary roots of maize (*Zea mays* L.). *Proteomics* 5:4885–4893.

Iseli, C., Jongeneel, C.V., and P. Bucher. 1999. ESTScan: A program for detecting, evaluating, and reconstructing potential coding regions in EST sequences. *Proc Int Conf Intell Syst Mol Biol* 138–148.

James, K.R., Cant, B., and T. Ryan. 2003. Responses of fresh water biota to rising salinity levels and implications for saline water management: A review. *Aust J Bot* 51:703–713.

Jang, Y.S., Park, J.M., Choi, S. et al. 2012. Engineering of microorganisms for the production of biofuels and perspectives based on systems metabolic engineering approaches. *Biotechnol Adv* 30:989–1000.

Joset, F., Jeanjean, R., and M. Hagemann. 1996. Dynamics of response of cyanobacteria to salt stress: Deciphering the molecular events. *Physiol Plant* 96:738–744.

Jude, F., Arpin, C., Brachet-Castang, C., Capdepuy, M., Caumette, P., and C. Quentin. 2004. TbtABM, a multidrug efflux pump associated with tributyltin resistance in *Pseudomonas stutzeri*. *FEMS Microbiol Lett* 232:7–14.

Jung, J.H., Fouad, W.M., Vermerris, W., Gallo, M., and F. Altpeter. 2012. RNAi suppression of lignin biosynthesis in sugarcane reduces recalcitrance for biofuel production from lignocellulosic biomass. *Plant Biotechnol J* 10:1067–1076.

Kaneko, T., Sato, S., Kotani, H. et al. 1996. Sequence analysis of the genome of the unicellular cyanobacterium *Synechocystis* sp. strain PCC6803. II. Sequence determination of the entire genome and assignment of potential protein-coding regions (supplement). *DNA Res* 3:185–209.

Kasuga, M., Liu, Q., Miura, S., Yamaguchi-Shinozaki, K., and K. Shinozaki. 1999. Improving plant drought, salt, and freezing tolerance by gene transfer of a single stress-inducible transcription factor. *Nat Biotechnol* 17:287–291.

Katavic, V., Agrawal, G.K., Hajduch, M., Harris, S.L., and J.J. Thelen. 2006. Protein and lipid composition analysis of oil bodies from two *Brassica napus* cultivars. *Proteomics* 6:4586–4598.

Khan, M.S., Ali, S., and J. Iqbal. 2011. Developmental and photosynthetic regulation of delta-endotoxin reveals that engineered sugarcane conferring resistance to 'dead heart' contains no toxins in cane juice. *Mol Biol Rep* 38:2359–2369.

Kim, C., Zhang, D., Auckland, S. et al. 2012. SSR-based genetic maps of *Miscanthus sinensis* and *M. sacchariflorus*, and their comparison to sorghum. *Theor Appl Genet* 124:1325–1338.

Koller, A., Washburn, M.P., Lange, B.M. et al. 2002. Proteomic survey of metabolic pathways in rice. *Proc Natl Acad Sci USA* 99:11969–11974.

Koziol, A.G., Marquez, B.K., Huebsch, M.P., Smith, J.C., and I. Altosaar. 2012. The starch granule associated proteomes of commercially purified starch reference materials from rice and maize. *J Proteom* 75:993–1003.

Krishnamurthy, L., Serraj, R., Hash, C.T., Dakheel, A.J., and B.V.S. Reddy. 2007. Screening sorghum genotypes for salinity tolerant biomass production. *Euphytica* 156:15–24.

Lan, E.I. and J.C. Liao. 2011. Metabolic engineering of cyanobacteria for 1-butanol production from carbon dioxide. *Metab Eng* 3:353–363.

Lan, E.I. and J.C. Liao. 2012. ATP drives direct photosynthetic production of 1-butanol in cyanobacteria. *Proc Natl Acad Sci USA* 109:6018–6023.

Lee, H.S., Vermaas, W.F., and B.E. Rittmann. 2010. Biological hydrogen production: Prospects and challenges. *Trends Biotechnol* 28:262–271.

Lembke, C.G., Nishiyama M.Y., Sato, P.M., de Andrade, R.F., and G.M. Souza. 2012. Identification of sense and antisense transcripts regulated by drought in sugarcane. *Plant Mol Biol* 79:461–477.

Li, J., Li, M.R., Wu, P.Z., Tian, C.E., Jiang, H.W., and G.J. Wu. 2008. Molecular cloning and expression analysis of a gene encoding putative beta-ketoacyl-acyl carrier protein synthase III (KAS III) from *Jatropha curcas*. *Tree Physiol* 28:921–927.

Lin, J., Jin, Y., Zhou, M., Zhou, X., and J. Wang. 2009. Molecular cloning, characterization and functional analysis of a 3-hydroxy-3-methylglutaryl coenzyme A reductase gene from *Jatropha curcas*. *Afr J Biotechnol* 8:3455–3462.

Lin, J., Li, Y.X., Zhou, X.W., Tang, K.X., and F. Chen. 2003a. Cloning and characterization of a curcin gene encoding a ribosome inactivating protein from *Jatropha curcas*. *DNA Sequence* 14:311–317.

Lin, J., Yan, F., Tang, L., and F. Chen. 2003b. Antitumor effects of curcin from seeds of *Jatropha curcas*. *Acta Pharmacol Sin* 24:241–246.

Lindberg, P., Park, S., and A. Melis. 2010. Engineering a platform for photosynthetic isoprene production in cyanobacteria, using *Synechocystis* as the model organism. *Metab Eng* 12:70–79.

Liu, X. and R. Curtiss. 2009. Nickel-inducible lysis system in *Synechocystis* sp. PCC 6803. *Proc Natl Acad Sci USA* 106:21550–21554.

Liu, X., Fallon, S., Sheng, J., and R. Curtiss. 2011. CO_2-limitation-inducible green recovery of fatty acids from cyanobacterial biomass. *Proc Natl Acad Sci USA* 108:6905–6908.

Lonosky, P.M., Zhang, X., Honavar, V.G., Dobbs, D.L., Fu, A., and S.R. Rodermel. 2004. A proteomic analysis of maize chloroplast biogenesis. *Plant Physiol* 134:560–574.

Lopez, J.L. 2007. Two-dimensional electrophoresis in proteome expression analysis. *J Chromatogr B Anal Technol Biomed Life Sci* 849:190–202.

Ma, B., Zhang, K., Hendrie, C. et al. 2003. PEAKS: Powerful software for peptide de novo sequencing by tandem mass spectrometry. *Rapid Commun Mass Spectrom* 17:2337–2342.

Ma, H.M., Schulze, S., Lee, S. et al. 2004. An EST survey of the sugarcane transcriptome. *Theor Appl Genet* 108:851–863.

Ma, W., Muthreich, N., Liao, C. et al. 2010. The mucilage proteome of maize (*Zea mays* L.) primary roots. *J Proteome Res* 9:2968–2976.

Macedo, I.C., Seabra, J.E.A., and J. Silva. 2008. Greenhouse gases emissions in the production and use of ethanol from sugarcane in Brazil: The 2005/2006 averages and a prediction for 2020. *Biomass Bioenerg* 32:582–595.

Majeran, W., Cai, Y., Sun, Q., and K.J. van Wijk. 2005. Functional differentiation of bundle sheath and mesophyll maize chloroplasts determined by comparative proteomics. *Plant Cell* 17:3111–3140.

Majeran, W., Friso, G., Asakura, Y. et al. 2012. Nucleoid enriched proteomes in developing plastids and chloroplasts from maize leaves: A new conceptual framework for nucleoid functions. *Plant Physiol* 158:156–189.

Majeran, W., Friso, G., Ponnala, L. et al. 2010. Structural and metabolic transitions of C4 leaf development and differentiation defined by microscopy and quantitative proteomics in maize. *Plant Cell* 22:3509–3542.

Majeran, W., Zybailov, B., Ytterberg, A.J., Dunsmore, J., Sun, Q., and K.J. van Wijk. 2008. Consequences of C4 differentiation for chloroplast membrane proteomes in maize mesophyll and bundle sheath cells. *Mol Cell Proteom* 7:1609–1638.

Manners, J. and R. Casu. 2011. Transcriptome analysis and functional genomics of sugarcane. *Trop Plant Biol* 4:9–21.

Margarido, G.R.A., Souza, A.P., and A.A.F. Garcia. 2007. One Map: Software for genetic mapping in outcrossing species. *Hereditas* 144:78–79.

May, P., Wienkoop, S., Kempa, S. et al. 2008. Metabolomics and proteomics-assisted genome annotation and analysis of the draft metabolic network of *Chlamydomonas reinhardtii*. *Genetics* 179:157–166.

McClintock, B. 1930. A cytological demonstration of the location of an interchange between two non-homologous chromosomes of *Zea mays*. *Proc Natl Acad Sci USA* 16:791–796.

McNeely, K., Xu, Y., Bennette, N., Bryant, D., and G. Dismukes. 2010. Redirecting reductant flux into hydrogen production via metabolic engineering of fermentative carbon metabolism in a cyanobacterium. *Appl Environ Microb* 76:5032–5038.

Mechin, V., Balliau, T., Chateau-Joubert, S. et al. 2004. A two-dimensional proteome map of maize endosperm. *Phytochemistry* 65:1609–1618.

Mendes, S.G., Berges, H., Bocs, S. et al. 2011. The sugarcane genome challenge: Strategies for sequencing a highly complex genome. *Trop Plant Biol* 4:145–156.

Menendez, R., Fernandez, S.I., Del-Rio, A. et al. 1994. Policosanol inhibits cholesterol biosynthesis and enhances low density lipoprotein processing in cultured human fibroblasts. *Biol Res* 27:199–203.

Menossi, M., Silva-Filho, M.C., Vincentz, M., Van Sluys, M.A., and G.M. Souza. 2008. Sugarcane functional genomics: Gene discovery for agronomic trait development. *Int J Plant Genomics* 2008:458732.

Merchant, S.S., Prochnik, S.E., Vallon, O. et al. 2007. The *Chlamydomonas* genome reveals the evolution of key animal and plant functions. *Science* 318:245–250.

Ming, R., Moore, P.H., Wu, K. et al. 2006. Sugarcane improvement through breeding and biotechnology. *Plant Breed Rev* 27:115–118.

Moore, P.M. 1995. Temporal and spatial regulation of sucrose accumulation in the sugarcane stem. *Aust J Plant Physiol* 22:661–679.

Murata, M., Fujimoto, H., Nishimura, K. et al. 2011. Molecular strategy for survival at a critical high temperature in *Escherichia coli*. *PLOS ONE* 6:e20063, 1–9.

Ndimba B.K and R. Ngara. 2013. Sorghum and sugarcane proteomics. In *Genomics of the Saccharinae, Plant Genetics and Genomics: Crops and Models*, ed. A.H. Paterson, pp. 141–168. New York: Springer.

Nelson, D.E., Shen, B., and H.J. Bohnert. 1998. Salinity tolerance mechanisms, models and the metabolic engineering of complex traits: Principles and methods. In *Genetic Engineering*, ed. J. Setlow, pp. 153–176. New York: Plenum Press.

Nestler, J., Schutz, W., and F. Hochholdinger. 2011. Conserved and unique features of the maize (*Zea mays* L.) root hair proteome. *J Proteom Res* 10:2525–2537.

Ngara, R. and B.K. Ndimba. 2011. Mapping and characterisation of the sorghum cell suspension culture secretome. *Afr J Biotechnol* 10:253–266.

Ngara, R., Ndimba, R., Borch-Jensen, J., Nørregaard, J.O., and B.K. Ndimba. 2012. Identification and profiling of salinity stress-responsive proteins in *Sorghum bicolor* seedlings. *J Proteom* 75:4139–4150.

Ngara, R., Rees, J., and B.K. Ndimba. 2008. Establishment of sorghum cell suspension culture system for proteomics studies. *Afr J Biotechnol* 7:744–749.

Nishiyama-Jr, M.Y., Vicente, F., Sato, P.M., Ferreira, S.S., Feltus, F.A., and G.M. Souza. 2013. Transcriptome analysis in the saccharinae. In *Genomics of the Saccharinae*, ed. A.H. Paterson, pp.121–140. New York: Springer.

Okamoto, T., Higuchi, K., Shinkawa, T. et al. 2004. Identification of major proteins in maize egg cells. *Plant Cell Physiol* 45:1406–1412.

Pandhal, J., Wright, P.C., and C.A. Biggs. 2008. Proteomics with a pinch of salt: A cyanobacterial perspective. *Saline Syst* 4:1–18.

Parker, R., Flowers, T.J., Moore, A.L., and N.V. Harpham. 2006. An accurate and reproducible method for proteome profiling of the effects of salt stress in the rice leaf lamina. *J Exp Bot* 57:1109–1118.

Parthibane, V., Rajakumari, S., Venkateshwari, V., Iyappan, R., and R. Rajasekharan. 2012. Oleo sin is bifunctional enzyme that has both monoacylglycerol acyltransferase and phospholipase activities. *J Biol Chem* 287:1946–1954.

Pastina, M.M., Malosetti, M., Gazaffi, R. et al. 2012. A mixed model QTL analysis for sugarcane multiple-harvest-location trial data. *Theor Appl Genet* 124:835–849.

Pastina, M.M., Pinto, L.R., Oliveira, K.M., Souza, A.P., and A.A. Garcia. 2010. Molecular mapping of complex traits. In *Genetics, Genomics and Breeding of Sugarcane*, eds. R. Henry, and C. Kole, pp. 117–148. Enfield, NH: Science Publishers.

Pechanova, O., Pechan, T., Ozkan, S., McCarthy, F.M., Williams, W.P., and D.S. Luthe. 2010. Proteome profile of the developing maize (*Zea mays* L.) rachis. *Proteomics* 10:3051–3055.

Pechanova, O., Takac, T., Samaj, J., and T. Pechan. 2013. Maize proteomics: An insight into the biology of an important cereal crop. *Proteomics* 13:637–662.

Penque, D. 2009. Two-dimensional gel electrophoresis and mass spectrometry for biomarker discovery. *Proteom Clin Appl* 3:155–172.

Popluechai, S., Froissard, M., Jolivet, P., and D. Breviario. 2011. *Jatropha curcas* oil body proteome and oleosins: L-form JcOle3 as a potential phylogenetic marker. *Plant Physiol Biochem* 49:352–356.

Porubleva, L., Velden, V.K., Kothari, S., Oliver, D.J., and P.R. Chitnis. 2001. The proteome of maize leaves: Use of gene sequence and expressed sequence tag data for identification of proteins with peptide mass fingerprints. *Electrophoresis* 92:1724–1738.

Prasad, S., Singh, A., Jain, N., and H.C. Joshi. 2007. Ethanol production from sweet sorghum syrup for utilization as automotive fuel in India. *Energ Fuel* 21:2415–2420.

Ramagopal, S. 1994. Protein variation accompanies leaf dedifferentiation in sugarcane (*Saccharum officinarum*) and is influenced by genotype. *Plant Cell Rep* 13:692–696.

Ramagopal, S. and J.B. Carr. 1991. Sugarcane proteins and messenger RNAs regulated by salt in suspension cells. *Plant Cell Environ* 14:47–56.

Reed, R.H. and W.D.P. Stewart. 1988. The responses of cyanobacteria to salt stress. In *Biochemistry of the Algae and Cyanobacteria*, ed. L.J. Rogers, pp. 217–231. Oxford, U.K.: Oxford Science Publisher.

Rittmann, B.E. 2008. Opportunities for renewable bioenergy using microorganisms. *Biotechnol Bioeng* 100:203–212.

Roberts, M.F. 2005. Organic compatible solutes of halotolerant and halophilic microorganisms. *Saline Syst* 1:1–5.

Ruffing, A.M. 2011. Engineered cyanobacteria: Teaching an old bug new tricks. *Bioeng Bugs* 2:136–149.

Schirmer, A., Rude, M.A., Li, X., Popova, E., and S.B. del Cardayre. 2010. Microbial biosynthesis of alkanes. *Science* 329:559–562.

Schnable, P.S., Ware, D., Fulton, R.S. et al. 2009. The B73 maize genome: Complexity, diversity, and dynamics. *Science* 326:1112–1115.

Sheng, J., Vannela, R., and B.E. Rittmann. 2011. Evaluation of methods to extract and quantify lipids from *Synechocystis* PCC 6803. *Bioresour Technol* 102:1697–1703.

Shigi, N., Sakaguchi, Y., Asai, S., Suzuki, T., and K. Watanabe. 2008. Common thiolation mechanism in the biosynthesis of tRNA thiouridine and sulphur containing cofactors. *EMBO J* 7:3267–3278.

Shinozaki, K., Yamaguchi-Shinozaki, K., and M. Seki. 2003. Regulatory network of gene expression in the drought and cold stress responses. *Curr Opin Plant Biol* 6:410–417.

Sobhanian, H., Aghaei, K., and S. Komatsu. 2011. Changes in the plant proteome resulting from salt stress: Toward the creation of salt-tolerant crops? *J Proteomics* 74:1323–1337.

Stal, L. 2007. Algae and cyanobacteria in extreme environments. In *Cellular Origin, Life in Extreme Habitats and Astrobiology*, ed. J. Seckbach, pp. 659–680. Springer, Netherlands.

Stancu, M.M. 2011. Effect of organic solvents on solvent-tolerant *Aeromonas hydrophila* IBBPo8 and *Pseudomonas aeruginosa* IBBPo10. *Indian J Biotechnol* 10:352–361.

Stauber, E.J. and M. Hippler. 2004. *Chlamydomonas reinhardtii* proteomics. *Plant Physiol Biochem* 42:989–1001.

Stone, R. 2006. Return to the inferno: Chernobyl after 20 years. *Science* 312:180–192.

Stone, R. 2011. Devastation in Japan. Fukushima cleanup will be drawn out and costly. *Science* 331:1507.

Strable, J. and M.J. Scanlon. 2009. Maize (*Zea mays*): A model organism for basic and applied research in plant biology. *Cold Spring Harb Protoc* 4:1–9.

Sugiharto, B., Ermawathi, N., Mori, H. et al. 2002. Identification and characterization of a gene encoding drought-inducible protein localizing in the bundle sheath cell of sugarcane. *Plant Cell Physiol* 43:350–354.

Sun, Q., Zybailov, B., Majeran, W., Friso, G., Olinares, P.D., and K.J. van Wijk. 2009. PPDB, the plant proteomics database at Cornell. *Nucleic Acids Res* 37:D969–D974.

Suokko, A., Poutanen, M., Savijoki, K., Kalkkinen, N., and P. Varmanen. 2008. ClpL is essential for induction of thermotolerance and is potentially part of the HrcA regulon in *Lactobacillus gasseri*. *Proteomics* 8:1029–1041.

Swami, A.K., Alam, S.I., Sengupta, N., and R. Sarin. 2011. Differential proteomic analysis of salt stress response in *Sorghum bicolor* leaves. *Environ Exp Bot* 71:321–328.

Swaminathan, K., Alabady, M., Varala, K. et al. 2010. Genomic and small RNA sequencing of *Miscanthus × giganteus* shows the utility of sorghum as a reference genome sequence for Andropogoneae grasses. *Genome Biol* 11:R12–R15.

Swaminathan, K., Chae, W., Mitros, T. et al. 2012. A framework genetic map for *Miscanthus sinensis* from RNA seq-based markers shows recent tetraploidy. *BMC Genomics* 13:142–147.

Takahama, K., Matsuoka, M., Nagahama, K., and T. Ogawa. 2003. Construction and analysis of a recombinant cyanobacterium expressing a chromosomally inserted gene for an ethylene-forming enzyme at the psbAI locus. *J Biosci Bioeng* 95:302–305.

Takeda, Y. 1982. Development study on *Jatropha curcas* (Sabu Dum) oil as a substitute for diesel engine oil in Thailand. *J Agric Assoc China* 120:1–8.

Tang, M.J., Sun, J.W., Liu, Y., Chen, F., and S.H. Shen. 2007. Isolation and functional characterization of the JcERF gene, a putative AP2/EREBP domain-containing transcription factor, in the woody oil plant *Jatropha curcas*. *Plant Mol Biol* 63:419–428.

Vemuri, G.N. and A.A. Aristidou. 2005. Metabolic engineering in the omics era: Elucidating and modulating regulatory networks. *Microbiol Mol Biol Rev* 69:197–216.

Vettore, A.L., Da Silva, F., Kemper, E.L. et al. 2003. Analysis and functional annotation of an expressed sequence tag collection for tropical crop sugarcane. *Genome Res* 13:2725–2735.

Vinnemeier, J., Kunert, A., and M. Hagemann. 1998. Transcriptional analysis of the isiAB operon in salt-stressed cells of the cyanobacterium *Synechocystis* sp. PCC6803. *FEMS Microbiol Lett* 169:323–330.

Waclawovsky, A.J., Sato, P.M., Lembke, C.G., Moore, P.H., and G.M. Souza. 2010. Sugarcane for bioenergy production: An assessment of yield and regulation of sucrose content. *Plant Biotechnol J* 8:263–276.

Waditee, R., Bhuiyan, M.N., Rai, V. et al. 2005. Genes for direct methylation of glycine provide high levels of glycine betaine and abiotic-stress tolerance in *Synechococcus* and *Arabidopsis*. *Proc Natl Acad Sci USA* 102:1318–1323.

Wakeel, A., Asif, A.R., Pitann, B., and S. Schubert. 2011. Proteome analysis of sugar beet (*Beta vulgaris* L.) elucidates constitutive adaptation during the first phase of salt stress. *J Plant Pathol* 168:519–526.

Woods, J. 2001.The potential for energy production using sweet sorghum in southern Africa. *Energ Sust Dev* 5:31–38.

Wu, K.K., Burnquist, W., Sorrells, M.E., Tew, T.L., Moore, P.H., and S.D. Tanksley. 1992. The detection and estimation of linkage in polyploids using single-dose restriction fragments. *Theor Appl Genet* 83:294–300.

Wu, L. and R.G. Birch. 2007. Doubled sugar content in sugarcane plants modified to produce a sucrose isomer. *Plant Biotechnol J* 5:109–117.

Wyn Jones, R.G. 1981. Salt tolerance. In *Physiological Processes Limiting Plant Productivity*, ed. C.B. Johnson, pp. 271–292. London, U.K.: Butterworths.

Yang, M.F., Liu, Y.J., Liu, Y., Chen, H., Chen, F., and S.H. Shen. 2009. Proteomic analysis of oil mobilization in seed germination and post germination development of *Jatropha curcas*. *J Proteome Res* 8:1441–1451.

Zhang, F., Rodriguez, S., and J.D. Keasling. 2011. Metabolic engineering of microbial pathways for advanced biofuels production. *Curr Opin Biotechnol* 22:775–783.

Zhu, J., Chen, S., Alvarez, S. et al. 2006. Cell wall proteome in the maize primary root elongation zone. I. Extraction and identification of water-soluble and lightly ionically bound proteins. *Plant Physiol* 140:311–325.

Zhu, J.K. 2001. Plant salt tolerance. *Trends Plant Sci* 6:66–71.

21

Biofuel Cells
Concepts and Perspectives for Implantable Devices

Mukesh Yadav, Nirmala Sehrawat,
Simran Preet Kaur, and Ram Sarup Singh

Contents

Abstract

Biofuel cells are devices capable of transforming chemical energy of a fuel into electrical energy. Biofuel cells can be abiotic, enzymatic, microbial, or mammalian type depending on the catalysts used for the oxidation and reduction reactions. The development and fabrication of biofuel cell are a multidisciplinary assignment that requires conceptual understanding of the metabolic pathways of microorganisms, catalysts, material sciences, fabrication, and bioelectronics. Presently, researchers are focusing on the development of biofuel cells that can be implanted in living organisms to power the medical devices or biosensors. Moreover, biofuel cells can also be better alternate for clean energy. This chapter focuses on the working principles for fuel cell, concepts of biofuel cell, types of biofuel cells, and application-based examples of biofuel cells for implantable devices.

21.1 Introduction

The demand for electric power is increasing continuously throughout the world. Globally, two major challenges that have forced various researchers to work on renewable sources of energy are ecological warming and limited amount of fossil fuels. Owing to an increasing demand of fuel-based electric power and control carbon emissions, major emphasis has been laid on providing sustainable sources of energy and more efficient use of that energy (Scott et al. 2012). Currently, scientists are focusing on the efficient conversion of solar energy, wind power, fuel cells, and biofuel cells. Hydrogen can also be used as fuel. Systems are under development that use wind or solar power to produce hydrogen by electrolysis (Steeb et al. 1985; Fischer 1986; Zhang et al. 2010; Joshi et al. 2011). Hydrogen can also be produced through solar thermochemical processes (Aldo 2005). Not only the hydrogen-based energy but also fuel cell research is under progress (Scott et al. 2012). Fuel cells are electrochemical devices that are used to convert chemical energy of a fuel into electrical energy. In recent years, biofuel cells have been developed, which are capable of using a wide range of carbon sources as fuels. Even wastewater and sludge can act as potential fuel for biofuel cells. Biofuel cells are defined (in broad sense) as devices capable of transforming chemical energy into electrical energy via electrochemical reactions involving biochemical pathways (Bullen et al. 2006). In a simple way, the fuel is being combusted in a simple reaction without the generation of heat (Scott et al. 2012). Heydorn and Gee (2004) have discussed several factors responsible for biofuel cell research including (1) the demand of clean energy from renewable sources, (2) demand for small, lightweight power sources to provide power in remote areas, and (3) demand for small implantable devices particularly medical devices. Moreover, biofuel cells are less expensive due to the nonrequirement of costly metal catalysts. The biofuel cells can have more effective configuration as compared to traditional fuel cells. The development of biofuel cell is a multidisciplinary task that demands the knowledge and conceptual understanding of the metabolic pathways of microbes, catalysts, material sciences, fabrication, and bioelectronics (Heydorn and Gee 2004). Biofuel cells also offer several advantages over traditional batteries, including the use of renewable and nontoxic components, reaction selectivity, fuel flexibility, and ability to operate at lower temperatures and near-neutral pH (Neto and De Andrade 2013).

In a typical biofuel cell, the electrochemical oxidation of hydrogen takes place at anode in the presence of biocatalytic component to generate hydrogen (Heydorn and Gee 2004; Neto and De Andrade 2013). The reducing biocatalyst reduces the oxygen at cathode. Depending on the biocatalyst, various types of biofuel cells have been described. Although biofuel cell in a living rat (Cinquin et al. 2010), snail (Halámková et al. 2012), and mammalian biofuel cell (MBFC) (Güven et al. 2013) has also been described, majority of research work is focused on enzymatic biofuel cell (EBFC) or microbial fuel cells. In an EBFC, oxidative enzymes are used at anode for the oxidation of fuel substrate, while reducing enzymes are used at cathode for oxygen reduction. Microbial fuel cell employs microorganisms as the biocatalytic component. Both direct electron transfer (DET) or mediated electron transfer (MET)-based enzymatic and microbial fuel cells have been described by various researchers. To improve the electron transfer mechanism, majority of microbial fuel cells use mediators to help the coupling of intracellular electron transfer process with electrochemical reactions at the electrodes of biofuel cells (Heydorn and Gee 2004).

Advances in the medical sciences have increased the number and type of implantable electrically operated devices (Güven et al. 2013). One of the problems with these medical devices is continuous requirement of power supply. Batteries are mostly used for this purpose. However, batteries have a limited lifetime and they need to be replaced. There is a continuous novel development in the area of implantable electronic devices in the diagnosis, management, and treatment of human disease (particularly cardiac diseases). Therefore, continuous increase in demand has been observed for devices having unlimited functional lifetimes and that can integrate seamlessly into their host biological systems (Sharma et al. 2011; Rapoport et al. 2012). Ideally, implanted devices would take advantage of the natural fuel substances found in the body and then continue to draw power for lifetime (Güven et al. 2013). Power recovery from body heat, breathing, blood pressure, typing, arm motion, pedaling, and walking have been considered as power supply methods (Starner and Paradiso 2004; Güven et al. 2013). Different physiological power sources such as thermal and pressure gradients or different movement-based energy storage devices have been considered as energy suppliers (Starner 1996; Starner and Paradiso 2004). Biofuel cells are an alternative energy source for electrically powered devices. Besides the medical devices, implantable biofuel cells (IBFCs) can have potent applications in environmental biosensors and therefore can be used for environmental monitoring. Biofuel cells can also be used to power other low-/high-power electric or electronic devices and can be promising in providing clean and sustainable energy form. This chapter describes the working principles for fuel cell, biofuel cell, various forms of biofuel cells, application areas of biofuel cells, and application examples of developed IBFCs.

21.2 Fuel Cells

Fuel cells are devices that convert chemical energy of a fuel into electrical energy. Fuel cells are scalable and, therefore, provide a wide range of possibilities for various application-based practices. Fuel cells can be used for both small- and large-scale devices. The basic principle of fuel cell involves a chemical reaction between hydrogen and oxygen that ultimately leads to the production of power and heat.

In a typical fuel cell configuration (Figure 21.1), fuel cell has an oxidative (anode) and a reductive electrode (cathode) connected externally via a circuit (Palmore 2008). Fuel is supplied continuously to the anode, while oxidant is supplied to the cathode in a similar manner (Scott et al. 2012). The fuel cell also contains an ion-selective membrane or barrier (commonly known as electrolyte membrane), and this selective barrier separates the anode and cathode internally (Palmore 2008). The membrane also restricts immediate crossing over and mixing of fuel and oxidant from one compartment to another. In a typical hydrogen/oxygen fuel cell, hydrogen (H_2) is oxidized at the anode (Equation 21.1), and both protons (H^+) and electrons (e^-) are released. The released protons migrate internally through the electrolyte membrane and electrons (Equation 21.2) migrate through the external circuit to form water (Duteanu 2008; Scott et al. 2012). The movement of electrons through the circuit creates electric current. At cathode, protons and electrons combine with oxygen and form water (Equation 21.2):

$$2H_2 = 4H^+ + 4e^- \tag{21.1}$$

$$O_2 + 4H^+ + 4e^- = 2H_2O \tag{21.2}$$

Fuel cells are usually much more efficient than other types of energy converters as they are capable of supplying electrical energy over longer time duration. As compared to

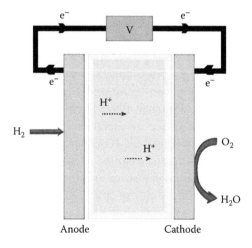

Figure 21.1 Schematic presentation of working principle of a typical hydrogen/oxygen fuel cell.

battery (which contains only limited amount of fuel and oxidant), a fuel cell will produce electricity as long as fuel and oxidant are supplied constantly (Cook 2002; Rayment and Sherwin 2003; Justin 2004).

Conventionally, fuel cells operate using relatively simple inorganic chemistries (Bullen et al. 2006), consuming fuels like hydrogen or methanol, and producing energy, water, and carbon dioxide (in the case of methanol). Other fuels are also used including lower-order alcohols and alkanes, but they are frequently reformed to produce hydrogen (Vielstich et al. 2003; Mitsos et al. 2004). Fuel cell also differs from internal combustion devices, where a fuel is burnt and gas is expanded to do work, converting chemical energy directly into electrical energy. The development of fuel cells offers many advantages over other types of power sources because they provide high efficiency, do not emit pollutants, and provide a truly sustainable power output if hydrogen is produced from a renewable energy source (Cook 2002; Justin 2004). Moreover, fuel cells require catalysts, which are expensive (Larminie and Dicks 2000; Bullen et al. 2006). Platinum is the most commonly used electrocatalyst in the fuel cells due to its efficiency of oxidizing hydrogen and capability of producing high currents in a fuel cell (Kannan et al. 2008). The major disadvantage of platinum is its cost and limited availability, making hydrogen fuel cells an expensive method of energy production. Moreover, platinum and other catalysts are also poisoned by carbon monoxide (CO) impurities (Atanassov et al. 2007; Kannan et al. 2008). Removal of CO adds to the cost of fuel cell system (Kannan et al. 2008). Biofuel cells offer several advantages over conventional fuel cells, and therefore, researchers have focused on the development of efficient biofuel cells particularly for biomedical applications and implantable devices.

21.3 Biofuel Cells

Biofuel cells can be defined as those fuel cells that use enzymes as the catalysts for the conversion of chemical energy into electrical energy (Scott et al. 2012) or at least for part of their activity (Palmore and Whitesides 1994). Biofuel cell works analogous to fuel cell, containing components similar as hydrogen/oxygen fuel cell. Biofuel cells use whole organism or enzyme, instead of inorganic catalysts, in at least one of their electrodes (Palmore and Whitesides 1994; Bertilsson et al. 1997; Kim et al. 2006; Davis and Higson 2007; Sarma et al. 2009; Fan et al. 2011). The biofuel cell can have fuels as simple as hydrogen/methane or as complicated as sugars or hydrocarbons. If glucose is used as fuel, it is oxidized at the anode (Equation 21.3) by specific enzyme or whole cell and oxygen is reduced at the enzyme/whole cell cathode (Equation 21.4). The electrochemical reaction relays electrons through the external electric circuit that produces electric current:

$$C_6H_{12}O_6 + 6H_2O = 6CO_2 + 24H^+ + 24e^- \tag{21.3}$$

$$24H^+ + 24e^- + 6O_2 = 12H_2O \tag{21.4}$$

Compared with conventional fuel cells, biofuel cells are more selective to a particular fuel and have the advantage of working at ambient temperature and physiological pH (Fan et al. 2011). These properties make biofuel cells an attractive development prospect for use in applications where it is difficult to generate high temperatures or where harsh reaction conditions are undesirable (Bullen et al. 2006). Therefore, biofuel cells are suitable for application in implantable devices and portable power supplies. Biofuel cells tend to operate under mild conditions (Bullen et al. 2006). Further, biofuel cells are not limited to these mild conditions because there are possibilities of exploitation of extremophile organisms or enzymes derived from them to be used under a wide variety of reaction conditions. Biocatalysts, either nonenzyme protein, enzyme, or whole organism, can be relatively cheaper to be used over expensive metallic catalysts, although this is not likely to be the case until the consumption of the enzyme is sufficient to merit large-scale production (Bullen et al. 2006). The choice of electrodes and biocatalysts are the main factors that affect the power output of the biofuel cell (Yang et al. 2013).

21.4 Abiotic Biofuel Cells

Abiotically catalyzed cells use abiotic catalysts like platinum or other noble metals to carry out the electrooxidation of biofuel (Kerzenmacher et al. 2010). Several catalysts such as silver and activated carbon can selectively catalyze oxygen reduction in the presence of glucose, while platinum alloys such as platinum–bismuth can catalyze glucose oxidation in deaerated solutions (Oncescu and Erickson 2011). Abiotic glucose fuel cell converts the chemical energy of glucose and oxygen into electric power using noble metals as the catalysts (Franks and Nevin 2010; Slaughter and Sunday 2014). The general electrode reactions of an abiotic biofuel cell involve oxidation of glucose to gluconic acid at a platinum-based anode catalyst and oxygen is reduced to water at the cathode. Released protons travel from the anode to the cathode through a proton-conducting membrane or electrolyte and generate electric power (Kerzenmacher et al. 2010). The biofuel may be glucose, methanol, and ethanol. Glucose is used more commonly as biofuel. Though in recent decade much attention has been paid to enzymatic, microbial, and whole cell/organism-based biofuel cells for implantable devices, abiotic biofuel cells also possess some advantages over biotic biofuel cells. Though the use of noble catalysts results in higher cost of developed biofuel cell system, abiotic biofuel cells promise tolerance to high temperatures during steam sterilization or wide range of pH. Abiotic cells also promise for long-term stability at operative and physiological conditions (Kerzenmacher et al. 2010; Slaughter and Sunday 2014). Moreover, abiotic biofuel cells may also perform better at physiological concentrations of glucose (Slaughter and Sunday 2014). Abiotic glucose fuel cell exhibits higher stability and longer life span as compared to EBFC. In the case of abiotic biofuel cells, more complex organic fuel molecules can be oxidized at relatively slow reaction rates and, therefore, result in low power densities. Further, an efficient separator for reactants is required to avoid mixed potential formation (Kerzenmacher et al. 2008, 2010; Slaughter and Sunday 2014). Recently, a membraneless single-compartment abiotic glucose fuel cell has been described. This abiotic biofuel cell was based on a combination of Al, Au, and ZnO via sputtering and hydrothermal methods (Slaughter and Sunday 2014). Anode was fabricated with Al/Au/ZnO, while platinum rod was used as cathode. The performance of the developed biofuel cell was investigated at physiological glucose and pH level. The Al/Au/ZnO anode selectively catalyzed the oxidation of glucose in the presence of oxygen and platinum specifically catalyzed the oxygen reduction at cathode. Membraneless single compartment was designed for testing the performance of developed abiotic biofuel cell. The developed biofuel cell under specified conditions possessed an open-circuit voltage of 840 mV. A peak power density of 16.2 μW cm^{-2} at cell voltage of 495 mV was obtained. The results were comparable to the complex system–based biofuel cells. The results provided a firm base using Al-/Au-/ZnO-based electrode to design a membraneless and single-compartment abiotic glucose cell providing stable electric power output for a long period of time. Earlier, an abiotic biofuel cell based on porous carbon paper (CP) support was described (Oncescu and Erickson 2011). The CP was used for the anodic catalyst layer to reduce the amount of metal catalysts required for fabricating high surface area electrodes. The CP is conductive and porous and has a high surface area; therefore, the platinum catalyst layer can be deposited directly on the membrane, thus requiring less platinum catalyst and less intermediate steps in order to obtain a high surface area electrode. A thin, enzyme-free fuel cell with high current density and good stability at a current density of 10 μA cm^{-2} has been described earlier (Oncescu and Erickson 2011). A nonenzymatic approach was preferred because of higher long-term stability. The fuel cell uses a stacked electrode design in order to achieve glucose and oxygen separation. This fuel cell was also developed without any electrode-separating membrane. The peak power output of the fuel cell is approximately 2 μW cm^{-2} and has a sustainable power density of 1.5 μW cm^{-2} at 10 μA cm^{-2}. These reports suggested the potential of abiotic biofuel cells and their possible applications for implantable devices.

21.5 Enzymatic Biofuel Cell

In EBFC, isolated enzymes are used for oxidation and reduction reactions at anode and cathode, respectively. The interest in EBFCs has increased due to implantable medical devices and biosensors for physiological substances (Scott et al. 2012). Besides the health-care applications, enzyme-based biofuel cells have also been used to power various portable and low-power devices (Ramanavicius et al. 2005; Liu and Dong 2007; Cracknell et al. 2008; Cinquin et al. 2010; Scott et al. 2012; Xu and Minteer 2012; Fujita et al. 2014; Song et al. 2014). Because of their specificity and selectivity, enzymes are preferred biocatalysts where mixed fuel or reactants are to be used. Enzymes also facilitate the construction of biofuel cells with impure substrates (Kerzenmacher et al. 2010). Practically, the lifetime of EBFCs is limited and researchers have focused on different possible solutions for long-term operational stability of EBFCs (Bullen et al. 2006; Kerzenmacher et al. 2010). These possibilities included enzyme immobilization, genetic engineering of enzymes, and process development to replenish the enzyme level at the electrodes. The operation of an EBFC resembles the functioning of conventional fuel cell (Neto and De Andrade 2013). The operating principle of EBFC or biobattery has also been described (Kannan et al. 2008). Biofuel cell generates electricity from carbohydrates (sugar) utilizing enzymes as the catalysts through the principles of power generation present in living organisms. The biobattery incorporates an anode consisting of carbohydrate-digesting enzymes and mediator and a cathode comprising oxygen-reducing enzymes and mediator on either side of a separator membrane. The anode extracts electrons and hydrogen ions from the sugar (glucose) through enzymatic oxidation. At cathode, the hydrogen ions and electrons combine with oxygen and produce water (Figure 21.2). It is interesting to note that the catalytic four-electron reduction of oxygen to water could take place at an enzyme electrode in a neutral solution. Due to the selective reactivity of the enzymes at each electrode, no cross-reaction occurs between the anode and the cathode (Taniguchi et al. 2006). In general, the EBFC could be classified into many types based on fuel containment, fuel and catalyst sources, origin of the catalytic enzymes, and method of electron transfer between the reaction site and the electrode (Bullen et al. 2006). Similarly, the enzymes generally used in biofuel cells can also be divided into three groups depending on the location of the enzyme active centers and type of electron transfer between the enzyme and the electrode (Heller 1992; Bullen et al. 2006; Scott et al. 2012). These groups are (1) enzymes having nicotinamide adenine dinucleotide (NADH/NAD$^+$) or nicotinamide adenine dinucleotide phosphate (NADPH/NADP$^+$) redox centers (these redox centers are weakly bound to the enzyme protein), (2) enzymes having redox centers at near or peripheral locations,

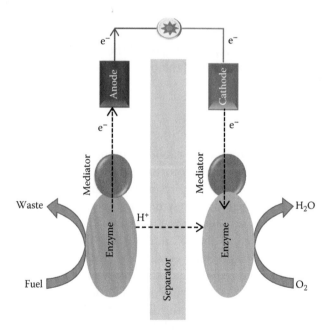

Figure 21.2 Schematic presentation of a typical enzymatic (MET) biofuel cell. (Modified from Kannan, A.M. et al., *J. Nanosci. Nanotechnol.*, 8, 1, 2008.)

and (3) enzymes having strongly bound redox centers or enzymes where redox centers are located deep in the protein or glycoprotein shell. The enzymes described in points (1) and (2) carry out DET between the electrode surface and enzyme active centers, while enzymes in group (3) are not able to perform the DET with the electrode surface (Heller 1990). Therefore, electron transfer to the electrode surface can be achieved by the use of electron transfer mediators. These mediator molecules (either electron donor or acceptor) can be accepted by the redox enzymes used in biofuel cells, and therefore, MET-based EBFCs were focused by different research groups worldwide. The working principle of a typical enzymatic (MET) biofuel cell has been presented in Figure 21.2.

21.6 Microbial Biofuel Cell

Microbial fuel cells are generally based on whole living microbial cells (Rabaey et al. 2004; Bullen et al. 2006; Kerzenmacher et al. 2010). These are used as catalysts at the electrode of biofuel cell. Some of the drawbacks (denaturation and loss of activity) to use enzymes as catalysts in biofuel cells can be circumvented by using microbial biofuel cells. The use of whole organism as a biofuel cell system allows multiple enzymes and, therefore, multiple substrates (or mixed substrates) to be used (Bullen et al. 2006). During evolution, living organisms have developed highly efficient electron control systems. Living cells can exchange electrons with the metabolic system inside the cell and the electrode outside it or with a mediator that can contact the membrane and collect electrons without penetrating the cell (Bullen et al. 2006). In MET, either indirect electron transfer systems with freely diffusing mediator molecules (diffusive MET) or indirect transfer systems with the mediator incorporated into the electrode or the cell membrane (nondiffusive MET) can be the appropriate approach (Bullen et al. 2006). In MET cases, it is desirable for a mediator to have a reaction potential close to that of the biological component to minimize potential losses (Bullen et al. 2006). In some organisms, DET can be achieved through the existence of electron-transferring capacity within the cell membrane (Kim et al. 1999, 2002; Bullen et al. 2006).

The concept of microbial biofuel cell has been discussed earlier (Rabaey et al. 2004). In a microbial-based biofuel cell, bacteria do not directly transfer the electrons that are produced by their terminal electron acceptor. The electrons are first relayed toward the anode followed by the cathode via a resistance or power using device. In this manner, the chemical energy of a bacterial reaction is converted into electrical energy (Rao et al. 1976).

In general, microbial fuel cells have been categorized into three groups: (1) photo-autotrophic-type biofuel cell (Tsujimura et al. 2001), (2) heterotrophic-type biofuel cell (Cooney et al. 1996), and (3) sediment-type biofuel cell (Bond et al. 2002). The important parameters for operational effectiveness include bacterial metabolism, bacterial electron transfer, efficiency or performance of proton exchange membrane, internal resistance of electrolyte, and efficiency of electrodes for electron transfer (Katz et al. 2003). Most of these parameters have direct influence on the potential losses due to their electron transfer resistance at the electrode, which are generally described as internal resistance of biofuel cells.

The most important step in biofuel process is the transfer of electrons from bacterial cell to electrode. Bacterial cell capable of expressing electron chain components in the outer wall is potentially adapted for use in the microbial fuel cell environment (Park and Zeikus 2002) since they provide easy access for electron transfer. In addition, electron mediators need to be added to the medium or to the electrode of biofuel cells in order to obtain higher currents and improved electron transfer (Ikeda and Kano 2001). It is also known that some bacteria are capable of producing compounds that act as electron shuttle between bacteria and electron acceptor (Hernandez and Newman 2001). The microbial consortia capable of self-MET have been reported (Rabaey et al. 2004). Diluted sludge was used as reactor inoculum, and the microbial consortium attached to electrode was isolated and enriched with repeated transfer. The bacterial community in suspended and attached form was studied for microbial fuel cell to generate higher power output. The bacterial species capable of producing power output were identified. The bacterial consortium harvested from the anode compartment of a biofuel cell, in which glucose was used as fuel, increased the output from an initial level of 0.6 W m^{-2} of electrode surface to a maximal level of 4.31 W m^{-2} (664 mV, 30.9 mA) when plain graphite electrodes were used. The result was obtained with an average loading rate of 1 g of glucose L^{-1} day^{-1} and corresponded to 81% efficiency for electron transfer from glucose to electricity. Results pointed out that the microbial consortium had either membrane-bound or excreted redox components that were not initially detected in the community. Six different bacterial species were identified including *Pseudomonas aeruginosa*, *Bacillus cereus*, *Alcaligenes faecalis*, *Ochrobactrum* sp., *Enterococcus* sp., and *Bacillus* sp. The community consisted mainly of facultative anaerobic bacteria (such as *A. faecalis* and *Enterococcus gallinarum*) capable of hydrogen production. For several isolates, electrochemical activity was mainly due to excreted redox mediators, and one of these mediators, pyocyanin produced by *P. aeruginosa*, could be characterized. Microorganisms selected through the enrichment procedure were capable of mediating the electron transfer either by direct bacterial transfer or by excretion of redox components. Both membrane-bound electron transfer and transfer through soluble redox mediators were observed in the analyzed bacterial species. The potential losses in an active biofuel cell, which decrease the output, are dependent on several factors: (1) the overpotentials at the anode, (2) the overpotentials at the cathode, and (3) the internal resistance of the fuel cell system. It was observed that bacteria facilitated the extracellular electron transfer and reduced the overpotential at the anode, thereby increasing the power output. The capability of bacteria to produce and use redox mediators as electron shuttles enables them to thrive in a microbial fuel cell. In a microbial fuel cell, the redox conditions are disadvantageous for anaerobic bacterial species, and therefore, bacterial species like *P. aeruginosa* can become dominant. Research is needed to further elucidate the microecological significance of these findings. If the mediators are produced in considerable quantities, other bacteria could use them as electron shuttles (Hernandez et al. 2004). Moreover, the more distant layers of the biofilm would be in contact with the anode besides the bacteria growing on a surface and using the surface as an electron acceptor. Results of this investigation suggest the potential use of such bacterial species for biofuel cell design and development that can self-mediate the electron transfer.

21.7 Mammalian Biofuel Cell

Biofuel cells are an alternative energy source for electrically powered devices. In 2013, MBFC was proposed (Güven et al. 2013). It was based on similar principle as microbial fuel cells. The MBFC was devised for harvesting the electrochemical energy spontaneously generated in a physiological process from mammalian cells. The practical and potential application for the MBFC is as power source for implantable medical devices such as glucose sensors, neural stimulators, and "smart" drug delivery chips. MBFC-powered device has the need for constant, reliable, long-term power provided by a physiological biofuel. Earlier, electrochemical response of leukocytes has been reported in various voltammetric experiments in simple three-electrode systems using bare graphite as working electrodes (Ci et al. 1998; Li et al. 1999). Preliminary studies have also explored the current response of human leukocytes/lymphocytes in fuel cell setups (Justin et al. 2004, 2005, 2011; Güven et al. 2006, 2013). The proposed power output of MBFC requires the supply of electrons from a mammalian cell–based source to be coupled with an electrode in a nondestructing way for the living organism (Güven et al. 2013). At the anode of the MBFC, intracellular electrons are generated in the cytosolic environment of the cell and subsequently released through extramembrane transport process. Either NADPH oxidase (present in leukocyte family of blood cells) are responsible for current production (Schrenzel et al. 1998) or electrochemically active molecules released from cells to the extracellular surroundings produce current after contact with the electrode (Guven et al. 2013).

Although the idea of MBFC generating power from various metabolic processes occurring in human cells is very challenging, the outcome of the preliminary study revealed the potential applicability of mammalian cells in the biofuel cell system as an alternative power production method for implantable medical devices (Güven et al. 2013).

21.8 Application-Based Examples of Biofuel Cells for Implantable Devices

Various reports for biofuel cells and their potential applications for implantable devices are available at present. Major research belongs to recent years. Examples of various biofuel cells, which have been designed and fabricated to provide energy to implantable devices, have been discussed in this section.

21.8.1 Refuelable Biofuel Cell to Power Walkman® Device

Recently, Fujita et al. (2014) have developed the refuelable biofuel cell based on a hierarchical porous carbon (HPC) electrode. They have immobilized multiple biocatalytic components on a carbon electrode by using a porous carbon material. The catalytic components were immobilized on HPC to fabricate an anode. The anode was coated with glutaraldehyde-cross-linked poly-L-lysine (PLL) to hold the HPC on the carbon fiber (CF). The authors fabricated the mediated bioanode composed of NAD-dependent glucose dehydrogenase (GDH), diaphorase (Di), NAD, and the mediator anthraquinone-2-sulfonate (AQ2S). In an earlier report, a similar Di-linked system was used but with 2-amino-1,4-naphthoquinone (ANQ) as an electron mediator (Sugiyama et al. 2010). But in this experimentation, they used AQ2S due to its more negative redox potential than ANQ31. The two-electron oxidation of glucose was catalyzed by GDH with NAD^+. The NADH was oxidized by AQ2S through Di followed by oxidation of $AQ2SH_2$ on the electrode surface. Cathode was fabricated using bilirubin oxidase (BOD). The developed biofuel cell demonstrated promising durability. They provided stable performance of mediated bioanode at a current level of $1.5\,mA\,cm^{-2}$ over 50 consecutive refueling cycles. The refuelable, mediated bioanode can be improved further to allow the oxidation of glucose to CO_2 since it utilizes NAD, which is a versatile cofactor in the glycolysis process in living cells (Palmore et al. 1998; Matsumoto et al. 2010; Tsujimura et al. 2012; Fujita et al. 2013). The immobilization of both AQ2S and Di in this bioanode facilitates the transfer of electrons to the electrode via multiple enzymatic reactions. The developed cell can be repeatedly refueled and can be used to generate sufficient electricity to continuously power the Walkman® device and connected loudspeakers simultaneously. Biofuel cells assembled with these bioanodes and BOD-based biocathodes can be repeatedly used to power a portable music player at $1\,mW\,cm^{-3}$ through 10 refueling cycles. The described technique is highly practical since it is simple, low cost, safe, and robust and offers good energy density. The biofuel cell has been achieved by the electrochemically and enzymatically active and robust immobilization of mediator, cofactor, and enzymes on a carbon electrode with porous carbon particles. The immobilization technique will likely have applications toward electrochemical devices, such as biofuel cells and biosensors, along with heterogeneous catalysis. This is particularly an important discovery for NAD since covalent immobilization techniques have often resulted in decreased activity with enzymes. The large quantity of immobilized NAD would be widely applicable for the design of reactors, columns, biosensors, diagnostic tools, and medical treatments.

21.8.2 Mammalian Biofuel Cell for Implantable Medical Devices

In 2013, MBFC was proposed (Güven et al. 2013). The working principle of this MBFC was analogous to microbial fuel cells and can harvest electrochemical energy generated spontaneously by physiological process from mammalian cells. Practically, the potential application of the MBFC includes its use as a power source for implantable medical devices such as glucose sensors, neural stimulators, and "smart" drug delivery chips. MBFC-powered device has the need for constant, reliable, long-term power provided by a physiological biofuel. The authors described power generation from human leukocytes/lymphocytes in MBFC. The electrochemical response of leukocytes in voltammetric experiments in simple three-electrode systems (using bare graphite as working electrodes) has been described earlier, but in these setups, the origin of anodic response was not clearly identified and the lifetime of that signal lasted few voltammograms (Ci et al. 1998; Li et al. 1999). Current response of human leukocytes/lymphocytes in fuel cell setups has also been studied earlier (Justin et al. 2004, 2011; Güven et al. 2006). Güven et al. (2013) characterized the obtained net power output for the proposed MBFC. The MBFC consists of two compartments (anode and cathode compartment) separated by a membrane. Carbon mesh electrode modified with PLL was used as a working electrode for anode characterization, and nonmodified carbon mesh electrode was used as a working electrode for cathode characterization in the MBFC setup. The PLL-modified anode was used to immobilize the human leukocytes for this setup. The electrochemical activity of the immobilized human leukocytes was investigated by cyclic voltammetry (CV). Oxidation peaks were observed in the presence of leukocytes but not in their absence, which confirmed the electrochemical activity of the leukocytes. The peak currents at 0.33 V versus Ag/AgCl for 2×10^6 and 1×10^6 leukocyte cells were found to be 1.92 and 0.58 µA, respectively. The peak current of 2×10^6 cells was increased from 1.92 to 3.62 µA, due to the addition of the activator. The increased peak current suggested an increased activity of the cells due to the activator. The MBFC required additional investigation and optimization prior to biomedical implementation. Miniaturization and biocompatibility of such a device should be studied before operating *in vivo*. Recent improvements in biofuel cell concepts in terms of carbon nanotube (CNT) compression and DET led to high open-circuit voltage, high power output, and stabilities over long period of time (Zebda et al. 2011). The MBFCs that produce significant level of energy at a single location could be utilized as the power source for implanted sensor devices dedicated to medical monitoring. Further optimization of MBFC could be expected to provide opportunities for other medical applications (Güven et al. 2013; Zebda et al. 2013). Future implantable medical devices such as cardiac defibrillators, deep brain neurostimulators, spinal cord stimulators, gastric stimulators, foot drop implants, cochlear implants, and insulin pumps powered by implanted MBFC extracting electrical energy directly from human white blood cells are possible (Zebda et al. 2013).

21.8.3 Abiotic Biofuel Cell Implanted in an Orange

Recently, abiotic biofuel cell has been reported by Holade et al. (2015). The biofuel cell contained catalytic electrodes modified with inorganic nanoparticles (NPs) deposited on carbon black (CB). The cathode was made of CP modified with Pt-NPs/CB, while the anode was formed with buckypaper modified with $Au_{80}Pt_{20}$-NPs/CB. The electrodes were implanted in orange fruit pulp capable of extracting power/energy from glucose and fructose content of the juice. The open-circuit voltage was found to be 0.36 V, while the maximum power obtained was 182 µW. The biofuel cell power was applied to wireless transmitter. Further research is required for the production of higher power output and its use in broader applications. This research will produce significant impact on biofuel research and will be important for the development of IBFCs to power electric/electronic devices.

21.8.4 Single-Compartment Enzymatic Biofuel Cell

Single-compartment biofuel cell without separator has been constructed. This EBFC was based on D-fructose dehydrogenase (FDH) from *Gluconobacter* sp. and laccase from *Trametes* sp. (TsLAC). The enzymatic fuel cell worked as DET type. FDH was used for two-electron oxidation of D-fructose and laccase in four-electron reduction of dioxygen. Although other multicopper oxidases were tested, the TsLAC was found as best for the DET-type biocatalysis of dioxygen reduction at pH 5. Ketjen black (KB) and carbon aerogel (CG) were used to modify the CP electrodes. For FDH and TsLAC adsorption, KB and CG particles were used, respectively. The FDH-adsorbed KB-modified CP electrode and TsLAC-adsorbed CG-modified CP electrode were combined to construct a single-compartment biofuel cell without separator. The open-circuit voltage was 790 mV. The maximum current density of 2.8 mA cm^{-2} and the maximum power density of 850 μW cm^{-2} were achieved at 410 mV of the cell voltage under stirring (Kamitaka et al. 2007).

21.8.5 Glucose Fuel Cell for Implantable Brain–Machine Interfaces

The cerebrospinal fluid represents a promising environment for an implantable fuel cell. It is virtually acellular, is under minimal immune surveillance, has a 100-fold lower protein content than the blood and other tissues, and is therefore less prone to induce the biofouling of implanted devices, and further its glucose levels are comparable to those of the blood and other tissues (Davson and Segal 1996). Rapoport et al. (2012) described a detailed model of a fuel cell able to consume glucose and oxygen in implanted condition in the subarachnoid space surrounding the human brain and analyzed the impact of such fuel cell on glucose and oxygen homeostasis. The fuel cell was manufactured using a novel approach that employed semiconductor fabrication technique. Glucose was used as fuel. Platinum and CNT-based anode and cathode were used, respectively. Glucose got oxidized at the nanostructured surface of activated platinum anode, while oxygen was reduced to water at CNT-based cathode. The electrodes were separated by Nafion membrane. The fuel cell was configured in a half-open geometry that protected the anode while exposing the cathode. This configuration resulted in the formation of oxygen gradient that favored oxygen reduction at the cathode. Glucose approached the protected anode by the process of diffusion through the nanotube mesh that does not catalyze glucose oxidation. The developed implantable biofuel device produced 3.4 μW cm^{-1} steady-state power when driving a load of 550 kΩ. The fuel cell generated an open-circuit voltage of 192 mV and transiently generated a peak power of more than 180 μW cm^{-2} when sourcing 1.5–1.85 mA cm^{-2}. The natural recirculation of cerebrospinal fluid around the human brain theoretically permits glucose energy harvesting at a rate on the order of at least 1 mW without any adverse physiological effect. Thus, low-power brain–machine interface can benefit from having their implanted units powered by glucose fuel cells (Rapoport et al. 2012).

21.8.6 Biofuel Cell Implanted in Living Snail

Whole organism–based biofuel cell has also been described (Halámková et al. 2012). Sustainable generation of electric power was achieved *in vivo* by implanting electrodes in a snail (*Neohelix albolabris*) and demonstrated that metabolically regenerated glucose can "recharge" the living battery for continuous production of electricity. Pyrroloquinoline quinone (PQQ)-dependent GDH was used for biocatalytic oxidation of glucose at the anode and laccase was used as a biocatalytic component at the cathode. The buckypaper electrodes were modified with PBSE linker, and then PQQ-GDH or laccase was immobilized to yield the biocatalytic anode or cathode, respectively. The biocatalytic electrodes were then implanted in a snail. Glucose is the major form of carbohydrate found in the hemolymph of most gastropods (Liebsch et al. 1978). The oxygen content in hemolymph depends on the physiological conditions and varies accordingly, but its concentration is always higher than the glucose concentration (Barnhart 1986). Therefore, the oxygen content would not be a limiting factor for the operation of an implanted biofuel cell in snail. The implantable electrodes were inserted into the snail through two holes cut in the shell and placed into the hemolymph between the body wall and the visceral mass. The implanted electrodes performing the bioelectrocatalytic reactions were connected through external circuit having a variable-load resistance. The open-circuit voltage of 530 mV and short-circuit current of 42.5 μA were achieved in the biofuel cell. The maximum power produced by the implanted biofuel cell was 7.45 μW on the optimum resistance of 20 kΩ. The obtained short-circuit current corresponds to an activity of 0.013 enzyme units, which suggested that only 6% of the total enzyme content was electrically wired onto the modified electrodes. The fact that a relatively small fraction of the immobilized enzymes contributed to the current generation provided an opportunity for further improvement of the biofuel cell operation through the optimization of the electrical wiring of the enzymes. The electric potential of electrified and biotechnological living device as biofuel cell was sufficient to power various types of bioelectric devices. According to the researchers, this investigation has demonstrated an actual implanted biofuel cell operating in a small creature living with the bioelectrodes for a long period of time (several months). Successful implantation of the biofuel cell and its significant operational stability (while using physiological glucose) in living snail has opened new perspectives for biofuel cells operating *in vivo* for many biotechnological applications.

21.8.7 Proton Exchange Membrane Fuel Cell for Possible Applications in Medical Devices

The biofuel cell developed by Ghaffari et al. (2013) is a type of proton exchange membrane fuel cell (PEMFC) that converts chemical energy into electrical energy. The chemical energy source can be glucose present in the blood, and the electrical energy produced can be used in pacemakers, insulin pumps, drug delivery systems, and defibrillators. The biofuel cell was fabricated using gold metal with a thickness of 700 µm as catalyst in anode. Cathode was fabricated using cloth carbon, and the membrane was made of PVA–PAA. The peak power density for developed biofuel cell (BFC) was 1.2 µW cm^{-2} between 115 and 125 mV (Ghaffari et al. 2013).

21.8.8 Implantable Biofuel Cell to Power Pacemakers or Robotic Artificial Urinary Sphincters

The development of functional and implantable glucose biofuel cell (GBFC) in retroperitoneal space of freely moving rats has been described by Cinquin et al. (2010). The GBFC-contained anode was fabricated using a mixture of graphite, ubiquinone, glucose oxidase (GOX), and catalase. The cathode contained quinhydrone, polyphenol oxidase (PPO) or urease, and graphite. Instead of covalent binding to electrodes, the enzymes and mediators were kept mechanically restricted through the use of dialysis bags and/or mechanical compression of graphite particles, enzymes, and redox mediators. The *in vitro* performance of the developed quinone–ubiquinone GBFC was analyzed in a glucose concentration and pH similar to extracellular fluid. The average maximum power from 30 to 40 days was about 1.65 µW (standard deviation 0.13 µW), reflecting an excellent operational stability. After demonstration of *in vitro* performance, the developed GBFC was surgically inserted in the rat's retroperitoneal space that enabled glucose and O_2 to flow from the ECF into the GBFC. The open-circuit voltage of the GBFC was recorded as 0.275 V. Maximum specific power of 24.4 µW mL^{-1} was obtained. The peak specific power provided by the developed GBFC is significant with respect to the requirement of a pacemaker. The results and feasibility of the GBFC have strengthened the way for the development of new generation implantable artificial organs, covering a wide range of medical applications.

Cinquin et al. (2013) have also reported the IBFC capable of producing electricity from the naturally occurring chemical energy in the body. The feasibility of developed IBFC has been investigated using rats as model. The IBFC working was based on the utilization of glucose and oxygen present in the physiological fluid. The feasibility of electrical energy production for other implanted medical devices was investigated. Researchers obtained a power to volume ratio of the order of magnitude of 1 µW µL^{-1}. Another type of IBFC involves a biomimetic approach. The biofuel cell generated electricity by imitating salt transport that occurs naturally in the cells of the body; electrolytes such as sodium are used to provide the fuel for power generation. The critical part of the device is a lipid bilayer containing transport proteins similar to the ones naturally occurring in a cellular membrane. Gradients of sodium are then transformed into voltage and power. Preliminary results show that a biomimetic membrane of 0.3 cm^2 can produce an output of 59 µW, which is encouraging since such membranes can be easily stacked in series and in parallel. The very interesting power to volume ratio opens newer perspectives of applications of IBFCs, for instance, to power implantable medical devices such as leadless pacemakers or robotic artificial urinary sphincters.

Southcott et al. (2013) have reported the development of IBFC operating under conditions similar to human blood circulatory system to power a pacemaker. The biocatalytic electrodes were made of buckypaper modified with PQQ-dependent GDH on the anode and with laccase on the cathode. The electrodes were assembled in a flow biofuel cell filled with serum solution mimicking the human blood circulatory system. The power generated by the IBFC was investigated to activate a pacemaker connected to the cell via a charge pump and a DC–DC converter interface circuit to adjust the voltage produced by the biofuel cell to the value required by the pacemaker. Successful and sustainable operation of the pacemaker was achieved with the system mimicking the human physiological conditions using a single biofuel cell. The activation of pacemaker by the physiologically produced electrical energy has also been demonstrated. Promising results show the use of electronic implantable medical devices powered by electricity harvested from the human body in the future.

21.8.9 Implanted Glucose Biofuel Cell to Harvest Energy from Mammal's Body Fluid

Zebda et al. (2013) described the development of implanted GBFC. The developed biofuel cell was capable to generate electric power sufficient for electric devices from the mammal's (rat) body fluid. The biofuel cell was based on CNT/enzyme electrode. The biocomponent included GOX and laccase for glucose oxidation and dioxygen reduction, respectively. GOX from *Aspergillus niger* and laccase from *Trametes versicolor* were used for oxidation/reduction reactions. Laccase was used at the cathode and GOX with catalase was used at the anode. Catalase degraded H_2O_2 and prevented enzyme deactivation. Electrodes were wrapped in dialysis membrane and placed in perforated silicon tube. Further, electrodes were placed in dialysis bag and sutured inside a Dacron® bag. The GBFC was implanted in the abdominal cavity of rat. The implanted biofuel cell produced an average open-circuit voltage of 0.57 V. GBFC delivered a power output of 38.7 µW corresponding to a power density of 193.5 µW cm^{-2} and a volumetric power of 161 µW mL^{-1}. This single implanted GBFC showed the capacity to power a light-emitting diode (LED) or a digital thermometer. Additionally, no sign of rejection or inflammation was observed even after 110 days of implantation in rat body.

21.8.10 Epidermal Biofuel Cell to Harvest Energy from Human Perspiration

An epidermal biofuel cell capable of harvesting the energy from human perspiration has been described by Jia et al. (2013). Lactate oxidase (LOX) was used at the anode for the utilization of lactate as biofuel. LOX converted lactate in pyruvate and extracted energy. Dioxygen was reduced to water at platinum-based cathode. It is an important report on electric power generation from human perspiration in a noninvasive and continuous pattern through the application of epidermal biofuel cell.

21.8.11 Enzymatic Biofuel Cell Implanted in Living Lobsters

MacVittie et al. (2013) described an enzyme-based biofuel cell implanted in living lobsters. The extraction of electric power sufficient to activate the electric watch from two lobsters having implanted biofuel cells connected in series has been investigated. The developed biofuel cell system generated an open-circuit voltage of up to 1.2 V. The authors also discussed biofuel cell designed as fluidic systems mimicking human blood circulation. This fluidic system was composed of five cells filled with human serum solution. These cells were connected in series and generated power sufficient for a pacemaker. The activation of pacemaker by physiologically produced electrical energy promised the development of advanced electronic implantable medical devices capable of generating electric power from the human body.

21.8.12 Efficient and Ultrathin Implantable Direct Glucose Fuel Cell

Three different types of direct glucose fuel cells with Pt anode have been described by Sharma et al. (2011). Mesoporous silica as functional membrane and reduced graphite were used for cathode fabrication for ultrathin implantable direct glucose fuel cell. Platinum thin film was deposited on silica substrate to fabricate the anode of implantable biofuel device. Graphene pressed on a stabilized steel mesh was used as the cathode. Results demonstrated that mesoporous silica thin film is capable of replacing the conventional polymer-based membranes with an improved power generation. Mesoporous silica possesses stable surface chemistry and long shelf life (Kaufhold et al. 2008). The developed biofuel cell showed significant power performance that indicated the contribution of reduced graphene oxide (as an alternative cathodic material to the conventionally used activated carbon) and hydrophilic mesoporous silica to increase the power output. The higher open-circuit potential of Pt/MPS/G (314.6 mV) compared to Pt/PP/G (181.4 mV) indicates that there is lower crossing of oxygen to the anode, which can be attributed to the hydrophilic nature of the mesoporous silica membrane. Further research is required to establish significant advantages of using mesoporous silica along with its optimization. As mesoporous silica and graphene processing are becoming semiconductor clean room friendly, the future holds great promise for the development of mass-producible, high-power-density, and ultrathin direct glucose fuel cells for biomedical implant applications.

Sharma et al. (2011) focused on strategies that can enable the creation of high-power direct glucose fuel cells by shifting the implant site from a subclavicular skin pouch to an intravenous stent mounted implant. Moving the implant site helps by increasing glucose diffusion rates. It is estimated that the glucose mass transfer rate in the case of intravenous implants is enhanced by a blood flow of 1–10 cm s^{-1} and is expected to be 1–2 mA cm^{-2} (Barton and Atanassov 2004), whereas the same implant in tissues in the absence of convection would account for glucose flux rates limited to only 0.2 mA cm^{-2} (Weidlich et al. 1976). However, reduction in the thickness of the present fuel cell is critical for a stent-based application. Hence, the motivation behind the use of ultrathin MPS as the membrane is twofold: (1) to provide well-controlled physicochemical properties of the nanopores (shape, size, distribution, and surface functionalization) for enhanced glucose diffusion and (2) to drastically reduce the thickness of direct glucose fuel cell. Since the problem associated with the use of blood glucose is diffusion in sufficient concentration, reduction in the membrane thickness would also help boost efficiency of direct glucose fuel cell by reducing the path length of the glucose molecules (Beltzer and Batzold 1969). Combining the strategies mentioned earlier, the authors have successfully assembled a highly efficient, ultrathin direct glucose fuel cell.

21.8.13 Biofuel Cell for Low-Power Implantable Devices and Large-Scale Devices

To power the low-power implantable devices, abiotic glucose-based single-layer biofuel cells have been reported by Kloke et al. (2008). Both anode and cathode were platinum based. Anode was fabricated with Pt–Zn, while Pt–Al was used for cathode preparation. Anode and cathode were placed side by side to assemble the biofuel cell. A maximum power density of 2.0 $\mu W\ cm^{-2}$ was achieved at a current density of 8.95 $\mu A\ cm^{-2}$. The value was about 40% less than the 3.3 $\mu W\ cm^{-2}$ as reported for a stacked activated carbon-based biofuel cell (Kerzenmacher et al. 2008). An improved oxygen tolerance was found for Pt–Zn anode. The open-circuit cell voltages were reduced by only 8%–25% for Pt–Zn anodes instead of 45% for state-of-the-art anodes, if oxygen saturation is increased from 0% to 7%. Therefore, Pt–Zn anodes can be suggested as tolerant for oxygen. It was demonstrated that layer stacking and the oxygen depletion are not essential at the anode for the operation of a glucose fuel cell. Stacking can be avoided by placing electrodes side by side, and it can thus facilitate fuel cell integration on implant capsules. The calculated maximum power densities with varying ratios of cathode to anode were also discussed by the authors. At ratio 1:1 (cathode to anode) and 9.5:1 (cathode to anode), the power output was comparable, while at ratio 3:1 (cathode to anode) overall maximum output was predicted. A power density of 3.0 $\mu W\ cm^{-2}$ was predicted for a cathode to anode area ratio of 3. The predicted power density was only 10% lower than reported for stacked fuel cells.

Sakai et al. (2008) described the fabrication of multistacked structured passive type of biofuel cell. The fabricated biofuel cell was based on NAD-dependent GDH that catalyzed two-electron oxidation of glucose by NAD^+. Di was used as a biocatalyst for NADH oxidation, while 2-methyl-1,4-naphthoquinone (VK_3) was used as a mediator to transfer electrons from Di to electrode. GDH, NADH, DI, and VK_3 were immobilized densely on glassy carbon (GC) plate electrode with polyion complex (PIC) method. To further enhance the current density, CF sheet was used instead of GC. BOD was used as a biocatalyst at the cathode. $Fe(CN)_6$ was used as an electron transfer mediator. Both BOD and $K_3(Fe[CN]_6)$ were immobilized on PLC-modified CF electrode. Three important techniques that were applied to develop the efficient biofuel cell included (1) dense entrapment of enzymes and mediators on CF electrodes without the loss of enzyme activity, (2) optimization of buffer concentrations for immobilized enzyme in electrolyte solution, and (3) designing of cathode structure for efficient O_2 supply. The developed passive-type biofuel cell unit generated a power over 100 mW/80 cm². The biofuel cell produced the maximum power density of 1.45 ± 0.24 mW cm^{-2} at 0.3 V in two-electron oxidation of glucose and four-electron reduction of O_2 at pH 7 under quiescent conditions. The demonstrated biofuel cell promised the application of biofuel cells for small- or large-scale devices with durability.

Glucose is widely used fuel in biofuel cells and in majority of the developed glucose biofuel systems, glucose is oxidized to gluconolactone (Xu and Minteer 2012), for the development of six-enzyme cascade-based bioanode for complete oxidation of glucose to CO_2. The enzymatic cascade included PQQ-dependent GDH and PQQ-dependent 2-gluconate dehydrogenase to oxidize the glucose to gluconolactone and then glucuronic acid. The glucuronic acid was used by aldolase and led to form glyceraldehyde and hydroxypyruvate. These molecules are intermediate of glucose oxidation pathway (Arechederra and Minteer 2009). Therefore, researchers have employed PQQ-dependent alcohol dehydrogenase, PQQ-dependent aldehyde dehydrogenase, and oxalate oxidase to completely oxidize these intermediate molecules. Anode was fabricated using a Toray paper electrode with tetrabutylammonium bromide–modified Nafion polymer. Platinum was used to fabricate the cathode. Glucose (^{13}C labeled) at a concentration of 100 mm was used as biofuel for the developed biofuel system. This six-enzyme cascade-based system resulted in a 46.8-fold increase in power density, while the current density was increased to 33-fold as compared to two-enzyme cascade-based system. Glucose was oxidized to CO_2. The glucose-based enzymatic bioanode coupled with air-breathing cathode yielded a maximum power density of 6.74 ± 1.43 μW cm^{-2}. The maximum current density at 0.001 V was 31.5 ± 6.5 μA cm^{-2} for the developed enzymatic cascade.

Among the glucose-oxidizing enzymes, GOX is the most common enzyme constituting the biocomponent of EBFCs (Song et al. 2014). GOX provides advantages due to its thermostability and specificity for glucose as substrate. GOX is an oxidoreductase that causes competition between GOX and laccase catalysis. Therefore, GDH is preferred over GOX. Song et al. (2014) described a computational model of EBFC based on a 3D interdigitated microelectrode array. To achieve the ideal working of EBFC, glucose should interact with the whole surface of electrode to fully utilize the enzyme immobilized on their surface. Glucose reacts immediately with the top surface portion of electrode and then diffused gradually down to the bottom and interacts with the electrode. Researchers observed this nonuniform concentration or uneven distribution along the surface of electrode. A gradual decrease in concentration of glucose along the vertical direction inside the well from top to bottom is generally observed. The effect of different designs and spatial distributions of the microelectrode arrays on mass transport of fuel, enzyme-mediated rate of reaction, open-circuit output potential, and current density was investigated. GDH was used for the biocomponent of the EBFC. GDH and laccase were immobilized on 3D microelectrode to form the anode and cathode, respectively. The authors suggested the optimal configuration for the EBFC, which included the ratio of height and well width of electrode to be 2:1. The maximum power density for 3D microelectrode reached 110 μW cm^{-2} at 0.44 V in voltage when the height of electrode was kept 200 μm and well width of 100 μm.

Recently, an improved fructose/dioxygen biofuel cell has been reported (So et al. 2014). The biofuel cell developed was a DET-type bioelectrochemical device. The biofuel contained BOD as cathode and FDH as anode. Improvement in cathode performance was observed after adsorption of bilirubin or related substances on electrode before BOD adsorption. The substrate modification method was applied successfully, and FDH adsorption conditions were also optimized. The developed one-compartment DET-type biofuel cell provided maximum power density of 2.6 mW cm^{-2} at 0.46 V of cell voltage under passive and atmospheric air conditions.

21.9 Conclusions

Continuous developments and advancements in implantable medical devices have forced various scientific groups to develop the continuous source of electric power to these medical devices. Biofuel cells can be the better option for implantable medical devices. Biofuel cells capable of generating electric power under the physiological conditions and, in some cases, physiological fluid have been used as a source of electric power. Recent research has shown efficient conversion capacity of biofuel cells in living organisms including mammals, lobsters, and snails. Biofuel cell has been shown to draw electric power not only in animals but also from the orange pulp. Successful implantation of biofuel cells in living organisms and the development of MBFCs have appeared as promising advancements in biofuel cell technology. These significant findings will be useful in the near future in the field of biofuel cells and will be considered as benchmark. Biofuel cells can be used for environmental sensors, medical sensors, implantable medical devices, and other low-power electric or electronic devices. The LED and audio devices have been powered successfully using biofuel cells. Implantable medical devices such as pacemakers, artificial urinary sphincters, and brain–machine interfaces can be powered by biofuel cells. Further, biofuel cell capable of extracting energy from human perspiration has also been demonstrated. The recent findings promise for the future advancements in the area and expand the area of applications of biofuel cells. Not only civilians but also IBFCs can be used for army and other defense services. Some improvements and advancements are required for continuous power supply to various devices. The practical approach of implantable devices may be increased with biofuel cells capable of utilizing different enzymes and microbial or mammalian cells. Nowadays, the choice for different enzymes, electrodes, electrode materials, fabrication techniques, and enzyme immobilization processes is available. Biofuel cells with higher power output capacity, conversion efficiency, stability, and durability are the requirement of energy-based sectors. Several challenges are needed to resolve for further advances in the biofuel technology. Major issue is to achieve the higher power output from biofuel cells along with stability and durability of the developed biofuel cell system. Designed or enzymes with improved specificity and efficiency can provide advantage for biofuel cell. Further, immobilization of the enzymes and microbes and use of the NPs can increase the enzyme stability and electron transfer. The biofuel cells can be good alternative approach to decrease the burden on traditional resources, fossil fuels, and coal-based electric power. Moreover, microbial biofuel cells can be useful in deriving electric power from wastewater and sludge. Sugars, alcohol, and organic acids can be used as renewable fuels for various types of biofuel cells. The possibility of using various fuels to draw electric power also makes the biofuel cells as preferred choice over the conventional batteries. The major limitation is that biofuel cell needs continuous supply of biofuel and higher energy requirement of some implantable devices. In the future, these limitations are required to be overcome for real ground applications of biofuel cells.

Acknowledgment

The authors acknowledge the help and support by Head, Department of Biotechnology, Maharishi Markandeshwar University, Mullana, Ambala, India.

References

Aldo, S. 2005. Solar thermochemical production of hydrogen review. *Sol Energ* 78:603–615.

Arechederra, R.L. and S.D. Minteer. 2009. Complete oxidation of glycerol in an enzymatic biofuel cell. *Fuel Cell* 9:63–69.

Atanassov, P., Apblett, C., Banta, S. et al. 2007. Enzymatic biofuel cells. *Electrochem Soc Interface* 16:28–31.

Barnhart, M.C. 1986. Respiratory gas tension and gas exchange in active and dormant land snails, *Otala lactea. Physiol Zool* 59:733–745.

Barton, S.C. and P. Atanassov. 2004. Enzymatic biofuel cells for implantable and micro-scale devices. *Preprint Papers Am Chem Soc Div Fuel Chem* 49:476–477.

Beltzer, M. and J.S. Batzold. 1969. Limitations of blood plasma as a fuel cell electrolyte. In Paper presented at *Proceeding Intersociety Energy Conversion Engineering Conference*, Washington, DC.

Bertilsson, L., Butt, H.J., Nelles, G., and D.D. Schlereth. 1997. Cibacron Blue F3G-A anchored monolayers with biospecific affinity for NAD(H)-dependent lactate dehydrogenase: Characterization by FTIR-spectroscopy and atomic force microscopy. *Biosens Bioelectron* 12:839–852.

Bond, D.R., Holmes, D.E., Tender, L.M., and D.R. Lovley. 2002. Electrode reducing microorganisms that harvest energy from marine sediments. *Science* 295:483–485.

Bullen, R.A., Arnot, T.C., Lakeman, J.B., and F.C. Walsh. 2006. Biofuel cells and their development. *Biosens Bioelectron* 21:2015–2045.

Ci, Y.-X., Li, H.-N., and J. Feng. 1998. Electrochemical method for determination of erythrocytes and leukocytes. *Electroanalysis* 10:921–925.

Cinquin, P., Cosnier, S., Belgacem, N., Cosnier, M.L., Dal Molin, R., and D.K. Martin. 2013. Implantable glucose biofuel cells for medical devices. *J Phys Confer Ser* 476:012063.

Cinquin, P. Gondran, C., Giroud, F. et al. 2010. A glucose biofuel cell implanted in rats. *PLOS ONE* 5:e10476.

Cook, B. 2002. Introduction to fuel cells and hydrogen technology. *Eng Sci Educ J* 11:205–216.

Cooney, M.J., Roschi, E., Marison, I.W., Comninellis, C., and U. vonStockar. 1996. Physiologic studies with the sulfate-reducing bacterium *Desulfovibrio desulfuricans*: Evaluation for use in a biofuel cell. *Enzyme Microb Technol* 18:358–365.

Cracknell, J.A., Vincent, K.A., and F.A. Armstrong. 2008. Enzymes as working or inspirational electrocatalysts for fuel cells and electrolysis. *Chem Rev* 108:2439–2461.

Davis, F. and S.P.J. Higson. 2007. Biofuel cells—Recent advances and applications. *Biosens Bioelectron* 22:1224–1235.

Davson, H. and M.B. Segal. 1996. *Physiology of the CSF and Blood-Brain Barriers*. Boca Raton, FL: CRC Press.

Duteanu, N. 2008. Pile de combustie directa a metanolului echipate cu electrolit polimer solid. Timisoara, Romania: Editura 'POLITEHNICA'.

Fan, M., Maréchal, M., Finn, A., Harrington, D.A., and A.G. Brolo. 2011. Layer-by-layer characterization of a model biofuel cell anode by (*in situ*) vibrational spectroscopy. *J Phys Chem C* 115:310–316.

Fischer, M. 1986. Review of hydrogen production with photovoltaic electrolysis systems. *Int J Hydro Energ* 11:495–501.

Franks, A.E. and K.P. Nevin. 2010. Microbial fuel cells, a current review. *Energies* 3:899–919.

Fujita, S., Matsumoto, R., Ogawa, K. et al. 2013. Bioelectrocatalytic oxidation of glucose with antibiotic channel-containing liposomes. *Phys Chem Chem Phys* 15:2650–2653.

Fujita, S., Yamanoi, S., Murata, K. et al. 2014. A repeatedly refuelable mediated biofuel cell based on a hierarchical porous carbon electrode. *Sci Rep* 4:4937.

Ghaffari, S., Asgarpour, A., Mousavi, R., and M. Salehieh. 2013. Fabrication and simulation of implantable glucose biofuel cell with gold catalyst. In Paper presented at *Medical Measurements and Applications Proceedings (MeMeA)*, Gatineau, QC, Canada, pp. 63–66.

Güven, G., Lozano-Sanchez, P., and A. Güven. 2013. Power generation from human leukocytes/lymphocytes in mammalian biofuel cell. *Int J Electrochem* 2013:706792.

Güven, G., Sanchez, P.L., Lenas, P., and I. Katakis. 2006. Towards self-powered implants: Electrical power from isolated human lymphocytes. In *Proceedings of the 11th Transfrontier Meeting on Sensors and Biosensors*, Girona, Spain, p. 15.

Halámková, L., Halámek, J., Bocharova, V., Szczupak, A., Alfonta, L., and E. Katz. 2012. Implanted biofuel cell operating in a living snail. *J Am Chem Soc* 134:5040–5043.

Heller, A. 1990. Electrical wiring of redox enzymes. *Acc Chem Res* 23:128–134.

Heller, A. 1992. Electrical connection of enzyme redox centers to electrodes. *J Phys Chem* 96:3579–3587.

Hernandez, M.E., Kappler, A., and D.K. Newman. 2004. Phenazines and other redox-active antibiotics promote microbial mineral reduction. *Appl Environ Microbiol* 70:921–928.

Hernandez, M.E. and D.K. Newman. 2001. Extracellular electron transfer. *Cell Mol Life Sci* 58:1562–1571.

Heydorn, B. and R. Gee. 2004. Biofuel cells: Can they fulfil their promise? *Fuel Cell Rev* 1:23–26.

Holade, Y., MacVittie, K., Conlon, T. et al. 2015. Wireless information transmission system powered by an abiotic biofuel cell implanted in an orange. *Electroanalysis* 27:276–280.

Ikeda, T. and K. Kano. 2001. An electrochemical approach to the studies of biological redox reactions and their applications to biosensors, bioreactors, and biofuel cells. *J Biosci Bioeng* 92:9–18.

Jia, W., Valdés-Ramírez, G., Bandodkar, A.J., Windmiller, J.R., and J. Wang. 2013. Epidermal biofuel cells: Energy harvesting from human perspiration. *Angew Chem Int Ed* 52:7233–7236.

Joshi, A.S., Dincer, I., and B.V. Reddy. 2011. Solar hydrogen production: A comparative performance assessment. *Int J Hydro Energ* 36:11246–11257.

Justin, G.A. 2004. Biofuel cells as a possible power source for implantable electronic devices. Master of Science thesis, University of Pittsburgh, Pittsburgh, PA.

Justin, G.A., Zhang, Y., Cui, X.T., Bradberry, C.W., Sun, M., and R.J. Sclabassi. 2011. A metabolic biofuel cell: Conversion of human leukocyte metabolic activity to electrical currents. *J Biol Eng* 5:5.

Justin, G.A., Zhang, Y., Sun, M., and R. Sclabassi. 2004. Biofuel cells: A possible power source for implantable electronic devices. In Paper presented at *Proceedings of the 26th Annual International Conference of the IEEE Engineering in Medicine and Biology Society*, San Francisco, CA, pp. 4096–4099.

Justin, G.A., Zhang, Y., Sun, M., and R. Sclabassi. 2005. An investigation of the ability of white blood cells to generate electricity in biofuel cells. In Paper presented at *Proceedings of the IEEE 31st Annual Northeast Bioengineering Conference*, Hoboken, NJ, pp. 277–278.

Kamitaka, Y., Tsujimura, S., Setoyama, N., Kajino, T., and K. Kano. 2007. Fructose/dioxygen biofuel cell based on direct electron transfer-type bioelectrocatalysis. *Phys Chem Chem Phys* 9:1793–1801.

Kannan, A.M., Renugopalakrishnan, V., Filipek, S., Li, P., Audette, G.F., and L. Munukutla. 2008. Bio-batteries and bio-fuel cells: Leveraging on electronic charge transfer proteins. *J Nanosci Nanotechnol* 8:1–13.

Katz, E., Shipway, A.N., and I. Willner. 2003. Biochemical fuel cells. In *Handbook of fuel cells—Fundamentals, Technology and Applications*, eds. W. Vielstich, H.A. Gasteiger, and A. Lamm, pp. 355–381. New York: John Wiley & Sons, Inc.

Kaufhold, S., Dohrmann, R., and C. Ulrichs. 2008. Shelf life stability of diatomites. *Appl Clay Sci* 41:158–164.

Kerzenmacher, S., Ducrée, J., Zengerle, R., and F. von Stetten. 2008. Energy harvesting by implantable abiotically catalyzed glucose fuel cells. *J Power Sources* 182:1–17.

Kerzenmacher, S., Rubenwolf, S., Kloke, A., Zengerle, R., and J. Gescher. 2010. Biofuel cells for the energy supply of distributed systems: State-of-the-Art and applications. Sensoren und Messsysteme, Nürnberg, Germany, pp. 562–565.

Kim, H.J., Hyun, M.S., Chang, I.S., and B.H. Kim. 1999. A microbial fuel cell type lactate biosensor using a metal-reducing bacterium, *Shewanella putrefaciens*. *J Microbiol Biotechnol* 9:365–367.

Kim, H.J., Park, H.S., Hyun, M.S., Chang, I.S., Kim, M., and B.H. Kim. 2002. A mediator-less microbial fuel cell using a metal reducing bacterium, *Shewanella putrefaciens*. *Enzyme Microb Technol* 30:145–152.

Kim, J., Jia, H., and P. Wang. 2006. Challenges in biocatalysis for enzyme-based biofuel cells. *Biotechnol Adv* 24:296–308.

Kloke, A., Biller, B., Kerzenmacher, S., Kräling, U., Zengerle, R., and F. von Stetten. 2008. A single layer biofuel cell as potential coating for implantable low power devices. In *Eurosensors*, Dresden, Germany.

Larminie, J. and A. Dicks. 2000. *Fuel Cell Systems Explained*. Chichester, U.K.: John Wiley & Sons, Inc.

Li, H.-N., Ci, Y-X., Feng, J., Cheng, K., Fu, S., and D.-B. Wang. 1999. The voltammetric behavior of bone marrow of leukaemia and its clinical application. *Bioelectrochem Bioenerg* 48:171–175.

Liebsch, M., Becker, W., and G. Gagelmann. 1978. An improvement of blood sampling technique for *Biomphalaria glabrata* using anaesthesia and long-term relaxation and the role of this method in studies of the regulation of hemolymph glucose. *Comp Biochem Physiol A* 59:169–174.

Liu, Y. and S. Dong. 2007. A biofuel cell harvesting energy from glucose-air and fruit juice-air. *Biosens Bioelectron* 23:593–597.

MacVittie, K., Halámek, J., Halámková, L. et al. 2013. From "cyborg" lobsters to a pacemaker powered by implantable biofuel cells. *Energ Environ Sci* 6:81–86.

Matsumoto, R., Kakuta, M., Sugiyama, T. et al. 2010. A liposome-based energy conversion system for accelerating the multi-enzyme reactions. *Phys Chem Chem Phys* 12:13904–13906.

Mitsos, A., Palou-Rivera, I., and P.I. Barton. 2004. Alternatives for micropower generation processes. *Ind Eng Chem Res* 43:74–84.

Neto, S.A. and A.R. De Andrade. 2013. New energy sources: The enzymatic biofuel cell. *J Braz Chem Soc* 24:1891–1912.

Oncescu, V. and D. Erickson. 2011. A microfabricated low cost enzyme-free glucose fuel cell for powering low-power implantable devices. *J Power Sources* 196:9169–9175.

Palmore, G.T.R. 2008. Biofuel cells. In *Bioelectrochemistry: Fundamentals, Experimental Techniques and Applications*, ed. P.N. Bartlett. New York: John Wiley & Sons, Inc.

Palmore, G.T.R., Bertschy, H., Bergens, S.H., and G.M. Whitesides. 1998. A methanol/dioxygen biofuel cell that uses NAD(1)-dependent dehydrogenases as catalysts: Application of an electro-enzymatic method to regenerate nicotinamide adenine dinucleotide at low overpotentials. *J Electroanal Chem* 443:155–161.

Palmore, G.T.R. and G.M. Whitesides. 1994. Microbial and enzymatic biofuel cells. In *Enzymatic Conversion of Biomass for Fuels Production*, eds. M.E. Himmel, J.O. Baker, and R.P. Overend, pp. 271–290. American Chemical Society. Washington, DC.

Park, D.H. and J.G. Zeikus. 2002. Impact of electrode composition on electricity generation in a single-compartment fuel cell using *Shewanella putrefaciens. Appl Microbiol Biotechnol* 59:58–61.

Rabaey, K., Boon, N., Siciliano, S.D., Verhaege, M., and W. Verstraete. 2004. Biofuel cells select for microbial consortia that self-mediate electron transfer. *Appl Environ Microbiol* 70:5373–5382.

Ramanavicius, A., Kausaite, A., and A. Ramanaviciene. 2005. Biofuel cell based on direct bioelectrocatalysis. *Biosens Bioelectron* 20:1962–1967.

Rao, J.R., Richter, G.J., Vonsturm, F., and E. Weidlich. 1976. Performance of glucose electrodes and characteristics of different biofuel cell constructions. *Bioelectrochem Bioenerg* 3:139–150.

Rapoport, B.I., Kedzierski, J.T., and R. Sarpeshkar. 2012. A glucose fuel cell for implantable brain-machine interfaces. *PLOS ONE* 7:e38436.

Rayment, C. and S. Sherwin. 2003. Introduction to fuel cell technology. University of Notre Dame, Notre Dame, IN.

Sakai, H., Nakagawa, T., and Y. Tokita. 2008. A high-power glucose/oxygen biofuel cell operating under quiescent conditions. *Energ Environ Sci* 2:133–138.

Sarma, A.K., Vatsyayan, P., Goswami, P., and S.D. Minteer. 2009. Recent advances in material science for developing enzyme electrodes. *Biosens Bioelectron* 24:2313–2322.

Schrenzel, J., Serrander, L., Bánfi, B. et al. 1998. Electron currents generated by the human phagocyte NADPH oxidase. *Nature* 392:734–737.

Scott, K., Yu, E.H., Ghangrekar, M.M., Erable, B., and N.M. Duţeanu. 2012. Biological and microbial fuel cells. In *Comprehensive Renewable Energy*, ed. A. Sayigh, Volume 4: Fuel cells and Hydrogen Technology. ed. A.J. Cruden, pp. 277–300. Elsevier.

Sharma, T., Hu, Y., Stoller, M. et al. 2011. Mesoporous silica as a membrane for ultra-thin implantable direct glucose fuel cells. *Lab Chip* 11:2460–2465.

Slaughter, G. and J. Sunday. 2014. A membraneless single compartment abiotic glucose fuel cell. *J Power Sources* 261:332–336.

So, K., Kawai, S., Hamano, Y. et al. 2014. Improvement of a direct electron transfer-type fructose/dioxygen biofuel cell with a substrate-modified biocathode. *Phys Chem Chem Phys* 16:4823–4829.

Song, Y., Penmatsa, V., and C. Wang. 2014. Modeling and simulation of enzymatic biofuel cells with three-dimensional microelectrodes. *Energies* 7:4694–4709.

Southcott, M., MacVittie, K., Halámek, J. et al. 2013. A pacemaker powered by an implantable biofuel cell operating under conditions mimicking the human blood circulatory system—Battery not included. *Phys Chem Chem Phys* 15:6278–6283.

Starner, T. 1996. Human-powered wearable computing. *IBM Syst J* 35:618–629.

Starner, T. and J.A. Paradiso. 2004. Human generated power for mobile electronics. In *Low Power Electronics Design*, ed. C. Piguet, pp. 1–35. Boca Raton, FL: CRC Press.

Steeb, H., Mehrmann, A., Seeger, W., and W. Schnurnberger. 1985. Solar hydrogen production: Photovoltaic/electrolyzer system with active power conditioning. *Int J Hydro Energ* 10:353–358.

Sugiyama, T., Goto, Y., Matsumoto, R., Sakai, H., Tokita, Y., and T. Hatazawa. 2010. A mediator-adapted diaphorase variant for a glucose dehydrogenase-diaphorase biocatalytic system. *Biosens Bioelectron* 26:452–457.

Taniguchi, I., Kishikawa, M., Ohtani, M., Tabata, D., and M. Tominaga. 2006. In Paper presented at *209th Electrochemical Society Meeting*, Abstract #580, Denver, CO.

Tsujimura, S., Fukuda, J., Shirai, O. et al. 2012. Micro-coulometric study of bioelectrochemical reaction coupled with TCA cycle. *Biosens Bioelectron* 34:244–248.

Tsujimura, S., Wadano, A., Kano, K., and T. Ikeda. 2001. Photosynthetic bioelectrochemical cell utilizing cyanobacteria and water-generating oxidase. *Enzyme Microb Technol* 29:225–231.

Vielstich, W., Lamm, A., and H.A. Gasteiger. 2003. *Handbook of Fuel Cells: Fundamentals, Technology, Applications*. Fundamentals and Survey of Systems. Chichester, U.K.: John Wiley & Sons, Inc.

Weidlich, E., Richter, G., von Sturm, F., Rao, J.R., Thorén, A., and H. Lagergren. 1976. Animal experiments with biogalvanic and biofuel cells. *Biomater Med Dev Artif Organs* 4:277–306.

Xu, S. and S.D. Minteer. 2012. Enzymatic biofuel cell for oxidation of glucose to CO_2. *ACS Catal* 2:91–94.

Yang, J., Ghobadian, S., Goodrich, P.J., Montazamia, R., and N. Hashemi. 2013. Miniaturized biological and electrochemical fuel cells: Challenges and applications. *Phys Chem Chem Phys* 15:14147–14161.

Zebda, A., Cosnier, S., Alcaraz, J.-P. et al. 2013. Single glucose biofuel cells implanted in rats power electronic devices. *Sci Rep* 3:1516.

Zebda, A., Gondran, C., Goff, A.L., Holzinger, M., Cinquin, P., and S. Cosnier. 2011. Mediatorless high-power glucose biofuel cells based on compressed carbon nanotube-enzyme electrodes. *Nat Comm* 2:370.

Zhang, X.-R., Yamaguchi, H., and Y. Cao. 2010. Hydrogen production from solar energy powered supercritical cycle using carbon dioxide. *Int J Hydro Energ* 35:4925–4932.

Index

A

Printed and bound by CPI Group (UK) Ltd, Croydon, CR0 4YY

01/11/2024

01782601-0013